Classics in Mathematics

Kunihiko Kodaira

Complex Manifolds and Deformation of Complex Structures

Kunihiko Kodaira was born on March 16, 1915 in Tokyo, Japan. He graduated twice from the University of Tokyo, with a degree in mathematics in 1938 and one in physics in 1941. From 1944 until 1949, Kodaira was an associate professor at the University of Tokyo but by this time his work was well known to mathematicians worldwide and in 1949 he accepted an invitation from H. Weyl to come to the Institute for Advanced Study. During his 12 years in Princeton, he was also Professor at Princeton University from 1952 to 1961. After a year at Harvard, he was then appointed in 1962 to the chair of mathematics at Johns Hopkins University, which he left in 1965 for a chair at Stanford University. Finally, after 2 years at Stanford, he returned to Japan to Tokyo University from 1967. He died in Kofu, Japan, in 1997.

Kodaira's work covers many topics, including applications of Hilbert space methods to differential equations and, importantly, the application of sheaves to algebraic geometry. Around 1960 he became involved in the classification of compact complex analytic spaces. One of the themes running through much of his work is the Riemann–Roch theorem, which played an important role in much of his research.

Kodaira received many honours for his outstanding research, in particular the Fields Medal in 1954.

Kunihiko Kodaira

Complex Manifolds and Deformation of Complex Structures

Reprint of the 1986 Edition

Originally published as Vol. 283 in the series
Grundlehren der mathematischen Wissenschaften

Library of Congress Control Number: 2004113281

Mathematics Subject Classification (2000): 32-01, 23C10, 58C10

ISSN 1431-0821
ISBN 3-540-22614-1 Springer Berlin Heidelberg New York

Springer is a part of Springer Science+Business Media
springeronline.com

Printed in Germany

Printed on acid-free paper 41/3142YL-5 4 3 2 1 0

Kunihiko Kodaira

Complex Manifolds and Deformation of Complex Structures

Translated by Kazuo Akao

With 22 Illustrations

Springer-Verlag
New York Berlin Heidelberg Tokyo

Kunihiko Kodaira
3-19-8 Nakaochiai
Shinjuku-Ku, Tokyo
Japan

Kazuo Akao (*Translator*)
Department of Mathematics
Gakushuin University
Tshima-ku, Tokyo
Japan

AMS Classifications: 32-01, 32C10, 58C10, 14J15

Library of Congress Cataloging in Publication Data
Kodaira, Kunihiko
Complex manifolds and deformation of complex structures.
(Grundlehren der mathematischen Wissenschaften; 283)
Translation of: Fukuso tayōtairon.
Bibliography: p. 459
Includes index.
1. Complex manifolds. 2. Holomorphic mappings.
3. Moduli theory. I. Title. II. Series.
QA331.K71913 1985 515.9′3 85-9825

Typeset by J. W. Arrowsmith Ltd., Bristol, England.
Printed and bound by R. R. Donnelley & Sons, Harrisonburg, Virginia.
Printed in the United States of America.

9 8 7 6 5 4 3 2 1

ISBN 0-387-96188-7 Springer-Verlag New York Berlin Heidelberg Tokyo
ISBN 3-540-96188-7 Springer-Verlag Berlin Heidelberg New York Tokyo

Dedicated to my esteemed colleague and friend
D. C. Spencer

Preface

This book is an introduction to the theory of complex manifolds and their deformations.

Deformation of the complex structure of Riemann surfaces is an idea which goes back to Riemann who, in his famous memoir on Abelian functions published in 1857, calculated the number of effective parameters on which the deformation depends. Since the publication of Riemann's memoir, questions concerning the deformation of the complex structure of Riemann surfaces have never lost their interest.

The deformation of algebraic surfaces seems to have been considered first by Max Noether in 1888 (M. Noether: *Anzahl der Modulen einer Classe algebraischer Flächen*, Sitz. Königlich. Preuss. Akad. der Wiss. zu Berlin, erster Halbband, 1888, pp. 123–127). However, the deformation of higher dimensional complex manifolds had been curiously neglected for 100 years. In 1957, exactly 100 years after Riemann's memoir, Frölicher and Nijenhuis published a paper in which they studied deformation of higher dimensional complex manifolds by a differential geometric method and obtained an important result. (A. Frölicher and A. Nijenhuis: A theorem on stability of complex structures, *Proc. Nat. Acad. Sci., U.S.A.*, **43** (1957), 239–241).

Inspired by their result, D. C. Spencer and I conceived a theory of deformation of compact complex manifolds which is based on the primitive idea that, since a compact complex manifold M is composed of a finite number of coordinate neighbourhoods patched together, its deformation would be a shift in the patches. Quite naturally it follows from this idea that an infinitesimal deformation of M should be represented by an element of the cohomology group $H^1(M, \Theta)$ of M with coefficients in the sheaf Θ of germs of holomorphic vector fields. However, there seemed to be no reason that any given element of $H^1(M, \Theta)$ represents an infinitesimal deformation of M. In spite of this, examination of familiar examples of compact complex manifolds M revealed a mysterious phenomenon that $\dim H^1(M, \Theta)$ coincides with the number of effective parameters involved in the definition of M. In order to clarify this mystery, Spencer and I developed the theory of deformation of compact complex manifolds. The process of the development was the most interesting experience in my whole mathematical life. It was similar to an experimental science developed by the interaction between experiments (examination of examples) and theory. In this book I have tried to reproduce this interesting experience; however I could not fully convey it. Such an experience may be a passing phenomenon which cannot be reproduced.

The theory of deformation of compact complex manifolds is based on the theory of elliptic partial differential operators expounded in the Appendix. I would like to express my deep appreciation to Professor D. Fujiwara who kindly wrote the Appendix and also to Professor K. Akao who spent the time and effort translating this book into English.

Tokyo, Japan
January, 1985

KUNIHIKO KODAIRA

Contents

Chapter 1

Holomorphic Functions

§1.1. Holomorphic Functions

(a) Holomorphic Functions

We begin by defining holomorphic functions of n complex variables. The n-dimensional complex number space is the set of all n-tuples $(z_1, \ldots, z_n)$ of complex numbers z_i, $i=1, \ldots, n$, denoted by $\mathbb{C}^n$. $\mathbb{C}^n$ is the Cartesian product of n copies of the complex plane: $\mathbb{C}^n = \mathbb{C} \times \cdots \times \mathbb{C}$. Denoting $(z_1, \ldots, z_n)$ by z, we call $z=(z_1, \ldots, z_n)$ a point of $\mathbb{C}^n$, and $z_1, \ldots, z_n$ the complex coordinates of z. Letting $z_j = x_{2j-1} + ix_{2j}$ by decomposing z_j into its real and imaginary parts (where $i=\sqrt{-1}$), we can express z as

$$z=(x_1, x_2, \ldots, x_{2n-1}, x_{2n}). \tag{1.1}$$

Thus $\mathbb{C}^n$ is considered as the $2n$-dimensional real Euclidean space $\mathbb{R}^{2n}$ equipped with the complex coordinates. $x_1, x_2, \ldots, x_{2n-1}, x_{2n}$ are called the real coordinates of z. Let $z=(z_1, \ldots, z_n)$ and $w=(w_1, \ldots, w_n)$ be points in $\mathbb{C}^n$. We define the linear combination $\lambda z + \mu w$ of z and w, viewed as vectors, by

$$\lambda z + \mu w = (\lambda z_1 + \mu w_1, \ldots, \lambda z_n + \mu w_n),$$

where λ and μ are complex numbers. This makes $\mathbb{C}^n$ a complex linear space. The length of $z=(z_1, \ldots, z_n)$ is defined by

$$|z| = \sqrt{|z_1|^2 + \cdots + |z_n|^2}. \tag{1.2}$$

Clearly we have

$$|\lambda z| = |\lambda||z|, \tag{1.3}$$

$$|z+w| \leqq |z| + |w|. \tag{1.4}$$

The distance of the two points $z, w \in \mathbb{C}^n$ is given by

$$|z-w| = \sqrt{|z_1 - w_1|^2 + \cdots + |z_n - w_n|^2}. \tag{1.5}$$

We introduce a topology on $\mathbb{C}^n$ by the identification with $\mathbb{R}^{2n}$ with the usual topology. Thus, for example, a subset $D \subset \mathbb{C}^n$ is a domain in $\mathbb{C}^n$ if D is a domain considered as a subset of $\mathbb{R}^{2n}$. Again, a complex-valued function $f(z)=f(z_1, \ldots, z_n)$ defined on a subset D in $\mathbb{C}^n$ is continuous if $f(z)$ is so as a function of the real coordinates $x_1, x_2, \ldots, x_{2n}$.

Now we consider a complex-valued function $f(z)=f(z_1, \ldots, z_n)$ of n complex variables $z_1, \ldots, z_n$ defined on a domain $D \subset \mathbb{C}^n$.

Definition 1.1. If $f(z)=f(z_1, \ldots, z_n)$ is continuous in $D \subset \mathbb{C}^n$, and holomorphic in each variable z_k, $k=1, \ldots, n$, separately, $f(z_1, \ldots, z_n)$ is said to be *holomorphic* in D. We also call $f(z)=f(z_1, \ldots, z_n)$ a *holomorphic function of n variables* $z_1, \ldots, z_n$.

Here, by saying that $f(z_1, \ldots, z_k, \ldots, z_n)$ is holomorphic in z_k separately, we mean that $f(z_1, \ldots, z_n)$ is a holomorphic function in z_k when the other variables $z_1, \ldots, z_{k-1}, z_{k+1}, \ldots, z_n$ are fixed.

The fundamental Cauchy integral formula with respect to a circle for holomorphic functions of one variable is extended to the case of holomorphic functions of n variables as follows.

Given a point $c=(c_1, \ldots, c_n) \in \mathbb{C}^n$ and positive real numbers $r_1, \ldots, r_n$, we put

$$U_r(c)=\{z \mid z=(z_1, \ldots, z_n) \mid |z_k - c_k| < r_k, k=1, \ldots, n\}, \tag{1.6}$$

where r denotes $(r_1, \ldots, r_n)$. Let $U_{r_k}(c_k)$ be the disk with centre c_k and radius r_k on the z_k-plane. Then we have

$$U_r(c)=U_{r_1}(c_1) \times \cdots \times U_{r_n}(c_n). \tag{1.7}$$

Thus we call $U_r(c)$ the polydisk with centre c. We denote by C_k the boundary of $U_{r_k}(c_k)$, that is, the circle of radius r_k with centre c_k on the z_k-plane. Of course C_k is represented by the usual parametrization $\theta_k \to \gamma(\theta_k)=c_k + r_k e^{i\theta_k}$ where $0 \leqq \theta_k \leqq 2\pi$. The product of $C_1, C_2, \ldots, C_n$

$$C^n = C_1 \times \cdots \times C_n \tag{1.8}$$

is called the determining set of the polydisk $U_r(c)$. C^n is an n-dimensional torus. Given a continuous function $\psi(\zeta)=\psi(\zeta_1, \ldots, \zeta_n)$, with $\zeta_1 \in C_1, \ldots, \zeta_n \in C_n$, we define its integral over C^n by

$$\int_{C^n} \psi(\zeta)\, d\zeta_1 \cdots d\zeta_n = \int_{C_1} \cdots \int_{C_n} \psi(\zeta)\, d\zeta_1 \cdots d\zeta_n$$

$$= \int_0^{2\pi} \cdots \int_0^{2\pi} \psi(\gamma_1(\theta_1), \ldots, \gamma_n(\theta_n)) \gamma_1'(\theta_1) \ldots \gamma_n'(\theta_n)\, d\theta_1 \ldots d\theta_n. \tag{1.9}$$

Theorem 1.1. *Let $f = f(z_1, \ldots, z_n)$ be a holomorphic function in a domain $D \subset \mathbb{C}^n$. Take a polydisk $U_r(c)$ with $[U_r(c)] \subset D$. Then for $z \in U_r(c)$, $f(z)$ is represented as*

$$f(z) = \left(\frac{1}{2\pi i}\right)^n \int_{C^n} \frac{f(\zeta_1, \ldots, \zeta_n)}{(\zeta_1 - z_1) \cdots (\zeta_n - z_n)} \, d\zeta_1 \cdots d\zeta_n, \tag{1.10}$$

where [ ] *denotes the closure.*

Proof. First we consider the case $n = 2$. In this case the right-hand side of (1.10) becomes

$$\left(\frac{1}{2\pi i}\right)^2 \int_0^{2\pi} \int_0^{2\pi} \frac{f(\gamma_1(\theta_1), \gamma_2(\theta_2))\gamma_1'(\theta_1)\gamma_2'(\theta_2)}{(\gamma_1(\theta_1) - z_1)(\gamma_2(\theta_2) - z_2)} \, d\theta_1 \, d\theta_2,$$

where the integrand is a continuous function of θ_1 and θ_2 for $(z_1, z_2) \in U_r(c)$. Hence by the formula of the iterated integral, this integral is equal to

$$\left(\frac{1}{2\pi i}\right)^2 \int_0^{2\pi} \frac{\gamma_1'(\theta_1)}{\gamma_1(\theta_1) - z_1} \, d\theta_1 \int_0^{2\pi} \frac{f(\gamma_1(\theta_1), \gamma_2(\theta_2))\gamma_2'(\theta_2)}{\gamma_2(\theta_2) - z_2} \, d\theta_2.$$

Therefore by the Cauchy integral formula, the right-hand side of (1.10) becomes

$$\begin{aligned} &\left(\frac{1}{2\pi i}\right)^2 \int_{C_1} \frac{1}{\zeta_1 - z_1} \, d\zeta_1 \int_{C_2} \frac{f(\zeta_1, \zeta_2)}{\zeta_2 - z_2} \, d\zeta_2 \\ &\quad = \frac{1}{2\pi i} \int_{C_1} \frac{f(\zeta_1, z_2)}{\zeta_1 - z_1} \, d\zeta_1 = f(z_1, z_2), \end{aligned}$$

which proves (1.10) in this case. Similarly for general n, by a repeated application of the Cauchy integral formula, the right-hand side of (1.10) becomes

$$\begin{aligned} &\left(\frac{1}{2\pi i}\right)^n \int_{C_1} \frac{d\zeta_1}{\zeta_1 - z_1} \cdots \int_{C_{n-1}} \frac{d\zeta_{n-1}}{\zeta_{n-1} - z_{n-1}} \int_{C_n} \frac{f(\zeta_1, \ldots, \zeta_{n-1}, \zeta_n)}{\zeta_n - z_n} \, d\zeta_n \\ &\quad = \left(\frac{1}{2\pi i}\right)^{n-1} \int_{C_1} \frac{d\zeta_1}{\zeta_1 - z_1} \cdots \int_{C_{n-1}} \frac{f(\zeta_1, \ldots, \zeta_{n-1}, z_n)}{\zeta_{n-1} - z_{n-1}} \, d\zeta_{n-1} \\ &\quad = \cdots = \frac{1}{2\pi i} \int_{C_1} \frac{f(\zeta_1, z_2, \ldots, z_n)}{\zeta_1 - z_1} \, d\zeta_1 = f(z_1, \ldots, z_n). \quad \blacksquare \end{aligned}$$

As in the case of holomorphic functions of one variable, we shall deduce the fundamental properties of holomorphic functions of n variables from the integral formula (1.10).

First let $\psi(\zeta_1, \ldots, \zeta_n)$ be a continuous function on $C^n = C_1 \times \cdots \times C_n$, and $m_1, \ldots, m_n$ natural numbers. Consider the integral

$$g(z) = \int_{C^n} \frac{\psi(\zeta_1, \ldots, \zeta_n)\, d\zeta_1 \cdots d\zeta_n}{(\zeta_1 - z_1)^{m_1} \cdots (\zeta_n - z_n)^{m_n}} \tag{1.11}$$

as a function of $z = (z_1, \ldots, z_n) \in U_r(c)$. Clearly $g(z)$ is continuous in $U_r(c)$. Then for fixed $z_2 \in U_{r_2}(c_1), \ldots, z_n \in U_{r_n}(c_n)$, put

$$\varphi(\zeta_1) = \int_{C_2} \cdots \int_{C_n} \frac{\psi(\zeta_1, \ldots, \zeta_n)\, d\zeta_2 \cdots d\zeta_n}{(\zeta_2 - z_2)^{m_2} \cdots (\zeta_n - z_n)^{m_n}}.$$

$\varphi(\zeta_1)$ is a continuous function of ζ_1 on C_1. Hence

$$g(z) = g(z_1, z_2, \ldots, z_n) = \int_{C_1} \frac{\varphi(\zeta_1)}{(\zeta_1 - z_1)^{m_1}} d\zeta_1 \tag{1.12}$$

is a holomorphic function of z_1 in $U_{r_1}(c_1)$. Similarly $g(z_1, \ldots, z_n)$ is a holomorphic function of each variable z_k, $k = 1, \ldots, n$, in $U_{r_k}(c_k)$. Hence $g(z) = g(z_1, \ldots, z_n)$ is a holomorphic function of n variables $z_1, \ldots, z_n$ in the polydisk $U_r(c)$. By (1.12) we have

$$\begin{aligned} \frac{\partial}{\partial z_1} g(z_1, \ldots, z_n) &= m_1 \int_{C_1} \frac{\varphi(\zeta_1)}{(\zeta_1 - z_1)^{m_1+1}} d\zeta_1 \\ &= m_1 \int_{C_1} \cdots \int_{C_n} \frac{\psi(\zeta_1, \ldots, \zeta_n)\, d\zeta_1 \cdots d\zeta_n}{(\zeta_1 - z_1)^{m_1+1} \cdots (\zeta_n - z_n)^{m_n}}. \end{aligned} \tag{1.13}$$

Thus $(\partial g/\partial z_1)(z_1, \ldots, z_n)$ is also holomorphic in $U_r(c)$. Similar results hold also for $\partial g/\partial z_k$.

By a repeated application of this result to the right-hand side of (1.10), we obtain the following theorem.

Theorem 1.2. *A holomorphic function $f(z) = f(z_1, \ldots, z_n)$ of n variables in a domain $D \subset \mathbb{C}^n$ is arbitrarily many times differentiable in $z_1, \ldots, z_k$ in D, and all its partial derivatives $\partial^{m_1+\cdots+m_n} f(z)/\partial z_1^{m_1} \cdots \partial z_n^{m_n}$ are holomorphic in D. Moreover taking a polydisk $U_r(c)$ such that $[U_r(c)] \subset D$, we have*

$$\begin{aligned} &\frac{\partial^{m_1+\cdots+m_n}}{\partial z_1^{m_1} \cdots \partial z_n^{m_n}} f(z_1, \ldots, z_n) \\ &\qquad = \frac{m_1! \cdots m_n!}{(2\pi i)^n} \int_{C^n} \frac{f(\zeta_1, \ldots, \zeta_n)\, d\zeta_1 \cdots d\zeta_n}{(\zeta_1 - z_1)^{m_1+1} \cdots (\zeta_n - z_n)^{m_n+1}} \end{aligned} \tag{1.14}$$

in $U_r(c)$. ∎

As in the case of functions of one variable we denote by $f^{(m_1\cdots m_n)}(z_1,\ldots,z_n)$ the partial derivative

$$\frac{\partial^{m_1+\cdots+m_n}}{\partial z_1^{m_1}\cdots\partial z_n^{m_n}}f(z_1,\ldots,z_n) \quad \text{of} \quad f(z)=f(z_1,\ldots,z_n).$$

Theorem 1.3. *Let $f(z)=f(z_1,\ldots,z_n)$ be a holomorphic function in a domain $D\subset\mathbb{C}^n$, and $c=(c_1,\ldots,c_n)\in D$. Then in a polydisk $U_{\rho(c)}\subset D$ with centre c, $f(z)$ has a power series expansion in $z_1-c_1,\ldots,z_n-c_n$,*

$$f(z)=\sum_{m_1,\ldots,m_n=0}^{\infty} a_{m_1\cdots m_n}(z_1-c_1)^{m_1}\cdots(z_n-c_n)^{m_n}, \tag{1.15}$$

which is absolutely convergent in $U_\rho(c)$. The coefficient $a_{m_1\cdots m_n}$ is given by

$$a_{m_1\cdots m_n}=\frac{1}{m_1!\cdots m_n!}f^{(m_1\cdots m_n)}(c_1,\ldots,c_n). \tag{1.16}$$

Proof. By replacing z_k by z_k-c_k, $k=1,\ldots,n$, we may assume that $c_1=c_2=\cdots=c_n=0$. For a point $z=(z_1,\ldots,z_n)\in U_\rho(0)$, $\rho=(\rho_1,\ldots,\rho_n)$, take $r=(r_1,\ldots,r_n)$ such that $|z_k|<r_k<\rho_k$ for $k=1,\ldots,n$. Then $z\in U_r(0)$, and $[U_r(0)]\subset U_\rho(0)\subset D$. Hence by (1.10)

$$f(z)=\left(\frac{1}{2\pi i}\right)^n\int_{C^n}\frac{f(\zeta_1,\ldots,\zeta_n)\,d\zeta_1\cdots d\zeta_n}{(\zeta_1-z_1)\cdots(\zeta_n-z_n)}. \tag{1.17}$$

Since $|\zeta_k|=r_k>|z_k|$, $k=1,\ldots,n$, we have

$$\frac{1}{\zeta_k-z_k}=\frac{1}{\zeta_k}\sum_{m_k=0}^{\infty}\left(\frac{z_k}{\zeta_k}\right)^{m_k}, \qquad \left|\frac{z_k}{\zeta_k}\right|=\frac{|z_k|}{r_k}<1, \qquad k=1,\ldots,n.$$

Substituting these into the right-hand side of (1.17), we obtain a power series expansion

$$f(z)=\sum_{m_1,\ldots,m_n=0}^{\infty} a_{m_1\cdots m_n}z_1^{m_1}\cdots z_n^{m_n},$$

$$a_{m_1\cdots m_n}=\left(\frac{1}{2\pi i}\right)^n\int_{C^n}\frac{f(\zeta_1,\ldots,\zeta_n)\,d\zeta_1\cdots d\zeta_n}{\zeta_1^{m_1+1}\cdots\zeta_n^{m_n+1}}.$$

Letting M be the maximum of $|f(\zeta_1,\ldots,\zeta_n)|$ on C^n, we have

$$|a_{m_1\cdots m_n}|\leq\frac{M}{r_1^{m_1}\cdots r_n^{m_n}},$$

which proves that the above power series is absolutely convergent in $U_r(0)$. From (1.14), it is clear that $a_{m_1 \cdots m_n} = f^{(m_1 \cdots m_n)}(0)/m_1! \cdots m_n!$. ∎

(b) Power Series

In this section we consider a power series

$$P(z) = P(z_1, \ldots, z_n) = \sum_{m_1, \ldots, m_n = 0}^{\infty} a_{m_1 \cdots m_n} z_1^{m_1} \cdots z_n^{m_n}$$

with centre 0. If $P(z)$ is a convergent at z, we denote its sum by the same notation $P(z)$.

Theorem 1.4. *Let $w = (w_1, \ldots, w_n)$ be such that $w_1 \neq 0, \ldots, w_n \neq 0$. If $P(z)$ is convergent at $z = w$, then $P(z)$ is absolutely convergent for $|z_1| < |w_1|, \ldots, |z_n| < |w_n|$, and its sum $P(z)$ is a holomorphic function of n variables $z_1, \ldots, z_n$ in $U_\rho(0)$ where $\rho = (|w_1|, \ldots, |w_n|)$.*

Proof. For simplicity we consider the case $n = 2$. The general case is proved similarly. Since, by hypothesis, $P(z)$ is convergent, there exists a constant M such that $|a_{m_1 m_2} w_1^{m_1} w_2^{m_2}| \leqq M < +\infty$. Hence

$$|a_{m_1 m_2}| \leqq \frac{M}{\rho_1^{m_1} \rho_2^{m_2}},$$

where $\rho_1 = |w_1|$ and $\rho_2 = |w_2|$. Therefore if $|z_1| < \rho_1$ and $|z_2| < \rho_2$,

$$\sum_{m_1, m_2 = 0}^{\infty} |a_{m_1 m_2} z_1^{m_1} z_2^{m_2}| \leqq M \sum_{m_1 = 0}^{\infty} \left(\frac{|z_1|}{\rho_1}\right)^{m_1} \sum_{m_2 = 0}^{\infty} \left(\frac{|z_2|}{\rho_2}\right)^{m_2} < +\infty,$$

namely, $P(z)$ is absolutely convergent. Moreover taking arbitrary r_1 and r_2 with $0 < r_1 < \rho_1$, $0 < r_2 < \rho_2$, we have

$$|a_{m_1 m_2} z_1^{m_1} z_2^{m_2}| \leqq |a_{m_1 m_2}| r_1^{m_1} m_2^{m_2}, \qquad \sum_{m_1, m_2 = 0}^{\infty} |a_{m_1 m_2}| r_1^{m_1} r_2^{m_2} < +\infty$$

for $z = (z_1, z_2)$ with $|z_1| < r_1$ and $|z_2| < r_2$. Therefore $P(z)$ is uniformly and absolutely convergent in $[U_r(0)]$ with $r = (r_1, r_2)$, hence continuous in $[U_r(0)]$. Since r_1 and r_2 are arbitrary real numbers with $0 < r_1 < \rho_1$, $0 < r_2 < \rho_2$, and $P(z)$ is clearly a holomorphic function in z_1 and z_2 separately, $P(z_1, z_2)$ is holomorphic in $U_\rho(0)$. ∎

Replacing the variables z_k by $z_k - c_k$, $k = 1, \ldots, n$, we obtain a power series

$$P(z-c) = P(z_1 - c_1, \ldots, z_n - c_n) = \sum_{m_1,\ldots,m_n=0}^{\infty} a_{m_1\cdots m_n}(z_1 - c_1)^{m_1} \cdots (z_n - c_n)^{m_n}$$

with centre $c = (c_1, \ldots, c_n)$.

Corollary. *If a power series $P(z-c)$ is convergent at $w = (w_1, \ldots, w_n)$ with $w_1 \neq c_1, \ldots, w_n \neq c_n$, $P(z-c)$ is absolutely convergent if $|z_k - c_k| < |w_k - c_k|$, $k = 1, \ldots, n$, and its sum $P(z-c)$ is a holomorphic function in $U_\rho(c)$, where $\rho = (|w_1 - c_1|, \ldots, |w_n - c_n|)$.* ∎

The region of convergence of a power series $P(z-c)$ is the union $D = \bigcup U_\rho(c)$ of all polydisks $U_\rho(c)$ where $P(z-c)$ is absolutely convergent. A region of convergence D is a domain if it is not empty. In case $n = 1$, the region of convergence of a power series is an empty set, an open disk, or the whole $\mathbb{C}$ itself, but in case $n \geqq 2$, the region of convergence of a power series may take various forms.

The next theorem follows immediately from this Corollary and Theorem 1.3.

Theorem 1.5. *A function $f(z) = f(z_1, \ldots, z_n)$ of n complex variables is holomorphic in a domain $D \subset \mathbb{C}^n$ if and only if for every point $c \in D$, $f(z)$ has a power series expansion $P(z-c)$ which is convergent in some neighbourhood of c.* ∎

(c) Cauchy–Riemann Equation

First consider a continuously differentiable function $f(z)$ of one complex variable z in a domain $D \subset \mathbb{C}$. Decompose z and $f(z)$ into their real and imaginary parts by writing $z = x + iy$ and $f(z) = u + iv$. Then u and v are continuously differentiable functions of the real coordinates x, y in D. Using z and $\bar{z}$, we have

$$x = \tfrac{1}{2}(z + \bar{z}), \qquad y = \frac{1}{2i}(z - \bar{z}).$$

Here z and $\bar{z}$ are *not* independent variables, but considering them as if they are independent, we define the partial derivatives of $f(z)$ with respect to z and $\bar{z}$ by

$$\frac{\partial f}{\partial z} = \frac{1}{2}\left(\frac{\partial f}{\partial x} - i\frac{\partial f}{\partial y}\right), \qquad \frac{\partial f}{\partial \bar{z}} = \frac{1}{2}\left(\frac{\partial f}{\partial x} + i\frac{\partial f}{\partial y}\right). \tag{1.18}$$

In terms of u and v, we have

$$\frac{\partial f}{\partial z}=\tfrac{1}{2}(u_x+v_y)+\frac{i}{2}(-u_y+v_x),$$
$$\frac{\partial f}{\partial \bar{z}}=\tfrac{1}{2}(u_x-v_y)+\frac{i}{2}(u_y+v_x). \tag{1.19}$$

Therefore by use of (1.18), the Cauchy-Riemann equation: $u_x=v_y$, $u_y=-v_x$, is written as

$$\frac{\partial f}{\partial \bar{z}}=0. \tag{1.20}$$

Thus a continuously differentiable function $f(z)$ is a holomorphic function of z in a domain D if and only if $\partial f/\partial \bar{z}=0$ identically in D. If $f(z)$ is holomorphic, $\partial f/\partial z=u_x+iv_x=f'(z)$ by (1.19), namely, for a holomorphic function $f(z)$, the partial derivative $(\partial f/\partial z)(z)$ is identical to the complex derivative $df(z)/dz$.

Next consider a function $f(z)=f(z_1,\ldots,z_n)$ of n complex variables. Put $f(z)=u+iv$ as above. $f(z)$ is said to be continuously differentiable, C^r, C^∞, etc. if u and v are continuously differentiable, C^r, C^∞, etc. in the real coordinates $x_1,\ldots,x_{2n}$.

Let $f(z)$ be a continuously differentiable function in a domain $W\subset\mathbb{C}^n$. Since

$$x_{2k-1}=\tfrac{1}{2}(z_k+\bar{z}_k),\qquad x_{2k}=\frac{1}{2i}(z_k-\bar{z}_k),\qquad k=1,\ldots,n,$$

we have

$$\frac{\partial f}{\partial z_k}=\frac{1}{2}\left(\frac{\partial f}{\partial x_{2k-1}}-i\frac{\partial f}{\partial x_{2k}}\right),\qquad \frac{\partial f}{\partial \bar{z}_k}=\frac{1}{2}\left(\frac{\partial f}{\partial x_{2k-1}}+i\frac{\partial f}{\partial x_{2k}}\right). \tag{1.21}$$

Since $f(z)=f(z_1,\ldots,z_n)$ is continuously differentiable, hence *a fortiori* continuous, $f(z)$ is a holomorphic function of n variables $z_1,\ldots,z_n$ if and only if $f(z)$ is holomorphic in each z_k separately. Therefore from the above results, we obtain the following theorem.

Theorem 1.6. *Let $f(z)=f(z_1,\ldots,z_n)$ be a continuously differentiable function of n complex variables $z_1,\ldots,z_n$ in a domain $D\subset\mathbb{C}^n$. Then $f(z)$ is holomorphic in D if and only if*

$$\frac{\partial f}{\partial \bar{z}_k}=0,\qquad k=1,\ldots,n. \ \blacksquare \tag{1.22}$$

As is clear from Theorem 1.2, a holomorphic function $f(z)=f(z_1,\ldots,z_n)$ in $D\subset\mathbb{C}^n$ is a C^∞ function in the real coordinates $x_1,\ldots,x_{2n}$.

A differential operator

$$\Delta=\frac{\partial^2}{\partial x_1^2}+\cdots+\frac{\partial^2}{\partial x_{2n}^2}$$

is called a Laplacian, and a C^2-real function $u=u(x_1,\ldots,x_{2n})$ defined on a domain in $\mathbb{R}^{2n}$ is called a harmonic function if it satisfies the Laplace equation

$$\Delta u=0.$$

Let $f(z)=u+iv$ be a holomorphic function of n complex variables $z_1,\ldots,z_n$. Then u and v are obviously harmonic functions since

$$\frac{\partial^2 u}{\partial x_{2k-1}^2}+\frac{\partial^2 u}{\partial x_{2k}^2}=0,\qquad \frac{\partial^2 v}{\partial x_{2k-1}^2}+\frac{\partial^2 v}{\partial x_{2k}^2}=0,\qquad k=1,\ldots,n.$$

Conversely in case $n=1$, let $u=u(x_1,x_2)$ be a harmonic function in a domain $D\subset\mathbb{C}$. Then for any $z\in D$, there is a holomorphic function $f(z)$ defined in some neighbourhood of z such that $u=\operatorname{Re} f(z)$ there. This does not hold in case $n\geqq 2$. For example, in case $n=2$, put $r^2=x_1^2+\cdots+x_4^2$. Then the function $u=1/r^2$ is harmonic, but $\partial^2u/\partial x_1^2+\partial^2u/\partial x_2^2\neq 0$, hence, u cannot be the real part of any holomorphic function. This already reveals the essential difference between holomorphic functions of one variable, and those of n variables with $n\geqq 2$.

(d) Analytic Continuation

Theorem 1.7. *Let $f(x)=f(z_1,\ldots,z_n)$ be a holomorphic function in $D\subset\mathbb{C}^n$. Unless $f(z)$ is identically zero in D, at each point z of D, among $f(z)$ and all its partial derivatives $f^{(m_1\cdots m_n)}(z)$ at least one does not vanish.*

Proof. Let D_0 be the set of points $z\in D$ such that $f(z)$ and all its partial derivatives vanish at z, and put $D_1=D-D_0$. Then, $D=D_0\cup D_1$, $D_1\cap D_0=\varnothing$ and D_0 is open by Theorem 1.3. Since D_1 is clearly open, and D is connected, either $D=D_0$ or $D=D_1$ holds. ∎

Corollary. *If two holomorphic functions $f(z)$ and $g(z)$ in a domain D coincide in some neighbourhood of a point $c\in D$, then $f(z)$ and $g(z)$ coincide in all of D.* ∎

As in the case of holomorphic functions of one complex variable, this Corollary implies the uniqueness of the analytic continuation.

First we must define the analytic continuation of a holomorphic function of n complex variables.

Let D_0 be a domain in $\mathbb{C}^n$, and $f_0(z)$ a holomorphic function in D_0.

Let D_1 be another domain in $\mathbb{C}^n$ with $D_0 \cap D_1 \neq \emptyset$. If there is a holomorphic function $f_1(z)$ defined on D_1 such that $f_1(z) = f_0(z)$ on $D_0 \cap D_1$, then $f_1(z)$ is said to be an *analytic continuation* of $f_0(z)$ *to* D_1. By the above Corollary, such $f_1(z)$ is unique if it exists.

Let $D_1, D_2, \ldots, D_m, \ldots$ be finitely or infinitely many domains in $\mathbb{C}^n$, and $f_k(z)$ a holomorphic function defined on each D_k. If for each $k \geqq 1$, $f_k(z)$ is an analytic continuation of $f_{k-1}(z)$, then every $f_k(z)$ is called an *analytic continuation* of $f_0(z)$. In this case putting $f(z) = f_k(z)$ for $z \in D_k$, we obtain a holomorphic function in $D = D_0 \cup D_1 \cup \cdots$. If this $f(z)$ is one-valued in D, then $f(z)$ is an analytic continuation of $f_0(z)$ to D in the above sense. In general, however, $f(z)$ is not necessarily one-valued, but in these cases too, we call $f(z)$ an *analytic continuation* of $f_0(z)$.

As in the case of functions of one variable, by saying simply a holomorphic function, we mean a one-valued holomorphic function, whereas a holomorphic function which may not be one-valued is called an analytic function.

Unlike in the case $n = 1$, in case $n \geqq 2$, there is a domain $D_0 \subset \mathbb{C}^n$ for which there exists a domain $D \supsetneqq D_0$ such that every holomorphic function defined on D_0 can be continued analytically to D. To see this, first consider the case $n = 2$. Let $\rho_1, \rho_2, \sigma_1, \sigma_2$ be real numbers with $0 < \sigma_1 < \rho_1$, $0 < \sigma_2 < \rho_2$, and let D be a polydisk with centre 0:

$$D = U_\rho(0) = \{(z_1, z_2) \mid |z_1| < \rho_1, |z_2| < \rho_2\}, \qquad \rho = (\rho_1, \rho_2).$$

Put

$$T(0) = \{(z_1, z_2) \mid \operatorname{Re} z_1 \geqq -\sigma_1, |z_2| \leqq \sigma_2\},$$

and $D_0 = D - T(0)$. If we write $z_1 = x_1 + ix_2$, and $z_2 = x_3 + ix_4$, the section of D_0 by the hyperplane $x_2 = 0$ is illustrated in Fig. 1.

Every holomorphic function defined on $D_0 = D - T(0)$ has an analytic continuation to D.

Proof. Let γ_r: $\theta \to \gamma_r(\theta) = (z_1, re^{i\theta})$, $0 \leqq \theta \leqq 2\pi$, be the circle of radius r with centre $(z_1, 0) \in D$, where $\sigma_2 < r < \rho_2$. Then $\gamma_r \subset D_0$. Consider the integral along γ_r

$$f_1(z_1, z_2) = \frac{1}{2\pi i} \int_{\gamma_r} \frac{f_0(z_1, \zeta)}{\zeta - z_2} d\zeta = \frac{1}{2\pi} \int_0^{2\pi} \frac{f_0(z_1, re^{i\theta})}{1 - r^{-1}e^{-i\theta}z_2} d\theta \tag{1.23}$$

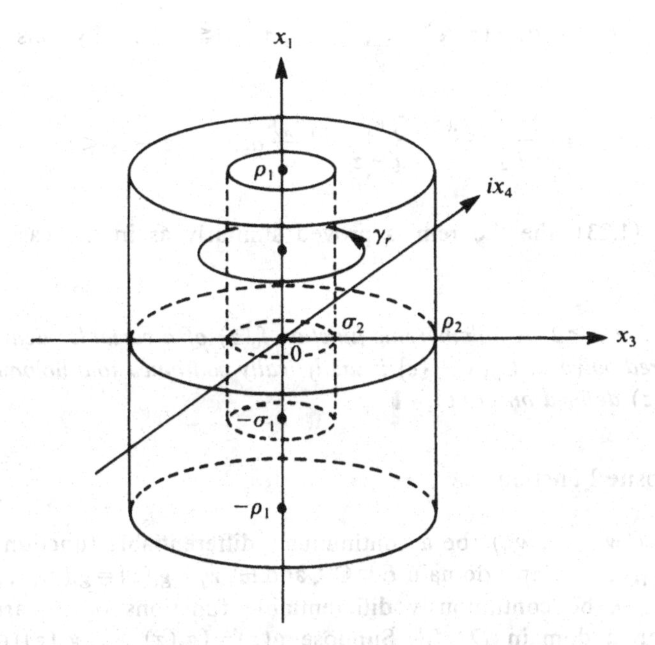

Figure 1

as a function of z_1 and z_2. Clearly $f(z_1, z_2)$ is continuous in $D_1 = \{(z_1, z_2) \in D \,|\, |z_2| < r\}$, and holomorphic in each variable separately there. Hence $f(z_1, z_2)$ is a holomorphic function of two variables z_1, z_2 in D_1. Since for each fixed z with $-\rho_1 < \operatorname{Re} z < \rho_1$, $(z_1, z_2) \in D_0$ if $|z_2| < \rho_2$, $f_0(z_1, z_2)$ is holomorphic with respect to z_2 in $|z_2| < \rho_2$. Therefore by the Cauchy integral formula $f_1(z_1, z_2) = f_0(z_1, z_2)$ if $|z_2| < r$. Namely, $f_1(z_1, z_2)$ and $f_0(z_1, z_2)$ coincide in a domain $E = \{z \in D_1 \,|\, \operatorname{Re} z_1 < \sigma_1\}$. Since $E \subset D_1 \cap D_0$, by the Corollary to Theorem 1.7, $f_1(z_1, z_2)$ and $f_0(z_1, z_2)$ coincide in $D_1 \cap D_0$. Hence putting $f(z_1, z_2) = f_1(z_1, z_2)$ for $(z_1, z_2) \in D_1$ and $f(z_1, z_2) = f_0(z_1, z_2)$ for $(z_1, z_2) \in D_0$, we obtain an analytic continuation $f(z_1, z_2)$ of $f_0(z_1, z_2)$ to $D = D_0 \cup D_1$. ∎

This result can be extended to the case $n \geqq 2$. For any $c = (c_1, \ldots, c_n) \in \mathbb{C}^n$, consider the polydisk $U_\rho(x)$, $\rho = (\rho_1, \ldots, \rho_n)$, with centre c. Let $0 < \sigma_1 < \rho_1$, and $0 < \sigma_2 < \rho_2$. Put

$$T(c) = \{z \in U_\rho(c) \,|\, \operatorname{Re}(z_1 - c_1) \geqq -\sigma_1, |z_2 - c_2| \leqq \sigma_2\}, \tag{1.24}$$

and $D_0 = U_\rho(c) - T(c)$, where $z = (z_1, \ldots, z_n)$.

Theorem 1.8 (Hartogs). *Every holomorphic function $f_0(z)$ defined on the domain $D_0 = U_\rho(c) - T(c)$ is continued analytically to a holomorphic function $f(z)$ defined on $U_\rho(z)$.*

Proof. Let $\gamma_r: \theta \to \gamma_r(\theta) = (z, re^{i\theta} + c_2, z_3, \ldots, z_n)$, $0 \leqq \theta \leqq 2\pi$. By considering the integral

$$f_1(z) = \frac{1}{2\pi i} \int_{\gamma_r} \frac{f_0(z_1, \zeta, z_3, \ldots, z_n)}{\zeta - z_2} d\zeta, \qquad |z_2 - c_2| < r$$

instead of (1.23), the theorem is proved similarly as in the case $n = 2$ above. ∎

Corollary. *Let $n \geqq 2$. A holomorphic function $f_0(z)$ of n variables defined on the punctured polydisk $U_\rho(x) - \{c\}$ is analytically continued to a holomorphic function $f(z)$ defined on $U_\rho(c)$.* ∎

(e) Composite Function

Let $f(w) = f(w_1, \ldots, w_m)$ be a continuously differentiable function of m variables $w_1, \ldots, w_m$ in a domain $E \subset \mathbb{C}^m$, and let $w_j = g_j(z) = g_j(z_1, \ldots, z_n)$, $j = 1, 2, \ldots, m$, be continuously differentiable functions of n variables $z_1, \ldots, z_n$ in a domain $D \subset \mathbb{C}^n$. Suppose $g(z) = (g_1(z), \ldots, g_n(z)) \in E$ if $z \in D$.

Let us consider the composite function $f(g(z)) = f(g_1(z), \ldots, g_n(z))$. Clearly $f(g(z))$ is a continuously differentiable function of $z = (z_1, \ldots, z_n)$ in D. The partial derivatives of $f(z)$ with respect to z_k and $\bar{z}_k$ are given by

$$\frac{\partial f(g(z))}{\partial \bar{z}_k} = \sum_{j=1}^{m} \left(\frac{\partial f(w)}{\partial w_j} \frac{\partial w_j}{\partial \bar{z}_k} + \frac{\partial f(w)}{\partial \bar{w}_j} \frac{\partial \bar{w}_j}{\partial \bar{z}_k} \right), \qquad w_j = g_j(z), \tag{1.25}$$

$$\frac{\partial f(g(z))}{\partial z_k} = \sum_{j=1}^{m} \left(\frac{\partial f(w)}{\partial w_j} \frac{\partial w_j}{\partial z_k} + \frac{\partial f(w)}{\partial \bar{w}_j} \frac{\partial \bar{w}_j}{\partial z_k} \right), \qquad w_j = g_j(z). \tag{1.26}$$

Proof. Decomposing w_j and z_k into their real and imaginary parts by writing $w_j = u_j + iv_j$, and $z_k = x_k + iy_k$, and writing $\partial/\partial \bar{z}_k$, $\partial/\partial z_k$, and $\partial/\partial \bar{w}_j$, $\partial/\partial w_j$ as linear combinations of $\partial/\partial x_k$, $\partial/\partial y_k$ and $\partial/\partial u_j$, $\partial/\partial v_j$ by (1.18), we obtain (1.25) and (1.26) from the usual chain rule for differentiation in the real variables. ∎

Theorem 1.9. *If $f(w) = f(w_1, \ldots, w_m)$ is holomorphic in $w_1, \ldots, w_m$, and $w_j = g_j(z) = g_j(z_1, \ldots, z_n)$, $j = 1, \ldots, m$, are holomorphic in $z_1, \ldots, z_n$, then the composite function $f(g(z))$ is a holomorphic function in $z_1, \ldots, z_n$. Its partial derivatives are given by*

$$\frac{\partial}{\partial z_k} f(g(z)) = \sum_{j=1}^{m} \frac{\partial f(w)}{\partial w_j} \frac{\partial w_j}{\partial z_k}, \qquad w_j = g_j(z). \tag{1.27}$$

Proof. Since by the assumption $f(w)$ is holomorphic in $w_1, \dots, w_m$, and $w_j = g_j(z)$ are holomorphic in $z_1, \dots, z_n$, by Theorem 1.6, $\partial f(w)/\partial \bar{w}_j = 0$, $\partial w_j/\partial \bar{z}_k = 0$. Therefore by (1.25) $\partial f(g(z))/\partial \bar{z}_k = 0$ for $k = 1, \dots, n$. Hence by Theorem 1.6, $f(g(z))$ is a holomorphic function in $z_1, \dots, z_n$. (1.27) follows from (1.26). ∎

(f) Weierstrass Preparation Theorem

In this section we consider functions which are holomorphic in a neighbourhood of $0 = (0, \dots, 0) \in \mathbb{C}^n$. For convenience sake we write $(w, z_2, \dots, z_n)$ instead of $(z_1, \dots, z_n)$. Let $f(w, z) = f(w, z_2, \dots, z_n)$ be a holomorphic function defined in a neighbourhood of $0 = (0, \dots, 0)$ such that $f(0, 0) = f(0, 0, \dots, 0) = 0$, and that $f(w, 0)$ does not vanish identically.

Let $a_1(z), \dots, a_s(z)$ be holomorphic functions in $(n-1)$ variables $(z_2, \dots, z_n)$ such that $a_1(0) = \cdots = a_s(0) = 0$. Then the polynomial in w

$$P(w, z) = w^s + a_1(z)w^{s-1} + \cdots + a_s(z)$$

is called a *distinguished polynomial.*

Theorem 1.10 (Weierstrass Preparation Theorem). *For any sufficiently small $\varepsilon > 0$, there exists $\delta = \delta(\varepsilon) > 0$ such that in the polydisk $U_{\varepsilon,\delta}(0) = \{(w, z) \mid |w| < \varepsilon, |z_2| < \delta, \dots, |z_n| < \delta\}$, $f(w, z)$ can be represented uniquely as the product of a distinguished polynomial $P(w, z)$, and a non-vanishing holomorphic function $u(w, z)$:*

$$f(w, z) = u(w, z)P(w, z). \tag{1.28}$$

Proof. By the assumption, $f(w, 0) = w^s(b_s + b_{s+1}w + \cdots)$ with $b_s \neq 0$, where s is a natural number. Hence, ε being taken sufficiently small, $f(w, 0) \neq 0$ for $0 < |w| \leqq \varepsilon$. Let μ be the minimum of $|f(w, 0)|$ on the circle $|w| = \varepsilon$. Then $|f(w, 0)| \geqq \mu > 0$ if $|w| = \varepsilon$. Therefore taking $\delta = \delta(\varepsilon)$ sufficiently small, we have

$$|f(w, z)| \geqq \mu/2 > 0 \quad \text{if } |w| = \varepsilon, \quad |z_2| < \delta, \dots, |z_n| < \delta.$$

Hence

$$\sigma_k(z) = \frac{1}{2\pi i} \int_{|w|=\varepsilon} \frac{f_w(w, z)}{f(w, z)} w^k \, dw, \qquad k = 0, 1, 2, \dots$$

are holomorphic functions of $z = (z_2, \dots, z_n)$ in $|z_2| < \delta, \dots, |z_n| < \delta$, where $\int_{|w|=\varepsilon}$ denotes the integration along the circle $\theta \to w = \varepsilon\, e^{i\theta}$, $0 \leqq \theta \leqq 2\pi$. For each z, $\sigma_0(z)$ is equal to the number of the zeros of $f(w, z)$ in the disk

$|w|<\varepsilon$. Therefore $\sigma_0(z)$ is a natural number independent of z. Since $\sigma_0(0)=s$, $\sigma_0(z)=s$. Let $\omega_1(z),\ldots,\omega_s(z)$ be the zeros of $f(w,z)$ in the disk $|w|<\varepsilon$. Then

$$\sigma_k(z)=\omega_1(z)^k+\cdots+\omega_s(z)^k.$$

Put

$$P(w,z)=w^s+a_1(z)w^{s-1}+\cdots+a_s(z)=\prod_{\nu=1}^{s}(w-\omega_\nu(z)).$$

The elementary symmetric functions $a_1(z),\ldots,a_s(z)$ of $\omega_1(z),\ldots,\omega_s(z)$ are polynomials of $\sigma_1(z),\ldots,\sigma_s(z)$: For example

$$a_1(z)=\sigma_1(z),$$

$$a_2(z)=\frac{1}{2!}(\sigma_1(z)^2-\sigma_2(z)),$$

$$a_3(z)=\frac{1}{3!}(\sigma_1(z)^3-3\sigma_1(z)\sigma_2(z)+2\sigma_3(z))\cdots.$$

Therefore $a_1(z),\ldots,a_s(z)$ are holomorphic functions of $z=(z_2,\ldots,z_n)$ in $|z_2|<\delta,\ldots,|z_n|<\delta$. Since $\omega_1(0)=\cdots=\omega_s(0)=0$, $a_1(0)=\cdots=a_s(0)=0$. Namely, $P(w,z)$ is a distinguished polynomial. Put

$$u(z,w)=\frac{f(w,z)}{P(w,z)},\qquad |z_2|<\delta,\ldots,|z_n|<\delta.$$

Taking $\varepsilon>0$ and $\delta>0$ sufficiently small, we may assume that $f(w,z)$ is defined on $|w|<3\varepsilon$, $|z_2|<\delta,\ldots,|z_n|<\delta$. Then, for any fixed z, $u(w,z)$ is a holomorphic function of w in $|w|<3\varepsilon$. Hence we have

$$u(w,z)=\frac{1}{2\pi i}\int_{|\zeta|=2\varepsilon}\frac{u(\zeta,z)}{\zeta-w}\,d\zeta,\qquad |w|<\varepsilon. \tag{1.29}$$

Moreover $u(w,z)$ does not vanish in $|w|\leqq\varepsilon$. On the other hand, since $|\omega_1(z)|<\varepsilon,\ldots,|\omega_s(z)|<\varepsilon$, $P(w,z)\neq 0$ for $|w|\geqq\varepsilon$. Hence $u(w,z)$ is a holomorphic function of n variables $w,z_2,\ldots,z_n$ in $\varepsilon<|w|<3\varepsilon$, $|z_2|<\delta,\ldots,|z_n|<\delta$. Therefore by (1.29) $u(w,z)$ is a holomorphic function of n variables $w,z_2,\ldots,z_n$ in $U_{\varepsilon,\delta}(0)$. Moreover $u(w,z)$ does not vanish there. ∎

(g) Power Series Ring

A power series $\sum_{m_1,\ldots,m_n=0}^{\infty}a_{m_1\cdots m_n}z_1^{m_1}\cdots z_n^{m_n}$ whose region of convergence is not empty is called a convergent power series. The sum $f=f(z)=\sum_{m_1,\ldots,m_n=0}^{\infty}a_{m_1\cdots m_n}z_1^{m_1}\cdots z_n^{m_n}$ of a convergent power series is a holomorphic function in a neighbourhood of 0. Conversely, a holomorphic function in a neighbourhood of 0 can be expanded into a convergent power series $\sum_{m_1,\ldots,m_n=0}^{\infty} a_{m_1\cdots m_n} z_1^{m_1}\cdots z_n^{m_n}$. If $f(z)$ and $g(z)$ are holomorphic in a neighbourhood of 0, $f(z)\pm g(z)$ and $f(z)g(z)$ are also holomorphic there. Therefore the set of all convergent power series forms a ring, which is called the power series ring and denoted by $\mathbb{C}\{z_1,\ldots,z_n\}$. If $f(z)$ and $g(z)$ are holomorphic, and $f(z)g(z)=0$ identically, then either $f(z)=0$ or $g(z)=0$ identically, hence the power series ring is an integral domain.

In general an element u of an integral domain R is called a *unit*, or an *invertible element* if there is an element $v\in R$ such that $uv=1$ where 1 denotes the identity of R. Let $f, g\in R$ be non-zero elements. If there is an element $h\in R$ such that $f=gh$, we say that g *divides* f or that f is *divisible by* g, and denote it by $g|f$. In this case g is called a *divisor* of f and f is called a *multiple* of g. If $g|f$ and $f|g$, $f=ug$ with a unit u. In this case f and g are said to be *associates*.

A non-zero element $f\in R$ is called *reducible* if it can be written as a product $f=gh$ where g, h are non-units. A non-zero element $p\in R$ is called *irreducible* if p is neither a unit nor a reducible element. If an element $f\in R$ is written as a product of finite number of irreducible elements as

$$f=p_1\cdots p_n, \tag{1.30}$$

f is said to be *factored into irreducible elements.* (1.30) is called an *irreducible factorization of* f. We say that f *has the unique irreducible factorization up to units* if the following condition is satisfied: Let $f=p_1\cdots p_m$ and $f=q_1\cdots q_l$ be two irreducible factorizations of f. Then we have $l=m$, and, after a suitable reordering, q_k and p_k are associates, that is, $q_k=u_kp_k$ with some unit u_k, for $k=1,\ldots,m$. An integral domain R is called a *unique factorization domain*, or simply a UFD, if every non-zero element $f\in R$, which is not a unit, is factored into irreducible elements uniquely up to units. Let R be a UFD, and $f=p_1\cdots p_m$ an irreducible factorization of $f\in R$. Then p_k, $k=1,\ldots,m$, is called an *irreducible factor* of f. Non-zero elements f and g are said to be *relatively prime* or *coprime* if f and g have no common irreducible factors. Let $f=p_1\cdots p_m$ and $g=q_1\cdots q_l$ be irreducible factorizations of f and g. Then f and g are relatively prime if no p_i and q_k are associates. This follows immediately from the fact that R is a UFD.

Let $R[w]$ be the polynomial ring in a variable w over an integral domain R. The following theorem is well known.

Theorem 1.11. *If R is a* UFD, *then $R[w]$ is also a* UFD.

We denote the power series ring $\mathbb{C}\{z_1,\ldots,z_n\}$ by $\mathcal{O}_0$ or $\mathcal{O}_0^n$:

If $f(z)\in\mathcal{O}_0$ is a unit, $f(0)\neq 0$ obviously. Conversely, if $f(0)\neq 0$ in a small neighbourhood of 0, then $1/f(z)$ is holomorphic, that is, $1/f(z)\in\mathcal{O}_0$. Thus $f(z)\in\mathcal{O}_0$ is a unit if and only if $f(0)\neq 0$.

Now we prove that the power series ring $\mathcal{O}_0 = \mathbb{C}\{z_1, \ldots, z_n\}$ is a UFD. We proceed by induction on n. We write w instead of z, and $f(w, z)$ instead of $f(z_1, z_2, \ldots, z_n)$ where $z = (z_2, \ldots, z_n)$. In case $n = 1$, any non-zero element $f(w) \in \mathbb{C}\{w\}$ which is not a unit can be written as $f(w) = w^s(b_s + b_{s+1}w + \cdots)$ where s is a natural number, and $b_s \neq 0$. Since w is irreducible and $b_s + b_{s+1}w + \cdots$ is a unit, we see immediately that $\mathbb{C}\{w\}$ is a UFD. Now consider the case $n > 1$. Put $R = \mathbb{C}\{z_2, \ldots, z_n\}$. Then $R \subset R[w]$, and the set of the units of $R[w]$ coincides with the set of the units of R. First consider an element $f(w, z) \in \mathcal{O}_0$ such that $f(0, 0) = 0$ and that $f(w, 0) \neq 0$. Then by the Weierstrass preparation theorem, $f(w, z)$ can be written uniquely as

$$f(w, z) = uP(w, z), \tag{1.31}$$

where u is a unit, and $P(w, z)$ is a distinguished polynomial. Suppose that $f(w, z)$ is written as a product

$$f(w, z) = \prod_{k=1}^{m} f_k(w, z), \tag{1.32}$$

where $f_k(w, z) \in \mathcal{O}_0$, and $f_k(0, 0) = 0$, $k = 1, \ldots, m$. Then, since $\prod_k f_k(w, 0) = f(w, 0) \neq 0$, we have $f_k(w, 0) \neq 0$, $k = 1, \ldots, m$, hence each $f_k(w, z)$ can also be written as

$$f_k(w, z) = u_k P_k(w, z), \tag{1.33}$$

where u_k is a unit and $P_k(w, z)$ is a distinguished polynomial. Then

$$f(w, z) = \prod_{k=1}^{m} u_k \prod_{k=1}^{m} P_k(w, z).$$

Since $\prod_k u_k$ is a unit, and $\prod_k P_k(w, z)$ is a distinguished polynomial, the uniqueness of the factorization (1.31) implies that

$$P(w, z) = \prod_{k=1}^{m} P_k(w, z). \tag{1.34}$$

Thus the factorization (1.32) of $f(w, z) = uP(w, z)$ corresponds to the factorization (1.34) of $P(w, z)$ in $R[w]$. Therefore if $P(w, z)$ is irreducible in $R[w]$, it is also irreducible in $\mathcal{O}_0$. In particular, putting $n = 1$ above, we see that $P(w, z)$ is irreducible in $\mathcal{O}_0$ if it is so in $R[w]$. Hence (1.32) is an irreducible factorization of $f(w, z)$ in $\mathcal{O}_0$ if and only if (1.34) is an irreducible factorization of $P(w, z)$ in $R[w]$. Therefore, in order to prove that $\mathcal{O}_0$ is a

UFD, it suffices to show that any distinguished polynomial $P(w, z)$ is factored uniquely into irreducible distinguished polynomials in $R[w]$. Since R is a UFD by the hypothesis of induction, $R[w]$ is also a UFD by Theorem 1.11. Put $P(w, z) = w^s + a_1(z)w^{s-1} + \cdots + a_s(z)$ with $a_1(0) = \cdots = a_s(0) = 0$, and let

$$P(w, z) = \prod_{k=1}^{m} Q_k(w, z), \qquad Q_k(w, z) = b_{k0}(z)w^{s_k} + \cdots + b_{ks_k}(z) \quad (1.35)$$

be the irreducible factorization of $P(w, z)$ in $R[w]$. Then since $\prod_{k=1}^{m} b_{k0}(z) = 1$, each $b_{k0}(z)$ is a unit in R. Consequently letting $P_k(w, z) = b_{k0}(z)^{-1}Q_k(w, z) = w^{s_k} + a_{k1}(z)w^{s_k-1} + \cdots + a_{ks_k}(z)$, we see that $P_k(w, z)$ is also irreducible in $R[w]$ and that

$$P(w, z) = \prod_{k=1}^{m} P_k(w, z). \tag{1.36}$$

Since $P(w, 0) = w^s$, $P_k(w, 0) = w^{s_k}$. Hence $P_k(w, z)$ is a distinguished polynomial. Since any unit in $R[w]$ is actually a unit in R, $P_k(w, z)$ is the unique distinguished polynomial associated with $Q_k(w, z)$ in $R[w]$. Thus we have proved the uniqueness of irreducible factorizations up to units for $f(w, z) \in \mathcal{O}_0$ such that $f(0, 0) = 0$ and that $f(w, 0) \neq 0$.

A convergent power series $f = f(z_1, \ldots, z_n)$ is said to be regular with respect to z_1 if $f(z_1, 0, \ldots, 0) \neq 0$. Thus we have proved that if $f \in \mathcal{O}_0$ is regular with respect to z_1, f is factored uniquely into irreducible elements up to units in $\mathcal{O}_0$. Next suppose that f is not regular with respect to z_1. Introduce new coordinates $z'_1, \ldots, z'_n$ by the linear transformation

$$z_k = \sum_{j=1}^{n} c_{jk} z'_j, \quad \text{with} \quad \det(c_{jk}) \neq 0. \tag{1.37}$$

By this change of coordinates a convergent power series $f(z_1, \ldots, z_n)$ in $z_1, \ldots, z_n$, becomes a convergent power series

$$f'(z'_1, \ldots, z'_n) = f\left(\sum_j c_{j1} z'_j, \ldots, \sum_j c_{jn} z'_j\right)$$

in $z'_1, \ldots, z'_n$. Hence $\mathcal{O}_0 = \mathbb{C}\{z_1, \ldots, z_n\} = \mathbb{C}\{z'_1, \ldots, z'_n\}$. We have $f'(z'_1, 0, \ldots, 0) = f(c_{11}z'_1, \ldots, c_{1n}z'_1)$. Therefore unless $f(z_1, \ldots, z_n) = 0$ identically, choosing a suitable linear change (1.37), we have $f'(z'_1, 0, \ldots, 0) \neq 0$, that is, $f'(z'_1, \ldots, z'_n)$ is regular with respect to z'_1. Thus we obtain the following theorem:

Theorem 1.12. *The power series ring* $\mathcal{O}_0 = \mathbb{C}\{z_1, \ldots, z_n\}$ *is a* UFD. ∎

Let $f, g \in \mathcal{O}_0$. Choosing, if necessary, a suitable linear change of coordinates, we assume that $f(z_1, \dots, z_n)$ and $g(z_1, \dots, z_n)$ are regular with respect to z_1. We write $w = z_1$, and $z = (z_2, \dots, z_n)$ as before.

Theorem 1.13. *$f = f(w, z)$ and $g = g(w, z)$ are relatively prime in $\mathcal{O}_0$ if and only if there exist $\alpha(w, z)$, $\beta(w, z) \in \mathcal{O}_0$ such that the equation*

$$\alpha(w, z)f(w, z) + \beta(w, z)g(w, z) = r(z), \qquad r(z) \neq 0, \tag{1.38}$$

holds, where $r(z) \in \mathbb{C}\{z_2, \dots, z_n\}$. Here $r(z) \neq 0$ means that $r(z)$ does not vanish identically.

Proof. If $f(w, z)$ and $g(w, z)$ have a common non-unit factor $h(w, z)$, then $h(w, 0) \neq 0$, and by (1.38) we obtain an equality

$$q(w, z)h(w, z) = r(z), \qquad q(w, z) \in \mathcal{O}_0.$$

From the proof of the Weierstrass preparation theorem given above, for every $z = (z_2, \dots, z_n)$ with $|z_2| < \delta, \dots, |z_n| < \delta$, the equation $h(w, z) = 0$ has at least one solution $w = \omega_1(z)$. Consequently $r(z) = q(\omega_1(z), z)h(\omega_1(z), z) = 0$ which is a contradiction. Thus the condition (1.38) is sufficient. Conversely, suppose that $f(w, z)$ and $g(w, z)$ are relatively prime. By the Weierstrass preparation theorem we can write

$$f(w, z) = uP(w, z), \qquad g(w, z) = vQ(w, z),$$

where u, v are units and $P(w, z)$, $Q(w, z)$ are distinguished polynomials. Then $P(w, z)$ and $Q(w, z)$ are relatively prime in $R[w]$, hence also in $K[w]$ where K is the quotient field of R. Consequently using the Euclidean algorithm, we obtain an equation

$$A(w)P(w, z) + B(w)Q(w, z) = 1, \qquad A(w), B(w) \in K[w].$$

By multiplying a suitable non-zero element $r(z) \in R$, we have $r(z)A(w)$, $r(z)B(w) \in R[w]$. Putting

$$\alpha(w, z) = u^{-1}r(z)A(w) \quad \text{and} \quad \beta(w, z) = u^{-1}r(z)B(w),$$

we obtain (1.38). ∎

The above results hold also for the power series ring $\mathbb{C}\{z_1 - c_1, \dots, z_n - c_n\}$ in $z_1 - c_1, \dots, z_n - c_n$, which we denote by $\mathcal{O}_c$, $c = (c_1, \dots, c_n)$.

Let $f(z) = f(z_1, \dots, z_n)$ be a holomorphic function in $F \subset \mathbb{C}^n$. Then by Theorem 1.3, for every $p \in D$, $f(z)$ has a convergent power series expansion in $z_1 - c_1, \dots, z_n - c_n$ in some neighbourhood of c:

$$f(z) = P(z - c) = P(z_1 - c_1, \dots, z_n - c_n).$$

We denote $P(z - c)$ by $f_c(z)$.

Theorem 1.14. *Let $f(z)$ and $g(z)$ be holomorphic functions in D, $0 \in D \subset \mathbb{C}^n$. Suppose that $f_0(z)$ and $g_0(z)$ are relatively prime in $\mathcal{O}_0$. Then if $\varepsilon > 0$ is small enough, $f_c(z)$ and $g_c(z)$ are relatively prime in $\mathcal{O}_c$ for any c, $|c| < \varepsilon$.*

Proof. First suppose $f(0) \neq 0$. Then for a sufficiently small $\varepsilon > 0$, $f(c) \neq 0$ in $|c| < \varepsilon$. Then $f_c(z)$ is a unit in $\mathcal{O}_c$, hence $f_c(z)$ and $g_c(z)$ are relatively prime. The case $g(0) \neq 0$ is proved similarly. Now suppose $f(0) = 0$ and $g(0) = 0$. Since $f_0(z)$ and $g_0(z)$ are relatively prime in $\mathcal{O}_0$, choosing coordinates $z_1, \ldots, z_n$ such that $f_0(z)$ and $g_0(z)$ are regular with respect to z_1, we see from Theorem 1.13 above that there exist $\alpha(z)$, $\beta(z) \in \mathcal{O}_0$ such that

$$\alpha(z) f_0(z) + \beta(z) g_0(z) = r(z) \neq 0.$$

Since for a sufficiently small $\varepsilon > 0$, $\alpha(z)$, $\beta(z)$ are holomorphic in $|z| < \varepsilon$, we have

$$\alpha(z) f(z) + \beta(z) g(z) = r(z_2, \ldots, z_n) \neq 0, \qquad |z| < \varepsilon.$$

Hence if $|c| < \varepsilon$

$$\alpha_c(z) f_c(z) + \beta_c(z) g_c(z) = r_c(z_2, \ldots, z_n) \neq 0,$$

where $r_c(z_2, \ldots, z_n)$ is the power series expansion of $r(z_2, \ldots, z_n)$ with centre $(c_2, \ldots, c_n)$. Hence Theorem 1.13 proves that $f_c(z)$ and $g_c(z)$ are relatively prime in $\mathcal{O}_0$. ∎

Definition 1.2. Let $f(z)$ and $g(z)$ be holomorphic in a domain $D \subset \mathbb{C}^n$. $f(z)$ and $g(z)$ are said to be *relatively prime in D* if $f_c(z)$ and $g_c(z)$ are relatively prime in $\mathcal{O}_c$ at every point $c \in D$.

By Theorem 1.14 above, if $f_c(z)$ and $g_c(z)$ are relatively prime in $\mathcal{O}_c$ at a point $c \in D$, $f(z)$ and $g(z)$ are relatively prime in a sufficiently small neighbourhood $U(c)$ of c.

(h) Analytic Hypersurface

Let $f(z)$ be a holomorphic function in a domain $D \subset \mathbb{C}^n$. The set $S = \{z \in D \mid f(z) = 0\}$ is called an *analytic hypersurface*. Let $0 \in S$ and choose coordinates $z_1, z_2, \ldots, z_n$ such that $f(z) = f(z_1, \ldots, z_n)$ is regular with respect to z_1. We write $w = z_1$, and $z = (z_2, \ldots, z_n)$ as before. Then by the Weierstrass preparation theorem, $f(w, z)$ is represented in a polydisk $U_{\varepsilon,\delta}(0) = \{(w, z) \mid |w| < \varepsilon, |z_2| < \delta, \ldots, |z_n| < \delta\}$, as

$$f(w, z) = uP(w, z), \qquad P(w, z) = w^s + a_1(z) w^{s-1} + \cdots + a_s(z),$$

where u does not vanish in $U_{\varepsilon,\delta}(0)$, and $P(w, z)$ is a distinguished polynomial. Consequently S is given in $U_{\varepsilon,\delta}(0)$ by the following algebraic equation of degree s in w:

$$w^s + a_1(z)w^{s-1} + \cdots + a_s(z) = 0.$$

If we denote the roots of this equation by $\omega_1(z), \ldots, \omega_s(z)$, then

$$S \cap U_{\varepsilon,\delta}(0) = \{(\omega_k(z), z) \mid k = 1, 2, \ldots, s, |z_2| < \delta, \ldots, |z_n| < \delta\}.$$

$\omega_1(z), \ldots, \omega_s(z)$ are continuous functions of $z = (z_2, \ldots, z_n)$, which is in general multi-valued if $s \geqq 2$. Note that $a_1(0) = \cdots = a_s(0) = 0$ implies that $\omega_1(0) = \cdots = \omega_s(0) = 0$. For each $z \in U_\delta(0) = \{z \mid |z_2| < \delta, \ldots, |z_n| < \delta\}$ the set $\{\omega_k(z) \mid k = 1, \ldots, s\}$ consists of at most k points. Thus $S \cap U_{\varepsilon,\delta}(0)$ is a hypersurface which is the orbit of these sets as z moves on the polydisk $U_\delta(0)$. The following theorem is an analogy of the Riemann extension theorem for holomorphic functions of one complex variable.

Theorem 1.15. *Let S be an analytic hypersurface in a domain $D \subset \mathbb{C}^n$. Any bounded holomorphic function $h(z_1, \ldots, z_n)$ defined on $D - S$ can be extended to a holomorphic function $\tilde{h}(z_1, \ldots, z_n)$ in D.*

Proof. We may assume that $0 \in S$ and $[U_{\varepsilon,\delta}(0)] \subset D$. We write $w = z_1$, and $z = (z_2, \ldots, z_n)$ as before. It suffices to show that $h(w, z)$ can be continued analytically to a holomorphic function $g(w, z)$ in $U_{\varepsilon,\delta}(0)$. For any fixed $z \in U_{\varepsilon,\delta}(0)$, $h(w, z)$ is a function of w which is holomorphic in $|w| < \varepsilon$ except at $\omega_1(z), \ldots, \omega_s(z)$, and by the assumption $h(w, z)$ is bounded. Hence $h(w, z)$ is extended to a holomorphic function $g(w, z)$ of w in $|w| < \varepsilon$. Let C be the circle of radius ε: $|w| = \varepsilon$ in the w-plane. By the Cauchy integral formula we obtain

$$g(w, z) = \frac{1}{2\pi i} \int_C \frac{h(\zeta - z)}{\zeta - z} \, d\zeta.$$

Hence $g(w, z)$ is a holomorphic function of n variables $w, z_2, \ldots, z_n$ in $U_{\varepsilon,\delta}(0)$. ∎

Let $S = \{f(w, z) = 0\}$ as before. S is called irreducible at 0 if the power series expansion $f_0(w, z)$ of $f(w, z)$ at 0 is irreducible in $\mathcal{O}_0 = \mathbb{C}\{w, z_2, \ldots, z_n\}$. In this case the equation $f_0(w, z) = 0$ is called an irreducible equation of S at 0. As stated in (g) above, the irreducibility of $f_0(w, z)$ in $\mathcal{O}_0$ is equivalent to that of $P(w, z)$ in $R[w]$. In the case $s = 1$, $P(w, z) = w + a_1(z)$ is irreducible, and $\omega_1(z) = -a_1(z)$ is holomorphic in $U_\delta(0)$.

Suppose $s \geqq 2$. If $P(w, z)$ is irreducible, $P(w, z)$ and $P_w(w, z) = sw^{s-1} + (s-1)a_1(z)w^{s-2} + \cdots + a_{s-1}(z)$ are relatively prime in $R[w]$. Consequently by Theorem 1.13 above there exist $\alpha(w, z)$, $\beta(w, z) \in R[w]$ such that

$$\alpha(w, z)P(w, z) + \beta(w, z)P_w(w, z) = r(z) \neq 0, \qquad r(z) \in R, \tag{1.39}$$

where $r(z) = r(z_2, \ldots, z_n)$ is a holomorphic function in $U_\delta(0)$ with $r(0) = 0$, and $r(z) \neq 0$ means that $r(z)$ does not vanish identically on $U_\delta(0)$. Thus $\Delta = \{z \in U_\delta(0) \mid r(z) = 0\}$ is an analytic hypersurface in $U_\delta(0)$. If $\omega_j(z)$ is a multiple root of the equation $P(w, z) = 0$, $P_w(\omega_j(z), z) = 0$, hence by (1.39), $r(z) = 0$. Therefore for $z \in U_\delta(0) - \Delta$, the s roots $\omega_1(z), \ldots, \omega_s(z)$ are all distinct: $\omega_j(z) \neq \omega_k(z)$ $(j \neq k)$. Hence $\omega_1(z), \ldots, \omega_s(z)$ are possibly multi-valued holomorphic functions of z in $U_\delta(0) - \Delta$. If z moves along a closed curve in $U_\delta(0) - \Delta$, $\omega_1(z)$ is continued analytically to one of $\omega_j(z)$'s. Thus any analytic continuation of $\omega_1(z)$ within $U_\delta(0) - \Delta$ is among $\omega_j(z)$, $j = 1, \ldots, s$. Conversely, $\omega_j(z)$, $j = 1, \ldots, s$, are all analytic continuations of $\omega_1(z)$. For suppose that $\omega_2(z), \ldots, \omega_t(z)$ are analytic continuations of $\omega_1(z)$, and that $\omega_{t+1}(z), \ldots, \omega_s(z)$ are not.

Put

$$P_1(w, z) = \prod_{j=1}^{t} (w - \omega_j(z)) = w^t + b_1(z)w^{t-1} + \cdots + b_t(z),$$

and

$$P_2(w, z) = \prod_{j=t+1}^{s} (w - \omega_j(z)) = w^{s-t} + c_1(z)w^{s-t-1} + \cdots + c_{s-t}(z).$$

Then $b_1(z), \ldots, b_t(z)$, $c_1(z), \ldots, c_{s-t}(z)$ are one-valued holomorphic functions in $U_\delta(0) - \Delta$, which are bounded, hence by Theorem 1.15 these are extended to holomorphic functions in $U_\delta(0)$, which are denoted again by $b_1(z), \ldots, b_t(z)$, $c_1(z), \ldots, c_{t-s}(z)$. Then $P_1(w, z)$ and $P_2(w, z)$ are distinguished polynomials in $R[w]$, and $P(w, z) = P_1(w, z)P_2(w, z)$ which contradicts the irreducibility of $P(w, z)$. Thus if S is irreducible at 0, $w_1(z), \ldots, w_s(z)$ are all branches in $U_\delta(0) - \Delta$ of one and the same holomorphic analytic function.

Theorem 1.16. *Let S be irreducible at* 0, *and $f_0(z_1, \ldots, z_n) = 0$ an irreducible equation of S at* 0. *If $g(z_1, \ldots, z_n) \in \mathcal{O}_0$ vanishes identically on S, $g = g(z_1, \ldots, z_n)$ is divisible by $f_0 = f_0(z_1, \ldots, z_n)$ in $\mathcal{O}_0$: $f_0 | g$.*

Proof. We may assume that f_0, and g are both regular with respect to z_1. We write $w = z_1$ and $z = (z_2, \ldots, z_n)$. Since f_0 is irreducible, f_0 and g are relatively prime if g is not divisible by f_0. This being the case, by Theorem

1.13 there exist $\alpha(w, z), \beta(w, z) \in \mathcal{O}_0$ such that

$$\alpha(w, z) f_0(w, z) + \beta(w, z) g(w, z) = r(z) \neq 0, \qquad r(z) \in R.$$

Denoting the roots of $f_0(w, z) = 0$ by $\omega_1(z), \omega_2(z), \ldots,$ we obtain $g(\omega_1(z), z) = 0$ since $(\omega_1(z), z) \in S$ for any $z \in U_\delta(0)$. Hence $r(z) = 0$ identically, which is a contradiction. ∎

Next we consider the general case in which the analytic hypersurface S given by the equation $f(z_1, \ldots, z_n) = 0$ is not necessarily irreducible at $0 \in S$.

In this case by Theorem 1.12, the power series expansion $f_0(z)$ of $f(z) = f(z_1, \ldots, z_n)$ with centre 0 is factored uniquely up to units into irreducible factors in $\mathcal{O}_0$. Let $p_1(z), \ldots, p_r(z)$ be distinct irreducible factors of $f_0(z)$. Then we have

$$f_0(z) = u(z) \prod_{\lambda=1}^{\nu} p_\lambda(z)^{m_\lambda},$$

where $u(z)$ is a unit in $\mathcal{O}_0$.

Put

$$\tilde{f}_0(z) = \prod_{\lambda=1}^{\nu} p_\lambda(z).$$

Then $f_0(z)$ and $\tilde{f}_0(z)$ vanish simultaneously, hence $\tilde{f}_0(z) = 0$ also gives an equation of S in a neighbourhood of 0, which is called a minimal equation of S.

Theorem 1.17. *If $g(z) \in \mathcal{O}_0$ vanishes identically on S, $g(z)$ is divisible by $\tilde{f}_0(z)$: $\tilde{f}_0(z) | g(z)$.*

Proof. By Theorem 1.16 above, $g = g(z)$ is divisible by each $p_\lambda = p_\lambda(z)$. Since $p_1(z), \ldots, p_\lambda(z)$ are mutually coprime, g is divisible by $\tilde{f}_0 = \prod_{\lambda=1}^{\nu} p_\lambda$. ∎

Corollary. *Let $f(z)$ and $g(z) \in \mathcal{O}_0$. If both $f(z) = 0$ and $g(z) = 0$ are minimal equations of S at $0 \in S$, $f(z)$ and $g(z)$ are associates.*

Theorem 1.18. *$f_0(z) = 0$ is a minimal equation of S at 0 if and only if $f_0(z)$ and at least one of its partial derivatives $f_{0z_k}(z) = \partial f_0(z)/\partial z_k$ are relatively prime in $\mathcal{O}_0$.*

Proof. Suppose $f_0(z)$ is not minimal. Then $f_0(z)$ has at least one multiple factor $p_1(z)$: $f_0(z) = p_1(z)^2 h(z)$. Therefore $f_{0z_k}(z) = p_1(z)^2 h_{z_k}(z) + 2p_{1z_k}(z) p_1(z) h(z)$ has a common factor $p_1(z)$ with $f_0(z)$. Suppose in turn that $f_0(z) = 0$ is minimal. Choose coordinates $z_1, \ldots, z_n$ such that

$f_0(z_1, \dots, z_n)$ is regular with respect to z_1, and write $w = z_1$, and $z = (z_2, \dots, z_n)$. Then by the Weierstrass preparation theorem, we have the factorization

$$f_0(w, z) = u(w, z)P(w, z),$$

where u is a unit in $\mathcal{O}_0$, and $P(w, z)$ is a distinguished polynomial. Let $P(w, z) = \prod_{k=1}^m P_k(w, z)$ be the factorization of $P(w, z)$ into irreducible distinguished polynomials in $R[w]$. Then

$$f_0(w, z) = u \prod_{k=1}^{m} P_k(w, z), \quad \text{with a unit} \quad u = u(w, z)$$

is an irreducible factorization of $f_0(z)$ in $\mathcal{O}_0$. Since $f_0(w, z) = 0$ is minimal, the irreducible polynomials $P_k(w, z)$, $k = 1, \dots, m$ are mutually coprime. Consequently

$$f_{0w}(w, z) = u_w \prod_k P_k(w, z) + u \sum_{j=1}^{m} P_{jw}(w, z) \prod_{k \neq j} P_k(w, z)$$

is not divisible by any $P_k(w, z)$. Hence $f_0(w, z)$ and $f_{0w}(w, z)$ are relatively prime. ∎

Corollary. *Let $f(z)$ be a holomorphic function in a domain of $\mathbb{C}^m$, and S the analytic hypersurface defined by the equation $f(z) = 0$. We assume $0 \in S$. Supppose $f_0(z) = 0$ is a minimal equation of S at 0. Then if $\varepsilon > 0$ is small enough, $f_c(z) = 0$ is a minimal equation of S at c for every $c \in S$ with $|c| < \varepsilon$.*

This follows immediately from Theorem 1.14. Let U be a domain where $f(z)$ is defined. $f(z) = 0$ is called a minimal equation of S in U if $f_c(z) = 0$ is a minimal equation of S at every $c \in S \cap U$. By the Corollary to Theorem 1.17, if $f(z) = 0$ and $g(z) = 0$ are both minimal equations of S in U, $u(z) = f(z)/g(z)$ is a non-vanishing holomorphic function in U.

§1.2. Holomorphic Map

In this section we consider a map $\Phi: z \to w = \Phi(z)$ of a domain $D \subset \mathbb{C}^n$ into $\mathbb{C}^m$.

Using complex coordinates, we can write Φ as follows:

$$\Phi: z = (z_1, \dots, z_n) \to (w_1, \dots, w_m) = (\varphi_1(z), \dots, \varphi_m(z)).$$

Φ is called continuous, continuously differentiable C^∞ in D if $\varphi_1(z), \ldots, \varphi_m(z)$ are continuous, continuously differentiable, C^∞ in D, respectively.

Definition 1.3. Φ is said to be *holomorphic in D* if $w_1 = \varphi_1(z), \ldots, w_m = \varphi_m(z)$ are holomorphic functions of n variables $z_1, \ldots, z_n$.

If a map $\Phi: z \to w = \Phi(z)$ is holomorphic, the matrix

$$\left(\frac{\partial w_j}{\partial z_k}\right)_{\substack{j=1,\ldots,m \\ k=1,\ldots,n}} = \begin{pmatrix} \dfrac{\partial w_1}{\partial z_1} & \cdots & \dfrac{\partial w_1}{\partial z_n} \\ \vdots & & \vdots \\ \dfrac{\partial w_m}{\partial z_1} & \cdots & \dfrac{\partial w_m}{\partial z_n} \end{pmatrix}$$

is called the *Jacobian matrix* of Φ, and denoted by $\partial(w_1, \ldots, w_m)/\partial(z_1, \ldots, z_n)$. In particular if $m = n$, the determinant of the Jacobian matrix of Φ:

$$J(z) = \det \frac{\partial(w_1, \ldots, w_n)}{\partial(z_1, \ldots, z_n)} \tag{1.40}$$

is called the *Jacobian of* Φ.

Introduce real coordinates by putting $z_k = x_{2k-1} + ix_{2k}$, and $w_j = u_{2j-1} + iu_{2j}$ and represent Φ in these terms as

$$\Phi: (x_1, \ldots, x_{2n}) \to (u_1, \ldots, u_{2n}) = \Phi(x_1, \ldots, x_{2n}).$$

Then we have

$$\det \frac{\partial(u_1, \ldots, u_{2n})}{\partial(x_1, \ldots, x_{2n})} = |J(z)|^2, \tag{1.41}$$

where the left-hand side of this equality represents the Jacobian of Φ with respect to $(x_1, \ldots, x_{2n})$ and $(u_1, \ldots, u_{2n})$.

Proof. We give proof for $n = 3$. Since by Theorem 1.16, $\partial w_j/\partial \bar{z}_k = (\partial \bar{w}_j/\partial z_k) = 0$, using an elementary calculation, we obtain

$$\begin{aligned} \det \frac{\partial(u_1, \ldots, u_6)}{\partial(x_1, \ldots, x_6)} &= \det \frac{\partial(w_1, w_2, w_3, \bar{w}_1, \bar{w}_2, \bar{w}_3)}{\partial(z_1, z_2, z_3, \bar{z}_1, \bar{z}_2, \bar{z}_3)} \\ &= \det \frac{\partial(w_1, w_2, w_3)}{\partial(z_1, z_2, z_3)} \cdot \frac{\partial(\bar{w}_1, \bar{w}_2, \bar{w}_3)}{\partial(\bar{z}_1, \bar{z}_2, \bar{z}_3)} = |J(z)|^2. \quad \blacksquare \end{aligned}$$

Let $\Phi: z \to w = \Phi(z)$ be a holomorphic map of $D \subset \mathbb{C}^n$ into $\mathbb{C}^m$, and $\Psi: w \to \zeta = \Psi(w)$ a holomorphic map of $E \subset \mathbb{C}^m$ into $\mathbb{C}^\nu$. If $\Phi(D) \subset E$, the composite $\Psi \circ \Phi: z \to \zeta = \Psi(\Phi(z))$ is a holomorphic map of D into $\mathbb{C}^\nu$. The Jacobian matrix of $\Psi \circ \Phi$ is the product of those of Ψ and Φ:

$$\frac{\partial(\zeta_1, \ldots, \zeta_\nu)}{\partial(z_1, \ldots, z_n)} = \frac{\partial(\zeta_1, \ldots, \zeta_\nu)}{\partial(w_1, \ldots, w_m)} \cdot \frac{\partial(w_1, \ldots, w_m)}{\partial(z_1, \ldots, z_n)}. \tag{1.42}$$

In particular if $\nu = m = n$,

$$\det \frac{\partial(\zeta_1, \ldots, \zeta_n)}{\partial(z_1, \ldots, z_n)} = \det \frac{\partial(\zeta_1, \ldots, \zeta_n)}{\partial(w_1, \ldots, w_n)} \cdot \det \frac{\partial(w_1, \ldots, w_n)}{\partial(z_1, \ldots, z_n)}.$$

If $\Psi = \Phi^{-1}: w \to z$,

$$\det \frac{\partial(z_1, \ldots, z_n)}{\partial(w_1, \ldots, w_n)} \cdot \det \frac{\partial(w_1, \ldots, w_n)}{\partial(z_1, \ldots, z_n)} = 1.$$

Thus if a holomorphic map $\Phi: z \to w$ has an inverse $\Phi^{-1}: w \to z$ which is holomorphic, then

$$\det \frac{\partial(w_1, \ldots, w_n)}{\partial(z_1, \ldots, z_n)} \neq 0.$$

Theorem 1.19. *Let $\Phi: z \to w = \Phi(w)$ be a holomorphic map of $D \subset \mathbb{C}^n$ into $\mathbb{C}^n$, and $J(z)$ its Jacobian. If $J(z^0) \neq 0$ at a point $z_0 \in D$, there exist a neighbourhood $U \subset D$ of z^0, and a neighbourhood W of $\Phi(z^0) = w^0$,* such that *Φ maps U bijectively on W. Moreover the inverse Φ^{-1} of Φ restricted to U is holomorphic on W.*

Proof. Consider Φ as a C^∞ map. Then by (1.41) the Jacobian of Φ as a C^∞ map is equal to $|J(z)|^2$. Since $|J(z^0)|^2 > 0$ by the assumption, the inverse mapping theorem for continuously differentiable maps shows that there exist a neighbourhood U of z^0 in D, and a neighbourhood W of $w^0 = \Phi(z^0)$ such that Φ maps U bijectively on W, and that the inverse Φ^{-1} of Φ restricted to U is continuously differentiable. Put

$$\Phi^{-1}: w \to (z_1, \ldots, z_n) = (\psi_1(w), \ldots, \psi_n(w)).$$

Then $z_h = \psi_h(\varphi_1(z), \ldots, \varphi_n(z))$. Since $\partial\varphi_j / \partial\bar{z}_k = 0$, taking the partial derivative with respect to $\bar{z}_k$, we obtain

$$0 = \sum_{j=1}^{n} \frac{\partial \bar{w}_j}{\partial \bar{z}_k} \frac{\partial \psi_h(w)}{\partial \bar{w}_j}, \qquad w_j = \varphi_j(z).$$

Consequently, since $\det(\partial \bar{w}_j/\partial z_k)_{j,k=1,\ldots,n} = \overline{J(z)} \neq 0$ on U, we have $\partial\psi_h(w)/\partial\bar{w}_j = 0, j = 1, \ldots, n$, hence $\psi_h(w)$ are holomorphic in $w_1, \ldots, w_n$. ∎

Corollary 1. *Let Φ be a holomorphic map of a domain $D \subset \mathbb{C}^n$ into $\mathbb{C}^n$. If $J(z)$ does not vanish in D, $\Phi(D)$ is a domain in $\mathbb{C}^n$.* ∎

Corollary 2. *Let Φ be a one-to-one holomorphic map of a domain $D \subset \mathbb{C}^n$ into $\mathbb{C}^n$. If $J(z)$ does not vanish in D, the inverse Φ^{-1} of Φ is a holomorphic map of the domain $E = \Phi(D)$ onto D.* ∎

If Φ maps a domain $D \subset \mathbb{C}^n$ bijectively onto a domain $E \subset \mathbb{C}^n$ and Φ^{-1} is also holomorphic, Φ is called a *biholomorphic map.* Two domains D and E are said to be *biholomorphic* if there exists a biholomorphic map Φ of D onto E.

Theorem 1.20. *Let $f_1(z), \ldots, f_m(z)$ be holomorphic in a domain of C^n. Suppose that*

$$\operatorname{rank} \frac{\partial(f_1(z), \ldots, f_m(z))}{\partial(z_1, \ldots, z_n)} = \nu$$

is independent of z. If z^0 is a point of this domain such that

$$\det \frac{\partial(f_1(z), \ldots, f_\nu(z))}{\partial(z_1, \ldots, z_\nu)} \neq 0 \quad at \quad z = z^0,.$$

then there exists a neightourhood $U(z^0)$ of z^0 such that $f_{\nu+1}(z), \ldots, f_m(z)$ are holomorphic functions of $f_1(z), \ldots, f_\nu(z)$ in $U(z^0)$.

Proof. Put $w_1 = f_1(z), \ldots, w_\nu = f_\nu(z)$. On a sufficiently small neighbourhood $U(z^0)$ of z^0,

$$\det \frac{\partial(w_1, \ldots, w_\nu, z_{\nu+1}, \ldots, z_n)}{\partial(z_1, \ldots, z_\nu, z_{\nu+1}, \ldots, z_n)} = \det \frac{\partial(w_1, \ldots, w_\nu)}{\partial(z_1, \ldots, z_\nu)} \neq 0.$$

Therefore by Theorem 1.19, $\Phi: (z_1, \ldots, z_\nu, z_{\nu+1}, \ldots, z_n) \to (w_1, \ldots, w_\nu, z_{\nu+1}, \ldots, z_n)$ is a biholomorphic map of $U(z^0)$ onto a neighbourhood $U(w^0)$ of $w^0 = \Phi(z^0)$. Put

$$g_j = g_j(w_1, \ldots, w_\nu, z_{\nu+1}, \ldots, z_n) = f_j(\Phi^{-1}(w_1, \ldots, w_\nu, z_{\nu+1}, \ldots, z_n)).$$

Then g_j are holomorphic functions of $w_1, \ldots, w_\nu, z_{\nu+1}, \ldots, z_n$ in $U(w^0)$, and

$$\operatorname{rank} \frac{\partial(g_1, \ldots, g_m)}{\partial(w_1, \ldots, w_\nu, z_{\nu+1}, \ldots, z_n)} = \nu.$$

Therefore, since $g_1 = w_1, \ldots, g_\nu = w_\nu$, we have

$$\frac{\partial g_{\nu+j}}{\partial z_{\nu+k}} = \det \frac{\partial(g_1, \ldots, g_\nu, g_{\nu+j})}{\partial(w_1, \ldots, w_\nu, z_{\nu+k})} = 0, \qquad j = 1, \ldots, m-\nu, \qquad k = 1, \ldots, n-\nu.$$

Hence $g_{\nu+j}(w_1, \ldots, w_\nu, z_{\nu+1}, \ldots, z_n)$ are holomorphic functions of $w_1, \ldots, w_\nu$, and do not depend on the variables $z_{\nu+1}, \ldots, z_n$:

$$g_{\nu+j}(w_1, \ldots, w_\nu, z_{\nu+1}, \ldots, z_n) = h_j(w_1, \ldots, w_\nu).$$

Consequently

$$f_{\nu+j}(z) = h_j(w_1, \ldots, w_\nu) = h_j(f_1(z), \ldots, f_\nu(z)). \quad \blacksquare$$

Chapter 2

Complex Manifolds

§2.1. Complex Manifolds

(a) Definition of Complex Manifolds

Recall that a Riemann surface $\mathcal{R}$ is a connected Hausdorff space Σ endowed with a system of local complex coordinates $\{z_1, z_2, \ldots, z_j, \ldots\}$. Each local complex coordinate z_j is a homeomorphism $z_j: p \to z_j(p)$ of a domain U_j in Σ onto a domain $\mathcal{U}_j \subset \mathbb{C}$ such that $\bigcup_j U_j = \Sigma$, and that for each pair of indices j, k with $U_j \cap U_k \neq \emptyset$, the map $\tau_{jk}: z_k(p) \to z_j(p)$, $p \in U_j \cap U_k$, is a biholomorphic map from the open set $\mathcal{U}_{kj} \subset \mathcal{U}_k$ onto $\mathcal{U}_{jk} \subset \mathcal{U}_j$. The concept of a complex manifold is a natural generalization of the concept of a Riemann surface. In case of a Riemann surface, the local complex coordinate of a point $p \in \Sigma$ is a complex number. Using an n-tuple of complex numbers $z_j(p) = (z_1(p), \ldots, z_n(p))$ instead, we obtain the concept of an n-dimensional complex manifold. More precisely, let Σ be a connected Hausdorff space, and $\{U_1, \ldots, U_j, \ldots\}$ an open covering of Σ consisting of at most countably many domains. Suppose that on each $U_j \subset \Sigma$, a homeomorphism

$$z_j: p \to z_j(p) = (z_j^1(p), \ldots, z_j^n(p)), \qquad p \in U_j,$$

is defined, which maps U_j onto a domain $\mathcal{U}_j \subset \mathbb{C}^n$. Then for each pair j, k with $U_j \cap U_k \neq \emptyset$, the map

$$\tau_{jk}: z_k(p) \to z_j(p), \qquad p \in U_j \cap U_k, \tag{2.1}$$

is a homeomorphism of the open set $\mathcal{U}_{kj} = \{z_k(p) \mid p \in U_j \cap U_k\} \subset \mathcal{U}_k$ in $\mathbb{C}^N$ onto the open set $\mathcal{U}_{jk} = \{z_j(p) \mid p \in U_j \cap U_k\} \subset \mathcal{U}_j$. If τ_{jk} is biholomorphic for any j, k such that $U_j \cap U_k \neq \emptyset$, each $z_j: p \to z_j(p)$ is called local complex coordinates defined on U_j, and the collection $\{z_1, \ldots, z_j, \ldots\}$ is called a system of local complex coordinates on Σ.

Definition 2.1. If a system of local complex coordinates $\{z_1, \ldots, z_j, \ldots\}$ is defined on a connected Hausdorff space Σ, we say that a complex structure is defined on Σ. A connected Hausdorff space is called a complex manifold if a complex structure is defined on it. We denote a complex manifold by the letters M, N, etc. The system of local complex coordinates $\{z_1, \ldots, z_j, \ldots\}$ which defines the complex structure of a complex manifold M is called the system of local complex coordinates of M. The dimension or complex dimension of M is defined to be n. We often denote M^n if we want to make explicit the dimension of M.

Thus the concept of complex manifolds is an obvious generalization of that of Riemann surfaces, and, in fact, a Riemann surface is nothing but a 1-dimensional complex manifold. If M is a complex manifold, and $\{z_1, \ldots, z_j, \ldots\}$ is the system of local complex coordinates of M, each domain U_j is called a coordinate neighbourhood. We call local complex coordinates simply local coordinates or a local coordinate system. The point $z_j(p) = (z_j^1(p), \ldots, z_j^n(p))$ of $\mathbb{C}^n$ is called the local (complex) coordinates of p. Let $p \in M$. Then if we choose a coordinate neighbourhood U_j with $p \in U_j$, p is determined *uniquely by its local coordinates* $z_j = (z_j^1, \ldots, z_j^n) = z_j(p)$. For $p \in U_j \cap U_k$, the coordinate transformation

$$\tau_{jk}\colon z_k \to z_j = (z_j^1, \ldots, z_j^n) = \tau_{jk}(z_k), \tag{2.2}$$

which transforms the local coordinates $z_k = (z_k^1, \ldots, z_k^n) = z_k(p)$ into the local coordinates $z_j = (z_j^1, \ldots, z_j^n) = z_j(p)$ is, by definition, a biholomorphic map.

Since $p \in U_j$ is determined uniquely by its local coordinates $z_j = z_j(p)$, identifying U_j with $\mathscr{U}_j$ via z_j, we can consider that a complex manifold M is obtained by glueing the domains $\mathscr{U}_1, \ldots, \mathscr{U}_j, \ldots$ in $\mathbb{C}^n$ via the isomorphisms $\tau_{jk}\colon \mathscr{U}_{kj} \to \mathscr{U}_{jk}$: $M = \bigcup_j \mathscr{U}_j$. Then $z_j \in \mathscr{U}_j$ and $z_k \in \mathscr{U}_k$ are the same point on M if and only if $z_j = \tau_{jk}(z_k)$.

Example 2.1. Any domain $\mathscr{U} \subset \mathbb{C}^n$ is a complex manifold. $M = \mathscr{U}$ has a system of local coordinates $\{z\}$ consisting of the single local coordinates $z \to z = (z^1, \ldots, z^n)$.

Example 2.2. For a point $(\zeta_0, \ldots, \zeta_n) \in \mathbb{C}^{n+1} - (0, \ldots, 0)$,

$$\zeta = \{(\lambda\zeta_0, \ldots, \lambda\zeta_n) \,|\, \lambda \in \mathbb{C}\}$$

is a complex line through $0 = (0, \ldots, 0)$. The collection of all complex lines through 0 is called the n-dimensional complex projective space, and denoted by $\mathbb{P}^n$. The Riemann sphere $\mathbb{S}$ is the 1-dimensional complex projective space $\mathbb{P}^1$. A point ζ of $\mathbb{P}^n$ represents a complex line

$$\zeta = \{(\lambda\zeta_0, \ldots, \lambda\zeta_n)\}.$$

$(\zeta_0, \ldots, \zeta_n)$ is called the homogeneous coordinates of $\zeta \in \mathbb{P}^n$, and denoted by $\zeta = (\zeta_0, \ldots, \zeta_n)$. The equality $\{\zeta'_0, \ldots, \zeta'_n) = (\zeta_0, \ldots, \zeta_n)$ means that

$(\zeta'_0, \ldots, \zeta'_n)$ and $(\zeta_0, \ldots, \zeta_n)$ are the homogeneous coordinates of the same point ζ, that is, $\zeta'_0 = \lambda\zeta_0, \ldots, \zeta'_n = \lambda\zeta_n$ for some $\lambda \neq 0$.

Put $U_j = \{\zeta \in \mathbb{P}^n \mid \zeta_j \neq 0\}$. $\zeta \in U_0$ is represented as $\zeta = (1, z^1, \ldots, z^n)$ where $z^\nu = \zeta_\nu/\zeta_0$. $(z^1, \ldots, z^n)$ is called the non-homogeneous coordinates of ζ. The map

$$z_0 \colon \zeta \to z_0(\zeta) = (z^1, \ldots, z^n)$$

gives local coordinates on U_0, where $\mathcal{U}_0 = z_0(U_0) = \mathbb{C}^n$. Similarly on U_j we define local coordinates

$$z_j \colon \zeta \to z_j(\zeta) = (z_j^0, \ldots, z_j^{j-1}, z_j^{j+1}, \ldots, z_j^n), \qquad z_j^\nu = \zeta_\nu/\zeta_j.$$

Then $\mathcal{U}_j = z_j(U_j) = \mathbb{C}^n$. Of course $z_0^\nu = z^\nu$. On $U_j \cap U_k$, we have

$$z_j^k = 1/z_k^j, \qquad z_j^\nu = z_k^\nu/z_k^j, \qquad \nu \neq j, \neq k. \tag{2.3}$$

Hence the coordinate transformations $\tau_{jk} \colon z_k \to z_j$ are biholomorphic. $\mathbb{P}^n$ is considered as the complex manifold obtained by glueing the $(n+1)$-copies of $\mathbb{C}^n$ via the isomorphisms (2.3).

(b) Holomorphic Functions and Holomorphic Maps

Let M be an n-dimensional complex manifold, $\{z_1, \ldots, z_j, \ldots\}$ the system of local complex coordinates, U_j the domain of z_j, and $\mathcal{U}_j = z_j(U_j)$. Let f be a real- or complex-valued function defined on a domain $D \subset M$. For $p \in D \cap U_j$, using the local coordinates $z_j = z_j(p)$, we define a function $f_j(z_j)$ by

$$f(p) = f_j(z_j), \tag{2.4}$$

then $f_j(z_j)$ is a function of n complex variables $(z_1, \ldots, z_n)$ defined in $\mathcal{D}_j = z_j(D \cap U_j) \subset \mathcal{U}_j$. By (2.1), if $z_j = z_{jk}(z_k)$, we have $f_j(z_j) = f_k(z_k)$. Since $p \to z_j = z_j(p)$ is a homeomorphism, f is continuous in D if and only if each $f_j(z_j)$ is continuous in $\mathcal{D}_j$ with respect to z_j.

Definition 2.2. f is said to be a continuously differentiable function, a C^r function, C^∞ function in $D \subset M$ if each f_j is a continuously differentiable function, a C^r function, a C^∞ function in $\mathcal{D}_j$ with respect to z_j respectively.

If we consider a complex manifold M as obtained by glueing the domains $\mathcal{U}_1, \ldots, \mathcal{U}_j, \ldots$ in $\mathbb{C}^n$: $M = \bigcup_j \mathcal{U}_j$, identifying $p \in M$ with the point $z_j = (z_j^1, \ldots, z_j^n) = z_j(p) \in \mathcal{U}_j$, the function $f(p)$ is written as $f(z_j)$. Since in this notation $z_j \in \mathcal{U}_j$ and $z_k \in \mathcal{U}_k$ are the same point of M if $z_j = \tau_{jk}(z_k)$, we have

$f_j(z_j) = f_k(z_k)$ if $z_j = \tau_{jk}(z_k)$. Note that here $f(z_k)$ *does not denote the function obtained from* $f(z_j)$ *by substituting* z_k *for* z_j. Also in this notation a function $f(p) = f(z_j)$ defined in a domain D of M is of class C^r (holomorphic) if and only if each $f(z_j)$ is of class C^r (holomorphic) with respect to z_j.

Similarly we define a holomorphic map from a complex manifold M to another complex manifold N. Let $\{w_1, \ldots, w_\lambda, \ldots\}$ be the system of local complex coordinates of N, W_λ the domain of w_λ, and $\mathcal{W}_\lambda = w_\lambda(W_\lambda) \subset \mathbb{C}^m$ where $m = \dim N$. Let $\Phi: p \to q = \Phi(p)$ be a continuous map from a domain $D \subset M$ into N. Since $z_j: p \to z_j(p)$ maps U_j homeomorphically onto $\mathcal{U}_j$, and $w_\lambda: q \to w_\lambda(q)$ maps W_λ homeomorphically onto $\mathcal{W}_\lambda$, for λ, j such that $\Phi^{-1}(W_\lambda) \cap U_j \neq \varnothing$,

$$\Phi_{\lambda j}: z_j(p) \to w_\lambda(q), \qquad q = \Phi(p), \qquad p \in \Phi^{-1}(W_\lambda) \cap U_j, \tag{2.5}$$

is a continuous map from the domain $\mathcal{U}_{j\lambda} = \{z_j(p) \mid p \in \Phi^{-1}(W_\lambda) \cap U_j\} \subset C^n$ into $\mathcal{W}_\lambda$:

$$\begin{array}{ccc} \Phi^{-1}(W_\lambda) \cap U_j & \xrightarrow{\Phi} & W_\lambda \\ {\scriptstyle z_j}\downarrow & & \downarrow{\scriptstyle w_\lambda} \\ \mathcal{U}_{j\lambda} & \xrightarrow[\Phi_{\lambda j}]{} & \mathcal{W}_\lambda. \end{array}$$

If $\Phi_{\lambda j}$ is of class C^r or holomorphic, Φ is said to be of class C^r or holomorphic where $r = 1, 2, \ldots, \infty$.

If there exists a biholomorphic map $w: p \to w(p)$ for a domain $W \subset M$ into $\mathbb{C}^n$, we can use this w as local complex coordinates. Thus if M^n is covered by at most countably many domains $W_1, \ldots, W_\lambda, \ldots$, and for each W_λ a biholomorphic map $w_\lambda: p \to w_\lambda(p)$ from W_λ into $\mathbb{C}^n$ is defined, then $\{w_1, \ldots, w_\lambda, \ldots\}$ makes a system of local complex coordinates of M^n. Hence there are infinitely many choices of systems of local complex coordinates for one and the same complex manifold M^n. In view of this fact we may define a complex manifold as follows: first let two systems of local complex coordinates $\{z_j\} = \{z_1, \ldots, z_j, \ldots\}$ and $\{w_\lambda\} = \{w_1, \ldots, w_\lambda, \ldots\}$ be given on a connected Hausdorff space Σ, U_j the domain of z_j, and W_λ the domain of w_λ. For j, λ such that $U_j \cap W_\lambda \neq \varnothing$,

$$w_\lambda z_j^{-1}: z_j(p) \to w_\lambda(p), \qquad p \in U_j \cap W_\lambda,$$

is a homeomorphism from the open set $\mathcal{U}_{j\lambda} = \{z_j(p) \mid p \in U_j \cap W_\lambda\}$ onto the open set $\mathcal{W}_{\lambda j} = \{w_j(p) \mid p \in W_\lambda \cap U_j\}$. Let M be a complex manifold defined by the system $\{z_j\}$. Then $w_\lambda: p \to w_\lambda(p)$ is biholomorphic if and only if $w_\lambda z_j^{-1}$ is biholomorphic for any j such that $U_j \cap W_\lambda \neq \varnothing$. Therefore if we say that

$\{z_j\}$ and $\{w_\lambda\}$ are holomorphically equivalent when $w_\lambda z_j^{-1}$ is biholomorphic for any pair j, λ such that $U_j \cap W_\lambda \neq \emptyset$, then $\{z_j\}$ and $\{w_\lambda\}$ are two systems of local complex coordinates of the same complex manifold if and only if they are holomorphically equivalent. Thus instead of the definition previously given, we define a complex manifold as follows

Definition 2.3. Let Σ be a connected Hausdorff space. A complex structure on Σ is defined as a holomorphic equivalence class of systems of local complex coordinates on Σ. A connected Hausdorff space endowed with a complex structure M is called a complex manifold and denoted by the same M. The complex structure M is called the complex structure of M, and a system of local complex coordinates belonging to M is called a system of local complex coordinates of the complex manifold M.

Two complex manifolds M and N are called complex analytically homeomorphic or biholomorphically equivalent if there is a biholomorphic map Φ from M onto N. In this case we consider M and N as the same complex manifold by identifying $p \in M$ and $q = \Phi(p) \in N$. In fact, since Φ is homeomorphic, M and N can be considered as the same Hausdorff space Σ. Next, let $\{z_j\}$ be a system of local complex coordinates of M and $\{w_\lambda\}$ a system of local complex coordinates of N. Then since Φ is biholomorphic, $\Phi_{\lambda j}$ defined in (2.5) is biholomorphic, hence $\{z_j\}$ and $\{w_\lambda\}$ are holomorphically equivalent systems of local complex coordinates on Σ.

If, as stated above, we consider $M = \bigcup_j \mathscr{U}_j$ and $N = \bigcup_\lambda \mathscr{W}_\lambda$, a map Φ from a domain $D \subset M$ into N is written as

$$\Phi\colon z_j \to w_\lambda = \Phi(z_j).$$

Thus we can dispense with the indices j, λ in $\Phi_{\lambda j}$ where $w_\lambda = \Phi_{\lambda j}(z_j)$.

(c) Locally Finite Coverings

Let Σ be a Hausdorff space. An open covering $\mathfrak{U}$ of Σ is said to be locally finite if for each point $p \in \Sigma$, there is a neighbourhood $U(p)$ such that $U(p) \cap U \neq \emptyset$ for only a finite number of the members $U \in \mathfrak{U}$. Let M^n be a complex manifold, and q a point of M. By local coordinates with centre q we mean a biholomorphic map $z_q\colon p \to z_q(p)$ of a domain $U(q)$ containing q onto a domain of $\mathbb{C}^n$ containing $(0, \ldots, 0)$ such that $z_q(q) = 0$. Let $z_q\colon p \to z_q(p) = (z_q^1(p), \ldots, z_q^n(p))$ be given local coordinates with centre q. By a coordinate polydisk $U_r(q)$ with centre q on M^n, we mean the inverse image

$$U_r(q) = z_q^{-1}(U_r(0)) = \{p \,|\, |z_q^1(p)| < r^1, \ldots, |z_q(p)| < r^n\}$$

by z_q of the polydisk with centre 0

$$U_r(0)=\{z\,|\,z=(z^1,\ldots,z_n),|z^1|<r^1,\ldots,|z^n|<r^n\},$$

such that $[U_r(0)]\subset z_q(U(q))$, where $r=(r^1,\ldots,r^n)$ and $r^i>0$ for all i.

Theorem 2.1. *Let M^n be a complex manifold. Suppose that for each point $q\in M$, local coordinates z_q with centre q and a coordinate polydisk $U_{R(q)}(q)$ are given. Then we can choose at most countably many coordinate polydisks*

$$U_j=U_{r(j)}(q_j)\subset U_{R(q)}(q_j),\qquad j=1,2,3,\ldots,$$

such that $U=\{U_j\,|\,j=1,2,3,\ldots\}$ is a locally finite open covering of M.

Proof. Omitted. ∎

(d) Submanifolds

Let M be a complex manifold. In the sequel we fix for each point $q\in M$ local coordinates $z_q\colon p\to z_q(p)$ with centre q arbitrarily.

Definition 2.4. Let S be a closed subset of M^n. S is called an analytic subset of M if for each $q\in S$, there are a finite number of holomorphic functions $f_q^1(p),\ldots,f_q^\nu(p)$, $\nu=\nu(q)$, defined in a neighbourhood $U(q)$ of q such that

$$S\cap U(q)=\{p\in U(q)\,|\,f_q^1(p)=\cdots=f_q^\nu(p)=0\}. \tag{2.6}$$

Thus S is a subset of M^n which is defined in a neighbourhood of each point $q\in S$ by a system of analytic equations $\{f_q^1(p)=\cdots=f_q^\nu(p)=0\}$. This system of equations is called a local equation of S at q. There are infinitely many choices of local equations for a given S. S is said to be smooth at q if a local equation $f_q^1(p)=\cdots=f_q^\nu(p)=0$ of S at q can be so chosen that

$$\operatorname{rank}\frac{\partial(f_q^1(p),\ldots,f_q^\nu(p))}{\partial(z_q^1(p),\ldots,z_q^n(p))}=\nu,\qquad p=q.$$

This being the case, we call $m=n-\nu$ the dimension of S at q. Otherwise we call $q\in S$ a singular point of S. If S is smooth at q, taking a sufficiently small neighbourhood $U(q)$ and an appropriate renumbering of $z_q^i(p)$'s, we may assume that

$$\det\frac{\partial(f_q^1(p),\ldots,f_q^\nu(p))}{\partial(z_q^{m+1}(p),\ldots,z_q^n(p))}\neq 0,\qquad m=n-\nu,$$

for $p \in U(q)$. Thus for a sufficiently small $U(q)$, the map

$$z_q(p) = (z_q^1(p), \ldots, z_q^m(p), f_q^1(p), \ldots, f_q^\nu(p))$$

is biholomorphic by Theorem 1.19. Therefore we may use

$$p \to (z_q^1(p), \ldots, z_q^m(p), f_q^1(p), \ldots, f_q^\nu(p)), \qquad m = n - \nu,$$

as local coordinates with centre q. In terms of these, we have

$$S \cap U(q) = \{p \in U(q) \mid z_q^{m+1}(p) = \cdots = z_q^n(p) = 0\}. \tag{2.7}$$

Definition 2.5. A connected analytic subset S of M^n without singular points is called a complex submanifold of M.

For each $q \in M$, choose a coordinate polydisk $U_{R(q)}(q)$ with respect to the local coordinates z_q: $p \to z_q(p)$ such that $U_{R(q)}(q) \cap S = \varnothing$ for $q \neq S$ and that $U_{R(q)} \subset U(q)$ for $q \in S$. Then by Theorem 2.1, we may choose coordinate polydisks

$$U_j = U_{r(j)}(q_j) \subset U_{R(q_j)}(q_j), \qquad j = 1, 2, \ldots,$$

such that $\mathfrak{U} = \{U_j \mid j = 1, 2, \ldots\}$ is a locally finite open covering. For simplicity we write z_j for z_{q_j}, $f_j^k(p)$ for $f_{q_j}^k(p)$ and ν_j for $\nu(q_j)$. Then either $S \cap U_j = \varnothing$, or $q_j \in S$, and if $q_j \in S$,

$$S \cap U = \{p \in U_j \mid f_j^1(p) = \cdots = f_j^{\nu_j}(p) = 0\}. \tag{2.8}$$

In particular if S is a complex submanifold, we have

$$S \cap U_j = \{p \in U_j \mid z_j^{m+1}(p) = \cdots = z_j^n(p) = 0\}, \tag{2.9}$$

where $m = n - \nu_j$ is independent of j. In fact if $S \cap U_j \cap U_k \neq \varnothing$, it is clear that $m_j = m_k$. Then the assertion follows from the connectedness of S. From this result, we see that S is itself a complex manifold. For, if we put $V_j = S \cap U_j$, $V = \{V_j \mid V_j \neq \varnothing\}$ is a locally finite open covering of S, and

$$z_{jS}: p \to z_{jS}(p) = (z_j^1(p), \ldots, z_j^m(p)))$$

is a homeomorphism of V_j onto a polydisk in $\mathbb{C}^m$. The coordinate transformation

$$\tau_{jkS}: z_{kS}(p) \to z_{jS}(p)$$

is biholomorphic since it is the restriction of τ_{jk}:

$$(z_k^1, \ldots, z_k^m, \ldots, z_k^n) \to (z_j^1, \ldots, z_j^m, \ldots, z_j^n) \quad \text{to} \quad \{z_k^{m+1} = \cdots = z_k^n = 0\}.$$

Consequently $\{z_{jS} \mid V_j \neq \varnothing\}$ forms a system of local complex coordinates, hence, defines a complex structure on S. Thus S is a complex manifold of dimension m.

An analytic subset S of M^n is called an analytic hypersurface if it is defined by a single equation $\{f_q(p) = 0\}$ in some neighbourhood of each $q \in S$. In this case, choosing U_j as in (2.8), we have

$$S \cap U_j = \{p \in U_j \mid f_j(p) = 0\}, \tag{2.10}$$

where $f_j(p)$ is holomorphic in U_j. If we identify U_j with $\mathcal{U}_j = z_j(U_j) \subset \mathbb{C}^n$ as usual,

$$S \cap \mathcal{U}_j = \{z_j \in \mathcal{U}_j \mid f_j(z_j) = 0\}$$

is an analytic hypersurface of $\mathcal{U}_j$, where $f_j(p) = f_j(z_j)$ is a holomorphic function of z_j in $\mathcal{U}_j$. Then if we take $U_{R(q)}(q)$ sufficiently small, by the Corollary to Theorem 1.18, we may choose $f_j(z_j)$ such that $f_j(z_j) = 0$ is a minimal equation of $S \cap \mathcal{U}_j$ in $\mathcal{U}_j$. In this case $f_j(p) = 0$ is called a minimal equation of S in U_j. Then S is smooth at $q \in S \cap U_j$ if and only if at least one of the partial derivatives $\partial f_j(z_j)/\partial z_j^k$ of $f_j(z_j)$ does not vanish at $z_j = z_j(q)$. Consequently the set S' of all smooth points of S is an open subset of S. Let $S'' = S - S'$ be the set of singular points of S. Then $S'' \cap \mathcal{U}_j$ is defined by the system of holomorphic equations

$$f_j(z_j) = \frac{\partial f_j(z_j)}{\partial z_j^1} = \cdots = \frac{\partial f_j(z_j)}{\partial z_j^n} = 0.$$

Thus S'' is an analytic subset of M^n. $M^n - S''$ is a complex manifold, and each connected component of $S' = S - S''$ is a submanifold of $M^n - S''$.

(e) Meromorphic Functions

Let M^n be a complex manifold. For each point $q \in M^n$, fix local coordinates $z_q \colon p \to (z_q^1, \ldots, z_q^n) = z_q(p)$ with centre q. A holomorphic function $h(p)$ defined in a domain $D \subset M^n$ is expanded into a convergent power series $h_q(z_q) = h_q(z_q^1, \ldots, z_q^n)$ of $z_q^1, \ldots, z_q^n$ in a neighbourhood of each point $q \in D$. Namely we have $h(p) = h_q(z_q(p))$ in a neighbourhood of q. Suppose given holomorphic functions $h(p)$ and $g(p)$ in D. $h(p)$ and $g(p)$ are called relatively prime at q if their power series expansions $h_q(z_q)$ and $g_q(z_q)$ are relatively prime in $\mathbb{C}\{z_q^1, \ldots, z_q^n\}$. $h_q(z_q)$ and $g_q(z_q)$ have a common divisor in $\mathbb{C}\{z_q^1, \ldots, z_q^n\}$ if and only if there are holomorphic functions d, h, and g_1 defined in a small neighbourhood of q such that

$$h(p) = h_1(p)d(p), \qquad g(p) = g_1(p)d(p), \quad \text{and} \quad d(q) = 0. \quad (2.11)$$

Consequently the above definition does not depend on the choice of local coordinates z_q. Since $\mathbb{C}\{z_q^1, \ldots, z_q^n\}$ is a UFD, we may choose $d(p)$ in (2.11) such that h_1 and g_1 are relatively prime at q. This being the case, we call $d(p)$ the greatest common divisor of $h(p)$ and $g(p)$ at q. We say that $h(p)$ and $g(p)$ are relatively prime in D if they are so at every point of D. By Theorem 1.14, if h and g are relatively prime at a point $q \in D$, they are relatively prime in a sufficiently small neighbourhood $U(q) \subset D$ of q.

A function $f(p)$ on M^n is called a meromorphic function if for each point $q \in M^n$, there are holomorphic functions $h_q(p)$ and $g_q(p)$ defined in some neighbourhood of q such that $f(p) = h_q(p)/g_q(p)$ there. From the above argument we may assume that $h_q(p)$ and $g_q(p)$ are relatively prime at q, hence, also in some neighbourhood $U(q)$ of q. Consequently by Theorem 2.1, as in (d) above, there is a locally finite open covering $\mathfrak{U} = \{U_j | j = 1, 2, \ldots\}$ such that each U_j is a coordinate polydisk, and that on each U_j, $f(p)$ is represented as the quotient of two relatively prime holomorphic functions f_j and g_j:

$$f(p) = \frac{h_j(p)}{g_j(p)} \qquad \text{for } p \in U_j. \qquad (2.12)$$

For each pair j, k with $U_j \cap U_k \neq \emptyset$, we have

$$h_j(p)g_k(p) = h_k(p)g_j(p) \quad \text{for } p \in U_j \cap U_k. \qquad (2.13)$$

Since $h_{jq}(z_q)$ and $g_{jq}(z_q)$ are relatively prime, and also $h_{kq}(z_q)$ and $g_{kq}(z_q)$ are relatively prime, $h_{jq}(z_q)$ and $h_{kq}(z_q)$ are associates, and $g_{jq}(z_q)$ and $g_{kq}(z_q)$ are also associates. Therefore

$$f_{jk}(p) = \frac{h_j(p)}{h_k(p)} = \frac{g_j(p)}{g_k(p)}, \qquad p \in U_j \cap U_k,$$

is a non-vanishing holomorphic function defined in $U_j \cap U_k$, and we have

$$h_j(p) = f_{jk}(p)h_k(p) \quad \text{and} \quad g_j(p) = f_{jk}(p)g_k(p), \qquad p \in U_j \cap U_k.$$

Let $f(p) = h_q(p)/g_q(p)$ be a meromorphic function, and suppose $g_q(q) = 0$. In case $n = 1$, since $h_q(p)$ and $g_q(p)$ are relatively prime, $g_q(q) = 0$ implies that $h_q(q) \neq 0$, hence

$$f(p) = \frac{a_{-m}}{z_q^m} + \cdots + \frac{a_{-1}}{z_q} + \cdots, \quad \text{with } z_q = z_q(p).$$

Thus q is a pole of $f(p)$ and $f(q)=\infty$. Considering $f: p \to f(p)$ as a map of the Riemann surface M^1 to the Riemann sphere $\mathbb{P}^1 = \mathbb{C} \cup \{\infty\}$, and letting (ζ_0, ζ_1) be the homogeneous coordinates of $\mathbb{P}^1$, we see that the map

$$f: p \to (\zeta_0, \zeta_1) = (g_q(p), h_q(p))$$

is holomorphic.

In case $n \geqq 2$, even if h_q and g_q are relatively prime, we may have $h_q(q) = g_q(q) = 0$. This being the case, the value $f(q)$ of $f(p) = h_q/g_q$ at q cannot be determined. For example, put $f(z_1, z_2) = z_2/z_1$, which is a meromorphic function in $\mathbb{C}^2$. Then $f(0,0)$ cannot be determined.

(f) Differentiable Manifolds

A connected Hausdorff space Σ is called a topological manifold if there is an open covering of Σ consisting of at most countably many domains $U_1, \ldots, U_j, \ldots$, such that each U_j is homeomorphic to a domain $\mathcal{U}_j$ in $\mathbb{R}^m$. In this case the homeomorphism of U_j onto $\mathcal{U}_j$:

$$x_j: p \to x_j(p) = (x_j^1(p), \ldots, x_j^m(p))$$

is called local coordinates or a local coordinate system defined on U_j. The collection of local coordinates $\{x_j\} = \{x_1, \ldots, x_j, \ldots\}$ is called a system of local coordinates on the topological manifold Σ. For j, k such that $U_j \cap U_k \neq \emptyset$,

$$\tau_{jk}: x_k(p) \to x_j(p), \qquad p \in U_j \cap U_k,$$

is a homeomorphism of the open set $\mathcal{U}_{kj} = \{x_k(p) \mid p \in U_k \cap U_j\} \subset \mathcal{U}_k$ onto the open set $\mathcal{U}_{jk} = \{x_j(p) \mid p \in U_j \cap U_k\}$. We call $\{x_j\}$ a system of local C^∞ coordinates if these τ_{jk} are all C^∞, which means that $x_j^1(p), \ldots, x_j^m(p)$ are C^∞ functions of $x_k^1(p), \ldots, x_k^m(p)$. Suppose given two systems of local C^∞ coordinates $\{x_j\}$ and $\{u_\lambda\}$ on Σ, and let U_j be the domain of x_j and W_λ the domain of u_λ. If for any pair j, λ with $U_j \cap W_\lambda \neq \emptyset$, the maps

$$x_j(p) \to u_\lambda(p) \quad \text{and} \quad u_\lambda(p) \to x_j(p)$$

are both C^∞ for $p \in U_j \cap W_\lambda$, $\{x_j\}$ and $\{u_\lambda\}$ are said to be C^∞ equivalent.

Definition 2.6. A C^∞ differentiable structure on a topological manifold Σ is defined to be an equivalence class of systems of local C^∞ coordinates on Σ. A topological manifold Σ endowed with a differentiable structure is called a differentiable manifold, whose differentiable structure is called the differentiable structure of the differentiable manifold Σ. A system of local C^∞ coordinates belonging to the differentiable structure of Σ is called a system of local C^∞ coordinates on the differentiable manifold Σ.

Let Σ be a differentiable manifold, $\{x_j\}$ a system of local C^∞ coordinates on Σ, and U_j the domain of x_j. A real- or complex-valued function $f(p)$ defined in a domain D of Σ is represented on each $D \cap U_j \neq \emptyset$ by a function of local coordinates $(x_j^1, \ldots, x_j^m) = x_j = x_j(p)$ as $f(p) = f_j(x_j)$. We call $f(p)$ continuously differentiable, C^r, C^∞ in D if each $f_j(x_j)$ is continuously differentiable, C^r, C^∞, respectively. The differentiability of a map of a domain $D \subset \Sigma$ to another differentiable manifold is defined similarly as follows. Let T be a differentiable manifold of dimension n, $\{u_\lambda\}$ a system of local C^∞ coordinates on T, W_λ the domain of u_λ, and $\Phi: p \to q = \Phi(p)$ a continuous map of D to T. For λ, j with $\Phi^{-1}(W_\lambda) \cap U_j \neq \emptyset$, the map

$$\Phi_{\lambda j}: x_j(p) \to u_\lambda(\Phi(p)), \qquad p \in \Phi^{-1}(W_\lambda) \cap U_j,$$

is a continuous map of the open set $\mathcal{U}_{j\lambda} = x_j(\Phi^{-1}(W_\lambda) \cap U_j) \subset \mathbb{R}^m$ into the domain $\mathcal{W}_\lambda = u_\lambda(W_\lambda) \subset \mathbb{R}^n$. If for any pair j, λ with $\Phi^{-1}(W_\lambda) \cap U_j \neq \emptyset$, $\Phi_{\lambda j}$ is C^r, Φ is called a C^r map where $1 \leqq r \leqq \infty$. If $m = n$, Φ maps the domain $D \subset \Sigma$ homeomorphically onto a domain $E \subset T$, and both Φ and Φ^{-1} are C^∞, then Φ is called a diffeomorphism, and D is said to be diffeomorphic to E. We may identify two mutually diffeomorphic differentiable manifolds.

Suppose that a complex structure M is defined on a connected Hausdorff space Σ. For a system of local complex coordinates $\{z_j\}$ belonging to M, the domain U_j of z_j is homeomorphic to the image $\mathcal{U}_j = z_j(U_j)$ in $\mathbb{C}^n = \mathbb{R}^{2n}$. Consequently Σ is a topological manifold, which is called the underlying topological manifold of M. For local complex coordinates

$$z_j = z_j(p) = (z_j^1(p), \ldots, z_j^n(p)),$$

putting $z_j^\nu(p) = x_j^{2\nu-1}(p) + i x_j^{2\nu}(p)$, $\nu = 1, 2, \ldots, n$, we introduce local real coordinates

$$x_j: p \to x_j(p) = (x_j^1(p), \ldots, x_j^{2n}(p)).$$

Then $\{x_j\}$ forms a system of local C^∞ coordinates on Σ. Consequently $\{x_j\}$ defines a C^∞ structure on Σ, which makes Σ a differentiable manifold. This is called the underlying differentiable manifold of the complex manifold M. Also we call M a complex structure on the differentiable manifold Σ.

Conversely, let Σ be a Hausdorff space on which a differentiable structure is given, which makes Σ a differentiable manifold. Then a system of local complex coordinates $\{z_j\}$ on the Hausdorff space Σ is a complex structure on the differentiable manifold Σ if and only if each z_j maps its domain U_j diffeomorphically onto a domain $\mathcal{U}_j = z_j(U_j) \subset \mathbb{C}^n = \mathbb{R}^{2n}$.

§2.2. Compact Complex Manifolds

A complex manifold M is said to be compact if its underlying topological manifold Σ is compact. In this book we mainly treat compact complex manifolds.

Let M be a compact complex manifold. Then since M is covered by a finite number of coordinate neighbourhoods, we may choose a system of local complex coordinates on M consisting of a finite number of local coordinates $\{z_1, \ldots, z_N\}$. Let U_j be the domain of $z_j: p \to z_j(p)$, and put $z_j(U_j) = \mathcal{U}_j \subset \mathbb{C}^n$. We have $M = \bigcup_j U_j$. Identifying U_j with $\mathcal{U}_j$ as usual, we may consider $M = \bigcup_j \mathcal{U}_j$. Thus a compact complex manifold M is obtained by glueing a finite number of domains $\mathcal{U}_1, \ldots, \mathcal{U}_N$ in $\mathbb{C}^n$ via the identification of $z_k \in \mathcal{U}_{kj} \subset \mathcal{U}_j$ with $z_j = \tau_{jk}(z_k) \in \mathcal{U}_{jk} \subset \mathcal{U}_k$.

A holomorphic function defined on a compact complex manifold M is a constant.

Proof. Suppose that $f(p)$ is holomorphic on all of M. Since M is compact, the continuous function $|f(p)|$ attains its maximum at some point $q \in M$. Let $q \in U_j$, and put $f(p) = f_j(z_j)$ on U_j where z_j is a local coordinate system on U_j. Then $f_j(z_j) = f_j(z_j^1, \ldots, z_j^n)$ is a holomorphic function on $\mathcal{U}_j = z_j(U_j)$. We may assume that $\mathcal{U}_j$ is a polydisk with centre $c_j = z_j(q)$. Put

$$g(w) = f_j(c_j^1 + w(z_j^1 - c_j^1), \ldots, c_j^n + w(z_j^n - c_j^n)).$$

Then for $(z_j^1, \ldots, z_j^n) \in U_j$, $g(w)$ is a holomorphic function of w on $|w| < 1 + \varepsilon$ if ε is sufficiently small, and $|g(w)|$ attains its maximum at $w = 0$. Consequently, by the maximum principle, $g(w)$ is a constant. Thus $f_j(p)$ is a constant on U_j, and, by the analytic continuation, one sees that $f(p)$ is a constant on all of M. ∎

In this section we give several examples of compact complex manifolds. A compact complex manifold is, theoretically, determined if a finite number of domains U_j and biholomorphic mappings τ_{jk} which glue them. But except for a few special cases as that of $\mathbb{P}^n$ (Example 2.2), this method of construction is very complicated, hence is not practical. In the following we shall explain various methods of construction of a new compact complex manifold from given ones.

(a) Submanifolds

First we take $\mathbb{P}^n$, and investigate submanifolds of $\mathbb{P}^n$. In the following we denote a point ζ of $\mathbb{P}^n$ by its homogeneous coordinates $(\zeta_0, \ldots, \zeta_n)$. Let $P(\zeta_0, \ldots, \zeta_n)$ be a homogeneous polynomial of degree m, which we often denote simply by $P(\zeta)$. Since $P(\lambda\zeta_0, \ldots, \lambda\zeta_n) = \lambda^m P(\zeta_0, \ldots, \zeta_n)$, the equation $P(\zeta_0, \ldots, \zeta_n) = 0$ gives a well-defined subset of $\mathbb{P}^n$. An algebraic subset S of $\mathbb{P}^n$ is, by definition, a subset defined by a system of algebraic equations $P_1(\zeta) = \cdots = P_\kappa(\zeta) = 0$, where $P_1(\zeta), \ldots, P_\kappa(\zeta)$ are homogeneous polynomials. As stated in Example 2.2, $\mathbb{P}^n$ is obtained by glueing $(n+1)$ copies $\mathcal{U}_j$ of $\mathbb{C}^n$, $j = 0, 1, \ldots, n$: $P^n = \bigcup_{j=0}^n \mathcal{U}_j$. The local coordinates on $\mathcal{U}_j$ are given by $(z_j^0, \ldots, z_j^{j-1}, \ldots, z_j^n)$ where $z_j^k = \zeta_k/\zeta_j$. Therefore $S \cap \mathcal{U}_j$ is defined by the system of algebraic equations:

$$P_\nu(z_j^0, \ldots, z_j^{j-1}, z_j^{j+1}, \ldots, z_j^n) = 0, \qquad \nu = 1, 2, \ldots, \kappa \tag{2.14}$$

on $\mathbb{C}^n = \mathcal{U}_j$. Thus an algebraic subset of $\mathbb{P}^n$ is an analytic subset of $\mathbb{P}^n$. An algebraic subset M which is a complex submanifold of $\mathbb{P}^n$ is called a projective algebraic manifold. In this case, $M \cap \mathcal{U}_j$ is a complex submanifold of $\mathbb{C}^n = \mathcal{U}_j$ defined by (2.14). In general a complex submanifold of $\mathbb{C}^n$ which is defined by a system of algebraic equations is called an affine algebraic manifold. Thus a projective algebraic manifold M is obtained by glueing a finite number of affine algebraic manifolds $M \cap \mathcal{U}_j = M_j$: $M = \bigcup_{j=1}^n M_j$. In this book, by an algebraic manifold we always mean a projective algebraic manifold unless otherwise mentioned. A (projective) algebraic manifold is obviously compact.

Let $P(\zeta) = P(\zeta_0, \ldots, \zeta_n)$ be a homogeneous polynomial of degree m. The algebraic subset S of $\mathbb{P}^n$ defined by a single equation $P(\zeta) = 0$ is called a hypersurface of degree m. Put $P_{\zeta_k}(\zeta) = \partial P(\zeta)/\partial \zeta_k$. Then we have

$$\zeta_0 P_{\zeta_0}(\zeta) + \zeta_1 P_{\zeta_1}(\zeta) + \cdots + \zeta_n P_{\zeta_n}(\zeta) = mP(\zeta).$$

Consequently, if for any $\zeta \in P^n$, at least one of $P_{\zeta_k}(\zeta)$ does not vanish, then at every point $(1, z^1, \ldots, z^n) \in S \cap \mathcal{U}_0$, at least one of $\partial P(1, z^1, \ldots, z^n)/\partial z^k$ does not vanish. Similar result holds for $S \cap \mathcal{U}_k$ with $k = 1, \ldots, n$. Therefore in this case S is non-singular. Moreover S is proved to be connected, hence, an algebraic manifold. Proof of the connectedness of S is omitted.

By Chow's theorem ([2]), any analytic subset of $\mathbb{P}^n$ is an algebraic subset. Thus any complex submanifold of $\mathbb{P}^n$ is an algebraic submanifold. An algebraic manifold of dimension 1 is said to be an algebraic curve, and an algebraic manifold of dimension 2 an algebraic surface. An algebraic curve is a compact Riemann surface.

Example 2.3. Let C be an algebraic curve in $\mathbb{P}^2$ defined by the equation $\zeta_0^m + \zeta_1^m + \zeta_2^m = 0$. We show that the genus g of C as a compact Riemann surface is given by $g = \frac{1}{2}m(m-3)+1$. Let (z_1, z_2) be the inhomogeneous coordinates of ζ, where $z_1 = \zeta_1/\zeta_0$, $z_2 = \zeta_2/\zeta_0$. Then C is given by $1 + z_1^m + z_2^m = 0$ on $\mathcal{U}_0$. If $1 + z_1^m \neq 0$, $z_2 = (-1 - z_1^m)^{1/m}$ is a (multi-valued) holomorphic function of z_1. Therefore we may use z_1 as a local coordinate on C if $1 + z_1^m \neq 0$. Similarly on a neighbourhood of $\mathfrak{p}_k = (1, e^{2\pi i k/m + \pi i m}, 0)$, $k =$

$1, 2, \ldots, m$, z_2 can be used as a local coordinate on C. C intersects with the line $\zeta_0 = 0$ at the m points $\mathfrak{q}_k = (0, e^{\pi i m}, e^{2\pi i k/m})$ with $k = 1, \ldots, m$. We use $w = \zeta_0/\zeta_1 = 1/z_1$ as a local coordinate on C in a neighbourhood of $\mathfrak{q}_k$. Consider the Abelian differential $\omega = dz_1$ on C. Since $1 + z_1^m + z_2^m = 0$, we have $z_1^{m-1}\, dz_1 + z_2^{m-1}\, dz_2 = 0$. Hence, $\omega = -z_2^{m-1} z_1^{1-m}\, dz_2$ in a neighbourhood of $\mathfrak{p}_k$. Thus $\mathfrak{p}_k$ is a zero of order $(m-1)$ of ω. In the neighbourhood of $\mathfrak{q}_k$, $\omega = -w^{-2}\, dw$, hence $\mathfrak{q}_k$ is a pole of order 2 of ω. Hence the divisor of $\mathfrak{k}$ of ω is given by

$$\mathfrak{k} = (\omega) = \sum_{k=1}^{m} (m-1)\mathfrak{p}_k - \sum_{k=1}^{m} 2\mathfrak{q}_k.$$

Thus $2g - 2 = \deg \mathfrak{k} = m(m-1) - 2m$, that is, $g = \frac{1}{2}m(m-3) + 1$.

Example 2.4. Let C be an algebraic curve in $\mathbb{P}^3$ defined by $\{\zeta_1\zeta_2 - \zeta_0\zeta_3 = \zeta_0\zeta_2 - \zeta_1^2 = \zeta_2^2 - \zeta_1\zeta_3 = 0\}$. Let (t_0, t_1) be the homogeneous coordinates of $\mathbb{P}^1$. Then the map

$$(t_0, t_1) \to (\zeta_0, \zeta_1, \zeta_2, \zeta_3) = (t_0^3, t_0^2 t_1, t_0 t_1^2, t_1^3)$$

maps $\mathbb{P}^1$ biholomorphically onto C. Therefore C is analytically isomorphic to $\mathbb{P}^1$.

Example 2.5. The equation $\zeta_0^m + \cdots + \zeta_3^m = 0$ defines an algebraic surface in $\mathbb{P}^3$ since at every point $\zeta \in \mathbb{P}^3$, at least one of $\partial(\zeta_0^m + \zeta_1^m + \zeta_2^m + \zeta_3^m)/\partial\zeta_k = m\zeta_k^{m-1}$, $k = 0, 2, 3$, does not vanish. We only give a calculation of the Euler number of S, putting aside various interesting properties of S. Let

$$\Phi: (\zeta_0, \zeta_1, \zeta_2, \zeta_3) \to (\zeta_0, \zeta_1, \zeta_2, 0)$$

be the projection. Then the restriction Φ_S of Φ to Σ is a holomorphic map of S onto the plane $\mathbb{P}^2$ defined by $\zeta_3 = 0$. Φ_S is m-to-one on $\zeta_0^m + \zeta_1^m + \zeta_2^m \neq 0$, and one-to-one on $\zeta_0^m + \zeta_1^m + \zeta_2^m = 0$. Let C be the algebraic curve in $\mathbb{P}^2$ defined by $\zeta_0^m + \zeta_1^m + \zeta_2^m = 0$. Then S is an m-fold branched covering of $\mathbb{P}^2$ with C as its branch locus of order $(m-1)$. Then denoting by $\chi(M)$ the Euler number of a manifold M, we have

$$\chi(S) = m\chi(\mathbb{P}^2) - (m-1)\chi(C).$$

Substituting $\chi(\mathbb{P}^2) = 3$ and $\chi(C) = 2 - 2g = m(3-m)$, we obtain

$$\chi(S) = m(m^2 - 4m + 6).$$

In general, let M^m be a complex submanifold of a complex manifold $W = W^n$. Then for given $q \in M$, we can choose local coordinates $w_q: p \to$

$w_q(p)=(w_q^1(p),\ldots,w_q^n(p))$ of W^n with centre q on a coordinate polydisk $U(q)$ such that

$$M\cap U(q)=\{p\in U(q)\,|\,w_q^{m+1}(p)=\cdots=w_q^n(p)=0\}.$$

Then the map $p\to(w_q^1(p),\ldots,w_q^m(p))$ gives local coordinates of M centred at q. Let $f(p)$ be a holomorphic function defined in a domain D of W^n. Then the restriction $f_M(p)$ of $f(p)$ to M is a holomorphic function in $M\cap D$. For, if we represent $f(p)$ by a holomorphic function of $(w_q^1,\ldots,w_q^n)$ in a neighbourhood of $q\in M\cap D$ as

$$f(p)=f_q(w_q^1,\ldots,w_q^n),$$

we have

$$f_M(p)=f_q(w_q^1,\ldots,w_q^m,0,\ldots,0).$$

Let $f(p)$ be a meromorphic function on W^n. For any $q\in W^n$, we can choose a sufficiently small $U(q)$ such that $f(p)=h_q(p)/g_q(p)$ on $U(q)$ where $h_q(p)$ and $g_q(p)$ are relatively prime holomorphic functions. If $U(q_1)\cap U(q_2)\neq\varnothing$, by (2.13) there is a non-vanishing holomorphic function $u(p)$ on $U(q_1)\cap U(q_2)$ such that $g_{q_1}(p)=u(p)g_{q_2}(p)$ there. If $q\in M$, the restrictions $h_{qM}(p)$, $g_{qM}(p)$ of $h_q(p)$, $g_q(p)$, respectively, to M are holomorphic in $M\cap U(q)$. If $g_{qM}(p)$ vanishes identically for some $q\in M$, then for all $q\in M$, $g_{qM}(p)$ vanishes identically. In fact, for $q_1, q_2\in M$ with $M\cap U(q_1)\cap U(q_2)\neq\varnothing$, we have $g_{q_1M}(p)=u_M(p)g_{q_2M}(p)$ on $M\cap U(q_1))\cap U(q_2))$ where $u_M(p)$ is non-vanishing. Therefore if $g_{q_1M}(p)=0$ identically, by the analytic continuation we obtain $g_{q_2M}(p)=0$ identically. If $g_{qM}(p)$ does not vanish identically, we have

$$f_M(p)=h_{qM}(p)/g_{qM}(p).$$

Thus *given a meromorphic function $f(p)=h_q(p)/g_q(p)$ on W^n, its restriction $f_M(p)$ to a submanifold $M\subset W^n$ is a meromorphic function on M unless $g_q(p)$ vanishes identically on M.*

Suppose given an algebraic manifold $M=M^m\subset\mathbb{P}^n$. Let $P(\zeta)$ and $Q(\zeta)$ be homogeneous polynomials of the same degree. We call $f(\zeta)=P(\zeta)/Q(\zeta)$ a rational function. Since a rational function $f(\zeta)$ is meromorphic on $\mathbb{P}^n$, *the restriction $f_M(\zeta)$ to M is a meromorphic function on M unless $Q(\zeta)\equiv 0$.* In particular $(\zeta_k/\zeta_j)_M$ is a meromorphic function on M unless $\zeta_j\equiv 0$ on M.

Let M be an algebraic manifold, and $q\in M$. *We may choose meromorphic functions $z_q^1(p),\ldots,z_q^m(p)$ on M such that $(z_q^1(p),\ldots,z_q^m(p))$ give local coordinates with centre q.*

Proof. We may assume that $M\cap\mathcal{U}_0\neq\varnothing$, and that $q\in M\cap\mathcal{U}_0$. Let $(z_q^1,\ldots,z_q^n)=(z_q^1(p),\ldots,z_q^n(p))$ be the non-homogeneous coordinates of

$p \in \mathscr{U}_0$. We choose local coordinates $w_q: p \to (w_q^1, \ldots, w_q^n) = w_q(p)$ such that

$$M \cap U(q) = \{p \mid w_q^{m+1} = \cdots = w_q^N = 0\}.$$

Since

$$\det \frac{\partial(z^1, \ldots, z^m, \ldots, z^n)}{\partial(w_q^1, \ldots, w_q^m, \ldots, w_q^n)} \neq 0,$$

after renumbering z_i if necessary, we may assume that

$$\det \frac{\partial(z^1, \ldots, z^m)}{\partial(w_q^1, \ldots, w_q^m)} \neq 0 \quad \text{at } q.$$

Put $z_q^k = z^k(p) - z^k(q)$ for $k = 1, \ldots, m$. Then since $(w_q^1, \ldots, w_q^m)$ are local coordinates on M with centre q, $p \to (z_q^1(p), \ldots, z_q^m(p))$ are also local coordinates on M with centre q, and each $z_q^k(p)$ is extended to a meromorphic function $(\zeta_k/\zeta_0)_M - z^k(q)$ on M. ∎

Thus there exist abundant meromorphic functions on an algebraic manifold.

(b) Quotient Space

Let W be a complex manifold. By an automorphism of W, we mean a biholomorphic map of W onto itself. In other words, an automorphism is a map which does not alter the complex structure of W. When we define the product of two automorphisms g_1, g_2 of W by their composite $g_1 g_2$, the set of all automorphisms of W forms a group, which we denote by $\mathscr{G}$. The unit of $\mathscr{G}$ is the identity of W, and the inverse of $g \in \mathscr{G}$ is the inverse map g^{-1} of g. Any subgroup of $\mathscr{G}$ is called a group of automorphisms of W. Let G be a group of automorphisms of W. For $p \in W$, the set $Gp = \{g(p) \mid g \in G\}$ is called the orbit of G through p. Two orbits Gp and Gq do not have a common element unless they coincide. Thus W is decomposed into the mutually disjoint orbits of G. The set of all orbits of G is called the quotient space of W by G, which we denote by W/G. We may consider that W/G is obtained from W by identifying $p \in W$ with $q \in W$ if there is an element $g \in G$ such that $q = g(p)$.

Example 2.6. Let $W = \mathbb{C}^{n+1} - (0, \ldots, 0)$, and $(\zeta_0, \ldots, \zeta_n)$ coordinates on $\mathbb{C}^{n+1}$. We denote by $\mathbb{C}^*$ the multiplicative group of all non-zero complex numbers. Any $g \in \mathbb{C}^*$ defines an automorphism of W via

$$g: (\zeta_0, \ldots, \zeta_n) \to (g\zeta_0, \ldots, g\zeta_n).$$

Thus $\mathbb{C}^*$ becomes a group of automorphisms of W. An orbit $\mathbb{C}^*(\zeta_0, \ldots, \zeta_n)$ of $\mathbb{C}^*$ on W is a complex line ζ on $\mathbb{C}^{n+1}$ with the origin deleted. Hence $W/\mathbb{C}^* = \mathbb{P}^n$.

In the sequel we explain a method of constructing a compact complex manifold as a quotient space of a given complex manifold. In general, the quotient space W/G of a complex manifold is not a complex manifold. In order that W/G may be a complex manifold, G must satisfy certain conditions. Let g be an automorphism of W. A point p of W is called a fixed point of g if $g(p) = p$. We say that G is fixed point free if any $g \in G$ except the identity has no fixed point.

Definition 2.7. An group G of automorphisms of W is called properly discontinuous if for any compact sets K_1, K_2 of W, there are only a finite number of elements $g \in G$ such that $g(k_1) \cap K_2 \neq \varnothing$.

If G is properly discontinuous, each orbit G_p is a discrete subset of W.

Theorem 2.2. *Let G be a group of automorphisms of a complex manifold W. If G is fixed point free and properly discontinuous, W/G has a canonical structure of a complex manifold induced from that of W.*

Proof. Put $\hat{W} = W/G$, and $\hat{p} = Gp$. For each $q \in W$, choose local coordinates $z_q\colon p \to z_q(p) = (z_q^1(p), \ldots, z_q^n(p))$ with centre q. Take $r > 0$ such that the closed polydisk $\{(z_q^1, \ldots, z_q^n) \mid |z_q^k| \leqq r, k = 1, \ldots, n\}$ is contained in the range of z_q, and put

$$U_r(q) = \{p \mid |z_q^1(p)| < r, \ldots, |z_1^n(p)| < r\}.$$

Given $q \in W$, if we choose r sufficiently small, we have $g(U_r(q)) \cap U_r(q) = \varnothing$ for any $g \in G$ except the identity. In fact, if otherwise, there is an element $g_n \in G$, $g_n \neq 1$, for each $n = 1, 2, \ldots$, such that $g_n(U_n) \cap U_n \neq \varnothing$, where $U_n = U_{r/n}(q)$. Then $g_n(U_1) \cap U_1 \neq \varnothing$ for any n. Since $[U_1]$ is compact, and G is properly discontinuous, $\{g_1, \ldots, g_n, \ldots\}$ must be a finite subset of G. Therefore we can find at least one g_i, say, g_1 such that $g_1(U_n) \cap U_n \neq \varnothing$ for infinitely many n. Then $g_1(q) = q$, hence, q is a fixed point of $g_1 \neq 1$, which contradicts the assumption. Consequently for sufficiently large n, $g(U_n) \cap U_n = \varnothing$ for any $g \in G$ with $g \neq 1$. Writing simply r instead of r/n, we have $g(U_r(q)) \cap U_r(q) = \varnothing$ as desired.

Thus for sufficiently small $r > 0$, $g(U_r(q)) \cap U_r(q) = \varnothing$ for $g \in G$, $g \neq 1$. Hence $Gp \cap U_r(q) = \{p\}$ for any $p \in U_r(q)$. Consequently the map $p \to \hat{p}$ is injective on $U_r(q)$. By the definition of the quotient topology, a subset $\hat{U}$ of $\hat{W}$ is open if and only if its inverse image by the above map $p \to \hat{p}$ is open in W. Thus $\hat{U}_r(\hat{q}) = \{\hat{p} \mid p \in U_r(q)\}$ is an open set, and the map $p \to \hat{p}$ maps $U_r(q)$ homeomorphically onto $\hat{U}_r(\hat{q})$. Thus $\hat{W}$ is a topological manifold.

For each point $q \in W$, fix a coordinate polydisk $U_r(q)$ with $r = r(q)$ satisfying the above condition. Then by Theorem 2.1, we can choose coordinate polydisks $U_j = U_j(q_j)$ with $0 < r_j \leqq r(q_j)$, $j = 1, 2, \ldots$, such that $\mathfrak{U} = \{U_j | j = 1, 2, \ldots\}$ forms a locally finite open covering of W. Then $\hat{W}$ is covered by $\hat{U}_j = \{\hat{p} | p \in U_j\}, j = 1, 2, \ldots$. Since the map $p \to \hat{p}$ maps U_j homeomorphically onto $\hat{U}_j$, the map $z_j : \hat{p} \to z_j(\hat{p})$ defined by

$$z_j(\hat{p}) = z_{q_j}(p) \quad \text{for } p \in U_j$$

is a homeomorphism of $\hat{U}_j$ into C^n. Suppose $\hat{U}_j \cap \hat{U}_k \neq \emptyset$, and take an arbitrary point $\hat{p}_0 \in \hat{U}_j \cap \hat{U}_k$ with $p_0 \in U_j$. Then there is a unique element $g \in G$ such that $gp_0 \in U_k$, and in a sufficiently small neighbourhood of $\hat{p}_0$, we have

$$z_j(\hat{p}) = z_{q_j}(p), \qquad z_k(\hat{p}) = z_{qk}(gp), \qquad p \in U_j.$$

Since g is biholomorphic, the map $z_j(\hat{p}) \to z_k(\hat{p})$ is also biholomorphic. Thus $\{z_1, z_2, \ldots\}$ forms a system of local complex coordinates on $\hat{W}$, and defines a complex structure on $\hat{W}$. ∎

Thus if G is properly discontinuous and fixed point free, the quotient space $\hat{W} = W/G$ becomes a complex manifold. The map $p \to \hat{p}$, which is represented as

$$z_{q_j}(p) \to z_j(\hat{p}) = z_{q_j}(p) \quad \text{on } U_j$$

in terms of the local coordinates defined above, is a holomorphic map of W onto $\hat{W}$, and biholomorphic on each coordinate neighbourhood U_j. We express this fact by saying that W is a covering manifold of $\hat{W}$. We call G the covering transformation group of W over $\hat{W}$.

If there is a compact subset K of W such that $\hat{W} = \hat{K} = \{\hat{p} | p \in K\}$, $\hat{W}$ is compact. Thus, taking a suitable group of automorphisms G of a complex manifold W, which is properly discontinuous and fixed point free, we can construct a new compact complex manifold W/G. Note that $\hat{K} = \hat{W}$ if and only if $W = \bigcup_{g \in G} g(K)$.

Example 2.7. Put $W = \mathbb{C}$ and fix a complex number ω with $\operatorname{Im} \omega > 0$. Let G be the group consisting of all parallel translations of the form $g_{mn}: z \to z + m\omega + n$, with $m, n \in \mathbb{Z}$. G is properly discontinuous and fixed point free on W. Let F be the parallelogram in $\mathbb{C}$ with vertices 0, 1, $\omega + 1$, and ω. Then $\bigcup g_{mn}(F) = \mathbb{C}$. Consequently $C = \mathbb{C}/G$ is a compact Riemann surface. C is obtained from F by identifying x, $0 \leqq x \leqq 1$, with $x + \omega$, and $t\omega$, $0 \leqq t \leqq 1$, with $t\omega + 1$. Hence C is diffeomorphic to a torus, and the genus of C is equal to 1.

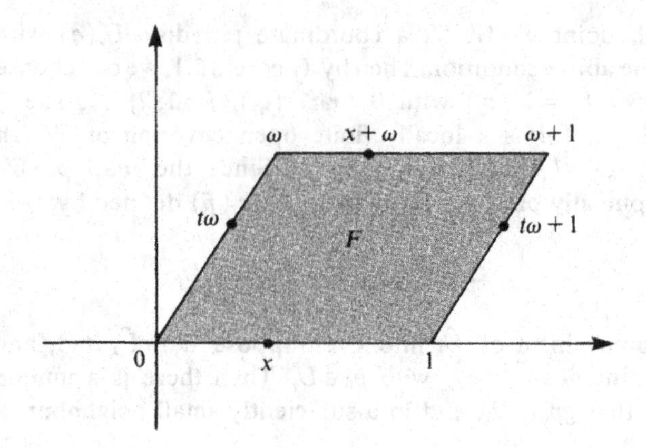

Figure 1

As is well known in the classical theory of elliptic functions, C is represented as a cubic curve in $\mathbb{P}^2$. We explain this briefly below (for details, see, for instance, [12], Absch. II, Kap. I). Put $\omega_{mn} = m\omega + n$, and define

$$\wp(z) = \frac{1}{z^2} + \sum_{(m,n)\neq(0,0)} \left(\frac{1}{(z-\omega_{mn})^2} - \frac{1}{\omega_{mn}^2} \right).$$

$\wp(z)$ is a meromorphic function on $\mathbb{C}$, called the Weierstrass $\wp$-function. $\wp(z)$ is G-invariant, that is, $\wp(z+\omega_{mn}) = \wp(z)$ for any ω_{mn}. Hence $\wp(z)$ is considered as a meromorphic function on $C = \mathbb{C}/G$. $\wp(z)$ has a pole of order 2 at each ω_{mn}, and holomorphic elsewhere. Therefore as a meromorphic function on C, $\wp(z)$ has a pole of order 2 at the point $\hat{0} \in C$ corresponding to $0 \in \mathbb{C}$, and holomorphic elsewhere. $\wp(z)$ satisfies the following differential equation:

$$\wp'(z)^2 - 4\wp(z)^3 + g_2\wp(z) + g_3 = 0, \tag{2.15}$$

where the coefficients g_2 and g_3 are given by

$$g_2 = 60 \sum_{(m,n)\neq 0} \frac{1}{\omega_{mn}^4}, \qquad g_3 = 140 \sum_{(m,n)\neq 0} \frac{1}{\omega_{mn}^6}.$$

In view of (2.15), put

$$P(\zeta) = \zeta_0\zeta_1^2 - 4\zeta_2^3 + g_2\zeta_0^2\zeta_2 + g_3\zeta_0^3.$$

Since $g_2^3 - 27g_3^2 \neq 0$, it is easy to see that for any $\zeta \in \mathbb{P}^2$, at least one of the partial derivatives of $P(\zeta)$ does not vanish. Hence $P(\zeta) = 0$ defines an algebraic curve Γ in $\mathbb{P}^2$. Consider the mapping Φ of C into $\mathbb{P}^2$ defined by

$$\Phi: z \to \zeta = \varphi(z) = (1, \wp'(z), \wp(z)).$$

Φ is clearly holomorphic except at $\hat{0}$. Since $\wp(z)$ has a pole of order 2 and $\wp'(z)$ has a pole of order 3 at $z=0$, writing $\Phi(z)$ as $\Phi(z) = (z^3, z^3\wp'(z), z^3\wp(z))$, we see that Φ is also holomorphic at $\hat{0}$. It is obvious from (2.15) that $\Phi(C) \subset \Gamma$. By more detailed investigation of the properties of $\wp(z)$, we can prove that Φ is a biholomorphic map of C onto Γ. Thus C may be identified with the cubic curve Γ in $\mathbb{P}^2$ as a Riemann surface. We call C an elliptic curve. If we put $\alpha = e^{2\pi i\omega}$, then $e^{2\pi i(m\omega+n)} = \alpha^m$. Therefore via the holomorphic map $z \to w = e^{2\pi iz}$, which maps $\mathbb{C}$ onto $\mathbb{C}^*$, g_{mn} induces an automorphism g^*_m of $\mathbb{C}^*$: $w \to \alpha^m w$. Put $G^* = \{g^*_m | m \in \mathbb{Z}\}$, which corresponds to G via the above map. Since $\operatorname{Im} \omega > 0$, $0 < |\alpha| < 1$. Therefore G^* is properly discontinuous and fixed point free on $\mathbb{C}^*$. Clearly $C = \mathbb{C}/G = \mathbb{C}^*/G^*$. Let F^* be the closed annulus $\{w \| \alpha | \leqq |w| \leqq 1, w \in \mathbb{C}^*\}$. Then C is obtained from F^* by identifying the points w and αw on the boundary of F^* where $|w| = 1$.

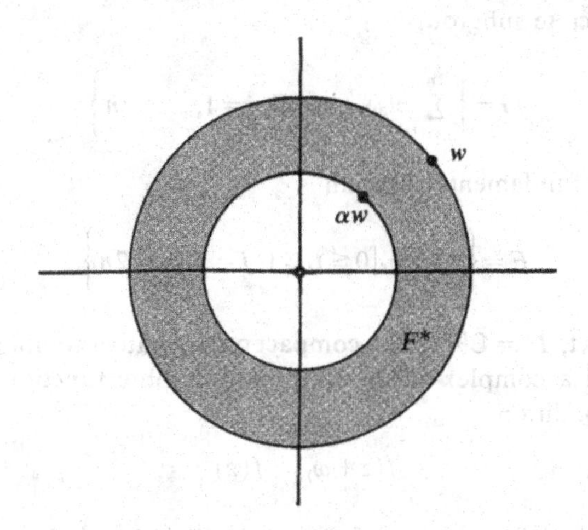

Figure 2

In general suppose given a group of automorphisms G of a complex manifold W, which is properly discontinuous and fixed point free. By a fundamental domain of G, we mean a closed domain $F \subset W$ satisfying the following conditions:

(i) $F = [(F)]$, where (F) denotes the interior of F.
(ii) If $p \in (F)$, $Gp \cap F = \{p\}$.
(iii) $\bigcup_{g \in G} g(F) = W$.

If F is a fundamental domain of G, the map $p \to \hat{p}$ maps (F) bijectively onto $\{\hat{p} | p \in (F)\}$, and F onto $\hat{W} = W/G$. Therefore if F is compact, $\hat{W}$ is also compact. In Example 2.7 above F and F^* are fundamental domains of G and G^* respectively.

A complex manifold W is called a complex Lie group if W is a group and the map of $W \times W$ to W defined by the group multiplication $(q, p) \to qp^{-1}$ is holomorphic. If W is a complex Lie group, identifying $q \in W$ with the automorphism $p \to qp$ of W, we see that W is a group of automorphisms of W itself. A subgroup G of W is called a discrete subgroup of W if it is a discrete subset of W. A discrete subgroup G acts on W in a properly discontinuous manner without fixed point. In fact, for a compact subsets K_1, K_2 of W, the set $\{g \in G | gK_1 \cap K_2 \neq \emptyset\}$ is contained in the compact subset $K_2K_1^{-1} = \{qp^{-1} | q \in K_2, p \in K_1\}$. Since clearly G is fixed point free, $\hat{W} = W/G$ is a complex manifold and has a group structure as the quotient group of W by G. Thus $\hat{W}$ is a complex Lie group.

Example 2.8. A complex vector space $\mathbb{C}^n$ is a complex Lie group with respect to the usual addition. Take $2n$ vectors $\omega_j = (\omega_j^1, \ldots, \omega_j^n) \in \mathbb{C}^n$ for $j = 1, \ldots, 2n$, such that these ω_j are linearly independent over $\mathbb{R}$. Then ω_j generate a discrete subgroup

$$G = \left\{ \sum_{j=1}^{2n} m_j \omega_j \,\middle|\, m_j \in \mathbb{Z}, j = 1, \ldots, 2n \right\}$$

of $\mathbb{C}^n$. Since a fundamental domain

$$F = \left\{ \sum_{j=1}^{2n} t_j \omega_j \,\middle|\, 0 \leqq t_j \leqq 1, j = 1, \ldots, 2n \right\}$$

of G is compact, $T^n = \mathbb{C}^n/G$ is a compact commutative complex Lie group, which we call a complex torus. If a meromorphic function $f(z)$ on $\mathbb{C}^n$ satisfies the condition

$$f(z + \omega_j) = f(z)$$

for $1 \leqq j \leqq 2n$ and for any $z \in \mathbb{C}^n$, $f(z)$ is called a periodic meromorphic function with the periods $\omega_1, \ldots, \omega_{2n}$. Such $f(z)$ gives a meromorphic function on T^n, which we denote also by $f(z)$. $\omega_1, \ldots, \omega_{2n}$ is called the periods of T^n, and the matrix

$$\Omega = \begin{pmatrix} \omega_1^1 & \cdots & \omega_1^n \\ \omega_2^1 & \cdots & \omega_2^n \\ \cdots & \cdots & \cdots \\ \omega_{2n}^1 & \cdots & \omega_{2n}^n \end{pmatrix}$$

is called the period matrix of T^n.

For $n=1$, as is stated in Example 2.7 above, $T^1 = C = \mathbb{C}/G$ is always an algebraic curve.

In case $n \geqq 2$, a complex torus $T^n = \mathbb{C}^n/G$ is not necessarily an algebraic manifold. Let J be an invertible alternating real $2n \times 2n$ matrix. Then $\sqrt{-1}\,{}^t\bar{\Omega}J^{-1}\Omega$ is a Hermitian matrix. By $\sqrt{-1}\,{}^t\bar{\Omega}J^{-1}\Omega > 0$, we mean that this matrix is positive definite. Ω is called a Riemann matrix if there exists a $2n \times 2n$ integral alternating matrix J satisfying the following conditions:

(i) ${}^t\Omega J^{-1}\Omega = 0$.
(ii) $\sqrt{-1}\,{}^t\bar{\Omega}J^{-1}\Omega > 0$.

T^n is an algebraic manifold if and only if its period matrix Ω is a Riemann matrix (*see* [28]). In this case we call T^n an Abelian variety. In general the period matrix of a complex torus T^n with $n \geqq 2$ is not a Riemann matrix. Moreover it is known that there exist no non-constant meromorphic functions on most general complex tori T^n. For example, the complex torus T^2 with the period matrix

$$\Omega = \begin{pmatrix} \sqrt{-5} & \sqrt{-7} \\ \sqrt{-2} & \sqrt{-3} \\ 1 & 0 \\ 0 & 1 \end{pmatrix}$$

has no non-constant meromorphic functions ([28], p. 104).

Example 2.9. T^n is an obvious generalization of an elliptic curve $C = \mathbb{C}/G$ given in Example 2.7 to the n-dimensional case. Here we give another generalization of C, considering C as $\mathbb{C}^*/G^*$.

Let $W = \mathbb{C}^n - \{0\}$, and G the infinite cyclic group generated by the automorphism

$$g\colon z = (z_1, \ldots, z_n) \to g(z) = (\alpha_1 z_1, \ldots, \alpha_n z_n)$$

of W, where $\alpha_1, \ldots, \alpha_n$ are constants with $|\alpha_1| > 1, \ldots, |\alpha_n| > 1$. G acts on W in a properly discontinuous manner without fixed point. The quotient space $M = W/G$ is called a Hopf manifold ([11]). *M is diffeomorphic to the product $S^1 \times S^{2n-1}$*. We give a proof of this only for $n=2$ below, but the generalization is straightforward. Let

$$S^3 = \{(\zeta_1, \zeta_2) \in \mathbb{C}^2 \,|\, |\zeta_1|^2 + |\zeta_2|^2 = 1\}.$$

Putting $\alpha_1 = e^{\beta_1}$ and $\alpha_2 = e^{\beta_2}$, consider a C^∞ map Φ of $\mathbb{R} \times S^3$ to W defined by

$$\Phi\colon (t, \zeta_1, \zeta_2) \to (z_1, z_2) = (\zeta_1 e^{t\beta_1}, \zeta_2 e^{t\beta_2}).$$

Then Φ is bijective. In fact, since $|\alpha_1|>1$ and $|\alpha_2|>1$, $r_1=\operatorname{Re}\beta_1>0$ and $r_2=\operatorname{Re}\beta_2>0$. Put

$$N(t)=|z_1\,e^{-t\beta_1}|^2+|z_2\,e^{-t\beta_2}|^2=|z_1|^2\,e^{-2r_1t}+|z_2|^2\,e^{-2r_2t}.$$

Then $N(t)$ is a monotonously decreasing function of t. Moreover $N(t)\to 0$ if $t\to\infty$, and $N(t)\to+\infty$ if $t\to-\infty$. Hence there is a unique t such that $N(t)=1$. Putting $\zeta_1=z_1\,e^{-t\beta_1}$, and $\zeta_2=z_2\,e^{-t\beta_2}$ for this t, we get the unique solution (t,ζ_1,ζ_2) of the system of equations $\zeta_1\,e^{t\beta_1}=z_1$, $\zeta_2\,e^{t\beta_2}=z_2$ with $|\zeta_1|^2+|\zeta_2|^2=1$. Clearly

$$\Phi^{-1}\colon(z_1,z_2)\to(t,\zeta_1,\zeta_2)\in\mathbb{R}\times S^3$$

is C^∞. Hence W is diffeomorphic to $\mathbb{R}\times S^3$. The automorphism g^m, $m\in\mathbb{Z}$, of W corresponds via Φ to that of $\mathbb{R}\times S^3$ given by

$$(t,\zeta_1,\zeta_2)\to(t+m,\zeta_1,\zeta_2).$$

Therefore we obtain a desired diffeomorphism $W/G\simeq\mathbb{R}/\mathbb{Z}\times S^3=S^1\times S^3$.

Thus a Hopf manifold $M=W/G$ is diffeomorphic to $S^1\times S^{2n-1}$, hence, M is compact, and its first Betti number is equal to 1. From the theory of harmonic differential forms we know that the first Betti number of an algebraic manifold is even (see [14], p. 346). Consequently *a Hopf manifold is not an algebraic manifold.* A complex torus is, in general, not an algebraic manifold, but it has the same topological structure as algebraic complex tori, while in the case of Hopf manifolds, even their topological structures are different from those of algebraic manifolds.

In the sequel we denote by $\hat{z}$ the point of $M=W/G$ corresponding to $z\in W$. For a meromorphic function $f(\hat{z})$ on M, putting $f(z)=f(\hat{z})$, we obtain a G-invariant meromorphic function $f(z)$ on W: $f(gz)=f(z)$ for $g\in G$. By Levi's theorem ([27], Band II, p. 220), any meromorphic function on W extends to a meromorphic function on all of $\mathbb{C}^n$. Thus $f(z)$ extends to a meromorphic function on $\mathbb{C}^n$, which we denote also by $f(z)$. From (e) of the preceding section, there are a neighbourhood $U_\varepsilon(0)$ of 0 and relatively prime holomorphic functions $\varphi(z)$ and $\psi(z)$ defined in $U_\varepsilon(0)$ such that $f(z)=\varphi(z)/\psi(z)$ in $U_\varepsilon(0)$. Using this fact we can determine all G-invariant meromorphic functions on W. For simplicity, we only treat the case $n=2$ below.

Since $f(z_1,z_2)=f(z)$ is G-invariant, $f(z)=f(g^{-m}(z))$ for any integer m. Since $|\alpha_1|>1$, $|\alpha_2|>1$ and $g^{-m}(z)=(\alpha_1^{-m_1}z_1,\alpha_2^{-m_2}z_2)$, for any given z, if we take m sufficiently large, we have $g^{-m}(z)\in U_\varepsilon(0)$. Hence

$$f(z)=f(g^{-m}(z))=\lim_{m\to+\infty}f(g^{-m}(z))=\lim_{m\to+\infty}\frac{\varphi(g^{-m}(z))}{\psi(g^{-m}(z))}. \tag{2.16}$$

Let

$$\varphi(z)=\sum_{h,k=0}^{+\infty} b_{hk}z_1^h z_2^k \quad\text{and}\quad \psi(z)=\sum_{h,k=0}^{+\infty} c_{hk}z_1^h z_2^k$$

be their power series expansions. Then

$$f(z)=\lim_{m\to+\infty}\frac{\sum(\alpha_1^h\alpha_2^k)^{-m}b_{hk}z_1^h z_2^k}{\sum(\alpha_1^h\alpha_2^k)^{-m}c_{hk}z_1^h z_2^k}.$$

Let μ be the minimum of $|\alpha_1^h\alpha_2^k|$ for all pairs (h,k) with $h,k=0,1,2,\ldots,$ such that not both b_{hk} and c_{hk} are not zero. Then

$$f(z)=\lim_{m\to+\infty}\frac{\sum(\alpha_1^h\alpha_2^k/\mu)^{-m}b_{hk}z_1^h z_2^k}{\sum(\alpha_1^h\alpha_2^k/\mu)^{-m}c_{hk}z_1^h z_2^k}.$$

Since $\lim_{m\to+\infty}(\alpha_1^h\alpha_2^k/\mu)^{-m}=0$ for $|\alpha_1^h\alpha_2^k|>\mu$, we have

$$f(z)=\lim_{m\to+\infty}\frac{\sum_{h,k}^{(\mu)}(\alpha_1^h\alpha_2^k/\mu)^{-m}b_{hk}z_1^h z_2^k}{\sum_{h,k}^{(\mu)}(\alpha_1^h\alpha_2^k/\mu)^{-m}c_{hk}z_1^h z_2^k},$$

where the summation $\sum_{h,k}^{(\mu)}$ is taken over all pairs (h,k) with $|\alpha_1^h\alpha_2^k|=\mu$. Thus $|\alpha_1^h\alpha_2^k/\mu|=1$ for all $\alpha_1^h\alpha_2^k/\mu$ appeared in the right-hand side of this equality. Let $e_\lambda=e^{i\theta_\lambda}$ with $0\leqq\theta_\lambda<2\pi$, and $\lambda=1,\ldots,\nu$, be all the distinct numbers among these $\alpha_1^h\alpha_2^k/\mu$, and put

$$P_\lambda(z)=\sum_{\alpha_1^h\alpha_2^k=\mu e_\lambda} b_{hk}z_1^h z_2^k \quad\text{and}\quad Q_\lambda(z)=\sum_{\alpha_1^h\alpha_2^k=\mu e_\lambda} c_{hk}z_1^h z_2^k.$$

Then

$$f(z)=\lim_{m\to+\infty}\frac{\sum_{\lambda=1}^{\nu}e^{-im\theta_\lambda}P_\lambda(z)}{\sum_{\lambda=1}^{\nu}e^{-m\theta_\lambda}Q_\lambda(z)}. \tag{2.17}$$

Since $f(z)$ is meromorphic, some $Q_\lambda(z)$ are not identically zero. Suppose, say, $Q_1(z)$ is not identically zero. Then from (2.17) we obtain

$$f(z)=\frac{P_1(z)}{Q_1(z)}.$$

We say that α_1 and α_2 are independent if $\alpha_1^h\neq\alpha_2^k$ for any pair (h,k) of natural numbers.

(i) The case that α_1 and α_2 are independent. In this case since there is a unique pair (h, k) with $a_1^h a_2^k = \mu e_1$, we have $P_1(z) = b_{hk} z_1^h z_2^k$ and $Q_1(z) = c_{hk} z_1^h z_2^k$ for this (h, k). Hence $f(z) = b_{hk}/c_{hk}$ is a constant. Thus there is no non-constant meromorphic function on $M = W/G$.

(ii) The case that α_1 and α_2 are not independent. Let (p, q) be the pair of natural numbers such that $\alpha_1^p = \alpha_2^q$ and that p is minimal among such pairs. Then if $\alpha_1^h = \alpha_2^k$, $(h, k) = (mp, mq)$ for some natural number m. Therefore letting (h_0, k_0) be the pair such that $\alpha_1^{h_0} \alpha_2^{k_0} = \mu e_1$ and that h_0 is minimal among those pairs satisfying the same condition, we see that any such (h, k) is written as $h = h_0 + mp$, $k = k_0 - mq$ for some non-negative integer m. Then

$$f(z) = \frac{P_1(z)}{Q_1(z)} = \frac{\sum_m b_m (z_1^p / z_2^q)^m}{\sum_m c_m (z_1^p / z_2^q)^m},$$

where we denote b_m for b_{hk} and c_m for c_{hk}. Since $\alpha_1^p = \alpha_2^q$, z_1^p / z_2^q is G-invariant, hence, gives a meromorphic function on $M = W/G$, and $f(z)$ is a rational function of z_1^p / z_2^q.

Thus a meromorphic function on a Hopf manifold $M = W/G$ of dimension 2 is either a constant if α_1 and α_2 are independent, or a rational function of z_1^p / z_2^q if α_1 and α_2 are not independent. From this result we see again that M is not algebraic. For $n \geqq 3$, too, if $\alpha_1, \ldots, \alpha_n$ are independent, that is, $\alpha_1^{k_1} \cdots \alpha_n^{k_n} \neq 1$ for any n-tuple $(k_1, \ldots, k_n) \neq (0, \ldots, 0)$ of natural numbers, there are no non-constant meromorphic functions on $M = W/G$.

(c) Surgery

We explain another method to construct complex manifolds.

Let M be a complex manifold, and $S \subset M$ a compact submanifold of M. We construct a new complex manifold $\tilde{M} = (M - S) \cup \tilde{S}$ by replacing S by another compact complex manifold $\tilde{S}$ as follows:

Take domains W and W_1 such that $S \subset W_1 \subset [W_1] \subset W \subset M$, where $[W]$ is assumed to be compact. Let $\tilde{S}$ be a compact complex submanifold of a complex manifold $\tilde{W}$ such that there are a domain $\tilde{W}_1$ with $\tilde{S} \subset \tilde{W}_1 \subset [\tilde{W}_1] \subset \tilde{W}$, and a biholomorphic map Φ of $\tilde{W} - \tilde{S}$ onto $W - S$ such that $\Phi(\tilde{W}_1 - \tilde{S}) = W_1 - S$. Let $\tilde{M}$ be the manifold obtained by glueing $M - [W_1]$ and $\tilde{W}$ by identifying $p \in W - [W_1]$ with $\tilde{p} = \Phi^{-1}(p) \in \tilde{W} - [\tilde{W}_1]$ via Φ:

$$\tilde{M} = (M - [W_1]) \cup \tilde{W}.$$

Since Φ is biholomorphic, $\tilde{M}$ becomes a complex manifold. *Thus $\tilde{M}$ is a complex manifold obtained from M by replacing W by $\tilde{W}$:*

$$\tilde{M} = (M - W) \cup \tilde{W}.$$

Identifying $p \in \tilde{W} - \tilde{S}$ with $p = \Phi(\tilde{p}) \in W - S$, we may identify $\tilde{W} - \tilde{S}$ and $\tilde{M} - \tilde{S}$ with $W - S$ and $M - S$, respectively. Then $\tilde{M}$ is considered to be a complex manifold obtained from M by replacing S by $\tilde{S}$.

$$\tilde{M} = (M - S) \cup \tilde{S}.$$

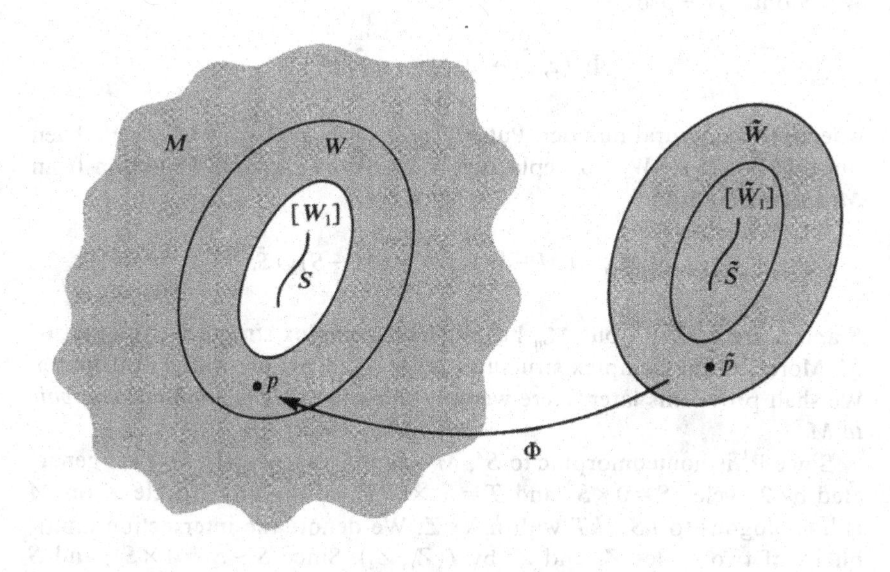

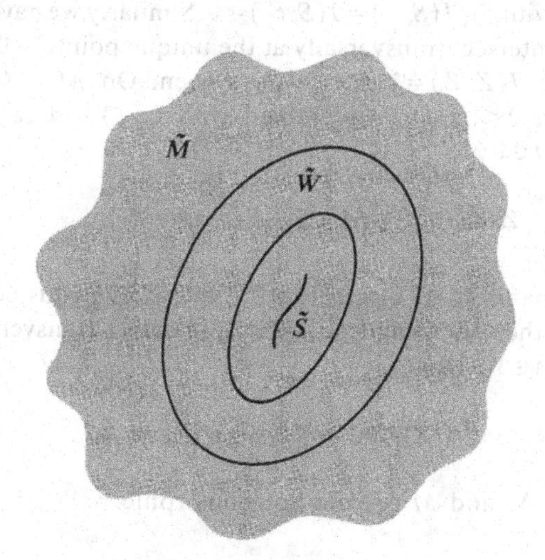

Figure 3

This method of construction is called a *surgery* or a *modification* of M. If M is compact, $\tilde{M}$ is also compact.

Example 2.10. Let $\zeta = \zeta_1/\zeta_0$ be the inhomogeneous coordinates on $\mathbb{P}^1$. Putting $\zeta = \infty$ for $\zeta_0 = 0$, we consider $\mathbb{P}^1 = \mathbb{C} \cup \infty$. Let $M = \mathbb{P}^1 \times \mathbb{P}^1 = \{(z, \zeta) | z \in \mathbb{P}^1, \zeta \in \mathbb{P}^1\}$, $S = 0 \times \mathbb{P}^1 \subset M$, and $W = U_\varepsilon \times \mathbb{P}^1$ where $U_\varepsilon = \{z | |z| < \varepsilon\}$. Further let $\tilde{W} = U_\varepsilon \times \mathbb{P}^1$, and $\tilde{S} = 0 \times \mathbb{P}^1 \subset \tilde{W}$. We denote a point of $\tilde{W}$ by $(z, \tilde{\zeta})$ where $z \in U_\varepsilon$, and $\tilde{\zeta} \in \mathbb{P}^1$. Define a biholomorphic map Φ of $\tilde{W} - \tilde{S}$ onto $W - S$ by

$$\Phi: (z, \tilde{\zeta}) \to (z, \zeta) = (z, \tilde{\zeta}/z^m),$$

where m is a natural number. Put $W_1 = U_{\varepsilon/2} \times \mathbb{P}^1$ and $\tilde{W}_1 = U_{\varepsilon/2} \times \mathbb{P}^1$. Then since $\Phi(\tilde{W}_1 - \tilde{S}) = W_1 - S$, replacing W by $\tilde{W}$ by surgery, we obtain from M a new manifold

$$\tilde{M}_m = (M - W) \cup \tilde{W} = (M - S) \cup \tilde{S}.$$

S and $\tilde{S}$ are both P^1, but $\tilde{M}_m$ has different complex structure from that of M. Moreover the complex structures of $\tilde{M}_m$ and $\tilde{M}_n$ are different if $m \neq n$. We shall prove this later. Here we only show that $\tilde{M}_1$ *is not homeomorphic to* M.

Since $\mathbb{P}^1$ is homeomorphic to S^2, $M = S^2 \times S^2$. Hence $H_2(M, \mathbb{Z})$ is generated by 2-cycles $S = 0 \times S^2$ and $T = S^2 \times 0$. Therefore any 2-cycle Z on M is homologous to $hS + kT$ with $h, k \in \mathbb{Z}$. We denote the intersection multiplicity of two cycles Z_1 and Z_2 by $I(Z_1, Z_2)$. Since $S \sim S_1 = 1 \times S^2$, and S does not intersect with S_1, $I(S, S) = I(S, S_1) = 0$. Similarly we have $I(T, T) = 0$. Since S and T intersect transversally at the unique point 0×0, $I(S, T) = I(T, S) = 1$. Hence $I(Z, Z) = 2hk$ is always even. On $\tilde{M} = (M - S) \cup \tilde{W}$, $(z, \zeta) \in M - S$ with $0 < |z| < \varepsilon$ is identified with $(z, \tilde{\zeta}) = (z, z\zeta) \in \tilde{W}$. Consequently for any $t \in \mathbb{C}$,

$$Z_t = \{(z, t) \in M - S\} \cup \{(z, zt) \in \tilde{W}\}$$

is a complex submanifold of dimension 1. Since Z_t depends continuously on t, $Z_t \sim Z_0$. On the other hand, Z_t and Z_0 intersect transversally at the unique point $(0, 0) \in \tilde{W}$. Hence

$$I(Z_0, Z_0) = I(Z_t, Z_0) = 1,$$

which proves that $\tilde{M}$ and M are not homeomorphic.

Example 2.11. Let $C = \mathbb{C}^*/G^*$ be an elliptic curve given in Example 2.7 where $G^* = \{g_m^* | m \in \mathbb{Z}\}$. We denote a point G^*w on C corresponding to $w \in \mathbb{C}^*$ by $[w]$. Since $g_m^*(w) = \alpha^m w, [\alpha^m w] = [w]$. Let $M = \mathbb{P}^1 \times C$, $W = U_\varepsilon \times C$ with $U_\varepsilon = \{z | |z| < \varepsilon\}$, $S = 0 \times C$, $\tilde{W} = U_\varepsilon \times C$ and $\tilde{S} = 0 \times C$. We denote a point on M by $(z, [w])$, and a point on $\tilde{W}$ by $(z, [\tilde{w}])$. Define a biholomorphic map Φ of $\tilde{W}-\tilde{S}$ onto $W-S$ by

$$\Phi\colon (z,[\tilde{w}])\to(z,[w])=(z,[z\tilde{w}]).$$

Put $W_1=U_{\varepsilon/2}\times C$ and $\tilde{W}_1=U_{\varepsilon/2}\times C$. Then since $\Phi(\tilde{W}_1-\tilde{S})=W_1-S$, we obtain by surgery a complex manifold

$$\tilde{M}=(M-W)\cup\tilde{W}.$$

$\tilde{M}$ is a Hopf manifold given in Example 2.9. To be precise, letting G be the infinite cyclic group of automorphisms of $\mathbb{C}^2-\{0\}$ generated by $g\colon (z_1, z_2)\to(\alpha z_1, \alpha z_2)$, we have

$$\tilde{M}=(\mathbb{C}^2-\{0\})/G.$$

To see this, put $N=(\mathbb{C}^2-\{0\})/G$. We want to construct a biholomorphic map of N onto $\tilde{M}$. We denote by $[z_1, z_2]$ the point of N corresponding to $(z_1, z_2)\in\mathbb{C}^2-\{0\}$. Since z_1/z_2 is invariant under $g, f(z_1, z_2)=z_1/z_2$ is a meromorphic function on N, and

$$f\colon [z_1, z_2]\to z=f(z_1, z_2)=z_1/z_2$$

is a holomorphic map of N onto $\mathbb{P}^1=\mathbb{C}\cup\infty$. Let $V_1=N-f^{-1}(0)$, and $V_2=f^{-1}(U_\varepsilon)$. Then $N=V_1\cup V_2$. Since $z_1\neq 0$ on V_1, and $z_2\neq 0$ on V_2, if we put

$$\Psi_1\colon [z_1, z_2]\to(z,[w])=(z,[z_1]),$$
$$\Psi_2\colon [z_1, z_2]\to(z,[\tilde{w}])=(z,[z_2]),$$

Ψ_1 maps V_1 onto $M-S$, and Ψ_2 maps V_2 onto $\tilde{W}$ biholomorphically. For $0<|z|<\varepsilon$,

$$(z,[w])=(z,[z_1])=(z,[zz_2])=(z,[z\tilde{w}]),$$

hence,

$$\Psi_1([z_1, z_2])=\Phi(\Psi_2([z_1, z_2])),$$

which implies that $\Psi_1\colon V_1\to M-S\subset\tilde{M}$ and $\Psi_2\colon V_2\to\tilde{W}\subset\tilde{M}$ coincide on $V_1\cap V_2$. Therefore defining $\Psi=\Psi_1$ on V_1, and $\Psi=\Psi_2$ on V_2, we have a biholomorphic map Ψ of N onto $\tilde{M}$. Thus $\tilde{M}=N$ is a Hopf manifold.

Example 2.12. Let $M=\mathbb{C}^n$, $S=0=(0,\ldots,0)\in\mathbb{C}^n$, and $\tilde{S}=\mathbb{P}^{n-1}$. We define a complex structure on $\tilde{M}=(M-S)\cup\tilde{S}$ as a complex submanifold of $\mathbb{C}^n\times\mathbb{P}^{n-1}$ in the following manner. Since each point ζ of $\mathbb{P}^{n-1}$ is, by definition, a line ζ in $\mathbb{C}^n$ through 0, $\zeta\times\zeta\in\mathbb{C}^n\times\mathbb{P}^{n-1}$ is a line in $\mathbb{C}^n\times\zeta$ through the point $0\times\zeta$. Put

$$\tilde{M}=\bigcup_{\zeta}\zeta\times\zeta=\{(z,\zeta)\in\mathbb{C}^n\times\mathbb{P}^{n-1}\,|\,z\in\zeta\}.$$

Let $(z_1,\ldots,z_n)$ be the standard coordinates on $\mathbb{C}^n$, and $(\zeta_1,\ldots,\zeta_n)$ the homogeneous coordinates on $\mathbb{P}^{n-1}$. In terms of these coordinates, we denote a point (z,ζ) by $(z_1,\ldots,z_n,\zeta_1,\ldots,\zeta_n)$. Then $z\in\zeta$ if $\zeta_j z_k-z_j\zeta_k=0$ for $j,k=1,\ldots,n$. Therefore $\tilde{M}$ is an analytic subset of $\mathbb{C}^n\times\mathbb{P}^{n-1}$ defined by

$$\zeta_j z_k-\zeta_k z_j=0,\qquad j,k=1,\ldots,n. \tag{2.18}$$

Since $\mathbb{P}^{n-1}$ is covered by n pieces of coordinate neighbourhoods $U_j=\{\zeta\in\mathbb{P}^{n-1}\,|\,\zeta_j\neq 0\}, j=1,\ldots,n$, $\mathbb{C}^n\times\mathbb{P}^{n-1}$ is covered by $\mathbb{C}^n\times U_j's$. To verify that $\tilde{M}$ is a submanifold $\mathbb{C}^n\times\mathbb{P}^{n-1}$, consider $\tilde{M}_1=\tilde{M}\cap\mathbb{C}^n\times U_1$. Let $(w_2,\ldots,w_k,\ldots,w_n)$ be the inhomogeneous coordinates on U_1 where $w_k=\zeta_k/\zeta_1$. Then on $\mathbb{C}^n\times U_1$, (2.18) is equivalent to

$$z_k=w_k z_1,\qquad k=2,3,\ldots,n. \tag{2.19}$$

Thus $\tilde{M}_1$ is a submanifold of $\mathbb{C}^n\times U_1$. Similarly each $M_j=\tilde{M}\cap\mathbb{C}^n\times U_j$ is a submanifold of $\mathbb{C}^n\times U_j$. Therefore $\tilde{M}$ is a complex submanifold of $\mathbb{C}^n\times\mathbb{P}^{n-1}$.

Let Φ be the restriction to $\tilde{M}$ of the projection $(z,\zeta)\to z$ of $\mathbb{C}^n\times\mathbb{P}^{n-1}$ onto $\mathbb{C}^n$. Then Φ is a holomorphic map of $\tilde{M}$ onto $M=\mathbb{C}^n$, and $\Phi^{-1}(0)=0\times\mathbb{P}^{n-1}$. Consequently, to see that $\tilde{M}=(M-S)\cup\tilde{S}$ where $\tilde{S}=\Phi^{-1}(0)=0\times\mathbb{P}^{n-1}$ and $S=0$, it suffices to show that $\tilde{S}$ is a submanifold of $\tilde{M}$ and that Φ maps $\tilde{M}-\tilde{S}$ biholomorphically onto $M-S$. By (2.19), we have

$$\tilde{M}_1=\{(z_1,w_2z_1,\ldots,w_nz_1,1,w_2,\ldots,w_n)\,|\,(z_1,w_2,\ldots,w_n)\in\mathbb{C}^n\}.$$

Therefore $(z_1,w_2,\ldots,w_n)$ gives a local coordinate system of $\tilde{M}$ defined on the coordinate neighbourhood $\tilde{M}_1$. We define $\tilde{M}_i$ similarly. Then $\tilde{M}$ is covered by these $\tilde{M}_1,\ldots,\tilde{M}_n$. In terms of these local coordinates, $\tilde{S}\cap\tilde{M}_1$ is a submanifold of $\tilde{M}_1$ defined by the equation $z_1=0$. Similar results hold for $\tilde{S}\cap M_2,\ldots,\tilde{S}\cap M_n$. Thus $\tilde{S}$ is a submanifold of $\tilde{M}$.

Let $(z,\zeta)=(z_1,\ldots,z_n,\zeta)\in\tilde{M}-\tilde{S}$. Since $z\in\zeta$, and $z\neq 0$, $\zeta=(z_1,\ldots,z_n)$. Hence on $\tilde{M}-\tilde{S}$, Φ is given by

$$\Phi\colon(z_1,\ldots,z_n,z_1,\ldots,z_n)\to(z_1,\ldots,z_n).$$

Then Φ maps $\tilde{M}-\tilde{S}$ biholomorphically onto $\mathbb{C}^n-\{0\}=M-S$. Therefore $\tilde{M}=(M-S)\cup\tilde{S}$ as required.

We have explained above how we construct $\tilde{M}$ from $M=\mathbb{C}^n$ by replacing $0\in M$ by $\tilde{S}=\mathbb{P}^{n-1}$. Let W_ε be the ε-neighbourhood of 0 in $\mathbb{C}^n$ with $\varepsilon>0$. Then the above procedure does not affect the complement of W_ε. Thus we have

$$\tilde{M}=(M-W_\varepsilon)\cup\tilde{W}_\varepsilon,\qquad \tilde{W}_\varepsilon=\Phi^{-1}(W_\varepsilon),$$

and $\tilde{W}_\varepsilon$ is a complex manifold obtained from W_ε by replacing $0\in W_\varepsilon$ by $\tilde{S}=\mathbb{P}^{n-1}$:

$$\tilde{W}_\varepsilon=(W_\varepsilon-\{0\})\cup\tilde{S}. \tag{2.20}$$

Given an arbitrary complex manifold M^n and any point $q\in M^n$, we can construct a new complex manifold $\tilde{M}^n$ from M^n, replacing $q\in M^n$ by $\mathbb{P}^{n-1}$ as follows. Let $p\to z_q(p)$ be local coordinates on M^n with centre q, and $W_\varepsilon(q)=\{p\,||z_q(p)|<\varepsilon\}$ where ε is sufficiently small. Then the map $p\to z_q(p)$ maps $W_\varepsilon(q)$ biholomorphically onto the ε-neighbourhood W_ε of 0 in $\mathbb{C}^n$, and $z_q(q)=0$. Thus identifying $W_\varepsilon(q)$ with W_ε via z_q, and putting $\tilde{W}_\varepsilon(q)=\tilde{W}_\varepsilon$, we obtain from (2.20)

$$\tilde{W}_\varepsilon(q)=(W_\varepsilon(q)-\{q\})\cup\tilde{S},$$

where $\tilde{S}=\mathbb{P}^{n-1}$. Thus letting

$$\tilde{M}^n=(M^n-W_\varepsilon(q))\cup\tilde{W}_\varepsilon(q),$$

we have

$$\tilde{M}^n=(M^n-\{q\})\cup\tilde{S}\quad\text{with}\quad\tilde{S}=\mathbb{P}^{n-1}$$

as required. Let Φ be the holomorphic map of $\tilde{W}_\varepsilon(q)=\tilde{W}_\varepsilon$ onto $W_\varepsilon(q)=W_\varepsilon$ defined above. Extending Φ by putting $\Phi(p)=p$ for $p\in M^n-W_\varepsilon(q)$, we obtain a holomorphic map Φ of $\tilde{M}^n$ onto M which maps $\tilde{S}$ to the point q and $M^n-\tilde{S}$ biholomorphically onto M^n-q. Thus $\Phi^{-1}(M^n)=\tilde{M}^n$ and $\Phi^{-1}(q)=\tilde{S}$. We denote Φ^{-1} by Q_q and call the *quadratic transformation with centre q*.

For example, let $M^2=\mathbb{P}^2$, and let (w_0, w_1, w_2) be its homogeneous coordinates and $q=(1,0,0)$. Put $\tilde{M}^2=Q_q(\mathbb{P}^2)$. We denote by $\mathbb{P}^1_\infty$ the projective line defined by $w_0=0$. Then

$$\mathbb{P}^2=\mathbb{C}^2\cup\mathbb{P}^1_\infty,\qquad \mathbb{C}^2=U_0=\{w\in P^2\,|\,w_0\neq 0\}.$$

As in the case of $\mathbb{P}^1=\mathbb{C}\cup\{\infty\}$, we call $\mathbb{P}^1_\infty$ the *line at infinity*, and any points $(0, w_1, w_2)$ on $\mathbb{P}^1_\infty$ a point at infinity. Let $(z_1, z_2)=(w_1/w_0, w_2/w_0)$ be the non-homogeneous coordinates on $\mathbb{C}^2=U_0$. Any line $\zeta=(\zeta_0,\zeta_1)$ on $\mathbb{C}^2$

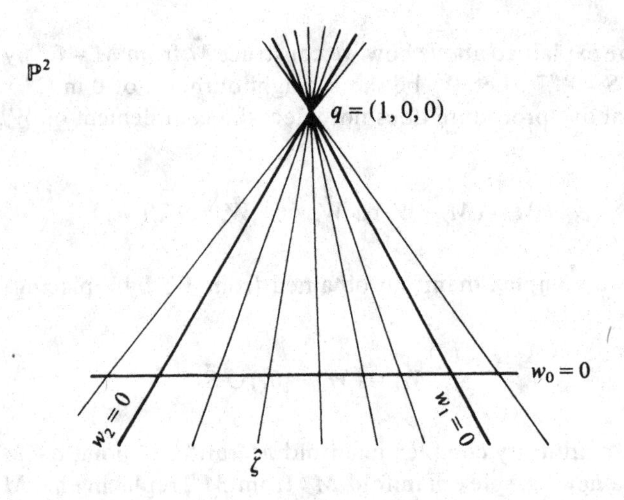

Figure 4

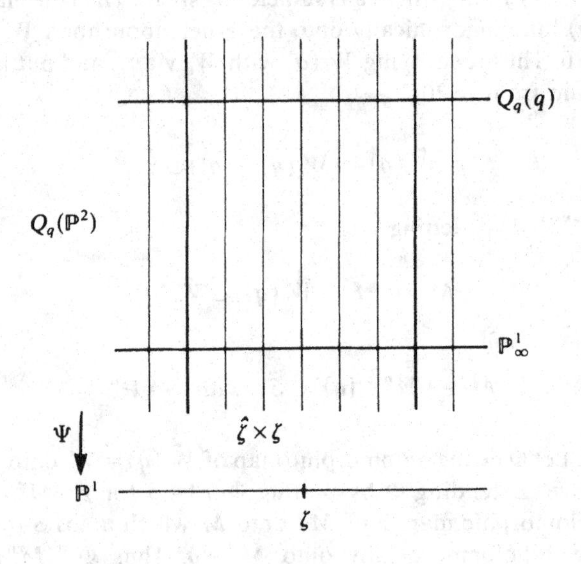

Figure 5

through $q=(0,0)$ extends to the projective line

$$\hat{\zeta}=\{(\lambda_0, \lambda_1\zeta_0, \lambda_1\zeta_1)\,|\,(\lambda_0, \lambda_1)\in\mathbb{P}^1\}$$

which passes through $(1,0,0)$ and $(0,\zeta_0,\zeta_1)$ on $\mathbb{P}^2$. Since $Q_q(\mathbb{P}^2)=Q_q(\mathbb{C}^2)\cup\mathbb{P}^1_\infty$, and $Q_q(\mathbb{C}^2)=\bigcup_\zeta \zeta\times\zeta$, we have

$$Q_q(\mathbb{P}^2)=\bigcup_\zeta \hat{\zeta}\times\zeta=\{(w,\zeta)\in\mathbb{P}^2\times\mathbb{P}^1\,|\,w\in\hat{\zeta}\}.$$

Thus $Q_q(\mathbb{P}^2)$ is a submanifold of $\mathbb{P}^2\times\mathbb{P}^1$. The restriction Ψ of the projection $\mathbb{P}^2\times\mathbb{P}^1\to\mathbb{P}^1$ to $Q_q(\mathbb{P}^2)$ maps $Q_q(\mathbb{P}^2)$ onto $\mathbb{P}^1$, and $\Psi^{-1}(\zeta)=\hat{\zeta}\times\zeta$. $Q_q(q)$ is a line on $Q_q(\mathbb{P}^2)$ which does not intersect $\mathbb{P}^1_\infty$.

It is easily verified that $Q_q(\mathbb{P}^2)$ is biholomorphic to $\tilde{M}_1$ given in Example 2.10.

§2.3. Complex Analytic Family

(a) Complex Analytic Family

In the definition of an elliptic curve C given in Example 2.7, there appears an arbitrary parameter ω on which the complex structure of C depends. Similarly the complex structure of a Hopf manifold given in Example 2.9 depends on the parameters $\alpha_1,\dots,\alpha_n$, while in the definition of $\mathbb{P}^n$ no arbitrary parameter appears. Thus the complex structure of a complex manifold M often varies as the parameters $t=(t_1,\dots,t_m)$ which appear in its definition varies. This being the case, we say that the complex structure of M depends on t. We write M_t for M if we want to express the dependence on t explicitly.

Let us consider how a compact complex manifold M_t depends on parameters t. First consider a complex-valued function $f(t)=f(t_1,\dots,t_n)$ defined in a domain $B\subset\mathbb{C}^m$. We may consider $f(t)$ as a complex number varying as t moves in B. If $f(t)$ is holomorphic, $f(t)$ is said to depend holomorphically on t. In this case the graph of $f(t)$

$$\mathscr{F}=\bigcup_{t\in B} f(t)\times t=\{(f(t),t)\in\mathbb{C}\times B\}$$

is a submanifold of $\mathbb{C}\times B$, hence, a complex manifold.

Similarly we may consider a complex manifold M_t depending on t as a "function" of $t\in B$, but unlike a function $f(t)$ above, there is no space containing all compact complex manifolds. Nevertheless, we can consider the set

$$\mathscr{M}=\bigcup_{t\in B} M_t\times t$$

corresponding to $\mathscr{F}=\bigcup_{t\ni B} f(t)\times t$ above. Thus we reach the following definition.

Definition 2.8. Suppose given a domain B in $\mathbb{C}^m$, and a set $\{M_t\,|\,t\in B\}$ of complex manifolds M_t depending on $t=(t_1,\dots,t_m)\in B$. We say that M_t depends holomorphically on t and that $\{M_t\,|\,t\in B\}$ is a *complex analytic family of compact complex manifolds* if there is a complex manifold $\mathscr{M}$ and a holomorphic map ϖ of $\mathscr{M}$ onto B satisfying the following conditions.

(i) $\varpi^{-1}(t)$ is a compact complex submanifold of $\mathcal{M}$.
(ii) $M_t = \varpi^{-1}(t)$.
(iii) The rank of the Jacobian of ϖ is equal to m at every point of $\mathcal{M}$.

Here in (iii) by the rank of the Jacobian of ϖ, we mean the rank of the Jacobian matrix in terms of local coordinates. Thus, let $(z_q^1, \ldots, z_q^n, z_q^{n+1}, \ldots, z_q^{n+m})$ be local coordinates on $\mathcal{M}$ and let

$$(t_1, \ldots, t_m) = \varpi(z_q^1, \ldots, z_q^n, z_q^{n+1}, \ldots, z_q^{n+m}).$$

Then (iii) implies that

$$\operatorname{rank} \frac{\partial(t_1, \ldots, t_m)}{\partial(z_q^1, \ldots, z_q^n, z_q^{n+1}, \ldots, z_q^{n+m})} = m.$$

Therefore by §2.1(d) we can choose a system of local complex coordinates $\{z_1, \ldots, z_j, \ldots\}$, $z_j : p \to z_j(p)$, and coordinate polydisks $\mathcal{U}_j$ with respect to z_j, satisfying the following conditions.

(i) $z_j(p) = (z_j^1(p), \ldots, z_j^n(p), t_1, \ldots, t_m)$, $(t_1, \ldots, t_m) = \varpi(p)$;
(ii) $\mathfrak{U} = \{\mathcal{U}_j \mid j = 1, 2, \ldots\}$ is locally finite.

Then

$$\{p \to (z_j^1(p), \ldots, z_j^n(p)) \mid \mathcal{U}_j \cap M_t \neq \emptyset\}$$

gives a system of local complex coordinates on M_t. In terms of these coordinates, ϖ is the projection given by

$$\varpi : (z_j^1, \ldots, z_j^n, t_1, \ldots, t_m) \to (t_1, \ldots, t_m).$$

For j, k with $U_j \cap U_k \neq \emptyset$, we denote the coordinate transformation from z_k to z_j by

$$f_{jk} : (z_k^1, \ldots, z_k^n, t) \to (z_j^1, \ldots, z_j^n, t) = f_{jk}(z_k^1, \ldots, z_k^n, t).$$

Note that $t_1, \ldots, t_m$ as part of local coordinates on $\mathcal{M}$ do not change under these coordinate transformations. Thus f_{jk} is given by

$$z_j^\alpha = f_{jk}^\alpha(z_k^1, \ldots, z_k^n, t_1, \ldots, t_m), \qquad \alpha = 1, \ldots, n. \tag{2.21}$$

In what follows by a complex analytic family, we mean a complex analytic family of compact complex manifolds unless otherwise mentioned. We sometimes denote a complex analytic family $\{M_t \mid t \in B\}$ by $(\mathcal{M}, B, \varpi)$, where $\mathcal{M}$, B and ϖ are as above. We call t its parameters, and B its *parameter space.*

Definition 2.8 *is obviously extended to the case B is an arbitrary complex manifold.*

Example 2.13. The set $\{C_\omega \mid \operatorname{Im}\omega > 0\}$ of elliptic curves $C_\omega = \mathbb{C}/G_\omega$ forms a complex analytic family where $G_\omega = \{m\omega + n \mid m, n \in \mathbb{Z}\}$. In fact, put $B = \{\omega \in \mathbb{C} \mid \operatorname{Im}\omega > 0\}$. Then a group of automorphisms of $\mathbb{C} \times B$ defined as

$$G = \{g_{mn}: (z, \omega) \to (z + m\omega + n) \mid m, n \in \mathbb{Z}\}$$

acts in a properly discontinuous manner without fixed point. Therefore $\mathcal{M} = \mathbb{C} \times B/G$ is a complex manifold. Since the projection $(z, \omega) \to \omega$ of $\mathbb{C} \times B$ to B commutes with g_{mn}, it induces a holomorphic map ϖ of $\mathcal{M}$ onto B. We have

$$\varpi^{-1}(\omega) = \mathbb{C} \times \omega/G = \mathbb{C}/G_\omega = C_\omega.$$

Using (z, ω) as local coordinates on $\mathcal{M}$, we easily see that the rank of the Jacobian matrix of ϖ is equal to 1. Thus $\{C_\omega \mid \omega \in B\}$ forms a complex analytic family.

Definition 2.9. Let M and N be two compact complex manifolds. N is called a *deformation* of M if M and N belong to the same complex analytic family, that is, if there is a complex analytic family $(\mathcal{M}, B, \varpi)$ with a complex manifold B as its parameter space such that $M = \varpi^{-1}(t_0)$ and $N = \varpi^{-1}(t_1)$ for some $t_0, t_1 \in B$.

Two complex analytic families $(\mathcal{M}, B, \varpi)$ and $(\mathcal{N}, B, \pi)$ are called *holomorphically equivalent* if there is a biholomorphic map Φ of $\mathcal{M}$ onto $\mathcal{N}$ such that $\varpi = \pi \circ \Phi$. This being the case, Φ maps $M_t = \varpi^{-1}(t)$ biholomorphically onto $N_t = \pi^{-1}(t)$, hence M_t and N_t are biholomorphic. As in the case of complex manifolds, we often identify two holomorphically equivalent complex analytic families.

Let M be a compact complex manifold, and B an arbitrary complex manifold. Then $(M \times B, B, \varpi)$ forms a complex analytic family where ϖ denotes the projection to the second factor. If a complex analytic family $(\mathcal{M}, B, \varpi)$ is holomorphically equivalent to $(M \times B, B, \varpi)$ above with $M = \varpi^{-1}(t_0)$ for some $t_0 \in B$, $(\mathcal{M}, B, \varpi)$ is called trivial. If $(\mathcal{M}, B, \varpi)$ is trivial, $M_t = \varpi^{-1}(t)$ is biholomorphic to M for all $t \in B$. In this case the complex structure of M_t is independent of t. A trivial family may be considered as an analogue of a constant function.

Let $(\mathcal{M}, B, \varpi)$ be a complex analytic family, and U a subdomain of B. Let $\mathcal{M}_U = \varpi^{-1}(U)$, and ϖ_U the restriction of ϖ to U. Then $(\mathcal{M}_U, U, \varpi_U)$ forms a complex analytic family, which we call the restriction of $(\mathcal{M}, B, \varpi)$ to U. If $(\mathcal{M}_U, U, \varpi_U)$ is trivial, we say that $(\mathcal{M}, B, \varpi)$ is trivial over U.

Theorem 2.3. *Let $(\mathcal{M}, B, \varpi)$ be a complex analytic family of compact complex manifolds, and t_0 any point of B. Then $M_t = \varpi^{-1}(t)$ is diffeomorphic to $M_{t_0} = \varpi^{-1}(t_0)$ for any $t \in B$.*

Thus *the differentiable structure of complex manifolds does not change under deformation.*

To prove this theorem, we need to prepare some results from the theory of differentiable manifolds. First let $\mathcal{M}$ be an arbitrary differentiable manifolds, $\{x_1, \ldots, x_j, \ldots\}$ a system of local coordinates on $\mathcal{M}$, and $\mathcal{U}_j$ the domain of $x_j\colon p \to (x_j^1, \ldots, x_j^m) = x_j(p)$. We assume that $\{\mathcal{U}_j \mid j = (1, 2, \ldots\}$ is a locally finite open covering of $\mathcal{M}$. For j, k with $\mathcal{U}_j \cap \mathcal{U}_k \neq \varnothing$, let

$$x_j^\alpha = f_{jk}^\alpha(x_j^1, \ldots, x_k^m)$$

be the coordinate transformation. Consider a smooth curve $\gamma\colon t \to x_j(t) = (x_j^1(t), \ldots, x_j^m(t))$ on $\mathcal{M}$. We denote $(d/dt)x_j^\alpha(t)$ by $\dot{x}_j^\alpha(t)$. Then

$$v_j = (v_j^1, \ldots, v_j^m) = (\dot{x}_j^1(t), \ldots, \dot{x}_j^m(t))$$

is the *tangent vector of* γ at $x_j(t)$.

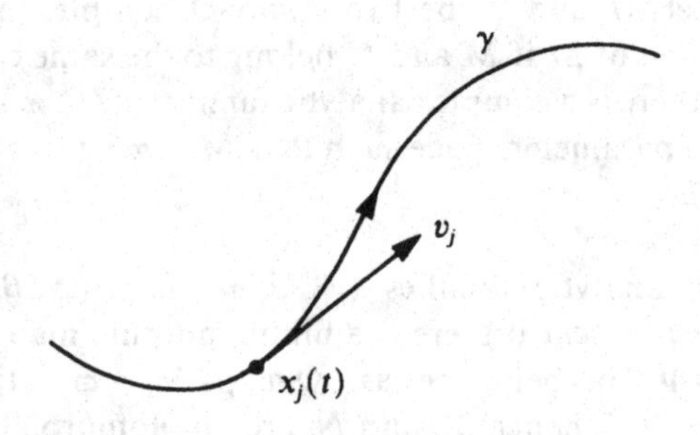

Figure 6

Under the coordinate transformation $x_j^\alpha = f_{jk}^\alpha(x_k)$, the tangent vector of γ is transformed as follows.

$$v_j^\alpha = \sum_{\beta=1}^{m} \frac{\partial f_{jk}^\alpha(x_k)}{\partial x_k^\beta} v_k^\beta = \sum_\beta \frac{\partial x_j^\alpha}{\partial x_k^\beta} v_k^\beta. \tag{2.22}$$

Let $\varphi(x_j)$ be a continuously differentiable function on M. Then we have

$$\frac{d}{dt}\varphi(x_j(t)) = \sum_{\alpha=1}^{m} v_j^\alpha \frac{\partial}{\partial x_j^\alpha}\varphi(x_j(t)).$$

Consequently the vector v_j corresponds to the *differential operator*

$$v = \sum_{\alpha=1}^{m} v_j^\alpha \frac{\partial}{\partial x_j^\alpha}.$$

Since by (2.22)

$$\sum_{\alpha=1}^{m} v_j^\alpha \frac{\partial}{\partial x_j^\alpha} = \sum_{\alpha=1}^{m} v_k^\alpha \frac{\partial}{\partial x_k^\alpha}, \tag{2.23}$$

the operator $\sum_\alpha v_j^\alpha(\partial/\partial x_j^\alpha)$ does not depend on the choice of local coordinates. If a tangent vector

$$v(x_j) = \sum_{\alpha=1}^{m} v_j^\alpha(x_j)\frac{\partial}{\partial x_j^\alpha}$$

is assigned to each point x_j of $\mathcal{M}$, we call $v(x_j)$ a *vector field.* If all $v_j^\alpha(x_j)$, $\alpha = 1, \ldots, m$, are C^∞, the vector field $v(x_j)$ is called a C^∞ vector field. Suppose given a C^∞ vector field $v(x_j)$ such that $v(x_j) \neq 0$ at every point of $\mathcal{M}$. Then *the system differential equations,*

$$\frac{d}{dt}x_j^\alpha = v_j^\alpha(x_j^1, \ldots, x_j^m), \qquad \alpha = 1, 2, \ldots, m, \tag{2.24}$$

has a unique solution $x_j^\alpha = x_j^\alpha(t)$ *under any given initial conditions*

$$x_j^\alpha(0) = \xi_i^\alpha, \qquad \alpha = 1, 2, \ldots, m.$$

We denote this solution by $x_j^\alpha(t, \xi_i)$. Then $x_j^\alpha(t, \xi_i)$ is a C^∞ function of t, $\xi_i^1, \ldots, \xi_i^m$. Since *the system of equations* (2.24) *are invariant under coordinate transformations by* (2.23), *its solution* $x_j^\alpha(t, \xi_i)$ *gives a smooth curve* $t \to x_j(t, \xi_i) = (x_j^1(t, \xi_i), \ldots, x_j^m(t, \xi_i))$ *on M starting from the point* ξ_i *of M.*

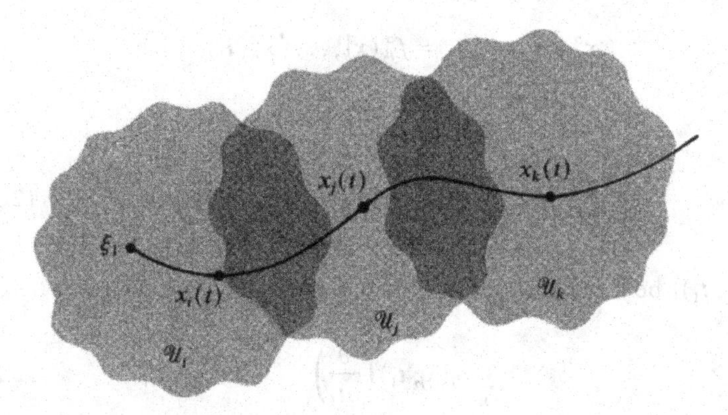

Figure 7

Next, let $\mathcal{M}$ be a differentiable manifold, B a domain of $\mathbb{R}^m$, and ϖ a C^∞ map of $\mathcal{M}$ onto B. We assume that $(\mathcal{M}, B, \varpi)$ satisfies the following conditions.

(i) $M_t = \varpi^{-1}(t)$ is compact for any $t \in B$.
(ii) the rank of the Jacobian of ϖ is equal to m at every point of $\mathcal{M}$.

Then, as in the case of a complex analytic family, we can choose a system of local coordinates $\{x_1, \ldots, x_j, \ldots\}$, $x_j: p \to x_j(p)$, satisfying the following conditions.

(i) $x_j(p) = (x_j^1(p), \ldots, x_j^n(p), t_1, \ldots, t_m)$, $\quad (t_1, \ldots, t_m) = \varpi(p)$.
(ii) $\{\mathcal{U}_j \mid j = 1, 2, \ldots\}$ forms a locally finite open covering of $\mathcal{M}$ where $\mathcal{U}_j$ is the domain of x_j.

Assume that $0 \in B$, and take an open cube $U = \{t \mid |t_1| < r, \ldots, |t_m| < r\}$, such that $[U] \subset B$. Let π be the projection of $M_0 \times U$ to the second factor.

Theorem 2.4. *There is a diffeomorphism Ψ of $M_0 \times U$ onto $\varpi^{-1}(U)$ such that $\varpi \circ \Psi = \pi$.*

Proof. We use induction on the dimension m of B.

(1°) The case $m = 1$. In this case B and U are open intervals in $\mathbb{R}$: $U = (-r, r)$, and $[-r, r] \subset B$. We denote $(x_j^1, \ldots, x_j^n, t_1)$ simply by (x_j, t_1). First let us construct a C^∞ vector field on $\mathcal{M}$ which is given on each $\mathcal{U}_j$ as

$$\sum_{\alpha=1}^{n} v_j^\alpha(x_j, t_1)\frac{\partial}{\partial x_j^\alpha} + \frac{\partial}{\partial t_1}. \tag{2.25}$$

We denote the vector field $\partial/\partial t_1$ on $\mathcal{U}_k$ by $(\partial/\partial t_1)_k$. Then with respect to the coordinate transformation

$$x_j^\alpha = f_{jk}^\alpha(x_k, t_1) = f_{jk}^\alpha(x_k^1, \ldots, x_k^n, t_1),$$

we have

$$\left(\frac{\partial}{\partial t_1}\right)_k = \sum_{\alpha=1}^{n} \frac{\partial}{\partial t_1} f_{jk}^\alpha(x_k, t_1)\frac{\partial}{\partial x_j^\alpha} + \left(\frac{\partial}{\partial t_1}\right)_j. \tag{2.26}$$

Let $\{\rho_j(x_j, t_1)\}$ be a partition of unity subordinate to $\{\mathcal{U}_j\}$. Then

$$\sum_j \rho_j(x_j, t_1)\left(\frac{\partial}{\partial t_1}\right)_j \tag{2.27}$$

is a C^∞ vector field on $\mathcal{M}$. On $\mathcal{U}_j$, we have

$$\sum_k \rho_k \left(\frac{\partial}{\partial t_1}\right)_k = \sum_{\alpha=1}^{n} \sum_{k \neq j} \rho_k \frac{\partial}{\partial t_1} f_{jk}^\alpha(x_k, t_1)\frac{\partial}{\partial x_j^\alpha} + \left(\frac{\partial}{\partial t_1}\right)_j,$$

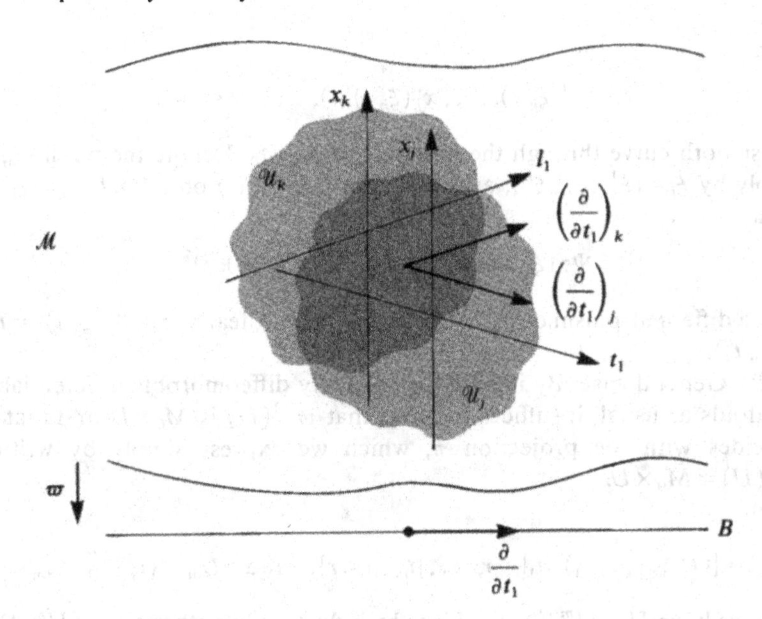

Figure 8

where we write ρ_k for $\rho_k(x_k, t_1)$. Therefore the vector field given by (2.27) has the required property (2.27) if we put

$$v_j^\alpha(x_j, t_1) = \sum_{k \neq j} \rho_k(x_k, t_1) \frac{\partial}{\partial t_1} f_{jk}^\alpha(x_k, t_1), \qquad x_k = f_{kj}(x_j, t_1).$$

Consider the system of differential equations for this vector field:

$$\begin{cases} \dfrac{dx_j^\alpha}{dt} = v_j^\alpha(x_j^1, \ldots, x_j^n, t_1), & \alpha = 1, \ldots, n, \\ \dfrac{dt_1}{dt} = 1. \end{cases} \tag{2.28}$$

For any point $(\xi_i, 0) = (\xi_i^1, \ldots, \xi_i^n, 0) \in M_0 = \varpi^{-1}(0)$, let

$$\begin{cases} x_j^\alpha = x_j^\alpha(\xi_i, t), & \alpha = 1, \ldots, n, \\ t_1(t) = t \end{cases}$$

be the solution of (2.28) satisfying the initial condition

$$\begin{cases} x_i^\alpha(0) = \xi_i^\alpha, & \alpha = 1, \ldots, n, \\ t_1(0) = 0. \end{cases}$$

Then

$$t \to (x_j^1(\xi_i, t), \ldots, x_j^n(\xi_i, t), t), \qquad -r < t < r,$$

is a smooth curve through the point $(\xi_i, 0)$ on M_0. Denote the point $(\xi_i, 0)$ simply by $\xi_i = (\xi_i^1, \ldots, \xi_i^n)$, and the point $((\xi_i, 0), t))$ on $M_0 \times U$ by (ξ_i, t). Then

$$\Psi: (\xi_i, t) \to (x_j^1(\xi_i, t), \ldots, x_j^n(\xi_i, t), t)$$

gives a diffeomorphism of $M_0 \times U$ onto $\varpi^{-1}(U)$. Clearly $\varpi \circ \Psi((\xi_i, t)) = t = \pi((\xi_i, t))$.

(2°) General case. By identifying mutually diffeomorphic differentiable manifolds as usual, it suffices to prove that $\varpi^{-1}(U)$ is $M_0 \times U$ and that ϖ coincides with the projection π, which we express simply by writing $\varpi^{-1}(U) = M_0 \times U$.

Put

$$U^{m-1} = \{(t_1, \ldots, t_{m-1}) \,|\, |t_1| < r, \ldots, |t_{m-1}| < r\}, \quad \text{and} \quad U_m = \{t_m \,|\, -r < t_m < r\}.$$

Then we have $U = U^{m-1} \times U_m$. Since by induction hypothesis $\varpi^{-1}(U^{m-1}) = M_0 \times U^{m-1}$, in order to prove that $\varpi^{-1}(U) = M_0 \times U$, it suffices to show that $\varpi^{-1}(U) = \varpi^{-1}(U^{m-1}) \times U_m$.

For $\varpi(p) = (t_1, \ldots, t_{m-1}, t_m)$ with $p \in M$, define ϖ_m by $\varpi_m(p) = t_m$. Then ϖ_m is a C^∞ map of M into $\mathbb{R}$ which maps $\varpi^{-1}(U)$ onto U_m. We denote by π_m the projection of $\varpi^{-1}(U^{m-1}) \times U_m$ onto U_m. As in (1°), using a partition of unity, we can construct a C^∞ vector field on M of the form

$$\sum_{\alpha=1}^{n} v_j^\alpha(x_j^1, \ldots, x_j^n, t_1, \ldots, t_m) \frac{\partial}{\partial x_j^\alpha} + \frac{\partial}{\partial t_m}.$$

By solving the corresponding system of differential equations

$$\begin{cases} \dfrac{dx_j^\alpha}{dt} = v_j^\alpha(x_j^1, \ldots, x_j^n, t_1, \ldots, t_m), & \alpha = 1, \ldots, n, \\[2ex] \dfrac{dt_\nu}{dt} = 0, & \nu = 1, \ldots, m-1, \\[2ex] \dfrac{dt_m}{dt} = 1, & \end{cases}$$

we obtain a diffeomorphism Ψ_m of $\varpi^{-1}(U^{m-1}) \times U_m$ onto $\varpi^{-1}(U)$ such that $\varpi_m \circ \Psi_m = \pi_m$. Thus we have $\varpi^{-1}(U) = \varpi^{-1}(U^{m-1}) \times U_m$. Hence we obtain

$$\varpi^{-1}(U) = \varpi^{-1}(U^{m-1}) \times U_m = M_0 \times U^{m-1} \times U_m = M_0 \times U$$

as desired. ∎

Applying Theorem 2.4 to a complex analytic family $(\mathcal{M}, B, \varpi)$ we obtain the following theorem.

Theorem 2.5. *Let $(\mathcal{M}, B, \varpi)$ be a complex analytic family, c an arbitrary point of B, and $M_c = \varpi^{-1}(c)$. For a sufficiently small coordinate polydisk $U(c)$ with centre c, there is a diffeomorphism Ψ_c of $M_c \times U(c)$ onto $\varpi^{-1}((U(c))$ such that $\varpi \circ \Psi_c = \pi_c$ where π_c is the projection of $M_c \times U(c)$ to $U(c)$.* ∎

The diffeomorphism Ψ_c maps each $M_c \times t$ onto $M_t = \varpi^{-1}(t)$ for $t \in U(c)$. Thus for $t \in U(c)$, M_t is diffeomorphic to M_c. Given any two points t_0, t of B, we can choose a series of coordinate polydisks $U(c_0), \ldots, U(c_l)$ as in Theorem 2.5 such that $t_0 \in U(c_0)$, $t \in U(c_l)$ and that $U(c_{j-1}) \cap U(c_j) \neq \varnothing$ for $j = 1, \ldots, l$. Therefore $M_t = \varpi^{-1}(t)$ is diffeomorphic to $M_{t_0} = \varpi^{-1}(t_0)$, which proves Theorem 2.3. ∎

Now consider holomorphic vector fields on a complex manifold. Let M be a complex manifold, $\{z_1, \ldots, z_j, \ldots\}$, $z_j \colon p \to (z_j^1, \ldots, z_j^n) = z_j(p)$, a system of local complex coordinates, and U_j the domain of z_j. A vector field on M is called a *holomorphic vector field* if it is represented on each coordinate neighbourhood U_j as

$$\sum_{\alpha=1}^{n} v_j^\alpha(p) \frac{\partial}{\partial z_j^\alpha},$$

with holomorphic functions $v_j^\alpha(p)$ on U_j. On $U_j \cap U_k \neq \varnothing$, we have

$$\sum_{\alpha=1}^{n} v_j^\alpha(p) \frac{\partial}{\partial z_j^\alpha} = \sum_{\alpha=1}^{n} v_k^\alpha(p) \frac{\partial}{\partial z_k^\alpha},$$

that is,

$$v_j^\alpha(p) = \sum_{\beta=1}^{n} \frac{\partial z_j^\alpha}{\partial z_k^\beta} v_k^\beta(p).$$

Given two holomorphic vector fields

$$v(p) = \sum_\alpha v_j^\alpha(p) \frac{\partial}{\partial z_j^\alpha} \quad \text{and} \quad w(p) = \sum_\alpha w_j^\alpha(p) \frac{\partial}{\partial z_j^\alpha},$$

we define their linear combination by

$$c_1 v(p) + c_2 w(p) = \sum_{\alpha=1}^{n} (c_1 v_j^\alpha(p) + c_2 w_j^\alpha(p)) \frac{\partial}{\partial z_j^\alpha} \quad \text{with } c_1, c_2 \in \mathbb{C}.$$

Then the set of all holomorphic vector fields on M forms a vector space.

(b) Examples

Let $\{M_t \mid t \in B\}$ be a complex analytic family. The structure of M_t may vary continuously or discontinuously as t varies continuously. We give some examples in this subsection.

Example 2.14 First consider the complex analytic family $\{C_\omega \mid \omega \in \mathbb{H}^+\}$ of elliptic curves where $\mathbb{H}^+ = \{\omega \in \mathbb{C} \mid \operatorname{Im} \omega > 0\}$. Let $C_\omega = \mathbb{C}/G_\omega$. There are infinitely many choices of generators of G_ω. Two elements $\omega_1 = a\omega + b$, and $\omega_2 = c\omega + d$ of G_ω with $a, b, c, d \in \mathbb{Z}$ generate G_ω if and only if $ad - bc = \pm 1$: $G_\omega = \{m\omega_1 + n\omega_2 \mid m, n \in \mathbb{Z}\}$. Interchanging ω_1 and ω_2 if necessary, we may assume that $ad - bc = 1$. By the coordinate transformation $z \to z' = z/\omega_2$, ω_1 and ω_2 are transformed into $\omega' = \omega_1/\omega_2$ and $1 = \omega_2/\omega_2$ respectively. Clearly $\operatorname{Im} \omega' > 0$. Conversely, if

$$\omega' = \frac{a\omega + b}{c\omega + d}, \qquad a, b, c, d \in \mathbb{Z} \quad \text{with} \quad ad - bc = 1, \tag{2.29}$$

$C_\omega = \mathbb{C}/G_\omega$ and $C_{\omega'} = \mathbb{C}/G_{\omega'}$ are biholomorphic to each other.

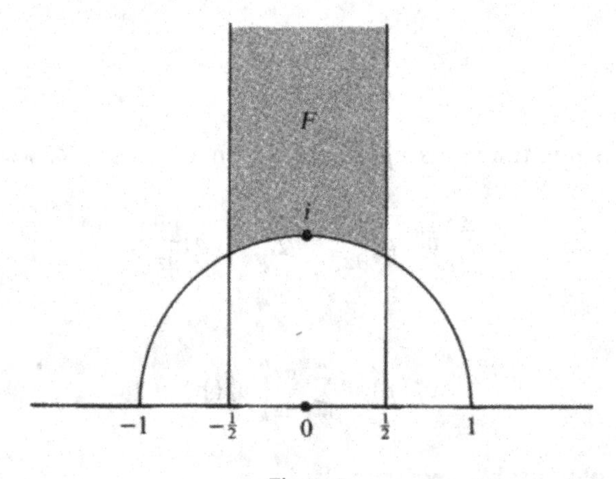

Figure 9

Conversely, if C_ω and $C_{\omega'}$ are complex analytically equivalent, a biholomorphic map of C_ω onto $C_{\omega'}$ is lifted to a linear map $z \to z' = \alpha z + \beta$ of C with $\alpha \neq 0$. Then $\alpha\omega$ and α must generate $G_{\omega'}$. Thus $\omega' = a\alpha\omega + b\alpha$, and $1 = c\alpha\omega + d\alpha$ for some $a, b, c, d \in \mathbb{Z}$ with $ad - bc = 1$. Therefore ω' is given by (2.29). Thus C_ω is biholomorphic to $C_{\omega'}$ if and only if (2.29) holds.

The set of all linear transformations of the form (2.29) forms a group of automorphisms of $\mathbb{H}^+$, which we denote by Γ. Γ acts in a properly discontinuous manner on $\mathbb{H}^+$. $F = \{\omega \mid -\frac{1}{2} \leqq \operatorname{Re} \omega \leqq \frac{1}{2}, \ |\omega| \geqq 1\}$ is a fundamental domain of Γ and H^+/Γ is biholomorphic to C. Consequently there is a Γ-invariant holomorphic function $J(\omega)$ defined on $\mathbb{H}^+$ which induces a biholomorphic map of $\mathbb{H}^+/\Gamma$ onto C. $J(\omega)$ is called the elliptic modular function. C_ω and $C_{\omega'}$ are biholomorphic if and only if $J(\omega)=J(\omega')$. Thus the complex structure of C_ω varies "continuously" as ω moves in $\mathbb{H}^+$.

Example 2.15. By a Hopf surface we mean a Hopf manifold of dimension 2. Let $W=\mathbb{C}^2-\{0\}$, and g_t an automorphism of W given by

$$g_t\colon (z_1, z_2)\to(\alpha z_1+tz_1, \alpha z_2),$$

where $0<|\alpha|<1$, and $t\in\mathbb{C}$. Let $G_t=\{g_t^m\,|\,m\in\mathbb{Z}\}$ be an infinite cyclic group generated by g_t. Then G_t acts on W in a properly discontinuous manner without fixed point. Put $M_t=W/G_t$. M_0 is a Hopf surface given in Example 2.9. M_t for $t\neq 0$ is also called a Hopf surface.

$\{M_t\,|\,t\in\mathbb{C}\}$ forms a complex analytic family. In fact, an automorphism of $W\times\mathbb{C}$ given by

$$g\colon (z_1, z_2, t)\to(\alpha z_1+tz_2, \alpha z_2, t)$$

generates an infinitely cyclic group G, which is properly discontinuous and fixed point free. Hence $\mathcal{M}=W\times\mathbb{C}/G$ is a complex manifold. Since the projection of $W\times\mathbb{C}$ to $\mathbb{C}$ commutes with g, it induces a holomorphic map ϖ of $\mathcal{M}$ to $\mathbb{C}$. Clearly the rank of the Jacobian matrix of ϖ is equal to 1. Thus $(\mathcal{M}, \mathbb{C}, \varpi)$ is a complex analytic family with $\varpi^{-1}(t)=W/G_t=M_t$.

Apparently the complex structure of M_t seems to vary as t moves in $\mathbb{C}$, but this is not true. Let $U=\mathbb{C}-\{0\}$. Then the restriction $(\mathcal{M}_U, U, \varpi_U)$ of $(\mathcal{M}, \mathbb{C}, \varpi)$ to U is proved to be trivial. This follows immediately from the equality

$$\begin{pmatrix}1 & 0\\ 0 & t\end{pmatrix}\begin{pmatrix}\alpha & t\\ 0 & \alpha\end{pmatrix}\begin{pmatrix}1 & 0\\ 0 & t^{-1}\end{pmatrix}=\begin{pmatrix}\alpha & 1\\ 0 & \alpha\end{pmatrix}.$$

In fact, introduce new coordinates $(w_1, w_2, t)=(z_1, tz_2, t)$ on $W\times U$. Then in terms of these coordinates, g is represented as

$$g\colon (w_1, w_2, t)\to(\alpha w_1, \alpha w_2+w_1, t).$$

Therefore

$$\mathcal{M}_U=W\times U/\mathcal{G}=W/G_1\times U=M_1\times U,$$

hence $(\mathcal{M}_U, U, \varpi_U)=(M_1\times U, U, \pi)$.

Thus M_t has the same complex structure as that of M_1 for $t\neq 0$. But the complex structure of M_0 is different from that of M_t with $t\neq 0$. To see this, consider holomorphic vector fields on M_t. A holomorphic vector field on

M_t is induced from a G_t-invariant holomorphic vector field on W. In what follows we write $z' = (z'_1, z'_2)$ instead of $(\alpha^m z_1 + m\alpha^{m-1} t z_2, \alpha^m z_2)$ for simplicity. In this notation we have

$$g_t^m : z = (z_1, z_2) \to z' = (z'_1, z'_2).$$

Let

$$v_1(z)\frac{\partial}{\partial z_1} + v_2(z)\frac{\partial}{\partial z_2} \tag{2.30}$$

be a G_t-invariant holomorphic vector field on W, where $v_1(z)$ and $v_2(z)$ are holomorphic functions on W. Since

$$\frac{\partial}{\partial z_1} = \alpha^m \frac{\partial}{\partial z'_1}, \qquad \frac{\partial}{\partial z_2} = m\alpha^{m-1} t \frac{\partial}{\partial z'_1} + \alpha^m \frac{\partial}{\partial z'_2},$$

the vector field (2.30) is transformed by g_t^m into the vector field

$$(\alpha^m v_1(z) + m\alpha^{m-1} t v_2(z))\frac{\partial}{\partial z'_1} + \alpha^m v_2(z)\frac{\partial}{\partial z'_2}.$$

Since (2.30) is G_t-invariant, we have

$$\begin{aligned} v_1(z') &= \alpha^m v_1(z) + m\alpha^{m-1} t v_2(z), \\ v_2(z') &= \alpha^m v_2(z). \end{aligned} \tag{2.31}$$

According to the Corollary to Theorem 1.8 (Hartog's theorem), holomorphic functions $v_1(z_1, z_2)$ and $v_2(z_1, z_2)$ on W are extended to holomorphic functions on all of $\mathbb{C}^2$. Therefore we may assume that $v_1(z_1, z_2)$ and $v_2(z_1, z_2)$ are holomorphic on all of $\mathbb{C}^2$. By (2.31), we have

$$v_2(z_1, z_2) = \frac{1}{\alpha^m} v_2(\alpha^m z_1 + m\alpha^{m-1} t z_2, \alpha^m z_2).$$

Consequently, since $0 < |\alpha| < 1$, letting

$$v_2(z_1, z_2) = \sum_{h,k=0}^{+\infty} c_{hk} z_1^h z_2^k$$

be the power series expansion of $v_2(z_1, z_2)$, we have

$$\begin{aligned} v_2(z_1, z_2) &= \lim_{m\to+\infty} \frac{1}{\alpha^m} \sum_{h,k} c_{hk} (\alpha^m z_1 + m\alpha^{m-1} t z_2)^h (\alpha^m z_2)^k \\ &= \lim_{m\to+\infty} \left(\frac{c_{00}}{\alpha^m} + c_{10} z_1 + \frac{mt}{\alpha} c_{10} z_2 + c_{01} z_2 \right). \end{aligned}$$

In order that this limit may exist for any z_1, z_2, c_{00} must be zero, and, if $t \neq 0$, c_{10} must also be zero. Thus we have

$$v_2(z_1, z_2) = c_{10}z_1 + c_{01}z_2,$$

where $c_{10} = 0$ if $t \neq 0$. Let

$$v_1(z_1, z_2) = \sum_{h,k=0}^{+\infty} b_{hk} z_1^h z_2^k.$$

Then by (2.31) we have

$$\begin{aligned} v_1(z_1, z_2) &= \lim_{m\to+\infty} \left(\frac{1}{\alpha^m} v_1(\alpha^m z_1 + m\alpha^{m-1} t z_2, \alpha^m z_2) - \frac{mt}{\alpha} v_2(z_1, z_2) \right) \\ &= \lim_{m\to+\infty} \left(\frac{b_{00}}{\alpha^m} + b_{10} z_1 + \frac{mt}{\alpha} b_{10} z_2 + b_{01} z_2 - \frac{mt}{\alpha}(c_{10}z_1 + c_{01}z_2) \right) \\ &= \lim_{m\to+\infty} \left(\frac{b_{00}}{\alpha^m} + \left[b_{10} - \frac{mt}{\alpha} c_{10} \right] z_1 + \frac{mt}{\alpha}[b_{10} - c_{01}] z_2 + b_{01} z_2 \right). \end{aligned}$$

Hence we have $b_{00} = 0$, and, if $t \neq 0$, we also have $b_{10} = c_{01}$. Therefore

$$v_1(z_1, z_2) = b_{10}z_1 + b_{01}z_2$$

holds where $b_{10} = c_{01}$ if $t \neq 0$.

To sum up, if we put $c_1 = b_{10}$, $c_2 = b_{01}$, $c_3 = c_{10}$, and $c_4 = c_{01}$, a holomorphic vector field on M_0 is given by

$$c_1 z_1 \frac{\partial}{\partial z_1} + c_2 z_2 \frac{\partial}{\partial z_1} + c_3 z_1 \frac{\partial}{\partial z_2} + c_4 z_2 \frac{\partial}{\partial z_2},$$

while a holomorphic vector field on M_t with $t \neq 0$ is given by

$$c_1 \left(z_1 \frac{\partial}{\partial z_1} + z_2 \frac{\partial}{\partial z_2} \right) + c_2 z_2 \frac{\partial}{\partial z_1}.$$

Consequently there are four linearly independent holomorphic vector fields on M_0, while on M_t with $t \neq 0$, there are only two such ones. Hence M_0 has a different complex structure from M_t with $t \neq 0$. Thus the complex structure of M_t "jumps" at $t = 0$.

Example 2.16. Consider deformations of the surface $\tilde{M}_m$ given in Example 2.10. Here by a surface we mean a compact complex manifold of dimension 2. Write $\mathbb{P}^1 = \mathbb{C} \cup \infty$ as $\mathbb{P}^1 = U_1 \cup U_2$ where $U_1 = C$ and $U_2 = \mathbb{P}^1 - \{0\}$. Let z_1

and z_2 be the non-homogeneous coordinates on U_1 and U_2 respectively. Then on $U_1 \cap U_2$, $z_1 z_2 = 1$ holds. In Example 2.10 we have defined the surface $\tilde{M}_m$ as

$$\tilde{M}_m = (M - S) \cup \tilde{W}, \qquad \tilde{W} = U_\varepsilon \times \mathbb{P}^1.$$

Since $M - S = U_2 \times \mathbb{P}^1$, putting $\varepsilon = \infty$, we have

$$\tilde{M}_m = U_2 \times \mathbb{P}^1 \cup U_1 \times \mathbb{P}^1,$$

where $(z_1, \zeta_1) \in U_1 \times \mathbb{P}^1$ and $(z_2, \zeta_2) \in U_2 \times \mathbb{P}^1$ are the same point on $\tilde{M}_m$ if

$$z_1 z_2 = 1 \quad \text{and} \quad \zeta_1 = z_2^m \zeta_2. \tag{2.32}$$

Note that here we write ζ_1, ζ_2, and $z_1 = 1/z_2$ for $1/\zeta$, $1/\tilde{\zeta}$, and z respectively in the notation given in Example 2.10.

We define a complex analytic family $\{M_t \mid t \in C\}$ with $M_0 = \tilde{M}_m$ as follows. Fix a natural number $k \leqq m/2$, and define M_t as

$$M_t = U_1 \times \mathbb{P}^1 \cup U_2 \times \mathbb{P}^1,$$

where $(z_1, \zeta_1) \in U_1 \times P^1$ and $(z_2, \zeta_2) \in U_2 \times \mathbb{P}^1$ are the same point of M_t if

$$z_1 z_2 = 1 \quad \text{and} \quad \zeta_1 = z_2^m \zeta_2 + t z_2^k. \tag{2.33}$$

Clearly M_t is a surface, and $\{M_t \mid t \in \mathbb{C}\}$ forms a complex analytic family.

For $t = 0$, $M_0 = \tilde{M}_m$ since in this case (2.33) becomes (2.32).

For $t \neq 0$, $M_t = \tilde{M}_{m-2k}$. To see this we introduce new coordinates (z_i, ζ_i') on $U_i \times \mathbb{P}^1$, $i = 1, 2$, as follows.

$$(z_1, \zeta_1') = \left(z_1, \frac{z_1^k \zeta_1 - t}{t \zeta_1} \right),$$

$$(z_2, \zeta_1') = \left(z_2, \frac{\zeta_2}{t z_2^{m-k} \zeta_2 + t^2} \right).$$

Since the determinants made by the coefficients of the linear transformations

$$\zeta_1' = \frac{z_1^k \zeta_1 - t}{t \zeta_1} \quad \text{and} \quad \zeta_2' = \frac{\zeta_2}{t z_2^{m-k} \zeta_2 + t^2}$$

are given by

$$\begin{vmatrix} z_1^k & -t \\ t & 0 \end{vmatrix} = t^2 \neq 0 \quad \text{and} \quad \begin{vmatrix} 1 & 0 \\ t z_2^{m-k} & t^2 \end{vmatrix} = t^2 \neq 0,$$

respectively, (z_i, ζ_i') actually define coordinates on $U_i\times\mathbb{P}^1$. By (2.33), we have

$$\zeta_1'=\frac{z_1^k\zeta_1-t}{t\zeta_1}=\frac{z_1^k(z_2^m\zeta_2+tz_2^k)-t}{t(z_2^m\zeta_2+tz_2^k)}$$
$$=\frac{z_2^{m-k}\zeta_2}{tz_2^m\zeta_2+t^2z_2^k}=z_2^{m-2k}\frac{\zeta_2}{tz_2^{m-k}+t^2}=z_2^{m-2k}\zeta_2'.$$

Thus in terms of these new coordinates, the relation (2.33) is given by

$$z_1\circ z_2=1 \quad\text{and}\quad \zeta_1'=z_2^{m-2k}\zeta_2',$$

hence, $M_t=\tilde{M}_{m-2k}$.

Thus *for any natural number* $k\leqq m/2$, $\tilde{M}_m$ *is a deformation of* $\tilde{M}_{m-2k}$. Hence by putting $k=m/2$ if m is even, and $k=m/2-\frac{1}{2}$ if m is odd, we see that $\tilde{M}_m$ *is a deformation of* $\tilde{M}_0=\mathbb{P}^1\times\mathbb{P}^1$ *if* m *is even, and a deformation of* $\tilde{M}_1$ *if* m *is odd.* Therefore by Theorem 2.3, $\tilde{M}_m$ *is diffeomorphic to* $\mathbb{P}^1\times\mathbb{P}^1$ *if* m *is even and to* $\tilde{M}_1$ *if* m *is odd.* We have already proved in Example 2.10 that $\tilde{M}_1$ and $\mathbb{P}^1\times\mathbb{P}^1$ are not diffeomorphic.

Thus $\tilde{M}_m$ and $\tilde{M}_n$ are diffeomorphic if $m\equiv n \pmod 2$, but *they are not biholomorphic,* if $m\neq n$. Consequently in the family $\{M_t\,|\,t\in C\}$ described above, $M_t=\tilde{M}_{m-2k}$ does not change its complex structure for all $t\neq 0$, and the complex structure of M_t jumps to that of $M_0=\tilde{M}_m$ at $t=0$. We show that $\tilde{M}_m$ is not biholomorphic to $\tilde{M}_n$ if $m\neq n$, by computing the number of linearly independent holomorphic vector fields on them.

First consider holomorphic vector fields on $\mathbb{P}^1=U_1\cup U_2$. A holomorphic vector field on $\mathbb{P}^1$ is represented as $v_1(z_1)(d/dz_1)$ on U_1, $v_2(z_2)(d/dz_2)$ on U_2, where $v_i(z_i)$ are entire functions of z_i on U_i for $i=1,2$. On $U_1\cap U_2$, they must coincide:

$$v_1(z_1)\frac{d}{dz_1}=v_2(z_2)\frac{d}{dz_2}. \tag{2.34}$$

Since $z_1=1/z_2$, we have

$$\frac{d}{dz_2}=\frac{dz_1}{dz_2}\frac{d}{dz_1}=-\frac{1}{z_2^2}\frac{d}{dz_1}=-z_1^2\frac{d}{dz_1}.$$

Hence, substituting this into (2.34), we obtain

$$v_1(z_1)=-z_1^2v_2\left(\frac{1}{z_1}\right).$$

Therefore, putting

$$v_1(z_1)=\sum_{n=0}^{\infty}c_{1n}z_1^n \quad\text{and}\quad v_2(z_2)=\sum_{n=0}^{\infty}c_{2n}z_2^n,$$

we have

$$c_{10}+c_{11}z_1+c_{12}z_1^2+\cdots=-c_{20}z_1^2-c_{21}z_1-c_{22}-\cdots.$$

Thus $v_1(z_1)$ must be a quadric $az_1^2+bz_1+c$ in z_1:

$$v_1(z_1)\frac{d}{dz_1}=(az_1^2+bz_1+c)\frac{d}{dz_1}. \tag{2.35}$$

Since a holomorphic function on $\mathbb{P}^1$ is a constant, a holomorphic vector field on $\tilde{M}_0=\mathbb{P}^1\times\mathbb{P}^1$ is given by

$$(a_1z_1^2+b_1z_1+c_1)\frac{\partial}{\partial z_1}+(\alpha_1\zeta_1^2+\beta_1\zeta_1+\gamma_1)\frac{\partial}{\partial \zeta_1}.$$

Consequently there are six linearly independent holomorphic vector fields on $\tilde{M}_0$.

Next we consider holomorphic vector fields on $\tilde{M}_m=U_1\times\mathbb{P}^1\cup U_2\times\mathbb{P}^1$, $m\geqq 1$. From (2.35) a holomorphic vector field on $U_1\times\mathbb{P}^1$ has the form

$$v_1(z_1)\frac{\partial}{\partial z_1}+(\alpha_1(z_1)\zeta_1^2+\beta_1(z_1)\zeta_1+\gamma_1(z_1))\frac{\partial}{\partial \zeta_1}, \tag{2.36}$$

where $v_1(z_1)$, $\alpha_1(z_1)$, $\beta_1(z_1)$, $\gamma_1(z_1)$ are entire functions of z_1. Similarly a holomorphic vector field on $U_2\times\mathbb{P}^1$ is given by

$$v_2(z_2)\frac{\partial}{\partial z_2}+(\alpha_2(z_2)\zeta_2^2+\beta_2(z_2)\zeta_2+\gamma_2(z_2))\frac{\partial}{\partial \zeta_2}, \tag{2.37}$$

where $v_2(z_2)$, $\alpha_2(z_2)$, $\beta_2(z_2)$, $\gamma_2(z_2)$ are entire functions of z_2.

For a holomorphic vector field on $\tilde{M}_m$ which has the form (2.36) on $U_1\times\mathbb{P}^1$, and the form (2.37) on $U_2\times\mathbb{P}^1$, we must have

$$\begin{aligned}v_1(z_1)\frac{\partial}{\partial z_1}&+(\alpha_1(z_1)\zeta_1^2+\beta_1(z_1)\zeta_1+\gamma_1(z_1))\frac{\partial}{\partial \zeta_1}\\ &=v_2(z_2)\frac{\partial}{\partial z_2}+(\alpha_2(z_2)\zeta_2^2+\beta_2(z_2)\zeta_2+\gamma_2(z_2))\frac{\partial}{\partial \zeta_2}\end{aligned} \tag{2.38}$$

on $U_1\times P^1\cap U_2\times P^1$. Since $z_1=1/z_2$ and $\zeta_1=z_2^m\zeta_2$ by (2.32), we have

$$\begin{cases}\dfrac{\partial}{\partial z_2}=-\dfrac{1}{z_2^2}\dfrac{\partial}{\partial z_1}+mz_2^{m-1}\zeta_2\dfrac{\partial}{\partial \zeta_1}=-z_1^2\dfrac{\partial}{\partial z_1}+mz_1\zeta_1\dfrac{\partial}{\partial \zeta_1},\\[2ex] \dfrac{\partial}{\partial \zeta_2}=z_2^m\dfrac{\partial}{\partial \zeta_1}=\dfrac{1}{z_1^m}\dfrac{\partial}{\partial \zeta_1}.\end{cases}$$

Substituting these into the right-hand side of (2.38), and comparing the coefficients, we have

$$\begin{cases} v_1(z_1) = -z_1^2 v_2\left(\dfrac{1}{z_1}\right), \\ \alpha_1(z_1) = z_1^m \alpha_2\left(\dfrac{1}{z_1}\right), \\ \beta_1(z_1) = m z_1 v_2\left(\dfrac{1}{z_1}\right) + \beta_2\left(\dfrac{1}{z_1}\right), \\ \gamma_1(z_1) = \dfrac{1}{z_1^m}\gamma_2\left(\dfrac{1}{z_2}\right). \end{cases}$$

From these equalities, we have $v_1(z_1) = az_1^2 + bz_1 + c$, $\alpha_1(z_1) = \sum_{k=0}^{n} c_k z_1^k$, $\beta_1(z_1) = -maz_1 + d$, and $\gamma_1(z_1) = 0$. Therefore on $\tilde{M}_m$ with $m \geqq 1$, there are $(m+5)$ linearly independent vector fields corresponding to the arbitrary constants a, b, c, $c_0, c_1, \ldots, c_m$ and d above.

Thus the number of linearly independent holomorphic vector fields on $\tilde{M}_m$ is 6 for $m = 0$, and $m+5$ for $m \geqq 1$. Hence if $m \equiv n \pmod 2$, and $m \neq n$, $\tilde{M}_m$ and $\tilde{M}_n$ are not biholomorphic.

Chapter 3

Differential Forms, Vector Bundles, Sheaves

As was stated in the Preface, the purpose of this book is to explain the development of the theory of deformations of complex structures since 1956. In this chapter we prepare various results about complex manifolds required for this purpose.

§3.1. Differential Forms

(a) Differential Forms on Differentiable Manifolds

Let Σ be a differential manifold, $\{x_1, \ldots, x_j, \ldots\}$, $x_j: p \to x_j(p) = (x_j^1, \ldots, x_j^m)$ a system of local C^∞ coordinates on Σ, and U_j the domain of x_j. We may assume that the open covering $\mathfrak{U} = \{U_j \mid j = 1, 2, \ldots\}$ of Σ is locally finite as in Theorem 2.1. Also as in §2.1(a), we may consider Σ as obtained by glueing the domains $\mathcal{U}_j = x_j(U_j) \subset \mathbb{R}^m$, by identifying $p \in U_j$ with its local coordinates $x_j = x_j(p)$: $\Sigma = \bigcup_j \mathcal{U}_j$. Thus in the sequel we write U_j for $\mathcal{U}_j$. Then we have $\Sigma = \bigcup_j U_j$ with $U_j \subset \mathbb{R}^m$.

Differential forms on Σ, their wedge products and exterior differentials are defined as follows. Let $\varphi_{j_1}, \ldots, \varphi_{j_m}$ be real or complex-valued functions on U_j. Suppose given on each U_j a linear combination of dx_j's

$$\sum_{\alpha=1}^{m} \varphi_{j\alpha}\, dx_j^\alpha = \varphi_{j1}\, dx_j^1 + \cdots + \varphi_{jm}\, dx_j^m.$$

If on $U_j \cap U_k$, the equality

$$\sum_{\alpha=1}^{m} \varphi_{j\alpha}\, dx_j^\alpha = \sum_{\alpha=1}^{m} \varphi_{k\alpha}\, dx_k^\alpha \tag{3.1}$$

holds, we call

$$\varphi = \varphi_{j1}\, dx_j^1 + \cdots + \varphi_{jm}\, dx_j^m \tag{3.2}$$

a differential form of degree 1 or simply a *1-form* on Σ. If $\varphi_{j\alpha}$, $\alpha = 1, \ldots, m$, are continuous, continuously differentiable, or C^∞, φ is called continuous, continuously differentiable, or C^∞, respectively. Equation (3.1) implies that by substituting

$$dx_j^\alpha = \sum_{\beta=1}^{m} \frac{\partial x_j^\alpha}{\partial x_k^\beta}\, dx_k^\beta,$$

we have

$$\varphi_{k\beta} = \sum_{\alpha=1}^{m} \frac{\partial x_j^\alpha}{\partial x_k^\beta} \varphi_{j\alpha}. \tag{3.3}$$

Also we mean by (3.2) that φ is given on each U_j by (3.2). In terms of arbitrary local coordinates $(x^1, \ldots, x^m)$, φ is given as

$$\varphi = \varphi_1\, dx^1 + \cdots + \varphi_m\, dx^m$$

where we put

$$\varphi_\alpha = \sum_{\beta=1}^{m} \frac{\partial x_j^\beta}{\partial x^\alpha} \varphi_{j\beta}.$$

Similarly we call

$$\varphi = \frac{1}{2} \sum_{\alpha,\beta=1}^{m} \varphi_{\alpha\beta}\, dx^\alpha \wedge dx^\beta = \frac{1}{2} \sum_{\alpha,\beta=1}^{m} \varphi_{j\alpha\beta}\, dx_j^\alpha \wedge dx_j^\beta$$

a 2-form on Σ, where $dx^\alpha \wedge dx^\beta = -dx^\beta \wedge dx^\alpha$. More generally, we call

$$\varphi = \frac{1}{r!} \sum \varphi_{\alpha_1 \cdots \alpha_r}\, dx^{\alpha_1} \wedge \cdots \wedge dx^{\alpha_r}$$

an *r-form* on Σ. If the coefficients $\varphi_{\alpha_1 \cdots \alpha_r}$ are continuous, continuously differentiable, or C^∞, φ is said to be continuous, continuously differentiable, or C^∞, respectively.

For $r = 2$, replacing $\varphi_{\alpha\beta}$ by $\frac{1}{2}(\varphi_{\alpha\beta} - \varphi_{\beta\alpha})$, we may assume that $\varphi_{\alpha\beta} = -\varphi_{\beta\alpha}$. More generally, we may assume that $\varphi_{\alpha_1 \cdots \alpha_r}$ are skew-symmetric in the indices $\alpha_1, \ldots, \alpha_r$. In this case we have

$$\varphi = \sum_{\alpha_1 < \cdots < \alpha_r} \varphi_{\alpha_1 \cdots \alpha_r}\, dx^{\alpha_1} \wedge \cdots \wedge dx^{\alpha_r}.$$

In the following, we use this convention unless otherwise mentioned. Writing $\alpha, \beta, \ldots, \gamma$ for $\alpha_1, \alpha_2, \ldots, \alpha_r$, we have

$$\varphi = \frac{1}{r!} \sum \varphi_{\alpha \cdots \gamma}\, dx^\alpha \wedge \cdots \wedge dx^\gamma.$$

We define a linear combination of r-forms φ, and ψ as

$$c_1\varphi + c_2\psi = \frac{1}{r!}\sum (c_1\varphi_{\alpha\cdots\gamma} + c_2\psi_{\alpha\cdots\gamma})\, dx^\alpha \wedge \cdots \wedge dx^\gamma.$$

The wedge product of an r-form φ and an s-form ψ is defined as follows:

$$\varphi \wedge \psi = \frac{1}{r!\, s!}\sum \varphi_{\alpha_1\cdots\alpha_r}\psi_{\beta_1\cdots\beta_s}\, dx^{\alpha_1} \wedge \cdots \wedge dx^{\alpha_r} \wedge dx^{\beta_1} \wedge \cdots \wedge dx^{\beta_s}. \tag{3.4}$$

For a continuously differentiable function f on Σ, its differential

$$df = \frac{\partial f}{\partial x^1}\, dx^1 + \cdots + \frac{\partial f}{\partial x^m}\, dx^m$$

is a continuous 1-form. For two such functions f and g, clearly

$$d(fg) = f\, dg + g\, df$$

holds. For a continuously differentiable r-form

$$\varphi = \frac{1}{r!}\sum_{\alpha,\ldots,\gamma} \varphi_{\alpha\cdots\gamma}\, dx^\alpha \wedge \cdots \wedge dx^\gamma,$$

we define its *exterior differential* $d\varphi$ by

$$d\varphi = \frac{1}{r!}\sum_{\alpha,\cdots,\gamma} d\varphi_{\alpha\cdots\gamma} \wedge dx^\alpha \wedge \cdots \wedge dx^\gamma. \tag{3.5}$$

Thus, for a 1-form φ we have

$$d\varphi = \sum_\alpha \sum_\beta \frac{\partial \varphi_\alpha}{\partial x^\beta}\, dx^\beta \wedge dx^\alpha = \frac{1}{2}\sum_{\alpha,\beta}\left(\frac{\partial \varphi_\beta}{\partial x^\alpha} - \frac{\partial \varphi_\alpha}{\partial x^\beta}\right) dx^\alpha \wedge dx^\beta,$$

and if φ is a 2-form, we have

$$d\varphi = \frac{1}{3!}\sum_{\alpha,\beta,\gamma}\left(\frac{\partial \varphi_{\beta\gamma}}{\partial x^\alpha} + \frac{\partial \varphi_{\gamma\alpha}}{\partial x^\beta} + \frac{\partial \varphi_{\alpha\beta}}{\partial x^\gamma}\right) dx^\alpha \wedge dx^\beta \wedge dx^\gamma.$$

In general for an r-form φ, if we write $d\varphi$ as

$$d\varphi = \frac{1}{(r+1)!}\sum (d\varphi)_{\alpha_0\alpha_1\cdots\alpha_r}\, dx^{\alpha_0} \wedge dx^{\alpha_1} \wedge \cdots \wedge dx^{\alpha_r},$$

we have

$$(d\varphi)_{\alpha_0\alpha_1\cdots\alpha_r} = \sum_{s=0}^{r} (-1)^s \frac{\partial \varphi_{\alpha_0\cdots\alpha_{s-1}\alpha_{s+1}\cdots\alpha_r}}{\partial x^{\alpha_s}}. \tag{3.6}$$

The definition of exterior differential d is independent of the choice of local coordinate systems. Namely, on $U_j \cap U_k$, we have

$$d\left(\frac{1}{r!}\sum \varphi_{j\alpha\cdots\gamma}\, dx_j^\alpha \wedge \cdots \wedge dx_j^\gamma\right) = d\left(\frac{1}{r!}\sum \varphi_{k\alpha\cdots\gamma}\, dx_k^\alpha \wedge \cdots \wedge dx_k^\gamma\right)$$

First we prove this for $r = 1$. Since by (3.3)

$$\varphi_{j\alpha} = \sum_\lambda \frac{\partial x_k^\lambda}{\partial x_j^\alpha} \varphi_{k\lambda},$$

we have

$$d\varphi_{j\alpha} = \sum_\lambda \varphi_{k\lambda}\, d\left(\frac{\partial x_k^\lambda}{\partial x_j^\alpha}\right) + \sum_\lambda \frac{\partial x_k^\lambda}{\partial x_j^\alpha}\, d\varphi_{k\lambda}.$$

But since $\partial^2 x_k^\lambda / \partial x_j^\beta\, \partial x_j^\alpha$ is symmetric, and $dx_j^\beta \wedge dx_j^\alpha$ is skew-symmetric in α and β, we have

$$\sum_\alpha d\left(\frac{\partial x_k^\lambda}{\partial x_j^\alpha}\right) \wedge dx_j^\alpha = \sum_{\beta,\alpha} \frac{\partial^2 x_k^\lambda}{\partial x_j^\beta\, \partial x_j^\alpha}\, dx_j^\beta \wedge dx_j^\alpha = 0, \tag{3.7}$$

hence

$$\sum_\alpha d\varphi_{j\alpha} \wedge dx_j^\alpha = \sum_\lambda d\varphi_{k\lambda} \wedge \sum_\lambda \frac{\partial x_k^\lambda}{\partial x_j^\alpha}\, dx_j^\alpha = \sum_\lambda d\varphi_{k\lambda} \wedge dx_k^\lambda.$$

Proof for general r also follows immediately from (3.7).

The wedge product $\varphi \wedge \psi$ is clearly bilinear in φ and ψ. If φ is an r-form, we denote it by φ^r explicitly. Then the following equality holds:

$$\varphi^r \wedge \psi^s = (-1)^{rs} \psi^s \wedge \varphi^r. \tag{3.8}$$

Clearly d is linear, that is

$$d(c_1\varphi + c_2\psi) = c_1\, d\varphi + c_2\, d\psi,$$

where c_1, c_2 are constant. Taking the exterior differentials of both sides of (3.4), we obtain

$$d(\varphi^r \wedge \psi^s) = d\varphi^r \wedge \psi^s + (-1)^r \varphi^r \wedge d\psi^s, \tag{3.9}$$

since

$$d\psi_{\beta_1\cdots\beta_s}\wedge dx^{\alpha_1}\wedge\cdots\wedge dx^{\alpha_r}=(-1)^r dx^{\alpha_1}\wedge\cdots\wedge dx^{\alpha_r}\wedge d\psi_{\beta_1\cdots\beta_s}.$$

If φ is C^2, then $\partial^2\varphi_{\alpha_1\cdots\alpha_r}/\partial x^\alpha\,\partial x^\beta$ is symmetric in α and β, hence by the definition of the exterior differential (3.5), we have

$$\begin{aligned}dd\varphi&=d\left(\frac{1}{r!}\sum\sum_{\alpha}\frac{\partial\varphi_{\alpha_1\cdots\alpha_r}}{\partial x^\alpha}\,dx^\alpha\wedge dx^{\alpha_1}\wedge\cdots\wedge dx^{\alpha_r}\right)\\&=\frac{1}{r!}\sum\sum_{\alpha}\sum_{\beta}\frac{\partial^2\varphi_{\alpha_1\cdots\alpha_r}}{\partial x^\alpha\,\partial x^\beta}\,dx^\beta\wedge dx^\alpha\wedge dx^{\alpha_1}\wedge\cdots\wedge dx^{\alpha_r}=0.\end{aligned}$$

Thus we obtain

$$dd\varphi=0. \tag{3.10}$$

Next we define the integral of a differential form. For this purpose first we define the orientation of a differentiable manifold. We call a differentiable manifold Σ *orientable* if we can choose a system of local C^∞ coordinates $\{x_1, x_2,\ldots\}$ such that on each $U_j\cap U_k\neq\emptyset$,

$$\det\left(\frac{\partial x_j^\alpha}{\partial x_k^\beta}\right)_{\alpha,\beta=1,\ldots,m}=\det\frac{\partial(x_j^1,\ldots,x_j^m)}{\partial(x_k^1,\ldots,x_k^m)}>0. \tag{3.11}$$

In case Σ is orientable, by choosing a system of local C^∞ coordinates $\{x_1, x_2,\ldots\}$ such that (3.11) holds on each $U_j\cap U_k\neq\emptyset$, we define an *orientation* on Σ. Let $\{u_1, u_2,\ldots\}$ be another system of local C^∞ coordinates satisfying (3.11) on each $W_\lambda\cap W_\nu\neq\emptyset$ where W_λ is the domain of u_λ. Then for all pairs (λ, j) of indices such that $W_\lambda\cap U_j\neq\emptyset$, $\det(\partial u_\lambda^\alpha/\partial x_j^\beta)$ are all simultaneously >0, or <0. In the former case we say that $\{u_\lambda\}$ defines the same orientation as $\{x_j\}$, and in the latter case the inverse orientation.

Let Σ be an orientable differentiable manifold oriented by a system of local C^∞ coordinates $\{x_1, x_2,\ldots\}$. Let ω be a continuous m-form with $m=\dim\Sigma$. Then, on each coordinate neighbourhood U_j, ω is written as

$$\omega=\frac{1}{m!}\sum\omega_{j\alpha_1\cdots\alpha_m}\,dx_j^{\alpha_1}\wedge\cdots\wedge dx_j^{\alpha_m}.$$

Since in this case $\{\alpha_1,\ldots,\alpha_m\}$ is a permutation of $\{1,\ldots,m\}$, we can write ω as

$$\omega=\omega_{1\cdots m}\,dx_j^1\wedge\cdots\wedge dx_j^m.$$

On $U_j\cap U_k\neq\emptyset$, we have

$$\omega_{j1\cdots m}\,dx_j^1\wedge\cdots\wedge dx_j^m=\omega_{k1\cdots m}\,dx_k^1\wedge\cdots\wedge dx_k^m.$$

Using $dx_k^\beta = \sum_{\alpha=1}^m (\partial x_k^\beta / \partial x_j^\alpha)\, dx_j^\alpha$, we see that

$$dx_k^1 \wedge \cdots \wedge dx_k^m = \sum_{\alpha_1,\ldots,\alpha_m} \frac{\partial x_k^1}{\partial x_j^{\alpha_1}} \cdots \frac{\partial x_k^m}{\partial x_j^{\alpha_m}}\, dx_j^{\alpha_1} \wedge \cdots \wedge dx_j^{\alpha_m}$$

$$= \det \frac{\partial(x_k^1, \ldots, x_k^m)}{\partial(x_j^1, \ldots, x_j^m)}\, dx_j^1 \wedge \cdots \wedge dx_j^m,$$

thus

$$dx_k^1 \wedge \cdots \wedge dx_k^m = \det \frac{\partial(x_k^1, x_k^2, \ldots, x_k^m)}{\partial(x_j^1, x_j^2, \ldots, x_j^m)}\, dx_j^1 \wedge \cdots \wedge dx_j^m. \tag{3.12}$$

Therefore we have

$$\omega_{j1\cdots m} = \det \frac{\partial(x_k^1, \ldots, x_k^m)}{\partial(x_j^1, \ldots, x_j^m)}\, \omega_{k1\cdots m}. \tag{3.13}$$

The integral of ω on Σ is defined as follows. Let $\{\rho_j | j = 1, \ldots\}$ be a partition of unity subordinate to a locally finite open covering $\mathfrak{U} = \{U_j | j = 1, \ldots\}$. Then we define

$$\int_\Sigma \omega = \sum_j \int_{U_j} \rho_j \omega_{1\cdots m}\, dx_j^1\, dx_j^2 \cdots dx_j^m. \tag{3.14}$$

If Σ is not compact, we must examine the convergence of the sum in the right-hand side of (3.14). Since by hypothesis $\det(\partial x_k^\alpha / \partial x_j^\beta) > 0$, it follows from (3.12) and (3.13) we have

$$|\omega_{j1\cdots m}|\, dx_j^1 \wedge \cdots \wedge dx_j^m = |\omega_{k1\cdots m}|\, dx_k^1 \wedge \cdots \wedge dx_k^m$$

on $U_j \cap U_k \neq \varnothing$. Hence, by putting

$$|\omega| = |\omega_{j1\cdots m}|\, dx_j^1 \wedge \cdots \wedge dx_j^m$$

on each U_j, we obtain a continuous m-form $|\omega|$ on Σ, and

$$\int_\Sigma |\omega| = \sum_j \int_{U_j} \rho_j |\omega_{j1\cdots m}|\, dx_j^1 \cdots dx_j^m.$$

The sum in the right-hand side of this equality either converges or diverges to $+\infty$. If $\int_\Sigma |\omega| < +\infty$, the right-hand side of (3.14) converges absolutely, hence $\int_\Sigma \omega$ exists. In this case we say that the integral $\int_\Sigma \omega$ converges absolutely. When $\int_\Sigma |\omega| = +\infty$, we don't define $\int_\Sigma \omega$.

If the orientation of the differentiable manifold Σ is fixed, the integral $\int_\Sigma \omega$ defined above is independent of the choice of systems of local C^∞

coordinates $\{x_1, x_2, \ldots\}$ and partitions of unity $\{\rho_j | j=1,\ldots\}$. In the definition of the integral we introduce a partition of unity, in order to, we must represent the integral of ω as a sum of continuous functions defined on domains of Σ. If we use the theory of Lebesgue integral, we can do without a partition of unity. Namely, if we choose Lebesgue measurable subsets $V_j \subset U_j, j=1,2,\ldots$ such that $V_j \cap V_k = \varnothing$ for $j \neq k$ and that $\bigcup_j V_j = \Sigma$, then

$$\int_\Sigma \omega = \sum_j \int_{V_j} \omega_{1\cdots m}\, dx_j^1 \cdots dx_j^m.$$

We say that an r-form

$$\varphi = \varphi(p) = \frac{1}{r!} \sum_{\alpha,\ldots,\gamma} \varphi_{j\alpha\cdots\gamma}\, dx_j^\alpha \wedge \cdots \wedge dx_j^\gamma$$

vanishes at $p \in \Sigma$ if all coefficients $\varphi_{j\alpha\cdots\gamma}$ vanish at p. We define the *support* of φ, denoted by supp φ, as

$$\operatorname{supp}\varphi = [\{p \in \Sigma | \varphi(p) \neq 0\}].$$

Theorem 3.1. *Let φ be a continuously differentiable $(m-1)$-form, with $m = \dim \Sigma$. If* supp φ *is compact, then*

$$\int_\Sigma d\varphi = 0. \tag{3.15}$$

Proof. Choose a system of local C^∞ coordinates $\{x_j\}$ such that each U_j is an interval $\{x_j \in \mathbb{R}^m \,|\, |x_j^1| < r_j^1, \ldots, |x_j^m| < r_j^m\}$. We may assume that all r_j^k are equal to 1. Since supp φ is compact, and $\mathfrak{U} = \{U_j\}$ is locally finite, there are only a finite number of U_j with $U_j \cap \operatorname{supp}\varphi \neq \varnothing$. Hence for a sufficiently large l we have

$$\int_\Sigma d\varphi = \int_\Sigma d\left(\sum_{j=1}^l \rho_j \varphi\right) = \sum_{j=1}^l \int_{U_j} d(\rho_j \varphi).$$

Therefore it suffices to prove $\int_{U_j} d(\rho_j\varphi) = 0$ for each U_j. We write

$$\rho_j \varphi = \sum_{\alpha=1}^m (-1)^{\alpha-1} \sigma_\alpha(x_j)\, dx_j^1 \wedge \cdots \wedge dx_j^{\alpha-1} \wedge dx_j^{\alpha+1} \wedge \cdots \wedge dx_j^m,$$

where $\sigma_\alpha = \sigma_\alpha(x_j)$ is a continuously differentiable function with $\operatorname{supp}\sigma_\alpha \subset U_j$. Since

$$d(\rho_j\varphi) = \sum_{\alpha=1}^m \frac{\partial \sigma_\alpha}{\partial x_j^\alpha}\, dx_j^1 \wedge \cdots \wedge dx_j^m,$$

we have

$$\int_{U_j} d(\rho_j\varphi) = \sum_{\alpha=1}^{m} \int_{U_j} \frac{\partial \sigma_\alpha}{\partial x_j^\alpha} dx_j^1 \cdots dx_j^m.$$

Then it is easy to see that each term of the right-hand side of this equality is 0. For example, the first term vanishes because

$$\int_{U_j} \frac{\partial \sigma_1}{\partial x_j^1} dx_j^1 \cdots dx_j^m = \int_{-1}^{1} \cdots \int_{-1}^{1} dx_j^2 \cdots dx_j^m \int_{-1}^{1} \frac{\partial \sigma_1}{\partial x_j^1} dx_j^1$$

$$= \int_{-1}^{1} \cdots \int_{-1}^{1} dx_j^2 \cdots dx_j^m [\sigma_1(1, x_j^2, \ldots, x_j^m)$$

$$- \sigma_1(-1, x_j^2, \ldots, x_j^m)] = 0. \quad \blacksquare$$

Corollary. *Let φ be a C^1 r-form on Σ and ψ a C^1 $(m-r-1)$-form. Suppose that at least one of* supp φ *and* supp ψ *is compact, then*

$$\int_\Sigma d\varphi \wedge \psi = (-1)^{r-1} \int_\Sigma \varphi \wedge d\psi. \tag{3.16}$$

Proof. By (3.9), $d(\varphi \wedge \psi) = d\varphi \wedge \psi + (-1)^r \varphi \wedge d\psi$. Then applying (3.15), we obtain (3.16). ∎

Any open subset W of Σ is itself a differentiable manifold. Hence the above argument applies also to W.

We call φ a *closed differential form* if φ is C^1 and $d\varphi = 0$. An r-form φ on a domain W is called an *exact differential form* if there is a C^1 $(r-1)$-form ψ on W such that $\varphi = d\psi$. If both φ and ψ are C^1, and $\varphi = d\psi$, then $d\varphi = 0$. This is obvious from (3.10) if ψ is C^2. In case $\psi = C^1$, this is proved as follows.

Proof. Since the problem is local, it suffices to see $d\varphi = dd\psi = 0$ on each U_j. For this it suffices to see that $\int_{U_j} dd\psi \wedge \eta = 0$ for any C^∞ $(m-r-1)$-form η on U_j with compact support. Since by (3.10) $dd\eta = 0$, using (3.16) we obtain

$$\int_{U_j} dd\psi \wedge \eta = (-1)^r \int_{U_j} d\psi \wedge d\eta = -\int_{U_j} \psi \wedge dd\eta = 0. \quad \blacksquare$$

Thus C^1 exact differential forms are closed. Locally its inverse also holds.

Theorem 3.2 (Poincaré's Lemma). *Suppose that a C^1 r-form φ, $r \geqq 1$, satisfies $d\varphi = 0$ on an interval $U = \{x \in \mathbb{R}^m \,|\, |x^\alpha| < 1,\ \alpha = 1, \ldots, m\}$. Then there is a C^1 $(r-1)$-form ψ on U such that $\varphi = d\psi$. Moreover if φ is C^∞, we may choose ψ to be C^∞.*

Proof. We fix r and prove the theorem by induction on m for $m \geqq r$. For simplicity we only treat the case $r=3$ below, but the generalization is straightforward.

(1°) The case $m=r=3$. Put $\varphi=\varphi_{312}\,dx^3\wedge dx^1\wedge dx^2$, and define $\psi=\psi_{12}\,dx^1\wedge dx^2$ by

$$\psi_{12}=\psi_{12}(x^1,x^2,x^3)=\int_0^{x^3}\varphi_{312}(x^1,x^2,x^3)\,dx^3.$$

Then ψ is a C^1 2-form on U and

$$d\psi=\frac{\partial\psi_{12}}{\partial x^3}\,dx^3\wedge dx^1\wedge dx^2=\varphi_{312}(x^1,x^2,x^3)\,dx^1\wedge dx^2\wedge dx^3=\varphi.$$

(2°) The case $m>r=3$. In what follows the indices α,β,γ represent numbers from 1 to $m-1$. Define

$$\psi=\frac{1}{2!}\sum_{\alpha,\beta}\psi_{\alpha\beta}\,dx^\alpha\wedge dx^\beta$$

by

$$\begin{aligned}\psi_{\alpha\beta}&=\psi_{\alpha\beta}(x^1,\ldots,x^{m-1},x^m)\\&=\int_0^{x^m}\varphi_{m\alpha\beta}(x^1,\ldots,x^{m-1},x^m)\,dx^m.\end{aligned}$$

Then ψ is a C^1 2-form. By (3.6) we have

$$\frac{\partial\varphi_{\alpha\beta\gamma}}{\partial x^m}-\frac{\partial\varphi_{m\beta\gamma}}{\partial x^\alpha}+\frac{\partial\varphi_{m\alpha\gamma}}{\partial x^\beta}-\frac{\partial\varphi_{m\alpha\beta}}{\partial x^\gamma}=(d\varphi)_{m\alpha\beta\gamma}=0,$$

hence

$$\begin{aligned}(d\psi)_{\alpha\beta\gamma}&=\int_0^{x^m}\left[\frac{\partial\varphi_{m\beta\gamma}}{\partial x^\alpha}-\frac{\partial\varphi_{m\alpha\gamma}}{\partial x^\beta}+\frac{\partial\varphi_{m\alpha\beta}}{\partial x^\gamma}\right]dx^m\\&=\int_0^{x^m}\frac{\partial\varphi_{\alpha\beta\gamma}}{\partial x^m}\,dx^m=\varphi_{\alpha\beta\gamma}(x)-\varphi_{\alpha\beta\gamma}(x^1,\ldots,x^{m-1},0),\end{aligned}$$

and

$$(d\psi)_{m\alpha\beta}=\frac{\partial\psi_{\alpha\beta}}{\partial x^m}=\varphi_{m\alpha\beta}(x).$$

Therefore putting

$$\varphi''=\frac{1}{3!}\sum_{\alpha,\beta,\gamma=1}^{m-1}\varphi_{\alpha\beta\gamma}(x^1,\ldots,x^{m-1},0)\,dx^\alpha\wedge dx^\beta\wedge dx^\gamma,$$

we obtain

$$d\psi = \varphi - \varphi''.$$

Since φ and φ'' are C^1, $d\psi$ is also C^1, hence, by the above result, $d\varphi'' = d\varphi - dd\psi = 0$. Since φ'' is a C^1 3-form of $m-1$ variables $x^1, \ldots, x^{m-1}$, by the induction hypothesis there is a C^1 2-form

$$\psi'' = \frac{1}{2!} \sum_{\alpha,\beta=1}^{m-1} \psi''_{\alpha\beta}(x^1, \ldots, x^{m-1})\, dx^\alpha \wedge dx^\beta$$

such that $d\psi'' = \varphi''$. Therefore $\varphi = d(\psi + \psi'')$.

If φ is C^∞, ψ and φ'' are also C^∞. Then since by induction hypothesis, we can choose ψ'' to be C^∞, $\psi + \psi''$ is also C^∞. ∎

From the above proof, one can see that Theorem 3.2 holds also for any domain in $\mathbb{R}^m$ satisfying the following condition (*):

(*) *If* $(x^1, \ldots, x^m) \in U$, *then for any* $\theta_\alpha : 0 \leqq \theta_\alpha \leqq 1$, $\alpha = 1, \ldots, m$, *we have* $(\theta_1 x^1, \ldots, \theta_m x^m) \in U$.

(b) Differential Forms on Complex Manifolds

Let M be a complex manifold, $\{z_1, \ldots, z_j, \ldots\}$, $z_j : p \to z_j(p) = (z_j^1, \ldots, z_j^n)$ a system of local complex coordinates, and U_j the domain of z_j. As stated in §2.1(a), by identifying $p \in U_j$ with its local coordinates $z_j = z_j(p)$, we may regard $M = \bigcup_j \mathcal{U}_j$, where $\mathcal{U}_j = z_j(U_j)$ is a domain in $\mathbb{C}^n$. In view of this fact, we write $\mathcal{U}_j$ also as U_j in the sequel. By Theorem 2.1, we may choose U_j such that $\mathfrak{U} = \{U_j \mid j = 1, 2, \ldots\}$ is locally finite, and that each U_j is a polydisk defined as

$$U_j = \{z_j \in \mathbb{C}^n \mid |z_j^1| < r_j^1, \ldots, |z_j^n| < r_j^n\}. \tag{3.17}$$

Taking z_j^α / r_j^α instead of z_j^α, we may assume that each U_j is a polydisk of radius 1:

$$U_j = \{z_j \in \mathbb{C}^n \mid |z_j^1| < 1, \ldots, |z_j^n| < 1\}. \tag{3.18}$$

Let Σ be the underlying differentiable manifold of M (see §2.1(f)). For the system of local complex coordinates $\{z_j\}$, $z_j = (z_j^1, \ldots, z_j^n)$, by decomposing z_j^α into the real and imaginary parts as $z_j^\alpha = x_j^{2\alpha-1} + i x_j^{2\alpha}$, $\alpha = 1, \ldots, n$, we obtain a system of local C^∞ coordinates $\{x_1, x_2, \ldots\}$, $x_j = (x_j^1, \ldots, x_j^{2n})$. Then by (1.41) we have

$$\det \frac{\partial(x_j^1, \ldots, x_j^{2n})}{\partial(x_k^1, \ldots, x_k^{2n})} = \left| \det \frac{\partial(z_j^1, \ldots, z_j^n)}{\partial(z_k^1, \ldots, z_k^n)} \right|^2 > 0. \tag{3.19}$$

Consequently Σ is orientable. We define the orientation of Σ by the above $\{x_j\}$, and call it the *orientation* of the complex manifold M.

Consider a differential form φ on M:

$$\varphi = \frac{1}{r!}\sum \varphi_{j\nu_1\cdots\nu_r}\, dx_j^{\nu_1}\wedge\cdots\wedge dx_j^{\nu_r}, \tag{3.20}$$

where the coefficient $\varphi_{j\nu_1\cdots\nu_r}$ is a complex-valued function. Since

$$x_j^{2\alpha-1} = \tfrac{1}{2}(z_j^\alpha + \bar{z}_j^\alpha) \quad \text{and} \quad x_j^{2\alpha} = \frac{1}{2i}(z_j^\alpha - \bar{z}_j^\alpha),$$

we have

$$dx_j^{2\alpha-1} = \tfrac{1}{2}(dz_j^\alpha + d\bar{z}_j^\alpha) \quad \text{and} \quad dx_j^{2\alpha} = \frac{1}{2i}(dz_j^\alpha - d\bar{z}_j^\alpha).$$

Substituting these into (3.20), we obtain

$$\varphi = \sum_{p+q=r} \frac{1}{p!\,q!}\sum \varphi_{j\alpha_1\cdots\alpha_p\bar{\beta}_1\cdots\bar{\beta}_q}\, dz_j^{\alpha_1}\wedge\cdots\wedge dz_j^{\alpha_p}\wedge d\bar{z}_j^{\beta 1}\wedge\cdots\wedge d\bar{z}_j^{\beta q}.$$

We assume that $\varphi_{j\alpha_1\cdots\alpha_p\bar{\beta}_1\cdots\bar{\beta}_q}$ are skew-symmetric in the indices $\alpha_1,\ldots,\alpha_p$ and $\beta_1,\ldots,\beta_q$ separately.

We call a differential form

$$\varphi = \frac{1}{p!\,q!}\sum \varphi_{j\alpha_1\cdots\alpha_p\bar{\beta}_1\cdots\bar{\beta}_q}\, dz_j^{\alpha_1}\wedge\cdots\wedge dz_j^{\alpha_p}\wedge d\bar{z}_j^{\beta_1}\wedge\cdots\wedge d\bar{z}_j^{\beta_q}$$

a *differential form of type* (p,q), or simply a (p,q)*-form*. Since the coordinate transformation: $z_k \to z_j$ is biholomorphic, we have

$$dz_j^\alpha = \sum_{\beta=1}^n \frac{\partial z_j^\alpha}{\partial z_k^\beta}\, dz_k^\beta \quad \text{and} \quad d\bar{z}_j^\alpha = \sum_{\beta=1}^n \frac{\partial \bar{z}_j^\alpha}{\partial \bar{z}_k^\beta}\, d\bar{z}_k^\beta.$$

Hence *the property that* φ *is a* (p,q)*-form does not depend on the choice of local complex coordinates.* If φ is a (p,q)-form, we often write it as $\varphi^{(p,q)}$ for the sake of distinction. An r-form φ^r is uniquely expressed as

$$\varphi^r = \varphi^{(r,0)} + \varphi^{(r-1,1)} + \cdots + \varphi^{(0,r)}.$$

Let f be a C^1 function in a domain of M. Then

$$df = \sum_{\alpha=1}^n \frac{\partial f}{\partial z_j^\alpha}\, dz_j^\alpha + \sum_{\alpha=1}^n \frac{\partial f}{\partial \bar{z}_j^\alpha}\, d\bar{z}_j^\alpha.$$

We define

$$\partial f=\sum_{\alpha=1}^{n}\frac{\partial f}{\partial z_j^\alpha}dz_j^\alpha \quad\text{and}\quad \bar\partial f=\sum_{\alpha=1}^{n}\frac{\partial f}{\partial \bar z_j^\alpha}d\bar z_j^\alpha.$$

Then we have

$$df=\partial f+\bar\partial f.$$

We can verify by easy calculation that ∂f and $\bar\partial f$ do not depend on the choice of complex coordinates. For a C^1 (p,q)-form

$$\varphi^{(p,q)}=\frac{1}{p!\,q!}\sum \varphi_{j\alpha_1\cdots\alpha_p\bar\beta_1\cdots\bar\beta_q}\,dz_j^{\alpha_1}\wedge\cdots\wedge dz_j^{\alpha_p}\wedge d\bar z_j^{\beta_1}\wedge\cdots\wedge d\bar z_j^{\beta_q},$$

we define

$$\partial\varphi^{(p,q)}=\frac{1}{p!\,q!}\sum \partial\varphi_{j\alpha_1\cdots\alpha_p\bar\beta_1\cdots\bar\beta_q}\wedge dz_j^{\alpha_1}\wedge\cdots\wedge dz_j^{\alpha_p}\wedge d\bar z_j^{\beta_1}\wedge\cdots\wedge d\bar z_j^{\beta_q},$$

and

$$\bar\partial\varphi^{(p,q)}=\frac{1}{p!\,q!}\sum \bar\partial\varphi_{j\alpha_1\cdots\alpha_p\bar\beta_1\cdots\bar\beta_q}\wedge dz_j^{\alpha_1}\wedge\cdots\wedge dz_j^{\alpha_p}\wedge d\bar z_j^{\beta_1}\wedge\cdots\wedge d\bar z_j^{\beta_q}.$$

Then we have

$$d\varphi^{(p,q)}=\partial\varphi^{(p,q)}+\bar\partial\varphi^{(p,q)},$$

where $\partial\varphi^{(p,q)}$ is a $(p+1,q)$-form, and $\bar\partial\varphi^{(p,q)}$ is a $(p,q+1)$-form. Hence the invariance of d under the coordinate transformation $z_k\to z_j$ implies those of ∂ and $\bar\partial$. As in (3.6), the coefficients of $\partial\varphi^{(p,q)}$ are given by

$$(\partial\varphi^{(p,q)})_{j\alpha_0\alpha_1\cdots\alpha_p\bar\beta_1\cdots\bar\beta_q}=\sum_{s=0}^{p}(-1)^s\frac{\partial\varphi_{j\alpha_0\alpha_1\cdots\alpha_{s-1}\alpha_{s+1}\cdots\alpha_p\bar\beta_1\cdots\bar\beta_q}}{\partial z_j^{\alpha_s}}.$$

Similar formulae hold for $\bar\partial\varphi^{(p,q)}$.

For an arbitrary C^1 form $\varphi=\sum_{p,q}\varphi^{(p,q)}$, we define

$$\partial\varphi=\sum_{p,q}\partial\varphi^{(p,q)}\quad\text{and}\quad\bar\partial\varphi=\sum_{p,q}\bar\partial\varphi^{(p,q)}.$$

Then clearly we have

$$d=\partial+\bar\partial. \tag{3.21}$$

Also the following formulae are immediately obtained from the definitions of ∂ and $\bar{\partial}$.

$$\partial(\varphi^r \wedge \psi) = \partial\varphi^r \wedge \psi + (-1)^r \varphi^r \wedge \partial\psi, \tag{3.22}$$

$$\bar{\partial}(\varphi^r \wedge \psi) = \bar{\partial}\varphi^r \wedge \psi + (-1)^r \varphi^r \wedge \bar{\partial}\psi. \tag{3.23}$$

Suppose that $\partial\varphi^{(p,q)}$ and $\bar{\partial}\varphi^{(p,q)}$ are both C^1. Then we have

$$\partial\partial\varphi^{(p,q)} + \bar{\partial}\partial\varphi^{(p,q)} + \partial\bar{\partial}\varphi^{(p,q)} + \bar{\partial}\bar{\partial}\varphi^{(p,q)} = dd\varphi^{(p,q)} = 0.$$

Considering the type of each term in this equality, we obtain

$$\partial\partial\varphi^{(p,q)} = 0, \qquad \bar{\partial}\partial\varphi^{(p,q)} + \partial\bar{\partial}\varphi^{(p,q)} = 0 \quad \text{and} \quad \bar{\partial}\bar{\partial}\varphi^{(p,q)} = 0,$$

hence the following equalities hold:

$$\partial\partial = 0, \qquad \bar{\partial}\bar{\partial} = 0 \quad \text{and} \quad \bar{\partial}\partial = -\partial\bar{\partial}. \tag{3.24}$$

A $(p, 0)$-form $\varphi = (1/p!) \sum \varphi_{j\alpha_1 \cdots \alpha_p} dz_j^{\alpha_1} \wedge \cdots \wedge dz_j^{\alpha_p}$ is called a *holomorphic p-form* if the coefficients $\varphi_{j\alpha_1 \cdots \alpha_p}$ are all holomorphic. For a C^1 $(p, 0)$-form φ,

$$\bar{\partial}\varphi = \frac{1}{p!} \sum \bar{\partial}\varphi_{j\alpha_1 \cdots \alpha_p} \wedge dz_j^{\alpha_1} \wedge \cdots \wedge dz_j^{\alpha_p} = 0$$

is equivalent to $\bar{\partial}\varphi_{j\alpha_1 \cdots \alpha_p} = 0$. Hence by Theorem 1.6, *$\varphi = \varphi^{(p,0)}$ is holomorphic if and only if $\bar{\partial}\varphi = 0$.* For a holomorphic p-form φ, have $d\varphi = \partial\varphi$.

The orientation of a complex manifold M is determined by the *volume element* $dx_j^1 \wedge \cdots \wedge dx_j^{2n}$. In terms of the local complex coordinates, we have

$$dx_j^1 \wedge \cdots \wedge dx_j^{2n} = \left(\frac{i}{2}\right)^n dz_j^1 \wedge d\bar{z}_j^1 \wedge \cdots \wedge dz_j^n \wedge d\bar{z}_j^n. \tag{3.25}$$

Since any domain W of M is itself a complex manifold, the above results hold also for differential forms on W.

In what follows we chiefly treat C^∞ differential forms.

A C^1 differential form φ is called a *$\bar{\partial}$-closed form* if $\bar{\partial}\varphi = 0$. Let $\varphi = \varphi^{(p,q)}$ be a C^∞ (p, q)-form on M or on a domain of M with $q \geqq 1$. If there is a C^∞ $(p, q-1)$-form ψ such that $\varphi = \bar{\partial}\psi$, then by (3.24), $\bar{\partial}\varphi = 0$. Locally its inverse is also true:

Theorem 3.3 (Dolbeault's Lemma). *If a C^∞ (p, q)-form $\varphi = \varphi^{(p,q)}$, $q \geqq 1$, defined on a polydisk $U_R = \{z \in \mathbb{C}^n \mid |z^1| < R, \ldots, |z^n| < R\}$, $0 \leqq R \leqq +\infty$, is $\bar{\partial}$-closed, there is a C^∞ $(p, q-1)$-form ψ on U_R such that $\varphi = \bar{\partial}\varphi$.*

In order to prove this theorem, for φ and ψ we put

$$\varphi_{\alpha_1\cdots\alpha_p}=\frac{1}{q!}\sum\varphi_{\alpha_1\cdots\alpha_p\bar\beta_1\cdots\bar\beta_q}\,d\bar z^{\beta_1}\wedge\cdots\wedge d\bar z^{\beta_q},\quad\text{and}$$

$$\psi_{\alpha_1\cdots\alpha_p}=\frac{1}{(q-1)!}\sum\psi_{\alpha_1\cdots\alpha_p\bar\beta_2\cdots\bar\beta_q}\,d\bar z^{\beta_2}\wedge\cdots\wedge d\bar z^{\beta_q}.$$

Then $\varphi_{\alpha_1\cdots\alpha_p}$ is a $(0,q)$-form, and $\psi_{\alpha_1\cdots\alpha_q}$ is a $(0,q-1)$-form. Moreover $\bar\partial\varphi=0$ is equivalent to $\bar\partial\varphi_{\alpha_1\cdots\alpha_p}=0$, and $\varphi=\bar\partial\psi$ is equivalent to $\varphi_{\alpha_1\cdots\alpha_p}=\bar\partial\psi_{\alpha_1\cdots\alpha_p}$. Hence it suffices to show Theorem 3.3 for a $(0,q)$-form.

First we show

Lemma 3.1. *Let $f(z)$ be a C^∞ function on $|z|<R$, and fix a real number ρ such that $0<\rho<R$. Put*

$$g(z)=\frac{1}{2\pi i}\iint_{|\zeta|\leqq\rho}\frac{f(\zeta)}{\zeta-z}\,d\zeta\wedge d\bar\zeta. \tag{3.26}$$

Then

$$\frac{\partial}{\partial\bar z}g(z)=f(z),\qquad |z|<\rho, \tag{3.27}$$

holds.

Proof. Take an arbitrary σ such that $0<\sigma<\rho$. We shall prove (3.27) for $|z|<\sigma$. Let $f_1(z)$ be a C^∞ function which coincides with f in $|z|\leqq\sigma$, and vanishes in $|z|\geqq\rho$, and let $f_2(z)=f(z)-f_1(z)$. Then $f_2(z)$ is a C^∞ function which vanishes in $|z|<\sigma$, and

$$f(z)=f_1(z)+f_2(z).$$

Then, putting

$$g_\nu(z)=\frac{1}{2\pi i}\iint_{|\zeta|\leqq\rho}\frac{f_\nu(\zeta)\,d\zeta\wedge d\bar\zeta}{\zeta-z},\qquad \nu=1,2,$$

we see that

$$g(z)=g_1(z)+g_2(z).$$

Since $f_2(\zeta)=0$ in $|\zeta|\leqq\sigma$, $g_2(z)$ is a holomorphic in $|z|<\sigma$, hence

$$\partial g_2(z)/\partial\bar z=0.$$

Defining $f_1(\zeta)=0$ in $|\zeta|\geqq R$, we extend $f_1(\zeta)$ on all of $\mathbb{C}$. Let $\zeta=z+w$, with $w=re^{i\theta}$. Then we have

$$g_1(z)=\frac{1}{2\pi i}\iint_{\mathbb{C}} f_1(z+w)\frac{dw\wedge d\bar{w}}{w}$$

$$=-\frac{1}{\pi}\iint f_1(z+re^{i\theta})e^{-i\theta}\,dr\,d\theta.$$

Consequently $g_1(z)$ is a C^∞ function of z, and we have

$$\frac{\partial g_1(z)}{\partial\bar{z}}=-\frac{1}{\pi}\iint\frac{\partial f_1(z+re^{i\theta})}{\partial\bar{z}}e^{-i\theta}\,dr\,d\theta$$

$$=\frac{1}{2\pi i}\iint_{\mathbb{C}}\frac{\partial f_1(z+w)}{\partial\bar{z}}\frac{dw\wedge d\bar{w}}{w}$$

$$=\frac{1}{2\pi i}\iint_{\mathbb{C}}\frac{\partial f_1(z+w)}{\partial\bar{w}}\frac{dw\wedge d\bar{w}}{w}$$

$$=-\frac{1}{2\pi i}\iint_{\mathbb{C}} d\left(f_1(z+w)\frac{dw}{w}\right).$$

Since $|z|<\sigma, f_1(z+w)=0$ for $|w|>\rho+\sigma$. Therefore by Green's theorem, the last integral is equal to

$$-\lim_{\varepsilon\to+0}\frac{1}{2\pi i}\iint_{\mathbb{C}-W_\varepsilon} d\left(f_1(z+w)\frac{dw}{w}\right)$$

$$=\lim_{\varepsilon\to+0}\frac{1}{2\pi i}\int_{\partial[W_\varepsilon]} f_1(z+w)\frac{dw}{w}=f_1(z),$$

where $W_\varepsilon=\{w\,|\,|w|<\varepsilon\}$. Thus for $|z|<\sigma$,

$$\frac{\partial}{\partial\bar{z}}g(z)=\frac{\partial}{\partial\bar{z}}g_1(z)=f_1(z)=f(z).$$

Since σ is arbitrary with $0<\sigma<\rho$, we are done. ∎

Proof of Theorem 3.3. (1°) Let $m\geqq q$, and let

$$\varphi=\sum_{\beta_1<\cdots<\beta_q\leqq m}\varphi_{\beta_1\cdots\beta_q}\,d\bar{z}^{\beta_1}\wedge\cdots\wedge d\bar{z}^{\beta_q},\qquad q\geqq 1,$$

be a $\bar{\partial}$-closed C^∞ $(0,q)$-form on U_R which does not involve the differentials $d\bar{z}^{m+1},\ldots,d\bar{z}^n$. Given an arbitrary real number ρ with $0<\rho<R$, we prove

by induction on m the existence of a C^∞ $(0, q-1)$-form

$$\psi = \sum_{\beta_1 < \cdots < \beta_{q-1} \leqq m-1} \psi_{\beta_1 \cdots \beta_{q-1}} d\bar{z}^{\beta_1} \wedge \cdots \wedge d\bar{z}^{\beta_{q-1}}$$

such that $\varphi = \bar{\partial}\psi$ on U_ρ. Since for $\nu \geqq m+1$, we have

$$\frac{\partial}{\partial \bar{z}^\nu} \varphi_{\beta_1 \cdots \beta_q}(z^1, \ldots, z^m, \ldots, z^n) = (\bar{\partial}\varphi)_{\nu\beta_1 \cdots \beta_q} = 0, \tag{3.28}$$

$\varphi_{\beta_1 \cdots \beta_q}(z^1, \ldots, z^m, z^{m+1}, \ldots, z^n)$ is holomorphic in $z^{m+1}, \ldots, z^n$, hence $\bar{\partial}$ does not affect on $z^{m+1}, \ldots, z^n$. The proof given below is an analogy of that of Poincaré's lemma. As before we consider only for the case $q=3$, but the generalization is straightforward.

Let

$$\varphi = \sum_{\alpha < \beta < \gamma \leqq m} \varphi_{\alpha\beta\gamma} d\bar{z}^\alpha \wedge d\bar{z}^\beta \wedge d\bar{z}^\gamma,$$

where $\varphi_{\alpha\beta\gamma}$ are holomorphic in $z^{m+1}, \ldots, z^n$. Take an arbitrary σ with $\rho < \sigma < R$, and define

$$\psi = \sum_{\alpha < \beta \leqq m-1} \psi_{\alpha\beta} d\bar{z}^\alpha \wedge d\bar{z}^\beta,$$

by

$$\begin{aligned} \psi_{\alpha\beta} &= \psi_{\alpha\beta}(z^1, \ldots, z^n) \\ &= \frac{1}{2\pi i} \iint_{|\zeta| \leqq \sigma} \frac{\varphi_{m\alpha\beta}(z^1, \ldots, z^{m-1}, \zeta, z^{m+1}, \ldots, z^n)}{\zeta - z^m} d\zeta \wedge d\bar{\zeta}. \end{aligned}$$

By Lemma 3.1, ψ is C^∞ in U_σ, and

$$\frac{\partial \psi_{\alpha\beta}}{\partial \bar{z}^m} = \varphi_{m\alpha\beta},$$

and by (3.28), we have

$$\frac{\partial \psi_{\alpha\beta}}{\partial \bar{z}^\nu} = 0, \quad \text{for } \nu = m+1, \ldots, n,$$

hence in case $m = q = 3$,

$$\bar{\partial}\psi = \bar{\partial}(\psi_{12} d\bar{z}^1 \wedge d\bar{z}^2) = \frac{\partial \psi_{12}}{\partial \bar{z}^3} d\bar{z}^3 \wedge d\bar{z}^1 \wedge d\bar{z}^2 = \varphi_{312} d\bar{z}^3 \wedge d\bar{z}^1 \wedge d\bar{z}^2 = \varphi.$$

In case $m>q=3$, we have

$$\bar{\partial}\psi = \sum_{\alpha<\beta\leqq m-1} \varphi_{m\alpha\beta}\, d\bar{z}^m \wedge d\bar{z}^\alpha \wedge d\bar{z}^\beta$$

$$+ \sum_{\alpha<\beta<\gamma\leqq m-1} \left(\frac{\partial\psi_{\beta\gamma}}{\partial\bar{z}^\alpha} - \frac{\partial\psi_{\alpha\gamma}}{\partial\bar{z}^\beta} + \frac{\partial\psi_{\alpha\beta}}{\partial\bar{z}^\gamma}\right) d\bar{z}^\alpha \wedge d\bar{z}^\beta \wedge d\bar{z}^\gamma.$$

Consequently putting $\varphi''=\varphi-\bar{\partial}\psi$, we obtain a $\bar{\partial}$-closed C^∞ $(0,3)$-form

$$\varphi'' = \sum_{\alpha<\beta<\gamma\leqq m-1} \varphi''_{\alpha\beta\gamma}\, d\bar{z}^\alpha \wedge d\bar{z}^\beta \wedge d\bar{z}^\gamma.$$

By induction hypothesis, there is a C^∞ $(0,2)$-form

$$\psi'' = \sum_{\alpha<\beta\leqq m-2} \psi''_{\alpha\beta}\, d\bar{z}^\alpha \wedge d\bar{z}^\beta$$

on U_ρ such that $\varphi''=\bar{\partial}\psi''$ on U_ρ. Hence on U_ρ, we have

$$\varphi = \bar{\partial}(\psi+\psi''),$$

which accomplishes the induction.

Thus, putting $m=n$, we see that for an arbitrary ρ with $0<\rho<R$, and for any given $\bar{\partial}$-closed C^∞ $(0,q)$-form φ on U_R, we can find a C^∞ $(0,q-1)$-form ψ on U_ρ such that $\varphi=\bar{\partial}\psi$ on U_ρ.

(2°) Let φ be a $\bar{\partial}$-closed C^∞ $(0,q)$-form, $q\geqq 1$. Fix a sequence $\{\rho_k\}$ such that $0<\rho_1<\rho_2<\cdots$, and that $\lim_{k\to\infty}\rho_k=R$. By (1°), we can find on each U_{ρ_k} a C^∞ $(0,q-1)$-form ψ_k such that $\bar{\partial}\psi_k=\varphi$ on U_{ρ_k}. We want to construct a C^∞ $(0,q-1)$-form ψ on U_R with $\bar{\partial}\psi=\varphi$ using $\psi_1,\psi_2,\ldots$. First let $q\geqq 2$. Put $|z|=\max_\alpha|z^\alpha|$ for $z=(z^1,\ldots,z^n)$. The method of construction of ψ is illustrated in Fig. 1.

Namely, first put $\tilde{\psi}_2=\psi_2$. Since on U_{ρ_2}, $\bar{\partial}\tilde{\psi}_2=\varphi=\bar{\partial}\psi_3$, we have $\bar{\partial}(\psi_3-\tilde{\psi}_2)=0$ on U_{ρ_2}. Hence by (1°), for any small $\varepsilon>0$, there exists a C^∞ $(0,q-2)$-form χ_2 on $U_{\rho_2-\varepsilon}$ such that $\bar{\partial}\chi_2=\psi_3-\tilde{\psi}_2$. Let $\lambda(r)$ be a C^∞ function of r with $r>0$ such that $\lambda(r)=1$ for $r\leqq\rho_1$, and that $\lambda(r)=0$ for $r\geqq\rho_2-2\varepsilon$, and define a C^∞ $(0,q-2)$-form $\tilde{\chi}_2=\tilde{\chi}_2(z)$ on U_{ρ_3} by

$$\tilde{\chi}_2(z)=\begin{cases}\chi_2(z)\lambda(|z|), & |z|<\rho_2-\varepsilon,\\ 0, & |z|\geqq\rho_2-\varepsilon.\end{cases}$$

Then we have

$$\tilde{\chi}_2(z)=\begin{cases}\chi_2(z), & |z|\leqq\rho_1,\\ 0, & |z|\geqq\rho_2-2\varepsilon.\end{cases}$$

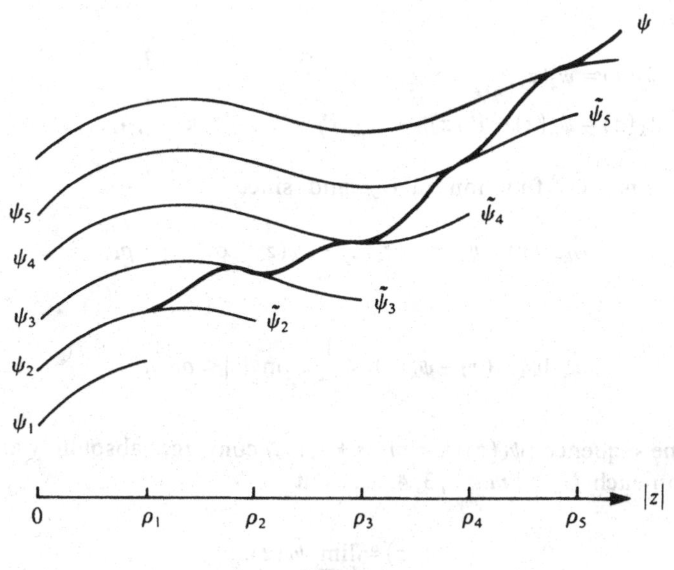

Figure 1

Put $\tilde{\psi}_3 = \psi_3 - \bar{\partial}\tilde{\chi}_2$. Then $\tilde{\psi}_3$ is a C^∞ $(0, q-1)$-form on U_{ρ_3} such that $\bar{\partial}\tilde{\psi}_3 = \bar{\partial}\psi_3 = \varphi$, and that $\tilde{\psi}_3 = \psi_3 - \bar{\partial}\tilde{\chi}_2 = \psi_3 - \bar{\partial}\chi_2 = \tilde{\psi}_2$ on U_{ρ_1}.

Similarly we can construct successively C^∞ $(0, q-1)$-forms $\tilde{\psi}_k$ on U_{ρ_k}, $k = 2, 3, \ldots$, such that $\bar{\partial}\tilde{\psi}_k = \varphi$ on U_{ρ_k}, and that $\tilde{\psi}_{k+1} = \tilde{\psi}_k$ on $U_{\rho_{k-1}}$. Let ψ be the C^∞ $(0, q-1)$-form on U_R obtained by putting $\psi = \tilde{\psi}_k$ on each U_{ρ_k}, for $k = 2, 3, \ldots$. Then $\bar{\partial}\psi = \varphi$ on U_R as desired.

In case $q = 1$, ψ_k is a C^∞ function on U_{ρ_k}, $k = 2, 3, \ldots$, and $\bar{\partial}(\psi_{k+1} - \psi_k) = 0$. Hence $f_k = \psi_{k+1} - \psi_k$ is holomorphic. Therefore, by Theorem 1.3, f_k is expanded into a convergent power series of $z^1, \ldots, z^n$:

$$f_k(z) = \sum_{m_1, \ldots, m_n = 0}^{\infty} a_{m_1 \ldots m_n} (z^1)^{m_1} \cdots (z^n)^{m_n}.$$

Since this series converges absolutely and uniformly on $U_{\rho_{k-1}}$, we can choose a sufficiently large l such that

$$P_k(z) = \sum_{m_1, \ldots, m_n = 0}^{l} a_{m_1 \cdots m_n} (z^1)^{m_1} \cdots (z^n)^{m_n}$$

satisfies

$$|f_k(z) - P_k(z)| < \frac{1}{2^k}, \qquad |z| < \rho_{k-1}. \tag{3.29}$$

If we put

$$\tilde{\psi}_2(z) = \psi_2(z),$$
$$\tilde{\psi}_k(z) = \psi_k(z) - P_2(z) - \cdots - P_{k-1}(z), \qquad k = 3, 4, \ldots,$$

then $\tilde{\psi}_k(z)$ is a C^∞ function on U_{ρ_k}, and, since

$$\tilde{\psi}_{k+1}(z) - \tilde{\psi}_k(z) = f_k(z) - P_k(z) \quad \text{on } |z| < \rho_k, \tag{3.30}$$

we have

$$|\tilde{\psi}_{k+1}(z) - \tilde{\psi}_k(z)| < \frac{1}{2^k} \quad \text{on } |z| < \rho_{k-1}.$$

Hence the sequence $\{\tilde{\psi}_k(z) \,|\, k = m, m+1, \ldots\}$ converges absolutely and uniformly on each $U_{\rho_{m-1}}$, $m = 2, 3, 4, \ldots$. Put

$$\psi(z) = \lim_{k \to \infty} \tilde{\psi}_k(z).$$

Then, by (3.30), we have

$$\psi(z) = \tilde{\psi}_m(z) + \sum_{k=m}^{\infty} (f_k(z) - P_k(z))$$

on $U_{\rho_{m-1}}$, where $\tilde{\psi}_m(z)$ is C^∞. Moreover since each $f_k(z) - P_k(z)$ is holomorphic, and $\sum_{k=m}^{\infty} (f_k(z) - P_k(z))$ converges absolutely and uniformly by (3.29), this sum is holomorphic on $U_{\rho_{m-1}}$. Consequently $\psi = \psi(z)$ is C^∞ and $\bar{\partial}\psi = \bar{\partial}\tilde{\psi}_m = \varphi$ on $U_{\rho_{m-1}}$. Therefore ψ is C^∞ and $\bar{\partial}\psi = \varphi$ on $U_R = \bigcup_{m=2}^{\infty} U_{\rho_{m-1}}$. ∎

§3.2. Vector Bundles

(a) Tangent Space

Let M be a C^∞ manifold, $\{x_1, \ldots, x_j, \ldots\}$, $x_j: p \to x_j(p) = (x_j^1, \ldots, x_j^m)$ a system of local C^∞ coordinates on it, and U_j the domain of x_j. Moreover we assume that $\mathfrak{U} = \{U_j\}$ is locally finite. As stated in §3.1(a), we identify $x_j = x_j(p)$ with p, and $\mathcal{U}_j = x_j(U_j)$ with U_j, respectively. As stated before (p. 62), for a smooth curve on M passing through p, its tangent vector at p as a linear differential operator is written as

$$v = \sum_{\alpha=1}^{m} v_j^\alpha \frac{\partial}{\partial x_j^\alpha}.$$

We call such v a *tangent vector of M at p*. Let $C^\infty(M)$ be the $\mathbb{R}$-vector space of all C^∞ functions on M. Then as a linear differential operator v defines an $\mathbb{R}$-linear map of $C^\infty(M)$ to $\mathbb{R}$ as follows:

$$v:\ \psi \to v(\psi) = \sum_{\alpha=1}^{m} v_j^\alpha \frac{\partial \psi}{\partial x_j^\alpha}(p), \qquad \psi \in C^\infty(M). \tag{3.31}$$

For $\psi, \varphi \in C^\infty(M)$, we have

$$v(\psi\varphi) = \psi(p)v(\varphi) + \varphi(p)v(\psi). \tag{3.32}$$

Conversely, *any $\mathbb{R}$-linear map v of $C^\infty(M)$ to $\mathbb{R}$ satisfying* (3.32) *is written as in* (3.31).

Proof. Replacing $(x_j^1, \ldots, x_j^m)$ by $(x_j^1 - x_j^1(p), \ldots, x_j^m - x_j^m(p))$, we may assume $x_j(p) = 0$. Moreover for $1 \in C^\infty(M)$, $v(1) = v(1.1) = 2v(1)$ by (3.32), hence $v(1) = 0$. Hence also $v(c) = 0$ for any constant c. Consequently $v(\psi) = v(\psi - \psi(p))$. Thus we may assume $\psi(p) = 0$.

Let $\psi = \psi(x_j^1, \ldots, x_j^m) = \psi(x_j)$ on U_j in terms of local coordinates. Then in a sufficiently small neighbourhood $U(p) \subset U_j$ of $p \in U_j$, we can write $\psi(x_j)$ as

$$\psi(x_j) = \sum_{\alpha=1}^{m} x_j^\alpha \Psi_\alpha(x_j),$$

where $\Psi_\alpha(x_j)$ is a C^∞ function of x_j. In fact, putting $tx_j = (tx_j^1, \ldots, tx_j^m)$, we obtain

$$\psi(x_j) = \int_0^1 \frac{d}{dt}\psi(tx_j)\, dt = \sum_{\alpha=1}^{m} x_j^\alpha \int_0^1 \frac{\partial \psi}{\partial x_j^\alpha}(tx_j)\, dt.$$

Hence we may put

$$\Psi_\alpha(x_j) = \int_0^1 \frac{\partial \psi}{\partial x_j^\alpha}(tx_j)\, dt,$$

which is clearly a C^∞ function of x_j. Take $\rho \in C^\infty(M)$ such that $\rho(p) = 1$, and that $\operatorname{supp} \rho \subset U(p)$. Since $\psi(p) = 0$, $v(\rho^2\psi) = v(\psi)$ by (3.32). On the other hand,

$$\rho^2\psi = \sum_{\alpha=1}^{m} \rho x_j^\alpha \rho \Psi_\alpha(x_j)$$

on U_j. We denote the natural extension of ρx_j^α and $\rho\Psi_\alpha(x_j)$ to all of M by the same notation ρx_j^α and $\rho\Psi_\alpha(x_j)$ respectively. Then by (3.32), we have

$$v(\psi)=v(\rho^2\psi)=\sum_{\alpha=1}^m \Psi_\alpha(0)v(\rho x_j^\alpha).$$

Since $\Psi_\alpha(0)=(\partial\psi/\partial x_j^\alpha)(p)$, putting $v_j^\alpha=v(\rho x_j^\alpha)$ we obtain (3.31). ∎

In view of this result, we may adopt the following "intrinsic" definition of a tangent vector.

Definition 3.1. A linear map v of $C^\infty(M)$ to R satisfying (3.32) is called a *tangent vector of M at p.*

This is the modern definition of a tangent vector, which does not involve local coordinates, but it is not so easy to understand at first glance.

For a tangent vector

$$v=\sum_{\alpha=1}^m v_j^\alpha \frac{\partial}{\partial x_j^\alpha}$$

at $p\in U_j\subset M$, $v_j^1,\ldots,v_j^m$ are called the *components* of v with respect to the local coordinates x_j. Suppose $p\in U_j\cap U_k$, and let $v_k^1,\ldots,v_k^m$ be the components of v with respect to x_k. Then the following equations hold:

$$v_j^\alpha=\sum_{\beta=1}^m \frac{\partial x_j^\alpha}{\partial x_k^\beta}v_k^\beta,\quad \text{where}\quad \frac{\partial x_j^\alpha}{\partial x_k^\beta}=\frac{\partial x_j^\alpha}{\partial x_k^\beta}(p). \tag{3.33}$$

This is the *transformation rule* between the components of a tangent vector with respect to x_j and x_k.

The $\mathbb{R}$-vector space consisting of all tangent vectors of M at p is called the *tangent space* of M at p, and is denoted by $T_p(M)$. $T_p(M)$ is an m-dimensional $\mathbb{R}$-vector space, and the map $v\to(v_j^1,\ldots,v_j^m)$ maps $T_p(M)$ isomorphically onto $\mathbb{R}^m$.

The union of all $T_p(M)$, $p\in M$, is called the *tangent bundle* of M and is denoted by $T(M)$: $T(M)=\bigcup_{p\in M}T_p(M)$. Since a tangent vector at $p\in U_j$ is written as $v_j=(v_j^1,\ldots,v_j^m)$ in terms of local coordinates, $T(U_j)=\bigcup_{p\in U_j}T_p(M)$ is the set of all pairs (p, v_j) with $p\in U_j$ and $v_j\in\mathbb{R}^m$:

$$T(U_j)=\{(p,v_j)\,|\,p\in U_j,\ v_j\in R^m\}=U_j\times\mathbb{R}^m.$$

Thus $T(U_j)$ *has the structure of the direct product* $U_j\times\mathbb{R}^m$. Hence we have

$$T(M)=\bigcup_j T(U_j)=\bigcup_j U_j\times\mathbb{R}^m, \tag{3.34}$$

where for $p \in U_j \cap U_k$, by (3.33) (p, v_j) and (p, v_k) are the same point of $T(M)$ if and only if

$$v_j^\alpha = \sum_{\beta=1}^{m} \frac{\partial x_j^\alpha}{\partial x_k^\beta}(p) v_k^\beta, \qquad \alpha = 1, \ldots, m. \tag{3.35}$$

Thus $T(M)$ may be regarded as *a differentiable manifold obtained from glueing up* $U_j \times \mathbb{R}^m$ *via the identification* (3.35). Each $U_j \times \mathbb{R}^m$ is a coordinate neighbourhood of $T(M)$ and $(x_j^1, \ldots, x_j^m, v_j^1, \ldots, v_j^m)$ gives the local C^∞ coordinate system on $U_j \times \mathbb{R}^m$.

Since the projections $U_j \times \mathbb{R}^m \to U_j$ and $U_k \times \mathbb{R}^m \to U_k$ coincide on $U_j \times \mathbb{R}^m \cap U_k \times \mathbb{R}^m$, they define the projection

$$\pi: T(M) \to M.$$

In terms of local coordinates, π is written as

$$\pi: (x_j^1, \ldots, x_j^m, v_j^1, \ldots, v_j^m) \to (x_j^1, \ldots, x_j^m),$$

hence π is C^∞. Moreover clearly $\pi^{-1}(p) = T_p(M)$.

(b) Vector Bundle

Definition 3.2. Let M be a differentiable manifold. A differentiable manifold F is called a *vector bundle* over M if it satisfies the following conditions:

(i) A C^∞ map $\pi: F \to M$ of F onto M is given.

(ii) For every $p \in M$, $\pi^{-1}(p)$ is a ν-dimensional $\mathbb{R}$-vector space: $\pi^{-1}(p) \cong \mathbb{R}^\nu$, where ν is independent of p.

(iii) For every $q \in M$, there exists a neighbourhood U, $q \in U \subset M$, such that $\pi^{-1}(U) = U \times \mathbb{R}^\nu$, and that for any $p \in U$, $p \times \mathbb{R}^\nu$ is isomorphic to $\pi^{-1}(p)$ as an $\mathbb{R}$-vector space: $\pi^{-1}(p) \cong p \times \mathbb{R}^\nu$.

We call $\pi^{-1}(p)$ the *fibre* of F over p, and denote it by F_p. ν is called the *rank* of F. M is called the *base space* of F, and π the *projection*. We also denote F by (F, M, π) if we want to make explicit its base space and projection.

Let $\mathfrak{U} = \{U_j \mid j = 1, 2, \ldots\}$ be a locally finite open covering of M, and (F, M, π) a vector bundle over M. If each U_j is taken sufficiently small, then by (iii) above, we have

$$\pi^{-1}(U_j) = U_j \times \mathbb{R}^\nu = \{(p, \xi_j) \mid p \in U_j, \xi_j = (\xi_j^1, \ldots, \xi_j^\nu) \in \mathbb{R}^\nu\}.$$

Since $p \times \mathbb{R}^\nu \simeq \pi^{-1}(p)$, $(\xi_j^1, \ldots, \xi_j^\nu)$ is a system of coordinates on the vector space $\pi^{-1}(p)$. For $p \in U_j \cap U_k$, $(\xi_j^1, \ldots, \xi_j^\nu)$ and $(\xi_k^1, \ldots, \xi_k^\nu)$ are two systems of coordinates on the same vector space $\pi^{-1}(p)$, hence there is an invertible linear transformation $(f^\alpha_{jk\beta})$ such that

$$\xi^\alpha_j = \sum_{\beta=1}^{\nu} f^\alpha_{jk\beta}(p)\xi^\beta_k, \qquad \alpha = 1, 2, \ldots, \nu. \tag{3.36}$$

Namely $(p, \xi_j) \in U_j \times \mathbb{R}^\nu$ and $(p, \xi_k) \in U_k \times \mathbb{R}^\nu$ are the same point on F if and only if (3.36) holds. We call $(\xi^1_j, \ldots, \xi^\nu_j)$ the *fibre coordinates* of $(p, \xi_j) \in F_p$ over U_j. Since the subdomain $\pi^{-1}(U_j)$ of M is, by (iii) above, isomorphic to $U_j \times \mathbb{R}^\nu$ as a differentiable manifold, these $\xi^1_j, \ldots, \xi^\nu_j$ are C^∞ functions on $\pi^{-1}(U_j)$. Consequently in (3.36), ξ^α_j is a C^∞ function of $(p, \xi^1_k, \ldots, \xi^\nu_k)$, hence $f^\alpha_{jk\beta}(p)$ is a C^∞ function on $U_j \cap U_k$. By using vector notation, (3.36) may be written simply as

$$\xi_j = f_{jk}(p) \cdot \xi_k, \tag{3.37}$$

where $f_{jk}(p)$ denotes the matrix $(f^\alpha_{jk\beta}(p))_{\alpha,\beta=1,2,\ldots,\nu}$. This is the transition relation between the fibre coordinates. $f_{jk} = f_{jk}(p)$ is called the *transition function.* Of course we have $\det f_{jk}(p) \neq 0$. *A vector bundle F is determined uniquely by giving f_{jk}, $j, k = 1, 2, \ldots$.* Hence we write $F = \{f_{jk}\}$.

Transition functions satisfy the following conditions. First we have

$$f_{jj}(p) = 1, \quad \text{where 1 denotes the unit matrix.} \tag{3.38}$$

Next, since for any $\xi_l \in \mathbb{R}^\nu$, $\xi_j = f_{jl}(p)\xi_l = f_{jk}(p)\xi_k = f_{jk}(p)f_{kl}(p)\xi_l$, we have

$$f_{jl}(p) = f_{jk}(p)f_{kl}(p), \qquad p \in U_j \cap U_k \cap U_l. \tag{3.39}$$

Putting $l = j$ in (3.39) we also obtain

$$f_{jk}(p)f_{kj}(p) = 1. \tag{3.40}$$

Conversely, given a locally finite covering $\mathfrak{U} = \{U_j\}$ and C^∞ functions $f_{jk}(p) = (f^\alpha_{jk\beta}(p))_{\alpha,\beta=1,\ldots,\nu}$ on each $U_j \cap U_k \neq \emptyset$ satisfying (3.39) and (3.40), we can construct a vector bundle $F = \bigcup_j U_j \times \mathbb{R}^\nu$ over M by identifying $(p, \xi_k) \in U_k \times \mathbb{R}^\nu$ with $(p, f_{jk}(p)\xi_k) \in U_j \times \mathbb{R}^\nu$ for $p \in U_j \cap U_k \neq \emptyset$.

Let F and G be vector bundles over M. If there is a diffeomorphism h of F onto G such that h maps each fibre F_p linearly onto G_p, F and G are called *equivalent*. In this case, take a locally finite open covering $\{U_j\}$ of M with sufficiently small U_j, and let $\xi_j = (\xi^1_j, \ldots, \xi^\nu_j)$ and $\eta_j = (\eta^1_j, \ldots, \eta^\nu_j)$ be the fibre coordinates of F and G over U_j respectively. Then h is written on $\pi^{-1}(U_j)$ as

$$h\colon (p, \xi_j) \to (p, \eta_j) = (p, h_j(p)\xi_j),$$

where each component of the matrix $h_j(p)=(h^\alpha_{j\beta}(p))_{\alpha,\beta=1,\ldots,\nu}$ is a C^∞ function of $p\in U_j$, and $\det h_j(p)\neq 0$. Let $f_{jk}(p)$ and $g_{jk}(p)$ be the transition functions of F and G respectively. Then as is easily seen, we have

$$g_{jk}(p)=h_j(p)f_{jk}(p)h_k(p)^{-1}, \qquad p\in U_j\cap U_k. \tag{3.41}$$

Thus we may consider that G is nothing but F with another fibre coordinates $\eta_j=(\eta^1_j,\ldots,\eta^\nu_j)$ obtained from $\xi_j=(\xi^1_j,\ldots,\xi^\nu_j)$ by applying the transformation $(p,\xi_j)\to(p,\eta_j)=(p,h_j(p)\xi_j)$. In this sense *we often identify F and G if they are equivalent.* $M\times\mathbb{R}^\nu$ is a vector bundle over M. If F is equivalent to $M\times\mathbb{R}^\nu$, F is called *trivial.*

Let F be a vector bundle over M, and π its projection. A continuous map $\sigma: p\to\sigma(p)$ of a domain $W\subset M$ into F is called a section of F over W if $\sigma(p)\in F_p$ for all $p\in W$. In terms of fibre coordinates,

$$\sigma(p)=(p,\sigma_j(p)), \sigma_j(p)=(\sigma^1_j(p),\ldots,\sigma^\nu_j(p)), \qquad p\in W\cap U_j,$$

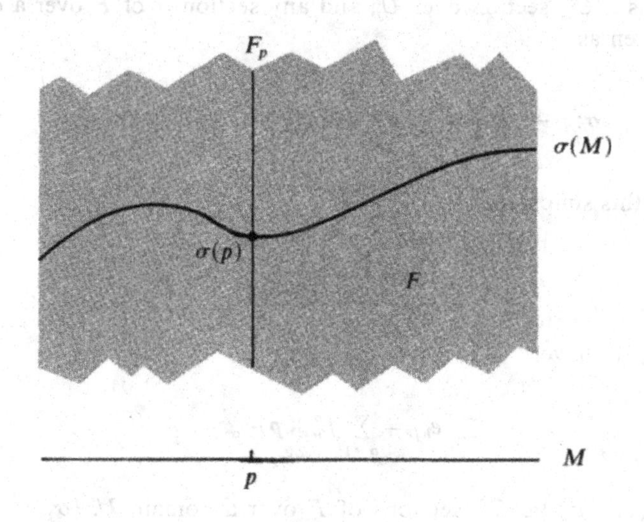

Figure 2

where $\sigma_j(p)$ are vector-valued continuous functions satisfying

$$\sigma_j(p)=f_{jk}(p)\cdot\sigma_k(p) \tag{3.42}$$

on $W\cap U_j\cap U_k$. If σ is continuously differentiable, C^∞, etc., that is, each function $\sigma_j(p)$ is continuously differentiable, C^∞, etc., the section σ is called continuously differentiable, C^∞, etc., respectively.

Comparing (3.33) and (3.42), we clearly see that the tangent bundle $T(M)$ of M is a vector bundle over M with the transition functions $f_{jk}(p)=(f^{\alpha}_{jk\beta}(p))$ where

$$f^{\alpha}_{jk\beta}(p)=\frac{\partial x^{\alpha}_j}{\partial x^{\beta}_k}(p).$$

A continuous vector field on M

$$v=\sum_{\alpha=1}^{m} v^{\alpha}_j(p)\frac{\partial}{\partial x^{\alpha}_j}$$

is nothing but a section of $T(M)$. Here each $\partial/\partial x^{\alpha}_j$ is a C^{∞} vector field on U_j, hence a C^{∞} section of $T(M)$ over U_j.

For an arbitrary vector bundle F, using the fibre coordinates over U_j, put

$$e_{j\alpha}: p\to e_{j\alpha}(p)=(p,(0,\ldots,0,1^{\alpha},0,\ldots,0))\in U_j\times\mathbb{R}^{\nu}.$$

Then $e_{j\alpha}$ is a C^{∞} section over U_j, and any section σ of F over a domain W is written as

$$\sigma: p\to\sigma(p)=\sum_{\alpha=1}^{\nu}\sigma^{\alpha}_j(p)e_{j\alpha}(p)\quad\text{for } p\in W\cap U_j.$$

We write this simply as

$$\sigma=\sum_{\alpha=1}^{\nu}\sigma^{\alpha}_j(p)e_{j\alpha}.\tag{3.43}$$

By (3.36), we have

$$e_{k\beta}=\sum_{\alpha=1}^{\nu}f^{\alpha}_{jk\beta}(p)e_{j\alpha}.\tag{3.44}$$

Let $\sigma_1,\ldots,\sigma_{\nu}$ be C^{∞} sections of F over a domain U. $\{\sigma_1,\ldots,\sigma_{\nu}\}$ is called a *basis of F over U* if for each $p\in U$, $\{\sigma_1(p),\ldots,\sigma_{\nu}(p)\}$ is a basis of $F_p=\pi^{-1}(p)$. For example, the above-mentioned $\{e_{j1},\ldots,e_{j\nu}\}$ is a basis of F over U_j, and $\{\partial/\partial x^1_j,\ldots,\partial/\partial x^m_j\}$ is a basis of $T(M)$ over U_j. Given a basis $\{\sigma_1,\ldots,\sigma_{\nu}\}$ of F over U, every point of $\pi^{-1}(U)$ is written uniquely as

$$\sum_{\alpha=1}^{\nu}\xi^{\alpha}\sigma_{\alpha}(p),\qquad p\in U,\qquad(\xi^1,\ldots,\xi^{\nu})\in\mathbb{R}^{\nu}.$$

By writing this point as (p,ξ), with $\xi=(\xi^1,\ldots,\xi^{\nu})$, we obtain an identification of $\pi^{-1}(U)$ with $U\times\mathbb{R}^{\nu}$. $(\xi^1,\ldots,\xi^{\nu})$ is called the fibre coordinates of

(p, ξ) with respect to the basis $\{\sigma_1, \ldots, \sigma_\nu\}$. In this notation we have

$$\sigma_\alpha(p) = (p, (0, \ldots, 0, 1^\alpha, 0, \ldots, 0)).$$

Replacing $\mathbb{R}$ by $\mathbb{C}$ and $\mathbb{R}^\nu$ by $\mathbb{C}^\nu$ in Definition 3.2, we obtain the definition of *a complex vector bundle.* In contrast to this, we call a vector bundle in the sense of Definition 3.2, *a real vector bundle.* For a complex vector bundle, its fibre coordinates $\zeta_j = (\zeta_j^1, \ldots, \zeta_j^\nu) \in \mathbb{C}^\nu$ form a complex vector, and the components of the transition function $f_{jk\beta}^\alpha(p)$ are complex-valued C^∞ functions.

Now consider vector bundles over a complex manifold. Let F be a complex vector bundle over a complex manifold M. If the transition functions $f_{jk}(p), j, k = 1, 2, \ldots$, are all holomorphic, F is called a *holomorphic vector bundle.* Here by saying that $f_{jk}(p) = (f_{jk\beta}^\alpha(p))$ is holomorphic, we mean that each component $f_{jk\beta}^\alpha(p)$ is a holomorphic function of $p \in U_j \cap U_k$. Then F is obtained by glueing up $U_j \times \mathbb{C}^\nu$ by identifying $(p, \zeta_k) \in U_k \times \mathbb{C}^\nu$ with

$$(p, \zeta_j) = (p, f_{jk}(p)\zeta_k) \in U_j \times \mathbb{C}^\nu: F = \bigcup_j U_j \times \mathbb{C}^\nu.$$

If $f_{jk}(p)$ is holomorphic, the map $(p, \zeta_k) \to (p, \zeta_j) = (p, f_{jk}(p)\zeta_k)$ is biholomorphic. Hence a holomorphic vector bundle $F = \bigcup_j U_j \times \mathbb{C}^\nu$ is a complex manifold, and the projection $\pi: F \to M$ is a holomorphic map.

Let F be a holomorphic vector bundle. A section σ of F over a domain $W \subset M$ is said to be holomorphic if it is holomorphic as a map of M into F. In terms of fibre coordinates, put $\sigma: p \to \sigma_j(p) = (\sigma_j^1(p), \ldots, \sigma_j^\nu(p)), p \in U_j$. Then $\sigma_j^\alpha(p)$ is a holomorphic function of p if σ is holomorphic. In particular,

$$e_{j\alpha}: p \to e_{j\alpha}(p) = (p, (0, \ldots, 0, 1^\alpha, 0, \ldots, 0))$$

is a holomorphic section of F over U_j, and (3.44) is also valid in this case.

Let $(z_j^1, \ldots, z_j^n)$ be a local coordinate system on M, and U_j the domain of z_j. We assume that U_j is a domain of M and that $\mathfrak{U} = \{U_j\}$ is a locally finite covering of M. Introduce local real coordinates $x_j = (x_j^1, \ldots, x_j^{2n})$ by putting $z_j^\alpha = x_j^{2\alpha-1} + ix_j^{2\alpha}$ as usual. Then $\{x_i, \ldots, x_j, \ldots\}$ becomes a system of local C^∞ coordinates of the underlying differentiable manifold of M.

The vector bundle of all tangent vectors of M which are written as

$$\sum_{\alpha=1}^n v_j^\alpha \frac{\partial}{\partial z_j^\alpha}, \qquad v_j^\alpha \in \mathbb{C},$$

in terms of local complex coordinates, is called the tangent bundle of the complex manifold M and is denoted by $T(M)$. The transition functions

$(f^{\alpha}_{jk\beta}(p))$ of $T(M)$ are given by

$$f^{\alpha}_{jk\beta}(p) = \frac{\partial z^{\alpha}_j}{\partial z^{\beta}_k}(p). \tag{3.45}$$

Hence $T(M)$ is a holomorphic vector bundle. A holomorphic vector field on M

$$v = \sum_{\alpha=1}^{n} v^{\alpha}_j(p) \frac{\partial}{\partial z^{\alpha}_j}$$

is nothing but a holomorphic section of $T(M)$. Note that $T(M)$ is different from the tangent bundle $T(\Sigma)$ of the underlying differentiable manifold Σ of M. We shall discuss the relation between them later.

(c) Dual Bundle, Tensor Product

In this paragraph, K denote either $\mathbb{R}$ or $\mathbb{C}$. Let E be a finite-dimensional vector space over K. As is well known, if $\{e_1, \ldots, e_\nu\}$ is a basis of E, any element $u \in E$ is uniquely represented as

$$u = \sum_{\alpha=1}^{\nu} \xi^{\alpha} e_{\alpha}, \qquad \xi^{\alpha} \in K, \tag{3.46}$$

and the map $u \to (\xi^1, \ldots, \xi^{\nu}) \in K^{\nu}$ gives an isomorphism of E onto K^{ν}: $E \cong K^{\nu}$.

The set of all linear functions $v = v(u)$ on E forms a vector space over K, which we call the *dual space* of E and denote by E^*. If $u \in E$ is written as in (3.46), we have

$$v(u) = v\left(\sum_{\alpha} \xi^{\alpha} e_{\alpha}\right) = \sum_{\alpha=1}^{\nu} \xi^{\alpha} v(e_{\alpha}).$$

Hence putting $\eta_{\alpha} = v(e_{\alpha})$ and representing v by $(\eta_1, \ldots, \eta_{\nu}) \in K^{\nu}$, we see that $E^* \cong K^{\nu}$ and that $v(u) = \sum_{\alpha=1}^{\nu} \xi^{\alpha}\eta_{\alpha}$. We write $\langle v, u\rangle$ instead of $v(u)$. Then

$$\langle v, u\rangle = v(u) = \sum_{\alpha=1}^{\nu} \xi^{\alpha}\eta_{\alpha}. \tag{3.47}$$

We call $\langle v, u\rangle$ the *inner product* of v and u. From (3.47) it is clear that $E^{**} = E$.

Definition 3.3. Let F be a real or complex vector bundle over a differentiable manifold M. A vector bundle G over M is called the *dual bundle* of F if it satisfies the following conditions:

(i) For every point $p \in M$, G_p is dual to F_p, i.e. the inner product $\langle v, u\rangle$ is defined between $v \in G_p$ and $u \in F_p$.

(ii) If we take a sufficiently fine locally finite open covering $\mathfrak{U} = \{U_j\}$ of M, then we can choose fibre coordinates $(\xi_j^1, \ldots, \xi_j^\nu)$ of F and $(\eta_j^1, \ldots, \eta_j^\nu)$ of G over U_j such that for $u = (p, \xi_j^1, \ldots, \xi_j^\nu) \in F_p$ and $v = (p, \eta_j^1, \ldots, \eta_j^\nu) \in G_p$, their inner product is given by

$$\langle v, u\rangle = \sum_{\alpha=1}^{\nu} \xi_j^\alpha \eta_{j\alpha}. \tag{3.48}$$

We denote the dual bundle of F by F^*. Let $f_{jk}(p) = (f_{jk\beta}^\alpha(p))$ be the transition functions, and $f_{jk}^*(p) = (f_{jk\alpha}^{*\beta}(p))$ those of F^*. Then we have

$$\xi_j^\alpha = \sum_{\beta=1}^{\nu} f_{jk\beta}^\alpha(p)\xi_k^\beta, \qquad \eta_{k\beta} = \sum_{\alpha=1}^{\nu} f_{kj\beta}^{*\alpha}(p)\eta_{j\alpha}.$$

Thus $\langle v, u\rangle = \sum_\alpha \xi_j^\alpha \eta_{j\alpha} = \sum_\beta \xi_k^\beta \eta_{k\beta}$ implies

$$\sum_\alpha \sum_\beta f_{jk\beta}^\alpha(p)\xi_k^\beta \eta_{j\alpha} = \sum_\beta \sum_\alpha f_{kj\beta}^{*\alpha}(p)\xi_k^\beta \eta_{j\alpha}.$$

Hence we have

$$f_{jk\alpha}^{*\beta}(p) = f_{kj\alpha}^\beta(p). \tag{3.49}$$

Therefore we obtain the following transition relation between fibre coordinates of F^*:

$$\eta_{j\alpha} = \sum_{\beta=1}^{\nu} f_{kj\alpha}^\beta(p)\eta_{k\beta}. \tag{3.50}$$

Note that we have $F^{**} = F$.

Now we review the definition of the tensor product of vector spaces. Let E and F be vector spaces over K, and let $\{e_1, \ldots, e_\mu\}$ and $\{f_1, \ldots, f_\nu\}$ bases of E and F respectively. Then the *tensor product* of E and F is the vector space with the set of all pairs (e_α, f_β) as its basis. We denote the tensor product of E and F by $E \otimes F$. Thus $E \otimes F$ consists of all elements of the form

$$w = \sum_{\alpha=1}^{\mu} \sum_{\beta=1}^{\nu} \zeta^{\alpha\beta}(e_\alpha, f_\beta), \qquad \zeta^{\alpha\beta} \in K.$$

For $u = \sum_\alpha \xi^\alpha e_\alpha \in E$, and $v = \sum_\beta \eta^\beta f_\beta \in F$, we define $u \otimes v$ by

$$u \otimes v = \sum_\alpha \sum_\beta \xi^\alpha \eta^\beta (e_\alpha, f_\beta).$$

Then we have $(e_\alpha, f_\beta) = e_\alpha \otimes f_\beta$, and any element w of $E \otimes F$ is represented as

$$w = \sum_{\alpha=1}^{\mu} \sum_{\beta=1}^{\nu} \zeta^{\alpha\beta} e_\alpha \otimes f_\beta, \qquad \zeta^{\alpha\beta} \in K. \tag{3.51}$$

The map $E \times F \to E \otimes F$ given by $(u, v) \to u \otimes v$ is bilinear.

Considering the dual space we verify that *the tensor product* $E \otimes F$ *does not depend on the choice of bases used in the above definition.* In fact for any $t \in (E \otimes F)^*$, $\langle t, u \otimes v\rangle$ is a bilinear function on $E \times F$. Conversely, any bilinear function $b(u, v)$ on $E \times F$ is written as $b(u, v) = \langle t, u \otimes v\rangle$ for some $t \in (E \otimes F)^*$. Thus we naturally identify $(E \otimes F)^*$ with the vector space of all bilinear functions on $E \times F$. Consequently $(E \otimes F)^*$, hence also $E \otimes F = (E \otimes F)^{**}$ do not depend on the choice of bases of E and F.

Definition 3.4. Let F and G be real or complex vector bundles over a differentiable manifold M. A vector bundle H over M is called the *tensor product* of F and G if the following two conditions are satisfied.

(i) For every $p \in M$, $H_p = F_p \otimes G_p$.

(ii) Take a sufficiently fine locally finite open covering $\mathfrak{U} = \{U_j\}$ of M and let $(\xi_j^1, \ldots, \xi_j^\mu)$ and $(\eta_j^1, \ldots, \eta_j^\nu)$ be fibre coordinates of F and G over U_j respectively. Then for $u = (p, \xi_j^1, \ldots, \xi_j^\mu) \in F_p$ and $v = (p, \eta_j^1, \ldots, \eta_j^\nu) \in G_p$, the fibre coordinates of $u \otimes v$ are given by $\zeta_j^{\alpha\beta} = \xi_j^\alpha \eta_j^\beta$, $\alpha = 1, \ldots, \mu$, $\beta = 1, \ldots, \nu$:

$$u \otimes v = (p, \xi_j^1 \eta_j^1, \ldots, \xi_j^\mu \eta_j^\nu). \tag{3.52}$$

We denote the tensor product of F and G by $F \otimes G$.

Let the transition functions of F and G be $f_{jk}(p) = (f_{jk\gamma}^\alpha(p))$ and $g_{jk}(p) = (g_{jk\delta}^\beta(P))$ respectively. Then since

$$\xi_j^\alpha \eta_j^\beta = \sum_{\gamma=1}^{\mu} \sum_{\delta=1}^{\nu} f_{jk\gamma}^\alpha(p) g_{jk\delta}^\beta(p) \xi_k^\gamma \eta_j^\delta,$$

the transition functions of $F \otimes G$ are given by

$$h_{jk\gamma\delta}^{\alpha\beta}(p) = f_{jk\gamma}^\alpha(p) g_{jk\delta}^\beta(p). \tag{3.53}$$

The matrix $h_{jk}(p) = (h_{jk\gamma\delta}^{\alpha\beta}(p))$ is called the *Kronecker product* of matrices $f_{jk}(p)$ and $g_{jk}(p)$.

For a section $\sigma: p \to \sigma(p)$ of F and a section $\tau: p \to \tau(p)$ of G over a domain $W \subset M$, $p \to \sigma(p) \otimes \tau(p)$ is a section of $F \otimes G$ by (3.52), which is denoted by $\sigma \otimes \tau$.

If M is a complex manifold, and F and G are holomorphic vector bundles, $F\otimes G$ is also a holomorphic vector bundle by (3.53). If σ and τ are holomorphic sections of F and G respectively, $\sigma\otimes\tau$ is a holomorphic section of $F\otimes G$.

Let $\{e_{j1},\ldots,e_{j\mu}\}$ and $\{f_{j1},\ldots,f_{j\nu}\}$ be bases of F and G over U_j respectively. Then $\{e_{j\alpha}\otimes f_{j\beta}\,|\,\alpha=1,\ldots,\mu,\beta=1,\ldots,\nu\}$ forms a basis of $F\otimes G$ over U_j, and a section ω of $F\otimes G$ over a domain $W\subset M$ is represented as

$$\omega=\sum_{\alpha=1}^{\mu}\sum_{\beta=1}^{\nu}\omega_j^{\alpha\beta}(p)\cdot e_{j\alpha}\otimes f_{j\beta},\qquad p\in W\cap U_j.$$

(d) Tensor Field

Let $\varphi=\sum_{\alpha=1}^{m}\varphi_{j\alpha}\,dx_j^\alpha$, $\varphi_{j\alpha}\in C$, be a 1-form on a differentiable manifold M at p, and

$$v=\sum_{\alpha=1}^{m}v_j^\alpha\frac{\partial}{\partial x_j^\alpha}$$

a tangent vector of M at p. Introduce the inner product of φ and v by

$$\langle\varphi,v\rangle=\sum_{\alpha=1}^{m}\varphi_{j\alpha}v_j^\alpha.$$

Then the set of all 1-forms at p forms the dual vector space $T_p^*(M)$ of $T_p(M)$ via this inner product. As in the case of $T(M)$, $\bigcup_{p\in M}T_p^*(M)$ has a natural structure of vector bundle over M. As is obviously seen from the transformation rule for 1-forms

$$\varphi_{j\alpha}=\sum_\beta\frac{\partial x_k^\beta}{\partial x_j^\alpha}\varphi_{k\beta}\tag{3.54}$$

with respect to the coordinate transformation $x_k\to x_j$, compared with (3.33), the vector bundle $T^*(M)=\bigcup_{p\in M}T_p^*(M)$ is the dual bundle of $T(M)$, and the inner product $\langle\varphi,v\rangle$ given above is invariant under coordinate changes. A section of $T^*(M)$ over a domain $W\subset M$ is a continuous 1-form on W. $\{dx_j^1,\ldots,dx_j^m\}$ forms a basis of $T^*(M)$ over U, and the equality

$$\left\langle dx_j^\alpha,\frac{\partial}{\partial x_j^\beta}\right\rangle=\delta_\beta^\alpha$$

holds.

A 1-form $\varphi=\sum_\alpha\varphi_{j\alpha}\,dx_j^\alpha\in T_p^*(M)$ at p is represented uniquely by the vector $(\varphi_{j1},\ldots,\varphi_{jm})$ in terms of the fixed local coordinates x_j, for $p\in U_j$.

The transformation rule (3.54) of this vector with respect to the coordinate change $x_k \to x_j$ is the same as that of the basis $\{\partial/\partial x_j^\alpha\}$ of $T(M)$. In view of this fact we call $(\varphi_{j1}, \dots, \varphi_{jm})$ a *covariant vector.* On the other hand, the transformation rule (3.33) of the components of a tangent vector $v = \sum_\alpha v_j^\alpha(\partial/\partial x_j^\alpha)$ is the inverse of that of the basis. Hence we call $(v_j^1, \dots, v_j^m)$ a *contravariant vector.*

Put

$$\overset{r}{\bigotimes} T^*(M) = \overbrace{T^*(M) \otimes T^*(M) \otimes \cdots \otimes T^*(M)}^{r \text{ times}}.$$

A C^∞ section σ of $\bigotimes^r T^*(M)$ is represented on each coordinate neighbourhood U_j as

$$\sigma = \sum_{\alpha,\beta,\dots,\gamma}^{m} \sigma_{j\alpha\beta\cdots\gamma}\, dx_j^\alpha \otimes dx_j^\beta \otimes \cdots \otimes dx_j^\gamma.$$

Here m^r coefficients $\sigma_{j\alpha\beta\cdots\gamma}$ are C^∞ functions on U_j, and for the coordinate change $x_j \to x_k$ on $U_j \cap U_k$, satisfy the following:

$$\sigma_{k\alpha\beta\cdots\gamma} = \sum_{\lambda,\mu,\dots,\nu=1}^{m} \frac{\partial x_j^\lambda}{\partial x_k^\alpha} \frac{\partial x_j^\mu}{\partial x_k^\beta} \cdots \frac{\partial x_j^\nu}{\partial x_k^\gamma} \sigma_{j\lambda\mu\cdots\nu}. \tag{3.55}$$

Given m^r C^∞ functions $\sigma_{j\alpha\cdots\gamma}$ on each U_j satisfying (3.55) on $U_j \cap U_k$, we call $\sigma_{j\alpha\cdots\gamma}$ a C^∞ *covariant tensor field* of rank r, and each $\sigma_{j\alpha\cdots\gamma}$ its component with respect to the local coordinates x_j. For arbitrary local coordinates $(x^1, \dots, x^m)$, the component of a covariant tensor field $\sigma_{\alpha\cdots\gamma}$ is given by

$$\sigma_{\alpha\cdots\gamma} = \sum_{\lambda,\dots,\nu} \frac{\partial x_j^\lambda}{\partial x^\alpha} \cdots \frac{\partial x_j^\nu}{\partial x^\gamma} \sigma_{j\lambda\cdots\nu}.$$

This is the classical definition of a covariant tensor field, which is, it is true, not *intrinsic*, but turns out to be very convenient. A C^∞ covariant tensor field of rank r is nothing but a C^∞ section of $\bigotimes^r T^*(M)$.

Similarly a section σ of $\bigotimes^r T(M) = \overbrace{T(M) \otimes T(M) \otimes \cdots \otimes T(M)}^{r \text{ times}}$ is represented on each U_j as

$$\sigma = \sum_{\alpha,\dots,\gamma} \sigma_j^{\alpha\cdots\gamma} \frac{\partial}{\partial x_j^\alpha} \otimes \cdots \otimes \frac{\partial}{\partial x_j^\gamma},$$

and satisfies on $U_j \cap U_k$ the relation

$$\sigma_k^{\alpha\cdots\gamma} = \sum_{\lambda,\dots,\nu} \frac{\partial x_k^\alpha}{\partial x_j^\lambda} \cdots \frac{\partial x_k^\gamma}{\partial x_j^\nu} \sigma_j^{\lambda\cdots\nu}. \tag{3.56}$$

We call $\sigma_j^{\alpha\cdots\gamma}$ a *contravariant tensor field* of rank r. With respect to arbitrary local coordinates $(x^1,\ldots,x^m)$, σ is written as

$$\sigma=\sum_{\alpha,\ldots,\gamma}\sigma^{\alpha\cdots\gamma}\frac{\partial}{\partial x^\alpha}\otimes\cdots\otimes\frac{\partial}{\partial x^\gamma}.$$

Further, for example, a section σ of $T(M)\otimes T^*(M)\otimes T^*(M)$ is written as

$$\sigma=\sum_{\alpha,\beta,\gamma}\sigma^\alpha_{\beta\gamma}\frac{\partial}{\partial x^\alpha}\otimes dx^\beta\otimes dx^\gamma.$$

$\sigma^\alpha_{\beta\gamma}$ is called a *mixed tensor field.* $\sigma^\alpha_{\beta\gamma}$ satisfies the following relations:

$$\sigma^\alpha_{k\nu\gamma}=\sum_{\lambda,\mu,\nu}\frac{\partial x^\alpha_k}{\partial x^\lambda_j}\frac{\partial x^\mu_j}{\partial x^\beta_k}\frac{\partial x^\nu_j}{\partial x^\gamma_k}\sigma^\lambda_{j\mu\nu}. \tag{3.57}$$

A covariant tensor field $\sigma_{\alpha\cdots\gamma}$ is called symmetric (or skew-symmetric) if it is symmetric (or skew-symmetric) in the indices $\alpha,\ldots,\gamma$. Thus a covariant tensor field $\sigma_{\alpha\beta}$ is symmetric if $\sigma_{\alpha\beta}=\sigma_{\beta\alpha}$ and skew-symmetric if $\sigma_{\alpha\beta}=-\sigma_{\beta\alpha}$. We define symmetric and skew-symmetric contravariant tensor fields similarly.

Let

$$\varphi=\frac{1}{r!}\sum_{\alpha,\beta,\ldots,\gamma}\varphi_{\alpha\beta\cdots\gamma}\,dx^\alpha\wedge dx^\beta\wedge\cdots\wedge dx^\gamma$$

be an r-form on M. Then $\varphi_{\alpha\beta\cdots\gamma}$ is a skew-symmetric covariant tensor field. Since

$$\sum_{\alpha,\ldots,\gamma}\varphi_{\alpha\cdots\gamma}\,dx^\alpha\otimes\cdots\otimes dx^\gamma$$

$$=\sum_{\alpha,\ldots,\gamma}\varphi_{\alpha\cdots\gamma}\frac{1}{r!}\sum_{\lambda,\ldots,\nu}\operatorname{sgn}\begin{pmatrix}\alpha\cdots\gamma\\ \lambda\cdots\nu\end{pmatrix}dx^\gamma\otimes\cdots\otimes dx^\nu,$$

by identifying

$$dx^\alpha\wedge\cdots\wedge dx^\gamma=\sum_{\lambda,\ldots,\nu}\operatorname{sgn}\begin{pmatrix}\alpha\cdots\gamma\\ \lambda\cdots\nu\end{pmatrix}dx^\lambda\otimes\cdots\otimes dx^\nu,$$

φ is regarded as a section of $\bigotimes^r T^*(M)$, where the summation $\sum_{\lambda,\ldots,\nu}$ is taken over all the permutations of $\alpha,\ldots,\gamma$, and $\operatorname{sgn}\binom{\alpha\cdots\gamma}{\lambda\cdots\nu}$ denote the signature of permutation.

A vector bundle G over M is called a *subbundle* of F if the following conditions (i), (ii) are satisfied:

(i) For every $p \in M$, G_p is a linear subspace of F_p.
(ii) Take a sufficiently fine locally finite open covering $U = \{U_j\}$. Then there exist μ C^∞ sections $\sigma_{j1}, \ldots, \sigma_{j\mu}$ of F over each U_j, such that $\{\sigma_{j1}(p), \ldots, \sigma_{j\mu}(p)\}$ forms a basis of G_p for any $p \in U_j$.

In this case $\{\sigma_{j1}, \ldots, \sigma_{j\mu}\}$ forms a basis of G over U_j, and, if we represent an element of G_p as $\sum_{\alpha=1}^{\mu} \xi_j^\alpha \sigma_{j\alpha}(p)$, $(\xi_j^1, \ldots, \xi_j^\mu)$ gives the fibre coordinates of G over U_j.

If vector bundles F and G over M are subbundles of a vector bundle H, and $H_p = F_p + G_p$ for every point $p \in M$, H is called the *Whitney sum* of F and G, and denoted by $F \oplus G$. In this case if $\{e_{j1}, \ldots, e_{j\mu}\}$ and $\{f_{j1}, \ldots, f_{j\nu}\}$ are bases of F and G over U_j respectively, $\{e_{j1}, \ldots, e_{j\mu}, f_{j1}, \ldots, f_{j\nu}\}$ forms a basis of $F \oplus G$ over U_j.

The set of all r-forms at each point of M forms a subbundle of $\bigotimes^r T^*(M)$, which we denote by

$$\bigwedge^r T^*(M) = \overbrace{T^*(M) \wedge \cdots \wedge T^*(M)}^{r \text{ times}}.$$

$\binom{m}{r}$ C^∞ sections $dx_j^\alpha \wedge \cdots \wedge dx_j^\gamma$, $1 \leqq \alpha < \cdots < \gamma \leqq m$, of $\bigotimes^r T^*(M)$ over U_j form a basis of $\bigwedge^r T^*(M)$ over U_j. A C^∞ r-form on a domain $W \subset M$ is a C^∞ section of $\bigwedge^r T^*(M)$ over W.

For a complex vector bundle F over M with the transition functions $f_{jk}(p) = (f_{jk\beta}^\alpha(p))$, $\bar{F}$ denotes a vector bundle over M with the transition functions $\bar{f}_{jk}(p) = (\bar{f}_{jk\beta}^\alpha(p))$, where $^-$ denotes the complex conjugation.

Suppose that M is a complex manifold. Let $\{z_1, \ldots, z_j, \ldots\}$, $z_j = (z_j^1, \ldots, z_j^n)$ be a system of local complex coordinates of M. Let $T(M)$ be the tangent bundle of M, which consists of all tangent vectors of the form $\sum_\alpha v_j^\alpha(\partial/\partial z_j^\alpha)$ with $v_j^\alpha \in \mathbb{C}$. Then $\overline{T(M)}$ is the set of all tangent vectors of the form $\sum_\alpha v_j^\alpha(\partial/\partial \bar{z}_j^\alpha)$ with $v_j^\alpha \in \mathbb{C}$, which we denote by $\bar{T}(M)$. The dual bundle $T^*(M)$ of $T(M)$ consists of all (0, 1)-forms $\sum_\alpha \varphi_{j\alpha}\, dz_j^\alpha$ at each point of M, and the dual bundle $\bar{T}^*(M)$ of $\bar{T}(M)$ consists of all (0, 1)-forms $\sum_\alpha \varphi_{j\alpha}\, d\bar{z}_j^\alpha$. $T(M)$ and $T^*(M)$ are holomorphic vector bundles.

Let Σ be the underlying differentiable manifold of M, and $T_{\mathbb{C}}(\Sigma)$ the vector bundle over M consisting of all tangent vectors of Σ with complex coefficients $\sum_{\alpha=1}^{2n} v_j^\alpha(\partial/\partial x_j^\alpha)$ with $v_j^\alpha \in C$. Since $z_j^\alpha = x_j^{2\alpha-1} + i x_j^{2\alpha}$, we have

$$\sum_{\alpha=1}^{2n} v_j^\alpha \frac{\partial}{\partial x_j^\alpha} = \sum_{\alpha=1}^{n} (v_j^{2\alpha-1} + i v_j^{2\alpha}) \frac{\partial}{\partial z_j^\alpha} + \sum_{\alpha=1}^{n} (v_j^{2\alpha-1} - i v_j^{2\alpha}) \frac{\partial}{\partial \bar{z}_j^\alpha}.$$

Hence we have

$$T_{\mathbb{C}}(\Sigma) = T(M) \oplus \bar{T}(M). \tag{3.58}$$

Similarly we have

$$T_{\mathbb{C}}^*(\Sigma) = T^*(M) \oplus \bar{T}^*(M). \tag{3.59}$$

A C^∞ section of the tensor product of arbitrary times of $T(M)$, $T^*(M)$, $\bar{T}(M)$ and $\bar{T}^*(M)$ is a C^∞ tensor field on M. $\bigotimes^r T(M)$ is a holomorphic vector bundle, and its holomorphic section is a holomorphic contravariant tensor field. Similar results hold for $\bigotimes^r T^*(M)$.

We denote by

$$\overset{p}{\bigwedge} T^*(M) \overset{q}{\bigwedge} \bar{T}^*(M) = \overbrace{T^*(M) \wedge \cdots \wedge T^*(M)}^{p \text{ times}} \wedge \overbrace{\bar{T}^*(M) \wedge \cdots \wedge \bar{T}^*(M)}^{q \text{ times}}$$

the subbundle of $\bigwedge^r T_{\mathbb{C}}(M)$, $r = p+q$, which has

$$\{\overbrace{dz_j^\alpha \wedge \cdots \wedge dz_j^\beta}^{p \text{ times}} \wedge \overbrace{d\bar{z}_j^\gamma \wedge \cdots \wedge d\bar{z}_j^\delta}^{q \text{ times}} \mid 1 \leqq \alpha < \cdots < \beta \leqq n, 1 \leqq \gamma < \cdots < \delta \leqq n\}$$

as a basis over U_j. Then a C^∞ (p,q)-form is a C^∞ section of $\bigwedge^p T^*(M) \bigwedge^q \bar{T}^*(M)$. $\bigwedge^p T^*(M)$ is a holomorphic vector bundle, and its holomorphic section is a holomorphic p-form.

§3.3. Sheaves and Cohomology

(a) Sheaves

First let M be a differentiable manifold. By a local C^∞ function of M we mean a C^∞ function defined on a domain $U \subset M$. We denote by $D(f)$ the domain of f. f is called a *local C^∞ function* at p if $p \in D(f)$. We say that two local C^∞ functions f and g at p are *equivalent at p* if $f = g$ on some neighbourhood of p. We denote it by writing $f \sim_p g$. It is obvious that $\sim_p$ is an equivalence relation. By a *germ of a C^∞ function* at p we mean an equivalence class of local functions at p. The equivalence class to which f belongs is called *the germ of f* at p, and denoted by f_p. The germ f_p can be regarded as a C^∞ function defined on an "infinitely" small neighbourhood of p. Let $\mathscr{A}_p$ be the set of germs of all complex-valued C^∞ functions at p. Let $\varphi, \psi \in \mathscr{A}_p$, and suppose that $\varphi = f_p$ and $\psi = g_p$. Then we define a linear combination of φ and ψ by putting

$$c_1\varphi + c_2\psi = (c_1 f + c_2 g)_p, \qquad c_1, c_2 \in \mathbb{C}.$$

Clearly this definition is independent of the choice of f and g. Thus $\mathscr{A}_p$ becomes a vector space over $\mathbb{C}$. Moreover we define the product of φ and ψ by

$$\varphi \cdot \psi = (f \cdot g)_p, \qquad f_p = \varphi, \quad g_p = \psi,$$

which makes $\mathscr{A}_p$ a ring.

Let

$$\mathscr{A} = \bigcup_{p \in M} \mathscr{A}_p$$

be the union of all $\mathscr{A}_p$, $p \in M$. We define a topology on $\mathscr{A}$ as follows: For a point $\varphi \in \mathscr{A}_p \subset \mathscr{A}$, choose any local C^∞ function f with $f_p = \varphi$, and an arbitrary open set U with $p \in U \subset D(f)$, and put

$$\mathscr{U}(\varphi; f, U) = \{f_q \mid q \in U\}.$$

Of course $\varphi = f_p \in U(\varphi; f, U)$. We introduce a topology on $\mathscr{A}$ by taking the family

$$\{\mathscr{U}(\varphi; f, U) \mid f_p = \varphi, p \in U \subset D(f)\}$$

consisting of all $\mathscr{U}(\varphi; f, U)$ with f and U satisfying the above conditions as the system of neighbourhood of $\varphi \in \mathscr{A}$. Namely, a subset $\mathscr{U} \subset \mathscr{A}$ is defined to be open if and only if U contains some $\mathscr{U}(\varphi; f, U)$ for any $\varphi \in U$. We denote by $\mathfrak{O}$ the family of open sets of $\mathscr{A}$. This definition makes sense. In fact, let $\mathscr{U}$, $\mathscr{V} \in \mathfrak{O}$ and $\varphi \in \mathscr{U} \cap \mathscr{V}$. Then there are neighbourhood $\mathscr{U}(\varphi; f, U)$ and $\mathscr{U}(\varphi; g, V)$ such that $\mathscr{U}(\varphi; f, U) \subset \mathscr{U}$ and $\mathscr{U}(\varphi; g, V) \subset \mathscr{V}$. Since $f_p = g_p = \varphi$, $f = g$ in some neighbourhood $W \subset U \cap V$, hence $\mathscr{U}(\varphi; f, W) \subset \mathscr{U}(\varphi; f, U) \cap \mathscr{U}(\varphi; g, V) \subset \mathscr{U} \cap \mathscr{V}$. Hence $\mathscr{U} \cap \mathscr{V} \subset \mathfrak{O}$. Thus $\mathscr{A}$ is a topological space with $\mathfrak{O}$ as the system of open sets. $\mathscr{U}(\varphi; f, U)$ is clearly open. For a distinct two points φ and $\psi \in \mathscr{A}$, there is a neighbourhood $\mathscr{U}(\varphi; f, U)$ of φ such that $\psi \notin \mathscr{U}(\varphi; f, U)$. In fact, suppose $\varphi \in \mathscr{A}_p$ and $\psi \in \mathscr{A}_q$. if $p \neq q$, then $q \notin U$ for some neighbourhood of p, and if $p = q$, $\mathscr{A}_p \cap \mathscr{U}(\varphi; f, U) = \{\varphi\}$. But $\mathscr{A}$ is not Hausdorff.

Example 3.1. Let $M = \mathbb{R}$ be the real line, and $u(x)$ and $v(x)$ C^∞ functions on $\mathbb{R}$ such that $u(x) = v(x)$ for $x \geqq 0$ and that $u(x) < v(x)$ for $x < 0$. Put $\varphi = u_0$ and $\psi = v_0$. Then $\varphi_0 \neq \psi_0$. But $\mathscr{U}(\varphi; f, U) \cap \mathscr{U}(\psi; g, V) \neq \varnothing$ for any f, g, U and V. In fact, $f_0 = \varphi = u_0$, and $g_0 = \psi = v_0$ mean that for a sufficient small $\delta > 0$, $f(x) = u(x)$ and $g(x) = v(x)$ on $|x| < \delta$. Hence for $\varepsilon \in U \cap V$, $0 < \varepsilon < \delta$, $f_\varepsilon = u_\varepsilon = v_\varepsilon = g_\varepsilon$.

Define ϖ: $\mathscr{A} \to M$ by $\varpi(\mathscr{A}_p) = p$. Then ϖ satisfies the following conditions:

(i) ϖ is a *local homeomorphism*, that is, for any $\varphi \in \mathscr{A}$ there exists a $\mathscr{U}(\varphi; f, U)$ such that ϖ: $\mathscr{U}(\varphi; f, U) \to U$ is a homeomorphism.

(ii) $\varpi^{-1}(p) = A_p$.

(iii) The linear combination $c_1\varphi + c_2\psi$ depends continuously on φ and ψ, $\varpi(\varphi) = \varpi(\psi)$.

Proof. (i) Since, restricted on $\mathscr{U}(\varphi; f, U)$, ϖ is represented by

$$\varpi: f_q \to q, \qquad q \in U,$$

ϖ maps certainly $\mathcal{U}(\varphi; f, U)$ bijectively onto U. For any $\psi = f_q \in \mathcal{U}(\varphi; f, U)$, a sufficiently small $\mathcal{U}(\psi; f, V)$, $q \in V \subset U$, is mapped bijectively by ϖ to the neighbourhood V of q. Hence ϖ is a local homeomorphism.

(ii) is clear.

(iii) Let $\chi = c_1\varphi + c_2\psi$ with $\varpi(\varphi) = \varpi(\psi) = p$. It suffices to prove that, for any $\mathcal{U}(\chi; f, U)$, there are sufficiently small $\mathcal{U}(\varphi; g, V)$ and $\mathcal{U}(\psi; h, V)$ such that $c_1\xi + c_2\eta \in \mathcal{U}(\chi; f, U)$ for $\xi \in \mathcal{U}(\varphi; g, V)$ and $\eta \in \mathcal{U}(\psi; h, V)$ with $\varpi(\xi) = \varpi(\eta)$. Since

$$c_1 g_p + c_2 h_p = c_1\varphi + c_2\psi = \chi = f_p,$$

f coincides with $c_1 g + c_2 h$ on some neighbourhood $V \subset U$ of p. Put $q = \varpi(\xi) = \varpi(\eta) \in V$. Then we have

$$c_1\xi + c_2\eta = c_1 g_q + c_2 h_q = f_q \in \mathcal{U}(\chi; f, U). \quad \blacksquare$$

$\mathcal{A}$ is called the *sheaf of germs of C^∞ functions over M.*

Let M be a complex manifold. Then using holomorphic functions instead of C^∞ functions, we obtain the definition of the *sheaf of germs of holomorphic functions* $\mathcal{O} = \bigcup_{p \in M} \mathcal{O}_p$. Let $z_p = (z_p^1, \ldots, z_p^n)$ be local complex coordinates centred at p of M. Then a local holomorphic function at p, that is, a holomorphic function defined on some neighbourhood of p, has a convergent power series expansion:

$$\begin{aligned} f = P(z_p) &= P(z_p^1, \ldots, z_p^n) \\ &= \sum_{m_1, \ldots, m_n = 0}^{\infty} a_{m_1 \cdots m_n} (z_p^1)^{m_1} \cdots (z_p^n)^{m_n}. \end{aligned}$$

f and g are equivalent if and only if f and g have the same power series expansion $P(z_p)$. Thus a germ of holomorphic functions at p corresponds in one-to-one manner to a convergent power series in $z_p^1, \ldots, z_p^n$. Hence, by identifying a germ of holomorphic functions with the corresponding convergent power series, we may identify $\mathcal{O}_p$ with the convergent power series ring $\mathbb{C}\{z_p^1, \ldots, z_p^n\}$:

$$\mathcal{O}_p = \mathbb{C}\{z_p^1, \ldots, z_p^n\}. \tag{3.60}$$

If in the definition of $U(\varphi; f, U) = \{f_q \mid q \in U\}$, $\varphi \in \mathcal{O}_p$, U is assumed to be connected, $U(\varphi; f, U)$ is uniquely determined by φ and U. For, by Corollary to Theorem 1.7, $f = g$ on U if $f_p = \varphi = g_p$ since U is connected. Moreover if $\varphi, \psi \in \mathcal{O}_p$ and $\varphi \neq \psi$,

$$U(\varphi; f, U) \cap U(\psi; g, U) = \emptyset$$

if U is connected. In fact, if $f_q = g_q$ at some $q \in U$, $f = g$ on all of U, hence $\varphi = f_p = g_p = \psi$. Therefore the sheaf of germs of holomorphic functions $\mathcal{O}$ is Hausdorff.

Define $\varpi: \mathcal{O} \to M$ by $\varpi(\mathcal{O}_p) = p$. Then, as in the case of $\mathcal{A}$, ϖ is a local homeomorphism, and the linear combination $c_1\varphi + c_2\psi$, $\varpi(\varphi) = \varpi(\psi)$, depends continuously on φ and ψ.

Let M be a differentiable or a complex manifold. In what follows, we denote by K one of $\mathbb{C}$, $\mathbb{R}$, and $\mathbb{Z}$.

Definition 3.5. A topological space $\mathcal{S}$ is called a *sheaf* over M if the following conditions are satisfied:

(i) A local homeomorphism ϖ of $\mathcal{S}$ onto M is defined.
(ii) For any $p \in M$, $\varpi^{-1}(p)$ is a K-module.
(iii) For $c_1, c_2 \in K$, the linear combination $c_1\varphi + c_2\psi$, of $\varphi, \psi \in S$ and $\varpi(\varphi) = \varpi(\psi)$, depends continuously on φ and ψ.

We call $\mathcal{S}_p = \varpi^{-1}(p)$ the *stalk* of S over p, and ϖ the *projection* of $\mathcal{S}$.

The above-mentioned $\mathcal{A}$ and $\mathcal{O}$ are typical examples of sheaves. We define the sheaf $\mathcal{A}^r$ of germs of C^∞ r-forms on a differentiable manifold M, and the sheaf Θ of germs of holomorphic vector fields on a complex manifold M similarly. More generally, given a complex vector bundle F over a differentiable manifold M, we define the sheaf of germs of all C^∞ sections of F as follows. By a local C^∞ section of F at p we mean a C^∞ section of F defined in a neighbourhood of p. Two local C^∞ sections σ and τ are called equivalent if $\sigma = \tau$ in some neighbourhood of p. We denote $\sigma \sim_p \tau$ if σ and τ are equivalent. Then $\sim_p$ is an equivalence relation. The equivalence class of local C^∞ sections is called a germ of C^∞ sections of F. The class to which σ belongs is called the germ of σ at p, and denoted by σ_p. We denote the set of all local C^∞ sections of F at p by $\mathcal{A}_p(F)$.

Put $\mathcal{A}(F) = \bigcup_{p \in M} \mathcal{A}_p(F)$. Then $\mathcal{A}(F)$ is a sheaf over M just as $\mathcal{A}$. $\mathcal{A}(F)$ is called the *sheaf of germs of C^∞ sections of F*. For example, taking $\bigwedge^r T^*(M)$ as F, we have $\mathcal{A}(\bigwedge^r T^*(M)) = \mathcal{A}^r$. For a complex manifold M, $\mathcal{A}^{p,q} = \mathcal{A}(\bigwedge^p T^*(M) \bigwedge^q \bar{T}^*(M))$ is the sheaf of germs of C^∞ (p, q)-forms on M.

In case F is a holomorphic vector bundle over a complex manifold M, the sheaf $\mathcal{O}(F) = \bigcup_{p \in M} \mathcal{O}_p(F)$ of germs of holomorphic sections of F is defined similarly. For example $\Theta = \mathcal{O}(T(M))$ is the sheaf of germs of holomorphic vector fields on M, and $\mathcal{O}(\bigwedge^r T^*(M))$ is the sheaf of germs of holomorphic r-forms on M. Θ plays an essential role in the theory of deformations of complex structures.

By a *locally constant function* we mean a local C^∞ function f with $f = c \in C$ on $D(f)$. *The sheaf of germs of locally constant functions* is identified with the direct product $M \times \mathbb{C}$ with the topology defined by the system of open sets $\mathfrak{O} = \{U \times c \mid U \subset M, c \in \mathbb{C}\}$, where U is an open set of M. In fact if we identify a locally constant function f with its graph $D(f) \times c \subset M \times \mathbb{C}$, then its germ f_p at $p \in D(f)$ is identified with $(p, c) \in M \times \mathbb{C}$, and $U(f_p; f, U) = \{f_q \mid q \in U\}$ with $U \times c \subset M \times \mathbb{C}$. We denote the sheaf $M \times \mathbb{C}$ of germs of locally constant functions simply by $\mathbb{C}$. Similarly we denote by $\mathbb{Z}$ the sheaf $M \times \mathbb{Z}$ of germs of $\mathbb{Z}$-valued locally constant functions.

Definition 3.6. Let $\mathscr{S}$ be a sheaf over M, and W an arbitrary subset of M. A *section* σ of $\mathscr{S}$ over W is a continuous map $\sigma: p \to \sigma(p)$ of W into $\mathscr{S}$ with $\sigma(p) \in \mathscr{S}_p$.

As an example, let us consider a section of the sheaf of germs of C^∞ functions over an open subset. Let $W \subset M$ be an open subset and f *a* C^∞ *function on* W. Then clearly *the map* $p \to f_p$, $p \in W$, *defines a section of* $\mathscr{A}$ *over* W. Conversely, *any section* $\sigma: p \to \sigma(p)$ *of* $\mathscr{A}$ *over* W *is given by some* C^∞ *function* f *on* W *as* $\sigma(p) = f_p$.

Proof. Each $\sigma(p)$ is the germ of some local C^∞ function $g^{[p]}$: $\sigma(p) = (g^{[p]})_p$. Define f by putting $f = f(p) = g^{[p]}(p)$. Then f is well defined. Since σ is continuous, for a neighbourhood $U(\sigma(p); g^{[p]}, U)$ of $\sigma(p)$, there is a neighbourhood $U(p) \subset U$ of p such that for $q \in U(p)$,

$$\sigma(q) \in U(\sigma(p); g^{[p]}, U) = \{(g^{[p]})_q \mid q \in U\},$$

which means that $\sigma(q) = (g^{[p]})_q$. Hence $f(q) = g^{[p]}(q)$. Thus for any $p \in W$, f coincides with a C^∞ function $g^{[p]}$ on $U(p)$. Hence f is C^∞, and $\sigma(p) = (g^{[p]})_p = f_p$. ∎

Similarly for the sheaf $\mathcal{O}$ of germs of holomorphic functions over a complex manifold M, a *section* σ of $\mathcal{O}$ *over an open subset* W *corresponds in a one-to-one manner to a holomorphic function* f *defined on* W *by* $\sigma(p) = f_p$. In this sense we can identify a section of $\mathcal{O}$ over an open set with a holomorphic function defined there. Note, however, that for a section σ, $\sigma(p)$ is a power series $P(z_p)$ whereas $f(p)$ is a complex number for a holomorphic function f. Similarly we can identify a *section* of $\mathscr{A}$ *over an open set with a* C^∞ *function defined there.*

Similar results hold for $\mathscr{A}(F)$ or $\mathcal{O}(F)$, where F denotes a complex vector bundle over a C^∞ manifold or a holomorphic vector bundle over a complex manifold. Namely, let s be a section of F over an open set W. Then by putting $\sigma(p) = s_p$, $p \in W$, we obtain a *one-to-one correspondence between the sections* σ *of* $\mathscr{A}(F)$ *over* W *and the sections* s *of* F *over* W. Thus we may identify them. Similarly a section of $\mathcal{O}(F)$ over an open set may be regarded as a holomorphic section of F defined there.

Let $\mathscr{S}$ be a sheaf over M. For a *section* $\sigma: p \to \sigma(p)$ *over an open set* $W \subset M$, $\mathscr{W} = \sigma(W) = \{\sigma(p) \mid p \in W\}$ *is an open subset of* $\mathscr{S}$, *and* $\sigma: W \to \mathscr{W}$ *is a homeomorphism.*

Proof. Since $\varpi\colon S\to M$ is locally homeomorphic, for any $q\in W$, there is a neighbourhood $\mathscr{V}$ of $\sigma(q)$ in $\mathscr{S}$ such that $\varpi\colon \mathscr{V}\to V=\varpi(\mathscr{V})$ is a homeomorphism. Here V is clearly a neighbourhood of q. Since σ is continuous, there is a neighbourhood $U\subset V$ of q such that $\mathscr{U}=\sigma(U)\subset \mathscr{V}$. $\varpi(\sigma(p))=p$ implies that $\varpi(\mathscr{U})=U$, and $\mathscr{U}$ is an open set of $\mathscr{V}$ since ϖ is homeomorphic. Thus for any $q\in W$, there is a neighbourhood $U\subset W$ such that $\mathscr{U}=\sigma(U)\subset \mathscr{W}$ is an open set of $\mathscr{S}$. Hence $\mathscr{W}$ is open in $\mathscr{S}$. Since $\varpi\colon \mathscr{U}\to U$ is homeomorphic for each $\mathscr{U}$, $\varpi\colon \mathscr{W}\to W$ is a homeomorphism. Hence its inverse $\sigma\colon W\to \mathscr{W}$ is also a homeomorphism. ∎

Conversely, *an open set $\mathscr{W}$ which intersects with each $\mathscr{S}_p$, $p\in \mathscr{W}$, at a single point $\sigma(p)$, defines a section σ over $W=\varpi(\mathscr{W})$.*

We denote by 0_p the identity of the abelian group $\mathscr{S}_p$. Then the map $p\to 0_p$ is a *section* of $\mathscr{S}$ over M.

Proof. It suffices to show that for any $q\in M$, $p\to 0_p$ is continuous in some neighbourhood $U(q)$ of q. Since ϖ is a local homeomorphism, for a sufficiently small $U(q)$ there is a neighbourhood $\mathscr{U}$ of 0_q such that $\varpi\colon \mathscr{U}\to U(q)$ is homeomorphic. Consequently its inverse $p\to\sigma(p)\in\mathscr{U}$ is continuous, hence by (iii) of Definition 3.5, $p\to 0_p=\sigma(p)-\sigma(p)$ is continuous on $U(q)$. ∎

Thus $p\to 0_p$ is a section of $\mathscr{S}$ over M, hence by the above argument, $\{0_p\,|\,p\in M\}$ is *an open subset* of $\mathscr{S}$.

Let W be an arbitrary subset of M with $W\neq\varnothing$. We denote the restriction of the map $p\to 0_p$ to W, which is a section over W, by 0. For sections σ, τ of $\mathscr{S}$ over W, we define its linear combination by

$$c_1\sigma+c_2\tau\colon p\to c_1\sigma(p)+c_2\tau(p),\qquad c_1, c_2\in K.$$

Then the *set of all sections of $\mathscr{S}$ over W forms a K-module.*

Definition 3.7. We denote by $\Gamma(W,\mathscr{S})$ the K-module of all sections of $\mathscr{S}$ over W.

The identity of the module $\Gamma(W,\mathscr{S})$ is the section $0\colon p\to 0_p$, $p\in W$. If $\Gamma(W,\mathscr{S})$ contains no element other than 0, we denote $\Gamma(W,\mathscr{S})=0$.

If W is open, we see from the above argument that $\Gamma(W,\mathscr{A})$ may be identified with the vector space of all C^∞ functions on W. Similarly $\Gamma(W,\mathscr{O})$, $\Gamma(W,\mathscr{A}(F))$ and $\Gamma(W,\mathscr{O}(F))$ are identified with the vector space of all holomorphic functions, all C^∞ sections of F over W and all holomorphic sections of F over W respectively.

Let $U\subset W$, and σ a section of $\mathscr{S}$ over W. Restricting σ to U, we obtain a section of $\mathscr{S}$ over U, which is called the restriction of σ to U, and denoted by $r_U\sigma$. $r_U\colon \sigma\to r_U\sigma$ is a homomorphism of $\Gamma(W,\mathscr{S})$ into $\Gamma(U,\mathscr{S})$.

Suppose given a finite number of sections $\sigma_\lambda \in \Gamma(W_\lambda, \mathscr{S})$, $\lambda = 1, 2, \ldots, \nu$. If $U = \bigcap_{\lambda=1}^{\nu} W_\lambda \neq \emptyset$, for any $c_\lambda \in K$, $\lambda = 1, \ldots, \nu$, the map $p \to \sum_{\lambda=1}^{\nu} c_\lambda \sigma_\lambda(p)$, $p \in U$, is a section of $\mathscr{S}$ over U, which we denote by $\sum_{\lambda=1}^{\nu} c_\lambda \sigma_\lambda$:

$$\sum_{\lambda=1}^{\nu} c_\lambda \sigma_\lambda = \sum_{\lambda=1}^{\nu} c_\lambda r_U \sigma_\lambda. \tag{3.61}$$

If $\bigcap_{\lambda=1}^{\nu} W_\lambda = \emptyset$, we do not define $\sum_{\lambda=1}^{\nu} c_\lambda \sigma_\lambda$. In what follows, if we write $\sum_{\lambda=1}^{\nu} c_\lambda \sigma_\lambda$, we always assume that $\bigcap_{\lambda=1}^{\nu} W_\lambda \neq \emptyset$.

(b) Cohomology Group

Let M be a differentiable manifold, $\mathscr{S}$ a sheaf over M, and $\mathfrak{U} = \{U_1, \ldots, U_j, \ldots\}$ an arbitrary locally finite open covering of M. First we shall define the cohomology group of M with coefficients in $\mathscr{S}$ with respect to $\mathfrak{U}$.

A *0-cochain* c^0 with respect to $\mathfrak{U}$ is a set $0 = \{\sigma_j\}$ of sections where $\sigma_j \in \Gamma(U_j, \mathscr{S})$ for each j. A 1-cochain $c^1 = \{\sigma_{jk}\}$ with respect to $\mathfrak{U}$ is a set of sections $\sigma_{jk} \in \Gamma(U_j \cap U_k, \mathscr{S})$ for all indices j, k with $U_j \cap U_k \neq \emptyset$ such that $\sigma_{jk} = -\sigma_{kj}$. In general, a q-cochain $c^q = \{\sigma_{k_0 \cdots k_q}\}$ with respect to $\mathfrak{U}$ is a set of sections

$$\sigma_{k_0 \cdots k_q} \in \Gamma(U_{k_0} \cap \cdots \cap U_{k_q}, \mathscr{S}),$$

for all $(q+1)$-tuples of indices $k_0, \ldots, k_q$ with $U_{k_0} \cap \cdots \cap U_{k_q} \neq \emptyset$ which are skew-symmetric in the indices $k_0, \ldots, k_q$. Given two q-cochains $\{\sigma_{j \cdots l}\}$, $\{\tau_{j \cdots l}\}$, we define their linear combination by

$$c_1\{\sigma_{j \cdots l}\} + c_2\{\tau_{j \cdots l}\} = \{c_1 \sigma_{j \cdots l} + c_2 \tau_{j \cdots l}\}, \quad \text{with } c_1, c_2 \in K.$$

Then *the set of all q-cochains with respect to* $\mathfrak{U}$ *becomes a K-module*, which we denote by $C^q(U, \mathscr{S})$ or simply, by $C^q(\mathfrak{U})$.

For a 0-cochain $c^0 = \{\sigma_j\}$, the 1-cochain $c^1 = \{\tau_{jk}\}$ defined by $\tau_{jk} = \sigma_k - \sigma_j$ is called the *coboundary* of c_0 and denoted by δc^0:

$$\delta\{\sigma_j\} = \{\tau_{jk}\}, \qquad \tau_{jk} = \sigma_k - \sigma_j.$$

Note that by $\sigma_k - \sigma_j$ we mean the linear combination of them in the sense of (3.61), hence by so writing, *we assume also* $U_j \cap U_k \neq \emptyset$. For a 1-cochain $\{\sigma_{jk}\}$, we define its coboundary by

$$\delta\{\sigma_{jk}\} = \{\tau_{jkl}\}, \qquad \tau_{jk} = \sigma_{kl} - \sigma_{jl} + \sigma_{jk} = \sigma_{kl} + \sigma_{lj} + \sigma_{jk}.$$

Clearly τ_{jkl} is skew-symmetric in j, k, l. In general for a $(q-1)$-cochain $\{\sigma_{k_1\cdots k_q}\}$ with respect to $\mathfrak{U}$, we define its coboundary by

$$\delta\{\sigma_{k_1\cdots k_q}\} = \{\tau_{k_0\cdots k_q}\}, \quad \text{where}$$

$$\begin{aligned}\tau_{k_0\cdots k_q} &= \sum_{s=1}^{q} (-1)^s \sigma_{k_0\cdots k_{s-1}k_{s+1}\cdots k_q} \\ &= \sum_{s=1}^{q} (-1)^s \sigma_{k_0\cdots \not{k}_s\cdots k_q},\end{aligned} \tag{3.62}$$

where / means that we omit this index. δ is a homomorphism of $C^{q-1}(\mathfrak{U})$ into $C^q(\mathfrak{U})$. If $\{\tau_{k_0\cdots k_q}\} = \delta\{\sigma_{k_1\cdots k_q}\}$, then we have $\delta\{\tau_{k_0\cdots k_q}\} = 0$. In fact

$$\begin{aligned}&\sum_{s=0}^{q+1} (-1)^s \tau_{k_0\cdots \not{k}_s\cdots k_q} \\ &\quad = \sum_{s=0}^{q+1} (-1)^s \left(\sum_{t=0}^{s-1} (-1)^t \sigma_{\cdots \not{k}_t\cdots \not{k}_s\cdots} + \sum_{t=s+1}^{q+1} (-1)^{t-1} \sigma_{\cdots \not{k}_s\cdots \not{k}_t\cdots} \right) \\ &\quad = \sum_{t<s} (-1)^{s+t} \sigma_{\cdots \not{k}_t\cdots \not{k}_s\cdots} + \sum_{s<t} (-1)^{t+s-1} \sigma_{\cdots \not{k}_s\cdots \not{k}_t\cdots} = 0.\end{aligned}$$

Hence we have

$$\delta\delta = 0. \tag{3.63}$$

If $\delta c^q = 0$, we call $c^q \in C^q(\mathfrak{U}, \mathscr{S})$ a *q-cocycle* with respect to $\mathfrak{U}$. We denote the set of q-cocycles with respect to $\mathfrak{U}$ by $Z^q(\mathfrak{U}, \mathscr{S})$ or $Z^q(\mathfrak{U})$.

$$Z^q(\mathfrak{U}, \mathscr{S}) = \{c^q \in C^q(\mathfrak{U}, \mathscr{S}) \mid \delta c^q = 0\}$$

is a submodule of $C^q(\mathfrak{U}, \mathscr{S})$. For $q \geqq 1$, $\delta\delta C^{q-1}(\mathfrak{U}, \mathscr{S}) = 0$ by (3.63), hence $\delta C^{q-1}(\mathfrak{U}, \mathscr{S}) \subset Z^q(\mathfrak{U}, \mathscr{S})$. We put

$$H^q(\mathfrak{U}, \mathscr{S}) = Z^q(\mathfrak{U}, \mathscr{S})/\delta C^{q-1}(\mathfrak{U}, \mathscr{S}), \qquad q \geqq 1,$$

which we call the *cohomology group with coefficients in* $\mathscr{S}$ with respect to $\mathfrak{U}$. For $q = 0$, we define

$$H^0(\mathfrak{U}, \mathscr{S}) = Z^0(\mathfrak{U}, \mathscr{S}).$$

If $\{\sigma_j\} \in Z^0(\mathfrak{U}, \mathscr{S})$, then $\sigma_j = \sigma_k$ on $U_j \cap U_k \neq \varnothing$, hence it defines an element $\sigma \in \Gamma(M, \mathscr{S})$ with $\sigma = \sigma_j$ on U_j. We have then $\sigma_j = r_{U_j}\sigma$. Thus we may identify $Z^0(\mathfrak{U}, \mathscr{S})$ with $\Gamma(M, \mathscr{S})$, that is,

$$H^0(\mathfrak{U}, \mathscr{S}) = \Gamma(M, \mathscr{S}). \tag{3.64}$$

Thus we define $H^q(\mathfrak{U}, \mathscr{S})$ for any locally finite open covering $\mathfrak{U}$ of M. $H^0(\mathfrak{U}, \mathscr{S}) = \Gamma(\mathfrak{U}, \mathscr{S})$ does not depend on the choice of $\mathfrak{U}$. For $q \geqq 1$, however, $H^q(\mathfrak{U}, \mathscr{S})$ does depend on $\mathfrak{U}$. Therefore we define the qth cohomology group $H^q(M, \mathscr{S})$ of M with coefficients in $\mathscr{S}$ as the "limit" of $H^q(\mathfrak{U}, \mathscr{S})$,

$$H^q(M, \mathscr{S}) = \lim_{\mathfrak{U}} H^q(\mathfrak{U}, \mathscr{S})$$

as each U_j of $\mathfrak{U}$ becomes "infinitely small". The limiting process will be explained now.

We say that the open covering $\mathfrak{B} = \{V_\lambda\}$ of M is a *refinement* of $\mathfrak{U}$ if each V_λ is contained in some member U_k of $\mathfrak{U}$. We write $\mathfrak{B} < \mathfrak{U}$ or $\mathfrak{U} > \mathfrak{B}$ if $\mathfrak{B}$ is a refinement of $\mathfrak{U}$. For each V_λ, we fix an arbitrary U_k with $V_\lambda \subset U_k$, and write it as $U_{k(\lambda)}$: $V_\lambda \subset U_{k(\lambda)}$. We define a homomorphism of $C^q(\mathfrak{U})$ into $C^q(\mathfrak{B})$

$$\Pi^{\mathfrak{U}}_{\mathfrak{B}}: \{\sigma_{k_0 \cdots k_q}\} \to \{\tau_{\lambda_0 \cdots \lambda_q}\} \tag{3.65}$$

by putting

$$\tau_{\lambda_0 \cdots \lambda_q} = r_V \sigma_{k(\lambda_0) \cdots k(\lambda_q)}, \quad \text{where} \quad V = \bigcap_{s=0}^{q} V_{\lambda_s} \neq \varnothing.$$

Since it is easily verified that

$$\delta \Pi^{\mathfrak{U}}_{\mathfrak{B}} = \Pi^{\mathfrak{U}}_{\mathfrak{B}} \delta,$$

$\Pi^{\mathfrak{U}}_{\mathfrak{B}}$ maps $Z^q(\mathfrak{U})$ into $Z^q(\mathfrak{B})$ and $\delta C^{q-1}(\mathfrak{U})$ into $\delta C^{q-1}(\mathfrak{B})$. Hence $\Pi^{\mathfrak{U}}_{\mathfrak{B}}$ defines a homomorphism of $H^q(\mathfrak{U}, \mathscr{S}) = Z^q(\mathfrak{U})/\delta C^{q-1}(\mathfrak{U})$ into $H^q(\mathfrak{B}, \mathscr{S}) = Z^q(\mathfrak{B})/\delta C^{q-1}(\mathfrak{B})$, which is also denoted by $\Pi^{\mathfrak{U}}_{\mathfrak{B}}$. $\Pi^{\mathfrak{U}}_{\mathfrak{B}}: C^q(\mathfrak{U}) \to C^q(\mathfrak{B})$ depend on the choice of $U_{k(\lambda)} \supset V_\lambda$ for each V_λ, but $\Pi^{\mathfrak{U}}_{\mathfrak{B}}: H^q(\mathfrak{U}, \mathscr{S}) \to H^q(\mathfrak{B}, \mathscr{S})$ turns out to be independent of this choice. In fact we have the following

Lemma 3.2. *The homomorphism* $\Pi^{\mathfrak{U}}_{\mathfrak{B}}: H^q(\mathfrak{U}, \mathscr{S}) \to H^q(\mathfrak{B}, \mathscr{S})$ *is uniquely determined by* $\mathfrak{U}$ *and* $\mathfrak{B}$.

Proof. For each λ choose an arbitrary $j(\lambda)$ with $U_{j(\lambda)} \supset V_\lambda$, and put

$$\eta_{\lambda_0 \cdots \lambda_q} = r_V \sigma_{j(\lambda_0) \cdots j(\lambda_q)}, \quad \text{where} \quad V = \bigcap_{s=1}^{q} V_{\lambda_s} \neq \varnothing.$$

It suffices to show that for $\{\sigma_{k_0 \cdots k_q}\} \in Z^q(\mathfrak{U})$, we have

$$\{\eta_{\lambda_0 \cdots \lambda_q}\} - \{\tau_{\lambda_0 \cdots \lambda_q}\} \in \delta C^{q-1}(\mathfrak{B}).$$

Put

$$\kappa_{\nu_1\cdots\nu_q}=\sum_{t=1}^{q}(-1)^{t-1}r_W\sigma_{k(\nu_1)\cdots k(\nu_t)j(\nu_t)\cdots j(\nu_q)},$$

where $W=\bigcap_{t=1}^{q}V_{\nu_t}\neq\varnothing$. Though $\kappa_{\nu_1\cdots\nu_q}$ is not necessarily skew-symmetric in the indices, we define

$$(\delta\kappa)_{\lambda_0\cdots\lambda_q}=\sum_{s=0}^{q}(-1)^s\kappa_{\lambda_0\cdots\not\lambda_s\cdots\lambda_q}.$$

Then we see by an easy calculation that

$$\eta_{\lambda_0\cdots\lambda_q}-\tau_{\lambda_0\cdots\lambda_q}=(\delta\kappa)_{\lambda_0\cdots\lambda_q}. \tag{3.66}$$

In fact if we write h_s for $k(\lambda_s)$ and j_s for $j(\lambda_s)$, we have

$$\begin{aligned}(\delta\kappa)_{\lambda_0\cdots\lambda_q}&=\sum_{s=0}^{q}(-1)^s\kappa_{\lambda_0\cdots\not\lambda_s\cdots\lambda_q}\\&=\sum_{s=0}^{q}(-1)^s\Bigg(\sum_{t=0}^{s-1}(-1)^t r_V\sigma_{h_0\cdots h_tj_t\cdots\not j_s\cdots j_q}\\&\qquad+\sum_{t=s+1}^{q}(-1)^{t-1}r_V\sigma_{h_0\cdots\not h_s\cdots h_tj_t\cdots j_q}\Bigg)\\&=\sum_{t=0}^{q}(-1)^{t+1}\Bigg(\sum_{s=0}^{t-1}(-1)^s r_V\sigma_{h_0\cdots\not h_s\cdots h_tj_t\cdots j_q}\\&\qquad+\sum_{s=t+1}^{q}(-1)^{s+1}r_V\sigma_{h_0\cdots h_tj_t\cdots\not j_s\cdots j_q}\Bigg).\end{aligned}$$

On the other hand by the hypothesis, we have,

$$\sum_{s=0}^{t}(-1)^s r_V\sigma_{h_0\cdots\not h_s\cdots h_tj_t\cdots j_q}+\sum_{s=t}^{q}(-1)^{s+1}r_V\sigma_{h_0\cdots h_tj_t\cdots\not j_s\cdots j_q}=0.$$

Hence we have

$$\begin{aligned}(\delta\kappa)_{\lambda_0\cdots\lambda_q}&=\sum_{t=0}^{q}r_V\sigma_{h_0\cdots h_{t-1}j_t\cdots j_q}-\sum_{t=0}^{q}r_V\sigma_{h_0\cdots h_tj_{t+1}\cdots j_q}\\&=r_V\sigma_{j_0\cdots j_q}-r_V\sigma_{h_0\cdots h_q}=\eta_{\lambda_0\cdots\lambda_q}-\tau_{\lambda_0\cdots\lambda_q}.\end{aligned}$$

Thus (3.66) holds.

The above $\kappa_{\nu_1\cdots\nu_q}$ is not necessarily skew-symmetric in $\nu_1, \ldots, \nu_q$, so we put

$$\tilde{\kappa}_{\nu_1\cdots\nu_q} = \frac{1}{q!}\sum \operatorname{sgn}\begin{pmatrix}\nu_1 & \cdots & \nu_q \\ \mu_1 & \cdots & \mu_q\end{pmatrix}\kappa_{\mu_1\cdots\mu_q},$$

where the summation is taken over all the permutations $\binom{\nu_1\cdots\nu_q}{\mu_1\cdots\mu_q}$ of $\nu_1, \ldots, \nu_q$. Then $\{\tilde{\kappa}_{\nu_1\cdots\nu_q}\} \in C^{q-1}(\mathfrak{V})$. By (3.66)

$$\begin{aligned}
&\eta_{\lambda_0\cdots\lambda_q} - \tau_{\lambda_0\cdots\lambda_q} \\
&\quad = \frac{1}{(q+1)!}\sum \operatorname{sgn}\begin{pmatrix}\lambda_0 & \cdots & \lambda_q \\ \nu_0 & \cdots & \nu_q\end{pmatrix}(\delta\kappa)_{\nu_0\cdots\nu_q} \\
&\quad = \frac{1}{(q+1)!}\sum \operatorname{sgn}\begin{pmatrix}\lambda_0 & \cdots & \lambda_s & \cdots & \lambda_q \\ \nu_0 & \cdots & \nu_s & \cdots & \nu_q\end{pmatrix}\sum_{s=0}^{q}(-1)^s\kappa_{\nu_0\cdots\hat{\nu}_s\cdots\nu_q} \\
&\quad = \frac{1}{(q+1)!}\sum \operatorname{sgn}\begin{pmatrix}\lambda_0 & \cdots & \lambda_s & \cdots & \lambda_q \\ \nu_0 & \cdots & \nu_s & \cdots & \nu_q\end{pmatrix}\sum_{s=0}^{q}(-1)^s\tilde{\kappa}_{\nu_0\cdots\hat{\nu}_s\cdots\nu_q}.
\end{aligned}$$

Since $\sum_s (-1)^s\tilde{\kappa}_{\nu_0\cdots\hat{\nu}_s\cdots\nu_q}$ is skew-symmetric in $\nu_0, \ldots, \nu_q$, we have

$$\eta_{\lambda_0\cdots\lambda_q} - \tau_{\lambda_0\cdots\lambda_q} = \sum_{s=0}^{q}(-1)^s\tilde{\kappa}_{\nu_0\cdots\hat{\nu}_s\cdots\nu_q},$$

namely,

$$\{\eta_{\lambda_0\cdots\lambda_q}\} - \{\tau_{\lambda_0\cdots\lambda_q}\} = \delta\{\tilde{\kappa}_{\nu_1\cdots\nu_q}\} \in \delta C^{q-1}(\mathfrak{V}). \quad \blacksquare$$

We define $\mathfrak{U} < \mathfrak{V}$ if $\mathfrak{V}$ is a refinement of $\mathfrak{U}$. Then $<$ defines a partial order on the set $\mathfrak{S} = \{\mathfrak{U}, \mathfrak{V}, \ldots\}$ of all locally finite open coverings of M, which makes $\mathfrak{S}$ a *directed set* since for any $\mathfrak{U}, \mathfrak{V} \in \mathfrak{S}$, there is a $\mathfrak{W} \in \mathfrak{S}$ such that $\mathfrak{V} < \mathfrak{W}$ and $\mathfrak{U} < \mathfrak{W}$. Write $H^q(\mathfrak{U})$ for $H^q(\mathfrak{U}, S)$ for simplicity. Then by Lemma 3.2, for $\mathfrak{U}, \mathfrak{V} \in S$ with $\mathfrak{U} > \mathfrak{V}$, a homomorphism

$$\Pi^{\mathfrak{U}}_{\mathfrak{V}}: H^q(\mathfrak{U}) \to H^q(\mathfrak{V})$$

is defined. Clearly $\Pi^{\mathfrak{U}}_{\mathfrak{U}}$ is the identity, and, if $\mathfrak{U} > \mathfrak{V} > \mathfrak{W}$, we have

$$\Pi^{\mathfrak{U}}_{\mathfrak{W}} = \Pi^{\mathfrak{V}}_{\mathfrak{W}} \cdot \Pi^{\mathfrak{U}}_{\mathfrak{V}}. \tag{3.67}$$

Now we define $\lim_{\mathfrak{U}} H^q(\mathfrak{U})$ as follows. First we *define* $g > 0$ *for* $g \in H^q(\mathfrak{U})$ *if there exists* $\mathfrak{V} < \mathfrak{U}$, $\mathfrak{V} \in \mathfrak{S}$, *such that* $\Pi^{\mathfrak{U}}_{\mathfrak{V}} g = 0$. Put $N^q(\mathfrak{U}) = \{g \in H^q(\mathfrak{U}) \mid g > 0\}$, which is a *subgroup* of $H^q(\mathfrak{U})$. In fact if $\Pi^{\mathfrak{U}}_{\mathfrak{V}} g = 0$, and

$\Pi_{\mathfrak{W}}^{\mathfrak{U}}h=0$, then taking $\mathfrak{X}$ with $\mathfrak{V}>\mathfrak{X}$ and $\mathfrak{U}>\mathfrak{X}$, we obtain by (3.67),

$$\Pi_{\mathfrak{X}}^{\mathfrak{U}}(c_1g+c_2h)=c_1\Pi_{\mathfrak{X}}^{\mathfrak{V}}\Pi_{\mathfrak{V}}^{\mathfrak{U}}g+c_2\Pi_{\mathfrak{X}}^{\mathfrak{W}}\Pi_{\mathfrak{W}}^{\mathfrak{U}}h=0, \qquad c_1, c_2\in K.$$

Put

$$\bar{H}^q(\mathfrak{U},\mathscr{S})=\bar{H}^q(\mathfrak{U})=H^q(\mathfrak{U})/N^q(\mathfrak{U}),$$

and let $\Pi^{\mathfrak{U}}$: $g\to\bar{g}$ be the canonical homomorphism of $H^q(\mathfrak{U})=H^q(\mathfrak{U},\mathscr{S})$ onto $\bar{H}^q(\mathfrak{U},\mathscr{S})$.

For $\mathfrak{U}>\mathfrak{V}$, we have $\Pi_{\mathfrak{V}}^{\mathfrak{U}}N^q(\mathfrak{U})\subset N^q(\mathfrak{V})$. In fact for $g\in N^q(\mathfrak{U})$, there is a $\mathfrak{W}<\mathfrak{U}$ with $\Pi_{\mathfrak{W}}^{\mathfrak{U}}g=0$. For $\mathfrak{X}\in\mathfrak{S}$ with $\mathfrak{V}<\mathfrak{X}$ and $\mathfrak{U}<\mathfrak{X}$, we have

$$\Pi_{\mathfrak{X}}^{\mathfrak{V}}\Pi_{\mathfrak{V}}^{\mathfrak{U}}g=\Pi_{\mathfrak{X}}^{\mathfrak{U}}g=\Pi_{\mathfrak{X}}^{\mathfrak{W}}\Pi_{\mathfrak{W}}^{\mathfrak{U}}g=0,$$

hence $\Pi_{\mathfrak{V}}^{\mathfrak{U}}g\in N^q(\mathfrak{V})$. Consequently $\Pi_{\mathfrak{V}}^{\mathfrak{U}}$: $H^q(\mathfrak{U})\to H^q(\mathfrak{V})$ *induces the homomorphism*

$$\bar{\Pi}_{\mathfrak{V}}^{\mathfrak{U}};\ \bar{H}^q(\mathfrak{U})=H^q(\mathfrak{U})/N^q(\mathfrak{U})\to\bar{H}^q(\mathfrak{V})=H^q(\mathfrak{V})/N^q(\mathfrak{V}).$$

This $\bar{\Pi}_{\mathfrak{V}}^{\mathfrak{U}}$ is an *injection.* For if $\bar{\Pi}_{\mathfrak{V}}^{\mathfrak{U}}\bar{g}=0$ with $\bar{g}=\Pi^{\mathfrak{U}}g$, $g\in H^q(\mathfrak{U})$, then $\Pi_{\mathfrak{V}}^{\mathfrak{U}}g\in N^q(\mathfrak{V})$, hence there is a $\mathfrak{W}<\mathfrak{V}$ with $\Pi_{\mathfrak{W}}^{\mathfrak{V}}\Pi_{\mathfrak{V}}^{\mathfrak{U}}g=0$. Hence $\Pi_{\mathfrak{W}}^{\mathfrak{U}}g=0$, which means $g\in N^q(\mathfrak{U})$, that is $\bar{g}=0$. Thus for $\mathfrak{U}<\mathfrak{V}$, $\bar{H}^q(\mathfrak{U})$ *may be identified with a submodule* $\bar{\Pi}_{\mathfrak{V}}^{\mathfrak{U}}\bar{H}^q(\mathfrak{U})$ *via the inclusion* $\bar{\Pi}_{\mathfrak{V}}^{\mathfrak{U}}$:

$$\bar{H}^q(\mathfrak{U},\mathscr{S})\hookrightarrow\bar{H}^q(\mathfrak{V},\mathscr{S}), \qquad \mathfrak{U}>\mathfrak{V}. \tag{3.68}$$

Taking *the union* $\bigcup_{\mathfrak{U}\in\mathfrak{S}}\bar{H}^q(\mathfrak{U},\mathscr{S})$ for $\mathfrak{U}\in\mathfrak{S}$, and considering it as $\lim_{\mathfrak{U}}H^q(\mathfrak{U},\mathscr{S})$, we define *the cohomology group of M with coefficients in* $\mathscr{S}$ *by*

$$H^q(M,\mathscr{S})=\bigcup_{\mathfrak{U}\in\mathfrak{S}}\bar{H}^q(\mathfrak{U},\mathscr{S}). \tag{3.69}$$

For any $\mathfrak{U}\in\mathfrak{S}$, $\Pi^{\mathfrak{U}}$: $H^q(\mathfrak{U},\mathscr{S})\to\bar{H}^q(\mathfrak{U},\mathscr{S})\subset H^q(M,\mathscr{S})$ is a *homomorphism* of $H^q(\mathfrak{U},\mathscr{S})$ into $H^q(M,\mathscr{S})$.

A q-cocycle $c^q\in Z^q(\mathfrak{U},\mathscr{S})$ defines an element

$$g\in H^q(\mathfrak{U},\mathscr{S})=Z^q(\mathfrak{U},\mathscr{S})/\delta C^{q-1}(\mathfrak{U},\mathscr{S}),$$

which in turn defines an element $\bar{g}=\Pi^{\mathfrak{U}}g\in H^q(M,\mathscr{S})$. We call $\bar{g}$ *the cohomology class of a q-cocycle* c^q, *and denote it by* $[c^q]$: $[c^q]=\Pi^{\mathfrak{U}}g\in H^q(M,\mathscr{S})$.

It is clear from (3.64) that

$$H^0(M,\mathscr{S})=\Gamma(M,\mathscr{S}). \tag{3.70}$$

For $q=1$, we have the following

Theorem 3.4. $\Pi^{\mathfrak{U}}: H^1(\mathfrak{U}, \mathscr{S}) \to H^1(M, \mathscr{S})$ *is injective.*

Proof. First we shall prove that $N^1(\mathfrak{U})=0$. Let $g \in N^1(\mathfrak{U})$. Then there is $\mathfrak{V} < \mathfrak{U}$, $\mathfrak{V} \in \mathfrak{S}$, such that $\Pi^{\mathfrak{U}}_{\mathfrak{V}} g = 0$. We let $\mathfrak{U} = \{U_j\}$, $\mathfrak{V} = \{V_\lambda\}$ and put $\mathfrak{W} = \{W_{j\lambda} \mid W_{j\lambda} = U_j \cap V_\lambda \neq \varnothing\}$. Since $W_{j\lambda} \subset V_\lambda$, we have $\mathfrak{V} > \mathfrak{W}$, hence $\Pi^{\mathfrak{U}}_{\mathfrak{W}} g = \Pi^{\mathfrak{V}}_{\mathfrak{W}} \Pi^{\mathfrak{U}}_{\mathfrak{V}} g = 0$. Here by Lemma 3.2, we may assume that $\Pi^{\mathfrak{U}}_{\mathfrak{W}}$ is defined via the inclusion $W_{j\lambda} \subset U_j$.

Suppose that g is represented by a 1-cocycle $\{\sigma_{jk}\} \in Z^1(\mathfrak{U})$. If we define $\Pi^{\mathfrak{U}}_{\mathfrak{W}}: Z^1(\mathfrak{U}) \to Z^1(\mathfrak{W})$ via the inclusion $W_{j\lambda} \subset U_j$, then

$$\Pi^{\mathfrak{U}}_{\mathfrak{W}}\{\sigma_{jk}\} = \{\tau_{j\lambda\, k\nu}\}, \qquad \tau_{j\lambda\, k\nu} = r_{W_{j\lambda} \cap W_{k\nu}} \sigma_{jk}, \quad \text{with} \quad W_{j\lambda} \cap W_{\kappa\nu} \neq \varnothing.$$

Since $\Pi^{\mathfrak{U}}_{\mathfrak{W}} g = 0$, $\{\tau_{j\lambda\, k\nu}\} \in \delta C^0(\mathfrak{W})$, hence

$$\tau_{j\lambda\, k\nu} = \tau_{k\nu} - \tau_{j\lambda}$$

for some $\{\tau_{j\lambda}\} \in C^0(\mathfrak{W})$. Since $\sigma_{jj} = 0$, $\tau_{j\nu} - \tau_{j\lambda} = \tau_{j\lambda\, j\nu} = 0$, which implies that $\tau_{j\nu} = \tau_{j\lambda}$ on $W_{j\lambda} \cap W_{j\nu} \neq \varnothing$. Therefore we define a section $\tau_j \in \Gamma(U_j, \mathscr{S})$ by putting $\tau_j = \tau_{j\lambda}$ on $U_j \cap W_\lambda = W_{j\lambda}$. On $U_j \cap U_k \cap W_\lambda$, $\sigma_{jk} = \tau_{j\lambda\, k\lambda} = \tau_k - \tau_j$, hence $\sigma_{jk} = \tau_k - \tau_j$ on $U_j \cap U_k$, that is, $\{\sigma_{jk}\} = \delta\{\tau_j\} \in \delta C^0(\mathfrak{U})$. Hence $g = 0$. Thus we obtain $N^1(\mathfrak{U}) = 0$, which implies

$$\bar{H}^1(\mathfrak{U}, \mathscr{S}) = H^1(\mathfrak{U}, \mathscr{S}). \tag{3.71}$$

Therefore $\Pi^{\mathfrak{U}}: H^1(\mathfrak{U}, \mathscr{S}) \to \bar{H}^1(\mathfrak{U}, \mathscr{S}) \subset H^1(M, \mathscr{S})$ is injective. ∎

Corollary. $H^1(M, \mathscr{S}) = \bigcup_{\mathfrak{U} \in \mathfrak{S}} H^1(\mathfrak{U}, \mathscr{S})$. ∎

Theorem 3.5. *If* $H^1(U_j, \mathscr{S}) = 0$ *for every* $U_j \in \mathfrak{U}$, *then* $H^1(\mathfrak{U}, \mathscr{S}) = H^1(M, \mathscr{S})$.

Proof. By the Corollary above, we have $H^1(M, \mathscr{S}) = \bigcup_{\mathfrak{V} \in \mathfrak{S}} H^1(\mathfrak{V})$, hence it suffices to prove $H^1(\mathfrak{V}) \subset H^1(\mathfrak{U})$ for any $\mathfrak{V} \in \mathfrak{S}$. Let $\mathfrak{U} = \{U_j\}$ and $\mathfrak{V} = \{V_\lambda\}$, and put $\mathfrak{W} = \{W_{j\lambda} \mid W_{j\lambda} = U_j \cap V_\lambda \neq \varnothing\}$. Then by (3.71) and (3.68), $H^1(\mathfrak{V}) \subset H^1(\mathfrak{W})$, hence we have only to show that $H^1(\mathfrak{W}) \subset H^1(\mathfrak{U})$. For this purpose it suffices to prove that for any 1-cocycle $\{\tau_{j\lambda\, k\nu}\} \in Z^1(\mathfrak{W})$, there are a 1-cocycle $\{\sigma_{jk}\} \in Z^1(\mathfrak{U})$ and a 0-cochain $\{\tau_{j\lambda}\} \in C^0(\mathfrak{W})$ such that

$$\tau_{j\lambda\, k\nu} = r_{W_{j\lambda} \cap W_{k\nu}} \sigma_{jk} + \tau_{k\nu} - \tau_{j\lambda}. \tag{3.72}$$

In fact (3.72) means that $\{\tau_{j\lambda\, k\nu}\} = \Pi^{\mathfrak{U}}_{\mathfrak{W}}\{\sigma_{jk}\} + \delta\{\tau_{j\lambda}\}$, which implies $H^1(\mathfrak{W}) \subset H^1(\mathfrak{U})$.

For a fixed U_j, $\mathfrak{W}_j = \{W_{j\lambda}\}$ is a locally finite open covering of U_j. Since by hypothesis $H^1(U_j, \mathscr{S}) = 0$, we have $Z^1(\mathfrak{W}_j)/\delta C^0(\mathfrak{W}_j) = H^1(\mathfrak{W}_j, \mathscr{S}) = 0$ by Theorem 3.4. Hence $\{\tau_{j\lambda\, k\nu}\} \in Z^1(\mathfrak{W}_j) = \delta C^0(\mathfrak{W}_j)$, i.e., there is a $\{\tau_{j\lambda}\} \in C^0(\mathfrak{W}_j)$

such that

$$\tau_{j\lambda\, j\nu} = \tau_{j\nu} - \tau_{j\lambda}.$$

Fixing such $\{\tau_{j\lambda}\}$ for each U_j, and putting

$$\sigma_{j\lambda\, k\nu} = \tau_{j\lambda\, k\nu} - \tau_{k\nu} + \tau_{j\lambda}, \tag{3.73}$$

we have $\{\sigma_{j\lambda\, k\nu}\} \in Z^1(\mathfrak{W})$, and $\sigma_{j\lambda\, j\nu} = 0$. Since $\sigma_{j\lambda\, j\nu} + \sigma_{j\nu\, k\mu} + \sigma_{k\mu\, j\lambda} = 0$, and $\sigma_{j\lambda\, k\mu} = -\sigma_{k\mu\, j\lambda}$, we have $\sigma_{j\lambda\, k\mu} = \sigma_{j\nu\, k\mu}$. Similarly we have $\sigma_{j\lambda\, k\mu} = \sigma_{j\lambda\, k\nu}$. Hence by putting $\sigma_{j\, k\mu} = \sigma_{j\lambda\, k\mu}$ on $U_j \cap U_k \cap V_\mu \cap V_\lambda \neq \emptyset$, we define a section $\sigma_{j\, k\mu} \in \Gamma(U_j \cap U_k \cap V_\mu)$. $\sigma_{j\lambda\, k\mu} = \sigma_{j\lambda\, k\nu}$ implies that $\sigma_{j\, k\mu} = \sigma_{j\, k\nu}$ on $U_j \cap U_k \cap V_\mu \cap V_\nu \neq \emptyset$. Hence again by putting $\sigma_{jk} = \sigma_{j\, k\mu}$ on $U_j \cap U_k \cap V_\mu \neq \emptyset$, we obtain a section $\sigma_{jk} \in \Gamma(U_j \cap U_k)$. $\{\sigma_{jk}\}$ is a 1-cocycle since

$$\sigma_{jk} + \sigma_{kl} + \sigma_{lj} = \sigma_{j\lambda\, k\lambda} + \sigma_{k\lambda\, l\lambda} + \sigma_{l\lambda\, j\lambda} = 0$$

on each $U_j \cap U_k \cap U_l \cap V_\lambda \neq \emptyset$. Then (3.72) follows immediately from (3.73). ∎

Theorem 3.6. *If $H^1(U_j, \mathscr{S}) = 0$ for each $U_j \in \mathfrak{U}$, then $\Pi^{\mathfrak{U}}: H^2(\mathfrak{U}, \mathscr{S}) \to H^2(M, \mathscr{S})$ is injective.*

Proof. It suffices to prove $N^2(\mathfrak{U}) = 0$. Let $g \in N^2(\mathfrak{U})$, then there is a $\mathfrak{B} < \mathfrak{U}$ such that $\Pi^{\mathfrak{U}}_{\mathfrak{B}} g = 0$. Let $\mathfrak{U} = \{U_j\}$ and $\mathfrak{B} = \{V_\lambda\}$ and put $\mathfrak{W} = \{W_{j\lambda} \mid W_{j\lambda} = U_j \cap V_\lambda \neq \emptyset\}$. Then $\Pi^{\mathfrak{U}}_{\mathfrak{W}} g = 0$. Suppose that g is represented by a 2-cocycle $\{\sigma_{ijk}\} \in Z^2(\mathfrak{U})$. If we define $\Pi^{\mathfrak{U}}_{\mathfrak{W}}: Z^2(\mathfrak{U}) \to Z^2(\mathfrak{W})$ via the inclusion $W_{j\lambda} \subset U_j$, we have

$$\Pi^{\mathfrak{U}}_{\mathfrak{W}}\{\sigma_{ijk}\} = \{\tau_{i\lambda\, j\mu\, k\nu}\}, \qquad \tau_{i\lambda\, j\mu\, k\nu} = r_{W_{i\lambda} \cap W_{j\mu} \cap W_{k\nu}} \sigma_{ijk}.$$

Since $\Pi^{\mathfrak{U}}_{\mathfrak{W}} g = 0$, there is a 1-cochain $\{\tau_{i\lambda\, k\nu}\} \in C^1(\mathfrak{W})$ such that

$$\tau_{i\lambda\, j\mu\, k\nu} = \tau_{j\mu\, k\nu} + \tau_{k\nu\, i\lambda} + \tau_{i\lambda\, j\mu}. \tag{3.74}$$

Since for each j, $\sigma_{jjj} = 0$, we have

$$\tau_{j\mu\, j\nu} + \tau_{j\nu\, j\lambda} + \tau_{j\lambda\, j\mu} = \tau_{j\lambda\, j\mu\, j\nu} = 0,$$

which means $\{\tau_{j\lambda\, j\nu}\} \in Z^1(\mathfrak{W}_j)$ with $\mathfrak{W}_j = \{\mathfrak{W}_{j\lambda}\}$. $\mathfrak{W}_j$ is a locally finite open covering of U_j. Since $H^1(U_j, \mathscr{S}) = 0$, there is a 0-chain $\{\tau_{j\lambda}\} \in C^0(\mathfrak{W}_j)$ such that

$$\tau_{j\lambda\, j\nu} = \tau_{j\nu} - \tau_{j\lambda}.$$

For each U_j fix such a $\{\tau_{j\lambda}\}$, and put

$$\sigma_{j\lambda\, k\nu} = \tau_{j\lambda\, k\nu} - \tau_{k\nu} + \tau_{j\lambda},$$

then by (3.74) we have

$$\tau_{i\lambda\, j\mu\, k\nu} = \sigma_{j\mu\, k\nu} + \sigma_{k\nu\, i\lambda} + \sigma_{i\lambda\, j\mu}. \tag{3.75}$$

Putting $i=j$, we have $\tau_{j\lambda\, j\mu\, k\nu}=0$ since $\sigma_{jjk}=0$. On the other hand, since $\sigma_{j\lambda\, j\mu}=0$, we have $\sigma_{j\lambda\, k\nu}=\sigma_{j\mu\, k\nu}$. Similarly we have $\sigma_{j\lambda\, k\mu}=\sigma_{j\lambda\, k\nu}$. Hence as in the proof of Theorem 3.5, we define a 0-cochain $\{\sigma_{jk}\}\in C^0(\mathfrak{U})$ by putting $\sigma_{jk}=\sigma_{j\lambda\, k\nu}$ on $U_j\cap U_k\cap V_\lambda\cap V_\nu\neq\emptyset$. Putting $\mu=\nu=\lambda$, by (3.75) we see that $\sigma_{ijk}=\sigma_{jk}+\sigma_{ki}+\sigma_{ij}$ on each $U_i\cap U_j\cap U_k\cap V_\lambda$. Hence $\{\sigma_{ijk}\}=\delta\{\sigma_{jk}\}\in C^1(\mathfrak{U})$, which means $g=0$. ∎

(c) Exact Sequence

Let $\mathscr{S}$ be a sheaf over a differentiable manifold M, and $\varpi\colon \mathscr{S}\to M$ its projection.

Definition 3.8. A subset $\mathscr{S}'\subset\mathscr{S}$ is called a *subsheaf* of $\mathscr{S}$ if the following conditions are satisfied.

(i) $\mathscr{S}'$ is an open subset of $\mathscr{S}$.
(ii) $\varpi(\mathscr{S}')=M$.
(iii) For any $p\in M$, $\mathscr{S}'_p=\varpi^{-1}(p)\cap\mathscr{S}'$ is a K-submodule of $\mathscr{S}_p$.

Clearly $\mathscr{S}'$ itself is a sheaf over M.
Let $\mathscr{S}''$ be a sheaf over M with the projection ϖ''.

Definition 3.9. A *homomorphism* h of $\mathscr{S}$ into $\mathscr{S}''$ is defined to be a continuous map of $\mathscr{S}$ into $\mathscr{S}''$ satisfying the following conditions:

(i) $\varpi''\circ h=\varpi$.
(ii) For any $p\in M$, $h\colon \mathscr{S}_p\to\mathscr{S}''_p$ is a K-homomorphism.

From (i) it follows that $h(\mathscr{S}_p)\subset\mathscr{S}''_p$. Since ϖ and ϖ'' are local homeomorphisms, h is also a local homeomorphism by (i).

Let $h\colon\mathscr{S}\to\mathscr{S}''$ be a homomorphism. By $h(\varphi)=0$, $\varphi\in\mathscr{S}_p$, we mean that $h(\varphi)=0''_p$, where $0''_p$ is the zero of $\mathscr{S}''_p$. Then $\mathscr{S}'=\{\varphi\in\mathscr{S}\mid h(\varphi)=0\}$ is a subsheaf of $\mathscr{S}$. *Proof.* $\mathscr{S}'$ obviously satisfies (ii) and (iii) in Definition 3.8. Moreover since h is continuous, and $0''=\{0''_p\mid p\in M\}$ is open in $\mathscr{S}''$, $\mathscr{S}'=h^{-1}(0'')$ is open in $\mathscr{S}$. ∎ We call this $\mathscr{S}'$ the kernel of h and denote by $\ker h$.

$$\ker h=\{\varphi\in\mathscr{S}\mid h(\varphi)=0\}.$$

$h(\mathscr{S})$ is a subsheaf of $\mathscr{S}''$. In fact, $h(\mathscr{S})$ is open in $\mathscr{S}''$ since h is a local homomorphism. The other conditions are obviously satisfied.

A homomorphism $h: \mathscr{S} \to \mathscr{S}''$ is injective if and only if $\ker h = \{0_p \mid p \in M\}$, which we denote by $\ker h = 0$.

Definition 3.10. $h: \mathscr{S} \to \mathscr{S}''$ is called an *isomorphism* of $\mathscr{S}$ onto $\mathscr{S}''$ if h is a homomorphism of $\mathscr{S}$ onto $\mathscr{S}''$ and if, for any $p \in M$, $h: \mathscr{S}_p \to \mathscr{S}''_p$ is a K-isomorphism of $\mathscr{S}_p$ onto $\mathscr{S}''_p$. If there exists an isomorphism of $\mathscr{S}$ onto $\mathscr{S}''$, $\mathscr{S}$ and $\mathscr{S}''$ are said to be *isomorphic*, and is denoted by $\mathscr{S} \cong \mathscr{S}''$.

We often identify $\mathscr{S}$ with $\mathscr{S}''$ if $\mathscr{S} \cong \mathscr{S}''$.

Since a homomorphism $h: \mathscr{S} \to \mathscr{S}''$ is a local homeomorphism, $h: \mathscr{S} \to h(\mathscr{S})$ is an isomorphism if $\ker h = 0$. In this case if we identify $h(\mathscr{S}) \subset \mathscr{S}''$ with $\mathscr{S}$ via h, then h can be regarded as the identity $\mathscr{S} \to \mathscr{S} \subset \mathscr{S}''$. Thus being regarded, h is called the *inclusion* map, and often denoted by ι.

Let $\mathscr{S}' \subset \mathscr{S}$ be a subsheaf of $\mathscr{S}$. Let $\mathscr{Q}_p = \mathscr{S}_p / \mathscr{S}'_p$ be the quotient group, and let $\mathscr{Q} = \bigcup_{p \in M} \mathscr{Q}_p$. Define $\varpi: \mathscr{Q} \to M$ by $\varpi(\mathscr{Q}_p) = p$. Let h_p be the natural map of $\mathscr{S}_p$ onto $\mathscr{Q}_p = \mathscr{S}_p / \mathscr{S}'_p$, and define $h: \mathscr{S} \to \mathscr{Q}$ by putting $h(\varphi) = h_p(\varphi)$ for $\varphi \in \mathscr{S}_p$. We give $\mathscr{Q}$ a topology by saying that $\mathscr{W}$ is open if and only if $h^{-1}(\mathscr{W})$ is open in $\mathscr{S}$. *Then $\mathscr{Q}$ is a sheaf over M. Proof.* First we prove that if for $\alpha, \beta \in \mathscr{S}_p$, we take sufficiently small neighbourhoods $\mathscr{U}$ and $\mathscr{V}$ with $\alpha \in \mathscr{U} \subset \mathscr{S}$, $\beta \in \mathscr{V} \subset \mathscr{S}$, such that $\varpi(U) = \varpi(\mathscr{V})$, then

$$\mathscr{U} + \mathscr{V} = \{\varphi + \psi \mid \varphi \in \mathscr{U}, \psi \in \mathscr{V}, \varpi(\varphi) = \varpi(\psi)\}$$

is *a neighbourhood of* $\alpha + \beta$. For, if we put $U = \varpi(\mathscr{U}) = \varpi(\mathscr{V})$, for sufficiently small $\mathscr{U}, \mathscr{V}$, $\varpi: \mathscr{U} \to U$, and $\varpi: \mathscr{V} \to U$ are both homeomorphisms. Their inverses $\sigma: p \to \varphi = \sigma(p)$ and $\tau: p \to \psi = \tau(p)$ are sections of $\mathscr{S}$ over U. Hence $\sigma + \tau: p \to \varphi + \psi = \sigma(p) + \tau(p)$ is also a section of $\mathscr{S}$ over U. Hence $\mathscr{U} + \mathscr{V} = \{\sigma(p) + \tau(p) \mid p \in U\}$ is open.

Since by definition $\mathscr{S}'$ is an open subset of $\mathscr{S}$, this implies that for any open $\mathscr{W} \subset \mathscr{S}$,

$$h^{-1}(h(\mathscr{W})) = \{\varphi + \psi \mid \varphi \in \mathscr{W}, \psi \in \mathscr{W}, \varpi(\varphi) = \varpi(\psi)\}$$

is open. Thus $h(\mathscr{W})$ is open in $\mathscr{Q}$, that is, h is an open map.

For any point $h(\alpha)$, $\alpha \in \mathscr{S}$, of $\mathscr{Q}$, take a neighbourhood $\mathscr{U}$ of α such that $\varpi: \mathscr{U} \to U$ is a homeomorphism. Then by the above argument $h(\mathscr{U})$ is an open subset of $\mathscr{Q}$. Moreover for any open subset $\mathscr{V} \subset \mathscr{U}$, $h(\mathscr{V})$ is an open subset of $h(\mathscr{U})$. Conversely, if $\bar{\mathscr{V}}$ is an open subset of $h(\mathscr{U})$, $\mathscr{V} = h^{-1}(\bar{\mathscr{V}}) \cap \mathscr{U}$ is an open subset of $\mathscr{U}$, and $h(\mathscr{V}) = \bar{\mathscr{V}}$. Hence $h: \mathscr{U} \to h(\mathscr{U})$ is a homeomorphism. Therefore $h: \mathscr{S} \to \mathscr{Q}$ is a local homeomorphism. Since $\varpi: \mathscr{U} \to U$ is a homeomorphism, and $\pi \circ h = \varpi$, $\varpi: h(\mathscr{U}) \to \mathscr{U}$ is also a homeomorphism. Hence $\pi: \mathscr{Q} \to M$ is a local homeomorphism. The continuity of $c_1\bar{\varphi} + c_2\bar{\psi}$ with $\pi(\bar{\varphi}) = \pi(\bar{\psi})$ with respect to $\bar{\varphi}, \bar{\psi} \in \mathscr{Q}$, follows from the continuity of $c_1\varphi + c_2\psi$ for $\varphi, \psi \in \mathscr{S}$. Hence $\mathscr{Q}$ is a sheaf over M with the projection π. ∎

Definition 3.11. $\mathcal{Q}$ is called the *quotient sheaf* of $\mathcal{S}$ by $\mathcal{S}'$ and is denoted by $\mathcal{S}/\mathcal{S}'$.

$h: \mathcal{S} \to \mathcal{Q}$ is a *surjective homomorphism* of $\mathcal{S}$ onto $\mathcal{Q}$.

For a homomorphism $h: \mathcal{S} \to \mathcal{S}''$, put $\mathcal{S}' = \ker h$. Then h maps $\mathcal{S}/\mathcal{S}'$ isomorphically onto $h(\mathcal{S})$, hence $h(\mathcal{S}) \cong \mathcal{S}/\mathcal{S}'$.

A homomorphism $h: \mathcal{S} \to \mathcal{S}''$ of sheaves over M defines homomorphisms of cohomology groups $H^q(M, \mathcal{S}) \to H^q(M, \mathcal{S}'')$ as follows: First for a section $\sigma: p \to \sigma(p)$ of S over a subset $W \subset M$,

$$h\sigma: p \to h(\sigma(p))$$

is a section of $\mathcal{S}''$ over W, and $h: \sigma \to h\sigma$ is a K-homomorphism of $\Gamma(W, \mathcal{S})$ into $\Gamma(W, \mathcal{S}'')$. Next let $\mathfrak{U} = \{U_j\}$ be a locally finite open covering of M. For any q-cochain $c^q = \{\sigma_{k_0 \cdots k_q}\} \in C^q(\mathfrak{U}, \mathcal{S})$, by defining $hc^q = \{h\sigma_{k_0 \cdots k_q}\}$, we obtain a homomorphism $h: c^q \to hc^q$ of $C^q(\mathfrak{U}, \mathcal{S})$ into $C^q(\mathfrak{U}, \mathcal{S}'')$. Since $h \circ \delta = \delta \circ h$ follows immediately from the definition (3.62) of the coboundary, h maps $Z^q(\mathfrak{U}, \mathcal{S})$ into $Z^q(\mathfrak{U}, \mathcal{S}'')$ and $\delta C^{q-1}(\mathfrak{U}, \mathcal{S})$ into $\delta C^{q-1}(\mathfrak{U}, \mathcal{S}'')$, hence h induces a homomorphism

$$h: H^q(\mathfrak{U}, \mathcal{S}) \to H^q(\mathfrak{U}, \mathcal{S}'').$$

Let $\mathfrak{V} = \{V_\lambda\}$ be an arbitrary refinement of $\mathfrak{U} = \{U_j\}$: $\mathfrak{U} > \mathfrak{V}$. For any V_λ, choose an arbitrary member $U_{k(\lambda)}$ of U with $U_{k(\lambda)} \supset V_\lambda$, and define $\Pi^{\mathfrak{U}}_{\mathfrak{V}}: C^q(\mathfrak{U}, \mathcal{S}) \to C^q(\mathfrak{V}, \mathcal{S})$ and $\Pi^{\mathfrak{U}}_{\mathfrak{V}}: C^q(\mathfrak{U}, \mathcal{S}'') \to C^q(\mathfrak{V}, \mathcal{S}'')$ by (3.65). Then clearly $h \circ \Pi = \Pi \circ h$ holds. Hence also for $\Pi^{\mathfrak{U}}_{\mathfrak{V}}: H^q(\mathfrak{U}, \mathcal{S}) \to H^q(\mathfrak{V}, \mathcal{S})$ and $\Pi^{\mathfrak{U}}_{\mathfrak{V}}: H^q(\mathfrak{U}, \mathcal{S}'') \to H^q(\mathfrak{V}, \mathcal{S}'')$, we have

$$h \circ \Pi^{\mathfrak{U}}_{\mathfrak{V}} = \Pi^{\mathfrak{U}}_{\mathfrak{V}} \circ h.$$

Therefore if $\Pi^{\mathfrak{U}}_{\mathfrak{V}} g = 0$ for $g \in H^q(\mathfrak{U}, \mathcal{S})$, then $\Pi^{\mathfrak{U}}_{\mathfrak{V}} hg = 0$. Thus $g > 0$ implies that $hg > 0$, that is, h maps $N^q(\mathfrak{U}, \mathcal{S}) = \{g \in H^q(\mathfrak{U}, \mathcal{S}) \mid g > 0\}$ into $N^q(\mathfrak{U}, \mathcal{S}'') = \{g'' \in H^q(\mathfrak{U}, \mathcal{S}'') \mid g'' > 0\}$. Consequently h induces a homomorphism

$$h_{\mathfrak{U}}: \bar{H}^q(\mathfrak{U}, \mathcal{S}) \to \bar{H}^q(\mathfrak{U}, \mathcal{S}'').$$

By (3.68) if $\mathfrak{U} > \mathfrak{V}$, we have $\bar{H}^q(\mathfrak{U}, \mathcal{S}) \subset \bar{H}^q(\mathfrak{V}, \mathcal{S})$, and $\bar{H}^q(\mathfrak{U}, \mathcal{S}'') \subset \bar{H}^q(\mathfrak{V}, \mathcal{S}'')$. It follows immediately from $h \circ \Pi^{\mathfrak{U}}_{\mathfrak{V}} = \Pi^{\mathfrak{U}}_{\mathfrak{V}} \circ h$ that $h_{\mathfrak{V}}: \bar{H}^q(\mathfrak{V}, \mathcal{S}) \to \bar{H}^q(\mathfrak{V}, \mathcal{S}'')$ coincides with $h_{\mathfrak{U}}$ on $\bar{H}^q(\mathfrak{U}, \mathcal{S})$. Since

$$H^q(M, \mathcal{S}) = \bigcup_{\mathfrak{U} \in \mathfrak{S}} \bar{H}^q(\mathfrak{U}, \mathcal{S}), \qquad H^q(M, \mathcal{S}'') = \bigcup_{\mathfrak{U} \in \mathfrak{S}} \bar{H}^q(\mathfrak{U}, \mathcal{S}''),$$

by putting $h = h_{\mathfrak{U}}$ on each $\bar{H}^q(\mathfrak{U}, \mathcal{S})$, we obtain a homomorphism

$$h: H^q(M, \mathcal{S}) \to H^q(M, \mathcal{S}'').$$

We write $\mathcal{S} \xrightarrow{h} \mathcal{S}''$ for a homomorphism $h: \mathcal{S} \to \mathcal{S}''$.

Definition 3.12. A sequence of sheaves and sheaf homomorphisms over M

$$\mathcal{S}_0 \xrightarrow{h_0} \cdots \to \mathcal{S}_{m-1} \xrightarrow{h_{m-1}} \mathcal{S}_m \xrightarrow{h_m} \cdots \to \mathcal{S}_l \tag{3.76}$$

is said to be *exact* if

$$h_{m-1}(\mathcal{S}_{m-1}) = \ker h_m, \qquad m = 1, 2, \ldots, l-1.$$

Similarly a sequence of K-modules and K-homomorphisms

$$H_0 \xrightarrow{h_0} \cdots \to H_{m-1} \xrightarrow{h_{m-1}} H_m \xrightarrow{h_m} \cdots$$

is said to be *exact* if

$$h_{m-1}(H_{m-1}) = \ker h_m, \qquad m = 1, 2, \ldots.$$

Suppose given an exact sequence of sheaves over M

$$0 \to \mathcal{S}' \xrightarrow{\iota} \mathcal{S} \xrightarrow{h} \mathcal{S}'' \to 0. \tag{3.77}$$

We investigate the relation between the cohomology groups $H^q(M, \mathcal{S}')$, $H^q(M, \mathcal{S})$, and $H^q(M, \mathcal{S}'')$. The exactness of (3.77) implies that $\ker \iota = 0$, $\ker h = \iota(\mathcal{S}')$, and $\mathcal{S}'' = h(\mathcal{S})$. Hence $\mathcal{S}' = \iota(\mathcal{S}) \subset \mathcal{S}$ is a subsheaf and ι is the inclusion map. Hence also $\mathcal{S}'' \cong \mathcal{S}/\mathcal{S}'$. We write $H^q(\mathcal{S}')$ for $H^q(M, \mathcal{S}')$ etc., for simplicity. The homomorphism $\mathcal{S}' \xrightarrow{\iota} \mathcal{S}$ induces a homomorphism $H^q(\mathcal{S}') \xrightarrow{\iota} H^q(\mathcal{S})$, and $\mathcal{S} \xrightarrow{h} \mathcal{S}''$ induces $H^q(\mathcal{S}) \xrightarrow{h} H^q(\mathcal{S}'')$. The following theorem is fundamental.

Theorem 3.7. *The exact sequence of sheaves* $0 \to \mathcal{S}' \xrightarrow{i} \mathcal{S} \xrightarrow{h} \mathcal{S}'' \to 0$ *induces the exact sequence of cohomology groups*

$$0 \to H^0(\mathcal{S}') \xrightarrow{i} H^0(\mathcal{S}) \xrightarrow{h} H^0(\mathcal{S}'') \xrightarrow{\delta^*} H^1(\mathcal{S}') \xrightarrow{\iota} \cdots$$

$$\xrightarrow{\delta^*} H^q(\mathcal{S}') \xrightarrow{\iota} H^q(\mathcal{S}) \xrightarrow{h} H^q(\mathcal{S}'') \xrightarrow{\delta^*} H^{q+1}(\mathcal{S}') \xrightarrow{\iota} \cdots. \tag{3.78}$$

Before we give the proof, we explain how δ^* is defined. First for $q = 0$, let us consider how δ^* must be defined in order that the sequence

$$H^0(\mathcal{S}) \xrightarrow{h} H^0(\mathcal{S}'') \xrightarrow{\delta^*} H^1(\mathcal{S}') \tag{3.79}$$

may be exact. The exactness of (3.79) means that for a given $\sigma'' \in H^0(\mathcal{S}'') = \Gamma(M, \mathcal{S}'')$, *there is a* $\sigma \in H^0(\mathcal{S}) = \Gamma(M, \mathcal{S})$ *such that* $h\sigma = \sigma''$ *if and only if* $\delta^*\sigma'' = 0$. Since the projections ϖ, ϖ'' of $\mathcal{S}$, $\mathcal{S}''$ are local homeomorphisms, and h is also a local homeomorphism, for a given σ'', if we take a sufficiently fine locally finite open covering $\mathfrak{U}=\{U_j\}$, there is a section $\sigma_j \in \Gamma(U_j, \mathscr{S})$ for each j such that $h\sigma_j = r_{U_j}\sigma''$. By (3.64), the 0-cocycle $\{r_{U_j}\sigma''\}$ gives the section σ''. Hence if we put $c^0=\{\sigma_j\}\in C^0(\mathfrak{U}, \mathscr{S})$, $hc^0=\{h\sigma_j\}=\{r_{U_j}\sigma''\}=\sigma''$. Thus for $\sigma''\in H^0(\mathfrak{U}, \mathscr{S}'')$, there exists a $c^0\in C^0(\mathfrak{U}, \mathscr{S})$ with $hc^0=\sigma''$. Note that c^0 may not be a cocycle. Take any such c^0, and put $\delta c^0=\{\tau'_{jk}\}$, $\tau'_{jk}=\sigma_k-\sigma_j$. Since $h\tau'_{jk}=h\sigma_k-h\sigma_j=0$, and $\ker h=\mathscr{S}'$, we see that $\tau'_{jk}\in\Gamma(U_j\cap U_k, \mathscr{S}')$. Since $\delta\delta c^0=0$, $\delta c^0\in Z^1(\mathfrak{U}, \mathscr{S}')$. We define $\delta^*\sigma''\in H^1(M, \mathscr{S}')$ *as the cohomology class of the* 1-*cocycle* δc^0: $\delta^*\sigma''=[\delta c^0]$. Since by Theorem 3.4, $H^1(\mathfrak{U}, \mathscr{S}')\subset H^1(M, \mathscr{S}')$, we may consider

$$\delta^*\sigma''=[\delta c^0]\in H^1(\mathfrak{U}, \mathscr{S}')=Z^1(\mathfrak{U}, \mathscr{S}')/\delta C^0(\mathfrak{U}, \mathscr{S}').$$

$\delta^*\sigma''$ *is determined uniquely by* σ''. For, if $\sigma''=0$, then $h\sigma_j=0$, hence $c^0=\{\sigma_j\}\in C^0(\mathfrak{U}, \mathscr{S}')$. $\delta^*\sigma''$ is also independent of the choice of an open covering.

If there is a $\sigma\in\Gamma(M, \mathscr{S})$ with $h\sigma=\sigma''$, we have $\delta^*\sigma''=0$. In fact, if we put $c^0=\{r_{U_j}\sigma\}$, then $hc^0=\sigma''$, hence $\delta c^0=0$. Conversely, assume $\delta^*\sigma''=0$. Then $\delta c^0=\{\tau'_{jk}\}\in\delta C^0(\mathfrak{U}, \mathscr{S}')$, that is, $\sigma_k-\sigma_j=\tau'_{jk}=\sigma'_k-\sigma'_j$ with $\sigma'_j\in\Gamma(\mathfrak{U}, \mathscr{S}')$. For $U_j\cap U_k\neq\varnothing$, $\sigma_j-\sigma'_j=\sigma_k-\sigma'_k$, hence putting $\sigma=\sigma_j-\sigma'_j$ on each U_j, we obtain a section $\sigma\in\Gamma(M, \mathscr{S})$. Since $h\sigma'_j=0$ and $h\sigma_j=r_{U_j}\sigma''$, we have $h\sigma=\sigma''$. Hence (3.79) is exact.

Now we assume $q\geqq 1$. We must define δ^* so that

$$H^q(\mathscr{S})\xrightarrow{h} H^q(\mathscr{S}'')\xrightarrow{\delta^*} H^{q+1}(\mathscr{S}') \tag{3.80}$$

may be exact.

Let M be a complex manifold. Suppose given for every p a local complex coordinate system $z_p=(z_p^1,\dots,z_p^n)$ centred at p, and a coordinate polydisk

$$U_{\varepsilon(p)}(p)=\{z_p\,||z_p^1|<\varepsilon(p),\dots,|z_p^n|<\varepsilon(p)\},\qquad \varepsilon(p)>0.$$

Then we may choose, by Theorem 2.1, at most countably many coordinate polydisks U_λ with

$$U_\lambda=U_{\varepsilon_\lambda}(p_\lambda)\subset U_{\varepsilon(p_\lambda)}(p_\lambda),\qquad 0<\varepsilon_\lambda\leqq\varepsilon(p_\lambda),\quad \lambda=1,2,\dots,$$

such that $\mathfrak{U}=\{U_\lambda\}$ is a locally finite open covering of M. Similarly for a differentiable manifold, using local C^∞ coordinates $(x_p^1,\dots,x_p^m)$ centred at p and an open set

$$U_\varepsilon(p)=\{x_p\,||x_p^1|<\varepsilon,\dots,|x_p^m|<\varepsilon\}\quad\text{with } \varepsilon>0,$$

we obtain the same result. Here we assume that the closure $[U_\varepsilon(p)]$ is a compact subset of the domain of local C^∞ coordinates $(x_p^1,\dots,x_p^m)$. We call $U_\varepsilon(p)$ a *coordinate* (*multi*)*interval* of radius ε. For a complex manifold

M, too, we may consider coordinate multi-intervals, regarding M as a differentiable manifold as usual.

Let $\mathfrak{U}=\{U_j\}$ be a locally finite open covering of M, and suppose $[U_j]$ is compact. Then we may choose an open subset W_j of each U_j such that $[W_j]\subset U_j$ and that $\mathfrak{W}=\{W_j\}$ is already a covering of M. *Proof.* Since for $p\in M$, there are only a finite number of U_j with $p\in U_j$, taking $\varepsilon(p)>0$ sufficiently small, we may assume that $[U_{\varepsilon(p)}(p)]\subset U_j$ if $p\in U_j$. Choose such $U_{\varepsilon(p)}(p)$ for each $p\in M$, and take

$$V_\lambda = U_{\varepsilon_\lambda}(p_\lambda)\subset U_{\varepsilon(p_\lambda)}(p_\lambda), \qquad 0<\varepsilon_\lambda \leqq \varepsilon(p_\lambda), \quad \lambda=1,2,\ldots,$$

so that $\{V_\lambda \mid \lambda=1,2,\ldots\}$ is a locally finite open covering of M. Then p_λ belongs to one of U_j, and if $p_\lambda\in U_j$, then $[V_\lambda]\subset[U_{\varepsilon(p_\lambda)}(p_\lambda)]\subset U_j$. Thus for each λ, $[V_\lambda]$ is contained in some U_j: $[V_\lambda]\subset U_j$. Then if we put

$$W_j = \bigcup_{[V_\lambda]\subset U_j} V_\lambda,$$

we have $\bigcup_j W_j=\bigcup_j V_j = M$, hence $\mathfrak{W}=\{W_j\}$ forms an open covering of M. Since $[U_j]$ is compact, and $\{V_\lambda\}$ is locally finite, there are only a finite number of V_λ with $[V_\lambda]\subset U_j$. Hence $[W_j]=\bigcup_{[V_\lambda]\subset U_j}[V_\lambda]\subset U_j$. ∎

In what follows we denote by $\mathfrak{U}, \mathfrak{V}, \mathfrak{W},\ldots$ locally finite open coverings of M.

Lemma 3.3. *Given a q-chain $c''^q\in C^q(\mathfrak{U},\mathscr{S}'')$, we can find a locally finite refinement $\mathfrak{V}$ of $\mathfrak{U}$ and $c^q\in C^q(\mathfrak{V},\mathscr{S})$ such that $\Pi^{\mathfrak{U}}_{\mathfrak{V}}c''^q = hc^q$.*

Proof. Put $\mathfrak{U}=\{U_j\}$. We give proof only for the case $q=2$, but the generalization is straightforward. Since for $\mathfrak{V}<\mathfrak{W}<\mathfrak{U}$, $\Pi^{\mathfrak{U}}_{\mathfrak{V}}=\Pi^{\mathfrak{W}}_{\mathfrak{V}}\Pi^{\mathfrak{U}}_{\mathfrak{W}}$, replacing $\mathfrak{U}$ by its appropriate refinement if necessary, we may assume that the closure $[U_j]$ of each U_j is compact. Then there is an open covering $\mathfrak{W}=\{W_j\}$ with $[W_j]\subset U_j$.

First let $c''^2=\{\sigma''_{ijk}\}\in C^2(\mathfrak{U},\mathscr{S}'')$, $\sigma''_{ijk}\in\Gamma(U_i\cap U_j\cap U_k,\mathscr{S}'')$. Take an arbitrary point $p\in M$, and denote by $U_\varepsilon(p)$ the coordinate multi-interval of radius $\varepsilon>0$ with centre p. Suppose $p\in U_i\cap U_j\cap U_k$, then since ϖ, ϖ'' and h are local homeomorphisms, for a sufficiently small $U_\varepsilon(p)\subset U_i\cap U_j\cap U_k$, there is an element $\tau\in\Gamma(U_\varepsilon(p),\mathscr{S})$ such that $h\tau = r_{U_\varepsilon(p)}\sigma''_{ijk}$. Since $\mathfrak{U}=\{U_j\}$ is locally finite, there are only a finite number of $U_i\cap U_j\cap U_k$ with $p\in U_i\cap U_j\cap U_k$ for a fixed p. Hence we can find for each p a sufficiently small $U_\varepsilon(p)$ satisfying the following conditions:

(i) If $p\in U_i\cap U_j\cap U_k$, $U_\varepsilon(p)\subset U_i\cap U_j\cap U_k$, and there is a $\tau\in\Gamma(U_\varepsilon(p),\mathscr{S})$ with $r_{U_\varepsilon(p)}\sigma''_{ijk}=h\tau$.

(ii) If $p\in W_j$, then $U_\varepsilon(p)\subset W_j$.

(iii) If $[W_j]\cap U_\varepsilon(p)\neq\varnothing$, then $U_\varepsilon(p)\subset U_j$.

In fact (ii) follows from the local finiteness of $\mathfrak{W}=\{W_j\}$. We can verify (iii) as follows. Since for a sufficiently small $U_{\varepsilon_0}(p)$, there are only a finite number of U_j with $U_{\varepsilon_0}(p)\cap U_j\neq\emptyset$, we can choose a sufficiently small ε with $0<\varepsilon<\varepsilon_0$ such that for $p\in U_j$, $U_\varepsilon(p)\subset U_j$, and that for $p\supset U_j$, $[W_j]\cap U_\varepsilon(p)=\emptyset$. Note that here we have $[W_j]\subset U_j$.

We fix $U_\varepsilon(p)=U_{\varepsilon(p)}(p)$ satisfying (i), (ii) and (iii) above for each $p\in M$. Choose coordinate multi-intervals

$$V_\lambda=U_{\varepsilon_\lambda}(p_\lambda)\subset U_{\varepsilon(p_\lambda)}(p_\lambda),\qquad 0<\varepsilon_\lambda\leqq\varepsilon(p_\lambda),\quad \lambda=1,2,\ldots,$$

such that $\mathfrak{V}=\{V_\lambda\}$ is a locally finite open covering of M. By the condition (ii), each V_λ is contained in some W_j. For each λ, choose an arbitrary $W_{j(\lambda)}$ with $V_\lambda\subset W_{j(\lambda)}$. Then $V_\lambda\subset W_{j(\lambda)}\subset U_{j(\lambda)}$, hence $\mathfrak{V}<\mathfrak{U}$. Define

$$\Pi^{\mathfrak{U}}_{\mathfrak{V}}\colon C^q(\mathfrak{U},\mathscr{S}'')\to C^q(\mathfrak{V},\mathscr{S}'')$$

via the inclusion $V_\lambda\subset U_{j(\lambda)}$. For $c''^2=\{\sigma''_{ijk}\}$, $\Pi^{\mathfrak{U}}_{\mathfrak{V}}c''^2=\{\tau''_{\lambda\mu\nu}\}$ is given by

$$\tau''_{\lambda\mu\nu}=r_V\sigma''_{j(\lambda)j(\mu)j(\nu)},\quad\text{where}\quad V=V_\lambda\cap V_\mu\cap V_\nu\neq\emptyset.$$

We must prove the existence of $\tau_{\lambda\mu\nu}\in\Gamma(V_\lambda\cap V_\mu\cap V_\nu,\mathscr{S})$ such that $h\tau_{\lambda\mu\nu}=\tau''_{\lambda\mu\nu}$ for each $\tau''_{\lambda\mu\nu}$. Put $i=j(\lambda)$, $j=j(\mu)$ and $k=j(\nu)$. Then $V_\lambda\cap W_j\neq\emptyset$ since $V_\lambda\subset W_i$, $V_\mu\subset W_j$, $V_\nu\subset W_k$ and $V_\lambda\cap V_\mu\cap V_\nu\neq\emptyset$. Therefore we have

$$W_j\cap U_{\varepsilon(p_\lambda)}(p_\lambda)\supset W_j\cap V_\lambda\neq\emptyset,$$

hence by (iii), $U_{\varepsilon(p_\lambda)}\subset U_j$. Similarly we have $U_{\varepsilon(p_\lambda)}(p_\lambda)\subset U_k$, while by (ii), $U_{\varepsilon(p_\lambda)}(p_\lambda)\subset W_i\subset U_i$. Hence we have

$$V_\lambda\subset U_{\varepsilon(p_\lambda)}(p_\lambda)\cup U_i\cap U_j\cap U_k.$$

Consequently by (i), there is a $\tau\in\Gamma(V_\lambda,\mathscr{S})$ with

$$h\tau=r_{V_\lambda}\sigma''_{ijk}.$$

Put $\tau_{\lambda\mu\nu}=r_V\tau$ with $V=V_\lambda\cap V_\mu\cap V_\nu$. Then we have

$$h\tau_{\lambda\mu\nu}=r_V\sigma''_{ijk}=\tau''_{\lambda\mu\nu}$$

as desired. ∎

For a sheaf $\mathscr{S}$, we denote the cohomology class with respect to $\mathfrak{U}$ of a q-cocycle $c^q\in Z^q(\mathfrak{U},\mathscr{S})$ by

$$[c_q]_{\mathfrak{U}}\in H^q(\mathfrak{U},\mathscr{S})=Z^q(\mathfrak{U},\mathscr{S})/\delta C^{q-1}(\mathfrak{U},\mathscr{S}).$$

Then the cohomology class $[c^q]$ of c^q is given by

$$[c^q]=\Pi^{\mathfrak{U}}[c^q]_{\mathfrak{U}}\in H^q(M,\mathscr{S}).$$

For $\mathfrak{U}>\mathfrak{V}$, $\Pi^{\mathfrak{U}}_{\mathfrak{V}}: H^q(\mathfrak{U},\mathscr{S})\to H^q(\mathfrak{V},\mathscr{S})$ induces an inclusion $\bar{H}^q(\mathfrak{U},\mathscr{S})\hookrightarrow \bar{H}^q(\mathfrak{V},\mathscr{S})$. Hence

$$\Pi^{\mathfrak{U}}[c^q]_{\mathfrak{U}}=\Pi^{\mathfrak{V}}\Pi^{\mathfrak{U}}_{\mathfrak{V}}[c^q]_{\mathfrak{U}},$$

hence in view of $\Pi^{\mathfrak{U}}_{\mathfrak{V}}[c^q]_{\mathfrak{U}}=[\Pi^{\mathfrak{U}}_{\mathfrak{V}}c^q]_{\mathfrak{V}}$, we see that

$$[c^q]=[\Pi^{\mathfrak{U}}_{\mathfrak{V}}c^q], \tag{3.81}$$

which means that *the cohomology class of c^q does not depend on the choice of refinements of a locally finite covering.*

Lemma 3.4. *For any $g''\in H^q(M,\mathscr{S}'')$, by a suitable choice of $\mathfrak{U}$, we can find a q-cocycle c''^q and a q-cochain c^q such that*

$$g''=[c''^q],\qquad hc^q=c''^q\in Z^q(\mathfrak{U},\mathscr{S}''),\quad\text{and}\quad c^q\in C^q(\mathfrak{U},\mathscr{S}). \tag{3.82}$$

Proof. Let g'' be given as $g''=[b''^q]$ by some q-cocycle $b''^q\in Z^q(\mathfrak{V},\mathscr{S}'')$. By Lemma 3.3, taking a suitable refinement $\mathfrak{U}<\mathfrak{V}$, we can find $c^q\in C^q(\mathfrak{U},\mathscr{S})$ such that $hc^q=\Pi^{\mathfrak{V}}_{\mathfrak{U}}b''^q$. Put $c''^q=\Pi^{\mathfrak{V}}_{\mathfrak{U}}b''^q\in Z^q(\mathfrak{U},\mathscr{S}'')$. Then by (3.81), we see that (3.82) holds. ∎

For any $g''\in H^q(M,\mathscr{S}'')$, take $c''^q\in Z^q(\mathfrak{U},\mathscr{S}'')$ and $c^q\in C^q(\mathfrak{U},\mathscr{S})$ as in Lemma 3.4. Then since

$$\delta\delta c^q=0,\qquad h\delta c^q=\delta hc^q=\delta c''^q=0,$$

and $\ker h=S'$, we have $\delta c^q\in Z^{q+1}(\mathfrak{U},\mathscr{S}')$. We define δ^*g'' by

$$\delta^*g''=[\delta c^q]\in H^{q+1}(M,\mathscr{S}'). \tag{3.83}$$

Then *δ^*g'' is determined uniquely by g'' and independent of the choice of $\mathfrak{U}$, c''^q, and c^q used above.* *Proof.* Take another $b''^q\in Z^q(\mathfrak{V},S'')$ and $b^q\in C^q(\mathfrak{V},\mathscr{S})$ such that $g''=[b''^q]$ and $hb^q=b''^q$. We must show $[\delta b^q]=[\delta c^q]$ in $H^{q+1}(M,\mathscr{S}')$. Take an arbitrary $\mathfrak{W}$ with $\mathfrak{W}<\mathfrak{U}$, and $\mathfrak{W}<\mathfrak{V}$. Then by (3.81), $g''=[\Pi^{\mathfrak{V}}_{\mathfrak{W}}b''^q]$, $[\delta b^q]=[\delta\Pi^{\mathfrak{V}}_{\mathfrak{W}}b^q]$, and $h\Pi^{\mathfrak{V}}_{\mathfrak{W}}b^q=\Pi^{\mathfrak{V}}_{\mathfrak{W}}b''^q$. Writing b''^q, c''^q, b^q and c^q simply instead of $\Pi^{\mathfrak{V}}_{\mathfrak{W}}b''^q$, $\Pi^{\mathfrak{V}}_{\mathfrak{W}}c''^q$, $\Pi^{\mathfrak{V}}_{\mathfrak{W}}b^q$ and $\Pi^{\mathfrak{V}}_{\mathfrak{W}}c^q$ respectively, we have

$$g''=[c''^q]=[b''^q],\qquad hc^q=c''^q,\qquad hb^q=b''^q.$$

If we put $a''=b''^q-c''^q$, and $a=b^q-c^q$, then we have

$$[a'']=0, \qquad ha=a''\in Z^q(\mathfrak{W},\mathcal{S}''), \qquad a\in C^q(\mathfrak{W},\mathcal{S}).$$

We must prove that $[\delta a]=0$ in $H^{q+1}(M,\mathcal{S}')$. Since $[a'']=0$, taking a suitable $\mathfrak{X}<\mathfrak{W}$, we can find $e''\in C^{q-1}(\mathfrak{X},\mathcal{S}'')$ such that $\Pi^{\mathfrak{W}}_{\mathfrak{X}}a''=\delta e''$. By Lemma 3.3, we can find $e\in C^{q-1}(\mathfrak{Y},\mathcal{S})$ such that $he=\Pi^{\mathfrak{X}}_{\mathfrak{Y}}e''$, where $\mathfrak{Y}$ is a suitable refinement of $\mathfrak{X}$. Since $\Pi^{\mathfrak{W}}_{\mathfrak{Y}}=\Pi^{\mathfrak{X}}_{\mathfrak{Y}}\Pi^{\mathfrak{W}}_{\mathfrak{X}}$, we have

$$h\Pi^{\mathfrak{W}}_{\mathfrak{Y}}a=\Pi^{\mathfrak{W}}_{\mathfrak{Y}}a''=\Pi^{\mathfrak{X}}_{\mathfrak{Y}}\delta e''=\delta\Pi^{\mathfrak{X}}_{\mathfrak{Y}}e''=\delta he=h\delta e,$$

hence putting $a'=\Pi^{\mathfrak{W}}_{\mathfrak{Y}}a-\delta e$, we have $ha'=0$. Since $\ker h=\mathcal{S}'$, we have $a'\in C^q(\mathfrak{Y},\mathcal{S}')$, hence

$$\Pi^{\mathfrak{W}}_{\mathfrak{Y}}\delta a=\delta\Pi^{\mathfrak{W}}_{\mathfrak{Y}}a=\delta a'\in C^q(\mathfrak{Y},\mathcal{S}').$$

Therefore $[\delta a]=[\Pi^{\mathfrak{W}}_{\mathfrak{Y}}\delta a]=[\delta a']=0$ in $H^{q+1}(M,\mathcal{S}')$. ∎

Next we shall prove the exactness of (3.80). We must prove that for $g''\in H^q(\mathcal{S}'')$, there exists $g\in H^q(\mathcal{S})$ with $g''=hg$ if and only if $\delta^*g''=0$. First suppose $g''=hg$, $g\in H^q(M,\mathcal{S})$. Then there exists a $c^q\in Z^q(\mathfrak{U},\mathcal{S})$ with $[c^q]=g$. If we put $hc^q=c''^q$, $g''=hg=$-$[c''^q]$. Hence by (3.83) we have $\delta^*g''=[\delta c^q]=0$.

Conversely, suppose $\delta^*g''=0$. Let $g''=[c''^q]$, and $hc^q=c''^q$ with $c^q\in C^q(\mathfrak{U},\mathcal{S})$ and $c''^q\in Z^q(\mathfrak{U},\mathcal{S}'')$. Then $[\delta c^q]=\delta^*g''=0$ in $H^{q+1}(M,\mathcal{S}')$, in other words, for a suitable $\mathfrak{B}<\mathfrak{U}$ and a suitable $c'^q\in C^q(\mathfrak{B},\mathcal{S}')$ we have $\Pi^{\mathfrak{U}}_{\mathfrak{B}}\delta c^q=\delta c'^q$. Then since

$$\delta(\Pi^{\mathfrak{U}}_{\mathfrak{B}}c^q-c'^q)=\Pi^{\mathfrak{U}}_{\mathfrak{B}}\delta c^q-\delta c'^q=0,$$

we see that $\Pi^{\mathfrak{U}}_{\mathfrak{B}}c^q-c'^q\in Z^q(\mathfrak{B},\mathcal{S})$. Hence if we put

$$g=[\Pi^{\mathfrak{U}}_{\mathfrak{B}}c^q-c'^q]\in H^q(M,\mathcal{S}),$$

since $hc'^q=0$, we have

$$hg=[\Pi^{\mathfrak{U}}_{\mathfrak{B}}hc^q]=[\Pi^{\mathfrak{U}}_{\mathfrak{B}}c''^q]=[c''^q]=g''.$$

Therefore (3.80) is exact.

Proof of Theorem 3.7. We may assume that $\mathcal{S}'\xrightarrow{\iota}\mathcal{S}$ is the inclusion. Then $H^0(\mathcal{S}')=\Gamma(M,\mathcal{S}')$ is a submodule of $H^0(\mathcal{S})=\Gamma(M,\mathcal{S})$. Hence the exactness of

$$0\to H^0(\mathcal{S}')\xrightarrow{\iota}H^0(\mathcal{S})\xrightarrow{h}H^0(\mathcal{S}'')\xrightarrow{\delta^*}H^1(\mathcal{S}')$$

follows immediately from the exactness of (3.79). Moreover since (3.80) is exact, in order to prove the exactness of (3.78), it suffices to verify the exactness of

$$H^{q-1}(\mathscr{S}'') \xrightarrow{\delta^*} H^q(\mathscr{S}') \xrightarrow{\iota} H^q(\mathscr{S}), \tag{3.84}$$

and

$$H^q(\mathscr{S}') \xrightarrow{\iota} H^q(\mathscr{S}) \xrightarrow{h} H^q(\mathscr{S}'') \tag{3.85}$$

for each $q = 1, 2, \ldots$.

Take an arbitrary $g' \in H^q(\mathscr{S}')$. If $g' = \delta^* g''$ with $g'' \in H^{q-1}(\mathscr{S}'')$, then by the definition (3.83) of δ^*, $g' = [\delta c^{q-1}] \in H^q(\mathscr{S}')$ for some $c^{q-1} \in C^{q-1}(\mathfrak{U}, \mathscr{S})$. Since ι is the inclusion $\mathscr{S}' \hookrightarrow \mathscr{S}$, $\iota g' = [\delta c^{q-1}]$ is the cohomology class of δc^{q-1} regarded as an element of $Z^q(\mathfrak{U}, \mathscr{S})$. Hence $\iota g' = 0$. Conversely, suppose $\iota g' = 0$. Let $g' = [c'^q]$ with $c'^q \in Z^q(\mathfrak{U}, \mathscr{S}')$. Then since $[c'^q] = \iota g' = 0$ means that $[c'^q] = 0$ as an element of $H^q(M, S)$, there are a refinement $\mathfrak{V} < \mathfrak{U}$ and an element $c^{q-1} \in C^{q-1}(\mathfrak{V}, \mathscr{S})$ such that $\Pi^{\mathfrak{U}}_{\mathfrak{V}} c'^q = \delta c^{q-1}$. Putting $c''^{q-1} = hc^{q-1}$, we obtain $\delta c''^{q-1} = h\delta c^{q-1} = \Pi^{\mathfrak{U}}_{\mathfrak{V}} hc'^q = 0$. Hence $c''^q \in Z^{q-1}(\mathfrak{U}, \mathscr{S}'')$. Therefore, putting $g'' = [c''^{q-1}] \in H^{q-1}(\mathscr{S}'')$, we obtain from (3.83)

$$\delta^* g'' = [\delta c^{q-1}] = [\Pi^{\mathfrak{U}}_{\mathfrak{V}} c'^q] = [c'^q] = g'.$$

Thus (3.84) is exact.

Now we proceed to show that (3.85) is exact. Take any $g \in H^q(\mathscr{S})$. If $g = \iota g'$ with $g' \in H^q(\mathscr{S}')$, clearly $hg = 0$. Let $g = [c^q]$ with $c^q \in Z^q(U, \mathscr{S})$. If $[hc^q] = hg = 0$, there are a refinement $\mathfrak{V} < \mathfrak{U}$ and an element $c''^{q-1} \in C^{q-1}(\mathfrak{V}, \mathscr{S}'')$ such that $\Pi^{\mathfrak{U}}_{\mathfrak{V}} hc^q = \delta c''^{q-1}$. By Lemma 3.3, taking a suitable refinement $\mathfrak{W} < \mathfrak{V}$ we can find an element $c^{q-1} \in C^{q-1}(\mathfrak{W}, \mathscr{S})$ for c''^{q-1} such that $hc^{q-1} = \Pi^{\mathfrak{V}}_{\mathfrak{W}} c''^{q-1}$. Hence we have

$$\Pi^{\mathfrak{U}}_{\mathfrak{W}} hc^q = \Pi^{\mathfrak{V}}_{\mathfrak{W}} \delta c''^{q-1} = \delta \Pi^{\mathfrak{V}}_{\mathfrak{W}} c''^{q-1} = \delta hc^{q-1},$$

that is, $h(\Pi^{\mathfrak{U}}_{\mathfrak{W}} c^q - \delta c^{q-1}) = 0$. Therefore $\Pi^{\mathfrak{U}}_{\mathfrak{W}} c^q - \delta c^{q-1} \in Z^q(\mathfrak{W}, \mathscr{S}')$. Consequently putting $g' = [\Pi^{\mathfrak{U}}_{\mathfrak{W}} c^q - \delta c^{q-1}] \in H^q(\mathscr{S}')$, we obtain

$$\iota g' = [\Pi^{\mathfrak{U}}_{\mathfrak{W}} c^q - \delta c^{q-1}] = [\Pi^{\mathfrak{U}}_{\mathfrak{W}} c^q] = [c^q] = g$$

since $[\delta c^{q-1}] = 0$ in $H^q(\mathscr{S})$. Thus (3.85) is exact. ∎

For given K-modules $H_1, H_2, \ldots, H_\nu, \ldots, L_1, L_2, \ldots, L_\nu, \ldots$ and K-homomorphisms h_ν: $H_\nu \to H_{\nu+1}$, ν: $L_\nu \to L_{\nu+1}$, φ_ν: $H_\nu \to L_\nu$, the diagram

$$\begin{array}{ccccccccc}
H_1 & \xrightarrow{h_1} & H_2 \to \cdots \to & H_\nu & \xrightarrow{h_\nu} & H_{\nu+1} \to \cdots \\
\downarrow \psi_1 & & \downarrow \psi_2 & \downarrow \psi_\nu & & \downarrow \psi_{\nu+1} \\
L_1 & \xrightarrow{k_1} & L_2 \to \cdots \to & L_\nu & \xrightarrow{k_\nu} & L_{\nu+1} \to \cdots
\end{array} \tag{3.86}$$

is called *commutative* if

$$k_\nu \circ \psi_\nu = \psi_{\nu+1} \circ h_\nu, \qquad \nu = 1, 2, 3, \ldots .$$

If (3.86) is commutative, and each row is exact, we call (3.86) an *exact commutative diagram.* Similarly we define an exact commutative diagram of sheaves and sheaf homomorphisms. For example,

$$\begin{array}{ccccccccc} 0 \to & \mathcal{S}' & \xrightarrow{\iota} & \mathcal{S} & \xrightarrow{h} & \mathcal{S}'' & \to 0 \\ & \downarrow{\scriptstyle \psi'} & & \downarrow{\scriptstyle \psi} & & \downarrow{\scriptstyle \psi''} & \\ 0 \to & \mathcal{T}' & \xrightarrow{\iota} & \mathcal{T} & \xrightarrow{k} & \mathcal{T}'' & \to 0 \end{array}$$

is an exact commutative diagram if each row is exact, $\iota \circ \psi' = \psi \circ \iota$ and $k \circ \psi = \psi'' \circ h$.

Theorem 3.8. *An exact commutative diagram of sheaves*

$$\begin{array}{ccccccccc} 0 \to & \mathcal{S}' & \xrightarrow{\iota} & \mathcal{S} & \xrightarrow{h} & \mathcal{S}'' & \to 0 \\ & \downarrow{\scriptstyle \psi'} & & \downarrow{\scriptstyle \psi} & & \downarrow{\scriptstyle \psi''} & \\ 0 \to & \mathcal{T}' & \xrightarrow{\iota} & \mathcal{T} & \xrightarrow{k} & \mathcal{T}'' & \to 0 \end{array}$$

induces an exact commutative diagram of cohomology groups

$$\begin{array}{ccccccccccc} 0 \longrightarrow & H^0(\mathcal{S}') & \xrightarrow{\iota} & H^0(\mathcal{S}) & \xrightarrow{h} & H^0(\mathcal{S}'') & \xrightarrow{\delta^*} & H^1(\mathcal{S}') & \xrightarrow{\iota} & \cdots \\ & \downarrow{\scriptstyle \psi'} & & \downarrow{\scriptstyle \psi} & & \downarrow{\scriptstyle \psi''} & & \downarrow{\scriptstyle \psi'} & & \\ 0 \longrightarrow & H^0(\mathcal{T}') & \xrightarrow{\iota} & H^0(\mathcal{T}) & \xrightarrow{k} & H^0(\mathcal{T}'') & \xrightarrow{\delta^*} & H^1(\mathcal{T}') & \xrightarrow{\iota} & \cdots \end{array} \tag{3.87}$$

$$\begin{array}{ccccccccccc} \cdots \xrightarrow{\iota} & H^{q-1}(\mathcal{S}) & \xrightarrow{h} & H^{q-1}(\mathcal{S}'') & \xrightarrow{\delta^*} & H^q(\mathcal{S}') & \xrightarrow{\iota} & H^q(\mathcal{S}) & \xrightarrow{h} & \cdots \\ & \downarrow{\scriptstyle \psi} & & \downarrow{\scriptstyle \psi''} & & \downarrow{\scriptstyle \psi'} & & \downarrow{\scriptstyle \psi} & & \\ \cdots \xrightarrow{\iota} & H^{q-1}(\mathcal{T}) & \xrightarrow{k} & H^{q-1}(\mathcal{T}'') & \xrightarrow{\delta^*} & H^q(\mathcal{T}') & \xrightarrow{\iota} & H^q(\mathcal{T}) & \xrightarrow{k} & \cdots . \end{array}$$

Proof. It suffices to verify the commutativity of (3.87). Since it is clear that $k \circ \psi = \psi'' \circ h$ and $\iota \circ \psi' = \psi \circ \iota$, we have only to show that the diagram

$$\begin{array}{ccc} H^q(\mathcal{S}'') & \xrightarrow{\delta^*} & H^{q+1}(\mathcal{S}') \\ \downarrow{\scriptstyle \psi''} & & \downarrow{\scriptstyle \psi'} \\ H^q(\mathcal{T}'') & \xrightarrow{\delta^*} & H^{q+1}(\mathcal{T}') \end{array}$$

is commutative for $q=0,1,\ldots$. For any $g''\in H^q(\mathscr{S}'')$, by Lemma 3.4 there are $c''^q\in Z^q(\mathfrak{U},\mathscr{S}'')$ and $c^q\in C^q(\mathfrak{U},\mathscr{S})$ with $g''=[c''^q]$, $hc^q=c''^q$ such that $\delta^*g''=[\delta c^q]$ where $\delta c^q\in Z^{q+1}(\mathfrak{U},\mathscr{S}')$. Hence $\psi'\delta^*g''=[\psi'\delta c^q]$. Since ι is the inclusion, $\iota\circ\psi'=\psi\circ\iota$ means that $\psi=\psi'$ on $\mathscr{S}'\subset\mathscr{S}$. Therefore $\psi'\delta c^q=\psi\delta c^q$, hence $\psi'\delta^*g''=[\psi\delta c^q]$. On the other hand, we have $\psi''g''=[\psi''c''^q]$ with $\psi''c''^q\in Z^q(\mathfrak{U},\mathscr{T}'')$. Since $\psi''\circ h=k\circ\psi$, we also have $\psi''c''^q=\psi''hc^q=k\psi c^q$. Hence by (3.83), we have $\delta^*\psi''g''=[\delta\psi c^q]$, which implies that $\delta^*\psi''g''=\psi'\delta^*g''$. Since g'' is an arbitrary element of $H^q(\mathscr{S}'')$, we obtain $\delta^*\circ\psi''=\psi'\circ\delta^*$ as desired. ∎

§3.4. de Rham's Theorem and Dolbeault's Theorem

(a) Fine Sheaves

Let $h\colon\mathscr{S}\to\mathscr{S}''$ be a homomorphism of sheaves over M. The closure of the set of points $p\in M$ with $h(\mathscr{S}_p)\neq 0$ is called the support of h and is denoted by supp h:

$$\operatorname{supp} h=[\{p\in M\mid h(\mathscr{S}_p)\neq 0\}].$$

Definition 3.13. Let $\mathscr{S}$ be a sheaf over M. M is called a *fine sheaf* if for any locally finite open covering $\mathfrak{U}=\{U_j\}$, there is a family of homomorphisms $\{h_j\}$: $h_j\colon\mathscr{S}\to\mathscr{S}$, such that

(i) $\operatorname{supp} h_j\subset U_j$,
(ii) $\sum_j h_j=$ identity, i.e. $\sum_j h_j(s)=s$ for any $s\in\mathscr{S}$.

Note that by the condition (i), $h_j(s)\neq 0$ for $s\in\mathscr{S}_p$ implies $p\in U_j$. Consequently the summation in (ii) makes sense since $\mathfrak{U}$ is locally finite.

Example 3.2. $\mathscr{A}$ is a finite sheaf. *Proof.* Let $\mathfrak{U}=\{U_j\}$ be a locally finite open covering of M, and $\{\rho_j\}$ a partition of unity subordinate to U. For $s=f_p\in\mathscr{A}_p$, with f a local C^∞ function at p, define h_js by

$$h_js=(\rho_jf)_p,\qquad s=f_p.$$

h_js is independent of the choice of f. In order to see that h_j is a homomorphism, it suffices to verify that h_j is continuous. Given a neighbourhood $U(h_js;g,U)$ of h_js, since $g_p=h_js=(\rho_jf)_p$, there is a neighbourhood $V\subset U$ such that $g=\rho_jd$ on V. Hence for $t\in\mathscr{U}(s;f,V)$, $q=\varpi(t)\in V$,

$$h_js=h_jf_q=(\rho_jf)_q=g_q\in\mathscr{U}(h_js;g,U).$$

Thus h_j is continuous, hence a homomorphism. It is clear that supp $h_j =$ supp $\rho_j \subset U_j$. Moreover $\sum_j h_j =$ identity follows from $\sum_j \rho_j = 1$. ∎. The essential point of the above argument is in the fact that if f is a local C^∞ function, so is $\rho_j f$. Consequently by a similar argument, we see that $\mathscr{A}^r$, $\mathscr{A}^{p,q}$ and $\mathscr{A}(F)$ are fine sheaves where F is a C^∞ vector bundle over M.

$\mathcal{O}$ is, however, not fine. In fact, if otherwise, for $\{h_j\}$ in Definition 3.13, $h_j 1 \in \Gamma(M, \mathcal{O})$ is holomorphic, where 1 is a holomorphic function which is identically one on M. Since $h_j 1$ is equal to 0 on $M -$ supp h_j, by the analytic continuation, $h_j 1 = 0$ on all of M, which contradicts $\sum_j h_j 1 = 1$.

Theorem 3.9. *If S is a fine sheaf over M,*

$$H^q(M, \mathscr{S}) = 0 \quad \text{for } q \geqq 1.$$

Proof. Since $H^q(M, \mathscr{S}) = \bigcup_{\mathfrak{U} \in \mathfrak{S}} \tilde{\Pi}^{\mathfrak{U}} H^q(\mathfrak{U}, \mathscr{S})$ by (3.69), it suffices to prove that

$$H^q(\mathfrak{U}, \mathscr{S}) = Z^q(\mathfrak{U}, \mathscr{S}) / \delta C^{q-1}(\mathfrak{U}, \mathscr{S}) = 0, \qquad q \geqq 1,$$

for any locally finite open covering $\mathfrak{U}$. We give the proof for the case $q = 2$, but the generalization is straightforward. Since $\mathscr{S}$ is fine, there is a family $\{h_j\}$, $h_j: \mathscr{S} \to \mathscr{S}$ of homomorphisms satisfying the above (i) and (ii). Each h_i induces a homomorphism $h_i: \Gamma(W, \mathscr{S}) \to \Gamma(W, \mathscr{S})$ for any open set $W \subset M$.

Let $c^2 = \{\sigma_{jkl}\} \in Z^2(\mathfrak{U}, \mathscr{S})$ be a 2-cocycle. For $U_j \cap U_k \cap U_i \neq \emptyset$, $h_i\sigma_{ijk}: p \to (h_i\sigma_{ijk})(p)$ is a section of S over $U_j \cap U_k \cap U_i$, and if $p \notin$ supp h_i, then $(h_i\sigma_{ijk})(p) = 0$. Since supp h_i is closed in U_i, putting

$$\tilde{\sigma}_{ijk}(p) = \begin{cases} (h_i\sigma_{ijk})(p), & p \in U_j \cap U_k \cap U_i, \\ 0, & p \in U_j \cap U_k - \operatorname{supp} h_i, \end{cases}$$

we get a section $\tilde{\sigma}_{ijk}: p \to \tilde{\sigma}_{ijk}(p)$ of S on $U_j \cap U_k$. $\tilde{\sigma}_{ijk}$ is an extension of $h_i\sigma_{ijk}$. For a fixed $p \in U_j \cap U_k$, $\tilde{\sigma}_{ijk}(p) \neq 0$ for only finite i's. Consequently putting

$$\tau_{jk} = \sum_i \tilde{\sigma}_{ijk},$$

we obtain $\tau_{jk} \in \Gamma(U_j \cap U_k, \mathscr{S})$ with $\tau_{kj} = -\tau_{jk}$, that is, $c^1 = \{\tau_{jk}\} \in C^1(\mathfrak{U}, \mathscr{S})$.

It suffices then to check $c^2 = \delta c^1$. Since $\delta c^2 = 0$, we have

$$\sigma_{jkl} = \sigma_{ikl} - \sigma_{ijl} + \sigma_{ijk} \quad \text{on} \quad U_i \cap U_j \cap U_k \cap U_l \neq \emptyset.$$

Hence

$$h_i\sigma_{jkl} = h_i\sigma_{ikl} - h_i\sigma_{ijl} + h_i\sigma_{ijk}.$$

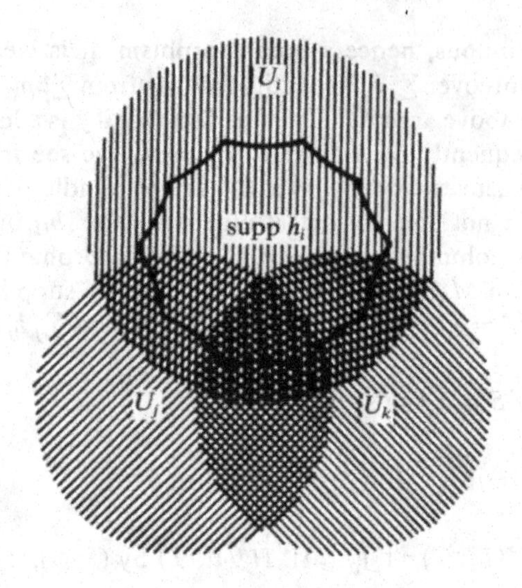

Figure 3

Since $\tilde{\sigma}_{ikl} = h_i\sigma_{ikl}$, $\tilde{\sigma}_{ijl} = h_i\sigma_{ijl}$, $\tilde{\sigma}_{ijk} = h_i\sigma_{ijk}$ on $U_j \cap U_k \cap U_l \cap U_i$, and $h_i\sigma_{jkl} = \tilde{\sigma}_{ikl} = \tilde{\sigma}_{ijl} = \tilde{\sigma}_{ijk} = 0$ on $U_j \cap U_k \cap U_l - U_i$, we have on $U_j \cap U_k \cap U_l$,

$$\sigma_{jkl} = \tau_{kl} - \tau_{jl} + \tau_{jk},$$

hence $c^2 = \delta c^1$. ∎

Now consider the sheaf $\mathscr{A}^r$ over a C^∞ manifold M. For a germ $s = \varphi_p$ of a local C^∞ r-form φ at p, we define its exterior differential by

$$ds = (d\varphi)_p, \qquad s = \varphi_p.$$

$d: s \to ds$ gives a homomorphism $\mathscr{A}^r \xrightarrow{d} \mathscr{A}^{r+1}$ of sheaves, where we put $\mathscr{A}^0 = \mathscr{A}$. The sheaf of locally constant functions $\mathbb{C}$ is a subsheaf of $\mathscr{A}^0$: $\mathbb{C} \hookrightarrow \mathscr{A}^0$. Then the sequence

$$0 \to \mathbb{C} \xrightarrow{\iota} \mathscr{A}^0 \xrightarrow{d} \mathscr{A}^1 \xrightarrow{d} \cdots \xrightarrow{d} \mathscr{A}^m \to 0 \tag{3.88}$$

is exact where $m = \dim M$, and ι is the inclusion. *Proof.* For $s = f_p \in \mathscr{A}^0$, $ds = (df)_p = 0$ implies that $df = 0$ on some neighbourhood of p, i.e., f is a constant there. Thus

$$0 \to \mathbb{C} \xrightarrow{\iota} \mathscr{A}^0 \xrightarrow{d} \mathscr{A}^1$$

is exact. The exactness of

$$\mathscr{A}^{r-1} \xrightarrow{d} \mathscr{A}^r \xrightarrow{d} \mathscr{A}^{r+1}, \qquad r \geqq 1, \tag{3.89}$$

follows from Poincaré's lemma (Theorem 3.2). In fact, take any $s = \varphi_p \in \mathscr{A}^r$, where φ is a local C^∞ r-form at p. $ds = (d\varphi)_p = 0$ implies that $d\varphi = 0$ on some neighbourhood U of p. We may assume that U is a coordinate multi-interval $U = \{x_p \,|\, |x_p^1| < \varepsilon, \ldots, |x_p^m| < \varepsilon\}$ with centre p. Then by Poincaré's lemma there is a C^∞ $(r-1)$-form ψ on U such that $\varphi = d\psi$. Putting $t = \psi_p \in \mathscr{A}^{r-1}$, we obtain $s = \varphi = (d\psi)_p = dt$. Thus if $ds = 0$, there is a $t \in \mathscr{A}^{r-1}$ with $dt = s$. Conversely, if $s = dt$ with $t \in \mathscr{A}^{r-1}$, $ds = d(d\psi)_p = (dd\psi)_p = 0$. Consequently (3.89) is exact. ∎

Definition 3.14. An exact sequence of sheaves over M

$$0 \to \mathscr{S} \xrightarrow{\iota} \mathscr{B}^0 \xrightarrow{d} \mathscr{B}^1 \xrightarrow{d} \mathscr{B}^2 \xrightarrow{d} \cdots \tag{3.90}$$

is called a *fine resolution* of $\mathscr{S}$ if each $\mathscr{B}^i$ is fine.

Thus Example 3.2 says that (3.88) is a fine resolution of $\mathbb{C}$.

Let $d: \mathscr{B} \to \mathscr{B}''$ be a homomorphism of sheaves over M. Put $\mathscr{S} = \ker d$. $\mathscr{S}$ is a subsheaf of $\mathscr{B}$. We denote the inclusion $\mathscr{S} \to \mathscr{B}$ by ι. Then

$$0 \to \mathscr{S} \xrightarrow{\iota} \mathscr{B} \xrightarrow{d} d\mathscr{B} \to 0$$

is exact. From this sequence we obtain

Lemma 3.5. *If $\mathscr{B}$ is fine,*

$$H^1(M, \mathscr{S}) \cong H^0(M, d\mathscr{B})/dH^0(M, \mathscr{B}),$$

$$H^q(M, \mathscr{S}) \cong H^{q-1}(M, d\mathscr{B}), \qquad q = 2, 3, \ldots.$$

Proof. By Theorem 3.7, we get the exact sequence of cohomology groups

$$0 \to H^0(\mathscr{S}) \xrightarrow{\iota} H^0(\mathscr{B}) \xrightarrow{d} H^0(d\mathscr{B}) \xrightarrow{\delta^*} H^1(\mathscr{S}) \xrightarrow{\iota} H^1(\mathscr{B}) \xrightarrow{d} \cdots$$

$$\to H^{q-1}(\mathscr{B}) \xrightarrow{d} H^{q-1}(d\mathscr{B}) \xrightarrow{\delta^*} H^q(\mathscr{S}) \xrightarrow{\iota} H^q(\mathscr{B}) \xrightarrow{d} \cdots.$$

Since $\mathscr{B}$ is fine, $H^q(\mathscr{B}) = 0$ for $q \geqq 1$ by Theorem 3.9. Therefore we obtain the exact sequences

$$0 \to H^0(\mathscr{S}) \to H^0(\mathscr{B}) \xrightarrow{d} H^0(d\mathscr{B}) \xrightarrow{\delta^*} H^1(\mathscr{S}) \to 0, \tag{3.91}$$

$$0 \to H^{q-1}(d\mathscr{B}) \xrightarrow{\delta^*} H^q(\mathscr{S}) \to 0, \qquad q = 2, 3, \ldots. \tag{3.92}$$

Hence

$$H^1(\mathscr{S}) \cong H^0(d\mathscr{B})/dH^0(\mathscr{B}), \qquad H^q(\mathscr{S}) \cong H^{q-1}(d\mathscr{B}), \qquad q \geqq 2. \quad \blacksquare$$

Theorem 3.10. *If*

$$0 \to \mathscr{S} \xrightarrow{\iota} \mathscr{B}^0 \xrightarrow{d} \mathscr{B}^1 \xrightarrow{d} \mathscr{B}^2 \xrightarrow{d} \cdots$$

is a fine resolution of $\mathscr{S}$ over M, then

$$H^q(M, \mathscr{S}) \cong \Gamma(M, d\mathscr{B}^{q-1})/d\Gamma(M, \mathscr{B}^{q-1}), \qquad q \geqq 1. \tag{3.93}$$

Proof. From the exact sequence

$$0 \to \mathscr{S} \xrightarrow{\iota} \mathscr{B}^0 \xrightarrow{d} \mathscr{B}^1 \xrightarrow{d} \dots,$$

we obtain the exact sequences

$$0 \to \mathscr{S} \xrightarrow{\iota} \mathscr{B}^0 \xrightarrow{d} d\mathscr{B}^0 \to 0$$

and

$$0 \to d\mathscr{B}^{r-1} \to \mathscr{B}^r \xrightarrow{d} d\mathscr{B}^r \to 0, \qquad r = 1, 2, 3, \dots.$$

By hypothesis each $\mathscr{B}^i$ is fine. Hence by Lemma 3.5, we have

$$H^1(\mathscr{S}) \cong H^0(d\mathscr{B}^0)/dH^0(\mathscr{B}^0),$$

$$H^q(\mathscr{S}) \cong H^{q-1}(d\mathscr{B}^0),$$

$$H^1(d\mathscr{B}^{r-1}) \cong H^0(d\mathscr{B}^r)/dH^0(d\mathscr{B}^{r-1}),$$

$$H^q(d\mathscr{B}^{r-1}) \cong H^{q-1}(d\mathscr{B}^r), \qquad q = 2, 3, 4, \dots.$$

Therefore we have

$$H^q(\mathscr{S}) \cong H^{q-1}(d\mathscr{B}^0) \cong H^{q-2}(d\mathscr{B}^1) \cong \cdots$$
$$\cong H^1(d\mathscr{B}^{q-2}) \cong H^0(d\mathscr{B}^{q-1})/dH^0(\mathscr{B}^{q-1}),$$

while by (3.70) we have $H^0(d\mathscr{B}^{q-1}) = \Gamma(M, d\mathscr{B}^{q-1})$ and $H^0(\mathscr{B}^{q-1}) = \Gamma(M, \mathscr{B}^{q-1})$. $\blacksquare$

(b) de Rham's Theorem

As stated earlier, the exact sequence (3.88)

$$0 \to \mathbb{C} \xrightarrow{\iota} \mathscr{A}^0 \xrightarrow{d} \mathscr{A}^1 \xrightarrow{d} \cdots \xrightarrow{d} \mathscr{A}^m \to 0$$

gives a fine resolution of $\mathbb{C}$ over a C^∞ manifold M of dimension m. Consequently by Theorem 3.10, we obtain the following theorem.

Theorem 3.11 (de Rham).

$$H^q(M, \mathbb{C}) \cong \Gamma(M, d\mathscr{A}^{q-1})/d\Gamma(M, \mathscr{A}^{q-1}), \; q \geqq 1. \quad \blacksquare \tag{3.94}$$

Note that $H^q(M, \mathbb{C})$ is the cohomology group of M with complex coefficients, depending only on the topology of M. As is well known, $H^q(M, \mathbb{C}) = 0$ for $q \geqq m+1$. This is also induced from (3.88) easily. In fact, from the proof of Theorem 3.10, $H^q(\mathbb{C}) \cong H^{q-m}(d\mathscr{A}^{m-1})$, where $d\mathscr{A}^{m-1} = \mathscr{A}^m$ is a fine sheaf. Hence by Theorem 3.9, $H^q(M, \mathbb{C}) \cong H^{q-m}(\mathscr{A}^m) = 0$ for $q \geqq m+1$.

Now we shall explain the meaning of the right-hand side of (3.94). Since for $q \geqq 1$, $0 \to d\mathscr{A}^{q-1} \to \mathscr{A}^q \xrightarrow{d} \mathscr{A}^{q+1}$ is exact, $d\mathscr{A}^{q-1}$ is the sheaf of local closed C^∞ q-forms. In fact $d\mathscr{A}^{q-1} = \ker d$ consists of germs $p \in A^q$ with $(d\varphi)_p = d(\varphi_p) = 0$, which means that $d\varphi = 0$ on some neighbourhood of p. Thus $\Gamma(M, d\mathscr{A}^{q-1})$ is the linear space of all d-closed C^∞ q-forms on M, while $d\Gamma(M, \mathscr{A}^{q-1})$ is the linear subspace of $\Gamma(M, d\mathscr{A}^{q-1})$ consisting of all exact r-forms on M. We call the quotient space

$$H^q_d(M) = \Gamma(M, d\mathscr{A}^{q-1})/d\Gamma(M, \mathscr{A}^{q-1}), \qquad q \geqq 1,$$

the *d-cohomology group of M*. Then de Rham's theorem:

$$H^q_d(M) \cong H^q(M, \mathbb{C}), \qquad q \geqq 1,$$

implies that the d-cohomology group $H^q_d(M)$, which *is defined with respect to the differentiable structure of M, actually depends only on the topology of M.*

(c) Dolbeault's Theorem

Let $M = M^n$ be an n-dimensional complex manifold, and $\mathscr{A}^{p,q}$ the sheaf of germs of C^∞ (p, q)-forms over M. The sheaf $\mathscr{O}$ of germs of holomorphic functions on M is a subsheaf of $\mathscr{A} = \mathscr{A}^{0,0}$. By Theorem 1.6, a local C^∞ function f is holomorphic if and only if $\bar{\partial} f = 0$. Thus

$$0 \to \mathscr{O} \xrightarrow{\iota} \mathscr{A}^{0,0} \xrightarrow{\bar{\partial}} \mathscr{A}^{0,1} \tag{3.95}$$

is exact, where $\iota: \mathcal{O} \to \mathcal{A}^{0,0}$ is the inclusion. Similarly Ω^p is a subsheaf of $\mathcal{A}^{p,0}$, and a local C^∞ $(p, 0)$-form φ is a holomorphic p-form if and only if $\bar\partial\varphi = 0$. Thus

$$0 \to \Omega^p \xrightarrow{\iota} \mathcal{A}^{p,0} \xrightarrow{\bar\partial} \mathcal{A}^{p,1} \tag{3.96}$$

is also exact.

By Dolbeault's lemma, a C^∞ (p, q)-form φ, $q \geqq 1$, on a polydisk is $\bar\partial$-closed if and only if there is a C^∞ $(p, q-1)$-form ψ such that $\varphi = \bar\partial\psi$. Consequently

$$\mathcal{A}^{p,q-1} \xrightarrow{\bar\partial} \mathcal{A}^{p,q} \xrightarrow{\bar\partial} \mathcal{A}^{p,q+1}, \qquad q \geqq 1, \tag{3.97}$$

is exact. Thus by (3.96) and (3.97) we obtain the exact sequence

$$0 \to \Omega^p \xrightarrow{\iota} \mathcal{A}^{p,0} \xrightarrow{\bar\partial} \mathcal{A}^{p,1} \xrightarrow{\bar\partial} \cdots \xrightarrow{\bar\partial} \mathcal{A}^{p,n} \to 0. \tag{3.98}$$

Since $\mathcal{A}^{p,q}$ is fine, (3.98) *gives a fine resolution of* Ω^p over M. For $p = 0$, we obtain a fine resolution of $\mathcal{O}$:

$$0 \to \mathcal{O} \xrightarrow{\iota} \mathcal{A}^{0,0} \xrightarrow{\bar\partial} \mathcal{A}^{0,1} \xrightarrow{\bar\partial} \cdots \xrightarrow{\bar\partial} \mathcal{A}^{0,n} \to 0.$$

Theorem 3.12 (Dolbeault's Theorem).

$$H^q(M, \Omega^p) \cong \Gamma(M, \bar\partial\mathcal{A}^{p,q-1})/\bar\partial\Gamma(M, \mathcal{A}^{p,q-1}), \qquad q \geqq 1. \tag{3.99}$$

Proof. Apply Theorem 3.10 to (3.98). ∎

In (3.99) $\bar\partial\mathcal{A}^{p,q-1}$ is the sheaf of germs of $\bar\partial$-closed C^∞ (p, q)-forms on M, $\Gamma(M, \bar\partial\mathcal{A}^{p,q-1})$ is the linear space of all $\bar\partial$-closed C^∞ (p, q)-forms on M, and $\bar\partial\Gamma(M, \mathcal{A}^{p,q-1})$ its subspace. The quotient space

$$H^{p,q}_{\bar\partial}(M) = \Gamma(M, \bar\partial\mathcal{A}^{p,q-1})/\bar\partial\Gamma(M, \mathcal{A}^{p,q-1})$$

is called the $\bar\partial$-*cohomology group* of M. Thus Dolbeault's theorem says that

$$H^q(M, \Omega^p) \cong H^{p,q}_{\bar\partial}(M). \tag{3.100}$$

In particular for $p = 0$, (3.100) reads

$$H^q(M, \mathcal{O}) \cong H^{0,q}_{\bar\partial}(M). \tag{3.101}$$

Ω^p is the sheaf of germs of holomorphic sections of $\bigwedge^p T^*(M)$ over M:

$$\Omega^p = \mathcal{O}\left(\bigwedge^p T^*(M)\right).$$

Dolbeault's theorem can be extended for the sheaf $\mathcal{O}(F)$ of germs of holomorphic sections of F over M, where F is a holomorphic vector bundle over M.

Each term $\mathscr{A}^{p,q}$ of (3.98) is the sheaf of germs of C^∞ sections of $\bigwedge^p T^*(M) \bigwedge^q \bar{T}^*(M)$, which we may identify with $\bigwedge^p T^*(M) \otimes \bigwedge^q \bar{T}^*(M)$. Thus

$$\mathscr{A}^{p,q} = \mathscr{A}\left(\overset{p}{\bigwedge} T^*(M) \otimes \overset{q}{\bigwedge} \bar{T}^*(M)\right).$$

Similarly we define $\mathscr{A}^{0,q}(F)$ by

$$\mathscr{A}^{0,q}(F) = \mathscr{A}\left(F \otimes \overset{q}{\bigwedge} \bar{T}^*(M)\right),$$

where $\mathscr{A}^{0,0}(F) = \mathscr{A}(F)$. We shall show that $\mathcal{O}(F)$ has a fine resolution

$$0 \to \mathcal{O}(F) \xrightarrow{\iota} \mathscr{A}^{0,0}(F) \xrightarrow{\bar\partial} \mathscr{A}^{0,1}(F) \xrightarrow{\bar\partial} \cdots \xrightarrow{\bar\partial} \mathscr{A}^{0,n}(F) \to 0. \tag{3.102}$$

First we investigate the structure of $\mathscr{A}^{0,q}(F)$ and define $\bar\partial$.

Let $\{e_{j1}, \ldots, e_{j\nu}\}$ be a basis of F over U_j where $\mathfrak{U} = \{U_j\}$ is a suitable locally finite open covering of M. We denote a point of M by z. Then each $e_{j\lambda}: z \to e_{j\lambda}(z)$ is a holomorphic section of F over U_j, and $\{e_{j1}(z), \ldots, e_{j\nu}(z)\}$ form a basis of F_z for $z \in U_j$. Let $f_{jk}(z) = (f^\lambda_{jk\mu}(z))_{\lambda,\mu=1,\ldots,\nu}$ be the transition function of F. Then on $U_j \cap U_k \neq \phi$,

$$e_{k\mu}(z) = \sum_{\lambda=1}^{\nu} f^\lambda_{jk\mu}(z) e_{j\lambda}(z), \tag{3.103}$$

where $f^\lambda_{jk\mu}(z)$ is holomorphic. Assume that each U_j is a coordinate neighbourhood and let $(z_j^1, \ldots, z_j^n)$ be the local complex coordinates on U_j. Then $\nu \times \binom{n}{q}$ sections

$$e_{j\lambda} \otimes \overbrace{d\bar z_j^\alpha \wedge \cdots \wedge d\bar z_j^\gamma}^{q \text{ times}}, \qquad \lambda = 1, \ldots, \nu, \quad 1 \leqq \alpha < \cdots < \gamma \leqq n,$$

form a basis of $F \otimes \bigwedge^q \bar{T}^*(M)$ over U_j. Consequently a C^∞ section φ of $F \otimes \bigwedge^q \bar{T}^*(M)$ over an open subset $W \subset M$ is given on each $U_j \cap W \neq \phi$ by

$$\varphi = \sum_{\lambda=1}^{\nu} \frac{1}{q!} \sum_{\alpha,\ldots,\gamma=1}^{n} \varphi^\lambda_{j\bar\alpha\cdots\bar\gamma}(z) e_{j\lambda} \otimes d\bar z_j^\alpha \wedge \cdots \wedge d\bar z_j^\gamma,$$

where $\varphi^\lambda_{j\bar\alpha\cdots\bar\gamma}(z)$ is a C^∞ function on $W \cap U_j$. Putting

$$\varphi^\lambda_j = \frac{1}{q!} \sum_{\alpha,\ldots,\gamma=1}^{n} \varphi^\lambda_{j\bar\alpha\cdots\bar\gamma}(z)\, d\bar z_j^\alpha \wedge \cdots \wedge d\bar z_j^\gamma,$$

we can write φ as

$$\varphi = \sum_{\lambda=1}^{\nu} e_{j\lambda} \otimes \varphi_j^\lambda,$$

where φ_j^λ is a C^∞ $(0, q)$-form on $W \cap U_j$. On $W \cap U_j \cap U_k \neq \varnothing$,

$$\varphi = \sum_{\lambda=1}^{\nu} e_{j\lambda} \otimes \varphi_j^\lambda = \sum_{\mu=1}^{\nu} e_{k\mu} \otimes \varphi_k^\mu$$

hence by (3.103) we have

$$\varphi_j^\lambda = \sum_{\mu=1}^{\nu} f_{jk\mu}^\lambda(z) \varphi_k^\mu. \tag{3.104}$$

If we denote $\sum_{\lambda=1}^{\nu} e_{j\lambda} \otimes \varphi_j^\lambda$ by the vector $(\varphi_j^1, \dots, \varphi_j^\nu)$, whose components are C^∞ $(0, q)$-forms, φ is represented on each $W \cap U_j$ by the vector

$$\varphi = (\varphi_j^1, \dots, \varphi_j^\lambda, \dots, \varphi_j^\nu),$$

which satisfies (3.104) on $W \cap U_j \cap U_k$. Conversely, suppose given on each $W \cap U_j \neq \varnothing$ a vector $(\varphi_j^1, \dots, \varphi_j^\nu)$ with φ_j^λ a C^∞ $(0, q)$-form, satisfying (3.104) on $W \cap U_j \cap U_k$. Then there is a C^∞ section φ of $F \otimes \bigwedge^q \bar{T}^*(M)$ over W such that $\varphi = \sum_{\lambda=1}^{\nu} e_{j\lambda} \otimes \varphi_j^\lambda$ on $W \cap U_j \neq \varnothing$. (3.104) gives the same transition relation as that of F:

$$\zeta_j^\lambda = \sum_{\mu=1}^{\nu} f_{jk\mu}^\lambda(z) \zeta_k^\mu.$$

Accordingly we call a C^∞ section $\varphi = (\varphi_j^1, \dots, \varphi_j^\nu)$ of $F \otimes \bigwedge^q \bar{T}^*(M)$ a *C^∞ $(0, q)$-form with coefficients in F*, and $\mathscr{A}^{0,q}(F)$ *the sheaf of germs of C^∞ $(0, q)$-forms with coefficients in F*. For $q = 0$, each φ_j^λ is a C^∞ function.

For a C^∞ $(0, q)$-form $\varphi = (\varphi_j^1, \dots, \varphi_j^\nu)$ with coefficients in F, we define

$$\bar{\partial}\varphi = (\bar{\partial}\varphi_j^1, \dots, \bar{\partial}\varphi_j^\nu). \tag{3.105}$$

Since $\bar{\partial} f_{jk\mu}^\lambda(z) = 0$, applying $\bar{\partial}$ to (3.104) we obtain

$$\bar{\partial}\varphi_j^\lambda = \sum_{\mu=1}^{\nu} f_{jk\mu}^\lambda(z) \bar{\partial}\varphi_k^\mu.$$

Therefore $\bar{\partial}\varphi$ is a C^∞ $(0, q+1)$-form with coefficients in F. Thus $\bar{\partial}$ gives a homomorphism

$$\bar{\partial}: \mathscr{A}^{0,q}(F) \to \mathscr{A}^{0,q+1}(F).$$

Since $\mathscr{A}^{0,q}(F)$ is fine, in order to show that (3.102) is a fine resolution, it suffices to prove the exactness of (3.102). For this it suffices to prove that at each point $z \in M$, the sequence of the stalks

$$0 \to \mathcal{O}(F)_z \xrightarrow{\iota} \mathscr{A}^{0,0}(F)_z \xrightarrow{\bar{\partial}} \mathscr{A}^{0,1}(F)_z \xrightarrow{\bar{\partial}} \cdots$$

is exact. Thus we have only to prove the exactness of (3.102) on each U_j. But on U_j, a local C^∞ $(0, q)$-form $\varphi = (\varphi_j^1, \ldots, \varphi_j^\nu)$ with coefficients in F may be regarded as a ν-tuple of C^∞ $(0, q)$-forms φ_j^λ, $\lambda = 1, \ldots, \nu$. Moreover $\bar{\partial}\varphi = (\bar{\partial}\varphi_j^1, \ldots, \bar{\partial}\varphi_j^\nu)$. Therefore the exactness of (3.102) follows from the exactness of

$$0 \to \mathcal{O} \xrightarrow{\iota} \mathscr{A}^{0,0} \xrightarrow{\bar{\partial}} \mathscr{A}^{0,1} \xrightarrow{\bar{\partial}} \cdots \xrightarrow{\bar{\partial}} \mathscr{A}^{0,n} \to 0.$$

Applying Theorem 3.10 to (3.102) we obtain immediately the following theorem.

Theorem 3.13.

$$H^q(M, \mathcal{O}(F)) \cong \Gamma(M, \bar{\partial}\mathscr{A}^{0,q-1}(F))/\bar{\partial}\Gamma(M, \mathscr{A}^{0,q-1}(F)), \qquad q \geqq 1. \quad \blacksquare \tag{3.106}$$

Let φ be a C^∞ section of $F \otimes \bigwedge^p T^*(M) \bigwedge^q \bar{T}^*(M)$ over an open set $W \subset M$. Then φ is represented on each $W \cap U_j \neq \varnothing$ by a vector $\varphi = (\varphi_j^1, \ldots, \varphi_j^\nu)$ where each φ_j^λ is a C^∞ (p, q)-form, and its transition relation on $W \cap U_j \cap U_k \neq \varnothing$ is given by (3.104). In view of this we call φ a C^∞ *(p, q)-form with coefficients in F*. We denote by $\mathscr{A}^{p,q}(F)$ the sheaf of germs of C^∞ (p, q)-forms with coefficients in F:

$$\mathscr{A}^{p,q}(F) = \mathscr{A}\left(F \otimes \bigwedge^p T^*(M) \bigwedge^q \bar{T}^*(M)\right).$$

If each φ_j^λ of $\varphi = (\varphi_j^1, \ldots, \varphi_j^\nu)$ is a holomorphic p-form, we call φ a *holomorphic p-form with coefficients in F*. We denote by $\Omega^p(F)$ the sheaf of germs of holomorphic p-forms with coefficients in F.

$$\Omega^p(F) = \mathcal{O}\left(F \otimes \bigwedge^p T^*(M)\right).$$

Substituting $F \otimes \bigwedge^p T^*(M)$ for F in (3.106), we obtain

$$H^q(M, \Omega^p(F)) \cong \Gamma(M, \bar{\partial}\mathscr{A}^{p,q-1}(F))/\bar{\partial}\Gamma(M, \mathscr{A}^{p,q-1}(F)), \qquad q \geqq 1, \tag{3.107}$$

where $\Gamma(M, \bar{\partial}\mathscr{A}^{p,q}(F))$ is the linear space of all $\bar{\partial}$-closed C^∞ (p, q)-forms with coefficients in F over M. The quotient space

$$H^{p,q}_{\bar\partial}(M, F) = \Gamma(M, \bar\partial\mathscr{A}^{p,q-1}(F))/\bar\partial\Gamma(M, \mathscr{A}^{p,q-1}(F))$$

is called *the $\bar\partial$-cohomology group of M with coefficients in F*. With this notation, (3.107) is written as

$$H^q(M, \Omega^p(F)) \cong H^{p,q}_{\bar\partial}(M, F). \tag{3.108}$$

§3.5. Harmonic Differential Forms

(a) Inner Product, Norm, Dual Form

Let M be an n-dimensional complex manifold, $\{z_1, \ldots, z_j, \ldots\}$, $z_j\colon p \to z_j(p) = (z_j^1, \ldots, z_j^n)$ local complex coordinates on M, and U_j the domain of z_j, and suppose that $\mathfrak{U} = \{U_j\}$ is locally finite. Suppose that $p \to z(p) = (z^1, \ldots, z^n)$ is an arbitrary local complex coordinates system on M. We represent a point of M by its local coordinates $z = z(p)$.

In the sequel, we represent a tensor field by its components with respect to local coordinates $(z^1, \ldots, z^n)$. Thus, for example, by a "tensor field $\sigma^\alpha_{\beta\bar\gamma}$", we mean a section

$$\sum_{\alpha,\beta,\gamma} \sigma^\alpha_{\beta\bar\gamma} \frac{\partial}{\partial z_\alpha} \otimes dz^\beta \otimes d\bar z^\gamma$$

of $T(M) \otimes T^*(M) \otimes \bar T^*(M)$. If we denote by $\sigma^\alpha_{j\beta\bar\gamma}$ the component of a tensor field $\sigma^\alpha_{b\bar\gamma}$ with respect to the local coordinates $(z_j^1, \ldots, z_j^n)$, we obtain

$$\sigma^\alpha_{j\beta\bar\gamma} = \sum_{\lambda,\mu,\nu=1}^n \frac{\partial z_j^\alpha}{\partial z^\lambda} \frac{\partial z^\mu}{\partial z_j^\beta} \overline{\left(\frac{\partial z^\nu}{\partial z_j^\gamma}\right)} \sigma^\lambda_{\mu\bar\nu}.$$

Definition 3.15. A C^∞ covariant tensor field $\sum_{\alpha,\beta=1}^n g_{\alpha\bar\beta}\, dz^\alpha \otimes d\bar z^\beta$, $g_{\alpha\bar\beta} = g_{\alpha\bar\beta}(z)$, is called a *Hermitian metric if* $g_{\beta\bar\alpha}(z) = \overline{g_{\alpha\bar\beta}(z)}$ and the Hermitian form $\sum_{\alpha,\beta=1}^n g_{\alpha\bar\beta}(z) \zeta^\alpha \bar\zeta^\beta$ with complex variables $\zeta^1, \ldots, \zeta^n$, is positive definite at every point of M.

$\sum g_{\alpha\bar\beta}\, dz^\alpha \otimes d\bar z^\beta$ is called a metric because by virtue of it we can define the length $\sqrt{\sum g_{\alpha\bar\beta} \zeta^\alpha \bar\zeta^\beta}$ of a tangent vector $\zeta = \sum_\alpha \zeta^\alpha (\partial/\partial z^\alpha) \in T_z(M)$.

We associate to the Hermitian metric $\sum_{\alpha,\beta=1}^n g_{\alpha\bar\beta}\, dz^\alpha \otimes d\bar z^\beta$ a (1, 1)-form

$$\omega = i \sum_{\alpha,\beta=1}^n g_{\alpha\bar\beta} dz^\alpha \wedge d\bar z^\beta, \qquad i = \sqrt{-1}.$$

Since

$$\bar{\omega} = -i \sum_{\alpha,\beta} \overline{g_{\alpha\bar{\beta}}}\, d\bar{z}^{\alpha} \wedge dz^{\beta} = i \sum_{\alpha,\beta} g_{\beta\bar{\alpha}}\, dz^{\beta} \wedge d\bar{z}^{\alpha} = \omega,$$

ω is a real $(1, 1)$-form.

Theorem 3.14. *Given any complex manifold M, we can introduce a Hermitian metric $\sum_{\alpha,\beta=1}^{n} g_{\alpha\bar{\beta}}\, dz^{\alpha} \otimes d\bar{z}^{\beta}$ on M.*

Proof. Let $\mathfrak{U} = \{U_j\}$ be a locally finite open covering of M, and $\{\rho_j\}$ a partition of unity subordinate to it. $\rho_j = \rho_j(z) \geqq 0$ is a C^{∞} function on M with supp $\rho_j \subset U_j$, and $\sum_j \rho_j(z) = 1$. Put

$$\omega_j(z) = i \sum_{\gamma=1}^{n} \rho_j(z)\, dz_j^{\gamma} \wedge d\bar{z}_j^{\gamma}.$$

Then $\omega_j = \omega_j(z)$ is a C^{∞} real $(1, 1)$-form on U_j, and $\omega_j(z) = 0$ on $U_j - \text{supp}\, \rho_j$. Consequently putting $\tilde{\omega}_j(z) = \omega_j(z)$ for $z \in U_j$, and $\tilde{\omega}_j(z) = 0$ for $z \notin U_j$, we can extend ω_j to a C^{∞} $(1, 1)$-form $\tilde{\omega}_j = \tilde{\omega}_j(z)$ on M. Put

$$\omega = \sum_j \tilde{\omega}_j.$$

Since supp $\tilde{\omega}_j = \text{supp}\, \rho_j \subset U_j$ and $\{U_j\}$ is locally finite, this summation makes sense. Therefore ω is a real C^{∞} $(1, 1)$-form on M with

$$\omega = i \sum_{\alpha,\beta=1}^{n} g_{\alpha\bar{\beta}}\, dz^{\alpha} \wedge d\bar{z}^{\beta}, \qquad g_{\beta\bar{\alpha}} = \overline{g_{\alpha\bar{\beta}}}.$$

We claim that $\sum_{\alpha,\beta=1}^{n} g_{\alpha\bar{\beta}}\, dz^{\alpha} \otimes d\bar{z}^{\beta}$ is a Hermitian metric. For this it suffices to see that $\sum_{\alpha,\beta=1}^{n} g_{\alpha\bar{\beta}}(z) \zeta^{\alpha} \bar{\zeta}^{\beta}$ is positive definite at each $z \in M$. By the definition of ω,

$$\sum_{\alpha,\beta=1}^{n} g_{\alpha\bar{\beta}}(z)\, dz^{\alpha} \wedge d\bar{z}^{\beta} = \sum_{U_j \ni z} \sum_{\gamma=1}^{n} \rho_j(z)\, dz_j^{\gamma} \wedge d\bar{z}_j^{\gamma}, \tag{3.109}$$

where the summation $\sum_{U_j \ni z}$ is taken over all j with $z \in U_j$. Consider $(\zeta^1, \ldots, \zeta^n)$ as the components of a tangent vector $\zeta \in T_z(M)$ with respect to $(z^1, \ldots, z^n)$, and denote by $(\zeta_j^1, \ldots, \zeta_j^n)$ the components of ζ with respect to $(z_j^1, \ldots, z_j^n)$. Then since the transition relation $\zeta_j^{\gamma} = \sum_{\alpha} (\partial z_j^{\gamma} / \partial z^{\alpha}) \zeta^{\alpha}$ of the components of a tangent vector is the same as that of the bases of $T^*(M)$: $dz_j^{\gamma} = \sum_{\alpha} (\partial z_j^{\gamma} / \partial z^{\alpha})\, dz^{\alpha}$, we obtain from (3.109) immediately

$$\sum_{\alpha,\beta=1}^{n} g_{\alpha\bar{\beta}}(z) \zeta^{\alpha} \bar{\zeta}^{\beta} = \sum_{U_j \ni z} \sum_{\gamma=1}^{n} \rho_j(z) |\zeta_j^{\gamma}|^2.$$

Therefore $\sum_{\alpha,\beta=1}^{n} g_{\alpha\bar{\beta}}(z)\, dz^{\alpha} \otimes d\bar{z}^{\beta}$ is positive definite. ∎

Let M be an n-dimensional complex manifold endowed with a Hermitian metric $\sum_{\alpha,\beta=1}^{n} g_{\alpha\bar\beta} dz^\alpha \otimes d\bar z^\beta$. Let

$$\omega = i \sum_{\alpha,\beta=1}^{n} g_{\alpha\bar\beta}\, dz^\alpha \wedge d\bar z^\beta$$

be its associated real (1, 1)-form. Put $g(z) = \det(g_{\alpha\bar\beta}(z))_{\alpha,\beta=1,\dots,n}$. We denote $\overbrace{\omega \wedge \cdots \wedge \omega}^{n\text{ fold}}$ by ω^n. Then by an elementary calculation we obtain

$$\omega^n = (i)^n n!\, g(z)\, dz^1 \wedge d\bar z^1 \wedge \cdots \wedge dz^n \wedge d\bar z^n. \tag{3.110}$$

For local real coordinates $(x^1, \dots, x^{2n})$ with $z^\alpha = x^{2\alpha-1} + ix^{2\alpha}$, we have

$$i\, dz^\alpha \wedge d\bar z^\alpha = 2\, dx^{2\alpha-1} \wedge dx^{2\alpha}, \qquad \alpha = 1, \dots, n.$$

Hence we have

$$\frac{\omega^n}{n!} = 2^n g(z)\, dx^1 \wedge \cdots \wedge dx^{2n}. \tag{3.111}$$

Since $\sum g_{\alpha\bar\beta}(z)\zeta^\alpha \bar\zeta^\beta$ is positive definite, $g(z) > 0$. Therefore using $\omega^n/n!$ as the *volume element*, we define *the integral of a continuous function* $f(z)$ *on* M *by*

$$\int_M f(z)\frac{\omega^n}{n!} = \int_M f(z) 2^n g(z)\, dx^1 \cdots dx^{2n}, \tag{3.112}$$

where $g(z)$ is the component $g_{12\cdots m}(z)$ of the skew-symmetric covariant tensor field of rank $2n$ with $m = 2n$.

We denote by $(g^{\bar\alpha\beta}(z))_{\alpha,\beta=1\dots,n}$ the inverse matrix of $(g_{\alpha\bar\beta}(z))_{\alpha=1,\dots,n}$.

$$\sum_\beta g^{\bar\alpha\beta}(z) g_{\beta\bar\gamma}(z) = \delta^\alpha_\gamma, \qquad \sum_\beta g_{\alpha\bar\beta}(z) g^{\bar\beta\gamma}(z) = \delta^\gamma_\alpha.$$

$g^{\bar\alpha\beta} = g^{\bar\alpha\beta}(z)$ gives a C^∞ contravariant tensor field on M. In fact with respect to the coordinate change $z_k \to z_j$, the transition relations of the component $g_{j\alpha\bar\beta}(z)$ of $g_{\alpha\bar\beta}(z)$ are given by

$$g_{j\alpha\bar\beta}(z) = \sum_{\lambda,\mu=1}^{n} \frac{\partial z_k^\lambda}{\partial z_j^\alpha} \overline{\left(\frac{\partial z_k^\mu}{\partial z_j^\beta}\right)} g_{k\lambda\bar\mu}(z), \tag{3.113}$$

and $(\partial z_j^\alpha / \partial z_k^\lambda)_{\lambda,\alpha=1,\dots,n} = (\partial z_k^\lambda / \partial z_j^\alpha)^{-1}_{\alpha,\beta=1,\dots,n}$. Consequently, putting $(g_j^{\alpha\bar\beta}(z)) = (g_{j\alpha\bar\beta}(z))^{-1}$ and $(g_k^{\bar\lambda\mu}(z)) = (g_{k\lambda\bar\mu}(z))^{-1}$, we obtain

$$g_j^{\bar\beta\alpha}(z) = \sum_{\mu,\lambda=1}^{n} \overline{\left(\frac{\partial z_j^\beta}{\partial z_k^\mu}\right)} \frac{\partial z_j^\alpha}{\partial z_k^\lambda} g_k^{\bar\mu\lambda}(z). \tag{3.114}$$

Let, for example, $\varphi=\frac{1}{2}\sum_{\alpha,\beta,\gamma=1}^{n}\varphi_{\alpha\beta\bar{\gamma}}(z)\,dz^{\alpha}\wedge dz^{\beta}\wedge d\bar{z}^{\gamma}$ be a C^{∞} (2, 1)-form on M. Then with respect to the coordinate change $z_k\to z_j$, the transition relations of $\varphi_{\alpha\beta\bar{\gamma}}$ are given by

$$\varphi_{j\alpha\beta\bar{\gamma}}(z)=\sum_{\lambda,\mu,\nu=1}^{n}\frac{\partial z_k^{\lambda}}{\partial z_j^{\alpha}}\frac{\partial z_k^{\mu}}{\partial z_j^{\beta}}\overline{\left(\frac{\partial z_k^{\nu}}{\partial z_j^{\gamma}}\right)}\varphi_{k\lambda\mu\bar{\nu}}(z).$$

From this and (3.114), it is clear that for C^{∞} (2, 1)-forms φ, ψ,

$$(\varphi,\psi)(z)=\tfrac{1}{2}\sum g^{\bar{\lambda}\alpha}(z)g^{\bar{\mu}\beta}(z)g^{\bar{\gamma}\nu}(z)\varphi_{\alpha\beta\bar{\gamma}}(z)\overline{\psi_{\lambda\mu\bar{\nu}}(z)}$$

is a C^{∞} function of z, independent of the choice of local coordinates. If we put

$$\bar{\psi}^{\alpha\beta\bar{\gamma}}(z)=\sum_{\lambda,\beta,\nu}g^{\bar{\lambda}\alpha}(z)g^{\bar{\mu}\beta}(z)g^{\bar{\gamma}\nu}(z)\overline{\psi_{\lambda\mu\bar{\nu}}(z)},$$

then we have

$$(\varphi,\psi)(z)=\frac{1}{2}\sum_{\alpha,\beta,\gamma=1}^{n}\varphi_{\alpha\beta\bar{\gamma}}(z)\bar{\psi}^{\alpha\beta\bar{\gamma}}(z).$$

Similarly *for* C^{∞} *(p, q)-forms*

$$\varphi=\frac{1}{p!\,q!}\sum\varphi_{\alpha_1\cdots\alpha_p\bar{\beta}_1\cdots\bar{\beta}_q}(z)\,dz^{\alpha_1}\wedge\cdots\wedge dz^{\alpha_p}\wedge d\bar{z}^{\beta_1}\wedge\cdots\wedge d\bar{z}^{\beta_q},$$

$$\psi=\frac{1}{p!\,q!}\sum\psi_{\alpha_1\cdots\alpha_p\bar{\beta}_1\cdots\bar{\beta}_q}(z)\,dz^{\alpha_1}\wedge\cdots\wedge dz^{\alpha_p}\wedge d\bar{z}^{\beta_1}\wedge\cdots\wedge d\bar{z}^{\beta_q},$$

we put

$$(\varphi,\psi)(z)=\frac{1}{p!\,q!}\sum\varphi_{\alpha_1\cdots\alpha_p\bar{\beta}_1\cdots\bar{\beta}_q}(z)\bar{\psi}^{\alpha_1\cdots\alpha_p\bar{\beta}_1\cdots\bar{\beta}_q}(z),\qquad(3.115)$$

where

$$\bar{\psi}^{\alpha_1\cdots\alpha_p\bar{\beta}_1\cdots\bar{\beta}_q}(z)=\sum g^{\bar{\lambda}_1\alpha_1}\cdots g^{\bar{\lambda}_p\alpha_p}g^{\bar{\beta}_1\mu_1}\cdots g^{\bar{\beta}_q\mu_q}\overline{\psi_{\lambda_1\cdots\lambda_p\bar{\mu}_1\cdots\bar{\mu}_q}(z)}.$$

Then $(\varphi,\psi)(z)$ *is a* C^{∞} *function of* z. We define *the inner product of* φ *and* ψ *by*

$$(\varphi,\psi)=\int_M(\varphi,\psi)(z)\frac{\omega^n}{n!}.\qquad(3.116)$$

The inner product satisfies the following properties:

(i) $(\psi, \varphi) = \overline{(\varphi, \psi)}$.
(ii) $(\varphi, \varphi) \geqq 0$. $(\varphi, \varphi) = 0$ if and only if $\varphi = 0$.

In fact for any $z_0 \in M$, choose local coordinates $(z^1, \ldots, z^n)$ such that $g_{\alpha\bar\beta}(z_0) = \delta_{\alpha\beta}$. Then $g^{\bar\beta\alpha}(z_0) = \delta_{\alpha\beta}$, and

$$(\varphi, \psi)(z_0) = \frac{1}{p!\,q!} \sum \varphi_{\alpha_1 \cdots \alpha_p \bar\beta_1 \cdots \bar\beta_q}(z_0) \overline{\psi_{\alpha_1 \cdots \alpha_p \bar\beta_1 \cdots \bar\beta_q}(z_0)},$$

$$(\varphi, \varphi)(z_0) = \frac{1}{p!\,q!} \sum |\varphi_{\alpha_1 \cdots \alpha_p \bar\beta_1 \cdots \bar\beta_q}(z_0)|^2.$$

From these equalities (i) and (ii) follow immediately. Also it is clear that (φ, ψ) *is linear in* φ. We call $\sqrt{(\varphi, \varphi)}$ the *norm* of φ and denote it by $\|\varphi\|$.

$$\|\varphi\| = \sqrt{(\varphi, \varphi)}.$$

For each (p, q)*-form* ψ, *we can find an* $(n-p, n-q)$*-form* $*\bar\psi$ *such that*

$$(\varphi, \psi)(z) \frac{\omega^n}{n!} = \varphi(z) \wedge *\bar\psi(z). \tag{3.117}$$

Proof. For simplicity we denote as follows.

$$A_p = \alpha_1 \cdots \alpha_p, \quad B_q = \beta_1 \cdots \beta_q \quad \text{with} \quad \alpha_1 < \cdots < \alpha_p, \quad \beta_1 < \cdots < \beta_q,$$

$$dz^{A_p} = dz_1^{\alpha_1} \wedge \cdots \wedge dz^{\alpha_p}, \qquad dz^{B_q} = dz_1^{\beta_1} \wedge \cdots \wedge dz_q^{\beta_q}.$$

Moreover for $A_p = \alpha_1 \cdots \alpha_p$, we put

$$A_{n-p} = \alpha_{p+1} \cdots \alpha_n,$$

where $\alpha_{p+1} < \cdots < \alpha_n$, and $\{\alpha_1, \ldots, \alpha_p, \alpha_{p+1}, \ldots, \alpha_n\}$ is a permutation of $\{1, \ldots, n\}$. Similarly we define B_{n-q} for a given $B_q = \beta_1 \cdots \beta_q$. Then with this notation we can write

$$\varphi = \sum_{A_p, B_q} \varphi_{A_p \bar B_q}(z)\, dz^{A_p} \wedge \overline{dz^{B_q}}.$$

Let

$$\operatorname{sgn}\begin{pmatrix} A_p & A_{n-p} \\ B_q & B_{n-q} \end{pmatrix} = \operatorname{sgn}\begin{pmatrix} \alpha_1 & \cdots & \alpha_p \alpha_{p+1} & \cdots & \alpha_n \\ \beta_1 & \cdots & \beta_q \beta_{q+1} & \cdots & \beta_n \end{pmatrix},$$

where sgn denotes the signature of the permutation. Then by (3.110)

$$\frac{\omega^n}{n!} = (i)^n(-1)^{n(n-1)/2} g(z)\, dz^1 \wedge \cdots \wedge dz^n \wedge d\bar{z}^1 \wedge \cdots \wedge d\bar{z}^n.$$

Since

$$\begin{aligned} dz^1 &\wedge \cdots \wedge dz^n \wedge d\bar{z}^1 \wedge \cdots \wedge d\bar{z}^n \\ &= \operatorname{sgn}\begin{pmatrix} \alpha_1 & \cdots & \alpha_n \\ \beta_1 & \cdots & \beta_n \end{pmatrix} dz^{\alpha_1} \wedge \cdots \wedge dz^{\alpha_n} \wedge d\bar{z}^{\beta_1} \wedge \cdots \wedge d\bar{z}^{\beta_n} \\ &= \operatorname{sgn}\begin{pmatrix} A_p & A_{n-p} \\ B_q & B_{n-q} \end{pmatrix} dz^{A_p} \wedge dz^{A_{n-p}} \wedge \overline{dz^{B_q} \wedge dz^{B_{n-q}}}, \end{aligned}$$

putting $k = \frac{1}{2}n(n-1) + (n-1)q$, we obtain

$$\frac{\omega^n}{n!} = (i)^n(-1)^k \operatorname{sgn}\begin{pmatrix} A_p & A_{n-p} \\ B_q & B_{n-q} \end{pmatrix} g(z)\, dz^{A_p} \wedge \overline{dz^{B_q}}\, dz^{A_{n-p}} \wedge \overline{dz^{B_{n-q}}}.$$

On the other hand, by (3.115)

$$(\varphi, \psi)(z) = \sum_{A_p, B_q} \varphi_{A_p\bar{B}_q}(z) \bar{\psi}^{A_p\bar{B}_q}(z).$$

Hence

$$\begin{aligned} (\varphi, \psi)(z)\frac{\omega^n}{n!} &= (i)^n(-1)^k \sum_{A_p, B_q} \varphi_{A_p\bar{B}_q}(z)\, dz^{A_p} \wedge \overline{dz^{B_q}} \\ &\qquad \wedge \operatorname{sgn}\begin{pmatrix} A_p & A_{n-p} \\ B_q & B_{n-q} \end{pmatrix} g(z) \bar{\psi}^{A_p\overline{B}_q}(z)\, dz^{A_{n-p}} \wedge \overline{dz^{B_{n-q}}}. \end{aligned}$$

Put $M_p = \mu_1 \cdots \mu_p$ with $1 \leqq \mu_1 < \mu_2 < \cdots < \mu_p \leqq n$. Then for $M_p \neq A_p$,

$$dz^{M_p} \wedge dz^{A_{n-p}} = dz^{\mu_1} \wedge \cdots \wedge dz^{\mu_p} \wedge dz^{\alpha_{p+1}} \wedge \cdots \wedge dz^{\alpha_n} = 0.$$

For in this case some μ_i coincides with one of $\alpha_{p+1}, \ldots, \alpha_n$. Similarly for $N_q \neq B_q$ with $N_q = \nu_1 \cdots \nu_q$, $1 \leqq \nu_1 < \cdots < \nu_q \leqq n$, $\overline{dz^{N_q} \wedge dz^{B_{n-q}}} = 0$. Consequently from the above result

$$\begin{aligned} (\varphi, \psi)(z)\frac{\omega^n}{n!} &= (i)^n(-1)^k \sum_{M_p, N_q} \varphi_{M_pN_q}(z)\, dz^{M_p} \wedge \overline{dz^{N_q}} \\ &\qquad \wedge \sum_{A_p, B_q} \operatorname{sgn}\begin{pmatrix} A_p & A_{n-p} \\ B_q & B_{n-q} \end{pmatrix} g(z) \bar{\psi}^{A_p\bar{B}_q}(z)\, dz^{A_{n-p}} \wedge \overline{dz^{B_{n-q}}} \\ &= \varphi \wedge (i)^n(-1)^k \sum_{A_p, B_q} \operatorname{sgn}\begin{pmatrix} A_p & A_{n-p} \\ B_q & B_{n-q} \end{pmatrix} g(z) \bar{\psi}^{A_p\overline{B}_q}(z)\, dz^{A_{n-p}} \wedge \overline{dz^{B_{n-1}}}. \end{aligned}$$

Hence defining $*\bar{\psi}$ by

$$*\bar{\psi}=(i)^n(-1)^k \sum_{A_p,B_q} \operatorname{sgn}\begin{pmatrix} A_p & A_{n-p} \\ B_q & B_{n-q}\end{pmatrix} g(z)\bar{\psi}^{A_p\overline{B_q}}(z)\, dz^{A_{n-p}}\wedge \overline{dz^{B_{n-q}}}, \tag{3.118}$$

with $k=\frac{1}{2}n(n-1)+(n-p)q$, we have (3.117). ∎

$*\bar{\psi}$ is independent of the choice of local coordinates used in (3.118), since so is $(\varphi,\psi)(z)\omega^n/n!$. Since ψ is a (p,q)-form,

$$\bar{\psi}=\frac{1}{p!\,q!}\overline{\sum \psi_{\alpha_1\cdots\alpha_p\bar{\beta}_1\cdots\bar{\beta}_q}(z)\, dz^{\alpha_1}\wedge\cdots\wedge dz^{\alpha_p}\wedge d\bar{z}^{\beta_1}\wedge\cdots\wedge d\bar{z}^{\beta_q}}$$

is a (q,p)-form. If we write

$$\bar{\psi}=\frac{1}{p!\,q!}\sum \bar{\psi}_{\beta_1\cdots\beta_q\bar{\alpha}_1\cdots\bar{\alpha}_p}(z)\, dz^{\beta_1}\wedge\cdots\wedge dz^{\beta_q}\wedge d\bar{z}^{\alpha_1}\wedge\cdots\wedge d\bar{z}^{\alpha_p},$$

then

$$\bar{\psi}_{\beta_1\cdots\beta_q\bar{\alpha}_1\cdots\bar{\alpha}_p}(z)=(-1)^{pq}\overline{\psi_{\alpha_1\cdots\alpha_p\bar{\beta}_1\cdots\bar{\beta}_q}}.$$

Hence putting

$$\bar{\psi}^{\bar{\beta}_1\cdots\bar{\beta}_q\alpha_1\cdots\alpha_p}(z)=\sum g^{\bar{\beta}_1\mu_1}\cdots g^{\bar{\beta}_q\mu_q}g^{\bar{\lambda}_1\alpha_1}\cdots g^{\bar{\lambda}_p\alpha_p}\bar{\psi}_{\mu_1\cdots\mu_q\bar{\lambda}_1\cdots\bar{\lambda}_p}(z),$$

we have

$$\bar{\psi}^{\alpha_1\cdots\alpha_p\bar{\beta}_1\cdots\bar{\beta}_q}(z)=(-1)^{pq}\bar{\psi}^{\bar{\beta}_1\cdots\bar{\beta}_q\alpha_1\cdots\alpha_p}(z).$$

Thus (3.118) is rewritten as

$$*\bar{\psi}=(i)^n(-1)^{n(n-1)/2+nq}\sum_{A_p,B_q}\operatorname{sgn}\begin{pmatrix} A_p & A_{n-p} \\ B_q & B_{n-q}\end{pmatrix}$$
$$\times g(z)\bar{\psi}^{\bar{B}_qA_p}(z)\, dz^{A_{n-p}}\wedge\overline{dz^{B_{n-q}}}.$$

Replacing $\bar{\psi}$ by ψ and interchanging A with B, we obtain

$$*\psi=(i)^n(-1)^{n(n-1)/2+np}\sum_{A_p,B_q}\operatorname{sgn}\begin{pmatrix} A_p & A_{n-q} \\ B_q & B_{n-q}\end{pmatrix}$$
$$\times g(z)\psi^{\overline{A_p}B_q}(z)\, dz^{B_{n-q}}\wedge\overline{dz^{A_{n-p}}}, \tag{3.119}$$

where

$$\psi^{\bar{A}_pB_q}(z)=\sum g^{\bar{\alpha}_1\lambda_1}\cdots g^{\bar{\alpha}_p\lambda_p}g^{\bar{\mu}_1\beta_1}\cdots g^{\bar{\mu}_q\beta_q}\psi_{\lambda_1\cdots\lambda_p\bar{\mu}_1\cdots\bar{\mu}_q}(z).$$

We call this $*\psi$ the *dual form* of a (p,q)-form ψ. $*\psi$ is an $(n-q,n-p)$-form.

Clearly the map $\psi \to *\psi$ is linear, namely, we have

$$*(c_1\varphi + c_2\psi) = c_1 * \varphi + c_2 * \psi \quad \text{for } c_1, c_2 \in \mathbb{C}.$$

We have

$$\overline{*\psi} = *\bar{\psi}. \tag{3.120}$$

Proof. It suffices to check this at any point $z_0 \in M$. Take local coordinates $(z^1, \ldots, z^n)$ with centre z_0 such that $g_{\alpha\bar{\beta}}(z_0) = \delta_{\alpha\beta}$. Then $g^{\bar{\alpha}\beta}(z_0) = \delta_{\alpha\beta}$, hence in (3.118) we obtain $\bar{\psi}^{A_p\bar{B}_q}(z_0) = \overline{\psi_{A_p\bar{B}_q}(z_0)}$. Similarly in (3.119) we have $\psi^{\bar{A}_pB_q}(z_0) = \psi_{A_p\bar{B}_q}(z_0)$. Moreover we have

$$\overline{dz^{B_{n-q}} \wedge d\overline{z^{A_{n-p}}}} = (-1)^{(n-q)(n-p)}\, dz^{A_{n-p}} \wedge d\overline{z^{B_{n-q}}}.$$

Substituting these equalities into (3.118) and (3.119), and comparing them, we obtain

$$\overline{*\psi(z_0)} = (-1)^{n+np+(n-q)(n-p)+(n-p)q} * \bar{\psi}(z_0) = *\bar{\psi}(z_0). \quad \blacksquare$$

We denote a (p, q)-form ψ by $\psi^{(p,q)}$. Then

$$**\psi^{(p,q)} = (-1)^{p+q}\psi^{(p,q)}. \tag{3.121}$$

Proof. Again we have only to check this pointwise. Fix $z_0 \in M$ and take local coordinates as in the proof of (3.120). Then since $g(z_0) = 1$, we have by (3.119)

$$(*\psi)_{B_{n-q}\bar{A}_{n-p}}(z_0) = (i)^n(-1)^{n(n-1)/2+np} \operatorname{sgn}\begin{pmatrix} A_p & A_{n-p} \\ B_q & B_{n-q} \end{pmatrix} \psi_{A_p\bar{B}_q}(z_0),$$

hence,

$$\begin{aligned}
(**\psi)_{A_p\bar{B}_q}(z_0) &= (i)^n(-1)^{n(n-1)/2+n(n-q)} \operatorname{sgn}\begin{pmatrix} B_{n-q} & B_q \\ A_{n-p} & A_p \end{pmatrix} (*\psi)_{B_{n-q}\bar{A}_{n-p}}(z_0) \\
&= (-1)^{n+np+n(n-q)} \operatorname{sgn}\begin{pmatrix} B_{n-q} & B_q \\ A_{n-p} & A_p \end{pmatrix} \operatorname{sgn}\begin{pmatrix} A_p & A_{n-p} \\ B_q & B_{n-q} \end{pmatrix} \psi_{A_p\bar{B}_q}(z_0) \\
&= (-1)^{n+np+n(n-q)+(n-q)q+(n-p)p} \psi_{A_p\bar{B}_q}(z_0) \\
&= (-1)^{p+q}\psi_{A_p\bar{B}_q}(z_0).
\end{aligned}$$

Hence $**\psi = (-1)^{p+q}\psi$. $\blacksquare$

By (3.117) *the inner product of φ and ψ is given by*

$$(\varphi, \psi) = \int_M \varphi \wedge *\bar{\psi}. \tag{3.122}$$

(b) Harmonic Differential Forms

Let M be a compact complex manifold of dimension n, and let $\mathscr{L}^{p,q} = \Gamma(M, \mathscr{A}^{p,q})$ be the linear space of C^∞ (p, q)-forms on M. $\partial: \varphi \to \partial\varphi$ is a *linear partial differential operator of order* 1 which maps $\mathscr{L}^{p,q}$ into $\mathscr{L}^{p+1,q}$, and $\bar{\partial}: \varphi \to \bar{\partial}\varphi$ is a linear partial differential operator of order 1 which maps $\mathscr{L}^{p,q}$ into $\mathscr{L}^{p,q+1}$. Define

$$\mathfrak{d} = -*\partial*. \tag{3.123}$$

Since $*: \psi \to *\psi$ maps $\mathscr{L}^{p,q}$ into $\mathscr{L}^{n-q,n-p}$, $\mathfrak{d}$ *maps* $L^{p,q}$, $q \geqq 1$ *into* $\mathscr{L}^{p,q-1}$. For $\psi \in \mathscr{L}^{p,0}$, $\mathfrak{d}\psi = 0$. $\mathfrak{d}$ is also a linear partial differential operator of order 1.

For $\varphi \in \mathscr{L}^{p,q-1}$ and $\psi \in \mathscr{L}^{p,q}$, we have

$$(\bar{\partial}\varphi, \psi) = (\varphi, \mathfrak{d}\psi). \tag{3.124}$$

Proof. Since M is compact, we have by Theorem 3.1,

$$\int_M d(\varphi \wedge *\bar{\psi}) = 0.$$

Since $\varphi \wedge *\bar{\psi}$ is $(n, n-1)$-form, $\partial(\varphi \wedge *\bar{\psi}) = 0$. Hence by (3.23) and by $d = \partial + \bar{\partial}$, we have

$$d(\varphi \wedge *\bar{\psi}) = \bar{\partial}(\varphi \wedge *\bar{\psi}) = \bar{\partial}\varphi \wedge *\bar{\psi} - (-1)^{p+q}\varphi \wedge \bar{\partial}*\bar{\psi}.$$

On the other hand by (3.120) and (3.121)

$$*\overline{\mathfrak{d}\psi} = -*(\overline{*\partial*\psi}) = -**\bar{\partial}*\psi = (-1)^{p+q}\bar{\partial}*\bar{\psi}.$$

Hence

$$d(\varphi \wedge *\bar{\psi}) = \bar{\partial}\varphi \wedge *\bar{\psi} - \varphi \wedge *\overline{\mathfrak{d}\psi}.$$

Hence

$$\int_M \bar{\partial}\varphi \wedge *\bar{\psi} - \int_M \varphi \wedge *\overline{\mathfrak{d}\psi} = 0.$$

Therefore by (3.122), $(\bar{\partial}\varphi, \psi) = (\varphi, \mathfrak{d}\psi)$. ∎ By (3.124), $\mathfrak{d}$ is the *formal adjoint operator of* $\bar{\partial}$. Taking the complex conjugate of (3.124), we obtain

$$(\mathfrak{d}\psi, \varphi) = (\psi, \bar{\partial}\varphi). \tag{3.125}$$

The coefficients $\psi_{\alpha_1\cdots\alpha_p\bar\beta_1\cdots\bar\beta_q}(z)$ of a C^∞ (p, q)-form ψ form a C^∞ covariant tensor field, while

$$\psi^{\bar\alpha_1\cdots\bar\alpha_p\beta_1\cdots\beta_q}(z)=\sum g^{\bar\alpha_1\lambda_1}\cdots g^{\bar\alpha_p\lambda_p}g^{\bar\mu_1\beta_1}\cdots g^{\bar\mu_q\beta_q}\psi_{\lambda_1\cdots\lambda_p\bar\mu_1\cdots\bar\mu_q}(z)$$

define a contravariant tensor field. Accordingly we call $\psi_{\alpha_1\cdots\alpha_p\bar\beta_1\cdots\bar\beta_q}(z)$ the *covariant component* of ψ, and $\psi^{\bar\alpha_1\cdots\bar\alpha_p\beta_1\cdots\beta_q}(z)$ the *contravariant component* of ψ. Since $\psi_{\alpha_1\cdots\alpha_p\bar\beta_1\cdots\bar\beta_q}(z)$ are skew-symmetric in the indices $\alpha_1,\cdots,\alpha_p$ and $\bar\beta_1,\ldots,\bar\beta_q$, so are $\psi^{\bar\alpha_1\cdots\bar\alpha_p\beta_1\cdots\beta_1}(z)$.

Theorem 3.15. *Put $A=\alpha_1\cdots\alpha_p$ with $1\leqq\alpha_1<\cdots<\alpha_p\leqq n$. Then*

$$(\mathfrak{d}\psi)^{\bar A\beta_2\cdots\beta_q}(z)=-\sum_{\beta=1}^{n}\frac{1}{g(z)}\frac{\partial}{\partial z^\beta}(g(z)\psi^{\beta\bar A\beta_2\cdots\beta_q}(z)). \tag{3.126}$$

Proof. Let $(z^1,\ldots,z^n)$ be local coordinates on a coordinate neighbourhood U. It suffices to prove (3.126) for U. Put the right-hand side of (3.126) as $\xi^{\bar A\beta_2\cdots\beta_q}(z)$, and let ξ be the $(p, q-1)$-form on U whose contravariant components are $\xi^{\bar A\beta_2\cdots\beta_q}(z)$. In order to prove that $\mathfrak{d}\psi=\xi$, it suffices to show that $(\mathfrak{d}\psi,\varphi)=(\xi,\varphi)$ for any C^∞ $(p, q-1)$-form φ with supp $\varphi\subset U$. For this, since $(\mathfrak{d}\psi,\varphi)=(\psi,\bar\partial\varphi)$ by (3.125) it suffices to verify

$$(\psi,\bar\partial\varphi)=(\xi,\varphi).$$

Put $\bar\varphi_{\bar A\beta_1\cdots\beta_q}(z)=\overline{\varphi_{A\bar\beta_1\cdots\bar\beta_q}(z)}$. We denote $\partial/\partial z^\beta$ by ∂_β for simplicity. Then we have

$$\overline{(\bar\partial\varphi)_{A\bar\beta_1\cdots\bar\beta_q}(z)}=\sum_{s=1}^{q}(-1)^{p+s-1}\partial_{\beta_s}\bar\varphi_{\bar A\beta_1\cdots\not\beta_s\cdots\beta_q}(z).$$

Hence

$$\begin{aligned}(\psi,\bar\partial\varphi)(z)&=\frac{1}{q!}\sum_{A,\beta_1,\ldots,\beta_q}\psi^{\bar A\beta_1\cdots\beta_q}(z)\overline{(\bar\partial\varphi)_{A\bar\beta_1\cdots\bar\beta_q}(z)}\\&=\frac{1}{q!}\sum\psi^{\bar A\beta_1\cdots\beta_q}(z)\sum_{s=1}^{q}(-1)^{p+s-1}\partial_{\beta_s}\bar\varphi_{\bar A\beta_1\cdots\not\beta_s\cdots\beta_q}(z)\\&=\frac{1}{(q-1)!}\sum\sum_{s=1}^{q}\psi^{\beta_s\bar A\beta_1\cdots\not\beta_s\cdots\beta_q}(z)\partial_{\beta_s}\bar\varphi_{\bar A\beta_1\cdots\not\beta_s\cdots\beta_q}(z)\\&=\frac{1}{(q-1)!}\sum_{A,\beta_2,\ldots,\beta_q}\sum_{\beta=1}^{n}\psi^{\beta\bar A\beta_2\cdots\beta_q}(z)\partial_\beta\bar\varphi_{\bar A\beta_2\cdots\beta_q}(z).\end{aligned}$$

Put $B=\beta_2\cdots\beta_q$, $1\leqq\beta_2<\cdots<\beta_q\leqq n$. Then

$$(\psi,\bar\partial\varphi)(z)=\sum_{A,B}\sum_{\beta=1}^{n}\psi^{\beta\bar AB}(z)\partial_\beta\bar\varphi_{\bar AB}(z).$$

Therefore by (3.112)

$$(\psi, \bar{\partial}\varphi) = \int_M \sum_{A,B} \sum_{\beta=1}^{n} \psi^{\beta\bar{A}B}(z)\partial_\beta\bar{\varphi}_{\bar{A}B}(z)2^n g(z)\, dx^1 \cdots dx^{2n}.$$

Since supp $\varphi \subset U$, and U is a polydisk, by integrating by part, the above integral is equal to

$$-\int_U \sum_{A,B} \sum_{\beta=1}^{n} \partial_\beta(g(z)\psi^{\beta\bar{A}B}(z))\bar{\varphi}_{\bar{A}B}(z)2^n g(z)\, dx^1 \cdots dx^{2n}$$

$$= -\int_U \sum_{A,B} \sum_{\beta=1}^{n} \frac{1}{g(z)}\partial_\beta(g(z)\psi^{\beta\bar{A}B}(x))\overline{\varphi_{A\bar{B}}(z)}\frac{\omega^n}{n!}.$$

Hence

$$(\psi, \bar{\partial}\varphi) = \int_U \sum_{A,B} \xi^{\bar{A}B}(z)\overline{\varphi_{A\bar{B}}(z)}\frac{\omega^n}{n!} = \int_U (\xi, \varphi)(z)\frac{\omega^n}{n!}.$$

Thus $(\psi, \bar{\partial}\varphi) = (\xi, \varphi)$. ∎

Definition 3.16. We define $\Box$ by

$$\Box = \bar{\partial}\mathfrak{d} + \mathfrak{d}\bar{\partial}. \tag{3.127}$$

$\Box$ is called the *complex Laplace-Beltram operator.* $\Box$: $\varphi \to \Box\varphi$ is a linear partial differential operator of rank 2. We denote the covariant components $\varphi_{\lambda_1\cdots\lambda_p\bar{\nu}_1\cdots\bar{\nu}_q}(z)$ of φ by $\varphi_{\Lambda\bar{N}}(z)$, the covariant components $(\Box\varphi)_{\alpha_1\cdots\alpha_p\bar{\beta}_1\cdots\bar{\beta}_q}(z)$ of $\Box\varphi$ by $(\Box\varphi)_{A\bar{B}}(z)$, $\partial/\partial z^\alpha$ by ∂_α and $\partial/\partial\bar{z}^\beta$ by $\bar{\partial}_\beta$. Since by (3.119)

$$\Box = -\bar{\partial}{*}\partial{*} - {*}\partial{*}\bar{\partial},$$

$(\Box\varphi)_{A\bar{B}}(z)$ is a linear combination of $\partial_\alpha\bar{\partial}_\beta\varphi_{\Lambda\bar{N}}(z)$, $\partial_\alpha\varphi_{\Lambda\bar{N}}(z)$, $\bar{\partial}_\beta\varphi_{\Lambda\bar{N}}(z)$ and $\varphi_{\Lambda\bar{N}}(z)$. The *principal part* of $\Box$ is defined to be the sum of all the terms involving $\partial_\alpha\bar{\partial}_\beta$.

Theorem 3.16. *The principal part of* $\Box$ *is given by*

$$-\textstyle\sum_{\alpha,\beta=1}^{n} g^{\bar{\beta}\alpha}(z)(\partial^2/\partial z^\alpha\, \partial\bar{z}^\beta).$$

More precisely we have

$$(\Box\varphi)_{A\bar{B}}(z) = -\sum_{\alpha,\bar{\beta}=1}^{n} g^{\bar{\beta}\alpha}\partial_\alpha\bar{\partial}_\beta\varphi_{A\bar{B}}(z)$$

$$+ \sum_{\Lambda,N}\left(\sum_{\alpha=1}^{n} a_{A\bar{B}}^{\Lambda\bar{N}\alpha}\partial_\alpha + \sum_{\beta=1}^{n} a_{A\bar{B}}^{\Lambda\bar{N}\bar{\beta}}\bar{\partial}_\beta + b_{A\bar{B}}^{\Lambda\bar{N}}\right)\varphi_{\Lambda\bar{N}}(z), \tag{3.128}$$

where $a_{A\bar{B}}^{\Lambda\bar{N}\alpha}$, $a_{A\bar{B}}^{\Lambda\bar{N}\bar{\beta}}$, $b_{A\bar{B}}^{\Lambda\bar{N}}$ *are polynomials of* $g_{\alpha\bar{\beta}}$, $g^{\bar{\beta}\alpha}$ *and their partial derivatives.*

Proof. Since $\square = -\bar{\partial}*\partial* - *\partial*\bar{\partial}$ and $*$ is an operator which makes a linear combination with coefficients in polynomials of $g_{\alpha\bar{\beta}}$ and $g^{\bar{\beta}\alpha}$, in order to determine the principal part of $\square$, it suffices to determine the terms of $(\square\varphi)_{A\bar{B}}(z)$ which contain no partial derivatives of $g_{\alpha\bar{\beta}}$, $g^{\bar{\beta}\alpha}$. For this purpose, we may assume that $g_{\alpha\bar{\beta}} = g_{\alpha\bar{\beta}}(z)$ are constants.

If $g_{\alpha\bar{\beta}}$ are assumed to be constants, $g^{\bar{\beta}\alpha}$ are also constants. Hence by (3.126)

$$(\mathfrak{d}\varphi)^{\bar{A}\beta_2\cdots\beta_q}(z) = -\sum_{\beta=1}^{n} \partial_\beta \varphi^{\beta\bar{A}\beta_2\cdots\beta_q}(z).$$

Therefore

$$\begin{aligned}(\mathfrak{d}\varphi)_{A\bar{\beta}_2\cdots\bar{\beta}_q}(z) &= -\sum_{\beta=1}^{n} \partial_\beta \varphi^{\beta}_{A\bar{\beta}_2\cdots\bar{\beta}_q}(z)\\ &= -\sum_{\gamma}\sum_{\beta} g^{\gamma\beta}\partial_\beta\varphi_{\bar{\gamma}A\bar{\beta}_2\cdots\bar{\beta}_q}(z)\\ &= (-1)^{p+1}\sum_{\beta,\gamma=1}^{n} g^{\bar{\gamma}\beta}\partial_\beta\varphi_{A\gamma\beta_2\cdots\bar{\beta}_q}(z).\end{aligned}$$

Hence

$$(\bar{\partial}\mathfrak{d}\varphi)_{A\bar{\beta}_1\cdots\bar{\beta}_q}(z) = \sum_{\beta,\gamma} g^{\bar{\gamma}\beta}\sum_{s=1}^{q}(-1)^s\bar{\partial}_{\beta_s}\partial_\beta\varphi_{A\gamma\beta_1\cdots\bar{\beta}_s\cdots\bar{\beta}_q}(z).$$

On the other hand

$$\begin{aligned}(\mathfrak{d}\bar{\partial}\varphi)_{A\bar{\beta}_1\cdots\bar{\beta}_q}(z) &= (-1)^{p+1}\sum_{\beta,\gamma} g^{\bar{\gamma}\beta}\partial_\beta(\bar{\partial}\varphi)_{A\bar{\gamma}\bar{\beta}_1\cdots\bar{\beta}_q}(z)\\ &= -\sum_{\beta,\gamma} g^{\gamma\bar{\beta}}(\partial_\beta\bar{\partial}_\gamma\varphi_{A\bar{\beta}_1\cdots\bar{\beta}_q}(z) + \sum_{s=1}^{q}(-1)^s\partial_\beta\bar{\partial}_{\beta_s}\varphi_{A\bar{\gamma}\bar{\beta}_1\cdots\bar{\beta}_s\cdots\bar{\beta}_q}(z)).\end{aligned}$$

Since $\bar{\partial}_{\beta_s}\partial_\beta = \partial_\beta\bar{\partial}_{\beta_s}$, we have

$$(\bar{\partial}\mathfrak{d}\varphi)_{A\bar{B}}(z) + (\mathfrak{d}\bar{\partial}\varphi)_{A\bar{B}}(z) = -\sum_{\beta,\gamma=1}^{n} g^{\bar{\gamma}\beta}\partial_\beta\partial_{\bar{\gamma}}\varphi_{A\bar{B}}(z).$$

Thus $(\square\varphi)_{A\bar{B}}(z) = -\sum_{\alpha,\beta} g^{\bar{\beta}\alpha}\partial_\alpha\bar{\partial}_\beta\varphi_{A\bar{B}}(z)$ if $g_{\alpha\bar{\beta}}$ are constants. Therefore the principal part of $\square$ is $-\sum_{\alpha,\beta=1}^{n} g^{\beta\bar{\alpha}}(z)(\partial^2/\partial z^\alpha\partial\bar{z}^\beta)$. ∎

Corollary. *The partial differential operator* $\Box$ *is strongly elliptic.* (For definition, see Definition 4.2 of the Appendix). ∎

$\Box$ is *formally self-adjoint*, that is,

$$(\Box\varphi, \psi) = (\varphi, \Box\psi) \tag{3.129}$$

holds for any $\varphi, \psi \in \mathcal{L}^{p,q}$. *Proof.* By (3.124) and (3.125)

$$\begin{aligned}(\Box\varphi, \psi) &= (\bar{\partial}\mathfrak{d}\varphi, \psi) + (\mathfrak{d}\bar{\partial}\varphi, \psi) = (\mathfrak{d}\varphi, \mathfrak{d}\psi) + (\bar{\partial}\varphi, \bar{\partial}\psi)\\ &= (\varphi, \bar{\partial}\mathfrak{d}\psi) + (\varphi, \mathfrak{d}\bar{\partial}\psi) = (\varphi, \Box\psi). \quad ∎\end{aligned}$$

Putting $\varphi = \psi$, we obtain

$$(\Box\varphi, \varphi) = \|\bar{\partial}\varphi\|^2 + \|\mathfrak{d}\varphi\|^2. \tag{3.130}$$

Here for $\varphi \in \mathcal{L}^{p,n}$, we have $(\Box\varphi, \varphi) = \|\mathfrak{d}\varphi\|^2$ since $\bar{\partial}\varphi = 0$. Similarly for $\varphi \in \mathcal{L}^{p,0}$, $(\Box\varphi, \varphi) = \|\bar{\partial}\varphi\|^2$.

Definition 3.17. $\varphi \in \mathcal{L}^{p,q}$ is called a *harmonic differential form*, or simply, a *harmonic form* if $\bar{\partial}\varphi = \mathfrak{d}\varphi = 0$.

If φ is harmonic, $\Box\varphi = \bar{\partial}\mathfrak{d}\varphi + \mathfrak{d}\bar{\partial}\varphi = 0$. Conversely, by (3.130) if $\Box\varphi = 0$, φ is harmonic. Thus *φ is harmonic if and only if $\Box\varphi = 0$*. We denote by $\mathbf{H}^{p,q}$ the linear subspace of $\mathcal{L}^{p,q}$ consisting of all harmonic (p, q)-forms.

$$\mathbf{H}^{p,q} = \{\varphi \in L^{p,q} \,|\, \Box\varphi = 0\}.$$

Since M is compact and $\Box$ is strongly elliptic, $\mathbf{H}^{p,q}$ *is finite dimensional.* (See Theorem 7.3 of Appendix.)

We say that *φ and ψ are orthogonal* if $(\varphi, \psi) = 0$. If $\varphi \in \mathbf{H}^{p,q}$, $(\varphi, \Box\psi) = (\Box\varphi, \psi) = 0$ for any $\psi \in \mathcal{L}^{p,q}$, hence, $\mathbf{H}^{p,q}$ and $\Box\mathcal{L}^{p,q}$ are orthogonal. *We denote the direct sum of mutually orthogonal subspaces by* $\oplus$. The following theorem is fundamental.

Theorem 3.17.

$$\mathcal{L}^{p,q} = \mathbf{H}^{p,q} \oplus \Box\mathcal{L}^{p,q}. \quad ∎ \tag{3.131}$$

For proof see Appendix (Corollary to Theorem 7.4 of Appendix).

Corollary.

$$\mathcal{L}^{p,q} = \mathbf{H}^{p,q} \oplus \bar{\partial}\mathcal{L}^{p,q-1} + \mathfrak{d}\mathcal{L}^{p,q+1}, \tag{3.132}$$

where we excise the term $\bar{\partial}\mathcal{L}^{p,q-1}$ for $q = 0$ and $\mathfrak{d}\mathcal{L}^{p,q+1}$ for $q = n$.

Proof. It is clear from (3.124) and (3.125) that $\mathbf{H}^{p,q}$, $\bar{\partial}\mathscr{L}^{p,q-1}$ and $\mathfrak{d}\mathscr{L}^{p,q+1}$ are mutually orthogonal. For any $\varphi \in \mathbf{H}^{p,q}$, by (3.131) there is an $h \in \mathbf{H}^{p,q}$ and $\psi \in \mathscr{L}^{p,q}$ such that

$$\varphi = h + \Box\psi.$$

Since $\Box\psi = \bar{\partial}\mathfrak{d}\psi + \mathfrak{d}\bar{\partial}\psi$, $\mathfrak{d}\psi \in \mathscr{L}^{p,q-1}$ and $\bar{\partial}\psi\Gamma \in \mathscr{L}^{p,q+1}$, we obtain (3.132). ∎

Consider the $\bar{\partial}$-cohomology $H_{\bar{\partial}}^{p,q}(M) = \Gamma(M, \bar{\partial}\mathscr{A}^{p,q-1})/\bar{\partial}\Gamma(M, \mathscr{A}^{p,q-1})$, $q \geqq 1$, of M. Then $\Gamma(M, \mathscr{A}^{p,q-1}) = \mathscr{L}^{0,q-1}$, and $\Gamma(M, \bar{\partial}\mathscr{A}^{p,q-1})$ is the linear subspace of $\mathscr{L}^{p,q}$ of all $\bar{\partial}$-closed $C^\infty(p, q)$-forms. From (3.132) we have

$$\Gamma(M, \bar{\partial}\mathscr{A}^{p,q-1}) = \mathbf{H}^{p,q} \oplus \bar{\partial}\mathscr{L}^{p,q-1}.$$

In fact for any $\varphi \in \mathscr{L}^{p,q}$ represent φ as $\varphi = h + \bar{\partial}\psi + \mathfrak{d}\eta$ by (3.132) where $h \in \mathbf{H}^{p,q}$, $\psi \in \mathscr{L}^{p,q-1}$ and $\eta \in \mathscr{L}^{p,q+1}$. Then if $\mathfrak{d}\eta = 0$, $\bar{\partial}\varphi = 0$ clearly. Conversely, if $\bar{\partial}\varphi = 0$,

$$(\mathfrak{d}\eta, \mathfrak{d}\eta) = (\bar{\partial}\mathfrak{d}\eta, \eta) = (\bar{\partial}\varphi, \eta) = 0.$$

Hence $\mathfrak{d}\eta = 0$. Thus we have the isomorphism

$$H_{\bar{\partial}}^{p,q}(M) \cong \mathbf{H}^{p,q}. \tag{3.133}$$

In other words *a harmonic form $\varphi \in \mathbf{H}^{p,q}$ gives a representative of a $\bar{\partial}$-cohomology class of (p, q)-forms.*

By Dolbeault's theorem (3.100), we have

$$H^q(M, \Omega^p) \cong H_{\bar{\partial}}^{p,q}(M), \qquad q \geqq 1.$$

Hence we obtain the following

Theorem 3.18 (Hodge-Dolbeault).

$$H^q(M, \Omega^p) \cong \mathbf{H}^{p,q}. \quad ∎ \tag{3.134}$$

For a $C^\infty(p, 0)$-form φ, $\bar{\partial}\varphi = 0$ means that φ is holomorphic. Consequently $\mathbf{H}^{p,0} = H^0(M, \Omega^p)$. Thus (3.134) holds also for $q = 0$. Since $\mathbf{H}^{p,q}$ is finite dimensional, we have

Corollary. *$H^q(M, \Omega^p)$ is a finite-dimensional vector space.*

(c) Harmonic Differential Forms with Coefficients in a Holomorphic Vector Bundle

Let M be a compact complex manifold of dimension n, and F a holomorphic vector bundle with the projection π. Choose a finite open covering $\mathfrak{U} = \{U_j\}$

such that $\pi^{-1}(U_j) \cong U_j \times \mathbb{C}^\nu$ for each U_j. Let $(\zeta_j^1, \ldots, \zeta_j^\nu)$ be the fibre coordinates of F over U_j, and $f_{jk}(z) = (f_{jk\mu}^\lambda(z))_{\lambda,\mu=1,\ldots,\nu}$ the transition function of F. Then on $U_j \cap U_k \neq \emptyset$,

$$\zeta_j^\lambda = \sum_{\mu=1}^{\nu} f_{jk\mu}^\lambda(z)\zeta_k^\mu. \tag{3.135}$$

Definition 3.18. *A Hermitian metric on the fibres of F* is defined by specifying on each U_j a positive definite form

$$\sum_{\lambda,\mu=1}^{\nu} a_{j\lambda\bar{\mu}}(z)\zeta_j^\lambda\bar{\zeta}_j^\mu,$$

such that $a_{j\lambda\bar{\mu}}(z)$ is C^∞ and that

$$\sum_{\lambda,\mu=1}^{\nu} a_{j\lambda\bar{\mu}}(z)\zeta_j^\lambda\bar{\zeta}_j^\mu = \sum_{\lambda,\mu=1}^{\nu} a_{k\lambda\bar{\mu}}(z)\zeta_k^\lambda\bar{\zeta}_k^\mu \tag{3.136}$$

for $z \in U_j \cap U_k$.

By (3.135), (3.136) is equivalent to

$$a_{k\lambda\bar{\mu}}(z) = \sum_{\alpha,\beta=1}^{\nu} f_{jk\lambda}^\alpha(z)\overline{f_{jk\mu}^\beta(z)}a_{j\alpha\bar{\beta}}(z). \tag{3.137}$$

Therefore, in view of the transition relations (3.50) of F^*, $a_{j\lambda\bar{\mu}}(z)$ represents a C^∞ section of $F^* \otimes \bar{F}^*$ where $\bar{F}^*$ is the conjugate of the dual bundle F^*.

Let $\mathscr{L}^{p,q}(F)$ be the linear space of all $C^\infty(p, q)$-forms with coefficients in F: $\mathscr{L}^{p,q}(F) = \Gamma(M, \mathscr{A}^{p,q}(F))$. $\varphi \in \mathscr{L}^{p,q}(F)$ is represented by a vector $\varphi = (\varphi_j^1, \ldots, \varphi_j^\nu)$ on U_j where φ_j^λ are $C^\infty(p, q)$-forms satisfying, on $U_j \cap U_k \neq \emptyset$,

$$\varphi_j^\lambda = \sum_{\mu=1}^{\lambda} f_{jk\mu}^\lambda(z)\varphi_k^\mu. \tag{3.138}$$

For $\varphi, \psi \in \mathscr{L}^{p,q}(F)$, $\sum_{\lambda,\mu=1}^{\nu} a_{j\lambda\bar{\mu}}(z)\varphi_j^\lambda \wedge *\bar{\psi}_j^\mu$ is a C^∞ $2n$-form on U_j. Comparing (3.138) with (3.135), we see from (3.136) that, on $U_j \cap U_k \neq \emptyset$,

$$\sum_{\lambda,\mu=1}^{\nu} a_{j\lambda\bar{\mu}}(z)\varphi_j^\lambda \wedge *\bar{\psi}_j^\mu = \sum_{\lambda,\mu=1}^{\nu} a_{k\lambda\bar{\mu}}(z)\varphi_k^\lambda \wedge *\bar{\psi}_k^\mu$$

holds. Consequently there is a $2n$-form on M which coincides with $\sum a_{j\lambda\bar{\mu}}(z)\varphi_j^\lambda \wedge *\bar{\psi}_j^\mu$ on each U_j. We denote it by $\sum_{\lambda,\mu} a_{\lambda\bar{\mu}}(z)\varphi^\lambda \wedge *\bar{\psi}^\mu$. The inner product of φ, $\psi \in \mathscr{L}^{p,q}(F)$ is defined by

$$(\varphi, \psi) = \int_M \sum_{\lambda,\mu=1}^{\nu} a_{\lambda\bar{\mu}}(z)\varphi^\lambda \wedge *\bar{\psi}^\mu. \tag{3.139}$$

It is easy to verify that (i) $(\psi, \varphi) = \overline{(\varphi, \psi)}$ and that (ii) $(\varphi, \varphi) \geqq 0$ and $(\varphi, \varphi) = 0$ implies $\varphi = 0$.

For $\varphi = (\varphi^1, \ldots, \varphi^\nu) \in \mathscr{L}^{p,q-1}(F)$, $\bar{\partial}\varphi = (\bar{\partial}\varphi^1, \ldots, \bar{\partial}\varphi^\nu) \in \mathscr{L}^{p,q}(F)$ by (3.105), that is, $(\bar{\partial}\varphi)^\lambda = \bar{\partial}\varphi^\lambda$. Put $(a^{\bar{\lambda}\mu}(z))_{\lambda,\mu=1,\ldots,\nu} = (a_{\lambda\bar{\mu}}(z))^{-1}_{\lambda,\mu=1,\ldots,\nu}$. For $\psi \in \mathscr{L}^{p,q}(F)$ define $\mathfrak{d}_a\psi \in \mathscr{L}^{p,q-1}(F)$ by

$$(\mathfrak{d}_a\psi)^\lambda = -\sum_{\tau=1}^{\nu} a^{\bar{\tau}\lambda}(z) * \partial\left(\sum_{\mu=1}^{\nu} a_{\mu\bar{\tau}}(z) * \psi^\mu\right). \tag{3.140}$$

$\mathfrak{d}_a$ is the adjoint of $\bar{\partial}$, that is,

$$(\bar{\partial}\varphi, \psi) = (\varphi, \mathfrak{d}_a\psi), \qquad \varphi \in L^{p,q-1}(F), \qquad \psi \in L^{p,q}(F). \tag{3.141}$$

In fact, writing $a_{\lambda\bar{\mu}}$ for $a_{\lambda\bar{\mu}}(z)$, we have

$$\begin{aligned}\sum_{\lambda,\mu} a_{\lambda\bar{\mu}}\varphi^\lambda \wedge *\overline{(\mathfrak{d}_a\psi)^\mu} &= \sum_\lambda \varphi^\lambda \wedge **\bar{\partial}\left(\sum_\mu a_{\lambda\bar{\mu}} * \bar{\psi}^\mu\right)\\ &= (-1)^{p+q} \sum_\lambda \varphi^\lambda \wedge \bar{\partial}\left(\sum_\mu a_{\lambda\bar{\mu}} * \overline{\psi^\mu}\right).\end{aligned}$$

Since $\sum_{\lambda,\mu} a_{\lambda\bar{\mu}}\varphi^\lambda \wedge *\overline{\psi^\mu}$ is an $(n, n-1)$-form on M, we have

$$\begin{aligned}d\left(\sum_{\lambda,\mu} a_{\lambda\bar{\mu}}\varphi^\lambda \wedge *\overline{\psi^\mu}\right) &= \bar{\partial}\left(\sum_{\lambda,\mu} a_{\lambda\bar{\mu}}\varphi^\lambda \wedge *\overline{\psi^\mu}\right)\\ &= \sum_{\lambda,\mu} a_{\lambda\bar{\mu}}\bar{\partial}\varphi^\lambda \wedge *\overline{\psi^\mu} + (-1)^{p+q-1}\sum_\lambda \varphi^\lambda \wedge \bar{\partial}\left(\sum_\mu a_{\lambda\bar{\mu}} * \overline{\psi^\mu}\right).\end{aligned}$$

Hence we have

$$d\left(\sum_{\lambda,\mu} a_{\lambda\bar{\mu}}\varphi^\lambda \wedge *\overline{\psi^\mu}\right) = \sum_{\lambda,\mu} a_{\lambda\bar{\mu}}\bar{\partial}\varphi^\lambda \wedge *\overline{\psi^\mu} - \sum_{\lambda,\mu} a_{\lambda\bar{\mu}} \wedge *\overline{(\mathfrak{d}_a\psi)^\mu}.$$

Integrating over M, we obtain $(\bar{\partial}\varphi, \psi) = (\varphi, \mathfrak{d}_a\psi)$.

$\mathfrak{d}_a$ is an analogy of $\mathfrak{d} = -*\partial*$. As in the case of $\mathfrak{d}$, we have

$$(\mathfrak{d}_a\psi, \varphi) = (\psi, \bar{\partial}\varphi).$$

By (3.140) we have

$$(\mathfrak{d}_a\psi)^\lambda = \mathfrak{d}\psi^\lambda - \sum_{\tau=1}^{\nu} a^{\bar{\tau}\lambda}(z) * \left(\sum_{\mu=1}^{\nu} \partial a_{\mu\bar{\tau}}(z) \wedge *\psi^\mu \right). \tag{3.142}$$

We define

$$\Box_a = \bar{\partial}\mathfrak{d}_a + \mathfrak{d}_a\bar{\partial} \tag{3.143}$$

analogously to $\Box$. It is clear from (3.142) that the principal part of $\Box_a$ coincides with that of $\Box$, hence is equal to $-\sum_{\alpha,\beta=1}^{n} g^{\bar{\beta}\alpha}(z)(\partial^2/\partial a^\alpha \partial \bar{z}^\beta)$. Since

$$(\Box_a\varphi, \psi) = (\varphi, \Box_a\psi), \tag{3.144}$$

$\Box_a$ is a self-adjoint strongly elliptic linear partial differential operator. $\varphi \in \mathscr{L}^{p,q}(F)$ is called *a harmonic form with coefficients in F* if $\bar{\partial}\varphi = \mathfrak{d}_a\varphi = 0$.

Since

$$(\Box_a\varphi, \varphi) = \|\bar{\partial}\varphi\|^2 + \|\mathfrak{d}_a\varphi\|^2, \tag{3.145}$$

$\varphi \in \mathscr{L}^{p,q}(F)$ is harmonic if and only if $\Box_a\varphi = 0$. We define $\mathbf{H}^{p,q}(F)$ by

$$\mathbf{H}^{p,q}(F) = \{\varphi \in \mathscr{L}^{p,q}(F) \mid \Box_a\varphi = 0\}.$$

As with $\mathbf{H}^{p,q}$, $\mathbf{H}^{p,q}(F)$ is finite dimensional. As with $\Box$, we have the following

Theorem 3.19.

$$\mathscr{L}^{p,q}(F) = \mathbf{H}^{p,q}(F) \oplus \Box_a \mathscr{L}^{p,q}(F). \quad \blacksquare \tag{3.146}$$

Corollary.

$$\mathscr{L}^{p,q}(F) = \mathbf{H}^{p,q}(F) \oplus \bar{\partial}\mathscr{L}^{p,q-1}(F) \oplus \mathfrak{d}_a\mathscr{L}^{p,q+1}(F). \quad \blacksquare \tag{3.147}$$

Consider the $\bar{\partial}$-cohomology with coefficients in F:

$$H_{\bar{\partial}}^{p,q}(F) = \Gamma(M, \bar{\partial}\mathscr{A}^{p,q-1}(F)) / \bar{\partial}\Gamma(M, \mathscr{A}^{p,q-1}(F)).$$

$\Gamma(M, \bar{\partial}\mathscr{A}^{p,q-1}(F))$ is the linear space of all $\varphi \in \mathscr{L}^{p,q}(F)$ with $\bar{\partial}\varphi = 0$. Hence by (3.147)

$$\Gamma(M, \bar{\partial}\mathscr{A}^{p,q-1}(F)) = \mathbf{H}^{p,q}(F) \oplus \bar{\partial}\mathscr{L}^{p,q-1}(F).$$

Since $\Gamma(M, \mathscr{A}^{p,q-1}(F)) \cong \mathscr{L}^{p,q-1}(F)$, we have

$$H_{\bar{\partial}}^{p,q}(M, F) \simeq \mathbf{H}^{p,q}(F).$$

Thus by (3.108) we obtain the following theorem.

Theorem 3.20.

$$H^q(M, \Omega^p(F)) \cong \mathbf{H}^{p,q}(F). \quad \blacksquare \tag{3.148}$$

Corollary. *$H^q(M, \Omega^p(F))$ is finite dimensional.* ∎

Let F^* be the dual bundle of F, and $(\eta_{j1}, \ldots, \eta_{j\nu})$ the fibre coordinates of F^* over U_j. Then the transition relations are, by (3.50),

$$\eta_{k,\mu} = \sum_{\lambda=1}^{\nu} f^{\lambda}_{jk\mu}(z)\eta_{j\lambda}, \tag{3.149}$$

where $(f^{\lambda}_{jk\mu}(z))$ is *the transition function of F*. For $\psi = (\psi^1_j, \ldots, \psi^\nu_j) \in \mathscr{L}^{p,q}(F)$, put

$$\psi^*_{j\lambda} = \sum_{\mu=1}^{\nu} a_{j\lambda\bar{\mu}}(z) * \bar{\psi}^{\mu}_j. \tag{3.150}$$

Then since on $U_j \cap U_k \neq \varnothing$, $\psi^{\mu}_j = \sum_{\tau=1}^{\nu} f^{\mu}_{jk\tau}(z)\psi^{\tau}_k$, we obtain from (3.137)

$$\psi^*_{k\mu} = \sum_{\lambda=1}^{\nu} f^{\lambda}_{jk\mu}(z)\psi^*_{j\lambda}.$$

Comparing this with (3.149), we see that $(\psi^*_{j1}, \ldots, \psi^*_{j\nu})$ is a C^∞ $(n-p, n-q)$-form ψ^* with coefficients in F^*: $\psi^* = (\psi^*_{j1}, \ldots, \psi^*_{j\nu}) \in \mathscr{L}^{n-p,n-q}(F^*)$. Solving (3.150) with respect to ψ^{μ}_j, since $**\psi^{\mu}_j = (-1)^{p+q}\psi^{\mu}_j$, we obtain

$$\psi^{\mu}_j = (-1)^{p+q} \sum_{\lambda=1}^{\nu} a^{\bar{\lambda}\mu}_j(z) * \overline{\psi^*_{j\lambda}}.$$

Thus the map $\psi \to \psi^*$ maps $\mathscr{L}^{p,q}(F)$ onto $\mathscr{L}^{n-p,n-q}(F^*)$ bijectively. *This map is* $\mathbb{R}$*-linear.* In terms of the components with respect to arbitrary fibre coordinates $(\eta_1, \ldots, \eta_\nu)$, $(\zeta^1, \ldots, \zeta^\nu)$ of F and F^* with $\sum_\lambda \zeta^\lambda \eta_\lambda = \sum_\lambda \zeta^\lambda_j \eta_{j\lambda}$, we write (3.150) as

$$\psi^*_\lambda = \sum_{\mu=1}^{\nu} a_{\lambda\bar{\mu}}(z) * \bar{\psi}^{\mu}.$$

For $\varphi \in \mathscr{L}^{p,q}(F)$ and $\psi \in \mathscr{L}^{n-p,n-q}(F^*)$, $\sum_{\lambda=1}^{\nu} \psi^*_{j\lambda} \wedge \varphi^\lambda_j$ is a C^∞ $2n$-form on U_j, and on $U_j \cap U_k \neq \varnothing$, we have

$$\sum_\lambda \psi^*_{j\lambda} \wedge \varphi^\lambda_j = \sum_\lambda \psi^*_{k\lambda} \wedge \varphi^\lambda_k,$$

which implies that $\sum_\lambda \psi^*_\lambda \wedge \varphi^\lambda$ is a C^∞ $2n$-form on M.

Theorem 3.21. *Define the inner product of $\psi^* \in \mathbf{H}^{n-p,n-q}(F^*)$ and $\varphi \in \mathbf{H}^{p,q}(F)$ by*

$$\langle \psi^*, \varphi \rangle = \int_M \sum_{\lambda=1}^{\nu} \varphi^\lambda \wedge \psi^*_\lambda.$$

Then with respect to this inner product $\mathbf{H}^{n-p,n-q}(F^*)$ is the dual of $\mathbf{H}^{p,q}(F)$.

Proof. Define the inner product of $\psi^* \in \mathscr{L}^{n-p,n-q}(F^*)$ and $\varphi \in \mathscr{L}^{p,q}(F)$ by

$$\langle \psi^*, \varphi \rangle = \int_M \sum_\lambda \varphi^\lambda \wedge \psi^*_\lambda.$$

Then since $\psi^*_\lambda = \sum_\mu a_{\lambda\mu}(z) * \bar{\psi}^\mu$, we have

$$\langle \psi^*, \varphi \rangle = (\varphi, \psi).$$

Introducing a Hermitian metric $\sum_{\lambda,\mu} a^{\bar{\mu}\lambda}(z)\eta_\lambda \bar{\eta}_\mu$ on the fibres of F^*, we define the inner product of $\psi^*, \varphi^* \in \mathscr{L}^{n-p,n-q}(F^*)$ by

$$(\psi^*, \varphi^*) = \int_M \sum_{\lambda,\mu} a^{\bar{\mu}\lambda}(z)\psi^*_\lambda \wedge * \overline{\varphi^*_\mu}.$$

Since

$$\sum_\mu a^{\bar{\mu}\lambda}(z) * \overline{\varphi^*_\mu} = \sum_\mu a^{\bar{\mu}\lambda}(z) \sum_\tau a_{\tau\bar{\mu}}(z) ** \varphi^\tau = (-1)^{p+q}\varphi^\lambda,$$

we have

$$(\psi^*, \varphi^*) = (-1)^{p+q} \int_M \psi^*_\lambda \wedge \varphi^\lambda = \int_M \sum_\lambda \varphi^\lambda \wedge \psi^*_\lambda,$$

Hence

$$(\psi^*, \varphi^*) = \langle \psi^*, \varphi \rangle = (\varphi, \psi). \tag{3.151}$$

For $\varphi \in \mathscr{L}^{p,q-1}(F)$ and $\psi^* \in \mathscr{L}^{n-p,n-q}(F^*)$, $\sum_\lambda \varphi^\lambda \wedge \psi^*_\lambda$ is a C^∞ $(n, n-1)$-form on M. Hence

$$\int_M \bar{\partial}\left(\sum_\lambda \varphi^\lambda \wedge \psi^*_\lambda\right) = \int_M d\left(\sum_\lambda \varphi^\lambda \wedge \psi^*_\lambda\right) = 0.$$

Consequently integrating the equality

$$\bar{\partial}\left(\sum_\lambda \varphi^\lambda \wedge \psi^*_\lambda\right) = \sum_\lambda \bar{\partial}\varphi^\lambda \wedge \psi^*_\lambda - (-1)^{p+q} \sum_\lambda \varphi^\lambda \wedge \bar{\partial}\psi^*_\lambda$$

over M, we obtain

$$\langle \psi^*, \bar{\partial}\mathfrak{d}\rangle = (-1)^{p+q}\langle \bar{\partial}\psi^*, \varphi\rangle. \tag{3.152}$$

Since $\mathfrak{d}_a$ is the adjoint of $\bar{\partial}$, we have, by (3.151)

$$\langle \psi^*, \bar{\partial}\varphi\rangle = (\bar{\partial}\varphi, \psi) = (\varphi, \mathfrak{d}_a\psi) = ((\mathfrak{d}_a\psi)^*, \varphi^*).$$

On the other hand, $\langle \bar{\partial}\psi^*, \varphi\rangle = (\bar{\partial}\psi^*, \varphi^*)$ holds, hence by (3.152),

$$(\bar{\partial}\psi^*, \varphi^*) = (-1)^{p+q}((\mathfrak{d}_a\psi)^*, \varphi^*).$$

Since $\varphi^* \in \mathscr{L}^{n-p,n-q}(F)$ is arbitrary, we obtain the equality

$$\bar{\partial}\psi^* = (-1)^{p+q}(\mathfrak{d}_a\psi)^*, \qquad \psi \in \mathscr{L}^{p,q}(F). \tag{3.153}$$

Similarly since $\langle \bar{\partial}\psi^*, \varphi^*\rangle = (\bar{\partial}\psi^*, \varphi) = (\psi^*, \mathfrak{d}_a\varphi^*)$, and $\langle \psi^*, \bar{\partial}\varphi\rangle = (\psi^*, (\bar{\partial}\varphi)^*)$, we have $(\psi^*, \mathfrak{d}_a\varphi^*) = (-1)^{p+q}(\psi^*, (\bar{\partial}\varphi)^*)$, hence $\mathfrak{d}_a\varphi^* = (-1)^{p+q}(\bar{\partial}\varphi)^*$. Replacing $\varphi \in \mathscr{L}^{p,q-1}(F)$ by $\psi \in \mathscr{L}^{p,q}(F)$, we obtain

$$\mathfrak{d}_a\psi^* = (-1)^{p+q+1}(\bar{\partial}\psi)^*, \qquad \psi \in \mathscr{L}^{p,q}(F). \tag{3.154}$$

(3.153) and (3.154) imply that *ψ^* is harmonic if and only if ψ is harmonic.* Thus $\psi \to \psi^*$ maps $\mathbf{H}^{p,q}(F)$ bijectively onto $\mathbf{H}^{n-p,n-q}(F^*)$.

Choose an orthonormal basis $\{e_1, \ldots, e_m\}$ of $\mathbf{H}^{p,q}(F)$ where $m = \dim \mathbf{H}^{p,q}(F)$. Then $\{e_1^*, \ldots, e_m^*\}$ is an orthonormal basis of $\mathbf{H}^{n-p,n-q}(F^*)$. For, by (3.151)

$$(e_i^*, e_k^*) = (e_k, e_i) = \delta_{ik},$$

and for any $\psi = \sum_{k=1}^m c_k e_k \in \mathbf{H}^{p,q}(F)$, we have $\psi^* = \sum_{k=1}^m \overline{c_k} e_k^*$. Since by (3.151)

$$\langle e_i^*, e_k\rangle = (e_k, e_i) = \delta_{ik},$$

we have

$$\langle \psi^*, \varphi\rangle = \sum_{k=1}^m v_k u_k$$

for $\varphi = \sum_{k=1}^m u_k e_k \in \mathbf{H}^{p,q}(F)$, $u_k \in \mathbb{C}$, and $\psi^* = \sum_{k=1}^m v_k e_k^*$, $v_k \in \mathbb{C}$. Thus $\mathbf{H}^{n-p,n-q}(F^*)$ is the dual space of $\mathbf{H}^{p,q}(F)$. ∎

Theorem 3.22. *Let M be a compact complex manifold M of dimension n, and F a holomorphic vector bundle over M. Then the equality*

$$\dim H^q(M, \Omega^p(F)) = \dim H^{n-q}(M, \Omega^{n-p}(F^*)) \tag{3.155}$$

holds where F^* is the dual bundle of F.

Proof. Since by Theorem 3.21, $\dim \mathbf{H}^{p,q}(F) = \dim H^{n-p,n-q}(F^*)$, (3.155) follows from (3.148) immediately. ∎

Let Θ be the sheaf of germs of holomorphic vector fields: $\Theta = \mathcal{O}(T(M))$, where $T(M)$ is the tangent bundle of M. Since $\mathcal{O}(T(M)) = \Omega^0(T(M))$, by (3.155)

$$\dim H^q(M, \Theta) = \dim H^{n-q}(M, \Omega^n(T^*(M)).$$

Moreover we have $\Omega^n = \mathcal{O}(\bigwedge^n T^*(M))$. Let $\{U_j\}$ be a finite open covering of M consisting of coordinate neighbourhoods U_j, and $(z_j^1, \ldots, z_j^n)$ the local complex coordinates on U_j. Then $dz_j^1 \wedge \cdots \wedge dz_j^n$ is a basis of $\bigwedge^n T^*(M)$ over U_j. $\bigwedge^n T^*(M)$ is a vector bundle of rank 1, and each fibre $\bigwedge^n T_z^*(M)$ at $z \in U_j$ is a 1-dimensional linear space $\mathbb{C}\, dz_j^1 \wedge \cdots \wedge dz_j^n$. Put

$$J_{jk}(z) = \det \frac{\partial(z_k^1, \ldots, z_k^n)}{\partial(z_j^1, \ldots, z_j^n)}$$

Then since

$$dz_k^1 \wedge \cdots \wedge dz_k^n = J_{jk}(z)\, dz_j^1 \wedge \cdots \wedge dz_j^n,$$

the transition relations for the fibre coordinates on $\bigwedge^n T^*(M)$ are given by

$$\zeta_j = J_{jk}(z)\zeta_k. \tag{3.156}$$

$\bigwedge^n T^*(M)$ is called the canonical bundle of M, and is denoted by K: $K = \bigwedge^n T^*(M)$. K plays an important role in the theory of complex manifolds and algebraic geometry. Since

$$\Omega^n(T^*(M)) = \mathcal{O}(K \otimes T^*(M)) = \mathcal{O}(T^*(M) \otimes K) = \Omega^1(K),$$

we obtain the following equality

$$\dim H^q(M, \Theta) = \dim H^{n-q}(M, \Omega^1(K)). \tag{3.157}$$

§3.6. Complex Line Bundles

(a) Complex Line Bundles

By a *complex line bundle* we mean a holomorphic vector bundle of rank 1 over a complex manifold. Let (F, M, π) be a complex line bundle. We choose a finite open covering $\mathfrak{U}=\{U_j\}$ such that $\pi^{-1}(U_j)\cong U_j\times\mathbb{C}$. Then $(z, \zeta_j)\in U_j\times\mathbb{C}$ and $(z, \zeta_k)\in U_k\times\mathbb{C}$ are the same point on F if

$$\zeta_j=f_{jk}(z)\zeta_k,$$

where the transition function $f_{jk}(z)$ is a *non-vanishing holomorphic* function on $U_j\cap U_k$. Moreover we have $f_{jj}(z)=1$, $f_{kj}(z)=f_{jk}(z)^{-1}$ by (3.40) and on $U_i\cap U_j\cap U_k\neq\emptyset$, by (3.39),

$$f_{ik}(z)=f_{ij}(z)f_{jk}(z). \tag{3.158}$$

Let $\mathcal{O}_p^*$ be the set of germs of non-vanishing holomorphic functions at $p\in M$. $\mathcal{O}_p^*$ forms a group with respect to the multiplication. Since $\mathcal{O}_p^*$ is an Abelian group, considering it as a $\mathbb{Z}$-module, we can define a sheaf $\mathcal{O}^*=\bigcup_{p\in M}\mathcal{O}_p^*$ of germs of non-vanishing holomorphic functions over M. $\mathcal{O}^*$ is an open subset of $\mathcal{O}$. Let $g=g(z)$ be a local holomorphic function on M. We denote $f=f(z)=e^{2\pi i g(z)}$ by $e(g)$. Then $e\colon g\to f=e(g)$ defines a homomorphism $g_p\to e(g)_p$ of $\mathcal{O}$ into $\mathcal{O}^*$, which we denote also by e. Since $e^{2\pi i g(z)}=1$ if and only if $g(z)$ is a $\mathbb{Z}$-valued locally constant function, $\ker e=\mathbb{Z}$, hence,

$$0\to\mathbb{Z}\xrightarrow{\iota}\mathcal{O}\xrightarrow{e}\mathcal{O}^*\to 0 \tag{3.159}$$

is exact. By Theorem 3.7 we obtain from this exact sequence the following exact sequence of cohomologies.

$$\cdots\to H^1(M,\mathcal{O})\to H^1(M,\mathcal{O}^*)\xrightarrow{\delta^*}H^2(M,\mathbb{Z})\to\cdots. \tag{3.160}$$

If we regard the transition function $f_{jk}=f_{jk}(z)$ of a complex line bundle $F=(F, M, \pi)$ as a section $f_{jk}\in\Gamma(U_j\cap U_k, \mathcal{O}^*)$ of $\mathcal{O}^*$ over $U_j\cap U_k\neq\emptyset$, $\{f_{jk}\}$ forms a 1-cocycle with respect to $\mathfrak{U}$ by (3.158): $\{f_{jk}\}\in Z^1(\mathfrak{U}, \mathcal{O}^*)$. Suppose that each U_j is a coordinate polydisk. Then as with (3.160), the sequence

$$\cdots\to H^1(U_j,\mathcal{O})\to H^1(U_j,\mathcal{O}^*)\xrightarrow{\delta^*}H^2(U_j,\mathbb{Z})\to\cdots$$

is exact. By Dolbeault's theorem (Theorem 3.12), we have

$$H^1(U_j,\mathcal{O})\cong\Gamma(U_j,\bar\partial\mathscr{A})/\bar\partial\Gamma(U_j,\mathscr{A})$$

where A is the sheaf of germs of C^∞ functions. By Dolbeault's lemma (Theorem 3.3), for any $\bar\partial$-closed $(0,1)$-form $\varphi \in \Gamma(U_j, \bar\partial \mathscr{A})$ on U_j, there is a $\psi \in \Gamma(U_j, \mathscr{A})$ with $\varphi = \bar\partial \psi$, hence $H^1(U_j, \mathcal{O}) = 0$. Since $H^2(U_j, Z) = 0$, we obtain $H^1(U_j, \mathcal{O}^*) = 0$. Therefore, by Theorem 3.5, we have

$$H^1(\mathfrak{U}, \mathcal{O}^*) = H^1(M, \mathcal{O}^*). \tag{3.161}$$

Thus 1-cocycles $\{f_{jk}\}$ and $\{g_{jk}\} \in Z^1(\mathfrak{U}, \mathcal{O}^*)$ belong to the same cohomology class if and only if there is a 0-cochain $\{h_j\} \in C^0(\mathfrak{U}, \mathcal{O}^*)$ such that $\{f_{jk} g_{jk}^{-1}\} = \delta\{h_j\}$, which means that $f_{jk} g_{jk}^{-1} = h_k h_j^{-1}$, i.e. $g_{jk} = h_j f_{jk} h_k^{-1}$. Consequently $\{f_{jk}\}$ and $\{g_{jk}\}$ belong to the same cohomology class if and only if f_{jk} and g_{jk} are transition functions of the same line bundle F (see p. 99). Thus *we identify a complex line bundle with the cohomology class of the* 1*-cocycle* $\{f_{jk}\}$:

$$F = [\{f_{jk}\}] \in H^1(M, \mathcal{O}^*).$$

Thus $H^1(M, \mathcal{O}^*)$ is *the group of complex line bundle over* M with tensor product as its group multiplication.

Definition 3.19. For a complex line bundle $F \in H^1(M, \mathcal{O}^*)$, $\delta^* F \in H^2(M, \mathbb{Z})$ is called *its Chern class*, and is denoted by $c(F)$: $c(F) = \delta^* F$.

Actually (3.161) holds for any sufficiently fine finite open covering $\mathfrak{U} = \{U_j\}$. In fact, take a finite open covering $\mathfrak{W} = \{W_\lambda\}$ of M such that each W_λ is a polydisk. Then if each U_j is sufficiently small, we may assume that $\mathfrak{U} = \{U_j\}$ is a refinement of $\mathfrak{W}$: $\mathfrak{U} < \mathfrak{W}$. Consequently by (3.71), (3.68) and Theorem 3.4, we have

$$H^1(\mathfrak{W}, \mathcal{O}^*) \subset H^1(\mathfrak{U}, \mathcal{O}^*) \subset H^1(M, \mathcal{O}^*).$$

Since by (3.161), $H^1(\mathfrak{W}, \mathcal{O}^*) = H(M, \mathcal{O}^*)$, we have the desired result.

Let $\mathfrak{U} = \{U_j\} < \mathfrak{W}$ be a finite open covering such that U_j and $U_j \cap U_k$ $(\neq \emptyset)$ are all simply connected, and that $U_i \cap U_j \cap U_k \neq \emptyset$ is connected. Let $F = [\{f_{jk}\}] \in H^1(M, \mathcal{O}^*)$ where $\{f_{jk}\} \in Z^1(\mathfrak{U}, \mathcal{O}^*)$. Then since $f_{jk}(z) = f_{jk}$ is a non-vanishing holomorphic function on a simply connected open set $U_j \cap U_k \neq \emptyset$, any branch of $\log f_{jk}(z)$ is a one-valued holomorphic function on $U_j \cap U_k$. Put

$$g_{jk}(z) = \frac{1}{2\pi\sqrt{-1}} \log f_{jk}(z), \qquad g_{kj}(z) = -g_{jk}(z).$$

Then by (3.158)

$$e(g_{jk}(z) - g_{ik}(z) + g_{ij}(z)) = f_{jk}(z) f_{ik}(z)^{-1} f_{ij}(z) = 1.$$

Hence

$$c_{ijk} = g_{jk}(z) - g_{ik}(z) + g_{ij}(z)$$

is a constant integer on the doman $U_i \cap U_j \cap U_k \neq \varnothing$. Since $e\{g_{jk}\} = \{f_{jk}\}$, and $\delta\{g_{jk}\} = \{c_{ijk}\} \in Z^2(\mathfrak{U}, \mathbb{Z})$, $\delta^* F$ is represented by the 2-cocycle $\{c_{ijk}\}$ by the definition of δ^* (see p. 130). Thus the Chern class $c(F) = \delta^* F$ of a complex line bundle $F = [\{f_{jk}\}]$ is given by

$$c(F) = [\{c_{ijk}\}], \qquad c_{ijk} = \frac{1}{2\pi\sqrt{-1}}(\log f_{jk} - \log f_{ik} + \log f_{ij}). \quad (3.162)$$

By a *divisor* on a compact complex manifold $M = M^n$ we mean a linear combination $D = \sum_\nu m_\nu S_\nu$ of a finite number of analytic hypersurface $S_1, \ldots, S_\nu, \ldots$, with coefficients $m_\nu \in \mathbb{Z}$. For a sufficiently fine finite open covering $\mathfrak{U} = \{U_j\}$, S_ν is given in U_j, with $U_j \cap S_\nu \neq \varnothing$ by its minimal equation $f_{\nu j}(z) = 0$:

$$S_\nu \cap U_j = \{z \in U_j \,|\, f_{\nu j}(z) = 0\}. \quad (3.163)$$

In case $U_j \cap S_\nu = \varnothing$, if we take as $f_{\nu j}(z)$ an arbitrary non-vanishing holomorphic function, e.g., a constant function 1, then (3.163) is also valid there. $f_{\nu j}(z)/f_{\nu k}(z)$ is a non-vanishing holomorphic function on $U_j \cap U_k \neq \varnothing$. Put

$$f_{jk}(z) = \prod_\nu \left(\frac{f_{\nu j}(z)}{f_{\nu k}(z)}\right)^{m_\nu}, \qquad z \in U_j \cap U_k. \quad (3.164)$$

Then $f_{jk}(z)$ is also non-vanishing on $U_j \cap U_k \neq \varnothing$. Hence $\{f_{jk}\} \in Z^1(\mathfrak{U}, \mathcal{O}^*)$. Thus there is a complex line bundle F with the transition function f_{jk}, which *we call the complex line bundle defined by D and denote it by* $[D]$: $[D] = F = [\{f_{jk}\}]$.

If S is non-singular and connected, S is a submanifold of M^n of codimension 1, i.e. of dimension $(n-1)$. In this case we call $D = S$ a *non-singular prime divisor.*

A divisor $D = \sum m_\nu S_\nu$ on M is a $(2n-2)$-dimensional cycle, hence determines a homology class in $H_{2n-2}(M, \mathbb{Z})$. We shall investigate the relation between the Chern class $c([D])$ of $[D]$ and the homology class of D. Consider a simplicial decomposition of M, and let $p_1, \ldots, p_k, \ldots$ be its vertices, s_{jk}, $j<k$, its 1-simplices, s_{ijk}, $i<j<k$, its 2-simplices, s_{ijkl}, $i<j<k<l$, its 3-cimplices and so on, where we suppose that $\partial s_{jk} = p_k - p_j$, $\partial s_{ijk} = s_{jk} - s_{ik} + s_{ij}$ and so on.

Let U_j be an interior of the union of all simplices with vertex p_j. Then a finite open covering $\mathfrak{U} = \{U_j\}$ has the following property:

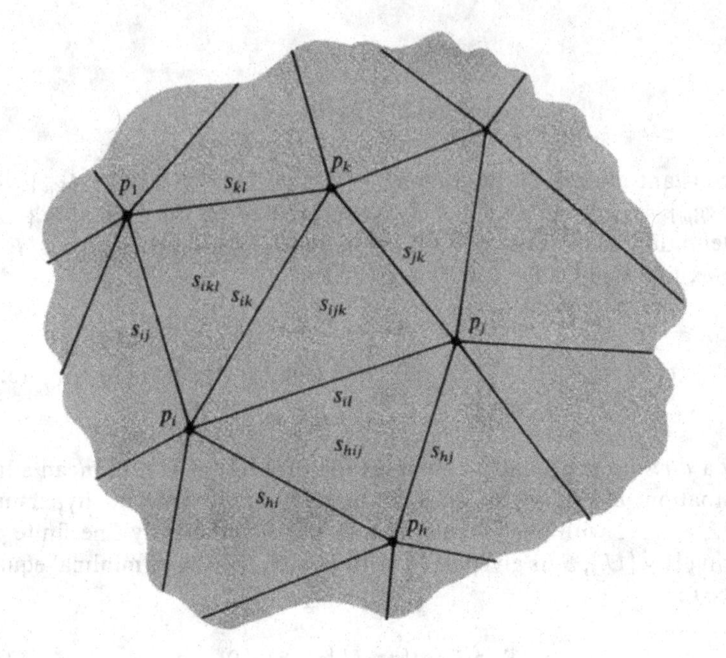

Figure 4

For $i<j<\cdots<k<\cdots$, $U_i \cap U_j \cap \cdots \cap U_k \cap \cdots \neq \varnothing$ if and only if $p_1, \ldots, p_k, \ldots$ are vertices of a simplex $s_{ij\ldots k\ldots}$. U_j, $U_i \cap U_j \neq \varnothing$, $U_i \cap U_j \cap U_k \neq \varnothing, \ldots$, are all simply connected.

Consequently $c(F)$ is given by (3.162) with respect to $\mathfrak{U}$. Taking a sufficiently fine simplicial decomposition, we may assume that $H^1(\mathfrak{U}, \mathcal{O}^*) = H^1(M, \mathcal{O}^*)$. Since $H^1(\mathfrak{U}, \mathbb{Z}) = 0$, we have $H^2(\mathfrak{U}, \mathbb{Z}) \subset H^2(M, \mathbb{Z})$ by Theorem 4.6. (Actually we have isomorphisms $H^q(\mathfrak{U}, \mathbb{Z}) = H^q(M, \mathbb{Z})$ for $q \geqq 0$.)

Let $\gamma = \sum m_{ijk} s_{ijk}$ be a 2-cycle where $m_{ijk} \in \mathbb{Z}$, and the summation is taken over all simplices s_{ijk}. Let $c = [\{c_{ijk}\}]$ be the coholomogy class of a 2-cocycle $\{c_{ijk}\} \in Z^2(\mathfrak{U}, \mathbb{Z})$. Then we define the product of c and γ by

$$\langle c, \gamma \rangle = \sum c_{ijk} m_{ijk}.$$

Since $\sum m_{ijk} \partial s_{ijk} = \partial \gamma = 0$, and $\partial s_{ijk} = s_{jk} - s_{ik} + s_{ij}$, we get

$$\sum m_{ijk} s_{jk} - \sum m_{ijk} s_{ik} + \sum m_{ijk} s_{ij} = 0, \tag{3.165}$$

which implies that $\langle c, \gamma \rangle$ does not depend on the choice of the representative $\{c_{ijk}\}$ of c.

For a divisor D on $M = M^n$ considered as a $(2n-2)$-cycle, we denote the intersection number of D and γ by $I(D, \gamma)$. Then we have

$$I(D, \gamma) = \langle c([D]), \gamma \rangle. \tag{3.166}$$

To see (3.166), we assume for simplicity that $D=S$ is a non-singular prime divisor and let $f_j(z)=0$ be a minimal equation of S on U_j. Since S is a submanifold of codimension 1, for any $q\in S$ we can choose local coordinates $(z_q^1,\ldots,z_q^n)$ centred at q such that

$$f_j(z)=z_q^1.$$

Then for $D=S$, (3.164) reads

$$f_{jk}(z)=\frac{f_j(z)}{f_k(z)}.$$

$[S]$ is a line bundle with transition function $f_{jk}(z)$.

Assume that S meets no 1-simplices s_{jk}, and that it meets 2-simplices s_{ijk} transversally if it does. Let $I(S, s_{ijk})$ be the number of intersection points of S and s_{ijk} counted with signature. Then

$$I(S,\gamma)=\sum m_{ijk}I(S, s_{ijk}).$$

Let p_{ijk} be the centre of mass of s_{ijk} and p_{ij} the middle point of s_{ij}, and decompose s_{ijk} into three cells e_{ijk}, e_{jki}, e_{kij} as illustrated below. If we assume that S does not meet any of three line segments joining p_{ijk} with p_{jk}, p_{ik} and p_{ij} respectively, then

$$I(S, s_{ijk})=I(S, e_{ijk})+I(S, e_{jki})+I(S, e_{kij}).$$

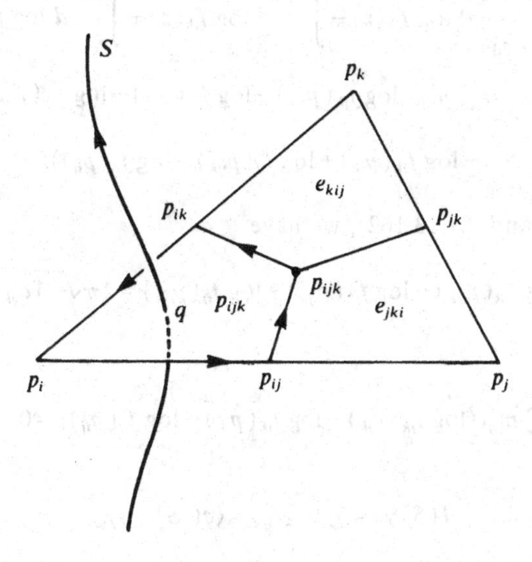

Figure 5

Since $f_j(z)=0$ is a minimal equation of S on U_j, and $e_{ijk}\subset U_i$,

$$I(S, e_{ijk})=\frac{1}{2\pi\sqrt{-1}}\int_{\partial e_{ijk}} d\log f_i(z)$$

by the residue theorem, where ∂e_{ijk} is the boundary of e_{ijk}. Hence

$$2\pi\sqrt{-1}I(S, s_{ijk})=\int_{\partial e_{ijk}} d\log f_i(z)+\int_{\partial e_{jki}} d\log f_j(z)+\int_{\partial e_{kij}} d\log f_k(z),$$

hence by (3.165)

$$\begin{aligned}2\pi\sqrt{-1}I(S, \gamma)=\sum m_{ijk}\Bigg(&\int_{p_{ij}}^{p_{ijk}} d\log f_i(z)-\int_{p_{ik}}^{p_{ijk}} d\log f_i(z)\\ &+\int_{p_{jk}}^{p_{ijk}} d\log f_j(z)-\int_{p_{ij}}^{p_{ijk}} d\log f_j(z)\\ &+\int_{p_{ik}}^{p_{ijk}} d\log f_k(z)-\int_{p_{jk}}^{p_{ijk}} d\log f_k(z)\Bigg).\end{aligned}$$

Since $f_i(z)/f_j(z)=f_{ij}(z)$, and so on, the right-hand side of this equality is equal to

$$\begin{aligned}&\sum m_{ijk}\left(\int_{p_{jk}}^{p_{ijk}} d\log f_{jk}(z)-\int_{p_{ik}}^{p_{ijk}} d\log f_{ik}(z)+\int_{p_{ij}}^{p_{ijk}} d\log f_{ij}(z)\right)\\ &\quad=\sum m_{ijk}(\log f_{jk}(p_{ijk})-\log f_{jk}(p_{jk})-\log f_{ik}(p_{ijk})\\ &\qquad+\log f_{ik}(p_{ik})+\log f_{ij}(p_{ijk})-\log f_{ij}(p_{ij})).\end{aligned}$$

On the other hand, by (3.162), we have

$$\log f_{jk}(p_{ijk})-\log f_{ik}(p_{ijk})+\log f_{ij}(p_{ijk})=2\pi\sqrt{-1}c_{ijk},$$

and by (3.165)

$$\sum m_{ijk}(\log f_{jk}(p_{jk})-\log f_{ik}(p_{ik})+\log f_{ij}(p_{ij}))=0.$$

Hence

$$I(S, \gamma)=\sum m_{ijk}c_{ijk}=\langle c([S]), \gamma\rangle.$$

(3.166) can also be proved similarly for general D.

(b) Submanifolds of Codimension 1

Let S be a submanifold of codimension 1 of a compact complex manifold $M=M^n$ where we assume $n \geqq 2$. We denote by $\mathcal{O}(-S)$ the sheaf over M of germs of holomorphic functions vanishing on S. $\mathcal{O}(-S)$ is a subsheaf of $\mathcal{O}$. We can identify $\mathcal{O}/(\mathcal{O}(-S))$ with $\mathcal{O}_S$ where $\mathcal{O}_S$ is the sheaf of germs of holomorphic functions on S. In fact, let $(z_p^1, \ldots, z_p^n)$ be local coordinates with centre p such that $z_p^1=0$ on S. Then we may regard $(z_p^2, \ldots, z_p^n)$ as local coordinates on S. In

$$(\mathcal{O}/\mathcal{O}(-S))_p = \mathcal{O}_p/\mathcal{O}(-S)_p,$$

$\mathcal{O}_p$ is the power series ring: $\mathcal{O}_p = \mathbb{C}\{z_p^1, \ldots, z_p^n\}$. If we write $\varphi(z_p) \in \mathcal{O}_p$ in the form

$$\varphi(z_p) = \varphi_0(z_p^2, \ldots, z_p^n) + \sum_{m=1}^{\infty} (z_p^1)^m \varphi_m(z_p^2, \ldots, z_p^n),$$

we have $\varphi(z_p) = \varphi_0(z_p^2, \ldots, z_p^n)$ on S. Therefore $\mathcal{O}(-S)_p$ is the ideal of $\mathcal{O}_p$ consisting of all power series of the form $\sum_{m=1}^{\infty} (z_p^1)^m \varphi_m(z_p^2, \ldots, z_p^n)$. Consequently $\mathcal{O}_p/\mathcal{O}(-S)_p$ is isomorphic to $\mathbb{C}\{z_p^2, \ldots, z_p^n\}$. Thus

$$(\mathcal{O}/\mathcal{O}(-S))_p \cong (\mathcal{O}_S)_p \quad \text{for } p \in S.$$

If $p \notin S$, $\mathcal{O}(-S)_p = \mathcal{O}_p$, and $(\mathcal{O}/\mathcal{O}(-S))_p = 0$. Thus we may consider

$$\mathcal{O}/\mathcal{O}(-S) = \mathcal{O}_S. \tag{3.167}$$

The homomorphism

$$r_S \colon \mathcal{O} \to \mathcal{O}_S = \mathcal{O}/\mathcal{O}(-S)$$

given by $r_S \colon \varphi(z_p) \to \varphi_0(z_p^2, \ldots, z_p^n)$ is called the *restriction map* to S, because $\varphi_0(z_p^2, \ldots, z_p^n) = \varphi(0, z_p^2, \ldots, z_p^n)$ is the restriction of $\varphi(z_p) = \varphi(z_p^1, \ldots, z_p^n)$ to S. r_S defines a homomorphism $r_S \colon \Gamma(W, \mathcal{O}) \to \Gamma(S \cap W, \mathcal{O}_S)$ for any open set $W \subset M$. For $\varphi \in \Gamma(W, \mathcal{O})$, $r_S\varphi$ is the local holomorphic function on S obtained by restricting φ to $W \cap S$.

Let $\mathfrak{U} = \{U_j\}$ be a sufficiently fine finite open covering and $f_j(z)=0$ a minimal equation of S on U_j. *Then we have*

$$\mathcal{O}(-S) \cong \mathcal{O}([S]^{-1}). \tag{3.168}$$

Proof. Let φ be a local section of $\mathcal{O}(-S)$ defined over an open set W. In other words $\varphi = \varphi(z)$ is a holomorphic function on W vanishing on S.

Consequently on each $W \cap U_j \neq \emptyset$, $\varphi(z) = f_j(z)\psi_j(z)$ with some local function $\psi_j(z)$ on $W \cap U_j$. Since on $W \cap U_j \cap U_k \neq \emptyset$, $f_j(z)\psi_j(z) = f_k(z)\psi_k(z)$, we have

$$\psi_j(z) = f_{jk}^{-1}(z)\psi_k(z). \tag{3.169}$$

Consequently putting $\psi(z) = (z, \psi_j(z))$ on each $W \cap U_j$, we obtain a holomorphic section $\psi: z \to \psi(z)$ of $[S]^{-1}$ over W. Conversely, if $\psi: z \to \psi(z) = (z, \psi_j(z))$ is a holomorphic section of $[S]^{-1}$ over W, $\psi_j(z)$ is a holomorphic function on $W \cap U_j$ such that $f_j(z)\psi_j(z) = f_k(z)\psi_k(z)$ on $W \cap U_j \cap U_k \neq \emptyset$. Consequently putting $\varphi(z) = f_j(z)\psi_j(z)$ on each $W \cap U_j \neq \emptyset$, we obtain a holomorphic function $\varphi = \varphi(z)$ on W vanishing on S. Thus we obtain an isomorphism $\mathcal{O}(-S) \cong \mathcal{O}([S]^{-1})$. ∎

The sequence

$$0 \to \mathcal{O}(-S) \xrightarrow{\iota} \mathcal{O} \xrightarrow{r_S} \mathcal{O}_S \to 0$$

is exact, where ι is the inclusion. Replacing $\mathcal{O}(-S)$ by $\mathcal{O}([S]^{-1})$ we obtain

$$0 \to \mathcal{O}([S]^{-1}) \to \mathcal{O} \xrightarrow{r_S} \mathcal{O}_S \to 0. \tag{3.170}$$

Let F be a complex line bundle over M, $\mathcal{O}(F)$ the sheaf of germs of holomorphic sections of F, $\mathcal{O}(F-S)$ the sheaf of germs of holomorphic sections of F vanishing on S, and $\mathcal{O}(F)_S$ the sheaf of germs of holomorphic section of F over S. Then similarly as in (3.167) we have

$$\mathcal{O}(F)/\mathcal{O}(F-S) = \mathcal{O}(F)_S.$$

Let F_S be the line bundle over S obtained from F by restricting the base space to S. We call F_S the *restriction* of F to S. Then $\mathcal{O}(F)_S$ is identified with $\mathcal{O}(F_S)$: $\mathcal{O}(F)_S = \mathcal{O}(F_S)$. Therefore we obtain as in (3.168)

$$\mathcal{O}(F-S) \cong \mathcal{O}(F \otimes [S]^{-1}). \tag{3.171}$$

Hence we obtain the exact sequence

$$0 \to \mathcal{O}(F \otimes [S]^{-1}) \to \mathcal{O}(F) \xrightarrow{r_S} \mathcal{O}(F_S) \to 0. \tag{3.172}$$

Putting $F = [S]$, we obtain from (3.172)

$$0 \to \mathcal{O} \to \mathcal{O}(F) \to \mathcal{O}(F_S) \to 0, \ F = [S], \tag{3.173}$$

since $F \otimes F^{-1} = M \otimes \mathbb{C}$. (3.172) holds also for holomorphic vector bundles of rank $\geqq 2$ over M.

Let K and $K(S)$ be the canonical bundles of M and S respectively. *Then we have*

$$K(S) = K_S \otimes [S]_S. \tag{3.174}$$

Proof. Let $\mathfrak{U} = \{U_j\}$ be a sufficiently fine finite open covering such that each U_j is a coordinate neighbourhood. Let $(z_j^1, \ldots, z_j^n)$ be the local coordinates on U_j and $f_j(z) = 0$ a minimal equation of S on U_j. Put $f_{jk}(z) = f_j(z)/f_k(z)$. Further assume that $f_j(z) = z_j^1$ for $U_j \cap S \neq \varnothing$. Then $(z_j^2, \ldots, z_j^n)$ may be regarded as local coordinates of S on $S \cap U_j$. The transition function of K is given by

$$J_{jk}(z) = \det \frac{\partial(z_k^1, z_k^2, \ldots, z_k^n)}{\partial(z_j^1, z_j^2, \ldots, z_j^n)}.$$

Put

$$J(S)_{jk}(z) = \det \frac{\partial(z_k^2, \ldots, z_k^n)}{\partial(z_j^2, \ldots, z_j^n)} \qquad \text{on} \quad S \cap U_j \cap U_k \neq \varnothing.$$

Then $J(S)_{jk}(z)$ is the transition function of $K(S)$. Thus it suffices to prove

$$J(S)_{jk}(z) = J_{jk}(z) f_{jk}(z) \qquad \text{on} \quad S \cap U_j \cap U_k \neq \varnothing.$$

For j, k with $S \cap U_j \cap U_k \neq \varnothing$,

$$z_k^1 = f_{kj}(z) z_j^1 \quad \text{on} \quad U_j \cap U_k.$$

Hence

$$dz_k^1 = f_{kj}(z)\, dz_j^1 + z_j^1\, df_{kj}(z).$$

On the other hand, on $U_j \cap U_k$, we have

$$dz_j^1 \wedge dz_k^2 \wedge \cdots \wedge dz_k^n = \det \frac{\partial(z_k^2, \ldots, z_k^n)}{\partial(z_j^2, \ldots, z_j^n)}\, dz_j^1 \wedge dz_j^2 \wedge \cdots \wedge dz_j^n.$$

Since $z_j^1 = 0$ on S, we have, on $S \cap U_j \cap U_k$

$$\begin{aligned} dz_k^1 \wedge dz_k^2 \wedge \cdots \wedge dz_k^n &= f_{kj}(z)\, dz_j^1 \wedge dz_k^2 \wedge \cdots \wedge dz_k^n \\ &= f_{kj}(z) J(S)_{jk}(z)\, dz_j^1 \wedge dz_j^2 \wedge \cdots \wedge dz_j^n. \end{aligned}$$

Thus $J_{jk}(z) = f_{kj}(z) J(S)_{jk}(z)$, which proves (3.174). ∎

(c) Projective Space

In this section we study complex line bundles on $\mathbb{P}^n$.

Let $\sum_{\alpha,\beta=1}^m g_{\alpha\bar\beta}\, dz^\alpha \otimes d\bar z^\beta$ be a Hermitian metric on a complex manifold M^n, and $\omega = \sqrt{-1}\sum_{\alpha,\beta=1} g_{\alpha\bar\beta}\, dz^\alpha \wedge d\bar z^\beta$ the associated real (1, 1)-form. We call $\sum_{\alpha,\beta=1} g_{\alpha\bar\beta}\, dz^\alpha \otimes d\bar z^\beta$ a *Kähler metric* if $d\omega = 0$. ω is then called its *Kähler form.* Any complex manifold M has a Hermitian metric, but does not always have a Kähler metric. A complex manifold endowed with a Kähler metric is called a *Kähler manifold.*

Let Ω^p be the sheaf of germs of holomorphic p-forms over a complex manifold $M = M^n$. Put

$$h^{p,q} = \dim H^q(M, \Omega^p) \quad \text{and} \quad b_r = \dim H^r(M, \mathbb{C}),$$

where b_r is the rth *Betti number* of M. If M is Kähler, the following equalities hold:

$$\sum_{p+q=r} h^{p,q} = b_r, \qquad h^{q,p} = h^{p,q}. \tag{3.175}$$

It follows from these equalities that b_r *is even for odd* r.

Thus let M^n, $n \geqq 2$, be a Hopf manifold given in Example 2.9. M^n is diffeomorphic to $S^1 \times S^{2n-1}$, hence cannot have a Kähler metric because its first Betti number is 1.

If N is a submanifold of a Kähler manifold M, restricting the Kähler metric of M to N, we obtain a Kähler metric of N. Thus a submanifold of a Kähler manifold is again Kähler.

Let $(\zeta_0, \ldots, \zeta_n)$ be the homogeneous coordinates on $\mathbb{P}^n$. Let $\mathfrak{U} = \{U_j\}$ be a finite open covering of $\mathbb{P}^n$, $(z_j^1, \ldots, z_j^n)$ local complex coordinates on U_j such that on each U_j, $\zeta_{\beta(j)} \neq 0$ for some $\beta(j)$, and that

$$(z_j^1, \ldots, z_j^n) = \left(\frac{\zeta_0}{\zeta_{\beta(j)}}, \ldots, \frac{\zeta_{\beta(j)-1}}{\zeta_{\beta(j)}}, \frac{\zeta_{\beta(j)+1}}{\zeta_{\beta(j)}}, \ldots, \frac{\zeta_n}{\zeta_{\beta(j)}}\right).$$

Put $a(z_j) = 1 + \sum_{\alpha=1}^n |z_j^\alpha|^2$. Then

$$\partial\bar\partial \log a(z_j) = \sum_{\alpha,\beta=1}^n g_{j\alpha\bar\beta}\, dz_j^\alpha \wedge d\bar z_j^\beta,$$

where

$$g_{j\alpha\beta} = \frac{1}{a(z_j)^2}(a(z_j)\delta_{\alpha\beta} - \bar z_j^\alpha z_j^\beta).$$

The Hermitian form $\sum_{\alpha,\beta} g_{j\alpha\bar\beta} w^\alpha \overline{w^\beta}$ is easily seen to be positive definite where $w^1, \ldots, w^n \in \mathbb{C}$. On the other hand, since $a(z_j) = \sum_{\alpha=0}^n |\zeta_\alpha/\zeta_{\beta(j)}|^2$, we obtain

$$a(z_j) = |e_{jk}(z)|^2 a(z_k) \quad \text{on} \quad U_j \cap U_k \neq \varnothing,$$

where $e_{jk}(z) = \zeta_{\beta(k)}/\zeta_{\beta(j)}$. $e_{jk}(z)$ is a non-vanishing holomorphic function. Hence

$$\partial\bar{\partial} \log a(z_j) = \partial\bar{\partial} \log a(z_k).$$

Thus $\sum_{\alpha,\beta} g_{j\alpha\beta}\, dz_j^\alpha \otimes d\bar{z}_j^\beta$ gives a Hermitian metric on $\mathbb{P}^n$. Its associated real (1, 1)-form is given by

$$\begin{aligned}\omega &= \sqrt{-1}\, \partial\bar{\partial} \log a(z_j) \\ &= \sqrt{-1}\, d\bar{\partial} \log a(z_j),\end{aligned}$$

since $d = \partial + \bar{\partial}$. Therefore $d\omega = 0$, and $\sum_{\alpha,\beta} g_{j\alpha\beta}\, dz_j^\alpha \otimes d\bar{z}_j^\beta$ is a Kähler metric on $\mathbb{P}^n$.

Thus $\mathbb{P}^n$ is a Kähler manifold. Consequently projective algebraic manifolds are all Kähler, being submanifolds of the Kähler manifold $\mathbb{P}^n$.

Since the rth Betti number of $\mathbb{P}^n$ is 1 for even r and 0 for odd r, where $r \geqq 2n$, we obtain from (3.175)

$$\dim H^q(\mathbb{P}^n, \Omega^p) = h^{p,q} = \begin{cases} 1, & p = q, \\ 0, & p \neq q. \end{cases} \tag{3.176}$$

In particular for $\mathcal{O} = \Omega^0$, we have

$$\dim H^q(\mathbb{P}^n, \mathcal{O}) = 0, \qquad q \geqq 1. \tag{3.177}$$

Hence by (3.160) we have

$$H^1(\mathbb{P}^n, \mathcal{O}^*) \cong H^2(\mathbb{P}^n, \mathbb{Z}) = \mathbb{Z}. \tag{3.178}$$

Thus the group $H^1(\mathbb{P}^n, \mathcal{O}^*)$ of complex line bundles over $\mathbb{P}^n$ is an infinite cyclic group.

Let P_∞ be the hyperplane $\zeta_0 = 0$, and $E = [P_\infty]$ the line bundle over $\mathbb{P}^n$ defined by the divisor P_∞. Since $\zeta_0/\zeta_{\beta(j)} = 0$ is a minimal equation of P_∞ on U_j, the transition function of E is given by

$$e_{jk}(z) = \frac{\zeta_0}{\zeta_{\beta(j)}} \bigg/ \frac{\zeta_0}{\zeta_{\beta(k)}} = \frac{\zeta_{\beta(k)}}{\zeta_{\beta(j)}}. \tag{3.179}$$

E is a generator of $H^1(\mathbb{P}^n, \mathcal{O}^*)$. *Proof.* Let G be a generator of $H^1(\mathbb{P}^n, \mathcal{O}^*)$, and suppose that $E = G^m$ where m is an integer. Let γ be a line joining a point of P_∞ with $(1, 0, \ldots, 0)$. Then $\gamma \cong P^1$ is a 2-cycle. Since $c(E) = mc(G)$, we have, by (3.166),

$$I(P_\infty, \gamma) = \langle c(E), \gamma \rangle = m\langle c(G), \gamma \rangle.$$

On the other hand, $I(P_\infty, \gamma) = 1$, and $\langle c(G), \gamma \rangle$ is an integer. Consequently $m = \pm 1$, and $E = G$ or $E = G^{-1}$. ∎ Thus any complex line bundle $F \in H^1(\mathbb{P}^n, \mathcal{O}^*)$ is written as $F = E^h$ where h is an integer.

Let $\Psi(\zeta) = \Psi(\zeta_0, \ldots, \zeta_n)$ be a homogeneous polynomial of degree h in $\zeta_0, \ldots, \zeta_n$. Put $\psi_j(z) = \Psi(\zeta)/\zeta^h_{\beta(j)}$ on each U_j. Then $\psi_j(z_j)$ is a holomorphic function of $(z_j^1, \ldots, z_n^n)$ on U_j, and on $U_j \cap U_k \neq \emptyset$,

$$\psi_j(z) = e_{jk}(z)^h \psi_k(z_k).$$

Therefore ψ: $z \to \psi(z) = (z, \psi_j(z))$ gives a holomorphic section of E^h over $\mathbb{P}^n$: $\psi \in \Gamma(\mathbb{P}^n, \mathcal{O}(E))$ where $\psi_j(z)$ is the fibre coordinate of the point $\psi(z)$ of E^h. We denote ψ by $\psi = \{\psi_j(z)\}$.

Conversely any holomorphic section $\psi \in \Gamma(\mathbb{P}^n, \mathcal{O}(E^h))$ with h a positive integer, is represented as $\psi = \{\psi_j(z)\}$ with $\psi_j(z) = \Psi(\zeta)/\zeta^h_{\beta(j)}$ by some homogeneous polynomial $\Psi(\zeta)$ of degree h in $\zeta_0, \ldots, \zeta_n$. *Proof.* Let P_∞ be the hyperplane $\zeta_0 = 0$. Then $\mathbb{P}^n - P_\infty = \mathbb{C}^n$, and $(z_1, \ldots, z_n)$ give complex coordinates on $\mathbb{C}^n$, where $z_\alpha = \zeta_\alpha/\zeta_0$. Since $\psi_j(z)$ is holomorphic on U_j and since

$$\psi_j(z) = \left(\frac{\zeta_{\beta(k)}}{\zeta_{\beta(j)}}\right)^h \psi_k(z) \quad \text{on } U_j \cap U_k \neq \emptyset,$$

we have on $\mathbb{C}^n \cap U_j \cap U_k \neq \emptyset$

$$\left(\frac{\zeta_{\beta(j)}}{\zeta_0}\right)^h \psi_j(z) = \left(\frac{\zeta_{\beta(k)}}{\zeta_0}\right)^h \psi_k(z).$$

Therefore we can define a holomorphic function $\varphi(z_1, \ldots, z_n)$ on $\mathbb{C}^n$ by putting

$$\varphi(z_1, \ldots, z_n) = \left(\frac{\zeta_{\beta(j)}}{\zeta_0}\right)^h \psi_j(z) \tag{3.180}$$

on each $\mathbb{C}^n \cap U_j \neq \emptyset$. Denote the homogeneous part of degree m of the power series expansion of φ by φ_m. We want to see that $\varphi_m(z_1, \ldots, z_n) = 0$ for $m > k$. For this, take any point $p_\infty = (0, \zeta_1, \ldots, \zeta_n)$ of P_∞. p_∞ is contained in some U_j. If $p_\infty \in U_j$, we have, by (3.180),

$$\sum_{m=0}^{\infty} \frac{\zeta_0^h}{\zeta_0^m} \varphi_m(\zeta_1, \ldots, \zeta_n) = \zeta^h_{\beta(j)} \psi_j(z). \tag{3.181}$$

For a sufficiently small $\varepsilon > 0$, the right-hand side of (3.181) is a holomorphic function in ζ_0 in $|\zeta_0| < \varepsilon$. Hence for $m > h$, $\varphi_m(\zeta_1, \ldots, \zeta_n) = 0$. Thus φ is a polynomial of degree at most h in $z_1, \ldots, z_n$. Consequently

$$\Psi(\zeta_0, \ldots, \zeta_n) = \zeta_0^h \varphi\left(\frac{\zeta_1}{\zeta_0}, \ldots, \frac{\zeta_n}{\zeta_0}\right)$$

is a homogeneous polynomial of degree h, and by (3.181) we obtain

$$\psi_j(z) = \Psi(\zeta)/\zeta_{\beta(j)}^h. \quad \blacksquare$$

Clearly $\Gamma(\mathbb{P}^n, \mathcal{O}(E^0)) = \Gamma(\mathbb{P}^n, \mathcal{O}) = \mathbb{C}$. For $h < 0$, $\Gamma(\mathbb{P}^n, \mathcal{O}(E^h)) = 0$ since in this case using (3.181) we obtain $\varphi_m = 0$ for $m \geqq 0$.

Recall that for a homogeneous polynomial $\Psi(\zeta) = \Psi(\zeta_0, \ldots, \zeta_n)$ of degree h in $\zeta_0, \ldots, \zeta_n$, the algebraic subset of $\mathbb{P}^n$ defined by $\Psi(\zeta) = 0$ is called a hypersurface of degree h.

Theorem 3.23. *Let S be a submanifold of codimension 1 of $\mathbb{P}^n$. Then $[S] = E^h$, where h is a positive integer, and S is a hypersurface of degree h.*

Proof. Let $f_j(z) = 0$ be a minimal equation of S on U_j. Then the transition function of $[S]$ is given by $f_{jk} = f_{jk}(z) = f_j(z)/f_k(z)$. Since $H^1(\mathbb{P}^n, \mathcal{O}^*) = \mathbb{Z}$, there is an integer h such that $[S] = E^h$, hence $[\{f_{jk}(z)\}] = [\{e_{jk}^h\}]$ with $e_{jk} = e_{jk}(z)$. Since by (3.161) $H^1(\mathbb{P}^n, \mathcal{O}^*) = H^1(\mathfrak{U}, \mathcal{O}^*)$, there is a 0-cochain $\{u_j\} \in C^0(\mathfrak{U}, \mathcal{O}^*)$ such that $f_{jk} = u_j^{-1} e_{jk}^h u_k$. As $u_j = u_j(z)$ is a non-vanishing holomorphic function on U_j, we have

$$u_j(z) f_j(z) = e_{jk}(z)^h u_k(z) f_k(z)$$

on $U_j \cap U_k \neq \emptyset$, which implies that $\{u_j(z) f_j(z)\} \in \Gamma(\mathbb{P}^n, \mathcal{O}(E^h))$. Since $\Gamma(\mathbb{P}^n, \mathcal{O}(E^h)) = 0$ for $h < 0$, and $\Gamma(\mathbb{P}^n, \mathcal{O}(E^0)) = \mathbb{C}$, h must be a positive integer. Consequently by the above result, $\{u_j(z) f_j(z)\}$ is represented as

$$u_j(z) f_j(z) = \frac{\Psi(\zeta)}{\zeta_{\beta(j)}^h}, \tag{3.182}$$

with a homogeneous polynomial $\Psi(\zeta)$ of degree h. On each U_j, $f_j(z) = 0$ is equivalent to $\Psi(\zeta) = 0$. Thus S is a hypersurface of degree h defined by $\Psi(\zeta) = 0$. $\blacksquare$

Note that this is a special case of Chow's theorem (see p. 40).

Let $P(\zeta) = P(\zeta_0, \ldots, \zeta_n)$ be a homogeneous polynomial of degree $n \geqq 2$ in $\zeta_0, \ldots, \zeta_n$. If at each point $\zeta \in \mathbb{P}^n$, at least one of the partial derivatives $\partial P(\zeta)/\partial \zeta_\alpha$ of $P(\zeta)$ does not vanish, the hypersurface S defined by $P(\zeta) = 0$ is non-singular. Moreover S is connected. For, if not, each component S_i, $i = 1, 2, \ldots,$ of S is a submanifold of codimension 1 of $\mathbb{P}^n$. Therefore by

Theorem 3.23, S_1 is a hypersurface defined by an algebraic equation $\Psi(\zeta) = 0$. Then $\Psi(\zeta)$ does not vanish on S_2. Therefore $\zeta_\alpha^h/\Psi(\zeta)$ with $h = \deg \Psi(\zeta)$ is a holomorphic function on S_2, hence, a constant. Consequently $c_\alpha = \zeta_\alpha/\Psi(\zeta)^{1/h}$ is also a constant, which implies that $(\zeta_0, \ldots, \zeta_n) = (c_0, \ldots, c_n)$ on S_2. This is a contradiction. Thus S is connected, hence, is a submanifold of codimension 1. Let $m = \deg P(\zeta)$, and put $f_j(\zeta) = P(\zeta)/\zeta_{\beta(j)}^m$. Then $f_j(z) = 0$ gives a minimal equation of S on U_j since at least one of the partial derivatives $\partial P(\zeta)/\partial\zeta_\alpha$ does not vanish at each point ζ. Therefore $f_j(z)/f_k(z) = (\zeta_{\beta(k)}/\zeta_{\beta(j)})^m$ gives the transition function of $[S]$. Thus $[S] = E^m$.

The canonical bundle K of $\mathbb{P}^n$ is given by

$$K = E^{-n-1}. \tag{3.183}$$

Proof. Let $U_j = \{\zeta \in \mathbb{P}^n \mid \zeta_j \neq 0\}$. Then $\mathfrak{U} = \{U_j\}$ is an open covering of $\mathbb{P}^n$, and

$$(z_j^0, \ldots, z_j^{j-1}, z_j^{j+1}, \ldots, z_j^n) \quad \text{with} \quad z^\alpha = \zeta_\alpha/\zeta_j$$

give local coordinates on U_j. Define J_{jk} by

$$dz_k^0 \wedge \cdots \wedge dz_k^{k-1} \wedge dz_k^{k+1} \wedge \cdots \wedge dz_k^n = J_{jk}\, dz_j^0 \wedge \cdots \wedge dz_j^{j-1} \wedge dz_j^{j+1} \wedge \cdots \wedge dz_j^n$$

on $U_j \cap U_k$. Then K is defined by $\{J_{jk}\}$. On the other hand, the transition function e_{jk} of E is given by $e_{jk} = \zeta_k/\zeta_j$. Since $z_k^\alpha = z_0^\alpha/z_0^k$, $\alpha = 1, \ldots, n$, and $z_k^0 = 1/z_0^k$, we have

$$\begin{aligned} &dz_k^0 \wedge dz_k^1 \wedge \cdots \wedge dz_k^{k-1} \wedge dz_k^{k+1} \wedge \cdots \wedge dz_k^n \\ &\quad = -\left(\frac{1}{z_0^k}\right)^{n+1} dz_0^k \wedge dz_0^1 \wedge \cdots \wedge dz_0^{k-1} \wedge dz_0^{k+1} \wedge \cdots \wedge dz_0^n \\ &\quad = (-1)^k \left(\frac{\zeta_0}{\zeta_k}\right)^{n+1} dz_0^1 \wedge \cdots \wedge dz_0^{k-1} \wedge dz_0^k \wedge \cdots \wedge dz_0^n. \end{aligned}$$

Thus $J_{0k} = (-1)^k (\zeta_0/\zeta_k)^{n+1}$, hence

$$J_{jk} = \frac{J_{0k}}{J_{0j}} = (-1)^{k-j}\left(\frac{\zeta_j}{\zeta_k}\right)^{n+1} = (-1)^{k-j} e_{jk}^{-n-1}.$$

Consequently $K = E^{-n-1}$. ∎

(d) Surfaces in $\mathbb{P}^3$

Let S be a non-singular surface of degree h in $\mathbb{P}^3$. By the results of the preceding subsection, $[S] = E^h$ where $E = [P_\infty]$.

By (3.183) $K = E^{-4}$ for $\mathbb{P}^3$. Therefore by (3.174) the canonical bundle $K(S)$ of S is given by

$$K(S) = E_S^{h-4}, \tag{3.184}$$

where E_S is the restriction of E to S.

The first Betti number b_1 of S vanishes. In fact we will let $\zeta_0^h, \zeta_0^{h-1}\zeta_1, \ldots, \zeta_0^{h_0}\zeta_1^{h_1}\zeta_2^{h_2}\zeta_e^{h_3}, \ldots, \zeta_2\zeta_3^{h-1}, \zeta_3^h$ be all the monomials of degree h in $\zeta_0, \zeta_1, \zeta_2, \zeta_3$. They are $\binom{h+3}{3}$ in number. Using these monomials we define a map of $\mathbb{P}^3$ into $\mathbb{P}^\mu$ where $\mu = \binom{h+3}{3} - 1$, as follows:

$$\iota: \zeta \to (\omega_0, \omega_1, \ldots, \omega_\mu) = (\zeta_0^h, \zeta_0^{h-1}\zeta_1, \ldots, \zeta_3^h),$$

where $(\omega_0, \omega_1, \ldots, \omega_\mu)$ is the homogeneous coordinates of $\mathbb{P}^\mu$. ι maps $\mathbb{P}^3$ biholomorphically into $\mathbb{P}^\mu$. Thus $\mathbb{P}^3 = \iota(\mathbb{P}^3) \subset \mathbb{P}^\mu$ is a submanifold of $\mathbb{P}^\mu$. Let $\Psi(\zeta) = 0$ be the defining equation of S. $\Psi(\zeta)$ is a linear combination of monomials of degree h:

$$\Psi(\zeta) = a_0\zeta_0^h + a_1\zeta_0^{h-1}\zeta_1 + \cdots + a_\nu\zeta_0^{h_0}\zeta_1^{h_1}\zeta_2^{h_2}\zeta_3^{h_3} + \cdots + a_\mu\zeta_3^h.$$

Therefore if H is a hyperplane of $\mathbb{P}^\mu$ defined by $a_0\omega_0 + \cdots + a_\mu\omega_\mu = 0$, then $S \subset \mathbb{P}^3 \subset \mathbb{P}^\mu$ is the section of $\mathbb{P}^3$ by H: $H \cap \mathbb{P}^3 = S$. Consequently by Lefschetz' theorem (see [23], also [24] for the proof using Morse theory),

$$H^1(S, \mathbb{C}) \cong H^1(\mathbb{P}^3, \mathbb{C}).$$

Since $H^1(\mathbb{P}^3, \mathbb{C}) = 0$, $b_1 = 0$ for S.

Let Ω^1 be the sheaf of germs of holomorphic 1-forms on S. Since S is Kähler, and $b_1 = 0$, we obtain from (3.175)

$$H^0(S, \Omega^1) = 0. \tag{3.185}$$

Let M be a 2-dimensional compact complex manifold which is not necessarily Kähler. Then we can prove that if $H^1(M, \mathbb{C}) = 0$, then $H^0(M, \Omega^1) = 0$ as follows. Let $\varphi \in H^0(M, \Omega^1)$ be a holomorphic 1-form on M. First we will prove that φ is d-closed. Let (z^1, z^2) be local complex coordinates on M and introduce local real coordinates (x^1, x^2, x^3, x^4) by putting $z^1 = x^1 + ix^2$, $z^2 = x^3 + ix^4$. If we put $d\varphi = \psi_{12}\, dz^1 \wedge dz^2$, then we have

$$\begin{aligned}\int_M d\varphi \wedge \overline{d\varphi} &= \int_M |\psi_{12}|\, dz^1 \wedge dz^2 \wedge d\bar{z}^1 \wedge d\bar{z}^2 \\ &= -\int_M |\psi_{12}|^2\, dz^1 \wedge d\bar{z}^1 \wedge dz^2 \wedge d\bar{z}^2 \\ &= 4\int_M |\psi_{12}|^2\, dx^1\, dx^2\, dx^3\, dx^4.\end{aligned}$$

On the other hand,

$$\int_M d\varphi \wedge \overline{d\varphi} = \int_M d\varphi \wedge d\bar{\varphi} = \int_M d(\varphi \wedge d\bar{\varphi}) = 0.$$

Hence $\psi_{12} = 0$. Thus $d\varphi = 0$.

Since by hypothesis $H^1(M, \mathbb{C}) = 0$, by de Rham's theorem $\Gamma(M, d\mathscr{A}^0) = d\Gamma(M, \mathscr{A}^0)$. Therefore for any $\varphi \in \Gamma(M, d\mathscr{A}^0)$, there is an $f \in \Gamma(M, \mathscr{A}^0)$ such that $\varphi = df$. Since $\varphi = df = \partial f + \bar{\partial} f$ is a (0, 1)-form, $\bar{\partial} f = 0$. Therefore f is holomorphic on M, hence a constant. Consequently $\varphi = df = 0$. Thus we obtain $H^0(M, \Omega^1) = 0$.

If $\dim M = 2$, $H^4(M, \mathbb{Z}) \cong \mathbb{Z}$. Therefore the cup product $cc' \in H^4(M, \mathbb{Z}) = \mathbb{Z}$ of any two cohomology classes $c, c' \in H^2(M, \mathbb{Z})$ is considered as an integer. In general, for a compact complex manifold $M = M^n$, the *first Chern class* c_1 of M is defined to be $-c(K)$ where K is the canonical bundle of M: $c_1 = -c(K)$. For $n = 2$, $c_1^2 = c_1 c_1$ is an integer. In this case we call $(1 + c_1^2)$ the linear genus of M.

Let S be a non-singular hypersurface of degree h in $\mathbb{P}^3$. Then

$$c(E_S)^2 = h. \tag{3.186}$$

Proof. Let P be a plane in general position in $\mathbb{P}^3$. Then P is not tangent to S, so the intersection $C = P \cap S$ is a non-singular algebraic curve on S. C, as a divisor on S, defines a line bundle $[C]$ on S. Then we have

$$E_S = [C]. \tag{3.187}$$

In fact, let $l(\zeta) = 0$ be the defining equation of P where $l(\zeta)$ is a linear form in $\zeta_0, \zeta_1, \zeta_2, \zeta_3$. Let $U = \{U_j\}$ be a finite open covering of $\mathbb{P}^3$ such that on each U_j, some $\zeta_{\beta(j)} \neq 0$. Then $l(\zeta)/\zeta_{\beta(j)} = 0$ gives a minimal equation of P on U_j. Therefore on $S \cap U_j \neq \varnothing$, $(l(\zeta)/\zeta_{\beta(j)})_S = 0$ gives a minimal equation of $C = P \cap S$, where $(l(\zeta)/\zeta_{\beta(j)})_S$ denotes the restriction of $l(\zeta)/\zeta_{\beta(j)}$ to S. Consequently the transition function of $[C]$ is given by

$$\frac{(l(\zeta)/\zeta_{\beta(j)})_S}{(l(\zeta)/\zeta_{\beta(j)})_S} = \left(\frac{\zeta_{\beta(k)}}{\zeta_{\beta(j)}}\right)_S.$$

Since $e_{jk} = \zeta_{\beta(k)}/\zeta_{\beta(j)}$ is the transition function of E, we obtain $[C] = E_S$ as desired.

Let P' be another plane in general position in $\mathbb{P}^3$, and put $C' = P' \cap S$. Then C and C' meet transversally on S. Then

$$c([C]) \cdot c([C']) = I(C, C'),$$

where $I(C, C')$ is the intersection number of C and C' (see [15]). Since by (3.187), $E_S = [C] = [C']$,

$$c(E_S)^2 = I(C, C').$$

On the other hand, $\gamma = P \cap P'$ is a line in $\mathbb{P}^3$, and the intersection points of $C = P \cap S$ with $C' = P' \cap S$ coincide with those of S with γ. Since a line γ in general position meets a hypersurface of degree h transversally in h distinct points, $I(C, C') = h$, hence $c(E_S)^2 = h$. ∎

Chapter 4

Infinitesimal Deformation

From now on we proceed to the main theme of this book.

§4.1. Differentiable Family

(a) Introduction

As stated in §2.1(a) in connexion with Definition 2.1 of complex manifolds, a complex manifold $M = M^n$ is obtained by glueing domains $\mathcal{U}_1, \ldots, \mathcal{U}_j, \ldots$ in $\mathbb{C}^n$: $M = \bigcup_j \mathcal{U}_j$. If we write the coordinate transformation by f_{jk} instead of τ_{jk} in (2.2),

$$f_{jk}: z_k \to z_j = (z_j^1, \ldots, z_j^n) = f_{jk}(z_k),$$

f_{jk} is a biholomorphic map of the open set $\mathcal{U}_{kj} \subset \mathcal{U}_k$ onto $\mathcal{U}_{jk} \subset \mathcal{U}_j$, and $z_j \in \mathcal{U}_j$ and $z_k \in \mathcal{U}_k$ are the same point of M if $z_j = f_{jk}(z_k)$. $z_j = f_{jk}(z_k)$ is written in detail as

$$z_j^\alpha = f_{jk}^\alpha(z_k) = f_{jk}^\alpha(z_k^1, \ldots, z_k^n), \qquad \alpha = 1, \ldots, n. \tag{4.1}$$

In the sequel we write U_j instead of $\mathcal{U}_j$ for simplicity: $U_j = \mathcal{U}_j$. By Theorem 2.1, we may assume that $\mathfrak{U} = \{U_j | j = 1, 2, \ldots\}$ is a locally finite open covering of M, and that each U_j is a polydisk:

$$U_j = \{z_j \in \mathbb{C}^n \,|\, |z_j^1| < r_j^1, \ldots, |z_j^n| < r_j^n\}. \tag{4.2}$$

Replacing z_j^α by z_j^α / r_j^α, we may assume further that U_j is of radius 1:

$$U_j = \{z_j \in \mathbb{C}^n \,|\, |z_j^1| < 1, \ldots, |z_j^n| < 1\}. \tag{4.3}$$

If M is compact, $\mathfrak{U} = \{U_j\}$ may be assumed to be a finite open covering. Thus a compact complex manifold M is obtained by glueing a finite number of polydisks $U_1, \ldots, U_j, \ldots, U_l$ by identifying $z_k \in U_k$ and $z_j = f_{jk}(z_k) \in U_j$: $M = \bigcup_{j=1}^l U_j$. *A deformation of M is considered to be the glueing of the same polydisks U_j via different identification.* In other words, replacing

$f_{jk}^{\alpha}(z_k)$ in (4.1) by the functions

$$f_{jk}^{\alpha}(z_k, t) = f_{jk}^{\alpha}(z_k, t_1, \ldots, t_m), \qquad f_{jk}^{\alpha}(z_k, 0) = f_{jk}^{\alpha}(z_k),$$

of z_k and the parameter $t = (t_1, \ldots, t_m)$, we obtain deformations M_t of $M = M_0$ by glueing the polydisks $U_1, \ldots, U_l$ by identifying $z_k \in U_k$ with $z_j = f_{jk}(z_k, t) \in U_j$. This is the fundamental idea of Spencer and the author on the deformation of complex structures [21]. We were, however, at first rather sceptic about this idea. For, the complex structure of $M_t = \bigcup_j U_j$ thus defined via the identification $z_j = f_{jk}(z_k, t)$ may not actually vary with t. For example, let $\zeta_j = (\zeta_j^1, \ldots, \zeta_j^n)$ be local coordinates of M, and choose holomorphic functions $z_j^{\alpha} = z_j^{\alpha}(\zeta_j, t)$, $\alpha = 1, \ldots, n$, of $(n+1)$ variables $(\zeta_j, t) = (\zeta_j^1, \ldots, \zeta_{j,t}^n)$, defined on $\zeta_j \in U_j$, $|t| < 1$, such that the map

$$z_j \colon \zeta_j \to z_j(\zeta_j, t) = (z_j^1(\zeta_j, t), \ldots, z_j^n(\zeta_j, t))$$

gives new local coordinates of M on U_j. Let

$$z_k \to z_j = f_{jk}(z_k, t)$$

be the coordinate transformation. Then since

$$z_j^{\alpha}(\zeta_j, t) = f_{jk}^{\alpha}(z_k(\zeta_k, t), t), \qquad \zeta_j = f_{jk}(\zeta_k),$$

$f_{jk}^{\alpha}(z_k, t)$ depend in general on t. But since $\{z_1, \ldots, z_j, \ldots\}$ is a system of local complex coordinates of the given complex manifold M, $M_t = \bigcup_j U_j$ obtained via the identification $z_j = f_{jk}(z_k, t)$ is just M, hence M_t does not vary with t. Thus *at a glance M_t appears to depend on t, but actually does not.* Therefore since, at first, we had no criterion for the acutual dependence of M_t on t, we were sceptic about the efficiency of our above-mentioned idea.

Later at a colloquium held in the Institute for Advanced Studies at Princeton, A. Frölicher reported his joint work with A. Nijenhuis on the deformation of complex structures, and explained their proof of the rigidity of $\mathbb{P}^n$ ([7]). Their theory depended on the differential geometric method developed by them. This encouraged us to develop our idea on the deformation anew.

(b) Differentiable Family

At first we did not take notice of the fact that the parameter appeared in the definition of a complex manifold is in general a complex one, and that therefore as a family of complex manifolds M_t depending on the parameter $t = (t_1, \ldots, t_m)$, it is more natural to consider a complex analytic family (Definition 2.8 in §2.3). Thus we at first considered the case that $f_{jk}^{\alpha}(z_k, t) =$

$f_{jk}^{\alpha}(z_k, t_1, \ldots, t_m)$ are C^∞ functions of the real parameters $t_1, \ldots, t_m$. For this we introduced the concept of a differentiable family of compact complex manifolds.

The definition of a differentiable family of compact complex manifolds is the C^∞ analogue of the definition of a complex analytic family $\{M_t \mid t \in B\} = (\mathcal{M}, B, \varpi)$ given in §2.3(a). We give the precise definition below.

Let $\mathcal{M}$ be a differentiable manifold, B a domain of $\mathbb{R}^m$, and ϖ a C^∞ map of $\mathcal{M}$ onto B. Suppose that the rank of the Jacobian matrix of ϖ is equal to m at every point of $\mathcal{M}$. Then, as stated in §2.3(a), we can choose a system of local C^∞ coordinates $\{z_1, \ldots, x_j, \ldots\}$, $x_j: p \to x_j(p)$ satisfying the following conditions:

(i) $x_j(p) = (x_j^1(p), \ldots, x_j^\nu(p), t_1, \ldots, t_m)$, $(t_1, \ldots, t_m) = \varpi(p)$,
(ii) $\{\mathcal{U}_j\}$ is a locally finite open covering of $\mathcal{M}$ where $\mathcal{U}_j$ is the domain of x_j.

Therefore for every point $t \in B$, if $\varpi^{-1}(t)$ is connected, $\varpi^{-1}(t)$ is a differentiable manifold whose system of local C^∞ coordinates is given by

$$\{p \to (x_j^1(p), \ldots, x_j^\nu(p) \mid \mathcal{U}_j \cap \varpi^{-1}(t) \neq \emptyset\}.$$

Moreover if $\varpi^{-1}(t)$ is compact, taking for each $t^0 \in B$ an open multi-interval I such that $t^0 \in I \subset [I] \subset B$, we see from Theorem 2.4 that $\varpi^{-1}(I) = \varpi^{-1}(t^0) \times I$.

Definition 4.1. Suppose given a compact complex manifold $M_t = M_t^n$ for each point t of a domain B of $\mathbb{R}^m$. $\{M_t \mid t \in B\}$ is called a *differentiable family of compact complex manifolds* if there are a differentiable manifold $\mathcal{M}$ and a C^∞ map ϖ of $\mathcal{M}$ onto B satisfying the following conditions:

(i) The rank of the Jacobian matrix of ϖ is equal to m at every point of $\mathcal{M}$.
(ii) For each $t \in B$, $\varpi^{-1}(t)$ is a compact connected subset of $\mathcal{M}$.
(iii) $\varpi^{-1}(t) = M_t$.
(iv) There are a locally finite open covering $\{\mathcal{U}_j \mid j = 1, 2, \ldots\}$ of $\mathcal{M}$ and complex-valued C^∞ functions $z_j^1(p), \ldots, z_j^n(p)$, $j = 1, 2, \ldots$, defined on $\mathcal{U}_j$ such that for each t

$$\{p \to (z_j^1(p), \ldots, z_j^n(p) \mid \mathcal{U}_j \cap \varpi^{-1}(t) \neq \emptyset\} \tag{4.4}$$

form a system of local complex coordinates of M_t.

By (i), (ii), each $\varpi^{-1}(t)$ is a compact differentiable manifold. The condition (iii) means that $\varpi^{-1}(t)$ is the underlying differentiable manifold of M_t. We call $t \in B$ the parameter of the differentiable family $\{M_t \mid t \in B\}$ and B its parameter space or base space.

We denote the differentiable family $\{M_t \mid t \in B\}$ by $(\mathcal{M}, B, \varpi)$ or simply by $\mathcal{M}$. If we write $z_j^\alpha(p) = x_j^{2\alpha-1}(p) + ix_j^{2\alpha}(p)$, $\alpha = 1, \ldots, n$, by dividing into the real and imaginary parts, and put

$$x_j(p) = (x_j^1(p), x_j^2(p), \ldots, x_j^{2n}(p), t_1, \ldots, t_m), \qquad (t_1, \ldots, t_m) = \varpi(p),$$

then by the condition (iv), $\{x_j \mid j = 1, 2, \ldots\}$, $x_j\colon p \to x_j(p)$ form a system of local C^∞ coordinates of the differentiable manifold $\mathcal{M}$. x_j maps $\mathcal{U}_j$ into $\mathbb{R}^{2n} \times B$ differentiably. Introducing complex coordinates in $\mathbb{R}^{2n}$ by

$$(z^1, \ldots, z^n) = (x^1 + ix^2, \ldots, x^{2n-1} + ix^{2n}),$$

we consider $\mathbb{R}^{2n}$ as $\mathbb{C}^n$, and x_j as a C^∞ map of $\mathcal{U}_j$ into $\mathbb{C}^n \times B$:

$$x_j\colon p \to x_j(p) = (z_j^1(p), \ldots, z_j^n(p), t_1, \ldots, t_m), \qquad (t_1, \ldots, t_m) = \varpi(p). \quad (4.5)$$

When we consider x_j as a map of $\mathcal{U}_j$ into $\mathbb{C}^n \times B$, we call $\{x_j \mid j = 1, 2, \ldots\}$ a system of local coordinates of the differentiable family $\mathcal{M}$. Since the coordinate transformation $x_k(p) \to x_j(p)$ for $\mathcal{U}_j \cap \mathcal{U}_k \neq \varnothing$ does not affect $t_1, \ldots, t_m$, it is written in the form

$$\begin{aligned}(z_k^1(p), \ldots, z_k^n(p), t) &\to (z_j^1(p), \ldots, z_j^n(p), t)\\ &= (f_{jk}^1(z_k(p, t), t), \ldots, f_{jk}^n(z_k(p, t), t), t), \qquad t = \varpi(p).\end{aligned}$$

Since by (iv), for a fixed t, $p \to (z_j^1(p), \ldots, z_j^n(p))$ gives local complex coordinates of the complex manifold M_t, the coordinate transformation on $\mathcal{U}_j \cap \mathcal{U}_k \cap \varpi^{-1}(t) \neq \varnothing$

$$(z_k^1(p), \ldots, z_k^n(p)) \to (z_j^1(p), \ldots, z_j^n(p))$$

is biholomorphic. Therefore C^∞ functions

$$f_{jk}^\alpha(z_k, t) = f_{jk}^\alpha(z_k^1, \ldots, z_k^n, t_1, \ldots, t_m), \qquad \alpha = 1, \ldots, n,$$

of $z_k^1, \ldots, z_k^n, t_1, \ldots, t_m$ are holomorphic in $z_k^1, \ldots, z_k^n$.

There are infinitely many choices of systems of local coordinates of the given differentiable family $\mathcal{M}$. Given a locally finite open covering $\{\mathcal{W}_\lambda \mid \lambda = 1, 2, \ldots\}$, and a diffeomorphism

$$u_\lambda\colon p \to u_\lambda(p) = (w_\lambda^1(p), \ldots, w_\lambda^n(p), t_1, \ldots, t_m),$$

of $\mathcal{W}_\lambda$ onto an open subset of $\mathbb{C}^n \times B$ for each λ, $\{u_1, \ldots, u_\lambda, \ldots\}$ form a system of local coordinates of the differentiable family $\mathcal{M}$ if and only if for each $t \in B$, $w_\lambda^\alpha(p)$, with $p \in M_t \cap \mathscr{W}_\lambda$ is a local holomorphic function on M_t. Thus if $\{u_\lambda\}$ is a system of local coordinates of the family $\mathscr{M}$, putting

$$w_\lambda^\alpha(p) = g_{\lambda j}^\alpha(z_j^1(p), \ldots, z_j^n(p), t_1, \ldots, t_m), \qquad (t_1, \ldots, t_m) = \varpi(p),$$

on each $\mathscr{W}_\lambda \cap \mathscr{U}_j$, we see that the C^∞ functions

$$g_{\lambda j}^\alpha(z_j^1, \ldots, z_j^n, t_1, \ldots, t_m), \qquad \alpha = 1, 2, \ldots, n, \tag{4.6}$$

of $z_j^1, \ldots, z_j^n, t_1, \ldots, t_m$, are holomorphic in $z_j^1, \ldots, z_j^n$.

Given a differentiable family $(\mathscr{M}, B, \varpi)$, by Theorem 2.1, we may choose its system of local coordinates $\{x_j\}$, $x_j\colon p \to x_j(p)$,

$$x_j(p) = (z_j^1(p), \ldots, z_j^n(p), t_1, \ldots, t_m), \qquad (t_1, \ldots, t_m) = \varpi(p),$$

such that for each j we have

$$x_j(\mathscr{U}_j) = U_j \times I_j, \tag{4.7}$$

where we put $U_j = \{z_j \in \mathbb{C}^n \,|\, |z_j^\alpha| < r_j^\alpha, \alpha = 1, \ldots, n\}$, and $I_j = \{t \in B \,|\, a_{j\nu} < t_\nu < b_{j\nu}, \nu = 1, \ldots, m\}$. Thus identifying $p \in \mathscr{U}_j$ with $x_j(p)$, we consider $\mathscr{M} = \bigcup_j U_j \times I_j$. From this viewpoint, $(z_j, t) \in U_j \times I_j$ and $(z_k, t) \in U_k \times I_k$ are the same point on $\mathscr{M}$ if $z_j = f_{jk}(z_k, t)$. In terms of the local coordinates (z_j, t), ϖ is written as

$$\varpi\colon (z_j^1, \ldots, z_j^n, t) \to t.$$

Consequently $\varpi(U_j \times I_j) = I_j$. Thus $M_t = \varpi^{-1}(t)$ is the union of all $U_j \times t$ with $t \in I_j$. Identifying $U_j \times t$ with U_j, we may consider

$$M_t = \bigcup_{I_j \ni t} U_j.$$

Then $z_j \in U_j$ and $z_k \in U_k$ are the same point on M_t if $z_j = f_{jk}(z_k, t)$. Thus *M_t is a compact manifold obtained by glueing polydisks U_j with $I_j \ni t$ by identifying $z_k \in U_k$ with $z_j = f_{jk}(z_k, t) \in U_j$.*

Since $\{\mathscr{U}_j \,|\, j = 1, 2, \ldots\}$ is locally finite, for any fixed $t^0 \in B$, there are only finitely many $\mathscr{U}_j$ such that $\mathscr{U}_j \cap M_{t^0} \neq \varnothing$. For simplicity we put $t^0 = 0$, and assume that $\mathscr{U}_j \cap M_0 \neq \varnothing$ for $j = 1, \ldots, l$, and that $\mathscr{U}_j \cap M_0 \neq \varnothing$ for $j \geqq l+1$. Then

$$M_0 \subset \bigcup_{j=1}^{l} U_j \times I_j.$$

Putting $I=\bigcap_{j=1}^{l} I_j$, and $M_I=\varpi^{-1}(I)$, we have

$$M_I=\varpi^{-1}(I)=\bigcup_{j=1}^{l} U_j\times I. \tag{4.8}$$

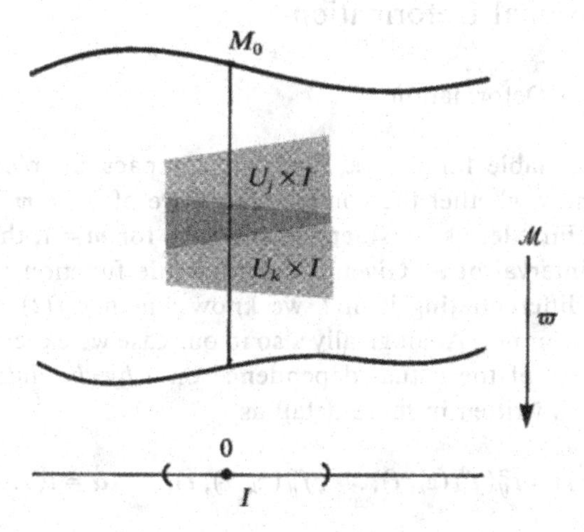

Figure 1

Of course I is an open multi-interval with $0\in I\subset B$. Therefore for each $t\in I$, $M_t=\bigcup_{j=1}^{l} U_j$ *is a complex manifold obtained by glueing* $U_1,\ldots,U_l$ *by identifying* $z_k\in U_k$ *with* $z_j=f_{jk}(z_k,t)\in U_j$.

Thus considering a differentiable family $(\mathcal{M},B,\varpi)$, we can get a clearer image of the deformation of compact complex manifolds. Namely, given a compact complex manifold M, if there is a differentiable family $(\mathcal{M},B,\varpi)$ with $\varpi^{-1}(t^0)=M$ for some $t^0\in B$, each $M_t=\varpi^{-1}(t)$ is called a deformation of M. By Theorem 2.4, *if* M_t *is a deformation of* M, M_t *and* M *are diffeomorphic.* Restricting the domain of the parameter t to a sufficiently small multi-interval I, we see that $M_t=\varpi^{-1}(t)=\bigcup_{j=1}^{l} U_j$ for $t\in I$, where *each* U_j *is a polydisk independent of* t, *and only the coordinate transformations* $z_k\to z_j=f_{jk}(z_k,t)$ *depend on* t. Thus *only the way of glueing* $U_1,\ldots,U_l$ *depends on* t. Each point of M_t belongs to one of U_j. Consequently the complex structure of a sufficiently small neighbourhood of each point of M_t does not vary under deformation. Note that at this point the deformation of complex manifolds is essentially different from the deformation of an elastic body. For, any small portion of an elastic body undergoes distortion under deformation. Let $p\in M_t=\bigcup_{j=1}^{l} U_j$, and suppose $p\in U_i\cap U_j\cap U_k$. Then putting $z_i=z_i(p)$, $z_j=z_j(p)$, and $z_k=z_k(p)$, we have

$$z_i=f_{ij}(z_j,t)=f_{ik}(z_k,t),\qquad z_j=f_{jk}(z_k,t).$$

Hence on $\mathscr{U}_i \cap \mathscr{U}_j \cap \mathscr{U}_k \neq \varnothing$, we obtain the following important equality.

$$f_{ik}(z_k, t) = f_{ij}(f_{jk}(z_k, t), t). \tag{4.9}$$

§4.2. Infinitesimal Deformation

(a) Infinitesimal Deformation

Given a differentiable family $(\mathscr{M}, B, \varpi)$ of compact complex manifolds, how can we know whether the complex structure of $M_t = \varpi^{-1}(t)$ actually depends on t? First let us consider this problem for $m = 1$, that is, in case B is an open interval of $\mathbb{R}$. Given a differentiable function $f(t)$ of a real variable t, by differentiating it in t, we know whether $f(t)$ is a constant independent of t or not. Analogically also in our case we expected to obtain a useful criterion of the actual dependence on t *by the differentiation of* (4.9) *in t*. (4.9) is written in more detail as

$$f_{ik}^{\alpha}(z_k, t) = f_{ij}^{\alpha}(f_{jk}^{1}(z_k, t), \ldots, f_{jk}^{n}(z_k, t), t), \qquad \alpha = 1, \ldots, n.$$

Putting $z_j^{\beta} = f_{jk}^{\beta}(z_k, t)$, we obtain

$$\frac{\partial f_{jk}^{\alpha}(z_k, t)}{\partial t} = \frac{\partial f_{ij}^{\alpha}(z_j, t)}{\partial t} + \sum_{\beta=1}^{n} \frac{\partial f_{ij}^{\alpha}(z_j, t)}{\partial z_j^{\beta}} \frac{\partial f_{jk}^{\beta}(z_k, t)}{\partial t}.$$

Also putting $z_i^{\alpha} = f_{ij}^{\alpha}(z_j, t)$, we have

$$\frac{\partial f_{ik}^{\alpha}(z_k, t)}{\partial t} = \frac{\partial f_{ij}^{\alpha}(z_j, t)}{\partial t} + \sum_{\beta=1}^{n} \frac{\partial z_i^{\alpha}}{\partial z_j^{\beta}} \frac{\partial f_{jk}^{\beta}(z_k, t)}{\partial t}.$$

Using holomorphic vector fields, we can rewrite these equalities in the following intrinsic form:

$$\sum_{\alpha=1}^{n} \frac{\partial f_{ik}^{\alpha}(z_k, t)}{\partial t} \frac{\partial}{\partial z_i^{\alpha}} = \sum_{\alpha=1}^{n} \frac{\partial f_{ij}^{\alpha}(z_j, t)}{\partial t} \frac{\partial}{\partial z_i^{\alpha}} + \sum_{\beta=1}^{n} \frac{\partial f_{jk}^{\beta}(z_k, t)}{\partial t} \frac{\partial}{\partial z_j^{\beta}},$$

where we use the equality

$$\frac{\partial}{\partial z_j^{\beta}} = \sum_{\alpha=1}^{n} \frac{\partial z_i^{\alpha}}{\partial z_j^{\beta}} \frac{\partial}{\partial z_i^{\alpha}}.$$

Introducing the vector fields

$$\theta_{jk}(t)=\sum_{\alpha=1}^{n}\frac{\partial f_{jk}^{\alpha}(z_k,t)}{\partial t}\frac{\partial}{\partial z_j^{\alpha}},\qquad z_k=f_{kj}(z_j,t),\tag{4.10}$$

we write the above equality as

$$\theta_{ik}(t)=\theta_{ij}(t)+\theta_{jk}(t).\tag{4.11}$$

Thus on each open set $U_j\cap U_k\neq\emptyset$ of $M_t=\bigcup_j U_j$, a holomorphic vector field $\theta_{jk}(t)$ is defined, and the equality (4.11) holds on $U_i\cap U_j\cap U_k\neq\emptyset$. (4.11) is also written as

$$\theta_{jk}(t)-\theta_{ik}(t)+\theta_{ij}(t)=0.\tag{4.12}$$

Putting $i=k$, since $f_{kk}^{\alpha}=z_k^{\alpha}$, we have $\theta_{kk}(t)=0$. Consequently from (4.11) we have

$$\theta_{kj}(t)=-\theta_{jk}(t).\tag{4.13}$$

From (4.12) and (4.13), we see that $\{\theta_{jk}(t)\}$ is a 1-cocycle. Namely, let Θ_t be the sheaf of germs of holomorphic vector fields over M_t. Then

$$\theta_{jk}(t)\in\Gamma(U_j\cap U_k,\Theta_t),\qquad \theta_{kj}(t)=-\theta_{jk}(t).$$

From (4.12), $\{\theta_{jk}(t)\}$ is a 1-cocycle with respect to the open covering $\mathfrak{U}_t=\{U_j\}$ on $M_t=\bigcup_j U_j$: $\{\theta_{jk}(t)\}\in Z^1(\mathfrak{U}_t,\Theta_t)$. We denote by $\theta(t)$ *the element of the cohomology group $H^1(M_t,\Theta_t)$ determined by the 1-cocycle $\{\theta_{jk}(t)\}$*. By the Corollary to Theorem 3.4, $H^1(\mathfrak{U}_t,\Theta_t)$ is a subgroup of $H^1(M_t,\Theta_t)$:

$$H^1(\mathfrak{U}_t,\Theta_t)\subset H^1(M_t,\Theta_t).\tag{4.14}$$

Thus $\theta(t)$ is considered to be an element of $H^1(M_t,\Theta_t)$. Then it will be obvious that $\theta(t)$ is *the cohomology class of* $\{\theta_{jk}(t)\}$. Since

$$\theta_{jk}(t)=\sum_{\alpha=1}^{n}\frac{\partial f_{jk}^{\alpha}(z_k,t)}{\partial t}\frac{\partial}{\partial z_j^{\alpha}},\qquad z_k=f_{kj}(z_j,t),$$

is obtained by differentiating $f_{jk}^{\alpha}(z_k,t)$ which define the complex structure of M_t, we may expect that $\theta(t)$ represents in some sense or another the "derivative" of the complex structure of M_t with respect to t. Hence we call $\theta(t)$ the *infinitesimal deformation of* M_t, and consider $\theta(t)$ as *the derivative of the complex structure of M_t with respect to t*. We denote it by

dM_t/dt:

$$\frac{dM_t}{dt}=\theta(t). \tag{4.15}$$

In the above argument, we assume that $x_j(\mathcal{U}_j)=U_j\times I_j$, but this assumption is redundant. Let $\{x_1,\ldots,x_j,\ldots\}$ be an arbitrary system of local coordinates of the differentiable family $\mathcal{M}$, and $\mathcal{U}_j$ the domain of x_j. Identifying $p\in\mathcal{U}_j$ with $(z_j,t)=(z_j^1,\ldots,z_j^n,t_1,\ldots,t_m)=x_j(p)$, we consider $\mathcal{U}_j$ with $(z_j,t)=\mathbb{C}^n\times B$, hence

$$\mathcal{U}_j\cap M_t=U_{jt}\times t,\qquad U_{jt}\subset\mathbb{C}^n.$$

Consequently identifying $U_{jt}\times t$ with U_{jt}, we obtain an open covering $\mathfrak{U}_t=\{U_{jt}\}$ of M_t. $\mathfrak{U}_t$ is a finite covering of M_t where we omit U_{jt} if it is empty. Let

$$(z_k,t)\to(z_j,t)=(f_{jk}(z_k,t),t)$$

be the coordinate transformation on $\mathcal{U}_j\cap\mathcal{U}_k$. Then we have the identity (4.9) on $\mathcal{U}_i\cap\mathcal{U}_j\cap\mathcal{U}_k\neq\emptyset$. Consequently defining $\theta_{jk}(t)$ by (4.10), we have $\{\theta_{jk}(t)\}\in Z^1(\mathfrak{U}_t,\Theta_t)$. We denote by $\theta(t)$ the *cohomology class of the* 1-cocycle $\{\theta_{jk}(t)\}$:

$$\theta(t)\in H^1(\mathfrak{U}_t,\Theta_t)\subset H^1(M_t,\Theta_t),$$

and define $\theta(t)$ as *the infinitesimal deformation of* M_t: $dM_t/dt=\theta(t)$. Thus the infinitesimal deformation is defined with respect to any system of local coordinates.

We must verify that $\theta(t)$ *does not depend on the choice of systems of local coordinates.* First we show that $\theta(t)$ does not change under the refinement of $\mathfrak{U}=\{\mathcal{U}_j\}$. Let $\mathfrak{V}=\{\mathcal{V}_\lambda\}$, $\mathcal{V}_\lambda\subset\mathcal{U}_{j(\lambda)}$, be any refinement of $\mathfrak{U}$, and define local coordinates $\hat{x}_\lambda$ on $\mathcal{V}_\lambda$ by $\hat{x}_\lambda: p\to\hat{x}_\lambda(p)=x_{j(\lambda)}(p)$. We must show that the infinitesimal deformation defined with respect to $\{\hat{x}\}$ coincides with $\theta(t)$ defined with respect to $\{x_j\}$. Since $\hat{x}_\lambda$ is the restriction of $x_{j(\lambda)}$ to $\mathcal{V}_\lambda$, the holomorphic vector field $\hat{\theta}_{\lambda\mu}(t)$ defined by (4.10) with respect to $\{\hat{x}_\lambda\}$, is the restriction of $\theta_{j(\lambda)j(\mu)}$ to $\mathcal{V}_\lambda\cap\mathcal{V}_\mu\cap\varpi^{-1}(t)$. Namely, putting $\mathfrak{V}_t=\{\mathcal{V}_{\lambda t}\}$, where $\mathcal{V}_{\lambda t}=\mathcal{V}_\lambda\cap M_t\neq\emptyset$, we have

$$\hat{\theta}_{\lambda\mu}(t)=r_V\theta_{j(\lambda)j(\mu)}(t),\qquad V=V_{\lambda t}\cap V_{\mu t}.$$

Hence $\{\hat{\theta}_{\lambda\mu}(t)\}=\Pi^{\mathfrak{U}_t}_{\mathfrak{V}_t}\{\theta_{jk}(t)\}$. Consequently by (3.81) the cohomology class of $\{\hat{\theta}_{\lambda\mu}(t)\}$ coincides with that of $\{\theta_{jk}(t)\}$. Thus the infinitesimal deformation does not change under the refinement of the open covering. Since any two locally finite open coverings have a common refinement, in order to prove the independence of $\theta(t)$ of the choice of systems of local coordinates, it remains to show the following: Given two local coordinates $x_j = (z_j, t)$ and $u_j = (w_j, t)$ on each U_j, the infinitesimal deformation $\eta(t)$ with respect to $\{u_j\}$ coincides with $\theta(t)$ defined with respect to $\{x_j\}$. Let

$$(w_k, t) \to (w_j, t) = (h_{jk}(w_k, t), t)$$

be the coordinate transformation of $\{u_j\}$ on $\mathcal{U}_j \cap \mathcal{U}_k \neq \emptyset$. Put as in (4.10)

$$\eta_{jk}(t) = \sum_{\alpha=1}^{n} \frac{\partial h_{jk}^{\alpha}(w_k, t)}{\partial t} \frac{\partial}{\partial w_j^{\alpha}}, \qquad w_k = h_{kj}(w_j, t).$$

Then $\eta(t)$ is the cohomology class of the 1-cocycle $\{\eta_{jk}(t)\} \in Z^1(\mathcal{U}_t, \Theta_t)$. By (4.6),

$$w_j^{\alpha} = g_j^{\alpha}(z_j^1, \ldots, z_j^n, t)$$

is a C^{∞} function of $z_j^1, \ldots, z_j^n, t$, which is holomorphic in $z_j^1, \ldots, z_j^n$. Since

$$g_j^{\alpha}(z_j, t) = w_j^{\alpha} = h_{jk}^{\alpha}(w_k, t) = h_{jk}^{\alpha}(g_k(z_k, t), t),$$

and $z_j = f_{jk}(z_k, t)$, we have

$$g_j^{\alpha}(f_{jk}(z_k, t), t) = h_{jk}^{\alpha}(g_k(z_k, t), t) \tag{4.16}$$

on $\mathcal{U}_k \cap \mathcal{U}_j \neq \emptyset$. Differentiating (4.16) in t, we obtain

$$\sum_{\beta=1}^{n} \frac{\partial g_j^{\alpha}}{\partial z_j^{\beta}} \frac{\partial f_{jk}^{\beta}}{\partial t} + \frac{\partial g_j^{\alpha}}{\partial t} = \sum_{\beta=1}^{n} \frac{\partial h_{jk}^{\alpha}}{\partial w_k^{\beta}} \frac{\partial g_k^{\beta}}{\partial t} + \frac{\partial h_{jk}^{\alpha}}{\partial t},$$

that is,

$$\sum_{\beta=1}^{n} \frac{\partial f_{jk}^{\beta}}{\partial t} \frac{\partial w_j^{\alpha}}{\partial z_j^{\beta}} + \frac{\partial g_j^{\alpha}}{\partial t} = \sum_{\beta=1}^{n} \frac{\partial g_k^{\beta}}{\partial t} \frac{\partial w_j^{\alpha}}{\partial w_k^{\beta}} + \frac{\partial h_{jk}^{\alpha}}{\partial t}.$$

Multiplying $\partial/\partial w_j^{\alpha}$ from the right, and taking the summation $\sum_{\alpha=1}^{n}$, we obtain the equality

$$\sum_{\beta=1}^{n} \frac{\partial f_{jk}^{\beta}}{\partial t} \frac{\partial}{\partial z_j^{\beta}} + \sum_{\alpha=1}^{n} \frac{\partial g_j^{\alpha}}{\partial t} \frac{\partial}{\partial w_j^{\alpha}} = \sum_{\beta=1}^{n} \frac{\partial g_k^{\beta}}{\partial t} \frac{\partial}{\partial w_k^{\beta}} + \sum_{\alpha=1}^{n} \frac{\partial h_{jk}^{\alpha}}{\partial t} \frac{\partial}{\partial w_j^{\alpha}}.$$

Putting

$$\theta_j(t) = \sum_{\alpha=1}^{n} \frac{\partial g_j^{\alpha}(z_j, t)}{\partial t} \frac{\partial}{\partial w_j^{\alpha}}, \qquad w_j^{\alpha} = g_j^{\alpha}(z_j, t),$$

we have

$$\theta_{jk}(t)-\eta_{jk}(t)=\theta_k(t)-\theta_j(t). \tag{4.17}$$

Since $\theta_j(t)$ is a holomorphic vector field on $U_{jt}=\mathcal{U}_j\cap M_t$, $\{\theta_j(t)\}\in C^0(\mathfrak{U}_t,\Theta_t)$ and (4.17) means that

$$\{\theta_{jk}(t)\}-\{\eta_{jk}(t)\}=\delta(\theta_j(t)\}.$$

Therefore $\eta(t)$ coincides with $\theta(t)$, and the infinitesimal deformation $\theta(t)$ does not depend on the choice of systems of local coordinates.

(b) Trivial Differentiable Family

The first problem in the theory of deformation was to see if we could justify the consideration that the infinitesimal deformation $\theta(t)$ represents the derivative of the complex structure of M_t with respect to t. For, this consideration seemed to the author too good to be true. (Spencer had, I think, a more optimistic view.)

If $\theta(t)$ is the derivative of the complex structure of M_t with respect to t, $\theta(t)=0$ *must hold if M_t does not vary with t*. For this we must give the precise definition for M_t not varying with t.

In general we define the equivalence of two differentiable families as follows.

Definition 4.2. Suppose given two differentiable families $(\mathcal{M}, B, \varpi)$ and $(\mathcal{N}, B, \pi)$ with the same base space $B\subset\mathbb{R}^n$. $\mathcal{M}$ and $\mathcal{N}$ are called *equivalent* if there is a diffeomorphism Φ of $\mathcal{M}$ onto $\mathcal{N}$ such that for each $t\in B$, Φ maps $M_t=\varpi^{-1}(t)$ biholomorphically onto $N_t=\pi^{-1}(t)$.

Suppose that $\mathcal{M}$ and $\mathcal{N}$ are equivalent. Let $\{x_j\}$, $x_j\colon p\to x_j(p)=(z_j(p),t)$, $t=\varpi(p)$, be the system of local coordinates of $\mathcal{M}$, and $\{u_\lambda\}$, $u_\lambda\colon q\to u_\lambda(q)=(w_\lambda(q),t)$, $t=\pi(q)$, that of $\mathcal{N}$. Then we have

$$u_\lambda(\Phi(p))=(w_\lambda(\Phi(p)),t).$$

If we represent $w_\lambda^\alpha(\Phi(P))$ by a C^∞ function of the local coordinates $(z_j(p),t)$ as

$$w_\lambda^\alpha=g_\lambda^\alpha(z_j(p),t),$$

then $g_\lambda^\alpha(z_j,t)$ is holomorphic in $z_j^1,\ldots,z_j^n$. In this case, identifying $p\in\mathcal{M}$ with $q=\Phi(p)\in\mathcal{N}$, we may consider that $\mathcal{M}$ and $\mathcal{N}$ are the same differentiable family. In fact, since Φ is diffeomorphic, we may identify $\mathcal{N}$ and $\mathcal{M}$ via Φ. Then since $g_\lambda^\alpha(z_j,t)$ is holomorphic in $z_j^1,\ldots,z_j^n$, the system of local coordinates $\{u_\lambda\}$, $u_\lambda=(w_\lambda,t)$, $w_\lambda=g_\lambda(z_j,t)$, of $\mathcal{N}$ can be regarded as a system of local coordinates of $\mathcal{M}$. Thus *if $\mathcal{M}$ and $\mathcal{N}$ are equivalent, we may consider $\mathcal{N}$ as the same differentiable family with $\mathcal{M}$ endowed with a new system of local coordinates.*

Let M be a compact complex manifold, and $\{w_\lambda\}$ a system of local complex coordinates. Let $\pi: M \times B \to B$ be the projection of $M \times B$ onto B. Then $(M \times B, B, \pi)$ is clearly a differentiable family of compact complex manifolds. A system of local coordinates of $M \times B$ is given by $\{u_\lambda\}$, where $u_\lambda = (w_\lambda, t)$. Of course the complex structure of $\pi^{-1}(t) = M \times t = M$ does not depend on t.

Definition 4.3. A differentiable family $(\mathcal{M}, B, \varpi)$ is called *trivial* if it is equivalent to $(M \times B, B, \pi)$ where $M = \varpi^{-1}(t^0)$ with $t^0 \in B$.

If the differentiable family $(\mathcal{M}, B, \varpi)$ is trivial, $\mathcal{M}$ coincides with $M \times B$ endowed with some system of local coordinates $\{x_j\}, x_j = (z_j, t) = (z_j^1, \ldots, z_j^n, t)$. Let $\mathcal{U}_j \subset M \times B$ be the domain of x_j, and $\{w_\lambda\}$ a system of local complex coordinates of $M = \varpi^{-1}(t^0)$. Then

$$z_j^\alpha = g_j^\alpha(w_\lambda, t), \qquad \alpha = 1, \ldots, n,$$

are C^∞ functions of $w_\lambda^1, \ldots, w_\lambda^n, t$ defined on $\mathcal{U}_j$ which are holomorphic in $w_\lambda^1, \ldots, w_\lambda^n$. On $\mathcal{U}_j \cap \mathcal{U}_k$,

$$z_j^\alpha = f_{jk}^\alpha(z_k, t)$$

are C^∞ functions of $z_j^1, \ldots, z_j^n, t$ which are holomorphic in $z_j^1, \ldots, z_j^n$. For any fixed t, $\{z_j | \mathcal{U}_j \cap M \times t\}$ form a system of local complex coordinates of $M_t = \varpi^{-1}(t)$. Since in general the coordinate transformation $z_k \to z_j = f_{jk}(z_k, t)$ does depend on t, the complex structure of M_t apparently depends on t.

Let $I \subset B$ be a subdomain of B. Then $(\mathcal{M}_I, I, \varpi)$ with $\mathcal{M}_I = \varpi^{-1}(I)$ is clearly a differentiable family, where the domain of ϖ is assumed to be restricted to $\mathcal{M}_I$.

Definition 4.4. A differentiable family $(\mathcal{M}, B, \varpi)$ is called *locally trivial* if for each $t \in B$, there is a subdomain I with $t \in I \subset B$ such that $(\mathcal{M}_I, I, \varpi)$ is trivial.

By saying that the complex structure of $M_t = \varpi^{-1}(t)$ of the differentiable family $(\mathcal{M}, B, \varpi)$ does not vary with t, we mean that $(\mathcal{M}, B, \varpi)$ is locally trivial. If $(\mathcal{M}, B, \varpi)$ is locally trivial, each $M_t = \varpi^{-1}(t)$ is biholomorphically equivalent to a fixed $M = \varpi^{-1}(t^0)$.

Let us consider the inverse. Namely, if each M_t is biholomorphic to a fixed M, is $(\mathcal{M}, B, \varpi)$ locally trivial? This problem was solved for complex analytic families affirmatively by Fischer and Grauert [4] later. (This paper was published in 1965, but Grauert explained the idea of proof already about 1960 at the "Nothing Seminar" held in Princeton University under the direction of Spencer.) At that time, however, nothing was known about it. We thought that the local triviality of $(\mathcal{M}, B, \varpi)$ would be stronger than the condition that each M_t is biholomorphic to a fixed M.

Now again let B be an open interval of $\mathbb{R}$. If $(\mathcal{M}, B, \varpi)$ is locally trivial, for any sufficiently small domain $I \subset B$, $(\mathcal{M}_I, I, \varpi)$ is equivalent to $(M \times I, I, \pi)$. Since the infinitesimal deformation $\theta(t)$ of $M_t = \varpi^{-1}(t)$ does not depend on the choice of systems of local coordinates, we can calculate $\theta(t)$ for $t \in I$ in terms of local coordinates $\{u_\lambda\}$, $u_\lambda = (w_\lambda, t)$ of $M \times I$, where $\{w_\lambda\}$ is a system of local coordinates of M. Let $w_\mu \to w_\lambda = h_{\lambda\mu}(w_\mu)$ be the coordinate transformation for $\{w_\lambda\}$. Then the coordinate transformation for $\{u_\lambda\}$ is given by

$$(w_\mu, t) \to (w_\lambda, t) = (h_{\lambda\mu}(w_\mu), t).$$

Since $h_{\lambda\mu}(w_\mu)$ is independent of t, we have

$$\theta_{\lambda\mu}(t) = \sum_{\alpha=1}^{n} \frac{\partial h_{\lambda\mu}^\alpha(w_\mu)}{\partial t} \frac{\partial}{\partial x_\lambda^\alpha} = 0,$$

hence $\theta(t) = 0$. Thus if *the complex structure of $M_t = \varpi^{-1}(t)$ does not vary with t in the sense that $(\mathcal{M}, B, \varpi)$ is locally trivial, we have $\theta(t) = 0$.*

If $\theta(t)$ is truly the derivative of the complex structure of M_t with respect to t, that $\theta(t) = 0$ identically must imply conversely the local triviality of $(\mathcal{M}, B, \varpi)$. We shall show below that this is true at least under a certain additional condition. (Cf. [21], Chap. II.)

For simplicity we consider the differentiable family

$$\mathcal{M}_I = \varpi^{-1}(I) = \bigcup_{j=1}^{l} U_j \times I$$

given by (4.8). Each $U_j \subset \mathbb{C}^n$ is a polydisk, and $(z_j, t) \in U_j \times I$ and $(z_k, t) \in U_k \times I$ are the same point on $\mathcal{M}_I$ if

$$z_j^\alpha = f_{jk}^\alpha(z_k, t), \qquad \alpha = 1, \ldots, n.$$

Let Θ_t be the sheaf of germs of holomorphic vector fields on $M_t = \varpi^{-1}(t)$, and $\mathfrak{U}_t = \{U_j\} = \{U_j \times t\}$.

Suppose $\theta(t) = 0$ identically. $\theta(t)$ is, by definition, the cohomology class of the 1-cocycle $\{\theta_{jk}(t)\} \in Z^1(\mathfrak{U}_t, \Theta_t)$ where $\theta_{jk}(t) = \sum_{\alpha=1}^n (\partial f_{jk}^\alpha / \partial t)(\partial / \partial z_j^\alpha)$. Since by (4.14),

$$\theta(t) \in H^1(\mathfrak{U}_t, \Theta_t) = Z^1(\mathfrak{U}_t, \Theta_t) / \delta C^0(\mathfrak{U}_t, \Theta_t),$$

$\theta(t)=0$ means that $\{\theta_{jk}(t)\}=\delta\{\theta_j(t)\}$, namely there is a 0-cochain $\{\theta_j(t)\}\in C^0(\mathfrak{U}_t, \Theta_t)$ such that

$$\theta_{jk}(t)=\theta_k(t)-\theta_j(t). \tag{4.18}$$

We want to prove that $(\mathcal{M}_I, I, \varpi)$ is equivalent to $(M\times I, I, \pi)$ with $M=\varpi^{-1}(0)$. In other words, we must show that there is a diffeomorphism Φ of $M\times I$ onto $\mathcal{M}_I=\bigcup_{j=1}^{l} U_j\times I$ such that Φ maps $M\times I$ biholomorphically onto $M_t=\varpi^{-1}(t)$ for each t. We proceed as in the proof of Theorem 2.4(1°). Let

$$\theta_j(t)=\sum_{\alpha=1}^{n}\theta_j^\alpha(z_j, t)\frac{\partial}{\partial z_j^\alpha}.$$

Then the equality (4.18) is written as

$$\sum_{\alpha=1}^{n}\frac{\partial f_{jk}^\alpha(z_k, t)}{\partial t}\frac{\partial}{\partial z_j^\alpha}=\sum_{\alpha=1}^{n}\theta_k^\alpha(z_k, t)\frac{\partial}{\partial z_k^\alpha}-\sum_{\alpha=1}^{n}\theta_j^\alpha(z_j, t)\frac{\partial}{\partial z_j^\alpha}.$$

We denote by $(\partial/\partial t)_k$ the vector field $\partial/\partial t$ on $U_k\times I\subset\mathcal{M}_I$. Then with respect to the coordinate transformation $(z_k, t)\to(z_j, t)$, $z_j^\alpha=f_{jk}^\alpha(z_k, t)$, $\alpha=1,\ldots,n$, we have as in (2.26)

$$\left(\frac{\partial}{\partial t}\right)_k=\sum_{\alpha=1}^{n}\frac{\partial f_{jk}^\alpha(z_k, t)}{\partial t}\frac{\partial}{\partial z_j^\alpha}+\left(\frac{\partial}{\partial t}\right)_j.$$

Consequently on each open subset $U_j\times I\cap U_k\times I\neq\varnothing$ of $\mathcal{M}_I$ we have

$$-\sum_{\alpha=1}^{n}\theta_j^\alpha(z_j, t)\frac{\partial}{\partial z_j^\alpha}+\left(\frac{\partial}{\partial t}\right)_j=-\sum_{\alpha=1}^{n}\theta_k^\alpha(z_k, t)\frac{\partial}{\partial z_k^\alpha}+\left(\frac{\partial}{\partial t}\right)_k.$$

We define the vector field v on $\mathcal{M}_I$, by putting on each $U_j\times I$

$$v=-\sum_{\alpha=1}^{n}\theta_j^\alpha(z_j, t)\frac{\partial}{\partial z_j^\alpha}+\frac{\partial}{\partial t}. \tag{4.19}$$

(4.19) is an analogy of (2.25). But in contrast to (2.25), it does not follow immediately that v is C^∞. In fact, though $\theta_{jk}(t)$ is a C^∞ vector field on $U_j\times I\cap U_k\times I$, $\theta_j(t)$ may not be differentiable in t. Moreover $\theta_j(t)$ is not uniquely determined by $\{\theta_{jk}(t)\}$. Hence there arises the following problem: Can we choose $\theta_j(t), j=1,\ldots,l$, such that each $\theta_j^\alpha(z_j, t)$ is C^∞ in $z_j^1,\ldots,z_j^n, t$? The following theorem is an answer to this problem.

Theorem 4.1. *If* $\dim H^1(M_t, \Theta_t)$ *is independent of* $t\in I$, *then we can choose* *a 0-cochain* $\{\theta_j(t)\}$ *with* $\delta\{\theta_j(t)\}=\{\theta_{jk}(t)\}$ *such that each* $\theta_j^\alpha(z_j, t)$ *is a* C^∞ *function of* $z_j^1, \ldots, z_j^n, t$ *where we put* $\theta_j(t)=\sum_{\alpha=1}^n \theta_j^\alpha(z_j, t)(\partial/\partial z_j^\alpha)$.

We give the proof of this theorem in §7.2(b). The proof depends on the method developed by Spencer and the author for the theory of variations of almost complex structures ([19]).

Now suppose that $\dim H^1(M_t, \Theta_t)$ does not depend on t. By Theorem 4.1, choose $\{\theta_j(t)\}$ such that each $\theta_j(t)$ is a C^∞ vector field on $U_j \times I \subset \mathcal{M}_I$. Then v in (4.19) is a C^∞ vector field on $\mathcal{M}_I$. Consider the simultaneous ordinary differential equations

$$\begin{cases} \dfrac{dz_j^\alpha}{ds} = -\theta_j^\alpha(z_j^1, \ldots, z_j^n, t), \qquad \alpha = 1, \ldots, n, \\ \\ \dfrac{dt}{ds} = 1, \end{cases} \tag{4.20}$$

for v as in (2.28). Take an arbitrary $p \in M = \varpi^{-1}(0)$, and assume $p \in U_i \times 0$. Let $(\zeta_i(p), 0)$ be the local coordinates of p. (4.20) has the unique solution

$$\begin{cases} z_j^\alpha(s) = z_j^\alpha(p, s), \\ \\ t = s, \end{cases}$$

under the initial conditions

$$\begin{cases} z_i^\alpha(0) = \zeta_i^\alpha(p), \qquad \alpha = 1, \ldots, n, \\ \\ t(0) = 0. \end{cases}$$

This solution gives a smooth curve on $\mathcal{M}_I$ passing through p:

$$\gamma_p \colon t \to \gamma_p(t) = (z_j(p, t), t), \qquad t \in I.$$

Thus we obtain a family of smooth curves $\{\gamma_p | p \in M\}$ on $\mathcal{M}_I$. By the uniqueness theorem of the solution of simultaneous ordinary differential equations, for each point $(z_j, t) \in \mathcal{M}_I$, there exists just one curve which passes through it. Consequently the map

$$\Phi \colon (p, t) \to (z_j, t) = (z_j(p, t), t)$$

is a diffeomorphism of $M \times I$ onto $\mathcal{M}_I$.

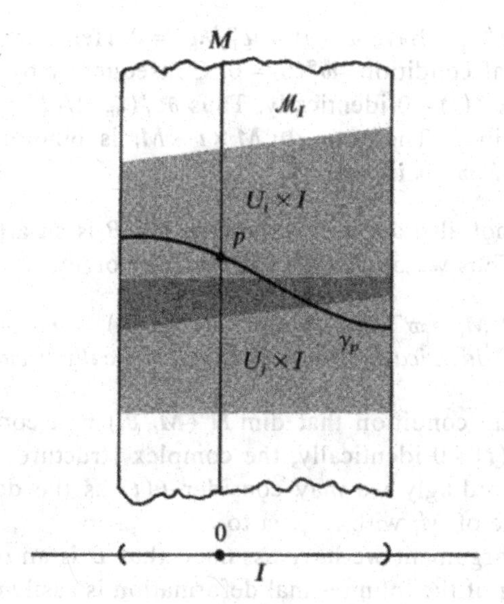

Figure 2

Since clearly Φ maps $M \times t$ onto M_t, in order to see that Φ: $M \times t \to M_t$ is biholomorphic, it suffices to show that $z_j(p, t)$ is holomorphic in p. For this we write $z_j(p, t)$ as a function of t and the local complex coordinates $\zeta_i^\lambda = \zeta_i^\lambda(p)$, $\lambda = 1, \ldots, n$, of p in the form

$$z_j^\alpha = z_j^\alpha(\zeta_i, t) = z_j^\alpha(\zeta_i^1, \ldots, \zeta_i^n, t).$$

Since $t = s$, by (4.20) we have

$$\frac{d}{dt} z_j^\alpha(\zeta_i, t) = -\theta_j^\alpha(z_j^1(\zeta_i, t), \ldots, z_j^n(\zeta_i, t), t).$$

By differentiating both sides of this equation in $\bar{\zeta}_i^\lambda$, since $\theta_j^\alpha(z_j, t)$ is holomorphic in $z_j^1, \ldots, z_j^n$, we obtain

$$\frac{d}{dt} \frac{\partial z_j^\alpha(\zeta_i, t)}{\partial \bar{\zeta}_i^\lambda} = -\sum_{\beta=1}^{n} \frac{\partial \theta_j^\alpha}{\partial z_j^\beta}(z_j(\zeta_i, t), t) \cdot \frac{\partial z_j^\beta(\zeta_i, t)}{\partial \bar{\zeta}_i^\lambda}.$$

Put $\omega_j^\alpha(t) = \partial z_j^\alpha(\zeta_i, t)/\partial \bar{\zeta}_i^\lambda$. Then $\omega_j^\alpha(t)$ is a solution of the simultaneous ordinary linear differential equations

$$\frac{d}{dt} \omega_j^\alpha(t) = -\sum_{\beta=1}^{n} \frac{\partial \theta_j^\alpha}{\partial z_j^\beta}(z_j(\zeta_i, t), t)\omega_j^\beta(t), \qquad \alpha = 1, \ldots, n.$$

Since $z_i^\alpha(\zeta_i, 0) = \zeta_i^\alpha$, we have $\omega_i^\alpha(0) = \partial\zeta_i^\alpha/\partial\bar{\zeta}_i^\lambda = 0$. Hence the solution $\omega_j^\alpha(t)$ satisfies the initial condition: $\omega_j^\alpha(0) = 0$. Consequently by the uniqueness of the solution, $\omega_j^\alpha(t) = 0$ identically. Thus $\partial z_j^\alpha(\zeta_i, t)/\partial\bar{\zeta}_i^\lambda = 0$, and $z_j^\alpha(p, t)$ is holomorphic in p. Therefore $\Phi: M \times t \to M_t$ is biholomorphic, which proves that $(\mathcal{M}_I, I, \varpi)$ is trivial. ∎

The above proof also applies to the case $t^0 \in B$ is an arbitrary point of B instead of 0. Thus we obtain the following theorem.

Theorem 4.2. *Let $M_t = \varpi^{-1}(t)$. If $\dim H^1(M_t, \Theta_t)$ is independent of t, and $\theta(t) = dM_t/dt = 0$ identically, then $(\mathcal{M}, B, \varpi)$ is locally trivial.*

Thus under the condition that $\dim H^1(M_t, \theta_t)$ is a constant, we have proved that if $\theta(t) = 0$ identically, the complex structure of M_t does not vary with t. Accordingly we may consider $\theta(t)$ as the derivative of the complex structure of M_t with respect to t.

In the above argument we have assumed that B is an open interval in $\mathbb{R}$. But the notion of the infinitesimal deformation is easily extended to the case B is an arbitrary domain in $\mathbb{R}^n$. Suppose given a differentiable family $(\mathcal{M}, B, \varpi)$ of compact complex manifolds where B is a domain of $\mathbb{R}^n$. Let $\{x_1, \ldots, x_j, \ldots\}$, $x_j = (z_j, t) = (z_j^1, \ldots, z_j^n, t_1, \ldots, t_m)$ be a system of local coordinates on $\mathcal{M}$, and $\mathcal{U}_j$ the domain of x_j. Suppose that on $\mathcal{U}_j \cap \mathcal{U}_k \neq \emptyset$, we have

$$z_j^\alpha = f_{jk}^\alpha(z_k, t) = f_{jk}^\alpha(z_j^1, \ldots, z_j^n, t_1, \ldots, t_m), \qquad \alpha = 1, \ldots, n.$$

As usual, identifying a point of $\mathcal{M}$ with its local coordinates, we may consider that $\mathcal{U}_j = x_j(\mathcal{U}_j) \subset \mathbb{C}^n \times B$, and that $\mathcal{M} = \bigcup_j \mathcal{U}_j$ is obtained by glueing the domains $\mathcal{U}_1, \ldots, \mathcal{U}_k, \ldots$ of $\mathbb{C}^n \times B$ by identifying (z_j, t) and (z_k, t) if $z_j = f_{jk}(z_k, t)$. $\mathfrak{U}_t = \{U_{jt} \mid U_{jt} = \mathcal{U}_j \cap M_t \neq \emptyset\}$ form a finite open covering of $M_t = \varpi^{-1}(t)$.

The tangent space $T_t(B)$ of B at t consists of all tangent vectors $\partial/\partial t = \sum_{\lambda=1}^m c_\lambda \partial/\partial t_\lambda$, $c_\lambda \in \mathbb{R}$. For $\partial/\partial t \in T_t(B)$, put

$$\theta_{jk}(t) = \sum_{\alpha=1}^m \frac{\partial f_{jk}^\alpha(z_k, t_1, \ldots, t_m)}{\partial t} \frac{\partial}{\partial z_j^\alpha}, \qquad z_k = f_{kj}(z_j, t). \tag{4.21}$$

Then as in the case $m = 1$, we have $\{\theta_{jk}(t)\} \in Z^1(\mathfrak{U}_t, \Theta_t)$. The cohomology class $\theta(t) \in H^1(M_t, \Theta_t)$ of the 1-cocycle $\{\theta_{jk}(t)\}$ is called the *infinitesimal deformation of M_t along $\partial/\partial t$*, and is denoted by $\partial M_t/\partial t$: $\partial M_t/\partial t = \theta(t)$. The infinitesimal deformation $\partial M_t/\partial t$ is independent of the choice of systems of local coordinates. *We define the map ρ_t by*

$$\rho_t: \frac{\partial}{\partial t} \to \rho_t\left(\frac{\partial}{\partial t}\right) = \frac{\partial M_t}{\partial t}. \tag{4.22}$$

ρ_t is an $\mathbb{R}$-linear map of $T_t(B)$ into $H^1(M_t, \Theta_t)$.

If $\rho_t(\partial/\partial t)=0$ for any $\partial/\partial t\in T_t(B)$, ρ_t is called the 0-map, and is denoted by $\rho_t=0$. If $(\mathcal{M}, B, \varpi)$ is trivial, we see immediately as in the case $m=1$ that $\rho_t=0$ identically, using the fact that $\theta(t)$ is independent of the choice of systems of local coordinates.

Theorem 4.3. *If* $\dim H^1(M_t, \Theta_t)$ *is independent of* t, *and* $\rho_t=0$ *identically, then* $(\mathcal{M}, B, \varpi)$ *is locally trivial.*

Proof. Consider the differentiable family

$$\mathcal{M}_I=\varpi^{-1}(I)=\bigcup_{j=1}^{l} U_j\times I$$

given in (4.8). For simplicity, we assume that $I=\{(t_1,\ldots,t_m)\,|\,|t_1|<r,\ldots,|t_m|<r\}$. We proceed by induction on m as in the proof of Theorem 2.4. Let

$$I^{m-1}=\{(t_1,\ldots,t_m)\,|\,|t_1|<r,\ldots,|t_{m-1}|<r\},\quad \text{and}$$

$$I_m=\{t_m\,|-r<t_m<r\}.$$

Then $I=I^{m-1}\times I_m$. Considering $I^{m-1}=I^{m-1}\times 0\subset I$, we see that $(\varpi^{-1}(I^{m-1}), I^{m-1}, \varpi)$ is a differentiable family. Since Theorem 4.3 is true for $m=1$, we may assume that $\varpi^{-1}(I^{m-1})$ is trivial. Define the map of $\varpi^{-1}(I^{m-1})\times I_m$ onto I by $(p, t_m)\to(\varpi(p), t_m)$ where $p\in\varpi^{-1}(I^{m-1})$, which makes $\varpi^{-1}(I^{m-1})\times I_m$ a differentiable family on the parameter space I. Since by hypothesis $\varpi^{-1}(I^{m-1})$ is equivalent to $M\times I^{m-1}$ where $M=\varpi^{-1}(0)$, $\varpi^{-1}(I^{m-1})\times I_m$ is equivalent to $M\times I$. Therefore in order to prove the triviality of $\mathcal{M}_I$, it suffices to verify that $\mathcal{M}_I$ is equivalent to $\varpi^{-1}(I^{m-1})\times I_m$. Put

$$\theta_{jk}(t)=\sum_{\alpha=1}^{n}\frac{\partial f_{jk}^{\alpha}(z_k, t_1,\ldots,t_m)}{\partial t_m}\frac{\partial}{\partial z_j^{\alpha}}.$$

Then the cohomology class $\theta(t)=\rho_t(\partial/\partial t_m)$ of the 1-cocycle $\{\theta_{jk}(t)\}$ is equal to 0 by the assumption. Consequently as in the case $m=1$, by (4.14) there exists a 0-cochain $\{\theta_j(t)\}\in C^0(\mathfrak{U}_t, \Theta_t)$ such that $\{\theta_{jk}(t)\}=\delta\{\theta_j(t)\}$. Since Theorem 4.1 remains true also for $m\geqq 2$ (see §7.2(b)), we may assume that the coefficients $\theta_j^{\alpha}(z_j, t_1,\ldots,t_m)$ in

$$\theta_j(t)=\sum_{\alpha=1}^{n}\theta_j^{\alpha}(z_j, t_1,\ldots,t_m)\frac{\partial}{\partial z_j^{\alpha}}\tag{4.23}$$

are C^∞ functions of $z_j^1,\ldots,z_j^n, t_1,\ldots,t_m$ in $U_j\times I$. Therefore as in the case $m=1$, we obtain a C^∞ vector field v on $\mathcal{M}_I$ such that on each $U_j\times I\subset\mathcal{M}_I$

$$v=-\sum_{\alpha=1}^{n}\theta_j^{\alpha}(z_j, t_1,\ldots,t_m)\frac{\partial}{\partial z_j^{\alpha}}+\frac{\partial}{\partial t_m}.\tag{4.24}$$

Consider the corresponding simultaneous ordinary differential equations

$$\begin{cases} \dfrac{dz_j^\alpha}{ds} = -\theta_j^\alpha(z_j, t_1, \ldots, t_m), & \alpha = 1, \ldots, n, \\[2ex] \dfrac{dt_\lambda}{ds} = 0, & \lambda = 1, \ldots, m-1, \\[2ex] \dfrac{dt_m}{ds} = 1. \end{cases} \tag{4.25}$$

Take any point $p \in \varpi^{-1}(I^{m-1}) = \bigcup_i U_i \times I^{m-1} \times 0$, and let $(\zeta_i(p), t_1, \ldots, t_{m-1}, 0)$ be its local coordinates where $(t_1, \ldots, t_{m-1}, 0) = \varpi(p)$. Then (4.25) has the unique solution

$$\begin{cases} z_j^\alpha(s) = z_j^\alpha(p, s) & \alpha = 1, \ldots, n, \\ t_\lambda(s) = t_\lambda, & \lambda = 1, \ldots, m-1, \\ t_m = s \end{cases}$$

under the initial conditions

$$\begin{cases} z_i^\alpha(0) = \zeta_i^\alpha(p), & \alpha = 1, \ldots, n, \\ t_\lambda(0) = t_\lambda, & \lambda = 1, \ldots, m-1, \\ t_m(0) = 0. \end{cases}$$

Thus we get a diffeomorphism Φ of $\varpi^{-1}(I^{m-1}) \times I_m$ onto $\mathcal{M}_I$:

$$\Phi: (p, t_m) \to (z_j(p, t_m), t_1, \ldots, t_{m-1}, t_m).$$

If, using the local coordinates $(\zeta_i^1, \ldots, \zeta_i^n, t_1, \ldots, t_{m-1}, 0)$, we write

$$z_j^\alpha(p, t_m) = z_j^\alpha(\zeta_i^1, \ldots, \zeta_i^n, t_1, \ldots, t_{m-1}, t_m),$$

then since $\theta_j^\alpha(z_j, t_1, \ldots, t_m)$ is holomorphic in $z_j^1, \ldots, z_j^n$, we see as in the case $m = 1$ that $z_j^\alpha(\zeta_i^1, \ldots, \zeta_i^n, t_1, \ldots, t_m)$ is holomorphic in $\zeta_i^1, \ldots, \zeta_i^n$. ∎

Theorem 4.4. ([19]). *For a differentiable family* $(\mathcal{M}, B, \varpi)$, $\dim H^1(M_t, \Theta_t)$ *is an upper-semicontinuous function of* t *where* $M_t = \varpi^{-1}(t)$.

By the upper-semicontinuity, we mean that for each $s \in B$,

$$\dim H^1(M_t, \Theta_t) \leqq \dim H^1(M_s, \Theta_s) \quad \text{if } |t-s| < \varepsilon$$

provided that ε is sufficiently small. We give proof of this theorem in §7.2(b).

From this theorem and Theorem 4.3, the Frölicher–Nijenhuis theorem ([7]) follows.

Theorem 4.5. *Let $(\mathcal{M}, B, \varpi)$ be a differentiable family of compact complex manifolds where B is a domain of $\mathbb{R}^m$, and $0 \in B$. If $H^1(M_0, \Theta_0) = 0$, with $M_0 = \varpi^{-1}(0)$, then for a sufficiently small open interval I with $0 \in I \subset B$, $(\mathcal{M}_I, I, \varpi)$ is trivial.*

A compact complex manifold M is called rigid if for any differentiable family $(\mathcal{M}, B, \varpi)$ with $0 \in B$ and $\varpi^{-1}(0) = M$, there is an open interval I with $0 \in I \subset B$ such that $(\mathcal{M}_I, I, \varpi)$ is trivial. If $(\mathcal{M}_I, I, \varpi)$ is trivial, $M_t = \varpi^{-1}(t)$ has the same complex structure with $M = M_0 = \varpi^{-1}(0)$ for $t \in I$. Thus if M is rigid, the complex structure of $M = M_0$ is invariant under small perturbation of t. By Theorem 4.5, *a compact complex manifold is rigid if $H^1(M, \Theta) = 0$*, where Θ is the sheaf of germs of holomorphic vector fields on M.

(c) The Case of Complex Analytic Families

Let $(\mathcal{M}, B, \varpi)$ be a complex analytic family of compact complex manifolds (Definition 2.8) where B is a domain of $\mathbb{C}^n$, and $\{(z_j, t)\}$ its system of local coordinates. Then each (z_j, t) is a local complex coordinate system of the complex manifold $\mathcal{M}$, and

$$z_j^\alpha = f_{jk}^\alpha(z_k^1, \ldots, z_k^n, t_1, \ldots, t_m), \qquad \alpha = 1, \ldots, n,$$

are holomorphic functions in $z_k^1, \ldots, z_k^n, t_1, \ldots, t_m$. As in the case of differentiable families, for a tangent vector $\partial/\partial t = \sum_{\lambda=1}^m c_\lambda (\partial/\partial t_\lambda)$, $c_\lambda \in \mathbb{C}$, of B, we put

$$\theta_{jk}(t) = \sum_{\alpha=1}^n \frac{\partial f_{jk}^\alpha(z_k, t)}{\partial t} \frac{\partial}{\partial z_j^\alpha}, \qquad z_k = f_{kj}(z_j, t). \tag{4.26}$$

The cohomology class $\theta(t) \in H^1(M_t, \Theta_t)$ of the 1-cocycle $\{\theta_{jk}(t)\}$ is called the infinitesimal deformation along $\partial/\partial t$ and is denoted by $\partial M_t/\partial t$. The infinitesimal deformation is independent of the choice of systems of local coordinates.

$$\rho_t \colon \frac{\partial}{\partial t} \to \rho_t\left(\frac{\partial}{\partial t}\right) = \frac{\partial M_t}{\partial t}$$

is a $\mathbb{C}$-linear map of $T_t(B)$ into $H^1(M_t, \Theta_t)$.

A complex analytic family $(\mathcal{M}, B, \varpi)$ *is called trivial if it is biholomorphically equivalent to* $(M \times B, B, \pi)$ *with* $M = \varpi^{-1}(t^0)$ where t^0 is some point of B (see p. 61). Similarly we define the local triviality of $(\mathcal{M}, B, \varpi)$. As in the case of differentiable families, $t = 0$ *identically if* $(\mathcal{M}, B, \varpi)$ *is locally trivial.*

By considering $B \subset \mathbb{C}^n = \mathbb{R}^{2n}$ as a domain of $\mathbb{R}^{2n}$, a complex analytic family $(\mathcal{M}, B, \varpi)$ may be regarded as a differentiable family. The tangent space of B at t considered as a subdomain of $\mathbb{R}^{2n}$ which consists of all tangent vectors with complex coefficients is $T_t(B) \oplus \overline{T_t(B)}$. Since $f_{jk}^{\alpha}(z_k, t_1, \ldots, t_m)$ are also holomorphic in $t_1, \ldots, t_m$, $\partial f_{jk}^{\alpha}(z_k, t)/\partial \bar{t} = 0$ for $\partial/\partial \bar{t} = \sum_{\lambda} c_{\lambda}(\partial/\partial \bar{t}_{\lambda}) \in \overline{T_t(B)}$, hence $\rho_t(\partial/\partial \bar{t}) = 0$. Thus *the map* ρ_t *for* $(\mathcal{M}, B, \varpi)$ *considered as a differentiable family coincides with that for* $(\mathcal{M}, B, \varpi)$ *considered as a complex analytic family.*

For a complex analytic family $(\mathcal{M}, B, \varpi)$ of compact complex manifolds, we have the following theorem similar to Theorem 4.3.

Theorem 4.6. *If* $\dim H^1(M_t, \Theta_t)$ *is independent of* $t \in B$, *and* $\rho_t = 0$ *identically, then the complex analytic family* $(\mathcal{M}, B, \varpi)$ *is locally trivial.*

Proof. Proof is similar to that of Theorem 4.3. Let $0 \in B$, and take a sufficiently small polydisk Δ with $0 \in \Delta \subset B$. Put $\mathcal{M}_{\Delta} = \varpi^{-1}(\Delta)$. Then as in (4.8), we have

$$\mathcal{M}_{\Delta} = \bigcup_{j=1}^{l} U_j \times \Delta, \tag{4.27}$$

where each U_j is a polydisk independent of t, and $(z_j, t) \in U_j \times \Delta$ and $(z_k, t) \in U_k \times \Delta$ are the same point on $\mathcal{M}_{\Delta}$ if $z_j^{\alpha} = f_{jk}^{\alpha}(z_k, t)$, $\alpha = 1, \ldots, n$. $f_{jk}^{\alpha}(z_k, t) = f_{jk}^{\alpha}(z_k^1, \ldots, z_k^n, t_1, \ldots, t_m)$ is a holomorphic function in the complex variables $z_k^1, \ldots, z_k^n, t_1, \ldots, t_m$.

We may assume that $\Delta = \{(t_1, \ldots, t_m) \in \mathbb{C}^n \mid |t_1| < r, \ldots, |t_m| < r\}$. We proceed by induction on m. For this we put

$$\Delta^{m-1} = \{(t_1, \ldots, t_m) \mid |t_1| < r, \ldots, |t_{m-1}| < r\}, \quad \text{and}$$

$$\Delta_m = \{t_m \mid |t_m| < r\}.$$

We consider $\Delta^{m-1} = \Delta^{m-1} \times 0 \subset \Delta^{m-1} \times \Delta_m = \Delta$. It suffices to prove that the complex analytic family $\varpi^{-1}(\Delta^{m-1}) \times \Delta_m$ is biholomorphically equivalent to $\mathcal{M}_{\Delta}$. Put

$$\theta_{jk}(t) = \sum_{\alpha=1}^{n} \frac{\partial f_{jk}^{\alpha}(z_k, t_1, \ldots, t_m)}{\partial t_m} \cdot \frac{\partial}{\partial z_j^{\alpha}}.$$

Since the cohomology class $\rho_t(\partial/\partial t_m)$ of the 1-cocycle $\{\theta_{jk}(t)\}$ is equal to 0, there exists a 0-cochain $\{\theta_j(t)\} \in C^0(\mathfrak{U}_t, \Theta_t)$ such that $\{\theta_{jk}(t)\} = \delta\{\theta_j(t)\}$ where we put $\mathfrak{U}_t = \{U_j \times t\}$. As was mentioned in the proof of Theorem 4.3

above, we may assume that each coefficient $\theta_j^\alpha(z_j, t_1, \ldots, t_m)$ in

$$\theta_j(t) = \sum_{\alpha=1}^{n} \theta_j^\alpha(z_j, t_1, \ldots, t_m)\frac{\partial}{\partial z_j^\alpha}$$

is a C^∞ function of $z_j^1, \ldots, z_j^n, t_1, \ldots, t_m$. In order to show that $\varpi^{-1}(\Delta^{m-1}) \times \Delta_m$ is biholomorphically equivalent to $\mathcal{M}_\Delta$, it suffices to verify that we can choose a 0-cochain $\{\theta_j(t)\}$ such that each coefficient $\theta_j^\alpha(z_j, t_1, \ldots, t_m)$ of $\theta_j(t)$ is holomorphic in $z_j^1, \ldots, z_j^n, t_1, \ldots, t_m$. In fact, if each $\theta_j^\alpha(z_j, t_1, \ldots, t_m)$ is holomorphic, the map $\Phi: \varpi^{-1}(\Delta^{m-1}) \times \Delta_m \to M_\Delta$ defined as in (4.25) by virtue of the solution of the simultaneous ordinary differential equations

$$\begin{cases} \dfrac{dz_j^\alpha}{ds} = -\theta_j^{\prime\alpha}(z_j, t_1, \ldots, t_m), & \alpha = 1, \ldots, n, \\[2ex] \dfrac{dt_\lambda}{ds} = 0, & \lambda = 1, \ldots, m-1, \\[2ex] \dfrac{dt_m}{ds} = 1, \end{cases}$$

is biholomorphic and Φ maps $\varpi^{-1}(\Delta^{m-1}) = \varpi^{-1}(\Delta^{m-1}) \times 0$ identically onto $\varpi^{-1}(\Delta^{m-1}) \subset \mathcal{M}_\Delta$. If we write

$$\theta_{jk}(t) = \sum_{\alpha=1}^{n} \theta_{jk}^\alpha(z_j, t)\frac{\partial}{\partial z_j^\alpha},$$

then $\theta_{jk}^\alpha(z_j, t) = (\partial/\partial t_m) f_{jk}^\alpha(z_k, t)$ with $z_k = f_{kj}(z_j, t)$ are holomorphic functions of $z_j^1, \ldots, z_j^n, t_1, \ldots, t_m$. The equality $\theta_{jk}(t) = \theta_k(t) - \theta_j(t)$ is written in the explicit form as

$$\sum_{\alpha=1}^{n} \theta_{jk}^\alpha(z_j, t)\frac{\partial}{\partial z_j^\alpha} = \sum_{\beta=1}^{n} \theta_k^\beta(z_k, t)\frac{\partial}{\partial z_k^\beta} - \sum_{\alpha=1}^{n} \theta_j^\alpha(z_j, t)\frac{\partial}{\partial z_j^\alpha},$$

namely,

$$\theta_{jk}^\alpha(z_k, t) = \sum_{\beta=1}^{n} \frac{\partial z_j^\alpha}{\partial z_k^\beta}\theta_k^\beta(z_k, t) - \theta_j^\alpha(z_j, t), \qquad \alpha = 1, \ldots, n.$$

Since $\partial z_j^\alpha/\partial z_k^\beta = \partial f_{jk}^\alpha(z_k, t)/\partial z_k^\beta$, $z_k = f_{kj}(z_j, t)$, and $\theta_{jk}^\alpha(z_j, t)$ are holomorphic in $t_1, \ldots, t_m$, differentiating both sides of the above equality in $\bar{t}_\lambda$, we obtain

$$\sum_{\beta=1}^{n} \frac{\partial z_j^\alpha}{\partial z_k^\beta}\frac{\partial \theta_k^\beta(z_k, t)}{\partial \bar{t}_\lambda} = \frac{\partial \theta_j^\alpha(z_j, t)}{\partial \bar{t}_\lambda},$$

namely,

$$\sum_{\beta=1}^{n} \frac{\partial \theta_k^\beta(z_k, t)}{\partial \bar{t}_\lambda} \cdot \frac{\partial}{\partial z_k^\beta} = \sum_{\alpha=1}^{n} \frac{\partial \theta_j^\alpha(z_j, t)}{\partial \bar{t}_\lambda} \cdot \frac{\partial}{\partial z_j^\alpha}.$$

Thus putting $\eta_\lambda(t) = \sum_\alpha (\partial\theta_j^\alpha(z_j, t)/\partial \bar{t}_\lambda) \cdot \partial/\partial z_j^\alpha$ on each $U_j \times t \subset M_t$, we obtain a holomorphic vector field $\eta_\lambda(t) \in H^0(M_t, \Theta_t)$ on M_t.

As was mentioned above, if we consider $(\mathcal{M}, B, \varpi)$ as a differentiable family, $\rho_t = 0$ also holds identically. Consequently by Theorem 4.3, $(\mathcal{M}, B, \varpi)$ is locally trivial as differentiable family. Therefore each $M_t = \varpi^{-1}(t)$ with $t \in B$ is biholomorphic to $M = \varpi^{-1}(0)$. Hence $\dim H^0(M_t, \Theta_t) = \dim H^0(M, \Theta)$. Note that we have $H^0(M, \Theta) = 0$ for many examples of compact complex manifolds. Such M as $\dim H^0(M, \Theta) \geqq 1$ is rather exceptional.

(1°) The case $\dim H^0(M, \Theta) = 0$. Then $H^0(M_t, \Theta_t) = 0$, too. Hence $\eta_\lambda(t) = 0$ identically. Therefore $\partial\theta_j^\alpha(z_j, t)/\partial \bar{t}_\lambda = 0$ for $\lambda = 1, \ldots, m$. Consequently $\theta_j^\alpha(z_j, t)$ are holomorphic in $z_j^1, \ldots, z_j^n, t_1, \ldots, t_m$.

(2°) The case $\dim H^0(M, \Theta) \geqq 1$. We use the following lemma.

Lemma 4.1. *For a complex analytic family $(\mathcal{M}, B, \varpi)$, suppose that $\dim H^0(M_t, \Theta_t) = d$ is independent of t. Then for a sufficiently small polydisk Δ with $\theta \in \Delta \subset B$, we can choose a basis $\{\varphi_1(t), \ldots, \varphi_d(t)\}$ of each linear space $H^0(M_t, \Theta_t)$ with*

$$\varphi_q(t) = \sum_{\alpha=1}^{n} \varphi_{qj}^\alpha(z_j, t) \frac{\partial}{\partial z_j^\alpha}$$

such that $\varphi_{qj}^\alpha(z_j, t)$ are holomorphic functions of $z_j^1, \ldots, z_j^n, t_1, \ldots, t_m$.

We give proof of this lemma in §7.2(b).

Put

$$\eta_\lambda(t) = \sum_{q=1}^{d} c_{q\lambda}(t) \varphi_q(t).$$

The coefficients $c_{q\lambda}(t)$ are C^∞ functions of $t_1, \ldots, t_m$. We denote $\bar{\partial}$ with respect to the variables $t_1, \ldots, t_m$ by $\bar{\partial}_t$. Then on each $U_j \times \Delta$, we have

$$\bar{\partial}_t \theta_j^\alpha(z_j, t) = \sum_{q=1}^{d} \varphi_{qj}^\alpha(z_j, t) \sum_{\lambda=1}^{m} c_{q\lambda}(t)\, d\bar{t}_\lambda. \tag{4.28}$$

Since $\bar{\partial}_t \varphi_{qj}^\alpha(z_j, t) = 0$ by the above lemma, we have

$$0 = \bar{\partial}_t \bar{\partial}_t \theta_j^\alpha(z_j, t) = \sum_{q=1}^{d} \varphi_{qj}^\alpha(z_j, t) \bar{\partial}_t \sum_{\lambda=1}^{m} c_{q\lambda}(t)\, d\bar{t}_\lambda.$$

Hence $\bar\partial_t \sum_{\lambda=1}^m c_{q\lambda}(t)\, d\bar t_\lambda = 0$. Since $\sum_{\lambda=1}^m c_{q\lambda}(t)\, d\bar t_\lambda$ is a (0, 1)-form on the polydisk Δ, by Dolbeault's lemma (Theorem 3.3), there exists a C^∞ function $c_q(t)$ on Δ, for each q such that

$$\bar\partial_t c_q(t) = \sum_{\lambda=1}^{m} c_{q\lambda}(t)\, d\bar t_\lambda. \tag{4.29}$$

Put

$$\tilde\theta_j(t) = \theta_j(t) - \psi(t), \qquad \psi(t) = \sum_{q=1}^{d} c_q(t)\varphi_q(t).$$

For each $t \in \Delta$, $\psi(t)$ is a holomorphic vector field on M_t, hence, $\tilde\theta_j(t)$ is a holomorphic vector field on $U_j \times t \subset M_t$. Clearly we have

$$\theta_{jk}(t) = \tilde\theta_j(t) - \tilde\theta_k(t).$$

Writing $\tilde\theta_j(t) = \sum_{\alpha=1}^n \tilde\theta_j^\alpha(z_j, t)\partial/\partial z_j^\alpha$, we have

$$\tilde\theta_j^\alpha(z_j, t) = \theta_j^\alpha(z_j, t) - \sum_{q=1}^{d} c_q(t)\varphi_{qj}^\alpha(z_j, t).$$

Therefore by (4.28) and (4.29), $\bar\partial_t \tilde\theta_j^\alpha(z_j, t) = 0$. Consequently $\tilde\theta_j^\alpha(z_j, t)$ are holomorphic functions of $z_j^1, \ldots, z_j^n, t_1, \ldots, t_m$. ∎

Corollary. *A complex analytic family* $(\mathcal{M}, B, \varpi)$ *is trivial if it is trivial as a differentiable family.*

Proofs of Theorems 4.1, 4.4 and Lemma 4.1 given in §7.2(b) are based on the theory of harmonic differential forms. Therefore the above theory of deformations of compact complex manifolds is also based on the theory of harmonic differential forms, hence cannot apply to the deformations of complex spaces. Grauert extended the above results to the case of complex spaces by developing the theory of deformations based on the theory of coherent analytic sheaves ([8]).

(d) Change of the Parameter

Suppose given a complex analytic family $\{M_t \mid M_t = \varpi^{-1}(t),\ t \in B\} = (\mathcal{M}, B, \varpi)$ of compact complex manifolds, where B is a domain of $\mathbb{C}^n$. Let D be a domain of $\mathbb{C}^r$, and $h: s \to t = h(s)$, $s \in D$, a holomorphic map of D into B. Then by changing the parameter from t to s, we obtain a complex analytic family $\{M_{h(s)} \mid s \in D\}$ on the parameter space D as follows.

We denote a point of $\mathcal{M}$ by p, and define the holomorphic map Π of $\mathcal{M}\times D$ to $B\times D$ by

$$\Pi\colon (p, s)\rightarrow (t, s)=(\varpi(p), s).$$

Then $(\mathcal{M}\times D, B\times D, \Pi)$ is a complex analytic family with the parameter space $B\times D$. We have $\Pi^{-1}(t, s)=M_t\times s$. The graph of h

$$G=\{(h(s), s)\in B\times D \mid s\in D\}$$

is a submanifold of $B\times D$, hence a complex manifold. Since the projection $P\colon B\times D\rightarrow D$ maps G biholomorphically onto D, we may identify G with D via P. $\mathcal{N}=\Pi^{-1}(G)$ is a submanifold of the complex manifold $\mathcal{M}\times D$. If we denote the restriction of Π to $\mathcal{N}$ again by Π, then $(\mathcal{N}, G, \Pi)$ is a complex analytic family over the parameter space G. Identifying G with D as mentioned above, we obtain the complex analytic family $(\mathcal{N}, D, \pi)$ where $\pi=P\circ\Pi$.

Definition 4.5. The complex analytic family $(\mathcal{N}, D, \pi)$ thus obtained is called *the complex analytic family induced from* $(\mathcal{M}, B, \varpi)$ *by the holomorphic map* $h\colon D\rightarrow B$.

Since $\Pi^{-1}(h(s), s)=M_{h(s)}\times s$, we may consider

$$\pi^{-1}(s)=\Pi^{-1}(h(s), s)=M_{h(s)}\times s=M_{h(s)}.$$

Thus $\{M_{h(s)} \mid s\in D\}$ form a complex analytic family $(\mathcal{N}, D, \pi)$.

To investigate the relation of the infinitesimal deformation of $M_{h(s)}$ and that of M_t, assume that $0\in B$, $0\in D$, and $h(0)=0$. Taking a sufficiently small coordinate polydisk Δ with $0\in\Delta\subset B$, we represent $\mathcal{M}_\Delta=\varpi^{-1}(\Delta)$ in the form of (4.27) as

$$\mathcal{M}_\Delta=\bigcup_{j=1}^{l} U_j\times\Delta,$$

where $(z_j, t)\in U_j\times\Delta$ and $(z_k, t)\in U_k\times\Delta$ are the same point of $\mathcal{M}_\Delta$ if $z_j=f_{jk}(z_k, t)$. Take a polydisk E with $0\in E\subset D$, and put

$$\mathcal{N}_E=\pi^{-1}(E)=\Pi^{-1}(G_E) \text{ where } G_E=\{(h(s), s) \mid s\in E\}.$$

$\mathcal{N}_E$ is a submanifold of $\mathcal{M}_\Delta\times E$. We have

$$\mathcal{M}_\Delta\times E=\bigcup_{j=1}^{l} U_j\times\Delta\times E,$$

where $(z_j, t, s) \in U_j \times \Delta \times E$ and $(z_k, t, s) \in U_k \times \Delta \times E$ are the same point on $\mathcal{M}_\Delta \times E$ if $z_j = f_{jk}(z_k, t)$. Consequently

$$\mathcal{N}_E = \Pi^{-1}(G_E) = \bigcup_{j=1}^{l} U_j \times G_E,$$

where $(z_j, h(s), s) \in U_j \times G_E$ and $(z_k, h(s), s) \in U_k \times G_E$ are the same point on $\mathcal{N}_E$ if $z_j = f_{jk}(z_k, h(s))$. Identifying G_E with $E = P(G_E)$ via P, we may consider

$$\mathcal{N}_E = \varpi^{-1}(E) = \bigcup_{j=1}^{l} U_j \times E.$$

Thus $\mathcal{N}_E$ is a complex manifold obtained by glueing $U_1 \times E, \ldots, U_l \times E$ by identifying $(z_k, s) \in U_k \times E$ and $(z_j, s) \in U_j \times E$ if $z_j = f_{jk}(z_k, h(s))$. Then the complex analytic family $(\mathcal{N}_E, E, \pi)$ induced from $(\mathcal{M}_\Delta, \Delta, \varpi)$ by the map $h\colon E \to \Delta$ is a complex analytic family obtained by substituting $t = h(s)$ into the functions $f_{jk}^\alpha(z_k, t)$ defining $\mathcal{M}_\Delta$.

Theorem 4.7. *For any tangent vector $\partial/\partial s \in T_s(D)$, the infinitesimal deformation of $M_{h(s)}$ along $\partial/\partial s$ is given by*

$$\frac{\partial M_{h(s)}}{\partial s} = \sum_{\lambda=1}^{m} \frac{\partial t_\lambda}{\partial s} \frac{\partial M_t}{\partial t_\lambda}, \qquad (t_1, \ldots, t_m) = h(s). \tag{4.30}$$

Proof. We put

$$\theta_{\lambda jk}(t) = \sum_{\alpha=1}^{n} \frac{\partial f_{jk}^\alpha(z_k, t_1, \ldots, t_m)}{\partial t_\lambda} \cdot \frac{\partial}{\partial z_j^\alpha},$$

$$\eta_{jk}(s) = \sum_{\alpha=1}^{n} \frac{\partial f_{jk}^\alpha(z_k, h(s))}{\partial s} \cdot \frac{\partial}{\partial z_j^\alpha}.$$

$\partial M_t/\partial t_\lambda$ is the cohomology class of the 1-cocycle $\{\theta_{\lambda jk}(t)\}$, and $\partial M_{h(s)}/\partial s$ is that of $\{\eta_{jk}(s)\}$. Since $h(s) = (t_1, \ldots, t_m)$, we have

$$\frac{\partial f_{jk}^\alpha(z_k, h(s))}{\partial s} = \sum_{\lambda=1}^{m} \frac{\partial t_\lambda}{\partial s} \cdot \frac{\partial f_{jk}^\alpha(z_k, t_1, \ldots, t_m)}{\partial t_\lambda}.$$

Hence we see that (4.30) holds. ∎

(4.30) is just the formula of the derivative of a composite function. A similar formula holds for differentiable families.

Our Theorems 4.2, 4.3, and 4.6 contain the assumption that $\dim H^1(M_t, \Theta_t)$ is independent of t. At first we did not know whether this assumption was essential or not. Since we might expect the local triviality of $(\mathcal{M}, B, \varpi)$ in case $\rho_t = 0$ if $\rho_t(\partial/\partial t) = \partial M_t/\partial t$ is truly the derivative of the complex structure of M_t, we suspected that we could get rid of this assumption. But the study of deformations of Hopf surfaces revealed the necessity of this assumption. We will explain this below.

Let $(\mathcal{M}, \mathbb{C}, \varpi)$ be the complex analytic family of Hopf surfaces $M_t = \pi^{-1}(t)$ given in Example 2.15. $(\mathcal{M}, \mathbb{C}, \varpi)$ has the following properties: (1°) If we put $U = \mathbb{C} - \{0\}$, $(\mathcal{M}_U, U, \varpi)$ is trivial; (2°) M_0 and M_t with $t \neq 0$ are not biholomorphically equivalent. Consider the complex analytic family $(\mathcal{N}, \mathbb{C}, \pi)$ induced from $(\mathcal{M}, \mathbb{C}, \varpi)$ by the holomorphic map $s \to t = s^2$. Then from (4.30) the infinitesimal deformation $\rho_s(d/ds)$ of $\pi^{-1}(s) = M_{s^2}$ is given by

$$\rho_s\left(\frac{d}{ds}\right) = \frac{dM_{s^2}}{ds} = \frac{dt}{ds}\frac{dM_t}{dt} = 2s\frac{dM_t}{dt}.$$

If $s \neq 0$, $t = s^2 \in U$, hence $dM_t/dt = 0$ because $(\mathcal{M}_U, U, \varpi)$ is trivial. Thus we have $\rho_s = 0$ identically. On the other hand, $(\mathcal{N}, \mathbb{C}, \pi)$ is not locally trivial. For, $\pi^{-1}(0) = M_0$ and $\pi^{-1}(s) = M_t$ with $t = s^2 \neq 0$ are not biholomorphically equivalent. Thus although $\rho_s = 0$ identically, $(\mathcal{N}, \mathbb{C}, \pi)$ is not locally trivial. Since $\rho_s = 0$ identically but $(\mathcal{N}, \mathbb{C}, \pi)$ is not locally trivial, $\dim H^1(M_{s^2}, \Theta_{s^2})$ must not be a constant. By making an explicit computation of $\dim H^1(M_t, \Theta_t)$ ([21]), we have

$$\dim H^1(M_t, \Theta_t) = \begin{cases} 4, & t = 0, \\ 2, & t \neq 0. \end{cases}$$

Chapter 5

Theorem of Existence

Let $(\mathcal{M}, B, \varpi)$ be a complex analytic family of compact complex manifolds where B is a domain of $\mathbb{C}$. Then the infinitesimal deformation dM_t/dt of $M_t = \varpi^{-1}(t)$ is an element of $H^1(M_t, \Theta_t)$. Consequently, given a compact complex manifold M, if $(\mathcal{M}, B, \varpi)$ with $0 \in B \subset \mathbb{C}$ is a complex analytic family such that $\varpi^{-1}(0) = M$, $(dM_t/dt)_{t=0} \in H^1(M, \Theta)$, where Θ is the sheaf of germs of holomorphic vector fields over M. Thus, if there is a complex analytic family $(\mathcal{M}, B, \varpi)$ with $\varpi^{-1}(0) = M$, the corresponding element $\theta = (dM_t/dt)_{t=0}$ of $H^1(M, \Theta)$ is determined.

Conversely, given a $\theta \in H^1(M, \Theta)$, does there exist a complex analytic family $(\mathcal{M}, B, \varpi)$ with $0 \in B \subset \mathbb{C}$ such that

$$\varpi^{-1}(0) = M, \qquad \left(\frac{dM_t}{dt}\right)_{t=0} = \theta \ ?$$

This was the next problem in the theory of deformation. In this chapter we explain the development of the theory of deformation in connexion with this problem.

§5.1. Obstructions

Lemma 5.1. *Let $U = \{(z^1, \ldots, z^n) \in \mathbb{C}^n \,|\, |z^1| < r, \ldots, |z^n| < r\}$ be a polydisk, and Θ the sheaf of germs of holomorphic vector fields over U. Then*

$$H^q(U, \Theta) = 0, \qquad q \geqq 1. \tag{5.1}$$

Proof. A germ of a holomorphic vector field $v \in \Theta_z$ at $z \in U$ is written in the form

$$v = v_1 \frac{\partial}{\partial z_1} + \cdots + v_n \frac{\partial}{\partial z_n},$$

where the coefficients v_α, $1 \leqq \alpha \leqq n$, are germs of holomorphic functions at z. Thus v can be represented as $v = (v_1, \ldots, v_n)$ with $v_\alpha \in \mathcal{O}_z$, where we denote the sheaf of germs of holomorphic functions over U by $\mathcal{O}$. Therefore

$$H^q(U,\Theta)=\underbrace{H^q(U,\mathcal{O})+\cdots+H^q(U,\mathcal{O})}_{n\text{ times}}.$$

By Dolbeault's theorem (Theorem 3.12), we have

$$H^q(U,\mathcal{O})\cong\Gamma(U,\bar\partial\mathcal{A}^{0,q-1})/\bar\partial\Gamma(U,\mathcal{A}^{0,q-1}),$$

while by Dolbeault's lemma (Theorem 3.3), for any $\bar\partial$-closed $(0,q)$-form $\varphi\in\Gamma(U,\bar\partial\mathcal{A}^{0,q-1})$ on a polydisk U, there is a $\psi\in\Gamma(U,\mathcal{A}^{0,q-1})$ such that $\varphi=\bar\partial\psi$. Therefore $H^q(U,\mathcal{O})=0$, hence $H^q(U,\Theta)=0$. ∎

Let $\mathfrak{U}=\{U_j\}$ be a finite open covering of a compact complex manifold M, and suppose that each U_j is a coordinate polydisk. Then by Lemma 5.1 above, we have $H^1(U_j,\Theta)=0$. Therefore, by Theorem 3.5 and 3.6, we obtain the following:

Lemma 5.2.

$$H^1(\mathfrak{U},\Theta)=H^1(M,\Theta), \tag{5.2}$$

$$H^2(\mathfrak{U},\Theta)\hookrightarrow H^2(M,\Theta). \tag{5.3}$$

Suppose given a compact complex manifold M, and let $(\mathcal{M},B,\varpi)$ with $0\in B\subset\mathbb{C}$ be a complex analytic family such that $\varpi^{-1}(0)=M$. Take a small disk Δ with centre 0 such that $0\in\Delta\subset B$, and represent $\mathcal{M}_\Delta=\varpi^{-1}(\Delta)$ in the form (4.27):

$$\mathcal{M}_\Delta=\bigcup_{j=1}^{l}U_j\times\Delta, \tag{5.4}$$

where each U_j is a polydisk, and $(z_j,t)\in U_j\times\Delta$ and $(z_k,t)\in U_k\times\Delta$ are the same point on $\mathcal{M}_\Delta$ if $z_j^\alpha=f_{jk}^\alpha(z_k,t)$ for $\alpha=1,\ldots,n$. Here each $f_{jk}^\alpha(z_k,t)=f_{jk}^\alpha(z_k^1,\ldots,z_k^n,t)$ is a holomorphic function of $z_k^1,\ldots,z_k^n,t$ defined on $U_k\times\Delta\cap U_j\times\Delta\neq\varnothing$.

The infinitesimal deformation $\theta(t)=dM_t/dt\in H^1(M_t,\Theta_t)$ is, by definition, the cohomology class of the 1-cocycle $\{\theta_{jk}(t)\}\in Z^1(\mathfrak{U}_t,\Theta_t)$ where $\mathfrak{U}_t=\{U_j\times t\}$, and the vector field

$$\theta_{jk}(t)=\sum_\alpha\theta_{jk}^\alpha(z_j,t)\frac{\partial}{\partial z_j^\alpha}$$

is given by

$$\theta_{jk}^\alpha(z_j,t)=\frac{\partial f_{jk}^\alpha(z_k,t)}{\partial t},\qquad z_k=f_{kj}(z_j,t). \tag{5.5}$$

Note that the functions $\theta^{\alpha}_{jk}(z_j, t)$ of $z^1_j, \ldots, z^n_j, t$ are obtained by *differentiating first the functions* $f^{\alpha}_{jk}(z_k, t)$ *of* $z^1_k, \ldots, z^n_k, t$ *with respect to* t, *and then substituting* $z^{\alpha}_k = f^{\alpha}_{kj}(z_j, t)$.

On $U_k \times \Delta \cap U_i \times \Delta \cap U_j \times \Delta \neq \varnothing$, we have the equalities

$$f^{\alpha}_{ik}(z_k, t) = f^{\alpha}_{ij}(f_{jk}(z_k, t), t), \qquad \alpha = 1, \ldots, n. \tag{5.6}$$

By differentiating both sides of these equalities in t, we have

$$\theta^{\alpha}_{ik}(z_i, t) = \theta^{\alpha}_{ij}(z_i, t) + \sum_{\beta=1}^{n} \frac{\partial z^{\alpha}_i}{\partial z^{\beta}_j} \theta^{\beta}_{jk}(z_j, t), \qquad \alpha = 1, \ldots, n. \tag{5.7}$$

As is stated in §4.2(a), these equalities imply that $\{\theta_{jk}(t)\}$ is a 1-cocycle.

Next, we consider the equalities obtained by differentiating both sides of (5.7) in t as functions of $z^1_j, \ldots, z^n_j, t$. We sometimes write $(\partial/\partial t)_j$ instead of $(\partial/\partial t)$ in order to make explicit that $\partial/\partial t$ denotes the differentiation of a function of $z^1_j, \ldots, z^n_j, t$ with respect to t. On $U_j \times \Delta \cap U_i \times \Delta \neq \varnothing$, we have the following equalities:

$$\begin{aligned} &\frac{\partial}{\partial z^{\alpha}_j} = \sum_{\beta=1}^{n} \frac{\partial z^{\beta}_i}{\partial z^{\alpha}_j} \frac{\partial}{\partial z^{\beta}_i}, \qquad \frac{\partial z^{\beta}_i}{\partial z^{\alpha}_j} = \frac{\partial f^{\beta}_{ij}(z_j, t)}{\partial z^{\alpha}_j}, \\ &\left(\frac{\partial}{\partial t}\right)_j = \sum_{\beta=1} \left(\frac{\partial z^{\beta}_i}{\partial t}\right)_j \frac{\partial}{\partial z^{\beta}_i} + \left(\frac{\partial}{\partial t}\right)_i, \qquad \left(\frac{\partial z^{\beta}_i}{\partial t}\right)_j = \frac{\partial f^{\beta}_{ij}(z_j, t)}{\partial t}. \end{aligned} \tag{5.8}$$

Consequently

$$\begin{aligned} \left(\frac{\partial}{\partial t}\right)_j \theta^{\alpha}_{ik}(z_i, t) &= \sum_{\beta=1}^{n} \frac{\partial f^{\beta}_{ij}(z_j, t)}{\partial t} \frac{\partial}{\partial z^{\beta}_i} \theta^{\alpha}_{ik}(z_i, t) + \frac{\partial \theta^{\alpha}_{ik}(z_i, t)}{\partial t} \\ &= \sum_{\beta=1}^{n} \theta^{\beta}_{ij}(z_i, t) \frac{\partial}{\partial z^{\beta}_i} \theta^{\alpha}_{ik}(z_i, t) + \frac{\partial \theta^{\alpha}_{ik}(z_i, t)}{\partial t}. \end{aligned}$$

Therefore putting

$$\dot{\theta}^{\alpha}_{ik}(z_i, t) = \frac{\partial \theta^{\alpha}_{ik}(z_i, t)}{\partial t},$$

we have

$$\left(\frac{\partial}{\partial t}\right)_j \theta^{\alpha}_{ik}(z_i, t) = \theta_{ij}(t) \cdot \theta^{\alpha}_{ik}(z_i, t) + \dot{\theta}_{ik}(z_i, t).$$

Similarly we have

$$\left(\frac{\partial}{\partial t}\right)_j \theta_{ij}^{\alpha}(z_i, t) = \theta_{ij}(t) \cdot \theta_{ij}^{\alpha}(z_i, t) + \dot{\theta}_{ij}^{\alpha}(z_i, t),$$

and

$$\left(\frac{\partial}{\partial t}\right)_j \sum_{\beta=1}^{n} \frac{\partial z_i^{\alpha}}{\partial z_j^{\beta}} \theta_{jk}^{\beta}(z_j, t)$$

$$= \sum_{\beta} \frac{\partial f_{ij}^{\alpha}(z_j, t)}{\partial z_j^{\beta}\, \partial t} \theta_{jk}^{\beta}(z_j, t) + \sum_{\beta} \frac{\partial z_i^{\alpha}}{\partial z_j^{\beta}} \dot{\theta}_{jk}^{\beta}(z_j, t)$$

$$= \sum_{\beta} \theta_{jk}^{\beta}(z_j, t) \frac{\partial}{\partial z_j^{\beta}} \theta_{ij}^{\alpha}(z_i, t) + \sum_{\beta} \frac{\partial z_i^{\alpha}}{\partial z_j^{\beta}} \dot{\theta}_{jk}^{\alpha}(z_j, t)$$

$$= \theta_{jk}(t) \cdot \theta_{ik}^{\alpha}(z_i, t) + \sum_{\beta} \frac{\partial z_i^{\alpha}}{\partial z_j^{\beta}} \dot{\theta}_{jk}^{\beta}(z_j, t).$$

Accordingly, differentiating (5.7) with respect to t, we obtain the following equalities:

$$\theta_{ij} \cdot \theta_{ik}^{\alpha}(z_i, t) + \dot{\theta}_{ik}^{\alpha}(z_i, t) = \theta_{ij}(t) \cdot \theta_{ij}^{\alpha}(z_i, t) + \dot{\theta}_{ij}^{\alpha}(z_i, t)$$
$$+ \theta_{jk}(t) \cdot \theta_{ij}^{\alpha}(z_i, t) + \sum_{\beta} \frac{\partial z_i^{\alpha}}{\partial z_j^{\beta}} \dot{\theta}_{jk}^{\beta}(z_j, t).$$

Since

$$\theta_{ik}^{\alpha}(z_i, t) - \theta_{ij}^{\alpha}(z_i, t) = \sum_{\beta} \frac{\partial z_i^{\alpha}}{\partial z_j^{\beta}} \theta_{jk}^{\beta}(z_j, t),$$

we can write the above equalities in the form

$$\begin{aligned} &\dot{\theta}_{ij}^{\alpha}(z_i, t) - \dot{\theta}_{ik}^{\alpha}(z_i, t) + \sum_{\beta=1}^{n} \frac{\partial z_i^{\alpha}}{\partial z_j^{\beta}} \dot{\theta}_{jk}^{\beta}(z_j, t) \\ &\quad = \theta_{ij}(t) \cdot \sum_{\beta} \frac{\partial z_i^{\alpha}}{\partial z_j^{\beta}} \theta_{jk}^{\beta}(z_j, t) - \theta_{jk}(t) \cdot \theta_{ij}^{\alpha}(z_i, t). \end{aligned} \tag{5.9}$$

In general we define the *bracket* of two local holomorphic vector fields $v = \sum_{\alpha=1}^{n} v_j^{\alpha}(\partial/\partial z_j^{\alpha})$ and $u = \sum_{\alpha=1}^{n} u_j^{\alpha}(\partial/\partial z_j^{\alpha})$ by

$$[v, u] = \sum_{\alpha=1}^{n} (v \cdot u_j^{\alpha} - u \cdot v_j^{\alpha}) \frac{\partial}{\partial z_j^{\alpha}}. \tag{5.10}$$

It can easily be verified that $[v, u]$ does not depend on the choice of coordinates. $[v, u]$ is bilinear in u and v, and clearly

$$[u, v] = -[v, u]$$

holds.

Putting $\dot{\theta}_{ij}(t) = \sum_{\alpha=1}^{n} \dot{\theta}_{ij}^{\alpha}(z_i, t)\, \partial/\partial z_i^{\alpha}$ and so on, and using the bracket, we rewrite (5.9) as

$$\dot{\theta}_{ij}(t) - \dot{\theta}_{ik}(t) + \dot{\theta}_{jk}(t) = [\theta_{ij}(t), \theta_{jk}(t)]. \tag{5.11}$$

Substituting $t = 0$, we obtain

$$\dot{\theta}_{ij}(0) - \dot{\theta}_{ik}(0) + \dot{\theta}_{jk}(0) = [\theta_{ij}(0), \theta_{jk}(0)]. \tag{5.12}$$

Since $M = M_0$ by assumption, identifying U_j with $U_j \times 0$, we may consider $\mathfrak{U} = \{U_j\}$ as a finite open covering of M. For a given $\theta \in H^1(M, \Theta)$, if $\theta = (dM_t/dt)_{t=0}$, θ is the cohomology class of the 1-cocycle $\{\theta_{jk}(0)\} \in Z^1(\mathfrak{U}, \Theta)$ above, hence (5.12) imposes a certain restriction on such θ. In order to make clear the meaning of this condition, we put for any 1-cocycle $\{\theta_{jk}\} \in Z^1(\mathfrak{U}, \Theta)$

$$\zeta_{ijk} = [\theta_{ij}, \theta_{jk}]$$

on $U_i \cap U_j \cap U_k \neq \varnothing$. ζ_{ijk} is a holomorphic vector field on $U_i \cap U_j \cap U_k$. $\{\zeta_{ijk}\}$ forms a 2-cocycle on U. In fact, first since

$$\zeta_{ikj} = [\theta_{ik}, \theta_{kj}] = -[\theta_{ij} + \theta_{jk}, \theta_{jk}] = -[\theta_{ij}, \theta_{jk}] = -\zeta_{ijk},$$

$$\zeta_{jik} = [\theta_{ji}, \theta_{ik}] = -[\theta_{ij}, \theta_{ij} + \theta_{jk}] = -[\theta_{ij}, \theta_{jk}] = -\zeta_{ijk},$$

ζ_{ijk} is skew-symmetric in i, j, k. Next, on $U_h \cap U_i \cap U_j \cap U_k \neq \varnothing$,

$$\begin{aligned} \zeta_{ijk} - \zeta_{hjk} + \zeta_{hik} - \zeta_{hij} &= [\theta_{ij} - \theta_{hj}, \theta_{jk}] + [\theta_{hi}, \theta_{ik} - \theta_{ij}] \\ &= -[\theta_{hi}, \theta_{jk}] + [\theta_{hi}, \theta_{jk}] = 0. \end{aligned}$$

Thus $\{\zeta_{ijk}\}$ is a 2-cocycle.

For arbitrary 1-cocycles $\{\theta_{jk}\}$ and $\{\eta_{jk}\} \in Z^1(\mathfrak{U}, \Theta)$, we put

$$\zeta_{ijk} = \tfrac{1}{2}([\theta_{ij}, \eta_{jk}] + [\eta_{ij}, \theta_{jk}]).$$

Then since

$$2\zeta_{ijk}=[\theta_{ij}+\eta_{ij},\ \theta_{jk}+\eta_{jk}]-[\theta_{ij},\ \theta_{jk}]-[\eta_{ij},\ \eta_{jk}],$$

$\{\zeta_{ijk}\}$ is also a 2-cocycle in $Z^2(\mathfrak{U}, \Theta)$, whose cohomology class ζ is uniquely determined by the cohomology classes θ of $\{\theta_{jk}\}$ and η of $\{\eta_{jk}\}$. In fact if $\theta_{jk}=\theta_k-\theta_j$, it follows from a simple calculation that

$$2\zeta_{ijk}=[\theta_j+\theta_k,\ \eta_{jk}]-[\theta_i+\theta_k,\ \eta_{ik}]+[\theta_i+\theta_j,\ \eta_{ij}],$$

namely, that $\{\zeta_{ijk}\}=\delta\{[\theta_j+\theta_k,\ \eta_{jk}]\}$. We define the bracket of θ, $\eta\in H^1(M,\Theta)$ by

$$[\theta,\ \eta]=\zeta. \tag{5.13}$$

$[\theta, \eta]$ is bilinear in θ and η, and the equality $[\theta, \eta]=-[\eta, \theta]$ holds.

Putting $i=k$ in (5.12), we obtain $\dot{\theta}_{kj}(0)+\dot{\theta}_{jk}(0)=0$ since $\dot{\theta}_{kk}(0)=0$. Thus $\{\dot{\theta}_{jk}(0)\}$ is a 1-cochain, and (5.12) can be rewritten as

$$\delta\{\dot{\theta}_{jk}(0)\}=\{\zeta_{ijk}\},\qquad \zeta_{ijk}=[\theta_{ij}(0),\ \theta_{jk}(0)].$$

Hence we obtain

$$[\theta(0),\ \theta(0)]=0. \tag{5.14}$$

From these results we obtain the following theorem immediately.

Theorem 5.1. *Suppose given a compact complex manifold M, and $\theta\in H^1(M,\Theta)$. In order that there may exist a complex analytic family $(\mathcal{M}, B, \varpi)$ such that $\varpi^{-1}(0)=M$, and that $(dM_t/dt)_{t=0}=\theta$, it is necessary that $[\theta, \theta]=0$ holds.*

In other words, if $[\theta, \theta]\neq 0$, there exists no deformation M_t with $M_0=M$ and $(dM_t/dt)_{t=0}=\theta$. In this sense we call $[\theta, \theta]\in H^2(M, \Theta)$ the *obstruction* to deformation of M.

The necessary condition $[\theta, \theta]=0$ is obtained by differentiating (5.7) with respect to t and putting $t=0$, while (5.7) is obtained by differentiating (5.6) with respect to t. Thus the condition $[\theta, \theta]=0$ is obtained from (5.6) by differentiating twice and then putting $t=0$. Similarly we obtain infinitely many necessary conditions by differentiating (5.6) m times and then putting $t=0$ for $m=3, 4, \ldots$. Thus we have infinitely many obstructions to deformation of M. In view of this fact we call $[\theta, \theta]$ the *primary obstruction.*

Theorem 5.1 above holds also for a differentiable family, as will be easily seen from its proof.

§5.2. Number of Moduli

(a) Wonder of the Number of Parameters

Let $(\mathcal{M}, B, \varpi)$ be a complex analytic family of compact complex manifolds where B is a domain of $\mathbb{C}^m$. Suppose that for any $t \in B$, the linear map

$$\rho_t: \frac{\partial}{\partial t} \to \rho_t\left(\frac{\partial}{\partial t}\right) = \frac{\partial M_t}{\partial t}, \qquad M_t = \varpi^{-1}(t),$$

of $T_t(B)$ to $H^1(M_t, \Theta_t)$ is injective. Then since if $\partial/\partial t = \sum_{\lambda=1}^m c_\lambda(\partial/\partial t_\lambda) \neq 0$, $\partial M_t/\partial t \neq 0$, we may consider that the complex structure of M_t actually varies with t. In this case we say that the parameter t is effective. More precisely,

Definition 5.1. If for every $t \in B$, ρ_t: $T_t(B) \to H^1(M_t, \Theta_t)$ is injective, we say that the parameter $t = (t_1, \ldots, t_m)$ of the complex analytic family $(\mathcal{M}, B, \varpi)$ is *effective*, and that $(\mathcal{M}, B, \varpi)$ is an *effectively parametrized complex analytic family.*

For a given compact complex manifold M, if there is an effectively parametrized complex analytic family $(\mathcal{M}, B, \varpi)$ where B is a domain of $\mathbb{C}^m$ containing 0, and $\varpi^{-1}(0) = M$, then since ρ_0: $T_0(B) \to H^1(M, \Theta)$ is injective, $\rho_0(T_0(B))$ is an m-dimensional linear subspace of $H^1(M, \Theta)$. Assume $\theta \in \rho_0(T_0(B))$. Then $\theta = \rho_0(\partial/\partial t)$ for some $\partial/\partial t = \sum_{\lambda=1}^m c_\lambda(\partial/\partial t_\lambda) \in T_0(B)$. Let $s \in \mathbb{C}$ be a variable and put

$$t(s) = (t_1(s), \ldots, t_m(s)) = (c_1 s, \ldots, c_m s).$$

Then if $\varepsilon > 0$ is sufficiently small, we have $t(s) \in B$ for $|s| < \varepsilon$. Consequently, considering the complex analytic family $\{M_{t(s)} \mid |s| < \varepsilon\}$ induced from $(\mathcal{M}, B, \varpi)$ by the holomorphic map $s \to t = t(s)$, we have $M_{t(0)} = M_0 = M$, and from (4.30),

$$\left(\frac{dM_{t(s)}}{ds}\right)_{s=0} = \sum_{\lambda=1}^m c_\lambda \left(\frac{\partial M_t}{\partial t_\lambda}\right)_{t=0} = \sum_\lambda c_\lambda \rho_0\left(\frac{\partial}{\partial t_\lambda}\right) = \rho_0\left(\frac{\partial}{\partial t}\right) = \theta.$$

Therefore by Theorem 5.1, θ satisfies the equality $[\theta, \theta] = 0$. Moreover, as was stated above, θ must satisfy infinitely many conditions. Thus, in general, $\rho_0(T_0(B)) \subsetneq H^1(M, \Theta)$, and we may expect that m is rather smaller than $\dim H^1(M, \Theta)$. Since $[\theta, \eta] = [\eta, \theta]$, if $\theta, \eta \in \rho_0(T_0(B))$, we have

$$[\theta, \eta] = \tfrac{1}{2}([\theta + \eta, \theta + \eta] - [\theta, \theta] - [\eta, \eta]) = 0.$$

If $m = \dim H^1(M, \Theta)$, we have $\rho_0(T_0(B)) = H^1(M, \Theta)$. In this case by considering the above $\{M_{t(s)} \| |s| < \varepsilon\}$, we see that for any $\theta \in H^1(M, \Theta)$, there is a complex analytic family $(\mathcal{M}, B, \varpi)$ with $0 \in B \subset \mathbb{C}$ such that $\varpi^{-1}(0) = M$ and that $(dM_t/dt)_{t=0} = \theta$.

In many examples of compact complex manifolds M, the definition of M itself contains several effective complex parameters, by use of which we can construct complex analytic family of deformations of M. For several simple cases, we compared the number m of parameters of M with $\dim H^1(M, \Theta)$ ([21], Chapter VI, §14). At first, however, since we had no precise definition of the effectivity of the parameter, we used as m the *number of the remaining parameters appeared in the definition of M after excluding the obviously ineffective ones.*

(i) Projective space $\mathbb{P}^n$. The definition of $\mathbb{P}^n$ contains no variable parameters. Hence $m = 0$. On the other hand, it is well known ([1]], see also p. 244) that $H^1(\mathbb{P}^n, \Theta) = 0$. Thus we have

$$m = \dim H^1(\mathbb{P}^n, \Theta).$$

(ii) Complex tori. Let $M = \mathbb{C}^n/G$ be a complex torus given in Example 2.8 with $G = \{\sum_{j=1}^{2n} m_j\omega_j \mid m_j \in \mathbb{Z}\}$, where $\omega_j = (\omega_j^1, \ldots, \omega_j^n) \in \mathbb{C}^n, j = 1, \ldots, 2n$, are linearly independent over $\mathbb{R}$. Among $\omega_1, \ldots, \omega_{2n}$, there are n linearly independent vectors over $\mathbb{C}$. We may assume that $\omega_{n+1}, \ldots, \omega_{2n}$ are such ones. If we take complex coordinates of $\mathbb{C}^n$ such that

$$\omega_{n+\alpha} = (0, \ldots, 0, 1^\alpha, 0, \ldots, 0), \qquad \alpha = 1, \ldots, n,$$

then the period matrix of M is given in terms of these coordinates by

$$\begin{bmatrix} \omega_1^1 & \omega_1^2 & \cdots & \omega_1^n \\ \cdots & \cdots & \cdots & \cdots \\ \omega_n^1 & \omega_n^2 & \cdots & \omega_n^n \\ 1 & 0 & \cdots & 0 \\ 0 & 1 & \cdots & 0 \\ \cdots & \cdots & \cdots & \cdots \\ 0 & 0 & \cdots & 1 \end{bmatrix}. \tag{5.15}$$

Therefore an n-dimensional complex torus M is determined by n^2 parameters $\omega_j^\alpha, j, \alpha = 1, \ldots, n$. Thus $m = n^2$. In this case *we get rid of the ineffective parameters through the coordinate change of* $\mathbb{C}^n$. On the other hand, we can easily compute $\dim H^1(M, \Theta)$ using the formula (3.148): $H^1(M, \Omega^p(F)) \cong \mathbf{H}^{p,q}(F)$. In our case since $\Theta = \mathcal{O}(T(M))$, we have

$$H^1(M, \Theta) \cong \mathbf{H}^{0,1}(T(M)). \tag{5.16}$$

G is a group of automorphisms of $\mathbb{C}^n$, and each element $\omega = (\omega^1, \dots, \omega^n) \in G$ represents the automorphism of $\mathbb{C}^n$ given by the translation

$$(z^1, \dots, z_n) \to (z^1 + \omega^1, \dots, z^n + \omega^n).$$

Vector fields, differential forms, etc., on $M = \mathbb{C}^n/G$ correspond in a one-to-one manner to G-invariant vector fields, differential forms, etc. on $\mathbb{C}^n$, respectively. Since holomorphic vector fields $\partial/\partial z^\alpha$, $\alpha = 1, \dots, n$, are G-invariant, they represent holomorphic vector fields on M, and

$$\left\{\frac{\partial}{\partial z^1}, \dots, \frac{\partial}{\partial z^n}\right\}$$

forms a basis of the tangent bundle $T(M)$ over M. Similarly

$$\{d\bar{z}^1, \dots, d\bar{z}^n\}$$

forms a basis of $\bar{T}^*(M)$ over M. Consequently any $\varphi \in \mathscr{L}^{0,1}(T(M)) = \Gamma(M, \mathscr{A}(\bar{T}^*(M) \otimes T(M))$ is represented as

$$\varphi = \sum_{\alpha=1}^{n} \sum_{\nu=1}^{n} \varphi^\alpha_{\bar{\nu}}\, d\bar{z}^\nu \otimes \frac{\partial}{\partial z^\alpha}.$$

Since the Hermitian metric $\sum_{\alpha=1}^n d\bar{z}^\alpha \otimes d\bar{z}^\alpha$ on $\mathbb{C}^n$ is G-invariant, it gives a Hermitian metric on M, which we write as $\sum_{\alpha,\beta=1}^n g_{\alpha\bar{\beta}}\, dz^\alpha \otimes d\bar{z}^\beta$ where $g_{\alpha\bar{\beta}} = \delta_{\alpha\beta}$. The complex Laplace–Beltrami operator $\Box$ for this metric is given by

$$\Box = -\sum_{\alpha=1}^{n} \frac{\partial^2}{\partial z^\alpha\, \partial \bar{z}^\alpha}$$

since $g^{\bar{\beta}\alpha} = \delta_{\alpha\beta}$, and each term of $a^{\Lambda \bar{N}\alpha}_{A\bar{B}}$, $a^{\Lambda \bar{N}\bar{\beta}}_{A\bar{B}}$ and $b^{\Lambda\bar{N}}_{A\bar{B}}$ in (3.128) contains at least one partial derivative of $g_{\alpha\bar{\beta}} = \delta_{\alpha\beta}$ or $g^{\bar{\beta}\alpha} = \delta_{\alpha\beta}$, hence, is equal to 0.

Writing a tangent vector $v \in T_z(M)$ in the form $v = \sum_{\alpha=1}^n \zeta^\alpha(\partial/\partial z^\alpha)$, we introduce the fibre coordinates $(\zeta^1, \dots, \zeta^n)$ of $T(M)$. We define a Hermitian metric on $T(M)$ by $\sum_\alpha \zeta^\alpha \bar{\zeta}^\alpha$. Then by (3.142) and (3.143), we have $\mathfrak{d}_a = \mathfrak{d}$ and

$$\Box_a = \Box = -\sum_{\alpha=1}^{n} \frac{\partial^2}{\partial z^\alpha\, \partial \bar{z}^\alpha}. \tag{5.17}$$

$$\mathrm{H}^{0,1}(T(M)) = \{\varphi \in \mathscr{L}^{0,1}(T(M)) \,|\, \Box_a \varphi = 0\},$$

while by (5.17) for $\varphi=\sum_\alpha\sum_\nu \varphi^\alpha_{\bar\nu}\, d\bar z^\nu\otimes(\partial/\partial z^\alpha)$, $\square_a\varphi=0$ means that

$$\square\varphi^\alpha_{\bar\nu}=-\sum_{\alpha=1}^{n}\frac{\partial^2}{\partial z^\alpha\,\partial\bar z^\alpha}\varphi^\alpha_{\bar\nu}=0,$$

for all $\varphi^\alpha_{\bar\nu}$. If $\square\varphi^\alpha_{\bar\nu}=0$, by (3.130)

$$\|\bar\partial\varphi^\alpha_{\bar\nu}\|^2=(\square\varphi^\alpha_{\bar\nu},\varphi^\alpha_{\bar\nu})=0,$$

hence $\bar\partial\varphi^\alpha_{\bar\nu}=0$, which means that $\varphi^\alpha_{\bar\nu}$ is a holomorphic function. Since any holomorphic function on a compact complex manifold is a constant, $\varphi^\alpha_{\bar\nu}=c^\alpha_{\bar\nu}=\text{constant}$. Thus we have

$$\mathbf{H}^{0,1}(T(M))=\left\{\varphi\;\middle|\;\varphi=\sum_{\alpha=1}^{n}\sum_{\nu=1}^{n}c^\alpha_{\bar\nu}\,d\bar z^\nu\otimes\frac{\partial}{\partial z^\alpha},\ c^\alpha_{\bar\nu}\in\mathbb{C}\right\}. \tag{5.18}$$

Consequently $\dim\mathbf{H}^{0,1}(T(M))=n^2$. Hence $\dim H^1(M,\Theta)=n^2$. Thus we have

$$m=\dim H^1(M,\Theta)$$

in this case.

(iii) Hypersurfaces in the projective space. Let $\zeta=(\zeta_0,\dots,\zeta_{n+1})$ be the homogeneous coordinates on $\mathbb{P}^{n+1}$. The number $(\mu+1)$ of the monomials $\zeta_0^{h_0}\cdots\zeta_{n+1}^{h_{n+1}}$ of degree $h=h_0+\cdots+h_{n+1}$ in $(n+2)$ variables $\zeta_0,\dots,\zeta_{n+1}$ is given by

$$\mu+1=\binom{n+1+h}{h}. \tag{5.19}$$

We associate with each point $t=(t_0,\dots,t_\mu)$ of the μ-dimensional projective space $\mathbb{P}^\mu$ the homogeneous polynomial of degree h given by

$$t(\zeta)=t_0\zeta_0^h+t_1\zeta_0^{h-1}\zeta_1+\cdots+t_\mu\zeta_{n+1}^h.$$

In order that the hypersurface $t(\zeta)=0$ has a singularity, $(t_0,\dots,t_\mu)$ must satisfy certain algebraic conditions. (See p. 42.) Therefore there is an algebraic subset $\mathfrak{S}\subsetneq\mathbb{P}^\mu$ such that the hypersurface M_t given by $t(\zeta)=0$ for $t\notin\mathfrak{S}$ is a submanifold of $\mathbb{P}^{n+1}$. Of course M_t is an n-dimensional algebraic manifold. We call M_t a non-singular hypersurface of *order* h in $\mathbb{P}^{n+1}$.

We shall treat the case $n=1$ later. The case $h=1$, $M_t=\mathbb{P}^n$, hence, it contains no parameter. In case $h=2$, M_t is a quadric hypersurface, which can be transformed into the normal form: $\zeta_0^2+\cdots+\zeta_{n+1}^2=0$ by a change of homogeneous coordinates. Thus it also contains no parameter. We consider now the cases $n \geqq 2$ and $h \geqq 3$.

First we must calculate the number of effective parameters. Put $B = \mathbb{P}^{\mu} - \mathfrak{S}$. Then $\{M_t \mid t \in B\}$ forms a complex analytic family, but its parameter space B is not effective. In fact, consider a projective transformation of $\mathbb{P}^{n+1}$:

$$\gamma: \zeta \to \zeta' = \gamma\zeta, \qquad \zeta'_\alpha = \sum_{\beta=0}^{n+1} \gamma_{\alpha\beta}\zeta_\beta, \qquad \det(\gamma_{\alpha\beta}) \neq 0.$$

Letting $(\gamma_{\alpha\beta})^{-1} = (\gamma^{-1}_{\alpha\beta})$ be the inverse matrix of $(\gamma_{\alpha\beta})$, we have $\zeta_\alpha = \sum_{\beta=0}^{n+1} \gamma^{-1}_{\alpha\beta}\zeta'_\beta$. Therefore any monomial of degree h in $\zeta_0, \ldots, \zeta_{n+1}$ is represented as a linear form of monomials of degree h in $\zeta'_0, \ldots, \zeta'_{n+1}$. Consequently if we let $t(\zeta) = t'(\zeta')$, $t \to t'$ is a projective transformation of $\mathbb{P}^{\mu}$, which we denote by γ:

$$\gamma: t \to t' = \gamma t, \qquad t(\zeta) = t'(\zeta').$$

Since γ is an automorphism of $\mathbb{P}^{n+1}$, γM_t and M_t are biholomorphically equivalent. Since $\zeta \in M_t$ if and only if $\zeta' \in \gamma M_t$, γM_t is a hypersurface given by $t'(\zeta) = 0$, namely,

$$\gamma M_t = M_{t'} = M_{\gamma t}.$$

Let $\mathfrak{G}$ be the projective transformation group of $\mathbb{P}^{n+1}$. Then since each element $\gamma \in \mathfrak{G}$ is determined by the ratio of $(n+2)^2$ components $\gamma_{\alpha\beta}$, $\alpha, \beta = 0, 1, \ldots, n+1$, of the matrix $(\gamma_{\alpha\beta})$, $\dim \mathfrak{G} = (n+2)^2 - 1$. For any point $s \in B$, $\mathfrak{G}s$ is a submanifold of dimension $(n+2)^2 - 1$, and as long as $t = \gamma s \in \mathfrak{G}s$, the complex structure of $M_t = \gamma M_s$ does not vary. Therefore among μ components of the local coordinates of B, $((n+2)^2 - 1)$ ones are ineffective parameters for M_t. Thus the number of the effective parameters of M_t is given by

$$m = \mu - (n+2)^2 + 1.$$

Hence by (5.19) we have

$$m = \binom{n+1+h}{h} - (n+2)^2. \tag{5.20}$$

On the other hand, calculating $\dim H^1(M_t, \Theta_t)$, we see that except in the case $n = 2$, $h = 4$, we have

$$\dim H^1(M_t, \Theta_t) = \binom{n+1+h}{h} - (n+2)^2. \tag{5.21}$$

Thus

$$m = \dim H^1(M_t, \Theta_t)$$

except in the case $n=2$, $h=4$. We shall give the calculation of $\dim H^1(M_t, \Theta_t)$ in (c) of this section.

In case $n=2, h=4$, that is, in case M_t is a quartic surface in $\mathbb{P}^3$, $\dim H^1(M_t, \Theta_t) = 20$ (see p. 308). Since by (5.19), $m = \binom{7}{4} - 4^2 = 19$, $m < \dim H^1(M_t, \Theta_t)$ in this case.

(iv) Quadratic transform of $\mathbb{P}^2$. Take ν distinct points $q_1, \ldots, q_\nu$ on $\mathbb{P}^2$, and let M be the algebraic surface obtained by applying quadratic transformations Q_{q_λ} at each q_λ for $\lambda = 1, \ldots, \nu$:

$$M = Q_{q_\nu} \cdots Q_{q_1}(\mathbb{P}^2). \tag{5.22}$$

We compute the number of effective parameters of this M when $q_1, \ldots, q_\nu$ move on $\mathbb{P}^2$. Choose the homogeneous coordinates on $\mathbb{P}^2$ such that

$$q_1 = (0,0,1), \qquad q_2 = (1,0,0), \qquad q_3 = (0,1,0), \qquad q_4 = (1,1,1).$$

Assume for simplicity that $q_5, \ldots, q_\nu$ do not lie on the line through q_2 and q_3, and put

$$q_\lambda = (\zeta_{\lambda 1}, \zeta_{\lambda 2}, 1), \qquad \lambda = 5, \ldots, \nu.$$

Them M is determined by $2(\nu - 4)$ parameters $\zeta_{51}, \zeta_{52}, \ldots, \zeta_{\nu 1}, \zeta_{\nu 2}$. In case $\nu \leqq 4$, M contains no parameters. Thus we have

$$m = \begin{cases} 0, & \nu \leqq 4, \\ 2\nu - 8, & \nu > 5, \end{cases} \tag{5.23}$$

for M.

Next we compute $\dim H^1(M, \Theta)$ using the Riemann-Roch-Hirzebruch theorem. For a while let M be an arbitrary algebraic surface. We call $q = \dim H^0(M, \Omega^1)$ the *irregularity* of M, $p_g = \dim H^0(M, \Omega^2)$ the *geometric genus* of M, and $p_a = p_g - q$ the *arithmetic genus* of M. By the Riemann-Roch-Hirzebruch theorem ([10]),

$$\dim H^0(M, \Theta) - \dim H^1(M, \Theta) + \dim H^2(M, \Theta) = \tfrac{1}{6}(7c_1^2 - 5c_2), \tag{5.24}$$

where $c_1 \in H^2(M, \mathbb{Z})$ is the first Chern class of M and c_2 is the Euler number of M. c_1^2 is an integer. Classical Noether's formula

$$12(p_a + 1) = c_1^2 + c_2 \tag{5.25}$$

also follows from the Riemann–Roch–Hirzebruch theorem. By (5.24) and (5.25) we obtain

$$\dim H^1(M,\Theta)=2c_2-14(p_a+1)+\dim H^0(M,\Theta)+\dim H^2(M,\Theta). \quad (5.26)$$

Again let M be the algebraic surface given by (5.22). If $M=\mathbb{P}^2$, $H^1(\mathbb{P}^2,\Theta)=0$ as stated in (i). By (3.157)

$$\dim H^2(\mathbb{P}^2,\Theta)=\dim H^0(M,\Omega^1(K)). \quad (5.27)$$

Let $\{(z_j^1, z_j^2)\}$ be a system of local complex coordinates on M. Then $\psi\in H^0(M,\Omega^1(K))$ is represented as

$$\psi=\frac{1}{2}\sum_{\alpha=1}^{2}\sum_{\beta,\gamma=1}^{2}\psi_{j\alpha\beta\gamma}\,dz_j^\alpha\otimes(dz_j^\beta\wedge dz_j^\gamma),$$

namely, ψ is a covariant tensor field $\psi_{\alpha\beta\gamma}$, $\psi_{\alpha\beta\gamma}=-\psi_{\alpha\gamma\beta}$ which is holomorphic on all of M. Consequently $H^2(\mathbb{P}^2,\Theta)=0$ follows from the fact that there is no non-zero holomorphic tensor field on $\mathbb{P}^2$, which is easily verified as follows.

Let $U_j=\{\zeta\in\mathbb{P}^1\,|\,\zeta_j\neq 0\}$, for $j=0,1,2$. Then each U_j is a copy of $\mathbb{C}^2$, and $\mathbb{P}^2=U_0\cup U_1\cup U_2$. Putting $z^1=\zeta_1/\zeta_0$ and $z^2=\zeta_2/\zeta_0$, we get the coordinates (z^1,z^2) on U_0. Similarly $(w^1,w^2)=(\zeta_0/\zeta_1,\zeta_2/\zeta_1)$ gives the coordinates on U_1. On $U_0\cap U_1$,

$$w^1=\frac{1}{z^1},\quad w^2=\frac{z^2}{z^1},\quad z^1=\frac{1}{w^1},\quad z^2=\frac{w^2}{w^1}. \quad (5.28)$$

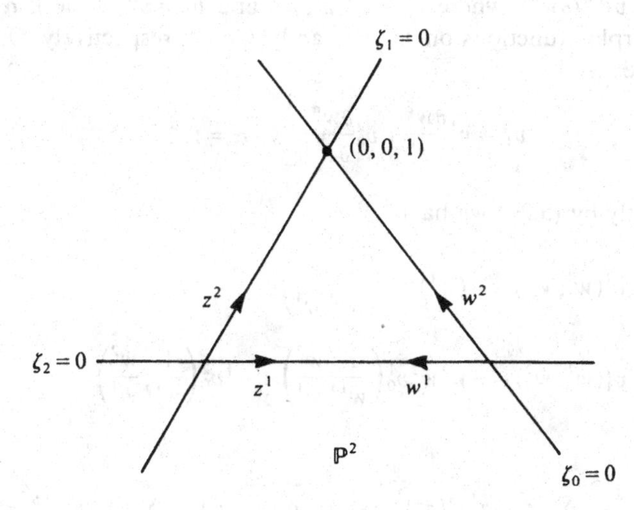

Figure 1

hold. A holomorphic covariant tensor field ψ of, say, rank 2 on $\mathbb{P}^2$ is written on U_0 as $\psi=\sum_{\alpha,\beta=1}^{2}\psi_{0\alpha\beta}\,dz^\alpha\otimes dz^\beta$ and on U_1 as $\psi=\sum_{\alpha,\beta=1}^{2}\psi_{1\alpha\beta}\,dw^\alpha\otimes dw^\beta$, where $\psi_{0\alpha\beta}$ and $\psi_{1\alpha\beta}$ are holomorphic functions on U_0 and U_1, respectively. On $U_0\cap U_1$, we have

$$\psi_{0\alpha\beta}=\sum_{\gamma,\delta=1}^{2}\psi_{1\gamma\delta}\frac{\partial w^\gamma}{\partial z^\alpha}\frac{\partial w^\delta}{\partial z^\beta},$$

while by (5.28) we have

$$\frac{\partial w^1}{\partial z^1}=-(w^1)^2,\qquad \frac{\partial w^2}{\partial z^1}=-w^1w^2,\qquad \frac{\partial w^1}{\partial z^2}=0,\qquad \frac{\partial w^2}{\partial z^2}=w^1. \tag{5.29}$$

Thus each $\partial w^\gamma/\partial z^\alpha$, α, $\gamma=1,2$, can be extended to a holomorphic function on U_1, which vanishes on the line $\zeta_0=0$. Therefore $\psi_{0\alpha\beta}$ is extended to a holomorphic function on $U_0\cup U_1=\mathbb{P}^2-\{(0,0,1)\}$ which vanishes on $\zeta_0=0$. This extended holomorphic function is again extended to all of $\mathbb{P}^2$ by Hartog's theorem (Theorem 1.8). Thus we have shown that $\psi_{0\alpha\beta}$ is a holomorphic function on all of $\mathbb{P}^2$, vanishing on the line $\zeta_0=0$. Since a holomorphic function on a compact complex manifold is a constant, $\psi_{0\alpha\beta}$ vanishes identically, hence, a covariant tensor field ψ is identically equal to 0.

A holomorphic differential form is a holomorphic skew-symmetric covariant tensor field. Consequently $H^0(\mathbb{P}^2,\Omega^1)=0$, and $H^0(\mathbb{P}^2,\Omega^2)=0$. Thus the *irregularity* q, the *geometric genus* p_g and *the arithmetic genus* p_a of $\mathbb{P}^2$ *are all equal to* 0.

Similarly we can determine holomorphic vector fields on $\mathbb{P}^2$. Let $v\in H^0(\mathbb{P}^2,\Theta)$. v is written as $v=v_0^1(\partial/\partial z^1)+v_0^2(\partial/\partial z^2)$ on U_0, and as $v=v_1^1(\partial/\partial w^1)+v_1^2(\partial/\partial w^2)$, where $v_0^\alpha=v_0^\alpha(z^1,z^2)$ and $v_1^\alpha=v_1^\alpha(w^1,w^2)$, $\alpha=1,2$, are holomorphic functions on $U_0=\mathbb{C}^2$ and $U_1=\mathbb{C}^2$, respectively. On $U_0\cap U_1$, we have

$$v_1^\alpha=v_0^1\frac{\partial w^\alpha}{\partial z^1}+v_0^2\frac{\partial w^\alpha}{\partial z^2},\qquad \alpha=1,2.$$

Consequently by (5.29) we have

$$v_1^1(w^1,w^2)=-(w^1)^2v_0^1\left(\frac{1}{w^1},\frac{w^2}{w^1}\right),$$

$$v_1^2(w^1,w^2)=-w^1w^2v_0^1\left(\frac{1}{w^1},\frac{w^2}{w^1}\right)+w^1v_0^2\left(\frac{1}{w^1},\frac{w^2}{w^1}\right).$$

Put

$$v_0^1(z^1,z^2)=\sum_{h,k=0}^{\infty}a_{hk}(z^1)^h(z^2)^k\quad\text{and}\quad v_0^2(z^1,z^2)=\sum_{h,k=0}^{\infty}b_{hk}(z^1)^h(z^2)^k.$$

Then by (5.28)

$$v_1^1(w^1, w^2) = -\sum_{h,k=0}^{\infty} \frac{a_{hk}(w^2)^k}{(w^1)^{h+k-2}},$$

$$v_1^2(w^1, w^2) = \sum_{h,k=0}^{\infty} \frac{b_{hk}(w^2)^k - a_{hk}(w^2)^{k+1}}{(w^1)^{h+k-1}}.$$

We must determine a_{hk}, b_{hk} such that these two power series contain no terms of negative exponent for w^1. Thus a holomorphic vector field v is written as

$$v = (c_1 + c_2 z^1 + c_3 z^2 + c_7 (z^1)^2 + c_8 z^1 z^2)\frac{\partial}{\partial z^1}$$

$$+ (c_4 + c_5 z^1 + c_6 z^2 + c_7 z^1 z^2 + c_8 (z^2)^2)\frac{\partial}{\partial z^2},$$

where $c_1, \ldots, c_8$ are arbitrary constants. Therefore we obtain

$$\dim H^0(\mathbb{P}^2, \Theta) = 8. \tag{5.30}$$

Since the Euler number of $\mathbb{P}^2$ is equal to 3, from (5.26) we have $H^1(\mathbb{P}^2, \Theta) = 0$.

In order to compute $\dim H^1(M, \Theta)$ for $M = Q_{q_\nu} \cdots Q_{q_1}(\mathbb{P}^2)$, it suffices to see how each term of (5.26) changes under a quadratic transformation. For this take an arbitrary point $q \in M$, and put $\tilde{M} = Q_q(M)$. As stated in Example 2.12, $\tilde{M}$ is a surface obtained from M by replacing q by the projective line $Q_q(q) = \tilde{S} = \mathbb{P}^1$:

$$\tilde{M} = (M - \{q\}) \cup \tilde{S}.$$

Therefore $\tilde{M} - \tilde{S} = M - \{q\}$.

First we investigate the correspondence between holomorphic tensor fields on $\tilde{M}$ and those on M. Given a holomorphic tensor field $\tilde{\psi}$ on $\tilde{M}$, by restricting $\tilde{\psi}$ to $\tilde{M} - \tilde{S}$, we get a holomorphic tensor field ψ' on $M - \{q\} = \tilde{M} - \tilde{S}$. By Hartog's theorem, ψ' is extended to a holomorphic tensor field ψ on M which coincides with $\tilde{\psi}$ on $M - \{q\} = \tilde{M} - \tilde{S}$. Thus to a holomorphic tensor field $\tilde{\psi}$ on $\tilde{M}$ corresponds a holomorphic tensor field ψ on M.

Conversely, given a holomorphic tensor field ψ on M and by restricting ψ to $M - \{q\}$, we obtain a holomorphic tensor field ψ' on $\tilde{M} - \tilde{S} = M - \{q\}$. We examine the possibility of extension of ψ' to all of $\tilde{M}$. Let (z_1, z_2) be local coordinates with centre q, and put $W = \{(z_1, z_2) \,|\, |z_1| < \varepsilon, |z_2| < \varepsilon\}$ for a sufficiently small $\varepsilon > 0$. Then

$$\tilde{W} = Q_q(W) = \{(z_1, z_2, \zeta) \in W \times \mathbb{P}^1 \,|\, \zeta_1 z_2 - \zeta_2 z_1 = 0\},$$

where (ζ_1, ζ_2) denotes the homogeneous coordinates of $\zeta \in \mathbb{P}^1$. Putting $U_1 = \{\zeta \mid \zeta_2 \neq 0\}$ and $U_2 = \{\zeta \mid \zeta_1 \neq 0\}$, we have $\mathbb{P}^1 = U_1 \cup U_2$. Therefore putting $\tilde{W}_\alpha = \tilde{W} \cap W \times U_\alpha$, $\alpha = 1, 2$, we have $\tilde{W} = \tilde{W}_1 \cup \tilde{W}_2$. Let $w_1 = \zeta_1/\zeta_2$ be the non-homogeneous coordinate of the point $\zeta \in U_1$. Then we have

$$\tilde{W}_1 = \{(z_1, z_2, w_1) \in W \times U_1 \mid z_1 = w_1 z_2\}.$$

Thus we may use (w_1, z_2) as coordinates on $\tilde{W}_1$. We write w_2 instead of z_2. Then $\tilde{W}_1 = \{(w_1, w_2) \in \mathbb{C}^2 \| w_2| < \varepsilon, |w_1 w_2| < \varepsilon\}$, and $Q_q^{-1}: \tilde{W}_1 \to W$ is given by

$$Q_q^{-1}: (w_1, w_2) \to (z_1, z_2) = (w_1 w_2, w_2).$$

$\tilde{S} \cap \tilde{W}_1$ is a submanifold of $\tilde{W}_1$ defined by $w_2 = 0$: $\tilde{S} \cap \tilde{W}_1 = \{(w_1, 0) \mid w_1 \in \mathbb{C}\}$. Therefore

$$\tilde{W}_1 - \tilde{S} = \{(w_1, w_2) \in \tilde{W}_1 \mid w_2 \neq 0\} \hookrightarrow W - \{q\},$$

where the inclusion map $\hookrightarrow$ is given by $(w_1, w_2) \to (z_1, z_2) = (w_1 w_2, w_2)$.

Let ψ be a covariant tensor field of, say, rank 2. Then on W, ψ is written as

$$\psi = \sum_{\alpha,\beta=1}^{2} \psi_{\alpha\beta}(z_1, z_2)\, dz_\alpha \otimes dz_\beta,$$

where $\psi_{\alpha\beta}(z_1, z_2)$ are holomorphic functions of z_1, z_2. Therefore if we represent ψ' on $\tilde{W}_1 - \tilde{S} \subset W - \{q\}$ as

$$\psi' = \sum_{\alpha,\beta=1}^{2} \psi'_{\alpha\beta}(w_1, w_2)\, dw_\alpha \otimes dw_\beta,$$

then

$$\psi'_{\alpha\beta}(w_1, w_2) = \sum_{\gamma,\delta=1}^{2} \psi_{\gamma\delta}(w_1 w_2, w_2) \frac{\partial z_\gamma}{\partial w_\alpha} \frac{\partial z_\delta}{\partial w_\beta},$$

while we have

$$\frac{\partial z_1}{\partial w_1} = w_2, \qquad \frac{\partial z_1}{\partial w_2} = w_1, \qquad \frac{\partial z_2}{\partial w_1} = 0, \qquad \frac{\partial z_2}{\partial w_2} = 1.$$

Consequently $\psi'_{\alpha\beta}(w_1, w_2)$ is extended to a holomorphic function on $\tilde{W}_1$, hence ψ' is extended to a holomorphic tensor field on $\tilde{W}_1$. Since similar result holds for $\tilde{W}_2$, ψ' is extended to a holomorphic tensor field $\tilde{\psi}$ on $\tilde{M}$.

Thus, if ψ is a holomorphic covariant tensor field, ψ' is extended to a covariant tensor field on $\tilde{M}$. Consequently *there is a one-to-one correspondence between the holomorphic covariant tensor fields* $\tilde{\psi}$ *on* $\tilde{M} = Q_q(M)$ *and those* ψ *on* M.

Next consider the case ψ is a holomorphic vector field on M. In this case ψ is represented on W as

$$\psi = \psi_1(z_1, z_2)\frac{\partial}{\partial z_1} + \psi_2(z_1, z_2)\frac{\partial}{\partial z_2},$$

where $\psi_1(z_1, z_2)$, $\psi_2(z_1, z_2)$ are holomorphic functions of z_1, z_2. Since $z_1 = w_1 w_2$, $z_2 = w_2$ and $w_1 = z_1/z_2$ on $\tilde{W}_1 - \tilde{S} \subset W - \{q\}$, we have

$$\frac{\partial}{\partial z_1} = \frac{1}{w_2}\frac{\partial}{\partial w_1}, \qquad \frac{\partial}{\partial z_2} = -\frac{w_1}{w_2}\frac{\partial}{\partial z_1} + \frac{\partial}{\partial z_2}.$$

Therefore we have

$$\psi' = \frac{\psi_1(w_1 w_2, w_2) - w_1 \psi_2(w_1 w_2, w_2)}{w_2}\frac{\partial}{\partial w_1} + \psi_2(w_1 w_2, w_2)\frac{\partial}{\partial w_2}.$$

Expanding $\psi_1(z_1, z_2)$, $\psi_2(z_1, z_2)$ into the power series in z_1, z_2, we see that the coefficient of $\partial/\partial w_1$ in ψ has the form

$$\frac{\psi_1(0,0) - w_1 \psi_2(0,0)}{w_2} + \text{a holomorphic function}$$

on $\tilde{W}_1$. Consequently ψ' is extended to a holomorphic vector field on $\tilde{W}_1$ if and only if $\psi_1(0,0) = \psi_2(0,0) = 0$, that is, ψ vanishes at q. Thus a holomorphic vector field $\tilde{\psi}$ on $\tilde{M} = Q_q(M)$ corresponds in a one-to-one manner to a holomorphic vector field ψ on M which vanishes at q.

$H^0(M, \Omega^1)$ is the vector space of all holomorphic covariant vector fields on M, $H^0(M, \Omega^2)$ is that of all holomorphic skew-symmetric covariant tensor fields of rank 2, and $H^0(M, \Omega^1(K))$ is that of all holomorphic covariant tensor fields $\psi_{\alpha\beta\gamma}$ of rank 3 which are skew-symmetric in the indices β, γ. Therefore from the above results, as to $M = Q_{q_\nu} \cdots Q_{q_1}(\mathbb{P}^2)$, we see that $H^0(M, \Omega^1)$, $H^0(M, \Omega^2)$ and $H^0(M, \Omega^1(K))$ are isomorphic to $H^0(\mathbb{P}^2, \Omega^1)$, $H^0(\mathbb{P}^2, \Omega^2)$ and $H^0(\mathbb{P}^2, \Omega^1(K))$ respectively. Consequently the irregularity q, the geometric genus p_g and $\dim H^2(M, \Theta) = \dim H^0(M, \Omega^1(K))$ all vanish, hence also the arithmetic genus p_a of M vanishes. Again from the above consideration, $H^0(M, \Theta)$ is isomorphic to the subspace of $H^0(\mathbb{P}^2, \Theta)$ consisting of holomorphic vector fields which vanishes at q_λ, $\lambda = 1, \ldots, \nu$. Therefore by (5.30) we have

$$\dim H^0(M, \Theta) = \begin{cases} 8 - 2\nu, & \nu \leqq 3, \\ 0, & \nu \geqq 4. \end{cases}$$

On the other hand, since the Euler number of q_λ is equal to 1, and that of $Q_{q_\lambda}(q_\lambda) = \mathbb{P}^1$ is equal to 2, the Euler number increases by one if we replace a point q_λ by $Q_{q_\lambda}(q_\lambda)$. Therefore the Euler number c_2 of $M = Q_{q_\nu} \cdots Q_{q_1}(P^2)$ is equal to $3+\nu$. Substituting these values into (5.26), we obtain

$$\dim H^1(M, \Theta) = \begin{cases} 0, & \nu \leqq 4, \\ 2\nu - 8, & \nu \geqq 5. \end{cases}$$

Consequently by (5.23) we have

$$m = \dim H^1(M, \Theta).$$

So far for several examples of compact complex manifolds M, we have compared $\dim H^1(M, \Theta)$ with the number of the parameters of M roughly computed. Except only one example—the quartic surfaces in $\mathbb{P}^3$—we have seen that the equality

$$m = \dim H^1(M, \Theta) \tag{5.31}$$

holds. *This seemed very strange* since at that time nothing was known concerning the existence of deformation of M. As stated at the head of this section, we had expected that in general the number m might be rather smaller than $\dim H^1(M, \Theta)$. The effort to explain this strange phenomenon by proving the theorem of existence for deformations contributed much to the subsequent development of the theory of deformation. In this sense *it may be said that the theory of deformation was at first an experimental science.*

A complex manifold of dimension 1 is just a Riemann surface. The idea of deformation of a Riemann surface goes back to Riemann. Let M be a compact Riemann surface of genus g. If $g=0$, M is $\mathbb{P}^1$, and if $g=1$, M is a 1-dimensional complex torus. Therefore we assume in the following that $g \geqq 2$. According to Riemann's well-known formula, the complex structure of a Riemann surface of genus $g \geqq 2$ depends on $3g-3$ parameters.

According to the theory of Teichmüller ([30]), the number of parameters of a compact Riemann surface M is given by

$$m = \dim H^0(M, \mathcal{O}(T^*(M) \otimes T^*(M))).$$

Let $\{z_j\}$ be the system of local complex coordinates, and U_j the domain of z_j. Then an arbitrary $\psi \in H^0(M, \mathcal{O}(T^*(M) \otimes T^*(M)))$ is written on each U_j in the form

$$\psi = \psi_j \, dz_j \otimes dz_j,$$

where ψ_j is a holomorphic function on U_j. Consequently ψ is called a holomorphic quadratic differential. There exist g linearly independent

Abelian differentials of the first kind on M. Let ω_0 be one of them, and $\mathfrak{k}=(\omega_0)$ the divisor of ω_0. The degree of $\mathfrak{k}$ is given by

$$\deg \mathfrak{k} = 2g-2.$$

Since we assume $g \geqq 2$, we have $\mathfrak{k}>0$. Since on each U_j, we can write $\omega_0=\omega_{0j}\,dz_j$ with holomorphic ω_{0j}, if we write a holomorphic quadratic differential ψ as

$$\psi = f\cdot \omega\otimes\omega,$$

f is a meromorphic function on M and we have $(f)+2\mathfrak{k}\geqq 0$. Thus we have $f\in\mathscr{F}(2\mathfrak{k})$. Conversely if $f\in\mathscr{F}(2\mathfrak{k})$, $\psi=f\cdot\omega\otimes\omega$ is a holomorphic quadratic differential on M. (For notation, see [17]). Therefore

$$H^0(M, \mathcal{O}(T^*(M)\otimes T^*(M)))\cong\mathscr{F}(2\mathfrak{k}).$$

By the Riemann–Roch formula, we have

$$\dim\mathscr{F}(2\mathfrak{k})-\dim\mathscr{F}(-\mathfrak{k})=\deg 2\mathfrak{k}-g+1.$$

On the other hand, since $\mathfrak{k}>0$, $\mathscr{F}(-\mathfrak{k})=0$, and $\deg\mathfrak{k}=2g-2$, hence $\dim\mathscr{F}(2\mathfrak{k})=3g-3$. Thus we obtain

$$m=\dim H^0(M, \mathcal{O}(T^*(M)\otimes T^*(M)))=3g-3.$$

By (3.157)

$$\dim H^1(M,\Theta)=\dim H^0(M,\Omega^1(K)),$$

while $K=T^*(M)$ in our case. Consequently $\Omega^1(K)=\mathcal{O}(T^*(M)\otimes T^*(M))$, hence

$$m=\dim H^1(M,\Theta).$$

Thus *if M is a compact Riemann surface,* $\dim H^1(M,\Theta)$ *coincides with the number of parameters of M already known long before.*

(b) The Number of Moduli

In the preceding section, we computed the number of parameters for several examples of compact complex manifolds, without giving their precise definition. In this section we shall give the precise definition. We begin with the definition of the completeness of a complex analytic family.

Definition 5.2. Let $(\mathcal{M}, B, \varpi)$ be a complex analytic family of compact complex manifolds, and $t^0 \in B$. $(\mathcal{M}, B, \varpi)$ is called *complete at* $t^0 \in B$ if for any complex analytic family $(\mathcal{N}, D, \pi)$ such that D is a domain of $\mathbb{C}^l$ containing 0 and that $\pi^{-1}(0) = \varpi^{-1}(t^0)$, there are a sufficiently small domain E with $0 \in E \subset D$, and a holomorphic map $h\colon s \to t = h(s)$ with $h(0) = 0$ such that $(\mathcal{N}_E, E, \pi)$ is the complex analytic family induced from $(\mathcal{M}, B, \varpi)$ by h where $\mathcal{N}_E = \pi^{-1}(E)$.

$N_s = \pi^{-1}(s)$ is a deformation of $N_0 = M_{t^0} = \varpi^{-1}(t^0)$. If $(\mathcal{N}_E, E, \pi)$ is induced from $(\mathcal{M}, B, \varpi)$ by h, $N_s = M_{h(s)} = \varpi^{-1}(h(s))$ for $s \in E$. Consequently *if* $(\mathcal{M}, B, \varpi)$ *is complete at* $t^0 \in B$, $(\mathcal{M}, B, \varpi)$ *contains any deformation* N_s *of* M_{t^0} *provided that* s *is sufficiently small. In this sense we may say that* $(\mathcal{M}, B, \varpi)$ *contains all sufficiently small deformations of* $M_{t^0} = \varpi^{-1}(t^0)$.

Definition 5.3. $(\mathcal{M}, B, \varpi)$ is called a *complete complex analytic family* if $(\mathcal{M}, B, \varpi)$ is complete at every point $t \in B$.

A complete complex analytic family $(\mathcal{M}, B, \varpi)$ contains all sufficiently small deformations of each $M_t = \varpi^{-1}(t)$ for $t \in B$.

Definition 5.4. Let M be a compact complex manifold. If there is an effectively parametrized and complete complex analytic family $(\mathcal{M}, B, \varpi)$ with $\varpi^{-1}(0) = M$ where B is a domain of $\mathbb{C}^m$ containing 0, we call $m = \dim B$ the *number of moduli* of M and denote it by $m(M)$.

We don't define the number of moduli of M if there exists no such family.

The number $m(M)$ of moduli of M is the number of effective parameters of the complex analytic family $(\mathcal{M}, B, \varpi)$ with $\varpi^{-1}(0) = M$ which contains all sufficiently small deformations of M. Thus the number of parameters of M in the precise sense is considered to be its number of moduli $m(M)$. If the number of moduli of M is not defined, we consider that the number of parameters of M cannot be determined. We will explain the reason of this later (p. 314).

We must verify that the number of moduli of M does not depend on the choice of complex analytic family used for the definition. For this, suppose given another effectively parametrized and complete complex analytic family $(\mathcal{N}, D, \pi)$ where D is a domain of $\mathbb{C}^l$ containing 0, and $\pi^{-1}(0) = M$. We must show $l = m$. Since $\pi^{-1}(0) = M = \varpi^{-1}(0)$, and $(\mathcal{M}, B, \varpi)$ is complete, there is a domain E with $0 \in E \subset D$ such that $(\mathcal{N}_E, E, \pi)$ with $N_E = \pi^{-1}(E)$ is the family induced from $(\mathcal{M}, B, \varpi)$ by a holomorphic map $h\colon E \to B$ with $h(0) = 0$. Put $M_t = \varpi^{-1}(t)$. Then $N_s = \pi^{-1}(s) = M_{h(s)}$. Since by assumption $s = (s_1, \ldots, s_l)$ is an effective parameter of $(\mathcal{N}, D, \pi)$, putting $t = h(s)$ for $s \in E$, we see that

$$\frac{\partial N_s}{\partial s_\mu} \in H^1(N_s, \Theta_s) = H^1(M_t, \Theta_t), \qquad \mu = 1, \ldots, l, \tag{5.32}$$

are linearly independent. On the other hand by (4.30), we have

$$\frac{\partial N_s}{\partial s_\mu} = \sum_{\lambda=1}^{m} \frac{\partial t_\lambda}{\partial s_\mu} \frac{\partial M_t}{\partial t_\lambda}, \qquad (t_1, \ldots, t_m) = t = h(s),$$

hence $l \leqq m$. Interchanging $\mathcal{M}$ and $\mathcal{N}$, we have also $m \leqq l$. Consequently $l = m$, and the number of moduli $m(M)$ of M does not depend on the choice of $(\mathcal{M}, B, \varpi)$ used for the definition.

Moreover since $l = m$, $\det(\partial t_\lambda / \partial s_\mu)_{\lambda,\mu=1,\ldots,m} \neq 0$ by (5.32). Therefore putting $\Delta = h(E)$, we see that Δ is a domain of B and that $h: E \to \Delta$ is holomorphic. If we take E sufficiently small, then Δ is also small, hence $\mathcal{M}_\Delta = \varpi^{-1}(\Delta)$ is represented in the form (4.27):

$$M_\Delta = \bigcup_j U_j \times \Delta.$$

Assume that $(z_j, t) \in U_j \times \Delta$ and $(z_k, t) \in U_k \times \Delta$ are the same point of $\mathcal{M}_\Delta$ if $z_j = f_{jk}(z_k, t)$. Then, as stated before (p. 207), we have

$$\mathcal{N}_E = \bigcup_j U_j \times E,$$

where $(z_j, s) \in U_j \times E$ and $(z_k, s) \in U_k \times E$ are the same point of $\mathcal{N}_E$ if $z_j = f_{jk}(z_k, h(s))$. Considering the biholomorphic map $h: s \to t = h(s)$ as the coordinate transformation of the parameter space, we see that $(\mathcal{N}_E, E, \pi)$ and $(\mathcal{M}_\Delta, \Delta, \varpi)$ may be considered as the same complex analytic family. In this sense, if for a given compact complex manifold M, an effectively parametrized and complete complex analytic family $(\mathcal{M}, B, \varpi)$ with $\varpi^{-1}(0) = M$ exists, then $(\mathcal{M}_\Delta, \Delta, \varpi)$ is uniquely determined by M provided that Δ is sufficiently small with $0 \in \Delta \subset B$.

For the examples of compact complex manifolds given in the preceding section, we have seen that the number of effective parameters m of M is equal to $\dim H^1(M, \Theta)$ except for one example. Since it is difficult to believe that for so many examples the equality $m = \dim H^1(M, \Theta)$ occurs only by accident, it might be reasonable to expect that the number m of parameters will coincide with the number $m(M)$ of moduli of M.

Let $t_1, \ldots, t_m$ be effective parameters contained in the definition of M, and put $t = (t_1, \ldots, t_m)$. Further let M_t be the manifold corresponding to the parameter t, and B the domain of t. Then $\{M_t | t \in B\}$ is a complex analytic family, which we denote by (M, B, π). In order to prove $m = m(M)$, we must prove that $(\mathcal{M}, B, \varpi)$ is complete and effectively parametrized.

For the completeness of a complex analytic family, we have the following theorem. Let $(\mathcal{M}, B, \varpi)$ be a complex analytic family of compact complex manifolds where B is a domain of $\mathbb{C}^m$, and $\rho_t: T_t(B) \to H^1(M_t, \Theta_t)$ the linear map defined in (4.22) for $M_t = \varpi^{-1}(t)$, that is,

$$\rho_t : \frac{\partial}{\partial t} \to \rho_t\left(\frac{\partial}{\partial t}\right) = \frac{\partial M_t}{\partial t}.$$

Theorem of Completeness ([20]). *If ρ_{t^0} is surjective at $t^0 \in B$, that is, $\rho_{t^0}(T_{t^0}(B)) = H^1(M_{t^0}, \Theta_{t^0})$, then $(\mathcal{M}, B, \varpi)$ is complete at t^0.*

We give the proof of this theorem in the next chapter. Using this theorem we see that if $(\mathcal{M}, B, \varpi)$ is effectively parametrized and $m = \dim H^1(M_t, \Theta_t)$ for every $t \in B$, then we have $m = m(M)$. In fact, if $(\mathcal{M}, B, \varpi)$ is effectively parametrized, ρ_t is injective, hence $\dim \rho_t(T_t(B)) = m$. Consequently $\rho_t(T_t(B)) = H^1(M_t, \Theta_t)$, and by the theorem of completeness, $(\mathcal{M}, B, \varpi)$ is complete.

Thus in order to prove that $m = m(M)$ for the examples given in the preceding section with $m = \dim H^1(M, \Theta)$, it suffices to verify that $(\mathcal{M}, B, \varpi) = \{M_t \mid t \in B\}$ is effectively parametrized.

(i) Projective space $\mathbb{P}^n$. Since $H^1(\mathbb{P}^n, \Theta) = 0$, by the theorem of completeness, the complex analytic family $\{\mathbb{P}^n\}$ consisting of only one member $\mathbb{P}^n$ is complete. Hence $m(\mathbb{P}^n) = 0$.

(ii) Complex tori. Let M be a complex torus of dimension n with the period matrix of the form (5.15):

$$\begin{bmatrix} t_1^1 & \cdots & t_1^n \\ \cdots & \cdots & \cdots \\ t_n^1 & \cdots & t_n^n \\ 1, 0 & \cdots & 0 \\ \cdots & \cdots & \cdots \\ 0 & \cdots & 1 \end{bmatrix}.$$

Put $t = (t_\alpha^\beta)_{\alpha,\beta=1,\ldots,n}$, and write M_t for M. Let

$$\omega_j(t) = (\omega_j^1(t), \ldots, \omega_j^n(t))$$

be the kth row of the period matrix. Namely

$$\omega_j^\beta(t) = \begin{cases} t_j^\beta, & j = 1, \ldots, n \\ \delta_{j-n}^\beta, & j = n+1, \ldots, 2n. \end{cases}$$

Then $M_t = \mathbb{C}^n / G_t$ where $G_t = \{\sum_{j=1}^{2n} m_j \omega_j(t) \mid m_j \in \mathbb{Z}\}$. $2n$ vectors $\omega_j(t), j = 1, \ldots, 2n$, are linearly independent over $\mathbb{R}$ if and only if $\det \operatorname{Im} t \neq 0$ where $\operatorname{Im} t = (\operatorname{Im} t_\alpha^\beta)_{\alpha,\beta=1,\ldots,n}$. *Let* $B = \{t \mid \det \operatorname{Im} t > 0\}$. Then $\{M_t \mid t \in B\}$ forms a complex analytic family. This is verified as in the case $n = 1$ given in Example

2.13. In fact, let $\mathscr{G}$ be the group of automorphisms of $\mathbb{C}^n \times B$ consisting of all automorphisms defined by

$$(z, t) \to \left(z + \sum_{j=1}^{2n} m_j \omega_j(t), t\right), \qquad m_j \in \mathbb{Z}, \quad j = 1, \ldots, 2n.$$

Then $\mathscr{G}$ acts properly discontinuously without fixed points, hence the quotient space $\mathscr{M} = \mathbb{C}^n \times B/\mathscr{G}$ is a complex manifold. The projection $(z, t) \to t$ of $\mathbb{C}^n \times B$ onto B induces a holomorphic map ϖ of $\mathscr{M}$ onto B. Then it is clear that $(\mathscr{M}, B, \varpi)$ is a complex analytic family and that $\varpi^{-1}(t) = \mathbb{C}^n \times t/G_t = M_t$.

In order to compute the infinitesimal deformation of M_t, we take the following system of coordinates on $\mathscr{M}$. Let Π be the canonical holomorphic map of $\mathbb{C}^n \times B$ onto $\mathscr{M} = \mathbb{C}^n \times B/\mathscr{G}$. Π is a locally biholomorphic map. Let $\{\mathscr{U}_k\}$ be a locally finite open covering of M where $\mathscr{U}_k$ is assumed to be a sufficiently small domain. Then $\Pi^{-1}(\mathscr{U}_k)$ consists of infinitely many mutually disjoint domains $\mathscr{U}_{k_1}, \mathscr{U}_{k_2}, \ldots$ of $\mathbb{C}^n \times B$. Choose one of them, say, $\mathscr{U}_{k_1}$, and let $(z_k^1, \ldots, z_k^n, t)$ be the restriction of the complex coordinates $(z^1, \ldots, z^n, t)$ of $\mathbb{C}^n \times B$ to $\mathscr{U}_{k_1}$. Since $\Pi\colon \mathscr{U}_{k_1} \to \mathscr{U}_k$ is biholomorphic, we may consider $(z_k^1, \ldots, z_k^n, t)$ as the complex coordinates on $\mathscr{U}_k$. We use these $(z_k, t) = (z_k^1, \ldots, z_k^n, t)$ as local complex coordinates of M on $\mathscr{U}_k$. On $\mathscr{U}_i \cap \mathscr{U}_k$ we have

$$z_i = z_k + \sum_{j=1}^{2n} m_{ik}^j \omega_j(t), \qquad m_{jk}^i \in \mathbb{Z},$$

namely

$$z_i^\beta = z_k^\beta + m_{ik}^{n+\beta} + \sum_{\alpha=1}^{n} m_{ik}^\alpha t_\alpha^\beta, \qquad \beta = 1, \ldots, n. \tag{5.33}$$

Put

$$f_{ik}^\beta(z_k, t) = z_k^\beta + m_{ik}^{n+\beta} + \sum_{\alpha=1}^{n} m_{ik}^\alpha t_\alpha^\beta,$$

and define $\theta_{ik}(t)$ as

$$\theta_{ik}(t) = \sum_{\gamma=1}^{n} \frac{\partial f_{ik}^\gamma(z_k, t)}{\partial t_\alpha^\beta} \frac{\partial}{\partial z_i^\gamma} = m_{ik}^\alpha \frac{\partial}{\partial z_i^\beta}. \tag{5.34}$$

Then the infinitesimal deformation $\partial M_t/\partial t_\alpha^\beta$ is the cohomology class $\theta(t) \in H^1(M_t, \Theta_t)$ of the 1-cocycle $\{\theta_{ik}(t)\}$.

Let $\Psi_t = \mathscr{A}(T(M_t))$. Ψ_t is the sheaf of germs of C^∞ vector fields of the form $\sum_\alpha \psi^\alpha(\partial/\partial z^\alpha)$ over M_t. By (3.106) we have

$$H^1(M_t, \Theta_t) \cong \Gamma(M_t, \bar\partial\Psi_t)/\bar\partial\Gamma(M_t, \Psi_t). \tag{5.35}$$

Since Ψ_t is a fine sheaf, from the exact sequence

$$0 \to \Theta_t \xrightarrow{\iota} \Psi_t \xrightarrow{\bar\partial} \bar\partial\Psi_t \to 0,$$

we obtain the exact sequence of cohomology groups

$$\cdots \to \Gamma(M_t, \Psi_t) \xrightarrow{\bar\partial} \Gamma(M_t, \bar\partial\Psi_t) \xrightarrow{\delta^*} H^1(M_t, \Theta_t) \to 0,$$

from which (5.35) is derived. Therefore a $\bar\partial$-closed vector (0, 1)-form $\varphi(t) \in \Gamma(M_t, \bar\partial\Psi_t)$ corresponds to $\theta(t) \in H^1(M_t, \Theta_t)$ if $\delta^*\varphi(t) = \theta(t)$. Let $\mathfrak{U}_t = \{U_{kt} \mid U_{kt} = \mathcal{U}_k \cap M_t \neq \varnothing\}$. If on each $U_{kt} \in \mathfrak{U}_t$, we can choose C^∞ vector field $\psi_k \in \Gamma(U_{kt}, \Psi_t)$ such that

$$\theta_{ik}(t) = \psi_k - \psi_i,$$

then since $\bar\partial\psi_i = \bar\partial\psi_k$ on each $U_{it} \cap U_{kt} \neq \varnothing$, we obtain a $\varphi(t) \in \Gamma(M_t, \bar\partial\Psi_t)$ which coincides with $\bar\partial\psi_k$ on each U_{kt}. Then by the definition of δ^*, $\delta^*\varphi(t) = \theta(t)$. For, if we put $c^0 = \{\psi_k\} \in C^0(\mathfrak{U}_t, \Psi_t)$, we have $\bar\partial c^0 = \varphi(t)$ and $\delta c^0 = \{\theta_{ik}(t)\}$.

Such a 0-cochain $\{\psi_k\} \in C^0(\mathfrak{U}_t, \Psi_t)$ as $\psi_k - \psi_i = \theta_{ik}(t)$ is obtained easily as follows. By (5.33),

$$\sum_{\alpha=1}^{n} m_{ik}^\alpha (t_\alpha^\beta - \bar t_\alpha^\beta) = \bar z_k^\beta - z_k^\beta - \bar z_i^\alpha + z_i^\alpha.$$

By assumption $\det(t_\alpha^\beta - \bar t_\alpha^\beta)_{\alpha,\beta=1,\dots,n} \neq 0$. Therefore letting $(u_\alpha^\beta)_{\alpha,\beta=1,\dots,n}$ be the inverse matrix of $(t_\alpha^\beta - \bar t_\alpha^\beta)_{\alpha,\beta=1,\dots,n}$, we have

$$m_{ik}^\alpha = \sum_{\gamma=1}^{n} (\bar z_k^\gamma - z_k^\gamma) u_\gamma^\alpha - \sum_{\gamma=1}^{n} (\bar z_i^\gamma - z_i^\gamma) u_\gamma^\alpha.$$

Consequently putting

$$\psi_k = \sum_{\gamma=1}^{n} (\bar z_k^\gamma - z_k^\gamma) u_\gamma^\alpha \frac{\partial}{\partial z^\beta},$$

since $\partial/\partial z_i^\beta = \partial/\partial z^\beta$, we obtain from (5.34)

$$\theta_{ik}(t) = \psi_k - \psi_i.$$

Since $\bar\partial \bar z_k^\gamma = d\bar z^\gamma$, and $\bar\partial z_k^\gamma = 0$, we have

$$\varphi(t) = \bar\partial\psi_k = \sum_{\gamma=1}^{n} u_\gamma^\alpha \, d\bar z^\gamma \frac{\partial}{\partial z^\beta}.$$

It is obvious from (5.18) that $\varphi(t)\in\Gamma(M_t, \bar{\partial}\Psi_t)$ is a harmonic vector (0, 1)-form: $\varphi(t)\in \mathbf{H}^{0,1}(T(M_t))$. Thus via the isomorphism (5.16):

$$H^1(M_t, \Theta_t)\cong \mathbf{H}^{0,1}(T(M_t)),$$

the infinitesimal deformation $\rho_t(\partial/\partial t_\alpha^\beta)=\partial M_t/\partial t_\alpha^\beta\in H^1(M_t, \Theta_t)$ corresponds to the harmonic vector (0, 1)-form $\sum_{\gamma=1}^n u_\gamma^\alpha\, d\bar{z}^\gamma\, \partial/\partial z^\beta$. Since $\det(u_\gamma^\alpha)\neq 0$, $\rho_t(\partial/\partial t_\alpha^\beta)$, $\alpha, \beta=1,\dots, n$, are linearly independent. Therefore n^2 parameters t_α^β, $\alpha, \beta=1,\dots, n$, are effective, and

$$m(M_t)=m=\dim H^1(M_t, \Theta_t).$$

However the situation for $n\geqq 2$ is somewhat different from that for $n\geqq 1$. In case $n=1$, M_t, with $\operatorname{Im} t>0$, is an elliptic curve, and, as stated in Example 2.14, in the complex analytic family $\{M_t \mid t\in \mathcal{H}^+\}$, $M_{t'}$ and M_t are biholomorphic if and only if

$$t'=\frac{at+b}{ct+d},\qquad a, b, c, d\in\mathbb{Z}\quad\text{and}\quad ad-bc=1.$$

Let Γ be the group of all linear transformations of the above form. Then Γ is properly discontinuous, and $\mathbb{H}^+/\Gamma\cong\mathbb{C}$. Thus up to the biholomorphic equivalence, an elliptic curve corresponds in a one-to-one manner to a point $\tau\in\mathbb{C}=\mathbb{H}^+/\Gamma$. We express this fact by saying that $\mathbb{C}$ is the space of moduli of elliptic curves. The number of moduli $m(M_t)=1$ is the dimension of the space of moduli $\mathbb{C}$.

Consider next the case $n\geqq 2$. Let $\mathbb{Z}^n$ be the set of all $n\times n$ integral matrices $a=(a_\alpha^\beta)$, $a_\alpha^\beta\in\mathbb{Z}$. Then in the above notation, M_t and $M_{t'}$ are biholomorphic if and only if

$$t'=(at+b)(ct+d)^{-1},\qquad a, b, c, d\in\mathbb{Z}_n,\qquad \det\begin{pmatrix} a & b\\ c & d\end{pmatrix}=1. \tag{5.36}$$

Let Γ be the group of automorphisms $\gamma\colon t\to t'$ of the form (5.36). Then for $n\geqq 2$, the group Γ of automorphisms of B is not properly discontinuous ([28]). Moreover in any neighbourhood U of any point $s\in B$ with $s\in U\subset B$, there is a $t\in U$ such that $\Gamma t\cap U$ is an infinite set. Consequently, in contrast to the case $n=1$, B/Γ is not even a Hausdorff space. Thus we cannot define B/Γ as the space of moduli. Let $\gamma_\nu t$, $\nu=1, 2, \dots$ be infinitely many points of $\Gamma t\cap U$. Then $M_{\gamma_1 t}, M_{\gamma_2 t},\dots$ are biholomorphic to one another. Thus *for any small neighbourhood U, there are infinitely many points $\gamma_1 t, \gamma_2 t,\dots$ of U such that $M_{\gamma_1 t}, M_{\gamma_2 t},\dots$ have the same complex structure* ([21]). *Nevertheless B is an effective parameter space.*

In view of this example, we suspected that there would exist a complex analytic family which is not locally trivial, but all members $M_t = \varpi^{-1}(t)$ of which are biholomorphic to one another. But, as already mentioned, Fischer and Grauert proved that this cannot happen.

(c) Complex Analytic Family of Hypersurfaces

In order to prove that for a non-singular hypersurface M_t of order h in $\mathbb{P}^{n+1}$ with $n \geqq 2$, $h \geqq 3$, the number $m(M_t)$ of moduli is defined and coincides with m given in (5.20) except for the case $n=2$, $h=4$, we need several results from the general theory on complex analytic families of non-singular hypersurfaces. First we explain them.

In general let S be a submanifold of a complex manifold W. By the codimension of S, we mean the difference of the dimensions of W and S. Then a non-singular hypersurface of order h in $\mathbb{P}^{n+1}$ is a submanifold of $\mathbb{P}^{n+1}$ of codimension 1.

Let W be a given compact complex manifold of dimension $n+1$. A complex analytic family $(\mathcal{M}, B, \varpi)$ of compact complex manifolds is called a complex analytic family of hypersurfaces in W if it satisfies the following conditions:

(i) Each $M_t = \varpi^{-1}(t)$ is a submanifold of codimension 1 of W for $t \in B$.
(ii) There is a holomorphic map Φ of $\mathcal{M}$ into W whose restriction to M_t coincides with the inclusion $M_t \hookrightarrow W$.

Since we are only interested in the infinitesimal deformations, it suffices to take a sufficiently small polydisk $\Delta \subset B$, and consider $(\mathcal{M}_\Delta, \Delta, \varpi)$ instead of $(\mathcal{M}, B, \varpi)$ where we put $\mathcal{M}_\Delta = \varpi^{-1}(\Delta)$. Moreover we may assume that $\mathcal{M}_\Delta$ has the form

$$\mathcal{M}_\Delta = \bigcup_{j=1}^{l} U_j \times \Delta,$$

where $(z_j, t) \in U_j \times \Delta$ and $(z_k, t) \in U_k \times \Delta$ are the same point of $\mathcal{M}_\Delta$ if $z_j^\alpha = f_{jk}^\alpha(z_k, t)$. Let $\{W_1, \ldots, W_l, \ldots\}$ be a finite open covering of W, where W_j is a coordinate neighbourhood, and let $w_j = (w_j^0, \ldots, w_j^n)$ be the local coordinates defined on W_j. Taking U_j and Δ sufficiently small, we may assume that $\Phi(U_j \times \Delta) \subset W_j$ for $j = 1, \ldots, l$, and that $\Phi(\mathcal{M}_\Delta) \cap [W_j] = \emptyset$ for $j \geqq l+1$. Φ is represented on each $U_j \times \Delta$ as

$$\Phi: (z_j, t) \to (w_j^0, \ldots, w_j^n) = (\varphi_j^0(z_j, t), \ldots, \varphi_j^n(z_j, t)).$$

Since Φ induces the inclusion $M_t \hookrightarrow W$,

$$\operatorname{rank} \frac{\partial(w_j^0, \ldots, w_j^n)}{\partial(z_j^1, \ldots, z_j^n)} = n.$$

We denote a point of $\mathcal{M}$ by p, and define the map $\tilde{\varphi}$ of $\mathcal{M}$ into $W\times B$ by

$$\tilde{\Phi}: p \to \tilde{\Phi}(p) = (\Phi(p), \pi(p)).$$

Then $\tilde{\Phi}$ is represented in terms of local coordinates as

$$(z_j, t) \to (w_j^0, \ldots, w_j^n, t) = (\varphi_j^0(z_j, t), \ldots, \varphi_j^n(z_j, t), t).$$

Therefore we have

$$\operatorname{rank} \frac{\partial(w_j^0 \ldots, w_j^n, t_1, \ldots, t_m)}{\partial(z_j^1, \ldots, z_j^n, t_1, \ldots, t_m)} = m + n = \dim \mathcal{M}.$$

Consequently $\tilde{\Phi}$ is a biholomorphic map of $\mathcal{M}_\Delta$ into $W\times\Delta$, and, identifying $\mathcal{M}_\Delta$ with $\tilde{\Phi}(\mathcal{M}_\Delta)$, we may consider that $\mathcal{M}_\Delta$ is a submanifold of codimension 1 of $W\times\Delta$. Then

$$\mathcal{M}_\Delta = \bigcup_{t\in\Delta} M_t \times t \subset W \times \Delta$$

is considered as the graph of $\varpi^{-1}: t \to M_t$.

If we take W_j and Δ sufficiently small, $\mathcal{M}_\Delta$ is given on each $W_j\times\Delta$ by a single minimal equation

$$S_j(w_j, t) = 0.$$

Here $S_j(w_j, t)$ is understood to be a non-vanishing holomorphic function if $W_j\times\Delta\cap\mathcal{M}_\Delta = \varnothing$. For a fixed t, $S_j(w_j, t) = 0$ gives a minimal equation of M_t in W_j. $S_j(w_j, t) = S_j(w_j^0, \ldots, w_j^n, t)$ is a holomorphic function of $w_j^0, \ldots, w_j^n, t_1, \ldots, t_m$, and at least one of the partial derivatives $\partial S_j(w_j, t)/\partial w_j^\lambda$, $\lambda = 1, \ldots, n$, does not vanish for each $w_j \in M_t$. On $W_j\times\Delta\cap W_k\times\Delta\neq\varnothing$, we have

$$S_j(w_j, t) = F_{jk}(w, t)\cdot S_k(w_k, t), \tag{5.37}$$

where $F_{jk}(w, t)$ is a non-vanishing holomorphic function defined on $(W_j\cap W_k)\times\Delta$, and w_j, w_k are local coordinates of $w\in W_j\cap W_k$. It is clear that

$$F_{ik}(w, t) = F_{ij}(w, t)\cdot F_{jk}(w, t) \quad \text{for} \quad w \in W_i\cap W_j\cap W_k. \tag{5.38}$$

Thus for every $t\in\Delta$, we can define a complex line bundle $\mathcal{F}_t$ over W with transition functions $F_{jk}(w, t)$. $\mathcal{F}_t$ is a complex line bundle over W defined by the divisor M_t: $\mathcal{F}_t = [M_t]$, and $S(t) = \{S_j(w_j, t)\}$ is a holomorphic section of $\mathcal{F}_t$ over W.

Fix $t\in\Delta$ for a while. Since $M_t=\bigcup_j U_j\times t$, if we put $U_j\times t=U_{jt}$, U_{jt} is a coordinate polydisk on M_t. We denote the point $z_j\times t\in U_j\times t=U_{jt}$ simply by z_j, and consider $z_j=(z_j^1,\dots,z_j^n)$ as the local coordinates on M_t. If we put

$$\varphi_j(z_j,t)=(\varphi_j^0(z_j,t),\dots,\varphi_j^n(z_j,t)),$$

the inclusion $\Phi\colon M_t\hookrightarrow W$ is represented on each $U_{jt}\subset M_t$ as

$$\Phi\colon z_j\to w_j=\varphi_j(z_j,t).$$

Put

$$F_{tjk}(z)=F_{jk}(\Phi(z),t),\qquad z\in U_{jt}\cap U_{kt}. \tag{5.39}$$

Then $F_{tjk}(z)$ is a non-vanishing holomorphic function on $U_{jt}\cap U_{kt}$, and from (5.38) we obtain immediately

$$F_{tik}(z)=F_{tij}(z)\cdot F_{tjk}(z),\qquad z\in U_{it}\cap U_{jt}\cap U_{kt}.$$

Therefore we can define a complex line bundle F_t on M_t with transition functions $F_{tjk}(z)$. F_t is the restriction of $\mathcal{F}_t$ to M_t.

In general when t moves, the complex structure of M_t as well as the position of M_t in W varies. The change of the position of M_t is called a displacement. On each neighbourhood W_j, M_t is given by the equation $S_j(w_j,t)=0$. In order to define the infinitesimal displacement along $\partial/\partial t$ for an arbitrary $\partial/\partial t\in T_t(\Delta)$, consider the partial derivative $\partial S_j(w_j,t)/\partial t$, and put

$$\sigma_j(z_j,t)=-\left(\frac{\partial S_j(w_j,t)}{\partial t}\right)_{w_j=\varphi_j(z_j,t)} \tag{5.40}$$

Applying $\partial/\partial t$ to both sides of (5.37), we have

$$\frac{\partial S_j(w_j,t)}{\partial t}=F_{jk}(w,t)\frac{\partial S_k(w_k,t)}{\partial t}+\frac{\partial F_{jk}(w,t)}{\partial t}\cdot S_k(w_k,t).$$

Restricting these equalities to M_t we obtain

$$\sigma_j(z_j,t)=F_{tjk}(z)\sigma_k(z_k,t),$$

since $S_k(w_k,t)=0$ on M_t. This means that $\sigma(t)=\{\sigma_j(z_j,t)\}$ is a holomorphic section of F_t over M_t.

Definition 5.5. $\sigma(t)\in H^0(M_t,\mathcal{O}(F_t))$ is called the *infinitesimal displacement* of the submanifold M_t of codimension 1 of W, and is denoted by

$$(\partial M_t\subset W)/\partial t.$$

We denote by $\rho_{d,t}$ the linear map $\partial/\partial t \to \sigma(t)$ of $T_t(\Delta)$ to $H^0(M_t, \mathcal{O}(F_t))$, namely,

$$\rho_{d,t}: \frac{\partial}{\partial t} \to \rho_{d,t}\left(\frac{\partial}{\partial t}\right) = \frac{\partial M_t \subset W}{\partial t} = \sigma(t). \tag{5.41}$$

Next we will consider the relation between the infinitesimal deformation and the infinitesimal displacement. Let $T(W)_t$ be the restriction of the tangent bundle $T(W)$ of W to M_t. $T(W)_t$ is a holomorphic vector bundle over M_t. Since the tangent space $T_z(M_t)$ at $z \in M_t$ consists of all tangent vectors of W at z which are tangent to M_t, the tangent bundle $T(M_z) = \bigcup_{z \in M_t} T_z(M_t)$ of M_t is a *subbundle* of $T(W)_t = \bigcup_{z \in M_t} T_z(W)$. A local holomorphic section ξ of $T(W)_t$ is represented in terms of local coordinates as

$$\xi = \sum_{\lambda=0}^{n} v_j^\lambda(z) \frac{\partial}{\partial w_j^\lambda},$$

where the coefficients $v_j^\lambda(z)$ are local holomorphic functions on M_t. Define $\tau_j(z_j, t)$ for this ξ by

$$\begin{aligned}\tau_j(z_j, t) &= \xi S_j(w_j, t)_{w_j = \varphi_j(z_j, t)} \\ &= \sum_{\lambda=0}^{n} v_j^\lambda(z) \left(\frac{\partial S_j(w_j, t)}{\partial w_j^\lambda}\right)_{w_j = \varphi_j(z_j, t)}\end{aligned}$$

Then $\tau_j(z_j, t)$ is a holomorphic function on $U_{jt} \subset M_t$, and from (5.37) we obtain

$$\tau_j(z_j, t) = F_{tjk}(z) \tau_k(z_k, t)$$

on $U_{jt} \cap U_{kt} \neq \emptyset$. Therefore $\tau(t) = \{\tau_j(z_j, t)\}$ is a holomorphic section of F_t over M_t, which we denote by $\xi S(t)$:

$$\xi S(t) = \tau(t). \tag{5.42}$$

$\xi S(t) = 0$, namely, $\xi S_j(w_j, t) = 0$ means that $\xi = \sum_\lambda v_j^\lambda(z)(\partial/\partial w_j^\lambda) \in T_z(M_t)$, namely, that ξ is a holomorphic vector field on M_t.

We denote by Ξ_t the sheaf of germs of holomorphic sections of $T(W)_t$: $\Xi_t = \mathcal{O}(T(W)_t)$. $\Theta_t = \mathcal{O}(T(M_t))$ is a subsheaf of Ξ_t. If we represent $\theta \in \Theta_t$ as $\theta = \sum_{\alpha=1}^n \theta_j^\alpha(\partial/\partial z_j^\alpha)$ where θ_j^α is a germ of a holomorphic function, the inclusion map $\iota: \Theta_t \hookrightarrow \Xi_t$ is given by

$$\iota: \theta = \sum_{\alpha=1}^{n} \theta_j^\alpha \frac{\partial}{\partial z_j^\alpha} \to \sum_{\lambda=0}^{n} \left(\sum_{\alpha=1}^{n} \theta_j^\alpha \frac{\partial w_j^\lambda}{\partial z_j^\alpha}\right) \frac{\partial}{\partial w_j^\lambda}, \tag{5.43}$$

where $\partial w_j^\lambda / \partial z_j^\alpha = \partial \varphi_j^\lambda(z_j, t)/\partial z_j^\alpha$.

If we put $\xi \in \Xi_t$ in (5.42), we have $\tau(t) = \xi S(t) \in \mathcal{O}(F_t)$. Consequently $\xi \to \tau(t) = \xi S(t)$ defines a homomorphism $\Xi_t \to \mathcal{O}(F_t)$ of sheaves, whose kernel is Θ_t as seen above. Thus we have the exact sequence

$$0 \to \Theta_t \to \Xi_t \to \mathcal{O}(F_t) \to 0. \tag{5.44}$$

For a fixed t, choose local coordinates $(z_j^1, \ldots, z_j^n)$ on M_t and $(w_j^0, \ldots, w_j^n)$ on W such that $\varphi_j^0(z_j, t) = 0$, $w_j^\alpha = \varphi_j^\alpha(z_j, t) = z_j^\alpha$ for $\alpha = 1, \ldots, n$, and put $S_j(w_j, t) = w_j^0$. If we write $\xi \in \Xi_t$ as

$$\xi = v_j^0 \frac{\partial}{\partial w_j^0} + \theta, \qquad \theta = \sum_{\lambda=1}^{n} v_j^\lambda \frac{\partial}{\partial w_j^\lambda},$$

then

$$\theta = \sum_{\lambda=1}^{n} v_j^\lambda \left(\frac{\partial}{\partial z_j^\lambda}\right) \in \Theta_t \quad \text{and} \quad \tau_j(z_j, t) = \xi S_j(w_j, t) = v_j^0.$$

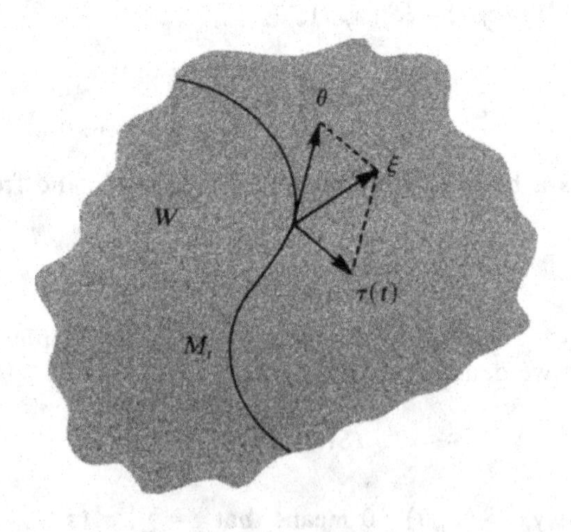

Figure 2

Thus $\tau(t) = \xi S(t)$ is the normal component of ξ to M_t. This is the meaning of (5.44). The exact sequence (5.44) gives the exact sequence of cohomology groups

$$\cdots \to H^0(M_t, \Xi_t) \to H^0(M_t, \mathcal{O}(F_t)) \xrightarrow{\delta^*} H^1(M_t, \Theta_t) \to \cdots. \tag{5.45}$$

Lemma 5.3. *The relation between the infinitesimal deformation and the infinitesimal displacement is given by the following.*

$$\rho_t = \delta^* \rho_{d,t}. \tag{5.46}$$

Proof. Since $\mathfrak{U}_t = \{U_{jt}\}$ is a finite covering of M_t, and each U_{jt} is a polydisk, we see from (5.2) that $H^1(M_t, \Theta_t) = H^1(\mathfrak{U}_t, \Theta_t)$. For any $\partial/\partial t \in T_t(\Delta)$, $\rho_t(\partial/\partial t)$ is the cohomology class of the 1-cocycle $\{\theta_{jk}(t)\} \in Z^1(\mathfrak{U}_t, \Theta_t)$, where

$$\theta_{jk}(t) = \sum_{\alpha=1}^{n} \theta_{jk}^{\alpha}(t)\frac{\partial}{\partial z_j^{\alpha}}, \qquad \theta_{jk}^{\alpha}(t) = \frac{\partial f_{jk}^{\alpha}(z_k, t)}{\partial t},$$

while by Definition 5.5 the infinitesimal displacement is defined as $\rho_{d,t}(\partial/\partial t) = \sigma(t) = \{\sigma_j(z_j, t)\}$, where $\sigma_j(z_j, t)$ is given by (5.40). Consequently in order to show that $\rho_t(\partial/\partial t) = \delta^* \rho_{d,t}(\partial/\partial t)$, by the definition of δ^*, it suffices to prove that there is a 0-cochain $\{\xi_j\}$ with $\xi_j \in \Gamma(U_{jt}, \Xi_t)$ such that

$$\xi_j S(t) = \sigma_j(z_j, t), \qquad \xi_k - \xi_j = \theta_{jk}(t).$$

Suppose that $w_j^{\lambda} = g_{jk}^{\lambda}(w_k)$ on $W_j \cap W_k$. Then since Φ: $z_j \to w_j = \varphi_j(z_j, t)$, we have on $U_{jt} \cap U_{kt}$

$$\varphi_j^{\lambda}(z_j, t) = g_{jk}^{\lambda}(\varphi_k(z_k, t)), \qquad z_j^{\alpha} = f_{jk}^{\alpha}(z_k, t).$$

Differentiating both sides of this equality with respect to t as functions of $z_k^1, \ldots, z_k^n$, t, we have

$$\sum_{\alpha=1}^{n} \frac{\partial w_j^{\lambda}}{\partial z_j^{\alpha}} \cdot \frac{\partial f_{jk}^{\alpha}(z_k, t)}{\partial t} + \frac{\partial \varphi_j^{\lambda}(z_j, t)}{\partial t} = \sum_{\nu=0}^{n} \frac{\partial w_j^{\lambda}}{\partial w_k^{\nu}} \cdot \frac{\partial \varphi_k^{\nu}(z_k, t)}{\partial t}.$$

Therefore putting

$$\xi_j = \sum_{\lambda=0}^{n} \frac{\partial \varphi_j^{\lambda}(z_j, t)}{\partial t} \frac{\partial}{\partial w_j^{\lambda}},$$

on each U_{jt}, we have

$$\sum_{\lambda=0}^{n} \left(\sum_{\alpha=1}^{n} \theta_{jk}^{\alpha}(t) \frac{\partial w_j^{\lambda}}{\partial z_j^{\alpha}} \right) \frac{\partial}{\partial w_j^{\lambda}} + \xi_j = \xi_k.$$

Consequently by (5.43)

$$\theta_{jk}(t) = \xi_k - \xi_j.$$

On the other hand, differentiating the equality $S_j(\varphi_j(z_j, t)) = 0$ with respect to t, we have

$$\sum_{\lambda=0}^{n} \frac{\partial \varphi_j^\lambda(z_j, t)}{\partial t} \cdot \frac{\partial S_j(w_j, t)}{\partial w_j^\lambda} + \frac{\partial S_j(w_j, t)}{\partial t} = 0, \qquad w_j = \varphi_j(z_j, t).$$

Thus we obtain

$$\xi_j S(t) = \sigma_j(z_j, t). \quad \blacksquare$$

We apply the above results to the case in which $W = \mathbb{P}^{n+1}$ with $n \geqq 2$, and M_t is a non-singular hypersurface of order h with $h \geqq 3$. We use the same notation as in (a)(iii) of this section. Let $(\zeta_0, \ldots, \zeta_{n+1})$ be the homogeneous coordinates on $W = \mathbb{P}^{n+1}$ and

$$t(\zeta) = t_0 \zeta_0^h + \cdots + t_\mu \zeta_{n+1}^h$$

a homogeneous polynomial of degree h in $\zeta_0, \ldots, \zeta_{n+1}$. M_t is defined by $t(\zeta) = 0$. Let Δ be a small polydisk in $B = \mathbb{P}^\mu - \mathfrak{S}$. We assume for simplicity that $t_0 = 1$ on Δ, and represent a point $t \in \Delta$ by its non-homogeneous coordinates $t_1, \ldots, t_\mu$. Thus we have

$$t(\zeta) = \zeta_0^h + t_1 \zeta_0^{h-1} \zeta_1 + \cdots + t_\mu \zeta_{n+1}.$$

If we consider $t(\zeta)$ as a function of $t_1, \ldots, t_\mu, \zeta_0, \ldots, \zeta_{n+1}$, the equation $t(\zeta) = 0$ defines a submanifold of codimension 1 on $W \times \Delta$, which we denote by $\mathcal{M}_\Delta$:

$$\mathcal{M}_\Delta = \{(\zeta, t) \in W \times \Delta \mid t(\zeta) = 0\}.$$

The projection $W \times \Delta \to \Delta$ induces a surjective holomorphic map $\varpi: \mathcal{M}_\Delta \to \Delta$. Then $(\mathcal{M}_\Delta, \Delta, \varpi)$ is a complex analytic family and for each $t \in \Delta$, $\varpi^{-1}(t) = M_t \times t = M_t$ is a non-singular hypersurface on W. The projection $W \times \Delta \to W$ induces a holomorphic map $\Phi: \mathcal{M}_\Delta \to W$. Φ restricted to $M_t = M_t \times t \subset \mathcal{M}_\Delta$ gives the inclusion $M_t \hookrightarrow W$. Thus $(\mathcal{M}_\Delta, \Delta, \varpi)$ is a complex analytic family of hypersurfaces on W.

Let $\{W_j\}$ be a finite open covering of $W = \mathbb{P}^{n+1}$, and $w_j = (w_j^0, \ldots, w_j^n)$ the local complex coordinates on W_j. We assume that $\{W_j\}$ and w_j satisfy the following: For each j, there is an index $\beta(j)$ such that $\zeta_{\beta(j)} \neq 0$ on W_j, and we have

$$w_j = (w_j^0, \ldots, w_j^n) = \left(\frac{\zeta_0}{\zeta_{\beta(j)}}, \ldots, \frac{\zeta_{\beta(j)-1}}{\zeta_{\beta(j)}}, \frac{\zeta_{\beta(j)+1}}{\zeta_{\beta(j)}}, \ldots, \frac{\zeta_{n+1}}{\zeta_{\beta(j)}} \right).$$

As above we assume that $\mathcal{M}_\Delta = \bigcup_{j=1}^l U_j \times \Delta$, and that $\Phi(U_j \times \Delta) \subset W_j$ for $j = 1, \ldots, l$ and $\Phi(\mathcal{M}_\Delta) \cap [W_j] = \varnothing$ for $j \geqq l+1$. On each $U_j \times \Delta$, Φ is assumed to be given by

$$\Phi: (z_j, t) \to w_j = \varphi_j(z_j, t).$$

Since $t(\zeta)$ is a homogeneous polynomial of degree h in $\zeta_0, \ldots, \zeta_{n+1}$, if we put

$$S_j(w_j, t) = \left(\frac{1}{\zeta_{\beta(j)}}\right)^h \cdot t(\zeta),$$

$S_j(w_j, t)$ is a polynomial of degree at most h in $w_j^0, \ldots, w_j^n$, and $S_j(w_j, t) = 0$ gives a minimal equation of M_t in W_j. Since

$$S_j(w_j, t) = F_{jk}(w) S_k(w_k, t), \qquad F_{jk}(w) = \left(\frac{\zeta_{\beta(k)}}{\zeta_{\beta(j)}}\right)^k, \tag{5.47}$$

the complex line bundle $\mathscr{F}_t = [M_t]$ depends only on the degree h and is independent of t. Let F be the complex line bundle over W with transition functions $F_{jk}(w)$. Then $F = [M_t]$. Further let E be the complex line bundle over $W = \mathbb{P}^{n+1}$ with transition functions $E_{jk}(w) = \zeta_{\beta(k)}/\zeta_{\beta(j)}$. Then

$$F = E^h = \overbrace{E \otimes \cdots \otimes E}^{h\,\text{times}}$$

Let P_∞ be the hyperplane of $W = \mathbb{P}^{n+1}$ defined by $\zeta_0 = 0$. Then $E = [P_\infty]$. Put

$$S_\lambda(\zeta) = \frac{\partial t(\zeta)}{\partial t_\lambda}, \qquad \lambda = 1, \ldots, \mu.$$

Each $S_\lambda(\zeta)$ is a monomial of degree h in $\zeta_0, \ldots, \zeta_{n+1}$. Therefore if we put

$$S_{\lambda j}(w_j) = \left(\frac{1}{\zeta_{\beta(j)}}\right)^h S_\lambda(\zeta),$$

$S_{\lambda j}(w_j)$ is a monomial of degree at most h in $w_j^0, \ldots, w_j^n$, and on $W_j \cap W_k \neq \varnothing$,

$$S_{\lambda j}(w_j) = F_{jk}(w) S_{\lambda k}(w_k) \tag{5.48}$$

holds. Consequently $S_\lambda = \{S_{\lambda j}(w_j)\}$ is a holomorphic section of F over W, namely, $S_\lambda \in H^0(W, \mathcal{O}(F))$.

We denote by F_t the restriction of F to $M_t \subset W$. By (5.39) F_t is defined by the transition functions $F_{tjk}(z) = F_{jk}(\Phi(z))$ with $z \in U_{jt} \cap U_{kt}$. Putting

$$\sigma_{\lambda j}(z_j, t) = S_{\lambda j}(\varphi_j(z_j, t)),$$

we have from (5.48)

$$\sigma_{\lambda j}(z_j, t) = F_{tjk}(z)\sigma_k(z_k, t)$$

on $U_{jt} \cap U_{kt} \neq \emptyset$. Consequently $\sigma_\lambda(t) = \{\sigma_{\lambda j}(z_j, t)\} \in H^0(M_t, \mathcal{O}(F_t))$. By Definition 5.5 of the infinitesimal displacement, it is clear that

$$\rho_{d,t}\left(\frac{\partial}{\partial t_\lambda}\right) = \frac{\partial M_t \subset W}{\partial t_\lambda} = -\sigma_\lambda(t), \qquad \lambda = 1, \ldots, \mu. \tag{5.49}$$

Lemma 5.4. $\{\sigma_1(t), \ldots, \sigma_\mu(t)\}$ *forms a basis of* $H^0(M_t, \mathcal{O}(F_t))$.

Proof. The sheaf $\mathcal{O}(F_t)$ is the restriction of $\mathcal{O}(F)$ to $M_t \subset W$. Denoting the restriction map by r_t, since $[M_t] = F$, we have $\ker r_t = \mathcal{O}(F \otimes [M_t]^{-1}) = \mathcal{O}$. Hence the sequence

$$0 \to \mathcal{O} \to \mathcal{O}(F) \xrightarrow{r_t} \mathcal{O}(F_t) \to 0$$

is exact. In the corresponding exact sequence of cohomology groups

$$0 \to H^0(W, \mathcal{O}) \to H^0(W, \mathcal{O}(F)) \xrightarrow{r_t} H^0(M_t, \mathcal{O}(F_t)) \to H^1(W, \mathcal{O}) \to \cdots,$$

$H^0(W, \mathcal{O}) = \mathbb{C}$, and $H^1(W, \mathcal{O}) = 0$. Thus we obtain the exact sequence

$$0 \to \mathbb{C} \to H^0(W, \mathcal{O}(F)) \xrightarrow{r_t} H^0(M_t, \mathcal{O}(F_t)) \to 0.$$

For $S_\lambda \in H^0(W, \mathcal{O}(F))$, we have

$$r_t S_\lambda = \sigma_\lambda(t), \qquad \lambda = 1, \ldots, \mu,$$

and for $S(t) = \{S_j(w_j, t)\} \in H^0(W, \mathcal{O}(F))$, $r_t S(t) = 0$. Therefore to prove Lemma 5.4, it suffices to show that $\{S(t), S_1, \ldots, S_\mu\}$ forms a basis of $H^0(W, \mathcal{O}(F))$. For this it suffices to verify that any $s = \{s_j(w_j)\} \in H^0(W, \mathcal{O}(F))$ is represented by some homogeneous polynomial $s(\zeta)$ of degree h in $\zeta_0, \ldots, \zeta_{n+1}$ as

$$s_j(w_j) = \left(\frac{1}{\zeta_{\beta(j)}}\right)^h \cdot s(\zeta). \tag{5.50}$$

For, $s(\zeta)$ is a linear combination of $t(\zeta), S_1(\zeta), \ldots, S_\mu(\zeta)$. Since $\{s_j(w_j)\}$ is a holomorphic section of F over W, each $s_j(w_j)$ is a holomorphic function on W_j, and on $W_j \cap W_k$

$$s_j(w_j) = \left(\frac{\zeta_{\beta(k)}}{\zeta_{\beta(j)}}\right)^h \cdot s_k(w_k) \tag{5.51}$$

holds. We have $\mathbb{P}^{n+1}-P_\infty=\mathbb{C}^{n+1}$, and putting $z_\alpha=\zeta_\alpha/\zeta_0$ for $\alpha=1,\dots,n+1$, we have complex coordinates $(z_1,\dots,z_{n+1})$ on $\mathbb{C}^{n+1}$. Since by (5.51) on $W_j\cap W_k\cap\mathbb{C}^{n+1}\neq\varnothing$, we have

$$\left(\frac{\zeta_{\beta(j)}}{\zeta_0}\right)^h\cdot s_j(w_j)=\left(\frac{\zeta_{\beta(k)}}{\zeta_0}\right)^h\cdot s_k(w_k),$$

there is a holomorphic function $g(z_1,\dots,z_{n+1})$ on $\mathbb{C}^{n+1}$ such that on each $W_j\cap\mathbb{C}^{n+1}$

$$g(z_1,\dots,z_{n+1})=\left(\frac{\zeta_{\beta(j)}}{\zeta_0}\right)^h\cdot s_j(w_j). \tag{5.52}$$

Let

$$g(z_1,\dots,z_{n+1})=\sum_{k=0}^{\infty}g_k(z_1,\dots,z_{n+1}),$$

where $g_k(z_1,\dots,z_{n+1})$ is the homogeneous part of degree k of the power series expansion of $g(z_1,\dots,z_{n+1})$. Take an arbitrary point $p=(0,\zeta_1,\dots,\zeta_{n+1})$ on P_∞. Then p belongs to one of W_j's. Fix $\zeta_1,\dots,\zeta_{n+1}$ for a while. If we take $\varepsilon>0$ sufficiently small, then for $0<|\zeta_0|<\varepsilon$, we have

$$\left(1,\frac{\zeta_1}{\zeta_0},\dots,\frac{\zeta_{n+1}}{\zeta_0}\right)=(\zeta_0,\dots,\zeta_{n+1})\in W_j.$$

Therefore by (5.52) we have

$$\zeta_0^h\sum_{k=0}^{\infty}\frac{g_k(\zeta_1,\dots,\zeta_{n+1})}{\zeta_0^k}=\zeta_{\beta(j)}^h\cdot s_j(w_j).$$

As a function of ζ_0, the right-hand side is holomorphic on $|\zeta_0|<\varepsilon$. Consequently we must have $g_k(\zeta_1,\dots,\zeta_{n+1})=0$ for $k\geqq h+1$. Since $\zeta_1,\dots,\zeta_{n+1}$ are arbitrary, $g(z_1,\dots,z_{n+1})$ is a polynomial of degree at most h in $z_1,\dots,z_{n+1}$. Therefore if we put

$$s(\zeta)=\zeta_0^h g\left(\frac{\zeta_1}{\zeta_0},\dots,\frac{\zeta_{n+1}}{\zeta_0}\right),$$

$s(\zeta)$ is a homogeneous polynomial of degree h in $\zeta_0,\dots,\zeta_{n+1}$, and (5.50) follows from (5.52). ∎

We consider the exact sequence (5.44):

$$0\to\Theta_t\to\Xi_t\to\mathcal{O}(F_t)\to 0$$

for $M_t \subset W = \mathbb{P}^{n+1}$. Let $\Xi = \mathcal{O}(T(W))$ be the sheaf of germs of holomorphic vector fields over W. Then Ξ_t is the restriction of Ξ to M_t. Let r_t be the restriction map. Then since $[M_t] = E^h$, $\ker r_t = \Xi \otimes E^{-h} = \mathcal{O}(T(W) \otimes E^{-h})$. Thus

$$0 \to \Xi \otimes E^{-h} \to \Xi \xrightarrow{r_t} \Xi_t \to 0 \tag{5.53}$$

is exact.

Lemma 5.5. *Except in the case $n=2$, $h=4$, we have*

$$H^q(\mathbb{P}^{n+1}, \Xi \otimes E^{-h}) = 0, \qquad q = 0, 1, 2.$$

Proof. Put $T = T(\mathbb{P}^{n+1})$. Then $\Xi \times E^{-h} = \mathcal{O}(T \otimes E^{-h})$, and the dual bundle of $T \otimes E^{-h}$ is $T^* \otimes E^h$. Hence by (3.155),

$$\dim H^q(\mathbb{P}^{n+1}, \Xi \otimes E^{-h}) = \dim H^{n+1-q}(\mathbb{P}^{n+1}, \Omega^{n+1}(T^* \otimes E^h)).$$

Putting $K = \bigwedge^{n+1} T^*$, we have

$$\Omega^{n+1}(T^* \otimes E^h) = \mathcal{O}(T^* \otimes K \otimes E^h) = \Omega^1(K \otimes E^h),$$

while the canonical bundle K of $\mathbb{P}^{n+1}$ is E^{-n-2}. Therefore

$$\dim H^q(\mathbb{P}^{n+1}, \Xi \otimes E^{-h}) = \dim H^{n+1-q}(\mathbb{P}^{n+1}, \Omega^1(E^{h-n-2})). \tag{5.54}$$

Since we assume that $n \geqq 2$ and $h \geqq 3$, Lemma 5.5 follows immediately from following Bott's theorem. ∎

Theorem 5.2 (Bott [1]). *$H^q(\mathbb{P}^{n+1}, \Omega^p(E^k))$ vanishes except for the following cases:*

(i) $p+q$ and $k=0$,
(ii) $q=0$ and $k>p$,
(iii) $q=n+1$ and $k<0-n-1$.

Thus except for the case $n=2$, $h=4$, in the exact sequence of cohomology groups induced from (5.53):

$$\begin{aligned} 0 \to H^0(W, \Xi \otimes E^{-h}) \to H^0(W, \Xi) &\xrightarrow{r_t} H^0(M_t, \Xi_t) \\ \to H^1(W, \Xi \otimes E^{-h}) &\to \cdots \end{aligned}$$

we have $H^q(W, \Xi \otimes E^{-h}) = 0$ for $q = 0, 1, 2$. Hence

$$H^0(W, \Xi) \overset{r_t}{\cong} H^0(M_t, \Xi_t),$$

and

$$H^1(W, \Xi) \cong H^1(M_t, \Xi_t). \tag{5.55}$$

Also from (5.54) and Theorem 5.2, we have $H^1(\mathbb{P}^{n+1}, \Xi) = 0$. Hence

$$H^1(M_t, \Xi_t) = 0.$$

An automorphism of $W = \mathbb{P}^{n+1}$ is a projective transformation, and the set $\mathfrak{G}$ of all of them forms a Lie transformation group of W with $\dim \mathfrak{G} = (n+2)^2 - 1$. Since $H^0(W, \Xi)$ is the Lie algebra consisting of the infinitesimal transformations of $\mathfrak{G}$, we have

$$\dim H^0(W, \Xi) = (n+2)^2 - 1.$$

Hence from (5.55) we have

$$\dim H^0(M_t, \Xi_t) = (n+2)^2 - 1. \tag{5.56}$$

In the exact sequence of cohomology groups induced from the exact sequence $0 \to \Theta_t \to \Xi_t \to \mathcal{O}(F_t) \to 0$, we have $H^1(M_t, \Xi_t) = 0$. Also we can prove that $H^0(M_t, \Theta_t) = 0$ (see [21], p. 406, Lemma 14.2). Therefore the sequence

$$0 \to H^0(M_t, \Xi_t) \to H^0(M_t, \mathcal{O}(F_t)) \to H^1(M_t, \Theta_t) \to 0 \tag{5.57}$$

is exact. By Lemma 5.4, $\dim H^0(M_t, \mathcal{O}(F_t)) = \mu$. Therefore by (5.56)

$$\dim H^1(M_t, \Theta_t) = m = \mu - (n+2)^2 + 1 = \binom{n+1+h}{h} - (n+2)^2.$$

Thus we obtain (5.21).

For a projective transformation $\gamma \in \mathfrak{G}$, define

$$(\gamma t)(\zeta) = t(\gamma^{-1}\zeta).$$

Then $\gamma\colon t \to \gamma t$ is a projective transformation of the parameter space $\mathbb{P}^\mu$. Since $S_j(w_j, t) = (1/\zeta_{\beta(j)})^h t(\zeta)$, if γ belongs to a sufficiently small neighbourhood of the identity of $\mathfrak{G}$, we have

$$S_j(w_j, \gamma t) = S_j(\gamma^{-1} w_j, t) \tag{5.58}$$

for $w_j \in W_j$ and $t = (1, t_1, \ldots, t_\mu) \in \Delta$. We denote by $\sum_{\lambda=1}^{\mu} \xi_\lambda(t)\, \partial/\partial t_\lambda$ the infinitesimal transformation of Δ determined by the infinitesimal transformation $\xi \in H^0(W, \Xi)$ of $\mathfrak{G}$. Since ξ is a holomorphic vector field on W, ξ is written on each W_j as $\xi = \sum_{\lambda=0}^{n+1} \xi_j^\alpha(w_j)\, \partial/\partial w_j^\alpha$. From (5.58) clearly we have

$$\sum_{\lambda=1}^{\mu} \xi_\lambda(t) \frac{\partial S_j(w_j, t)}{\partial t_\lambda} = -\xi S_j(w_j, t).$$

Substituting $w_j = \varphi_j(z_j, t)$ and denoting the restriction of ξ to M_t by $\xi_t \in H^0(M_t, \Xi_t)$, we obtain from (5.41) and (5.42) the equality

$$\sum_{\lambda=1}^{\mu} \xi_\lambda(t) \rho_{d,t}\left(\frac{\partial}{\partial t_\lambda}\right) = \xi_t S(t). \tag{5.59}$$

This equality implies that if $\sum_{\lambda=1}^{\mu} \xi_\lambda(t)\, \partial/\partial t_\lambda = 0$ for a point $t \in \Delta$, we have $\xi_t S(t) = 0$. On the other hand, $\xi_t \to \xi_t(S(t))$ gives the injection $H^0(M_t, \Xi_t) \to H^0(M_t, \mathcal{O}(F_t))$ in (5.57). Therefore $\xi_t S(t) = 0$ implies $\xi_t = 0$, hence by (5.55) we have $\xi = 0$. Thus if $\xi \neq 0$, $\sum_{\lambda=1}^{\mu} \xi_\lambda(t)\, \partial/\partial t_\lambda \neq 0$. This proves that *for any $t \in \Delta$, the map $\gamma \to \gamma t$ is a biholomorphic map on a sufficiently small neighbourhood of the unity of $\mathfrak{G}$.*

Fix an arbitrary point $t^0 \in \Delta$, and take a sufficiently small neighbourhood $\mathfrak{U}$ of the unity of $\mathfrak{G}$. Then $\gamma \to \gamma t^0$ maps $\mathfrak{U}$ biholomorphically onto $\mathfrak{U} t^0 \subset \Delta$. $\mathfrak{U} t^0$ is a domain of a submanifold of dimension $(n+2)^2 - 1$ of Δ containing t^0. Since $\mu = m + (n+2)^2 - 1$, rearranging $t_1, \ldots, t_\mu$ if necessary, we may assume that the coordinate plane $t_{m+1} = t^0_{m+1}, \ldots, t_\mu = t^0_\mu$ through t^0 intersects with $\mathfrak{U} t^0$ transversally at t^0. If we take a polydisk with centre t^0:

$$U = \{(t_1, \ldots, t_m, t^0_{m+1}, \ldots, t^0_\mu) \,|\, |t_1 - t^0_1| < \varepsilon, \ldots, |t_m - t^0_m| < \varepsilon\},$$

the map $(\gamma, t) \to \gamma t$ maps $\mathfrak{U} \times U$ biholomorphically onto the neighbourhood $\mathfrak{U} U \subset \Delta$ of t^0 provided that $\varepsilon > 0$ is sufficiently small.

Consider the restriction $(\mathcal{M}_U, U, \varpi)$ of $(\mathcal{M}_\Delta, \Delta, \varpi)$ to $U \subset \Delta$, where $\mathcal{M}_U = \varpi^{-1}(U)$. Since $\mathfrak{U} U$ is biholomorphically equivalent to $\mathfrak{U} \times U$, we have

$$T_t(\Delta) = T_t(\mathfrak{U} U) = T_t(\mathfrak{U} t) + T_t(U)$$

at each point $t \in U$. Let e be the unity of $\mathfrak{G}$. Then the tangent space $T_e(\mathfrak{U})$ of $\mathfrak{U}$ at $e \in \mathfrak{U}$ is $H^0(W, \Xi)$: $T_e(\mathfrak{U}) = H^0(W, \Xi)$. Therefore $T_t(\mathfrak{U} t)$ is isomorphic to $H^0(W, \Xi)$. The isomorphism $H^0(W, \Xi) \to T_t(\mathfrak{U} t)$ is given by

$$\eta : \xi \to \sum_{\lambda=1}^{\mu} \xi_\lambda(t) \frac{\partial}{\partial t_\lambda}.$$

By (5.59) we have

$$\rho_{d,t}(\eta \xi) = \xi_t S(t).$$

From this and (5.46), (5.57), we see that the diagram

$$\begin{array}{ccccccccc}
0 \longrightarrow & H^0(M_t, \Xi_t) & \longrightarrow & H^0(M_t, \mathcal{O}(F_t)) & \xrightarrow{\delta^*} & H^1(M_t, \Theta_t) & \longrightarrow 0 \\
& \uparrow r_t & & \uparrow \rho_{d,t} & \nearrow \rho_t & & \\
0 \longrightarrow & H^0(W, \Xi) & \xrightarrow{\eta} & T_t(\Delta) & & &
\end{array} \tag{5.60}$$

is an exact commutative diagram. In this diagram, $\rho_{d,t}$ is an isomorphism by (5.49) and Lemma 5.4, r_t is an isomorphism by (5.55) and $\eta H^0(W, \Xi) = T_t(\mathfrak{U}t)$. Consequently since $T_t(\Delta) = T_t(\mathfrak{U}t) + T_t(U)$, ρ_t maps $T_t(U)$ isomorphically onto $H^1(M_t, \Theta_t)$. Thus the complex analytic family $(\mathcal{M}_U, U, \varpi)$ is effectively parametrized and $\dim H^1(M_t, \Theta_t) = m$. Therefore the number of moduli $m(M_t) = m$ of each $M_t = \varpi^{-1}(t)$ for $t \in U$ is equal to $\dim H^1(M_t, \Theta_t)$.

Since t^0 is an arbitrary point of Δ, and $\Delta \subset B$ is an arbitrary sufficiently small polydisk in B, we have proved that except for the case $n = 2$, $h = 4$, the number of moduli $m(M_t)$ is defined for any non-singular hypersurface M_t of order h in $\mathbb{P}^{n+1}$, and that $m(M_t) = \dim H^1(M_t, \Theta_t)$.

In case $n = 2$, $h = 4$, as stated before, we have $m = 19$ and $\dim H^1(M_t, \Theta_t) = 20$. Also in this case we have the following exact commutative diagram similar to (5.60)

$$\begin{array}{ccccccccc}
0 \longrightarrow & H^0(M_t, \Xi_t) & \longrightarrow & H^0(M_t, \mathcal{O}(F_t)) & \longrightarrow & H^1(M_t, \Theta_t) & \longrightarrow & H^1(M_t, \Xi_t) & \longrightarrow 0 \\
& \uparrow & & \uparrow \rho_{d,t} & \nearrow \rho_t & & & & \\
0 \longrightarrow & H^0(W, \Xi) & \longrightarrow & T_t(\Delta) & & & & &
\end{array}$$

But $H^1(M_t, \Xi_t) \neq 0$ in this case. Since in the exact sequence of cohomology groups induced from (5.53) we have $H^1(\mathbb{P}^3, \Xi) = H^2(\mathbb{P}^3, \Xi) = 0$, we obtain $H^1(M_t, \Xi_t) \cong H^2(\mathbb{P}^3, \Xi \otimes E^{-4})$. Consequently by (5.54) $\dim H^1(M_t, \Xi_t) = \dim H^1(\mathbb{P}^3, \Omega^1)$, while $H^1(\mathbb{P}^3, \Omega^1)$ belongs to the exceptional cases (i) of Bott's theorem, and actually we have $\dim H^1(\mathbb{P}^3, \Omega^1) = 1$ (see p. 175). Therefore $\dim H^1(M_t, \Theta_t) = m + 1 = 20$.

At first we could not find the reason why the case $n = 2$, $h = 4$ is exceptional. Then S. Nakano, who stayed at the Institute for Advanced Study at Princeton at that time, pointed out that in this case the complex analytic family $(\mathcal{M}, B, \varpi)$ is not complete. Actually taking the hypersurface $\zeta_0^4 + \zeta_1^4 + \zeta_2^4 + \zeta_3^4 = 0$ of Fermat type as M_t, he showed that M_t is an elliptic surface and, using the theory of elliptic surfaces ([14]), proved that there is a complex analytic family $\{N_u \mid u \in \mathbb{C}\}$ with $N_0 = M_t$ of deformation of M_t consisting of elliptic surfaces such that for any given $\varepsilon > 0$, we can find a non-algebraic surface N_u with $|u| < \varepsilon$ in this family.

From this result, we considered that the reason why the number of effective parameters $m = 19$ of the complex analytic family $(\mathcal{M}, B, \varpi)$ for a quartic surface in $\mathbb{P}^3$ is less than $\dim H^1(M_t, \Theta_t) = 20$ would be that M does not contain all possible deformations of M_t, and so we expected that the number of moduli $m(M_t)$ of M_t could still be defined in this case and equal to $\dim H^1(M_t, \Theta_t) = 20$.

Next we studied hypersurfaces on Abelian varieties ([21], §14(δ)). We explain them briefly below. Let $A = A^{n+1}$ be an $(n+1)$-dimensional Abelian variety with $n \geqq 2$. A is a submanifold of some projective space $\mathbb{P}^N$. For a general hyperplane P of $\mathbb{P}^N$, $M = A \cap P$ is a non-singular hypersurface on A. Given a complex analytic family $\{M_t \mid t \in B\}$ of this hypersurface $M_0 = M$ with $0 \in B \subset \mathbb{C}^m$, there exists a complex analytic family of deformations $\{A_t \mid t \in \Delta\}$ of $A_0 = A$ with $M_t \subset A_t$ provided that Δ is a sufficiently small polydisk with $0 \in \Delta \subset B$. A_t is a complex torus, and $M_t \subset A_t$ implies that A_t is an Abelian variety. The number of moduli of $A_0 = A^{n+1}$ as a complex torus is equal to $(n+1)^2$, but if we only consider deformations of A_0 which are Abelian varieties, its number of moduli reduces to $(n+1)(n+2)/2$ since their period matrices must be Riemann matrices. In view of this fact we suspected that the number of moduli $m(M)$ of M would be smaller than $\dim H^1(M, \Theta)$, but the computation showed that $m(M) = \dim H^1(M, \Theta)$ also in this case. Of course the decrease of the number of moduli of A_t does not necessarily imply $m(M) < \dim H^1(M, \Theta)$, but at that time this result seemed surprising to us.

§5.3. Theorem of Existence

In the preceding section we have seen that for several examples of compact complex manifolds M, the number $m(M)$ of moduli of M is equal to $\dim H^1(M, \Theta)$. As stated in (a) of the preceding section, if $m(M) = \dim H^1(M, \Theta)$, for an arbitrary $\theta \in H^1(M, \Theta)$, there exists a complex analytic family $(\mathcal{M}, B, \varpi)$ with $0 \in B \subset \mathbb{C}$ satisfying the following conditions:

$$\varpi^{-1}(0) = M, \qquad \left(\frac{dM_t}{dt}\right)_{t=0} = \theta, \quad \text{where} \quad M_t = \varpi^{-1}(t). \tag{5.61}$$

Since at that time we could not find no example M with $m(M) \neq \dim H^1(M, \Theta)$, we expected that for any given compact complex manifold M and any $\theta \in H^1(M, \Theta)$, there would exist a complex analytic family $(\mathcal{M}, B, \varpi)$ satisfying (5.61) but for some exceptional cases.

(a) An Elementary Method

We first tried to prove the theorem of existence by an elementary method, using power series expansion. We state it in this section, for, although we did not succeed in proving the theorem by this method, this was an important step in the development of the theory of deformation.

In order to prove the theorem of existence, it suffices to take as B a disk $\Delta = \{t \in \mathbb{C} \mid |t| < r\}$ of sufficiently small radius $r > 0$, and construct a complex analytic family of the form (4.27)

$$\mathcal{M} = \bigcup_{j=1}^{l} U_j \times \Delta$$

satisfying (5.61). Here each U_j is a polydisk in $\mathbb{C}^n$ and $\mathcal{M}$ is the complex manifold obtained by glueing the domains $U_j \times \Delta, j=1,\ldots,l$, by identifying $(z_j, t) \in U_j \times \Delta$ and $(z_k, t) \in U_k \times \Delta$ if $z_j^\alpha = f_{jk}^\alpha(z_k, t)$. $f_{jk}^\alpha(z_k, t) = f_{jk}^\alpha(z_k^1, \ldots, z_k^n, t)$, $\alpha = 1, \ldots, n$, are holomorphic functions defined on the open subset $U_k \times \Delta \cap U_j \times \Delta$ of the domain $U_k \times \Delta$ of $\mathcal{M}$. We call these holomorphic functions $f_{jk}^\alpha(z_k, t)$ the *defining functions* of $\mathcal{M}$. These functions must satisfy the following compatibility conditions: On $U_k \times \Delta \cap U_j \times \Delta \neq \varnothing$,

$$f_{ik}^\alpha(z_k, t) = f_{ij}^\alpha(f_{jk}(z_k, t), t). \tag{5.62}$$

Since by (5.61) $M = \varpi^{-1}(0) = \bigcup_{j=1}^{l} U_j \times 0$, identifying U_j with $U_j \times 0$, U_j is considered to be a coordinate polydisk on M and $z_j = (z_j^1, \ldots, z_j^n)$ are considered to be local coordinates of M defined on U_j. If we expand $f_{jk}^\alpha(z_k, t)$ into the power series in t as

$$f_{jk}^\alpha(z_k, t) = \sum_{\nu=0}^{\infty} f_{jk|\nu}^\alpha(z_k) t^\nu, \tag{5.63}$$

the coefficients $f_{jk|\nu}^\alpha(z_k)$ *are holomorphic functions defined on the open subset* $U_k \cap U_j \neq \varnothing$ *of the coordinate polydisk* U_k *on* M. On $U_j \cap U_k = U_j \times 0 \cap U_k \times 0$, we have $z_j^\alpha = f_{jk|0}^\alpha(z_k)$. Thus $z_j^\alpha = f_{jk|0}^\alpha(z_k)$ gives the transformation of local coordinates of the complex manifold M. Therefore we may consider that *the coefficient* $f_{jk|0}^\alpha(z_k)$ *of* t^0 *in* (5.63) *is given with* M.

In order to construct a complex analytic family $\mathcal{M} = \bigcup_{j=1}^{l} U_j \times \Delta$ with $\varpi^{-1}(0) = \bigcup_{j=1}^{l} U_j \times 0 = M$, it suffices to give the defining functions $f_{jk}^\alpha(z_k, t)$ of $\mathcal{M}$. For this, since $f_{jk|0}^\alpha(z_k) = f_{jk}^\alpha(z_k, 0)$ are already given, we only have to determine the coefficients $f_{jk|\nu}^\alpha(z_k)$, $\nu = 1, 2, \ldots$, of (5.63). The defining functions of M satisfy *the equalities* (5.62):

$$f_{ik}^\alpha(z_k, t) = f_{ij}^\alpha(f_{jk}(z_k, t), t), \qquad \alpha = 1, \ldots, n.$$

Hence *the coefficients* $f_{jk|\nu}^\alpha(z_k)$, $\nu = 1, 2, \ldots$, *must be determined such as*

$$f_{jk}^\alpha(z_k, t) = f_{jk|0}^\alpha(z_k) + \sum_{\nu=1}^{\infty} f_{jk|\nu}^\alpha(z_k) t^\nu$$

satisfy these equalities.

For a given cohomology class $\theta \in H^1(M, \Theta)$, take an arbitrary 1-cocycle $\{\theta_{jk}\} \in Z^1(\{U_j\}, \Theta)$ belonging to θ, and write θ_{jk} in the form

$$\theta_{jk} = \sum_{\alpha=1}^{n} \theta_{jk}^\alpha(z_j) \frac{\partial}{\partial z_j^\alpha}.$$

Since

$$\sum_{\alpha=1}^{n}\left(\frac{\partial f_{jk}^{\alpha}(z_k, t)}{\partial t}\right)_{t=0}\frac{\partial}{\partial z_j^{\alpha}}=\sum_{\alpha=1}^{n} f_{jk|1}^{\alpha}(z_k)\frac{\partial}{\partial z_j^{\alpha}},$$

if we put

$$f_{jk|1}^{\alpha}(z_k)=\theta_{jk}^{\alpha}(z_k), \qquad z_j=f_{jk|0}(z_k),$$

the infinitesimal deformation of $M_t=\bigcup_{j=1}^{l} U_j\times t$ is by the definition (4.15),

$$\left(\frac{dM_t}{dt}\right)_{t=0}=\theta.$$

Therefore in order to construct a complex analytic family $\mathcal{M}=\bigcup_{j=1}^{l} U_j\times\Delta$ satisfying (5.61), it suffices to *put* $f_{jk|1}^{\alpha}(z_k)=\theta_{jk}^{\alpha}(z_j)$ *first and to determine* $f_{jk|\nu}^{\alpha}(z_k)$, $\nu=2, 3, \ldots$, *such that they satisfy* (5.62).

For simplicity, putting

$$f_{jk}(z_k, t)=(f_{jk}^{1}(z_k, t), \ldots, f_{jk}^{n}(z_k, t)),$$

and

$$f_{jk|\nu}(z_k)=(f_{jk|\nu}^{1}(z_k), \ldots, f_{jk|\nu}^{n}(z_k)),$$

and using vector notation, we write (5.63) in the form

$$f_{jk}(z_k, t)=\sum_{\nu=0}^{\infty} f_{jk|\nu}(z_k)t^{\nu}. \tag{5.64}$$

In general for a power series in t

$$P(t)=\sum_{\nu=0}^{\infty} P_{\nu}t^{\nu}=P_0+P_1t+\cdots,$$

we put

$$P^{\nu}(t)=P_0+P_1t+\cdots+P_{\nu}t^{\nu}.$$

Further for two power series $P(t)$ and $Q(t)$, we write $P(t)\equiv_{\nu} Q(t)$ if $P(t)\equiv Q(t) \pmod{t^{\nu+1}}$. Thus $P(t)\equiv_{\nu} Q(t)$ just means that $P^{\nu}(t)=Q^{\nu}(t)$.

The coefficient of t^{ν} in the power series expansion of $f_{ij}^{\alpha}(f_{jk}(z_k, t), t)$ in t is a polynomial of the functions $f_{ij|\mu}^{\alpha}(z_j)$, $f_{jk|\mu}^{\beta}(z_k)$, $\mu=1, \ldots, \nu$, $\beta=1, \ldots, n$, and their partial derivatives, while we have $z_j=f_{jk|0}(z_k)$. Hence the coefficient of t^{ν} is a holomorphic function defined on the open set $U_k\cap U_j\cap U_i\neq\varnothing$ on M, and (5.62) is reduced to the system of infinitely many congruences

$$f_{jk}^{\nu}(z_k, t)\underset{\nu}{\equiv} f_{ij}^{\nu}(f_{jk}^{\nu}(z_k, t), t), \qquad \nu=1, 2, \ldots. \tag{5.65$_\nu$}$$

As stated above, $f_{jk|0}(z_k)$ are given with M. Also we have put $f_{jk|1}(z_k)=\theta_{jk}(z_j)$ with $z_j=f_{jk|0}(z_k)$. Then we can verify $(5.65)_1$ easily as follows.

$$\begin{aligned} f_{ij}^1(f_{jk}^1(z_k,t),t) &\underset{1}{\equiv} f_{ij}^1(f_{jk|0}(z_k)+f_{jk|1}(z_k)t,t) \\ &\underset{1}{\equiv} f_{ij|0}(z_j+f_{jk|1}(z_k)t)+f_{ij|1}(z_j+f_{jk|1}(z_k)t)t \\ &\underset{1}{\equiv} f_{ij|0}(z_j)+\sum_{\beta=1}^{n}\frac{\partial}{\partial z_j^\beta}f_{ij|0}(z_j)\cdot f_{jk|1}^\beta(z_k)+f_{ij|1}(z_k)t, \end{aligned}$$

while $f_{ij|0}(z_j)=f_{ik|0}(z_k)$. Hence if we put $z_i^\alpha=f_{ij|0}^\alpha(z_j)$, $(5.65)_1$ is equivalent to the equality

$$f_{ik|1}^\alpha(z_k)=f_{ij|1}^\alpha(z_j)+\sum_{\beta=1}^{n}\frac{\partial z_i^\alpha}{\partial z_j^\beta}f_{jk|1}^\beta(z_k),$$

namely,

$$\sum_{\alpha=1}^{n}f_{ik|1}^\alpha(z_k)\frac{\partial}{\partial z_i^\alpha}=\sum_{\alpha=1}^{n}f_{ij|1}^\alpha(z_j)\frac{\partial}{\partial z_i^\alpha}+\sum_{\beta=1}^{n}f_{jk|1}^\beta(z_k)\frac{\partial}{\partial z_j^\beta}.$$

If we put $f_{jk|1}(z_k)=\theta_{jk}(z_j)$, the last equality follows from the fact that $\{\theta_{jk}(z_j)\}$ is a 1-cocycle.

We consider the vector $f_{jk|\nu}(z_k)=(f_{jk|\nu}^1(z_k),\ldots,f_{jk|\nu}^n(z_k))$ as the holomorphic vector field on $U_j\cap U_k$:

$$f_{jk|\nu}=\sum_{\alpha=1}^{n}f_{jk|\nu}^\alpha(z_k)\frac{\partial}{\partial z_j^\alpha}$$

if necessary. On each $U_k\cap U_j\neq\emptyset$, we determine $f_{jk|\nu}(z_k)$, $\nu=2,3,\ldots$, by induction on ν. For this assume that on each $U_k\cap U_j\neq\emptyset$,

$$f_{jk}^{\nu-1}(z_k,t)=f_{jk|0}(z_k)+\cdots+f_{jk|\nu-1}(z_k)t^{\nu-1}$$

is already determined in such a way as the congruence $(5.65)_{\nu-1}$

$$f_{ik}^{\nu-1}(z_k,t)\underset{\nu-1}{\equiv}f_{ij}^{\nu-1}(f_{jk}^{\nu-1}(z_k,t),t)$$

holds and consider $(5.65)_\nu$:

$$f_{ik}^\nu(z_k,t)\underset{\nu}{\equiv}f_{ij}^\nu(f_{jk}^\nu(z_k,t),t).$$

For simplicity, we write $f_{jk}^\nu(z_k,t)$, $f_{jk|\nu}(z_k)$, etc., as $f_{jk}^\nu(t)$, $f_{jk|\nu}$, etc., respectively. Then the right-hand side of $(5.65)_\nu$ is written as

$$\begin{aligned} f_{ij}^{\nu}(f_{jk}^{\nu}(t), t) &= f_{ij}^{\nu-1}(f_{jk}^{\nu}(t), t) + f_{ij|\nu}(f_{jk}^{\nu}(t))t^{\nu} \\ &= f_{ij}^{\nu-1}(f_{jk}^{\nu-1}(t) + f_{jk|\nu}t^{\nu}, t) + f_{ij|\nu}(f_{jk}^{\nu}(t))t^{\nu}. \end{aligned}$$

Putting $z_j + \zeta_j = (z_j^1 + \zeta_j^1, \ldots, z_j^n + \zeta_j^n)$ and expanding each component of the vector $f_{ij}^{\nu-1}(z_j + \zeta_j, t)$ into the power series in $\zeta_j^1, \ldots, \zeta_j^n$, we obtain

$$f_{ij}^{\nu-1}(z_j + \zeta_j, t) = f_{ij}^{\nu-1}(z_j, t) + \sum_{\beta=1}^{n} \frac{\partial}{\partial z_j^{\beta}} f_{ij}^{\nu-1}(z_j, t)\zeta_j^{\beta} + \cdots.$$

Putting $z_j = f_{jk}^{\nu-1}(t)$ and $\zeta_j = f_{jk|\nu}t^{\nu}$, since

$$\frac{\partial}{\partial z_j^{\beta}} f_{ij}^{\nu-1}(f_{jk}^{\nu-1}(t), t) f_{jk|\nu}^{\beta} t^{\nu} \underset{\nu}{\equiv} \frac{\partial}{\partial z_j^{\beta}} f_{ij|0}(f_{jk|0}) f_{jk|\nu}^{\beta} t^{\nu},$$

we have

$$f_{ij}^{\nu-1}(f_{jk}^{\nu-1}(t) + f_{jk|\nu}t^{\nu}, t) \underset{\nu}{\equiv} f_{ij}^{\nu-1}(f_{jk}^{\nu-1}(t), t) + \sum_{\beta=1}^{n} \frac{\partial}{\partial z_j^{\beta}} f_{ij|0}(f_{jk|0}) f_{jk|\nu}^{\beta} t^{\nu}.$$

Here by $(\partial/\partial z_j^{\beta}) f_{ij|0}(f_{jk|0})$ we mean the vector obtained by differentiating each component of $f_{ij|0}(z_j)$ first with respect to z_j^{β} and then substituting $z_j = f_{jk|0}(z_k)$. Putting $z_i = f_{ij|0}(z_j)$, we have

$$\frac{\partial}{\partial z_j^{\beta}} f_{ij|0}(f_{jk|0}) = \frac{\partial z_i}{\partial z_j^{\beta}} = \left(\frac{\partial z_i^1}{\partial z_j^{\beta}}, \ldots, \frac{\partial z_i^n}{\partial z_j^{\beta}}\right).$$

On the other hand, clearly we have

$$f_{ij|\nu}(f_{jk}^{\nu}(t))t^{\nu} \underset{\nu}{\equiv} f_{ij|\nu}(f_{jk|0})t^{\nu} = f_{ij|\nu}(z_j)t^{\nu}.$$

Therefore $(5.65)_\nu$ is reduced to

$$\begin{aligned} &f_{ik}^{\nu-1}(z_k, t) - f_{ij}^{\nu-1}(f_{jk}^{\nu-1}(z_k, t), t) \\ &\quad \underset{\nu}{\equiv} \sum_{\beta=1}^{n} \frac{\partial z_i}{\partial z_j^{\beta}} f_{jk|\nu}^{\beta}(z_k)t^{\nu} - f_{ik|\nu}(z_k)t^{\nu} + f_{ij|\nu}(z_j)t^{\nu}. \end{aligned}$$

Since by hypothesis $(5.65)_{\nu-1}$ holds, the power series expansion of the left-hand side of this congruence begins in the term of degree ν in t. If we put the coefficient of t^{ν} as $\Gamma_{ijk|\nu}(z_k)$, we have

$$f_{ik}^{\nu-1}(z_k, t) - f_{ij}^{\nu-1}(f_{jk}^{\nu-1}(z_k, t), t) \underset{\nu}{\equiv} \Gamma_{ijk|\nu}(z_k)t^{\nu}, \qquad (5.66)_\nu$$

hence, $(5.65)_\nu$ *is reduced to the equality*

$$\Gamma_{ijk|\nu}(z_k) = \sum_{\beta=1}^{n} \frac{\partial z_i}{\partial z_j^\beta} f_{jk|\nu}^\beta(z_k) - f_{ik|\nu}(z_k) + f_{ij|\nu}(z_j). \tag{5.67$_\nu$}$$

Here z_k, $z_j = f_{jk|0}(z_k)$ and $z_i = f_{ij|0}(z_j) = f_{ik|0}(z_k)$ are *local coordinates of one and the same point of M contained in* $U_k \cap U_j \cap U_i$. Each component $\Gamma_{ijk|\nu}^\alpha(z_k)$ of the vector $\Gamma_{ijk|\nu}(z_k)$ is a holomorphic function defined on the open set $U_i \cap U_j \cap U_k \neq \emptyset$ of M. The equality $(5.67)_\nu$ is written in terms of the components as

$$\Gamma_{ijk|\nu}^\alpha(z_k) = \sum_{\beta=1}^{n} \frac{\partial z_i^\alpha}{\partial z_j^\beta} f_{jk|\nu}^\beta(z_k) - f_{ik|\nu}^\alpha(z_k) + f_{ij|\nu}(z_j), \qquad \alpha = 1, \ldots, n.$$

Using vector field notation, we can write it in the form

$$\sum_\alpha \Gamma_{ijk|\nu}^\alpha(z_k)\frac{\partial}{\partial z_i^\alpha} = \sum_\beta f_{jk|\nu}^\beta(z_k)\frac{\partial}{\partial z_j^\beta} - \sum_\alpha f_{ik|\nu}^\alpha(z_k)\frac{\partial}{\partial z_i^\alpha} + \sum_\alpha f_{ij|\nu}^\alpha(z_k)\frac{\partial}{\partial z_i^\alpha}.$$

Therefore putting

$$\Gamma_{ijk|\nu} = \sum_\alpha \Gamma_{ijk|\nu}^\alpha(z_k)\frac{\partial}{\partial z_i^\alpha}, \qquad f_{jk|0} = \sum_\beta f_{jk|\nu}^\beta(z_k)\frac{\partial}{\partial z_j^\beta}, \ldots,$$

we can write $(5.67)_\nu$ as

$$\Gamma_{ijk|\nu} = f_{jk|\nu} - f_{ik|\nu} + f_{ij|\nu}. \tag{5.68$_\nu$}$$

Thus if $(5.65)_{\nu-1}$ holds, $(5.65)_\nu$ *is equivalent to* $(5.68)_\nu$.

If we consider $\{f_{jk|\nu}\}$ as a 1-cochain on the finite open covering $\mathfrak{U} = \{U_j\}$ of M, the right-hand side of $(5.68)_\nu$ is equal to its coboundary $\delta\{f_{jk|\nu}\}$. In our case, however, although $f_{jj|\nu} = 0$, $\nu \geqq 1$, since $f_{jj}(z_j, t) = z_j$ are independent of t, $f_{kj|\nu} = f_{jk|\nu}$ do not necessarily hold. In view of this *we extend the definitions of* 1-*cochains*, 2-*cocycles, etc., as follows.*

Let $\mathscr{S}$ be an arbitrary sheaf over M. Suppose that for each pair j, k of indices with $U_j \cap U_k \neq \emptyset$, a section $\sigma_{jk} \in \Gamma(U_j \cap U_k, \mathscr{S})$ is assigned. If $\sigma_{jj} = 0$ for every j, the set $\hat{c}^1 = \{\sigma_{jk}\}$ of these σ_{jk} is called a 1-cochain on $\mathfrak{U} = \{U_j\}$. Similarly suppose that each triple i, j, k with $U_i \cap U_j \cap U_k \neq \emptyset$, a section $\sigma_{ijk} \in \Gamma(U_i \cap U_j \cap U_k, S)$ is assigned. If $\sigma_{iik} = \sigma_{ikk} = 0$, and on each $U_h \cap U_i \cap U_j \cap U_k \neq \emptyset$

$$\sigma_{ijk} - \sigma_{hjk} + \sigma_{hik} - \sigma_{hij} = 0$$

holds, the set $\hat{c}^2 = \{\sigma_{ijk}\}$ is called a 2-cocycle on $\mathfrak{U}$. We define the coboundary of a 1-cochain $\hat{c}^1 = \{\sigma_{jk}\}$ as

$$\delta\hat{c}^1 = \{\tau_{ijk}\}, \qquad \tau_{ijk} = \sigma_{jk} = \sigma_{jk} - \sigma_{ik} + \sigma_{ij}.$$

It is easy to verify that *the coboundary* $\delta\hat{c}^1$ *is a* 2-*cocycle.* Let $\hat{C}^1(\mathfrak{U}, \mathscr{S})$ be the Abelian group of all 1-cochains $\hat{c}^1$ and $\hat{Z}^2(\mathfrak{U}, \mathscr{S})$ that of all 2-cocycles $\hat{c}^2$. Then $\delta\hat{C}^1(\mathfrak{U}, \mathscr{S})$ is a subgroup of $\hat{Z}^2(\mathfrak{U}, \mathscr{S})$. The quotient group

$$\hat{H}^2(\mathfrak{U}, \mathscr{S}) = \hat{Z}^2(\mathfrak{U}, \mathscr{S})/\delta\hat{C}^1(\mathfrak{U}, \mathscr{S}) \tag{5.69}$$

is called the 2-dimensional cohomology group of $\mathfrak{U}$ with coefficients in $\mathscr{S}$.

With this notation, $(5.68)_\nu$ is written as

$$\{\Gamma_{ijk|\nu}\} = \delta\{f_{jk|\nu}\}. \qquad (5.70)_\nu$$

Lemma 5.6. $\{\Gamma_{ijk|\nu}\}$ *is a* 2-*cocycle on* $\mathfrak{U}$:

$$\{\Gamma_{ijk|\nu}\} \in \hat{Z}^2(\mathfrak{U}, \Theta).$$

Proof. Using vector notation, we must prove that

$$\sum_\alpha \frac{\partial z_h}{\partial z_i^\alpha} \Gamma^\alpha_{ijk|\nu}(z_k) - \Gamma_{hjk|\nu}(z_k) + \Gamma_{hik|\nu}(z_k) - \Gamma_{hij|\nu}(z_j) = 0. \tag{5.71}$$

Here we put $z_h = (z_h^1, \ldots, z_h^n)$, and $z_h = f_{hk|0}(z_k)$. For simplicity we write $\Gamma_{hijk|\nu}$, $f_{hk}^{\nu-1}(t)$, etc., instead of $\Gamma_{hjk|\nu}(z_k)$, $f_{hk}^{\nu-1}(z_k, t)$, etc., respectively. By $(5.66)_\nu$ we have

$$\Gamma_{hik|\nu} t^\nu \underset{\nu}{\equiv} f_{hk}^{\nu-1}(t) - f_{hi}^{\nu-1}(f_{ik}^{\nu-1}(t), t),$$

$$\Gamma_{hjk|\nu} t^\nu \underset{\nu}{\equiv} f_{hk}^{\nu-1}(t) - f_{hj}^{\nu-1}(f_{jk}^{\nu-1}(t), t),$$

hence,

$$\Gamma_{hik|\nu} t^\nu - \Gamma_{hjk|\nu} t^\nu \underset{\nu}{\equiv} f_{hj}^{\nu-1}(f_{jk}^{\nu-1}(t), t) - f_{hi}^{\nu-1}(f_{ik}^{\nu-1}(t), t).$$

Substituting $z_j = f_{jk}^{\nu-1}(t)$ into the congruence

$$f_{hj}^{\nu-1}(z_j, t) - f_{hi}^{\nu-1}(f_{ij}^{\nu-1}(z_j, t), t) \equiv \Gamma_{hij|\nu}(z_j) t^\nu,$$

we obtain

$$\begin{aligned} &f_{hj}^{\nu-1}(f_{jk}^{\nu-1}(t), t) - f_{hi}^{\nu-1}(f_{ij}^{\nu-1}(f_{jk}^{\nu-1}(t), t), t) \\ &\qquad \underset{\nu}{\equiv} \Gamma_{hij|\nu}(f_{jk}^{\nu-1}(t)) t^\nu \underset{\nu}{\equiv} \Gamma_{hij|\nu}(f_{jk}(0)) t^\nu \underset{\nu}{\equiv} \Gamma_{hij|\nu}(z_j) t^\nu. \end{aligned}$$

Further since

$$f_{ik}^{\nu-1}(t) - f_{ij}^{\nu-1}(f_{jk}^{\nu-1}(t), t) \underset{\nu}{\equiv} \Gamma_{ijk|\nu} t^\nu,$$

we have

$$f_{hi}^{\nu-1}(f_{ik}^{\nu-1}(t), t) \underset{\nu}{\equiv} f_{hi}^{\nu-1}(f_{ij}^{\nu-1}(f_{jk}^{\nu-1}(t), t) + \Gamma_{ijk|\nu}t^\nu, t)$$

$$\underset{\nu}{\equiv} f_{hi}^{\nu-1}(f_{ij}^{\nu-1}(f_{jk}^{\nu-1}(t), t), t) + \sum_{\alpha=1}^{n} \frac{\partial z_h}{\partial z_i^\alpha}\Gamma_{ijk|\nu}^\alpha t^\nu,$$

where $z_h = f_{hi|0}(f_{ij|0}(f_{jk|0}(z_k))) = f_{hi|0}(z_i)$. Hence

$$f_{hj}^{\nu-1}(f_{jk}^{\nu-1}(t), t) - f_{hi}^{\nu-1}(f_{ik}^{\nu-1}(t), t) \underset{\nu}{\equiv} \Gamma_{hij|\nu}(z_j)t^\nu - \sum_{\alpha=1}^{n} \frac{\partial z_h}{\partial z_i^\alpha}\Gamma_{ijk|\nu}^\alpha t^\nu.$$

Therefore we obtain

$$\Gamma_{hik|\nu} - \Gamma_{hjk|\nu} = \Gamma_{hij|\nu}(z_j) - \sum_{\alpha=1}^{n} \frac{\partial z_h}{\partial z_i^\alpha}\Gamma_{ijk|\nu}^\alpha t^\nu.$$

Thus (5.71) holds. ∎

We denote the cohomology class of the 2-cocycle $\{\Gamma_{ijk|\nu}\} \in \hat{Z}^2(\mathfrak{U}, \Theta)$ by $\Gamma_\nu \in \hat{H}^2(\mathfrak{U}, \Theta)$. As stated above, when $f_{jk}^{\nu-1}(z_k, t)$ satisfy $(5.65)_{\nu-1}$, the system of congruences $(5.65)_\nu$ is equivalent to the equality $(5.70)_\nu$: $\{\Gamma_{ijk|\nu}\} = \delta\{f_{jk|\nu}\}$. Therefore in order to determine $f_{jk}^\nu(z_k, t)$ such that $(5.65)_\nu$ holds, it suffices to determine the 1-cochain $\{f_{jk|\nu}\} \in \hat{C}^1(\mathfrak{U}, \Theta)$ such that $\delta\{f_{jk|\nu}\} = \{\Gamma_{ijk|\nu}\}$. First suppose $\Gamma_\nu = 0$. Then if we take an arbitrary 1-cochain $\{\sigma_{jk}\} \in \hat{C}^1(\mathfrak{U}, \Theta)$ with $\{\Gamma_{ijk|\nu}\} = \delta\{\sigma_{jk}\}$, and put $f_{jk|\nu} = \sigma_{jk}$, $(5.65)_\nu$ holds. If $\Gamma_\nu \neq 0$, there exists no 1-cochain $\{\sigma_{jk}\}$ with $\{\Gamma_{ijk|\nu}\} = \delta\{\sigma_{jk}\}$, hence we cannot determine $\{f_{jk|\nu}\}$ such that $(5.65)_\nu$ holds. In this sense we call Γ_ν the *obstruction* to deformation of M.

We always have

$$\Gamma_{ijk|1} = f_{ik|0}(z_k) - f_{ij|0}(f_{jk|0}(z_k)) = 0,$$

hence $\Gamma_1 = 0$. Therefore we call Γ_2 the *first obstruction*, and $\Gamma_{\nu+1}$ *the* νth *obstruction.* Here note that Γ_ν is not defined unless $(5.65)_{\nu-1}$ holds, since $\{\Gamma_{ijk|\nu}\}$ is defined by the congruences $(5.66)_\nu$

$$f_{ik}^{\nu-1}(z_k, t) - f_{ij}^{\nu-1}(f_{jk}^{\nu-1}(z_k, t), t) \underset{\nu}{\equiv} \Gamma_{ijk|\nu}t^\nu.$$

In other words, unless $\Gamma_2 = 0, \ldots, \Gamma_\nu = 0$, $\Gamma_{\nu+1}$ is not defined. Moreover even if $\Gamma_2 = 0, \ldots, \Gamma_\nu = 0$, $\Gamma_{\nu+1}$ depends in general on the choice of $\{f_{jk|\nu}\}$ with $\delta|f_{jk|\nu}\} = \{\Gamma_{ijk|\nu}\}$, hence, it may be that $\Gamma_{\nu+1} \neq 0$ for one choice of $\{f_{jk|\nu}\}$, and $\Gamma_{\nu+1} = 0$ for another. Thus it is very difficult in general to see whether we can determine $\{f_{jk|\nu}\}$ such that the congruences $(5.65)_\nu$ hold for all $\nu = 2, 3 \ldots$

In case $\hat{H}^2(\mathfrak{U}, \Theta) = 0$, however, *since* $\hat{Z}^2(\mathfrak{U}, \Theta) = \delta\hat{C}^1(\mathfrak{U}, \Theta)$, *we can determine* 1*-cochains* $\{f_{jk|\nu}\}$ *successively for* $\nu = 2, 3, \ldots$ *in such a way as* $\delta\{f_{jk|\nu}\} = \{\Gamma_{ijk|\nu}\}$. Thus we can determine power series

$$f_{jk}^{\alpha}(z_k, t) = \sum_{\nu=0}^{\infty} f_{jk|\nu}^{\alpha}(z_k) t^{\nu}$$

satisfying the equalities (5.62):

$$f_{ik}^{\alpha}(z_k, t) = f_{ij}^{\alpha}(f_{jk}(z_k, t), t), \qquad \alpha = 1, \ldots, n.$$

If all these $f_{jk}^{\alpha}(z_k, t)$ converge, then on a sufficiently small disk $\Delta = \{t \in \mathbb{C} \,|\, |t| < r\}$, $f_{jk}^{\alpha}(z_k, t)$ are holomorphic functions on $(U_j \cap U_k) \times \Delta$, hence, the existence of a complex analytic family $(\mathcal{M}, B, \varpi)$ satisfying (5.61) is proved. But there are infinitely many choices of 1-cochains $\{f_{jk|\nu}\}$ with $\delta\{f_{jk|\nu}\} = \{\Gamma_{ijk|\nu}\}$ for each $\nu = 2, 3, \ldots$, and $f_{jk}^{\alpha}(z_k, t)$ do not converge in general for arbitrary choice of $\{f_{jk|\nu}\}$. Thus *we have to show that the power series*

$$f_{jk}^{\alpha}(z_k, t) = \sum_{\nu=0}^{n} f_{jk|\nu}^{\alpha}(z_k) t^{\nu}$$

converges if $\{f_{jk|\nu}\}$ are suitably chosen. But for all our efforts, we could not prove the convergence. Thus we did not succeed in proving the theorem of existence by an elementary method.

Although we failed to prove the theorem, we proved that, in case $\hat{H}^2(\mathfrak{U}, \Theta) = 0$, there exist formal power series

$$f_{jk}(z_k, t) = \sum_{\nu=1}^{\infty} f_{jk|\nu}(z_k) t^{\nu}$$

in t satisfying the fundamental equations $f_{ik}(z_k, t) = f_{ij}(f_{jk}(z_k, t), t)$. As will be shown below, $H^2(M, \Theta) = 0$ *implies* $\hat{H}^2(\mathfrak{U}, \Theta) = 0$, *hence, if* $H^2(M, \Theta) = 0$, *the fundamental equations* $f_{ik}(z_k, t) = f_{ij}(f_{jk}(z_k, t), t)$ *have formal power series solutions.* This gave some hope that, if $H^2(M, \Theta) = 0$, there might exist a complex analytic family $(\mathcal{M}, \Delta, \varpi)$ satisfying the conditions (5.61):

$$\varpi^{-1}(0) = M, \qquad \left(\frac{dM_t}{dt}\right)_{t=0} = \theta, \qquad M_t = \varpi^{-1}(t)$$

for any $\theta \in H^1(M, \Theta)$.

We now give proof of the fact that $H^2(M, \Theta) = 0$ implies $\hat{H}^2(\mathfrak{U}, \Theta) = 0$.

Proof. By a *local* C^{∞} *vector* $(0, q)$*-form* we mean a local C^{∞} section of the vector bundle $T(M) \otimes \bigwedge^q \bar{T}^*(M)$ over M where $T(M)$ is the tangent bundle of M. A C^∞ vector $(0, q)$-form ψ is written as $\psi = \sum_{\alpha=1}^n \psi_j^\alpha(\partial/\partial z_j^\alpha)$ where ψ_j^α are C^∞ $(0, q)$-forms. We denote by $\mathscr{A}^{0,q}(T(M))$ the sheaf of germs of C^∞ vector $(0, q)$-forms over M. Since $\Theta = \mathcal{O}(T(M))$, by (3.106) we have

$$H^2(M, \Theta) \cong \Gamma(M, \bar\partial \mathscr{A}^{0,1}(T(M)))/\bar\partial\Gamma(M, \mathscr{A}^{0,1}(T(M))). \tag{5.72}$$

Let $\{\rho_j\}$ be a partition of unity subordinate to $\mathfrak{U} = \{U_j\}$. Take an arbitrary 2-cocycle $\{\theta_{ijk}\} \in \hat{Z}^2(\mathfrak{U}, \Theta)$. By the definition of a 2-cocycle, we have

$$\theta_{ijk} = \theta_{hjk} - \theta_{hik} + \theta_{hij}.$$

Multiplying by ρ_h, and taking the summation with respect to h, we obtain

$$\theta_{ijk} = \xi_{jk} - \xi_{ik} + \xi_{ij}, \qquad \xi_{jk} = \sum_h \rho_h \theta_{hjk}.$$

Each ξ_{jk} is a C^∞ vector field on $U_j \cap U_k \neq \varnothing$. Since $\bar\partial\theta_{ijk} = 0$, we have

$$\bar\partial\xi_{jk} = \bar\partial\xi_{ik} - \bar\partial\xi_{ij},$$

hence

$$\bar\partial\xi_{jk} = \psi_k - \psi_j, \qquad \psi_j = \sum_i \rho_i \bar\partial\xi_{ij}.$$

Here each ψ_j is a C^∞ vector $(0, 1)$-form on U_j. Since $\bar\partial\psi_j = \bar\partial\psi_k$ on $U_j \cap U_k$, there is a $\bar\partial$-closed C^∞ vector $(0, 2)$-form $\varphi \in \Gamma(M, \bar\partial\mathscr{A}^{0,1}(T(M)))$ on M such that $\varphi = \bar\partial\psi_j$ on each U_j. Since $H^2(M, \Theta) = 0$ by assumption, by (5.72) there is a C^∞ vector $(0, 1)$-form $\psi \in \Gamma(M, \mathscr{A}^{0,1}(T(M)))$ such that $\varphi = \bar\partial\psi$. We have $\bar\partial(\psi_j - \psi) = 0$ on each U_j, while U_j is a polydisk. Therefore by Dolbeault's lemma (Theorem 3.3), there is a C^∞ vector field η_j on U_j with $\psi_j - \psi = \bar\partial\eta_j$. Then since

$$\bar\partial\xi_{jk} = \psi_k - \psi_j = \bar\partial\eta_k - \bar\partial\eta_j,$$

putting

$$\theta_{jk} = \xi_{jk} - \eta_k + \eta_j,$$

we have $\bar\partial\theta_{jk} = 0$, hence θ_{jk} is a holomorphic vector field on $U_j \cap U_k$. Moreover $\theta_{hjj} = 0$ implies $\xi_{jj} = 0$, hence $\theta_{jj} = 0$. Clearly we have

$$\theta_{ijk} = \xi_{jk} - \xi_{ik} + \xi_{ij} = \theta_{jk} - \theta_{ik} + \theta_{ik},$$

which implies that $\{\theta_{ijk}\} = \delta\{\theta_{jk}\}$, with $\{\theta_{jk}\} \in \hat{C}^1(\mathfrak{U}, \Theta)$. Thus $\hat{H}^2(\mathfrak{U}, \Theta) = 0$. ∎

In §5.1, we define the primary obstruction $[\theta, \theta]$. *We have* $[\theta, \theta] = 2\Gamma_2$.

Proof. Since $Z^2(\mathfrak{U}, \Theta) \subset \hat{Z}^2(\mathfrak{U}, \Theta)$, $C^1(\mathfrak{U}, \Theta) \subset \hat{C}^1(\mathfrak{U}, \Theta)$, and, as is easily seen,

$$Z^2(\mathfrak{U}, \Theta) \cap \delta\hat{C}^1(\mathfrak{U}, \Theta) = \delta C^1(\mathfrak{U}, \Theta),$$

we have

$$H^2(\mathfrak{U}, \Theta) \hookrightarrow \hat{H}^2(\mathfrak{U}, \Theta).$$

Consequently in order to prove that $[\theta, \theta] = 2\Gamma_2$, it suffices to show the existence of a 1-cochain $\{\tau_{jk}\} \in \hat{C}^1(\mathfrak{U}, \Theta)$ such that

$$2\Gamma_{ijk|2} - [\theta_{ij}, \theta_{jk}] = \tau_{jk} - \tau_{ik} + \tau_{ij}.$$

By $(5.66)_2$, we have

$$\Gamma_{ijk|2}(z_k)t^2 \underset{2}{\equiv} f^1_{ik}(z_k, t) - f^1_{ij}(f^1_{jk}(z_k, t), t),$$

hence $\Gamma_{ijk|2}(z_k)$ is the coefficient of t^2 in $-f^1_{ij}(f^1_{jk}(z_k, t), t)$. Since

$$f^1_{ij}(f^1_{jk}(t), t) = f_{ij|0}(f_{jk|0} + f_{jk|1}t) + f_{ij|1}(f_{jk|0} + f_{jk|1}t)t,$$

putting $z_i = f_{ij|0}(z_j)$, $z_j = f_{jk|0}(z_k)$, we have

$$\Gamma_{ijk|2} = -\frac{1}{2}\sum_{\beta,\gamma=1}^{n} \frac{\partial^2 z_i}{\partial z_j^\beta\, \partial z_j^\gamma} f^\beta_{jk|1} f^\gamma_{jk|1} - \sum_{\beta=1}^{n} \frac{\partial f_{ij|1}}{\partial z_j^\beta} f^\beta_{jk|1}.$$

Therefore, since $f_{jk|1} = \theta_{jk}$, we have

$$\begin{aligned} 2\Gamma_{ijk|2} &= -\sum_{\gamma,\beta} \theta^\gamma_{jk}\theta^\beta_{jk} \frac{\partial^2 z_i}{\partial z_j^\gamma\, \partial z_j^\beta} - 2\theta_{jk} \cdot \theta_{ij} \\ &= -\theta_{jk}\left(\sum_\beta \frac{\partial z_i}{\partial z_j^\beta}\theta^\beta_{jk}\right) + \sum_\beta \frac{\partial z_i}{\partial z_j^\beta}\theta_{jk} \cdot \theta^\beta_{jk} - 2\theta_{jk} \cdot \theta_{ij} \end{aligned}$$

while $\sum_\beta (\partial z_i/\partial z_j^\beta)\theta^\beta_{jk} = \theta_{ik} - \theta_{ij}$. Hence

$$2\Gamma_{ijk|2} = \sum_\beta \frac{\partial z_i}{\partial z_j^\beta}\theta_{jk} \cdot \theta^\beta_{jk} - \theta_{jk} \cdot \theta_{ik} - \theta_{jk} \cdot \theta_{ij}.$$

Since $[\theta_{ij}, \theta_{jk}] = [\theta_{ij}, \theta_{ik}] = \theta_{ij} \cdot \theta_{ik} - \theta_{ik} \cdot \theta_{ij}$, we have

$$2\Gamma_{ijk|2} - [\theta_{ij}, \theta_{jk}] = \sum_\beta \frac{\partial z_i}{\partial z_j^\beta}\theta_{jk} \cdot \theta^\beta_{jk} - \theta_{ik} \cdot \theta_{ik} + \theta_{ij} \cdot \theta_{ij}.$$

Thus putting $\tau_{jk} = \sum_\beta \theta_{jk} \cdot \theta^\beta_{jk}(\partial/\partial z_j^\beta)$ on each $U_j \cap U_k \neq \varnothing$, we obtain

$$2\Gamma_{ijk|2} - [\theta_{ij}, \theta_{jk}] = \tau_{jk} - \tau_{ik} + \tau_{ij}. \quad ∎$$

Let N be a 2-dimensional complex torus. Then the product $M=\mathbb{P}^1\times N$ gives an example for which there exists a $\theta\in H^1(M,\Theta)$ with $[\theta,\theta]\neq 0$. If the number of moduli in the sense of Definition 5.4 is defined for this $M=\mathbb{P}^1\times N$, we would have $m(M)<\dim H^1(M,\Theta)$ since there is a $\theta\in H^1(M,\theta)$ with $[\theta,\theta]\neq 0$. But for the number of moduli $m(M)$ to be defined for M, there must exist an effectively parametrized and complete analytic family $(\mathcal{M},B,\varpi)$ with $0\in B\subset\mathbb{C}^m$ such that $\varpi^{-1}(0)=M$. For $M=\mathbb{P}^1\times N$, such a family does not exist. This is verified as follows. Suppose that such a family $(\mathcal{M},B,\varpi)$ exists. The set of the infinitesimal deformations $\rho_0(\partial/\partial t)=(\partial M_t/\partial t)_{t=0}$ of $M_t=\varpi^{-1}(t)$ at $t=0$ forms a vector space $\rho_0(T_0(B))\subset H^1(M,\Theta)$. As stated in §5.2(a), if $\theta,\eta\in\rho_0(T_0(B))$, $[\theta,\eta]=0$. Consider two complex analytic families $(\mathcal{M}_\lambda,\Delta_r,\varpi_\lambda)$ with $\varpi_\lambda^{-1}(0)=M$ and $\Delta_r=\{s\in C\,||s|<r\}$ for $\lambda=1,2$, and let $\theta_\lambda=(d\varpi_\lambda^{-1}(s)/ds)_{s=0}$ be the corresponding infinitesimal deformations at $s=0$. Since $(\mathcal{M},B,\varpi)$ is assumed to be complete, taking a sufficiently small disk Δ_ε with $r>\varepsilon>0$, we may assume that both $(\varpi_\lambda^{-1}(\Delta_\varepsilon),\Delta_\varepsilon,\varpi_\lambda)$, $\lambda=1,2$, are complex analytic families induced from $(\mathcal{M},B,\varpi)$. Therefore by (4.30), $\theta_\lambda\in\rho_0(T_0(B))$ for $\lambda=1,2$, hence $[\theta_1,\theta_2]=0$. But for $M=\mathbb{P}^1\times N$, there exist complex analytic families $(\mathcal{M}_\lambda,\Delta_r,\varpi_\lambda)$, $\lambda=1,2$, with $\varpi_\lambda^{-1}(0)=M$ such that $[\theta_1,\theta_2]\neq 0$ ([21], §16). Thus the number of moduli $m(M)$ is not defined.

At that time we had no example M of compact complex manifolds whose number of moduli $m(M)$ is defined and yet $m(M)\neq\dim H^1(M,\Theta)$. In view of this we made the following conjecture.

Conjecture. *Let M be a compact complex manifold. If the number of moduli $m(M)$ of M is defined, $m(M)$ is equal to* $\dim H^1(M,\Theta)$.

Our paper "On deformations of complex analytic structures, I–II", often quoted already in this book, treats various results obtained up to that time centring on this conjecture. Given a compact complex manifold M, it is in general very difficult to compute the number of moduli $m(M)$ or $\dim H^1(M,\Theta)$. Consequently from the mere fact that $m(M)=\dim H^1(M,\Theta)$ always holds for several simple types of compact complex manifolds M for which we can compute $m(M)$ and $\dim H^1(M,\Theta)$, we cannot conclude the universal validity of this equality $m(M)=\dim H^1(M,\Theta)$. But the above conjecture has actually turned out to be very useful as a "working hypothesis" in our study of deformations.

(b) The Theorem of Existence

Since we did not succeed in proving the theorem of existence by an elementary method, we—Spencer and the author—tried to prove it by another method. Let $(\mathcal{M},B,\varpi)$ be a complex analytic family of compact complex manifolds, and put $M_t=\varpi^{-1}(t)$ where B is a domain of $\mathbb{C}^m$ containing the origin 0. Define $|t|=\max_\lambda|t_\lambda|$ for $t=(t_1,\ldots,t_m)\in\mathbb{C}^m$, and let $\Delta=\Delta_r=\{t\in\mathbb{C}^m \,|\, |t|<r\}$ the polydisk of radius $r>0$. If we take a sufficiently small $\Delta\subset B$, $\mathcal{M}_\Delta=\varpi^{-1}(\Delta)$ is represented in the form (4.27):

$$\mathcal{M}_\Delta=\bigcup_j U_j\times\Delta. \tag{5.73}$$

We denote a point of U_j by $\zeta_j=(\zeta_j^1,\ldots,\zeta_j^n)$. For simplicity we assume that $U_j=\{\zeta_j\in\mathbb{C}^m \,|\, |\zeta_j|<1\}$ where $|\zeta_j|=\max_\alpha|\zeta_j^\alpha|$. $(\zeta_j,t)\in U_j\times\Delta$ and $(\zeta_k,t)\in U_k\times\Delta$ are the same point on M_Δ if $\zeta_j^\alpha=f_{jk}^\alpha(\zeta_k,t)$, $\alpha=1,\ldots,n$, where $f_{jk}^\alpha(\zeta_k,t)$ is a holomorphic function of $\zeta_k^1,\ldots,\zeta_k^n,t_1,\ldots,t_m$ defined on $U_k\times\Delta\cap U_j\times\Delta$. So far we have considered that $M_t=\bigcup_j U_j$ is a compact complex manifold obtained by glueing a finite number of polydisks $U_1,\ldots,U_j,\ldots$ by identifying $\zeta_j\in U_j$ and $\zeta_k\in U_k$ if $\zeta_j=f_{jk}(\zeta_k,t)$, and that the complex structure of M_t varies since the manner of glueing varies with t. According to Theorem 2.3, every M_t is diffeomorphic to M_0, so the underlying differentiable manifold X of M_t does not depend on t. Namely, for any $t\in B$, the complex structure M_t is defined on one and the same differentiable manifold X. X does not, however, appear explicitly in the above-mentioned description of M_t.

Of course we can give a description of $(\mathcal{M}_\Delta,\Delta,\varpi)$ in which X appears explicitly. We shall explain it below. If we take a sufficiently small Δ, by Theorem 2.5, there is a diffeomorphism Ψ of $M\times\Delta$ onto $\mathcal{M}_\Delta$ such that $\varpi\circ\Psi$ is the projection $M\times\Delta\to\Delta$ where we put $M=M_0$. If we denote a point of M by z, we have

$$\varpi\circ\Psi(z,t)=t, \qquad t\in\Delta.$$

Moreover, as is clearly seen from the proof of Theorem 2.5, Ψ is the identify of $M=M\times 0$ onto $M=M_0\subset\mathcal{M}_\Delta$, namely $\Psi(z,0)=z$. Put $\Psi(z,t)=(\zeta,t)=(\zeta_j,t)$ for $\Psi(z,t)\in U_j\times\Delta$. Then each component $\zeta_j^\alpha=\zeta_j^\alpha(z,t)$, $\alpha=1,\ldots,n$, of $\zeta_j=(\zeta_j^1,\ldots,\zeta_j^n)$ is a C^∞ function:

$$\Psi(z,t)=(\zeta_j^1(z,t),\ldots,\zeta_j^n(z,t),t_1,\ldots,t_m).$$

If we identify $\mathcal{M}_\Delta=\Psi(M\times\Delta)$ with $M\times\Delta$ via Ψ, $\mathcal{M}_\Delta$ *is considered as a complex manifold with the complex structure defined on the* C^∞ *manifold* $M\times\Delta$ *by the system of local complex coordinates*

$$\{(\zeta_j,t)\,|\,j=1,2,3,\ldots\}, \qquad (\zeta_j,t)=(\zeta_j^1(z,t),\ldots,\zeta_j^n(z,t),t_1,\ldots,t_m).$$

Consequently for each t, M_t is *the complex structure of the differentiable manifold M defined by the system of local complex coordinates*:

$$\{z\to\zeta_j(z,t)\,|\,j=1,2,\ldots\}, \qquad \zeta_j(z,t)=(\zeta_j^1(z,t),\ldots,\zeta_j^n(z,t)).$$

Here by "the differentiable manifold M" we mean the underlying differentiable manifold of the complex manifold M.

If we describe $(\mathcal{M}_\Delta, \Delta, \varpi)$ as above, each component of local complex coordinates $\zeta_j(z, t)$ is represented as a local C^∞ function $\zeta_j^\alpha(z, t)$ on M depending on t, and the underlying C^∞ manifold M of M_t appears explicitly as the place on which $\zeta_j^\alpha(z, t)$ are defined.

M is itself a complex manifold. Let $(z^1, \ldots, z^n)$ be *arbitrary local complex coordinates of a point z of M.*

$$\zeta_j^\alpha(z, t) = \zeta_j^\alpha(z^1, \ldots, z^n, t_1, \ldots, t_m), \qquad \alpha = 1, \ldots, n,$$

are C^∞ *functions of the complex variables* $z^1, \ldots, z^n, t_1, \ldots, t_m$. Since for $t = 0$, both $(\zeta_j^1(z, 0), \ldots, \zeta_j^n(z, 0))$ and $(z^1, \ldots, z^n)$ are local complex coordinates on the complex manifold $M_0 = M$, $\zeta_j^\alpha(z, 0)$ are *holomorphic functions of* $z^1, \ldots, z^n$, and

$$\det\left(\frac{\partial \zeta_j^\alpha(z, 0)}{\partial z^\lambda}\right)_{\alpha, \lambda = 1, \ldots, n} \neq 0 \tag{5.74}$$

holds. On the open set $U_j \times t \cap U_k \times t$ of M_t, we have

$$\det\left(\frac{\partial \zeta_j^\alpha(z, t)}{\partial z^\lambda}\right) = \det\left(\frac{\partial \zeta_j^\alpha}{\partial \zeta_k^\beta}\right) \cdot \det\left(\frac{\partial \zeta_j^\beta(z, t)}{\partial z^\lambda}\right),$$

and $\det(\partial \zeta_j^\alpha / \partial \zeta_k^\beta) \neq 0$. From this and (5.74), if we take Δ sufficiently small, it follows that

$$\det\left(\frac{\partial \zeta_j^\alpha(z, t)}{\partial z^\lambda}\right)_{\alpha, \lambda = 1, \ldots, n} \neq 0 \tag{5.75}$$

for any $t \in \Delta$.

We denote the *tangent bundle of the complex manifold M by T*: $T = T(M)$. Also we denote by $\mathcal{L}^{0,q}(T)$ the linear space of all C^∞ vector $(0, q)$-forms, namely, C^∞ sections of $T \otimes \bigwedge^q \bar{T}^*$ over M: $\mathcal{L}^{0,q}(T) = \Gamma(M, \mathcal{A}^{0,q}(T))$. $\varphi \in \mathcal{L}^{0,q}(T)$ is represented in the form

$$\varphi = \sum_{\lambda=1}^n \varphi^\lambda \frac{\partial}{\partial z^\lambda}, \qquad \varphi^\lambda = \frac{1}{q!} \sum \varphi^\lambda_{\bar{\nu}_1 \cdots \bar{\nu}_q}(z)\, d\bar{z}^{\nu_1} \wedge \cdots \wedge d\bar{z}^{\nu_q},$$

where $\varphi^\lambda_{\bar{\nu}_1 \cdots \bar{\nu}_q}(z)$ are local C^∞ functions on M which are skew-symmetric in the indices $\nu_1, \ldots, \nu_q$.

For simplicity *we denote $\partial/\partial z^\lambda$ by ∂_λ, and $\partial/\partial \bar{z}^\lambda$ by $\bar{\partial}_\lambda$*. Consider $\bar{\partial} \zeta_j^\alpha(z, t) = \sum_\nu \bar{\partial}_\nu \zeta_j^\alpha(z, t)\, d\bar{z}^\nu$. The domain $\mathcal{U}_j = \Psi^{-1}(U_j \times \Delta)$ of $\zeta_j^\alpha(z, t)$ is a domain of $M \times \Delta$. Since by (5.75), $\det(\partial_\lambda \zeta_j^\alpha(z, t)) \neq 0$, there is a unique $(0, 1)$-form

$$\varphi_j^\lambda(z, t) = \sum_{\nu=1}^n \varphi_{j\bar{\nu}}^\lambda(z, t)\, d\bar{z}^\nu,$$

for each $\lambda = 1, \ldots, n$, such that

$$\bar{\partial}\zeta_j^\alpha(z, t) = \sum_{\lambda=1}^{n} \varphi_j^\lambda(z, t)\partial_\lambda\zeta_j^\alpha(z, t), \qquad \alpha = 1, \ldots, n. \tag{5.76}$$

The coefficients $\varphi_{j\bar{\nu}}^\alpha(z, t)$ are C^∞ functions on $\mathcal{U}_j$, and on $\mathcal{U}_j \cap \mathcal{U}_k$ we have

$$\sum_{\lambda=1}^{n} \varphi_j^\lambda(z, t)\partial_\lambda = \sum_{\lambda=1}^{n} \varphi_k^\lambda(z, t)\partial_\lambda. \tag{5.77}$$

Proof. Since $(z, t) \in \mathcal{U}_j \cap \mathcal{U}_k$, $(\zeta_j(z, t), t) \in U_j \times \Delta$ and $(\zeta_k(z, t), t) \in U_k \times \Delta$ are the same point on $\mathcal{M}_\Delta$, we have

$$\zeta_j^\alpha(z, t) = f_{jk}^\alpha(\zeta_k(z, t), t), \qquad \alpha = 1, \ldots, n. \tag{5.78}$$

Since $\zeta_j^\alpha = f_{jk}^\alpha(\zeta_k, t)$ are holomorphic functions of $\zeta_k^1, \ldots, \zeta_k^n, t_1, \ldots, t_m$, applying $\bar{\partial}$ to both sides of (5.78), we obtain

$$\bar{\partial}\zeta_j^\alpha(z, t) = \sum_{\mu=1}^{n} \frac{\partial \zeta_j^\alpha}{\partial \zeta_k^\beta} \cdot \bar{\partial}\zeta_k^\beta(z, t).$$

Hence, we have

$$\sum_\lambda \varphi_j^\lambda(z, t)\partial_\lambda\zeta_j^\alpha(z, t) = \sum_{\beta=1}^{n} \frac{\partial \zeta_j^\alpha}{\partial \zeta_k^\beta} \sum_\lambda \varphi_k^\lambda(z, t)\partial_\lambda\zeta_k^\beta(z, t).$$

On the other hand, applying ∂_λ to both sides of (5.78), we obtain

$$\partial_\lambda\zeta_j^\alpha(z, t) = \sum_{\beta=1}^{n} \frac{\partial \zeta_j^\alpha}{\partial \zeta_k^\beta}\partial_\lambda\zeta_k^\beta(z, t).$$

Hence

$$\sum_\lambda \varphi_j^\lambda(z, t)\partial_\lambda\zeta_j^\alpha(z, t) = \sum_\lambda \varphi_k^\lambda(z, t)\partial_\lambda\zeta_j^\alpha(z, t).$$

Therefore, since $\det(\partial_\lambda\zeta_j^\alpha(z, t)) \neq 0$, if we denote the domain of the local coordinates $(z^1, \ldots, z^n)$ by U, we have $\varphi_j^\lambda(z, t) = \varphi_k^\lambda(z, t)$ on $\mathcal{U}_j \cap \mathcal{U}_k \cap U \times \Delta \neq \varnothing$, hence

$$\sum_{\lambda=1}^{n} \varphi_j^\lambda(z, t)\partial_\lambda = \sum_{\lambda=1}^{n} \varphi_k^\lambda(z, t)\partial_\lambda$$

there. Since $(z^1, \ldots, z^n)$ are arbitrary local complex coordinates on M, we obtain (5.77). ∎

If for $(z, t) \in \mathcal{U}_j$, we define

$$\varphi(z, t) = \sum_{\lambda=1}^{n} \varphi_j^{\lambda}(z, t)\partial_{\lambda}.$$

From (5.77) $\varphi(t) = \varphi(z, t)$ *is a* C^∞ *vector* (0, 1)*-form on* M *for every* $t \in \Delta$. In terms of local complex coordinates we have

$$\varphi(z, t) = \sum_{\lambda=1}^{n} \varphi^{\lambda}(z, t)\partial_{\lambda}, \qquad \varphi^{\lambda}(z, t) = \sum_{\nu=1}^{n} \varphi_{\bar{\nu}}^{\lambda}(z, t)\, d\bar{z}^{\nu}.$$

In the sense that $\varphi_{\bar{\nu}}^{\lambda}(z, t)$ are all C^∞ functions of $z^1, \ldots, z^n, t_1, \ldots, t_m$, we say that $\varphi(t) = \varphi(z, t)$ *is* C^∞ *with respect to* t. By (5.76) we obtain

$$\bar{\partial}\zeta_j^{\alpha}(z, t) = \sum_{\lambda=1}^{n} \varphi^{\lambda}(z, t)\partial_{\lambda}\zeta_{\xi}(z, t), \qquad \alpha = 1, \ldots, n. \tag{5.79}$$

If we consider $\varphi(t) = \varphi(z, t) = \sum_{\lambda=1}^{n} \varphi^{l}(z, t)\partial_{\lambda}$ as a differential operator which associates to every local C^∞ function $f(z)$ a local (0, 1)-form $\sum_{\lambda=1}^{n} \varphi^{\lambda}(z, t)\partial_{\lambda} f(z)$, (5.79) can be rewritten as

$$(\bar{\partial} - \varphi(t))\zeta_j^{\alpha}(z, t) = 0, \qquad \alpha = 1, \cdots, n. \tag{5.80}$$

Since $\zeta_j^{\alpha}(z, 0)$ are holomorphic function of $z^1, \ldots, z^n$, $\bar{\partial}\zeta_j^{\alpha}(z, 0) = 0$. Hence by (5.79), we have

$$\varphi(0) = 0. \tag{5.81}$$

Theorem 5.3. *If we take a sufficiently small polydisk* Δ, *then, for* $t \in \Delta$, *a local* C^∞ *function* f *on* M *is holomorphic with respect to the complex structure* M_t *if and only if* f *satisfies the equation*

$$(\bar{\partial} - \varphi(t))f = 0. \tag{5.82}$$

Proof. Write $f = f(\zeta_j^1, \ldots, \zeta_j^n)$ as a C^∞ function of the local complex coordinates $\zeta_j^1 = \zeta_j^1(z, t), \ldots, \zeta_j^n = \zeta_j^n(z, t)$ on M_t. Then

$$\begin{aligned}(\bar{\partial} - \varphi(t))f &= (\bar{\partial} - \varphi(t))f(\zeta_j^1(z, t), \ldots, \zeta_j^n(z, t)) \\ &= \sum_{\alpha}^{n} (\bar{\partial} - \varphi(t))\zeta_j^{\alpha}(z, t)\frac{\partial f}{\partial \zeta_j^{\alpha}} + \sum_{\alpha=1}^{n} (\bar{\partial} - \varphi(t))\overline{\zeta_j^{\alpha}(z, t)}\frac{\partial f}{\partial \bar{\zeta}_j^{\alpha}}.\end{aligned}$$

From (5.80), $(\bar{\partial} - \varphi(t))\zeta_j^{\alpha}(z, t) = 0$, hence, denoting $\varphi_{\bar{\nu}}^{\alpha}(z, t)$ by $\varphi_{\bar{\nu}}^{\alpha}$, we obtain

$$(\bar{\partial} - \varphi(t))f = \sum_{\alpha=1}^{n}\sum_{\nu=1}^{n}\left(\overline{\partial_{\nu}\zeta_j^{\alpha}} - \sum_{\mu=1}^{n} \varphi_{\bar{\nu}}^{\mu}\partial_{\mu}\bar{\zeta}_j^{\alpha}\right) d\bar{z}^{\nu}\frac{\partial f}{\partial \bar{\zeta}_j^{\alpha}}.$$

Since $\bar{\partial}_\mu \zeta_j^\alpha = \sum_\lambda \varphi_{\bar{\mu}}^\lambda \partial_\lambda \zeta_j^\alpha$, we have

$$(\bar{\partial} - \varphi(t))f = \sum_\alpha \sum_\nu \sum_\lambda \left(\delta_\nu^\lambda - \sum_\mu \varphi_{\bar{\nu}}^\mu \overline{\varphi_{\bar{\mu}}^\lambda} \right) \overline{\partial_\lambda \zeta_j^\alpha} \, d\bar{z}^\nu \frac{\partial f}{\partial \bar{\zeta}_j^\alpha}.$$

Since the determinant

$$\det\left(\delta_\nu^\lambda - \sum_{\mu=1}^n \varphi_{\bar{\nu}}^\mu(z,t) \overline{\varphi_{\bar{\mu}}^\lambda(z,t)} \right)_{\lambda,\nu=1,\ldots,n}$$

is invariant under change of local coordinates $(z^1, \ldots, z^n)$, it is a C^∞ function on $M \times \Delta$. Since $\varphi_{\bar{\nu}}^\lambda(z,t) = 0$ at $t = 0$, we may assume that for any $t \in \Delta$,

$$\det\left(\delta_\nu^\lambda - \sum_{\mu=1}^n \varphi_{\bar{\nu}}^\mu(z,t) \overline{\varphi_{\bar{\mu}}^\lambda(z,t)} \right)_{\lambda,\nu=1,\ldots,n} \neq 0 \tag{5.83}$$

provided that Δ is sufficiently small. Therefore, since by (5.75) $\det(\partial_\lambda \zeta_j^\alpha) \neq 0$, $(\bar{\partial} - \varphi(t))f = 0$ if and only if $\partial f / \partial \bar{\zeta}_j^\alpha = 0$ for $\alpha = 1, \ldots, n$, which means that $f = f(\zeta_j^1, \ldots, \zeta_j^n)$ is a holomorphic function of the local complex coordinates $\zeta_j^1, \ldots, \zeta_j^n$ on M_t. ∎

This theorem implies that *the deformation M_t of the complex structure on M is represented by the vector* $(0,1)$*-form*

$$\varphi(t) = \sum_{\lambda=1}^n \varphi^\lambda(z,t) \partial_\lambda = \sum_{\lambda=1}^n \sum_{\nu=1}^n \varphi_{\bar{\nu}}^\lambda(z,t) \, d\bar{z}^\nu \partial_\lambda$$

on M.

$\varphi(t)$ satisfies the following conditions induced from (5.79). Denote $\zeta_j^\alpha(z,t)$ by ζ_j^α and $\varphi^\lambda(z,t)$ by φ^λ for simplicity. Then by (5.79) we have

$$0 = \bar{\partial}\bar{\partial}\zeta_j^\alpha = \bar{\partial} \sum_{\lambda=1}^n \varphi^\lambda \partial_\lambda \zeta_j^\alpha = \sum_{\lambda=1}^\nu \bar{\partial}\varphi^\lambda \partial_\lambda \zeta_j^\alpha - \sum_{\lambda=1}^n \varphi^\lambda \wedge \bar{\partial}\partial_\lambda \zeta_j^\alpha.$$

Hence

$$\sum_{\lambda=1}^n \bar{\partial}\varphi^\lambda \partial_\lambda \zeta_j^\alpha = \sum_{\mu=1}^n \varphi^\mu \wedge \bar{\partial}\partial_\mu \zeta_j^\alpha.$$

Since $\bar{\partial}_\nu \zeta_j^\alpha = \sum_\lambda \varphi_{\bar{\nu}}^\lambda \partial_\lambda \zeta_j^\alpha$ by (5.79), we have

$$\begin{aligned}
\bar{\partial}\partial_\mu{}_j^\alpha &= \sum_{\nu=1}^n \bar{\partial}_\nu \partial_\mu \zeta_j^\alpha \, d\bar{z}^\nu = \sum_\nu \partial_\mu \bar{\partial}_\nu \zeta_j^\alpha \, d\bar{z}^\nu \\
&= \sum_\nu \left(\partial_\mu \sum_\lambda \varphi_{\bar{\nu}}^\lambda \partial_\lambda \zeta_j^\alpha \right) d\bar{z}^\nu \\
&= \sum_\lambda \sum_\nu \partial_\mu \varphi_{\bar{\nu}}^\lambda \, d\bar{z}^\nu \partial_\lambda \zeta_j^\alpha + \sum_\lambda \varphi^\lambda \partial_\mu \partial_\lambda \zeta_j^\alpha.
\end{aligned}$$

Consequently putting

$$\partial_\mu \varphi^\lambda(z, t) = \sum_{\nu=1}^{n} \partial_\mu \varphi^\lambda_{\bar{\nu}}(z, t)\, d\bar{z}^\nu,$$

we obtain

$$\sum_\mu \varphi^\mu \wedge \bar{\partial}\partial_\mu \zeta_j^\alpha = \sum_\lambda \left(\sum_\mu \varphi^\mu \wedge \partial_\mu \varphi^\lambda \right) \partial_\lambda \zeta_j^\alpha + \sum_\lambda \sum_\mu \varphi^\mu \wedge \varphi^\lambda \partial_\mu \partial_\lambda \zeta_j^\alpha,$$

while, since $\varphi^\lambda \wedge \varphi^\mu = -\varphi^\mu \wedge \varphi^\lambda$, the last term of the above equality vanishes. Hence

$$\sum_{\lambda=1}^{n} \bar{\partial}\varphi^\lambda \partial_\lambda \zeta_j^\alpha = \sum_{\lambda=1}^{n} \left(\sum_{\mu=1}^{n} \varphi^\mu \wedge \partial_\mu \varphi^\lambda \right) \partial_\lambda \zeta_j^\alpha.$$

Since $\det(\partial_\lambda \zeta_j^\alpha) \neq 0$, we obtain

$$\bar{\partial}\varphi^\lambda(t) = \sum_{\mu=1}^{n} \varphi^\mu(t) \wedge \partial_\mu \varphi^\lambda(t). \tag{5.84}$$

Thus $\varphi(t) = \sum_{\lambda=1}^{n} \varphi^\lambda(t)\partial_\lambda$ satisfies (5.84).

For $\varphi = \sum_{\lambda=1}^{n} \varphi^\lambda \partial_\lambda \in \mathscr{L}^{0,p}(T)$ and $\psi = \sum_{\lambda=1}^{n} \psi^\lambda \partial_\lambda \in \mathscr{L}^{0,q}(T)$, we define their bracket by

$$[\varphi, \psi] = \sum_{\lambda=1}^{n} \sum_{\mu=1}^{n} (\varphi^\mu \wedge \partial_\mu \psi^\lambda - (-1)^{pq} \psi^\mu \wedge \partial_\mu \varphi^\lambda) \partial_\lambda, \tag{5.85}$$

where we put

$$\partial_\mu \psi^\lambda = \frac{1}{q!} \sum \partial_\mu \psi^\lambda_{\bar{\nu}_1 \cdots \bar{\nu}_q}(z)\, d\bar{z}^{\nu_1} \wedge \cdots \wedge d\bar{z}^{\nu_q}$$

for $\psi^\lambda = (1/q!) \sum \psi^\lambda_{\bar{\nu}_1 \cdots \bar{\nu}_q}(z)\, d\bar{z}^{\nu_1} \wedge \cdots \wedge d\bar{z}^{\nu_q}$. $[\varphi, \psi]$ *is independent of the choice of local complex coordinates* $(z^1, \ldots, z^n)$ *used for its definition,* and we have $[\varphi, \psi] \in \mathscr{L}^{0,p+q}(T)$.

By the definition (3.105), for $\varphi = \sum_{\lambda=1}^{n} \varphi^\lambda \partial_\lambda \in \mathscr{L}^{0,p}(T)$ we have

$$\bar{\partial}\varphi = \sum_{\lambda=1}^{n} \bar{\partial}\varphi^\lambda \partial_\lambda.$$

Therefore, using the bracket, we write (5.84) as

$$\bar{\partial}\varphi(t) = \tfrac{1}{2}[\varphi(t), \varphi(t)]. \tag{5.86}$$

For the bracket we can easily verify the following formulae: For $\varphi \in \mathscr{L}^{0,p}(T)$, $\psi \in \mathscr{L}^{0,q}(T)$ and $\tau \in \mathscr{L}^{0,r}(T)$,

$$[\psi, \varphi] = -(-1)^{pq}[\varphi, \psi], \tag{5.87}$$

$$\bar{\partial}[\varphi, \psi] = [\bar{\partial}\varphi, \psi] + (-1)^p[\varphi, \bar{\partial}\psi]. \tag{5.88}$$

$$(-1)^{pr}[[\varphi, \psi], \tau] + (-1)^{qp}[[\psi, \tau], \varphi] + (-1)^{rq}[[\tau, \varphi], \quad \psi = 0. \tag{5.89}$$

We call $\varphi = \mathscr{L}^{0,q}(T)$ a $\bar{\partial}$-closed vector $(0,1)$-form if $\bar{\partial}\varphi = 0$. We denote by $\mathscr{Z}^{0,q}_{\bar{\partial}}(T)$ the linear subspace of $\mathscr{L}^{0,q}(T)$ consisting of all $\bar{\partial}$-closed vector $(0,1)$-forms. Since $\mathscr{Z}^{0,q}_{\bar{\partial}}(T) = \Gamma(M, \bar{\partial}\mathscr{A}^{0,q-1}(T))$, we have by (3.106)

$$H^q(M, \Theta) \cong \mathscr{Z}^{0,q}_{\bar{\partial}}(T)/\bar{\partial}\mathscr{L}^{0,q-1}(T), \qquad q \geqq 1. \tag{5.90}$$

For the infinitesimal deformation $(\partial M_t/\partial t)_{t=0}$, we have the following theorem.

Theorem 5.4. $(\partial\varphi(t)/\partial t)_{t=0}$ *is a* $\bar{\partial}$*-closed* $(0,1)$*-form, and under the isomorphism*

$$H^1(M, \Theta) \cong \mathscr{Z}^{0,1}_{\bar{\partial}}(T)/\bar{\partial}\mathscr{L}^{0,0}(T), \tag{5.91}$$

$(\partial M_t/\partial t)_{t=0} \in H^1(M, \Theta)$ *corresponds to* $-(\partial\varphi(t)/\partial t)_{t=0} \in \mathscr{Z}^{0,1}_{\bar{\partial}}(T)$.

Proof. Put

$$\theta_{jk} = \sum_{\alpha=1}^{n} \left(\frac{\partial f^\alpha_{jk}(\zeta_k, t)}{\partial t}\right)_{t=0} \frac{\partial}{\partial \zeta^\alpha_j}. \tag{5.92}$$

The infinitesimal deformation $(\partial M_t/\partial t)_{t=0} \in H^1(M, \Theta)$ is the cohomology class of the 1-cocycle $\{\theta_{jk}\} \in Z^1(\{U_j\}, \Theta)$. We fix a tangent vector $\partial/\partial t \in T_0(\Delta)$, and for a C^∞ function $f(t)$ of $t \in \Delta$, denote $(\partial f(t)/\partial t)_{t=0}$ by $\dot{f}$. Differentiating (5.78):

$$\zeta^\alpha_j(z, t) = f^\alpha_{jk}(\zeta_k(z, t), t)$$

with respect to t, and putting $t = 0$, we obtain

$$\dot{\zeta}^\alpha_j = \sum_{\beta=1}^{n} \frac{\partial \zeta^\alpha_j}{\partial \zeta^\beta_k} \dot{\zeta}^\beta_k + \left(\frac{\partial f^\alpha_{jk}(\zeta_k, t)}{\partial t}\right)_{t=0},$$

where $\zeta^\alpha_j = \zeta^\alpha_j(z, 0)$ and $\zeta^\beta_k = \zeta^\beta_k(z, 0)$. Therefore putting

$$\xi_j = \sum_{\alpha=1}^{n} \dot{\zeta}^\alpha_j \frac{\partial}{\partial \zeta^\alpha_j}$$

for each j, we have

$$\theta_{jk} = \xi_j - \xi_k.$$

Since $\dot{\zeta}_j^\alpha = (\partial\zeta_j^\alpha(z,t)/\partial t)_{t=0}$ is a C^∞ function on U_j, ξ_j is a C^∞ vector field on U_j. Next, differentiating (5.79):

$$\bar{\partial}\zeta_j^\alpha(z,t) = \sum_{\lambda=1}^n \varphi^\lambda(z,t)\partial_\lambda\zeta_j^\alpha(z,t)$$

with respect to t, and putting $t=0$, we obtain

$$\bar{\partial}\dot{\zeta}_j^\alpha = \sum_{\lambda=1}^n \dot{\varphi}^\lambda\partial_\lambda\zeta_j^\alpha$$

since $\varphi^\lambda(z,0)=0$ by (5.81). Hence

$$\bar{\partial}\xi_j = \sum_\alpha \bar{\partial}\dot{\zeta}_j^\alpha \frac{\partial}{\partial\zeta_j^\alpha} = \sum_\lambda\sum_\alpha \dot{\varphi}^\lambda \frac{\partial\zeta_j^\alpha}{\partial\zeta_k^\beta}\frac{\partial}{\partial\zeta_j^\alpha} = \sum_\lambda \dot{\varphi}^\lambda\partial_\lambda = \dot{\varphi}.$$

The exact sequence of sheaves $0\to\Theta\to\mathscr{A}(T)\to\bar{\partial}\mathscr{A}(T)\to 0$ induces the exact sequence of cohomology groups

$$0\to H^0(M,\Theta)\to\mathscr{L}^{0,0}(T)\xrightarrow{\bar{\partial}}\mathscr{L}_{\bar{\partial}}^{0,1}(T)\xrightarrow{\delta^*}H^1(M,\Theta)\to 0,$$

from which the isomorphism $H^1(M,\Theta)\cong\mathscr{L}_{\bar{\partial}}^{0,1}(T)/\bar{\partial}\mathscr{L}^{0,0}(T)$ is induced. Thus for the 0-cochain $\{\xi_j\}\in C^0(\mathfrak{U},\mathscr{A}(T))$, we have $\bar{\partial}\{\xi_j\}=\dot{\varphi}$, and $\delta\{\xi_j\}=-\{\theta_{jk}\}$ from the above results. Consequently by the definition of δ^*, we have

$$\delta^*\dot{\varphi} = -\left(\frac{\partial M_t}{\partial t}\right)_{t=0}, \quad \text{where} \quad \dot{\varphi} = \left(\frac{\partial\varphi(t)}{\partial t}\right)_{t=0}. \quad \blacksquare$$

As stated above, a deformation M_t of M is represented by the vector $(0,1)$-form $\varphi(t)$ on M satisfying the condition (5.86): $\bar{\partial}\varphi(t)=\frac{1}{2}[\varphi(t),\varphi(t)]$. In order to clarify the meaning of this equality, fix t for a moment, and assume that z moves on the domain U of a single local complex coordinate system $(z^1,\ldots,z^n)$. We write $\varphi(t)$ as $\varphi=\sum_{\lambda=1}^n\varphi^\lambda\partial_\lambda$ where $\varphi^\lambda=\sum_{\nu=1}^n\varphi_{\bar{\nu}}^\lambda\,d\bar{z}^\nu$. Here $\varphi_{\bar{\nu}}^\lambda=\varphi_{\bar{\nu}}^\lambda(z)$ are C^∞ functions on U. If we introduce partial differential operators

$$L_\nu = \bar{\partial}_\nu - \sum_{\lambda=1}^n \varphi_{\bar{\nu}}^\lambda\partial_\lambda, \qquad \nu=1,\ldots,n,$$

the equation (5.82) $(\bar{\partial}-\sum_\lambda\varphi^\lambda\partial_\lambda)f=0$ is written in the form of simultaneous partial differential equations as

$$L_\nu f = 0, \qquad \nu = 1, \ldots, n. \tag{5.93}$$

From (5.83) $\det(\delta^\lambda_\nu - \sum_\mu \varphi^\mu_{\bar\nu}\overline{\varphi^\lambda_{\bar\mu}}) \neq 0$. It follows from this that $2n$ *partial differential operators* $L_1, \ldots, L_n, \bar{L}_1, \ldots, \bar{L}_n$ *are linearly independent.* In fact if

$$\sum_{\lambda=1}^n a_\lambda\Big(\bar\partial_\lambda - \sum_\mu \varphi^\mu_{\bar\lambda}\partial_\mu\Big) + \sum_{\mu=1}^n b_\mu\Big(\partial_\mu - \sum_\lambda \overline{\varphi^\lambda_{\bar\mu}}\bar\partial_\lambda\Big) = 0,$$

then $a_\lambda = \sum_\mu \overline{\varphi^\lambda_{\bar\mu}} b_\mu$ and $b_\mu = \sum_\nu \varphi^\mu_{\bar\nu} a_\nu$, hence $a_\lambda = \sum_{\mu,\nu} \varphi^\mu_{\bar\nu}\overline{\varphi^\lambda_{\bar\mu}} a_\nu$. Therefore by (5.83) $a_\lambda = 0$ for $\lambda = 1, \ldots, n$, hence also $b_\mu = 0$ for $\mu = 1, \ldots, n$.

(5.86) is written in the explicit form as

$$\bar\partial_\tau \varphi^\lambda_{\bar\nu} - \bar\partial_\nu \varphi^\lambda_{\bar\tau} = \sum_\mu \varphi^\mu_{\bar\tau}\partial_\mu \varphi^\lambda_{\bar\nu} - \sum_\mu \varphi^\mu_{\bar\nu}\partial_\mu \varphi^\lambda_{\bar\tau},$$

and is equivalent to the system of equations

$$L_\tau L_\nu - L_\nu L_\tau = 0, \qquad \tau, \nu = 1, \ldots, n. \tag{5.94}$$

As is clearly seen from the proof of (5.84), (5.94) *is a necessary condition for the system of partial differential equations to have n* C^∞ *solutions* $f = \zeta^\alpha_j = \zeta^\alpha_j(z)$, $\alpha = 1, \ldots, n$, *with* $\det(\partial_\lambda \zeta^\alpha_j) \neq 0$. For the sufficiency of (5.94), Newlander and Nirenberg proved the following theorem.

Theorem 5.5 (Newlander-Nirenberg [26]). *Let* $\varphi = \sum_{\lambda=1}^n \sum_{\nu=1}^n \varphi^\lambda_{\bar\nu}\, d\bar{z}^\nu \partial_\lambda$ *be a* C^∞ *vector* (0, 1)*-form defined on a domain U of* $\mathbb{C}^n$, *and put* $L_\nu = \bar\partial_\nu - \sum_\nu \varphi^\lambda_{\bar\nu}\partial_\lambda$. *Suppose that* $L_1, \ldots, L_n, \bar{L}_1, \ldots, \bar{L}_n$ *are linearly independent, and that they satisfy the condition* (5.94):

$$L_\tau L_\nu - L_\nu L_\tau = 0, \qquad \tau, \nu = 1, \ldots, n.$$

Then the system of partial differential equations

$$L_\nu f = 0, \qquad \nu = 1, \ldots, n,$$

has n linearly independent C^∞ *solutions* $f = \zeta^\alpha = \zeta^\alpha(z)$, $\alpha = 1, \ldots, n$, *in a sufficiently small neighbourhood of any point of U. Here the solutions* $\zeta^1, \ldots, \zeta^n$ *are said to be linearly independent if*

$$\det \frac{\partial(\zeta^1, \ldots, \zeta^n, \bar\zeta^1, \ldots, \bar\zeta^n)}{\partial(z^1, \ldots, z^n, \bar{z}^1, \ldots, \bar{z}^n)} \neq 0. \tag{5.95}$$

For solutions $f=\zeta^\zeta$, $\alpha=1,\ldots,n$, of the system of equations $L_\nu f=0$, $\nu=1,\ldots,n$, we obtain by a simple calculation the following equality:

$$\det\frac{\partial(\zeta^1,\ldots,\zeta^n,\bar{\zeta}^1,\ldots,\bar{\zeta}^n)}{\partial(z^1,\ldots,z^n,\bar{z}^1,\ldots,\bar{z}^n)}=\det\Big(\delta^\lambda_\nu-\sum_\mu\varphi^\mu_{\bar{\nu}}\overline{\varphi^\lambda_{\bar{\mu}}}\Big)\left|\det\frac{\partial(\zeta^1,\ldots,\zeta^n)}{\partial(z^1,\ldots,z^n)}\right|^2.$$

Therefore if the solutions $\zeta^1,\ldots,\zeta^n$ are linearly independent, we have $\det(\partial_\lambda\zeta^\alpha)\neq 0$. Consequently when $\det(\delta^\lambda_\nu-\sum_\mu\varphi^\mu_{\bar{\nu}}\overline{\varphi^\lambda_{\bar{\mu}}})\neq 0$, (5.94) gives a necessary and sufficient condition for the system of partial differential equations (5.93): $L_\nu f=0$, $\nu=1,\ldots,n$, to have n linearly independent C^∞ solutions $f=\zeta^\alpha$, $\alpha=1,\ldots,n$, in a neighbourhood of any point of U. (5.94) is called *the integrability condition* for (5.93). Thus for $\varphi(t)$ on M, the condition (5.86): $\bar{\partial}\varphi(t)=\frac{1}{2}[\varphi(t),\varphi(t)]$ is the integrability condition for (5.82): $(\bar{\partial}-\varphi(t))f=0$.

Let $\varphi=\sum_{\lambda=1}^n\varphi^\lambda\partial_\lambda$ be a C^∞ vector (0, 1)-form on a compact complex manifold M where $\varphi^\lambda=\sum_{\lambda=1}^n\varphi^\lambda_{\bar{\nu}}\,d\bar{z}^\nu$, and suppose $\det(\delta^\lambda_\nu-\sum_\mu\varphi^\mu_{\bar{\nu}}\overline{\varphi^\lambda_{\bar{\mu}}})\neq 0$. Assume that φ satisfies the integrability condition

$$\bar{\partial}\varphi=\tfrac{1}{2}[\varphi,\varphi].$$

Then, if we take a sufficiently fine finite open covering $\{U_j\}$ of M, by the Newlander-Nirenberg theorem, there are n linearly independent C^∞ functions $\zeta^\alpha_j=\zeta^\alpha_j(z)$, $\alpha=1,\ldots,n$, on each U_j such that

$$\Big(\bar{\partial}-\sum_{\lambda=1}^n\varphi^\lambda\partial_\lambda\Big)\zeta^\alpha_j=0. \tag{5.96}$$

Since

$$\det\frac{\partial(\zeta^1_j,\ldots,\zeta^n_j,\bar{\zeta}^1_j,\ldots,\bar{\zeta}^n_j)}{\partial(z^1,\ldots,z^n,\bar{z}^1,\ldots,\bar{z}^n)}\neq 0,$$

$\zeta_j\colon z\to\zeta_j(z)=(\zeta^1_j(z),\ldots,\zeta^n_j(z))$ gives complex coordinates on $U_j\subset M$ if M is considered as a differentiable manifold. By the proof of Theorem 5.3, a local C^∞ function f on U is a holomorphic function of $\zeta^1_j(z),\ldots,\zeta^n_j(z)$ if $(\bar{\partial}-\varphi(t))f=0$. Therefore on $U_j\cap U_k\neq\emptyset$, $\zeta^\alpha_j(z)$ are holomorphic functions of $\zeta^1_k(z),\ldots,\zeta^n_k(z)$, namely,

$$\zeta^\alpha_j(z)=f^\alpha_{jk}(\zeta_k(z)),\qquad \alpha=1,\ldots,n.$$

Consequently, $\{\zeta_1,\ldots,\zeta_j,\ldots\}$ is a system of local complex coordinates on the differentiable manifold M, hence, defines a complex structure on M, which we denote by M_φ. If φ is equal to $\varphi(t)$ defined by (5.79), it is clear that $M_{\varphi(t)}=M_t=\varpi^{-1}(t)$.

With these preparations, we prove the following theorem of existence.

Theorem 5.6 (Theorem of Existence [18]). *Let M be a compact complex manifold and suppose $H^2(M, \Theta)=0$. Then there exists a complex analytic family $(\mathcal{M}, B, \varpi)$ with $0 \in B \subset \mathbb{C}^m$ satisfying the following conditions:*

(i) $\varpi^{-1}(0)=M$.

(ii) $\rho_0: \partial/\partial t \to (\partial M_t/\partial t)_{t=0}$ *with* $M_t=\varpi^{-1}(t)$ *is an isomorphism of* $T_0(B)$ *onto* $H^1(M, \Theta)$: $T_0(B) \xrightarrow{\rho_0} H^1(M, \Theta)$.

Before we give its proof, we will explain the motivation for this theorem. According to the conjecture given at the end of the preceding section (a), we expect that $m(M)=\dim H^1(M, \Theta)$ holds if the number of moduli $m(M)$ of M is defined. If $m(M)=\dim H^1(M, \Theta)$ holds, by Definition 5.4 of the number of moduli, there is a complex analytic family $(\mathcal{M}, B, \varpi)$ satisfying the conditions (i) and (ii) of the above theorem. If there is a complex analytic family satisfying the conditions (i) and (ii), then from the consideration in the preceding section (a), we see that for an arbitrary $\theta \in H^1(M, \Theta)$, all obstructions $\Gamma_2=\frac{1}{2}[\theta, \theta]$, $\Gamma_3, \Gamma_4, \ldots$ must vanish. Since there exists a compact complex manifold M for which there is a $\theta \in H^1(M, \Theta)$ with $[\theta, \theta] \neq 0$, it is absurd to try to prove the existence of a complex analytic family satisfying (i) and (ii) for an arbitrary M. As stated in (a), if $H^2(M, \Theta)=0$, all the obstructions vanish, while we could find no other simple condition for all the obstructions to vanish. So we tried to prove the theorem of existence for M with $H^2(M, \Theta)=0$.

Assume that there is a complex analytic family $(\mathcal{M}, B, \varpi)$ satisying (i) and (ii). Take a sufficiently small Δ with $0 \in \Delta \subset B$, and define by (5.79) the C^∞ vector $(0, 1)$-form $\varphi(t)=\sum_{\lambda=1}^n \varphi^\lambda(z, t)\partial_\lambda$ on M corresponding to $(\mathcal{M}_\Delta, \Delta, \varpi)$. Then $\varphi(t)$ satisfies the integrability condition $\bar\partial\varphi(t)=\frac{1}{2}[\varphi(t), \varphi(t)]$, and we have $\varphi(0)=0$ by (5.81). If we put

$$\dot\varphi_\lambda=\left(\frac{\partial\varphi(t)}{\partial t_\nu}\right)_{t=0}, \qquad \lambda=1, \ldots, m,$$

$\dot\varphi_\lambda \in \mathscr{Z}_{\bar\partial}^{0,1}(T)$ by Theorem 5.4, and $\{\dot\varphi_1, \ldots, \dot\varphi_m\}$ forms a basis of $H^1(M, \Theta) \cong \mathscr{Z}_{\bar\partial}^{0,1}(T)/\bar\partial\mathscr{L}^{0,0}(T)$.

Conversely, given $\beta_\lambda \in \mathscr{Z}_{\bar\partial}^{0,1}(T)$ for $\lambda=1, \ldots, m$, such that $\{\beta_1, \ldots, \beta_m\}$ forms a basis of $H^1(M, \Theta)$, assume that there is a family $\{\varphi(t)\}|t \in \Delta\}$ of C^∞ vector $(0, 1)$-forms $\varphi(t)$ on M with $0 \in \Delta \subset C^m$, which satisfy the integrability condition

$$\bar\partial\varphi(t)=\tfrac{1}{2}[\varphi(t), \varphi(t)],$$

and the initial conditions

$$\varphi(0)=0, \qquad \left(\frac{\partial\varphi(t)}{\partial t_\lambda}\right)_{t=0}=\beta_\lambda, \qquad \lambda=1, \ldots, m. \tag{5.97}$$

Since Δ is assumed to be sufficiently small, we may assume that $\varphi(t)=$

$\sum_\lambda \sum_\nu \varphi^\lambda_{\bar\nu}(t)\, d\bar z^\nu \partial_\lambda$ satisfies the condition (5.83):

$$\det\left(\delta^\lambda_\nu - \sum_\mu \varphi^\mu_{\bar\nu}(t)\overline{\varphi^\lambda_{\bar\mu}(t)}\right) \neq 0.$$

Therefore, by the Newlander–Nirenberg theorem, each $\varphi(t)$ determines a complex structure $M_{\varphi(t)}$ on M. If the family $\{M_{\varphi(t)}\}$ thus obtained is a complex analytic family, it is clear that it satisfies the conditions (i) and (ii). First we will construct such a family $\{\varphi(t) \mid t \in \Delta\}$. The fact that $\{M_{\varphi(t)}\}$ forms actually a complex analytic family will be proved later.

We take as $\varphi(t)$ a power series in $t_1, \ldots, t_m$:

$$\varphi(t) = \sum \varphi_{\nu_1 \cdots \nu_m} t_1^{\nu_1} \cdots t_m^{\nu_m}, \qquad \varphi_{\nu_1 \cdots \nu_m} \in \mathscr{L}^{0,1}(T).$$

We shall fix some notation. For a power series $P(t)$ in $t_1, \ldots, t_m$, we denote by

$$P_\nu(t) = \sum_{\nu_1 + \cdots + \nu_m = \nu} P_{\nu_1 \cdots \nu_m} t_1^{\nu_1} \cdots t_m^{\nu_m}$$

its νth homogeneous part. Thus

$$P(t) = \sum_{\nu=0}^{\infty} P_\nu(0) = P_0(t) + \cdots + P_\nu(t) + \cdots.$$

Further we put

$$P^\nu(t) = P_0(t) + \cdots + P_\nu(t).$$

Given two power series $P(t)$ and $Q(t)$, if $P^\nu(t) = Q^\nu(t)$, we denote it by

$$P(t) \underset{\nu}{\equiv} Q(t).$$

We must find a power series $\varphi(t)$ satisfying the integrability condition and the initial conditions. If we put

$$\varphi_0 = 0, \qquad \varphi_1(t) = \sum_{\lambda=1}^{m} \beta_\lambda t_\lambda, \tag{5.98}$$

$$\varphi(t) = \sum_{\nu=1}^{\infty} \varphi_\nu(t) = \varphi_1(t) + \varphi_2(t) + \cdots$$

satisfies the initial conditions (5.97). Consequently it remains to determine $\varphi_2(t), \ldots, \varphi_\nu(t), \ldots$ such that

$$\bar\partial \varphi(t) = \tfrac{1}{2}[\varphi(t), \varphi(t)] \tag{5.99}$$

holds. Since $\beta_\lambda \in \mathscr{L}_{\bar\partial}^{0,1}(T)$, we have $\bar\partial\varphi_1(t)=0$, and $[\varphi_\mu(t), \varphi_\nu(t)]$ is a homogeneous polynomial of degree $\mu+\nu$ in $t_1, \ldots, t_m$. Therefore the equality (5.99) is reduced to the system of infinitely many congruences

$$\bar\partial\varphi^\nu(t) \underset{\nu}{\equiv} \tfrac{1}{2}[\varphi^{\nu-1}(t), \varphi^{\nu-1}(t)], \qquad \nu=2,3,\ldots. \tag{5.100$_\nu$}$$

Thus in order to obtain $\varphi(t)$, it suffices to determine $\varphi_\nu(t)$ satisfying $(5.100)_\nu$ by induction on ν.

First for $\nu=2$, $(5.100)_2$ is written explicitly as

$$\bar\partial\varphi_2(t)=\tfrac{1}{2}[\varphi_1(t), \varphi_1(t)]. \tag{5.101}$$

Putting

$$\psi_2(t)=\tfrac{1}{2}[\varphi_1(t), \varphi_1(t)],$$

since $\bar\partial\varphi_1(t)=0$, we obtain from (5.88) and (5.87)

$$\bar\partial\psi_2(t)=[\bar\partial\varphi_1(t), \varphi_1(t)]=0.$$

Therefore $\psi_2=\sum_{\lambda,\mu=1}^n \psi_{\lambda\mu}t_\lambda t_\mu$ with $\psi_{\lambda\mu}=\psi_{\mu\lambda}\in\mathscr{L}_{\bar\partial}^{0,2}(T)$. Since by the assumption

$$\mathscr{L}_{\bar\partial}^{0,2}(T)/\bar\partial\mathscr{L}^{0,1}(T)\cong H^2(M,\Theta)=0,$$

there is a $\varphi_{\lambda\mu}\in\mathscr{L}^{0,1}(T)$ such that $\psi_{\lambda\mu}=\bar\partial\varphi_{\lambda\mu}$. If we put $\varphi_2(t)=\sum\varphi_{\lambda\mu}t_\lambda t_\mu$, we have

$$\bar\partial\varphi_2(t)=\psi_2(t)=\tfrac{1}{2}[\varphi_1(t), \varphi_1(t)].$$

Thus $(5.110)_2$ hold.

Assume that already $\varphi^\nu(t)=\varphi_1(t)+\cdots+\varphi_n(t)$ is determined such that $(5.100)_\nu$ holds, and consider $(5.100)_{\nu+1}$. Since $\varphi^{\nu+1}(t)=\varphi^\nu(t)+\varphi_{\nu+1}(t)$, $(5.100)_{\nu+1}$ is written in the form

$$\bar\partial\varphi_{\nu+1}(t)\underset{\nu}{\equiv}\tfrac{1}{2}[\varphi^\nu(t), \varphi^\nu(t)]-\bar\partial\varphi^\nu(t). \tag{5.102}$$

Since $[\varphi^\nu(t), \varphi^\nu(t)]\equiv_\nu[\varphi^{\nu-1}(t), \varphi^{\nu-1}(t)]$, and $(5.100)_\nu$ holds by the hypothesis of induction, the right-hand side of (5.102) $\equiv_\nu 0$. Let $\psi_{\nu+1}(t)$ be the $(\nu+1)$th homogeneous part of the right-hand side of (5.102). Then we have

$$\tfrac{1}{2}[\varphi^\nu(t), \varphi^\nu(t)]-\bar\partial\varphi^\nu(t)\underset{\nu+1}{\equiv}\psi_{\nu+1}(t). \tag{5.103}$$

Thus the congruence (5.102) is reduced to the equation

$$\bar{\partial}\varphi_{\nu+1}(t) = \psi_{\nu+1}(t). \tag{5.104}$$

By (5.103), (5.88), and (5.87),

$$\bar{\partial}\psi_{\nu+1}(t) \underset{\nu+1}{\equiv} \tfrac{1}{2}\bar{\partial}[\varphi^{\nu}(t), \varphi^{\nu}(t)] = [\bar{\partial}\varphi^{\nu}(t), \varphi^{\nu}(t)]$$

while $\bar{\partial}\varphi^{\nu}(t) \equiv_{\nu} (1/2)[\varphi^{\nu}(t), \varphi^{\nu}(t)]$ as stated above. Since $\varphi^{\nu}(t) \equiv_0 0$. we have

$$[\bar{\partial}\varphi^{\nu}(t), \varphi^{\nu}(t)] \underset{\nu+1}{\equiv} \tfrac{1}{2}[[\varphi^{\nu}(t), \varphi^{\nu}(t)], \varphi^{\nu}(t)]$$

while by (5.89) $[[\varphi^{\nu}(t), \varphi^{\nu}(t)], \varphi^{\nu}(t)] = 0$. Hence $\bar{\partial}\psi_{\nu+1}(t) = 0$, namely,

$$\psi_{\nu+1}(t) = \sum_{\nu_1+\cdots+\nu_m=\nu+1} \psi_{\nu_1\cdots\nu_m} t_1^{\nu_1}\cdots t_m^{\nu_m}, \qquad \psi_{\nu_1\cdots\nu_m} \in \mathscr{L}_{\bar{\partial}}^{0,2}(T).$$

Consequently, since $H^2(M, \Theta) = 0$, by (5.90) we can find $\varphi_{\nu_1\cdots\nu_m} \in \mathscr{L}^{0,1}(T)$ such that $\psi_{\nu_1\cdots\nu_m}$. Thus there is a solution

$$\varphi_{\nu+1}(t) = \sum_{\nu_1+\cdots+\nu_m=\nu+1} \varphi_{\nu_1\cdots\nu_m} t_1^{\nu_1}\cdots t_m^{\nu_m}$$

of the equation (5.104): $\bar{\partial}\varphi_{\nu+1}(t) = \psi_{\nu+1}(t)$. Putting $\varphi^{\nu+1}(t) = \varphi^{\nu}(t) + \varphi_{\nu+1}(t)$, we see that $(5.100)_{\nu+1}$ also holds, which completes the induction.

Thus we construct a power series $\varphi(t) = \sum_{\nu=1}^{\infty} \varphi_{\nu}(t)$ satisfying $\bar{\partial}\varphi(t) = \frac{1}{2}[\varphi(t), \varphi(t)]$, and (5.97). Next, we must prove that this power series $\varphi(t)$ converges for sufficiently small $t_1, \ldots, t_m$. We will give proof of convergence below. Here we owe to Nirenberg the idea of using the Hölder norm for proof of convergence.

(c) Proof of Convergence

We introduce a Hermitian metric $\sum_{\lambda,\mu=1}^{n} g_{\lambda\bar{\nu}}\, dz^{\lambda} \otimes dz^{-\nu}$ on M, and define dual forms, inner products, norms, etc. of differential forms on M as in §3.5(a). For the tangent bundle T, we define its Hermitian metric on fibres by $\sum_{\lambda,\mu=1}^{n} g_{\lambda\bar{\nu}}\zeta^{\lambda}\bar{\zeta}^{\nu}$ for $\sum_{\lambda} \zeta^{\lambda}\partial_{\lambda} \in T$. Then the inner product of vector $(0, q)$-forms $\varphi = \sum_{\lambda} \varphi^{\lambda}\partial_{\lambda}$ and $\psi = \sum_{\lambda} \psi^{\lambda}\partial_{\lambda}$ on M is given by (3.139) as

$$(\varphi, \psi) = \int_M \sum_{\lambda,\nu=1}^{n} g_{\lambda\bar{\nu}}\varphi^{\lambda} \wedge *\bar{\psi}^{\nu}, \qquad \varphi, \psi \in \mathscr{L}^{0,q}(T).$$

We denote by $\mathfrak{d}$ the adjoint operator of $\bar{\partial}$ with respect to this inner product. Thus

$$(\bar{\partial}\varphi, \psi) = (\varphi, \mathfrak{d}\psi), \qquad \varphi \in \mathscr{L}^{0,q}(T), \qquad \psi \in \mathscr{L}^{0,q+1}(T).$$

Also we put

$$\square = \bar{\partial}\mathfrak{d} + \mathfrak{d}\bar{\partial}.$$

$\varphi \in \mathscr{L}^{0,q}(T)$ is called a *harmonic vector* $(0, q)$*-form* if $\bar{\partial}\varphi = \mathfrak{d}\varphi = 0$. $\varphi = \mathscr{L}^{0,q}(T)$ is harmonic if and only if $\square\varphi = 0$. The vector space

$$\mathbf{H}^{0,q}(T) = \{\varphi \in \mathscr{L}^{0,q}(T) \mid \square\varphi = 0\}$$

consists of all harmonic vector $(0, q)$-forms on M. By Theorem 3.19, we have

$$\mathscr{L}^{0,q}(T) = \mathbf{H}^{0,q}(T) \oplus \square\mathscr{L}^{0,q}(T). \tag{5.105}$$

From this it follows that

$$\mathscr{Z}^{0,q}_{\bar{\partial}}(T) = \mathbf{H}^{0,q}(T) \oplus \bar{\partial}\mathscr{L}^{0,q-1}(T). \tag{5.106}$$

Hence

$$H^q(M, \Theta) \cong \mathscr{Z}^{0,q}_{\bar{\partial}}(T)/\bar{\partial}L^{0,q-1}(T) \cong \mathbf{H}^{0,q}(T). \tag{5.107}$$

Next we define the Hölder norm. First let $f(x) = f(z^1, \ldots, x^{2n})$ be a complex-valued C^∞ function defined on a domain U of $\mathbb{R}^{2n}$. We denote any partial differential operator of order of order h with respect to $x^1, \ldots, x^{2n}$ by D^h:

$$D^h = \left(\frac{\partial}{\partial x^1}\right)^{h_1} \cdots \left(\frac{\partial}{\partial x^{2n}}\right)^{h_{2n}}, \qquad h_1 + \cdots + h_{2n} = h.$$

Let k be a non-negative integer, and α a real number with $0 < \alpha < 1$. Then the *Hölder norm* $|f|^U_{k+\alpha}$ of $f = f(x)$ is defined by

$$|f|^U_{k+\alpha} = \sum_{h=0}^{k} \sum_{D^h} \sup_{x \in U} |D^h f(x)| + \sum_{D^k} \sup_{x, y \in U} \frac{|D^k f(x) - D^k f(y)|}{|x - y|^\alpha},$$

where $|x - y| = \sqrt{\sum_\nu |x^\nu - y^\nu|^2}$, and the summation $\sum_{D^h}$ is taken over all partial differential operator D^h of order h.

For a C^∞ vector $(0, q)$-form $\varphi \in \mathscr{L}^{0,q}(T)$ on M, we define its Hölder norm $|\varphi|_{k+\alpha}$ as follows: Let $\{U_j\}$ be a finite open covering of M, where each U_j is a coordinate polydisk. Then by the definition of coordinate polydisks, there are local complex coordinates $\{z_j\}$ with $z_j = (z_j^1, \ldots, z_j^n)$ satisfying the following conditions:

(i) $[U_j]$ is contained in the domain of z_j.
(ii) $U_j = \{z_j \mid |z_j^1| < 1, \ldots, |z_j^n| < 1\}$.

Putting $z_j^\nu = x_j^{2\nu-1} + ix_j^{2\nu}$, we introduce the real coordinates $x_j = (x_j^1, \ldots, x_j^{2n})$. If we represent $\varphi = \mathscr{L}^{0,q}(T)$ on each U_j in the form

$$\varphi = \sum_\lambda \frac{1}{q!} \sum \varphi^\lambda_{j\bar\nu_1\cdots\bar\nu_q}(x_j)\, dz_j^{\bar\nu_1} \wedge \cdots \wedge dz_j^{\bar\nu_q} \frac{\partial}{\partial z_j^\lambda},$$

$\varphi^\lambda_{j\bar\nu_1\cdots\bar\nu_q} = \varphi^\lambda_{j\bar\nu_1\cdots\bar\nu_q}(x_j)$ are considered as C^∞ functions on $U_j \in \mathbb{R}^{2n}$. Define

$$|\varphi|_{k+\alpha} = \max_j \max_{\lambda,\nu_1\cdots\nu_q} |\varphi^\lambda_{j\bar\nu_1\cdots\bar\nu_q}|^{U_\alpha}_{k+\alpha}.$$

Since the the Hölder norm $|\varphi|_{k+\alpha}$ depends on the choice of $\{U_j\}$ and $\{z_j\}$, we fix $\{U_j\}$ and $\{z_j\}$ once for all below. Put

$$|\varphi|_0 = \max_j \max_{\lambda,\nu_1\cdots\nu_q} \sup_{x_j \in U_j} |\varphi^\lambda_{j\bar\nu_1\cdots\bar\nu_q}(x_j)|.$$

Theorem 5.7 ([3]). *For $k \geqq 2$, the inequality*

$$|\varphi|_{k+\alpha} \leqq c(|\square\varphi|_{k-2\oplus\alpha} + |\varphi|_0) \tag{5.108}$$

holds for $\varphi \in \mathscr{L}^{0,q}(T)$, where c is a constant depending only on k and α.

We give proof of this theorem in §4(b) of the Appendix. The fundamental inequality (5.108) is called the *a priori estimate.*

With these preparations, we consider $\mathscr{L}^{0,2}(T)$ for M with $H^2(M, \Theta) = 0$. In this case since $\mathbf{H}^{0,2}(T) = 0$ by (5.107), if $\square\psi = 0$ for $\psi \in \mathscr{L}^{0,2}(T)$, we have $\psi = 0$, and by (5.105)

$$\mathscr{L}^{0,2}(T) = \square\mathscr{L}^{0,2}(T).$$

Therefore the linear map $\square$ maps $\mathscr{L}^{0,2}(T)$ in a one-to-one manner onto itself. We denote its inverse by G: $G = \square^{-1}$. G is called the *Green operator.* It is clear that

$$\psi = \square G\psi, \qquad \psi \in \mathscr{L}^{0,2}(T). \tag{5.109}$$

If $\bar\partial\psi = 0$, we have

$$\psi = \bar\partial\mathfrak{d} G\psi, \qquad \psi \in \mathscr{L}^{0,2}_{\bar\partial}(T). \tag{5.110}$$

Proof. We have

$$\psi = \square G\psi = \bar\partial\mathfrak{d} G\psi + \mathfrak{d}\bar\partial G\psi.$$

Applying $\bar\partial$ to this equality, we obtain $\bar\partial\mathfrak{d}\bar\partial G\psi = \bar\partial\psi = 0$. Therefore

$$\|\mathfrak{d}\bar\partial G\psi\|^2 = (\mathfrak{d}\bar\partial G\psi, \mathfrak{d}\bar\partial G\psi) = (\bar\partial\mathfrak{d}\bar\partial G\psi, \bar\partial G\psi) = 0.$$

Thus $\mathfrak{d}\bar\partial G\psi = 0$. Hence $\psi = \bar\partial\mathfrak{d} G\psi$.

Lemma 5.7. *For $k \geqq 2$, the inequality*

$$|G\psi|_{k+\alpha} \leqq c_1 |\psi|_{k-2+\alpha} \tag{5.111}$$

holds for $\psi \in \mathscr{L}^{0,2}(T)$, where c_1 is a constant independent of ψ.

Proof. Since $\Box G\psi = \psi$, by the *a priori* estimate (5.108) we have

$$|G\psi|_{k+\alpha} \leqq c(|\psi|_{k-2+\alpha} + |G\psi|_0) \quad \text{for } \psi \in \mathscr{L}^{0,2}(T). \tag{5.112}$$

Therefore in order to prove (5.111) it suffices to show the inequality

$$|G\psi|_0 \leqq c_2 |\psi|_{k-1+\alpha} \quad \text{for } \psi \in \mathscr{L}^{0,2}(T), \tag{5.113}$$

where c_2 is a constant independent of ψ. Suppose that (5.113) does not hold for any c_2. Then for any natural number m, there exists $\psi^{(m)} \in \mathscr{L}^{0,2}(T)$ such that

$$|G\psi^{(m)}|_0 = 1, \qquad |\psi^{(m)}|_{k-2+\alpha} < \frac{1}{m}. \tag{5.114}$$

Putting $\varphi^{(m)} = G\psi^{(m)}$, we have from (5.112)

$$|\varphi^{(m)}|_{k+\alpha} \leqq 2c.$$

Thus, representing $\varphi^{(m)}$ on each U_j in the form

$$\varphi^{(m)} = \sum_{\lambda=1}^{n} \frac{1}{2} \sum_{\beta,\gamma=1}^{n} \varphi_{j\beta\gamma}^{(m)\lambda}(x_j)\, d\bar{z}_j^\beta \wedge d\bar{z}_j^\gamma \frac{\partial}{\partial z_j^\lambda},$$

we have

$$|\varphi_{j\beta\gamma}^{(m)\lambda}|_{k+\alpha}^{U_j} \leqq 2c.$$

Fix j, λ, β, and γ for a while, and put $f_m = f_m(x_j) = \varphi_{j\beta\gamma}^{(m)\lambda}(x_j)$. Denoting by D_j^h a partial differential operator of order h in $x_j^1, \ldots, x_j^{2n}$, since $|f_m|_{k+\alpha}^{U_j} \leqq 2c$, we have

$$|D_j^k f_m(x_j) - D_j^k f_m(y_j)| \leqq 2c|x_j - y_j|^\alpha,$$

and

$$|D_j^h f_m(x_j) - D_j^h f_m(y_j)| \leqq 2c|x_j - y_j| \quad \text{for } h < k.$$

Therefore for each D_j^h, $h \leqq k$, the sequence of functions $\{D_j^h f_m(x_j)\}$ is *equicontinuous* in U_j. Clearly $|D_j^h f_m(x_j)| \leqq 2c$ on U_j for $h \leqq k$. Consequently, by Ascoli's theorem, we can choose a subsequence $\{f_{m(\nu)}\}$ of $\{f_m\}$ with

$m(1) < m(2) < \cdots < m(\nu) < \cdots$ such that the sequence $\{D_j^h f_{m(\omega)}(x_j)\}$ converges uniformly on U_j for each $h \leqq k$. Writing $\{f_{m(1)}, f_{m(2)}, \ldots\}$ simply as $\{f_m\} = \{f_1, f_2, \ldots\}$, we may assume that $\{D_j^h f_m(x_j)\}$ converges uniformly on U_j for any D_j^h with $h \leqq k$. Put $f(x_j) = \lim_{m\to\infty} f_m(x_j)$. Then $f(x_j)$ is a C^k function on U_j, and $\{D_j^h f_m(x_j)\}$ converges uniformly to $D_j^h f(x_j)$ for any D_j^h with $h \leqq k$.

Since the above results hold for each quadruple of indices $j, \lambda, \beta, \gamma$, replacing $\{\psi^{(m)}\}$ by an appropriate subsequence, we may assume that there is a C^k vector $(0, 2)$-form φ such that for any D_j^h with $h \leqq k$, the sequence $\{D_j^h \varphi_{j\beta\gamma}^{(m)\lambda}(x_j)\}$ converges uniformly to $D_j^h \varphi_{j\beta\gamma}^{\lambda}(x_j)$ on U_j where $\varphi^{(m)} = G\psi^{(m)}$. Therefore $\varphi^{(m)}$ converges to φ uniformly on M if $m \to \infty$, and, since $k \geqq 2$, $\square\varphi^{(m)}$ also to $\square\varphi$ uniformly. Since $\varphi^{(m)} = G\psi^{(m)}$ and $\square\varphi^{(m)} = \square G\psi^{(m)}$, we have

$$|G\psi^{(m)} - \varphi|_0 \to 0 \qquad (m \to \infty),$$

$$|\psi^{(m)} - \square\varphi|_0 \to 0 \qquad (m \to \infty).$$

Therefore, since by (5.114) $|\psi^{(m)}|_0 \leqq |\psi^{(m)}|_{k-2+\alpha} \to 0$ $(m \to 0)$, we have $\square\varphi = 0$. Consequently for an arbitrary $\psi \in \mathscr{L}^{0,2}(T)$,

$$(\varphi, \psi) = (\varphi, \square G\psi) = (\square\varphi, G\psi) = 0,$$

which implies that $\varphi = 0$. Since by (5.114) $|G\psi^{(m)}|_0 = 1$, this contradicts the fact that $|G\psi^{(m)} - \varphi|_0 \to 0$. ∎

Suppose given a power series in $t_1, \ldots, t_m$

$$P(t) = \sum_{\nu_1, \ldots, \nu_m = 0}^{\infty} P_{\nu_1 \ldots \nu_m} t_1^{\nu_1} \cdots t_m^{\nu_m}, \qquad P_{\nu_1 \cdots \nu_m} \in \mathbb{C}.$$

A power series

$$a(t) = \sum_{\nu_1, \ldots, \nu_m = 0}^{\infty} a_{\nu_1 \cdots \nu_m} t_m^{\nu_1}, \qquad a_{\nu^1 \cdots \nu_m} \geqq 0,$$

is said to be a *majorant* of $P(t)$ if

$$|P_{\nu_1 \cdots \nu_m}| \leqq a_{\nu_1 \cdots \nu_m}, \qquad \nu_1, \ldots, \nu_m = 0, 1, \ldots,$$

and is denoted by

$$P(t) \ll a(t).$$

For a power series

$$\psi(t) = \sum_{\nu_1, \ldots, \nu_m = 0}^{\infty} \psi_{\nu \cdots \nu_m} t_1^{\nu_1} \cdots t_m^{\nu_m}, \qquad \psi_{\nu_1 \cdots \nu_m} \in \mathscr{L}^{0,q}(T),$$

we define $|\psi|_{k+\alpha}(t)$ by

$$|\psi|_{k+\alpha}(t)=\sum_{\nu_1,\ldots,\nu_m=0}^{\infty}|\psi_{\nu_1\cdots\nu_m}|_{k+\alpha}t_1^{\nu_1}\cdots t_m^{\nu_m}.$$

We write

$$|\psi|_{k+\alpha}(t)\ll a(t)$$

if

$$|\psi_{\nu_1\cdots\nu_m}|_{k+\alpha}\leqq a_{\nu_1\cdots\nu_m},\qquad \nu_1,\ldots,\nu_m=0,1,2,\ldots.$$

In order to prove the convergence of the power series $\varphi(t)=\sum_{\nu=1}^{\infty}\varphi_\nu(t)$ constructed in the preceding section (b) on some polydisk $\Delta_\varepsilon=\{t\in\mathbb{C}^m \,|\, |t_1|<\varepsilon,\ldots,|t_m|<\varepsilon\}$, $\varepsilon>0$, with respect to the Hölder norm, it suffices to show that there is a power series $a(t)$ which converges absolutely in Δ_ε such that

$$|\varphi|_{k+\alpha}(t)\ll a(t).$$

Since in the construction of $\varphi(t)=\sum_{\nu=1}^{\infty}\varphi_\nu(t)$ given above, $\bar\partial\varphi^\nu(t)$ in (5.103) is a polynomial of degree ν in $t_1,\ldots,t_m$, $\psi_{\nu+1}(t)$ in the right-hand side is the homogeneous part of degree $\nu+1$ of $\frac{1}{2}[\varphi^\nu(t),\varphi^\nu(t)]$, namely,

$$\psi_{\nu+1}(t)=\tfrac{1}{2}[\varphi^\nu(t),\varphi^\nu(t)]_{\nu+1}.$$

Since $\bar\partial\psi_{\nu+1}(t)=0$, if we put

$$\varphi_{\nu+1}(t)=\mathfrak{d}G\psi_{\nu+1}(t),$$

from (5.110) $\varphi_{\nu+1}(t)$ is a solution of the equation $\bar\partial\varphi_{\nu+1}(t)=\psi_{\nu+1}(t)$. Therefore if we put

$$\varphi_1(t)=\sum_{\lambda=1}^{m}\beta_\lambda t_\lambda$$

first, and define $\varphi_{\nu+1}(t)$ successively for $\nu=1,2,3,\ldots$ by

$$\varphi_{\nu+1}((t)=\tfrac{1}{2}\mathfrak{d}G[\varphi^\nu(t),\varphi^\nu(t)]_{\nu+1},\tag{5.115}$$

we obtain the power series $\varphi(t)=\sum_{\nu=1}^{\infty}\varphi_\nu(t)$.

We choose as $a(t)$ the power series

$$A(t)=\frac{b}{16c}\sum_{\nu=1}^{\infty}\frac{\psi^\nu(t_1+\cdots+t_m)^\nu}{\nu^2},$$

with $b>0$ and $c>0$. b and c will be determined later. As regards $A(t)$, we have the following inequality

$$A(t)^2 \ll \frac{b}{c} A(t). \tag{5.116}$$

Proof. Consider the power series $B(s)=\sum_{\nu=1}^{\infty} s^\nu/\nu^2$ in the variable s. Then

$$B(s)^2=\sum_{\lambda=1}^{\infty}\frac{s^\lambda}{\lambda^2}\sum_{\mu=1}^{\infty}\frac{s^\mu}{\mu^2}=\sum_{\nu=2}^{\infty}s^\nu\sum_{\lambda+\mu=\nu}\frac{1}{\lambda^2\mu^2},$$

while, since $\mu\geqq\nu/2$ if $\lambda+\mu=\nu$ and $\lambda\leqq\mu$, we have

$$\sum_{\lambda+\mu=\nu}\frac{1}{\lambda^2\mu^2}\leqq 2\sum_{\substack{\lambda+\mu=\nu\\ \lambda\leqq\mu}}\frac{1}{\lambda^2\mu^2}<2\sum_{\lambda=1}^{\infty}\frac{4}{\lambda^2\nu^2}=\frac{8}{\nu^2}\sum_{\lambda=1}^{\infty}\frac{1}{\lambda^2}.$$

As is well known, $\sum_{\lambda=1}^{\infty} 1/\lambda^2=\pi^2/6<2$. Therefore

$$B(s)^2 \ll 16\sum_{\nu=2}^{\infty}\frac{s^\nu}{\nu^2}=16B(s).$$

Since $A(t)=(b/16c)B(c(t_1+\cdots+t_m))$, (5.116) follows immediately from this inequality. ∎

Fix a natural number $k\geqq 2$. We will show by induction on ν that if we choose appropriate large b and c,

$$|\varphi^\nu|_{k+\alpha}(t)\ll A(t) \tag{5.117$_\nu$}$$

holds.

For $\nu=1$, $\varphi^1(t)=\beta_1 t_1+\cdots+\beta_m t_m$, and the linear term of $A(t)$ is $(b/16)$ $(t_1+\cdots+t_m)$. Therefore, if we choose b such that $|\beta_\lambda|_{k+\alpha}<b/16$ for $\lambda=1,\ldots,m$, $(5.117)_1$: $|\varphi^1|_{k+\alpha}(t)\ll A(t)$ holds.

Now we assume that $(5.117)_\nu$ holds, and make an estimate of $|\varphi_{\nu+1}|_{k+\alpha}(t)$. By (5.115)

$$\varphi_{\nu+1}(t)=\mathfrak{d}G\psi_{\nu+1}(t), \qquad \psi_{\nu+1}(t)=\tfrac{1}{2}[\varphi^\nu(t),\varphi^\nu(t)]_{\nu+1}.$$

By the definition of the Hölder norm, we have the inequalities

$$|\mathfrak{d}\psi|_{k+\alpha}\leqq K_1|\psi|_{k+1+\alpha}, \qquad \psi\in\mathscr{L}^{0,2}(T), \tag{5.118}$$

and

$$|[\varphi,\psi]|_{k-1+\alpha}\leqq K_2|\varphi|_{k+\alpha}|\psi|_{k+\alpha}, \qquad \varphi,\psi\in\mathscr{L}^{0,1}(T) \tag{5.119}$$

where K_1, K_2 are constants independent of ψ and φ. By (5.111) and (5.118),

$$|\mathfrak{d}G\psi|_{k+\alpha} \leqq K_1|G\psi|_{k+1+\alpha} \leqq K_1c_1|\psi|_{k-1+\alpha},$$

hence

$$|\varphi_{\nu+1}|_{k+\alpha}(t) \ll K_1c_1|\psi_{\nu+1}|_{k-1+\alpha}(t).$$

Since $\psi_{\nu+1}(t) = \frac{1}{2}[\varphi^\nu(t), \varphi^\nu(t)]_{\nu+1}$, by (5.119) we have

$$|\psi_{\nu+1}|_{k-1+\alpha}(t) \ll K_2|\varphi^\nu|_{k+\alpha}(t)|\varphi^\nu|_{k+\alpha}(t)$$

while by the hypothesis of induction $|\varphi^\nu|_{k+\alpha}(t) \ll A(t)$. Therefore using (5.116) we obtain

$$|\varphi_{\nu+1}|_{k+\alpha}(t) \ll K_1K_2c_1A(t)^2 \ll \frac{K_1K_2c_1b}{c}A(t).$$

Hence putting $c = K_1K_2c_1b$, we obtain

$$|\varphi_{\nu+1}|_{k+\alpha}(t) \ll A(t).$$

On the other hand, $\varphi^{\nu+1}(t) = \varphi^\nu(t) + \varphi_{\nu+1}(t)$ where $\varphi^\nu(t)$ is a polynomial of degree ν and $\varphi_{\nu+1}(t)$ is a homogeneous polynomial of degree $\nu+1$. Consequently since $|\varphi^\nu|_{k+\alpha}(t) \ll A(t)$, we have

$$|\varphi^{\nu+1}|_{k+\alpha}(t) \ll A(t).$$

Thus $(5.117)_{\nu+1}$ holds, which completes the induction. Therefore we have showed that

$$|\varphi|_{k+\alpha}(t) \ll A(t). \tag{5.120}$$

Since the radius of convergence of the power series $\sum_{\nu=1}^\infty s^\nu/\nu^2$ is equal to 1, $A(t)$ converges absolutely for $t \in \Delta_\varepsilon$ if $0 < \varepsilon \leqq 1/mc$. Hence by (5.120) $\varphi(t) = \sum_{\nu=1}^\infty \varphi_\nu(t)$ converges with respect to the Hölder norm $|\ |_{k+\alpha}$ for $t \in \Delta_\varepsilon$. Consequently $\varphi(t)$ is a C^k vector (0, 1)-form on $M \times \Delta_\varepsilon$, namely, if we write $\varphi(t)$ on each $U_j \times \Delta_\varepsilon$ in the form

$$\varphi(t) = \sum_\lambda \frac{1}{2} \sum_{\beta,\gamma} \varphi_{j\beta\gamma}^\lambda(z_j, t)\, d\bar{z}_j^\beta \wedge d\bar{z}_j^\gamma \frac{\partial}{\partial z_j^\lambda},$$

the coefficients $\varphi_{j\beta\gamma}^\lambda(z_j, t)$ are C^k functions of $z_k^1, \ldots, z_j^n, t_1, \ldots, t_m$. Although k is an arbitrary natural number not less than 2, we cannot deduce immediately from this result that $\varphi(t)$ is C^∞, because the constant $c = K_1K_2c_1b$ above does depend on k, hence it does not follow immediately that there is an $\varepsilon>0$ such that $\varepsilon \leqq 1/mc$ for all k. We must prove C^∞ differentiability of $\varphi(t)$ in another way.

By (5.107) we may consider that $\{\beta_1, \ldots, \beta_m\}$ above is a basis of $\mathbf{H}^{0,1}(T)$. Then for $\varphi_1(t)=\beta_1 t_1+\cdots+\beta_m t_m$, we have $\mathfrak{d}\varphi_1(t)=0$. Since $\varphi_\nu(t)=\mathfrak{d}G\psi_\nu(t)$ for $\nu=2,3,\ldots$, we have $\mathfrak{d}\varphi_\nu(t)=0$. Thus we have

$$\mathfrak{d}\varphi(t)=0,$$

hence, $\square\varphi(t)=\mathfrak{d}\bar{\partial}\varphi(t)$ while by the construction $\varphi(t)$ satisfies the integrability condition

$$\bar{\partial}\varphi(t)=\tfrac{1}{2}[\varphi(t), \varphi(t)].$$

Hence

$$\square\varphi(t)=\tfrac{1}{2}\mathfrak{d}[\varphi(t), \varphi(t)].$$

On the other hand, since $\varphi(t)$ is holomorphic in $t_1, \ldots, t_m$, $\partial\varphi(t)/\partial\bar{t}_\lambda=0$. Consequently $\varphi(t)$ is a solution of the partial differential equation of order 2

$$\left(-\sum_{\lambda=1}^{m}\frac{\partial^2}{\partial t_\lambda\,\partial\bar{t}_\lambda}+\square\right)\varphi(t)-\tfrac{1}{2}\mathfrak{d}[\varphi(t), \varphi(t)]=0. \tag{5.121}$$

According to (3.142) and (3.126) the principal part of $\mathfrak{d}[\varphi(t), \varphi(t)]$ is given by

$$\sum_{\lambda=1}^{n}\sum_{\beta,\gamma=1}^{n} g^{\bar{\gamma}\beta}\sum_{\mu=1}^{n}(\varphi^\mu(t)\partial_\beta\partial_\mu\varphi^\lambda_{\bar{\gamma}}(t)-\varphi^\mu_{\bar{\gamma}}(t)\partial_\beta\partial_\mu\varphi^\lambda(t))\partial_\lambda,$$

where we put $\partial_\beta\partial_\mu\varphi^\lambda(t)=\sum_\nu \partial_\beta\partial_\mu\varphi^\lambda_{\bar{\nu}}(t)\,d\bar{z}^\nu$. Since $\varphi(t)\to 0$ for $|t|\to 0$, taking a sufficiently small Δ_ε, we may assume that (5.121) is a quasi-linear elliptic partial differential equation $M\times\Delta_\varepsilon$. Therefore its solution $\varphi(t)$ is C^∞ on $M\times\Delta_\varepsilon$ ([3]). We shall give the proof of the C^∞ differentiability of the solution in §8 of the Appendix.

Thus we have constructed a family $\{\varphi(t)\}|t\in\Delta_\varepsilon\}$ of C^∞ vector (0, 1)-forms $\varphi(t)=\sum_{\lambda=1}^{n}\sum_{\nu=1}^{n}\varphi^\lambda_{\bar{\nu}}(z,t)\,d\bar{z}^\nu\partial_\lambda$ satisfying the integrability condition $\bar{\partial}\varphi(t)=\frac{1}{2}[\varphi(t), \varphi(t)]$, and the initial conditions (5.97), where $\varphi^\lambda_{\bar{\nu}}(t)$ are C^∞ functions of $z^1, \ldots, z^n, t_1, \ldots, t_m$.

As was already stated, each $\varphi(t)$ determines a complex structure $M_{\varphi(t)}$ on M. In order to show that $\{M_{\varphi(t)}\}|t\in\Delta_\varepsilon\}$ is a complex analytic family, we consider $\varphi=\varphi(t)$ as a vector (0, 1)-form on the complex manifold $M\times\Delta_\varepsilon$. Namely, we consider φ as

$$\varphi=\varphi(t)=\sum_{\lambda=1}^{n}\left(\sum_{\nu=1}^{n}\varphi^\lambda_{\bar{\nu}}\,d\bar{z}^\nu+\sum_{\mu=1}^{\nu}\varphi^\lambda_{n+\mu}\,d\bar{t}_\mu\right)\frac{\partial}{\partial z^\lambda}+\sum_{\mu=1}^{m}\varphi^{n+\mu}\frac{\partial}{\partial t_\mu},$$

with $\varphi^{n+\mu}=\varphi^{\lambda}_{\overline{n+\mu}}=0$ for $\mu=1,\dots,m$. Then since $\varphi^{\lambda}_{\bar\nu}=\varphi^{\lambda}_{\bar\nu}(z,t)$ are holomorphic in $t_1,\dots,t_m$, we have $\partial\varphi^{\lambda}_{\bar\nu}/\partial\bar t_\mu=0$ in

$$\bar\partial\varphi=\sum_{\lambda,\nu=1}^{n}\left(\sum_{\beta=1}^{n}\frac{\partial\varphi^{\lambda}_{\bar\nu}}{\partial\bar z^{\beta}}d\bar z^{\beta}+\sum_{\mu=1}^{m}\frac{\partial\varphi^{\lambda}_{\bar\nu}}{\partial\bar z_\mu}d\bar t_\mu\right)\wedge d\bar z^{\nu}\frac{\partial}{\partial z^{\lambda}}.$$

By $\bar\partial\varphi(t)$ we denote the exterior differential of $\varphi(t)$ as a vector $(0,1)$-form on M with t fixed. Then from the above consideration $\bar\partial\varphi$ coincides with $\bar\partial\varphi(t)$. Similarly we obtain $[\varphi,\varphi]=[\varphi(t),\varphi(t)]$. Therefore as a C^∞ vector $(0,1)$-form on $M\times\Delta_\varepsilon$, φ satisfies the integrability condition

$$\bar\partial\varphi=\tfrac{1}{2}[\varphi,\varphi].$$

If we put

$$L_\nu=\frac{\partial}{\partial\bar z_\nu}-\sum_{\lambda=1}^{n}\varphi^{\lambda}_{\bar\nu}(z,t)\frac{\partial}{\partial z^{\lambda}},$$

the partial differential equation $(\bar\partial-\varphi)f=0$ is reduced to the system of partial differential equations:

$$\begin{aligned} L_\nu f&=0, &&\nu=1,\dots,n,\\ \frac{\partial}{\partial\bar t_\mu}f&=0, &&\mu=1,\dots,m. \end{aligned}\tag{5.122}$$

$L_1,\dots,L_n,\bar L_1,\dots,\bar L_n,\partial/\partial\bar t,\dots,\partial/\partial\bar t_m,\partial/\partial t_1,\dots,\partial/\partial t_m$ are linearly independent. Consequently, by the Newlander–Nirenberg theorem, *φ defines a complex structure $\mathcal{M}$* on $M\times\Delta_\varepsilon$. If we choose a sufficiently fine locally finite open covering $\{U_j\}$ of M, and take a sufficiently small Δ_ε, the equation (5.122) has $n+m$ linearly independent solutions $f=\zeta^{\beta}_j(z,t)$, $\beta=1,\dots,m+n$ on each $U_j\times\Delta_\varepsilon$, and the map

$$\zeta_j\colon (z,t)\to(\zeta^1_j(z,t),\dots,\zeta^{n+m}_j(z,t))$$

gives local complex coordinates of $\mathcal{M}$ on $U_j\times\Delta_\varepsilon$. Since $f=t_\mu$ is clearly a solution of (5.122), we may assume that $\zeta^{n+\mu}_j(z,t)=t_\mu$ for $\mu=1,\dots,m$. Then we have

$$\zeta_j\colon (z,t)\to(\zeta^1_j(z,t),\dots,\zeta^{n}_j(z,t),t_1,\dots,t_m).$$

Therefore

$$\varpi\colon (\zeta^1_j(z,t),\dots,\zeta^{n}_j(z,t),t_1,\dots,t_m)\to(t_1,\dots,t_m)$$

is a holomorphic map of $\mathcal{M}$ onto Δ_ε. For each $t\in\Delta_\varepsilon$, $\varpi^{-1}(t)$ is a complex manifold whose system of local complex coordinates is given by

$\{(\zeta_j^1(z,t),\ldots,\zeta_j^n(z,t)\}$. Since $f=\zeta_j^\beta(z,t)$, $\beta=1,\ldots,n$, are linearly independent solutions of the equation $(\bar\partial-\varphi(t))f=0$ on U_j, $\varpi^{-1}(t)=M_{\varphi(t)}$. Thus $\{M_{\varphi(t)}\,|\,t\in\Delta_\varepsilon\}$ forms a complex analytic family $(\mathcal{M},\Delta_\varepsilon,\varpi)$.

Thus Theorem 5.6 (Theorem of Existence, [18]) is proved. Since each component $\zeta_j^\beta(z,t)$ of the local complex coordinates $(\zeta_j^1(z,t),\ldots,\zeta_j^n(z,t))$ of $M_{\varphi(t)}$ is a solution of (5.122), $\partial\zeta_j^\beta(z,t)/\partial\bar t_\mu=0$ for $\mu=1,\ldots,m$. Therefore C^∞ *functions* $\zeta_j^\beta(z,t)$ of $z^1,\ldots,z^n,t_1,\ldots,t_m$ *are holomorphic* in $t_1,\ldots,t_m$.

About the same time with us, Grauert proved this theorem of existence by an elementary method. He explained his remarkable idea at "Nothing Seminar" under the direction of D. C. Spencer. The main point of his idea is in replacing the fundamental equalities (5.62):

$$f_{jk}^\alpha(z_k,t)=f_{ij}^\alpha(f_{jk}(z_k,t),t)$$

by the inequalities

$$|f_{jk}^\alpha(z_k,t)-f_{ij}^\alpha(f_{jk}(z_k,t),t)|<\varepsilon.$$

By this method he succeeded in proving the convergence. Later he proved the most general theorem of existence of deformations for complex analytic spaces based on this idea ([9]).

Later another proof of Theorem 5.6 by an elementary method was given by Forster and Knorr ([5]).

We note that the theorem of completeness (p. 230) was not known at that stage of the development of the theory of deformations. However, since the proof of the theorem of completeness is elementary and independent of the existence theorem which is the main theme of this chapter, in verifying the equality $m=m(M)$ for concrete examples of M (pp. 230–248), we replaced for simplicity our original argument ([20], §14) by one using the theorem of completeness.

Chapter 6

Theorem of Completeness

§6.1. Theorem of Completeness

In this section we shall prove the theorem of completeness stated in §5.2(b).

Let $(\mathcal{M}, B, \varpi)$ be a complex analytic family, B a domain of $\mathbb{C}^n$ containing 0, and $\rho_t: T_t(B) \to H^1(M_t, \Theta_t)$, $M_t = \varpi^{-1}(t)$, the linear map defined by (4.22):

$$\rho_t: \frac{\partial}{\partial t} \to \rho_t\left(\frac{\partial}{\partial t}\right) = \frac{\partial M_t}{\partial t}.$$

Theorem 6.1 (Theorem of Completeness, [20]). *If $\rho_0: T_0(B) \to H^1(M_0, \Theta_0)$ is surjective, the complex analytic family $(\mathcal{M}, B, \varpi)$ is complete at $0 \in B$.*

Recall that we mean by the completeness of $(\mathcal{M}, B, \varpi)$ at 0 that given any complex analytic family $(\mathcal{N}, D, \pi)$ such that $0 \in D \subset \mathbb{C}^l$ and $\pi^{-1}(0) = M_0$, if we take a sufficiently small subdomain Δ with $0 \in \Delta \subset D$, and let $\mathcal{N}_\Delta = \pi^{-1}(\Delta)$, we can find a holomorphic map $h: s \to t = h(s)$, $h(0) = 0$, of Δ into B such that $(\mathcal{N}_\Delta, \Delta, \pi)$ is the complex analytic family induced from $(\mathcal{M}, B, \varpi)$ by h (see p. 228, Definition 5.2).

If $(\mathcal{N}_\Delta, \Delta, \pi)$ is the complex analytic family induced from $(\mathcal{M}, B, \varpi)$ by h, then for each $s \in \Delta$, $N_s = \pi^{-1}(s) = M_{h(s)} \times s$, and

$$\mathcal{N}_\Delta = \bigcup_{s \in \Delta} M_{h(s)} \times s \subset \mathcal{M} \times \Delta$$

is a submanifold of $\mathcal{M} \times \Delta$ (see p. 206). Let g be the restriction of the projection $\mathcal{M} \times \Delta \to \mathcal{M}$ to $\mathcal{N}_\Delta$. Then g is obviously a holomorphic map of $\mathcal{N}_\Delta$ into $\mathcal{M}$. Moreover g maps each N_s biholomorphically onto $M_{h(s)}$. For, writing a point of $N_s = M_{h(s)} \times s$ as (p, s), $p \in M_{h(s)}$, we have $g(p, s) = p$. Identifying $(p, 0) \in N_0 = M_0 \times 0$ with $p \in M_0$, we may consider $\pi^{-1}(0) = N_0 = M_0$. Then, if we denote the identity map $(p, 0) \to p$ by g_0, $g: \mathcal{N}_\Delta \to \mathcal{M}$ is an extension of $g_0: N_0 \to M_0$. Conversely, the following lemma holds.

Lemma 6.1. *If we can extend the identity $g_0: N_0 = M_0 \to M_0$ to a holomorphic map $g: \mathcal{N}_\Delta \to \mathcal{M}$ such that g maps each N_s, $s \in \Delta$, biholomorphically onto $M_{h(s)}$, then $(\mathcal{N}_\Delta, \Delta, \pi)$ is the complex analytic family induced from $(\mathcal{M}, B, \varpi)$ by h.*

Proof. Let $(\hat{\mathcal{N}}, \Delta, \hat{\pi})$ be the complex analytic family induced from $(\mathcal{M}, B, \varpi)$ by $h: \Delta \to B$. Then for each $s \in \Delta$, $\hat{\pi}^{-1}(s) = M_{h(s)} \times s$ and

$$\hat{\mathcal{N}} = \bigcup_{s\in\Delta} M_{h(s)} \times s \subset \mathcal{M} \times \Delta$$

is a submanifold of $\mathcal{M} \times \Delta$. Denoting a point of $\mathcal{N}_\Delta$ by q, we consider the holomorphic map $\Phi: q \to \Phi(q) = (g(q), \pi(q))$ of $\mathcal{N}_\Delta$ into $\mathcal{M} \times \Delta$. Since g maps each N_s biholomorphically onto $M_{h(s)}$, Φ maps N_s biholomorphically onto

$$\Phi(N_s) = M_{h(s)} \times s.$$

Hence Φ maps $\mathcal{N}_\Delta = \bigcup_{s\in\Delta} N_s$ biholomorphically onto $\hat{\mathcal{N}} = \bigcup_{s\in\Delta} M_{h(s)} \times s$. Moreover

$$\hat{\pi}\Phi(N_s) = \hat{\pi}(M_{h(s)} \times s) = s = \pi(N_s)$$

holds, i.e. $\pi = \hat{\pi}\Phi$. Hence $(\mathcal{N}_\Delta, \Delta, \pi)$ and $(\hat{\mathcal{N}}, \Delta, \hat{\pi})$ are biholomorphically equivalent (see p. 61), which proves that $(\mathcal{N}_\Delta, \Delta, \pi)$ is the complex analytic family induced from $(\mathcal{M}, B, \varpi)$ by h. ∎

By this lemma, in order to prove Theorem 6.1, it suffices to show that for any given complex analytic family $(\mathcal{N}, D, \pi)$ with $\pi^{-1}(0) = M_0$, if we take a sufficiently small domain Δ with $0 \in \Delta \subset D$, we can construct a holomorphic map $h: s \to t = h(s)$, $h(0) = 0$, of Δ into B, and a holomorphic map g of $\mathcal{N}_\Delta = \pi^{-1}(\Delta)$ into $\mathcal{M}$ satisfying the following condition: g is an extension of the identity $g_0: \pi^{-1}(0) = M_0 \to M_0$, and g maps each $N_s = \pi^{-1}(s)$ biholomorphically onto $M_{h(s)}$. In what follows we construct such h and g by an elementary method.

(a) Construction of Formal Power Series

In this subsection, for $t = (t_1, \ldots, t_m) \in \mathbb{C}^m$, $z = (z^1, \ldots, z^n) \in \mathbb{C}^n$, etc., we define $|t| = \max_\lambda |t_\lambda|$, $|z| = \max_\alpha |z^\alpha|$, etc. Since the problem is local with respect to B, we may assume that B is a polydisk: $B = \{t \in \mathbb{C}^m \,\|\, |t| < 1\}$, $\mathcal{M}$ is written in the form (4.27), namely,

$$\mathcal{M} = \bigcup_j \mathcal{U}_j, \qquad \mathcal{U}_j = \{(\zeta_j, t) \in \mathbb{C}^n \times B \,\|\, |\zeta_j| < 1\},$$

and $\varpi(\zeta_j, t) = t$. Of course $\bigcup_j$ is a finite union. If $U_j \cap U_k \neq \varnothing$, (ζ_j, t) and (ζ_k, t) are the same point of $\mathcal{M}$ if

$$\zeta_j = g_{jk}(\zeta_k, t) = (g^1_{jk}(\zeta_k, t), \ldots, g^n_{jk}(\zeta_k, t)), \tag{6.1}$$

where the $g^\alpha_{jk}(\zeta_k, t)$, $\alpha = 1, \ldots, n$, are holomorphic functions on $U_j \cap U_k$.

Similarly we assume that $D=\{s\in\mathbb{C}^l \,|\, |s|<1\}$,

$$\mathcal{N}=\bigcup_j \mathcal{W}_j, \qquad \mathcal{W}_j=\{(z_j,s)\in\mathbb{C}^n\times D \,|\, |z_j|<1\},$$

$\pi(z_j,s)=s$, and that for $W_j\cap W_k\neq\emptyset$, (z_j,s) and (z_k,s) are the same point of $\mathcal{N}$ if

$$z_j=f_{jk}(z_k,s)=(f_{jk}^1(z_k,s),\ldots,f_{jk}^n(z_k,s)). \tag{6.2}$$

Moreover, since by hypothesis, $N_0=M_0$, we assume $\mathcal{W}_j\cap N_0=\mathcal{U}_j\cap M_0$ and that *the local coordinates* $(\zeta_j,0)$ *and* $(z_j,0)$ *coincide on* $\mathcal{W}_j\cap N_0=\mathcal{U}_j\cap M_0$, namely, if $\zeta_j^1=z_j^1,\ldots,\zeta_j^n=z_j^n$, $(\zeta_j,0)$ and $(z_j,0)$ are the same point of $\mathcal{W}_j\cap N_0=\mathcal{U}_j\cap M_0$. Putting

$$b_{jk}(z_k)=f_{jk}(z_k,0), \tag{6.3}$$

from (6.1) and (6.2), we have

$$b_{jk}(\zeta_k)=g_{jk}(\zeta_k,0). \tag{6.4}$$

Thus we have

$$N_0=M_0=\bigcup_j U_j, \qquad U_j=\mathcal{W}_j\cap N_0=\mathcal{U}_j\cap M_0, \tag{6.5}$$

and $\{z_j\}$, $z_j=(z_j^1,\ldots,z_j^n)$, is a system of local complex coordinates of the complex manifold $N_0=M_0$ with respect to $\{U_j\}$. The coordinate transformation on $U_j\cap U_k$ is given by

$$z_j^\alpha=b_{jk}^\alpha(z_k), \qquad \alpha=1,\ldots,n.$$

Our purpose is to define a holomorphic map h: $s\to t=h(s)$ with $h(0)=0$ of $\Delta_\varepsilon=\{s\in D \,|\, |s|<\varepsilon\}$ into B for a sufficiently small $\varepsilon>0$, and at the same time to extend the identity g_0: $N_0\to M_0=N_0$ to a holomorphic map g: $\pi^{-1}(\Delta_\varepsilon)\to\mathcal{M}$ such that $\varpi\circ g=h\circ\pi$. The condition $\varpi\circ g=h\circ\pi$ means that the holomorphic map g maps each $N_s=\pi^{-1}(s)$, $s\in\Delta_\varepsilon$, into $M_{h(s)}=\varpi^{-1}(h(s))$. But then, since g is an extension of the identity g_0, it is easy to see that, if we take a sufficiently small $\delta<\varepsilon$, *g maps N_s biholomorphically onto $M_{h(s)}$* for $|s|<\delta$. Replacing ε by such δ, we may assume that g maps N_s biholomorphically onto $M_{h(s)}$ for $s\in\Delta_\varepsilon$. In what follows, we write $\Delta=\Delta_\varepsilon$ and $\mathcal{N}_\Delta=\pi^{-1}(\Delta)$ for simplicity.

Suppose for the moment that we can construct such h and g. Then since $g(U_j)=U_j\subset\mathcal{U}_j$ for each $U_j\subset N_0$, $g^{-1}(\mathcal{U}_j)\subset\mathcal{N}_\Delta$ is an open set containing U_j. Therefore we can choose a continuous function $\varepsilon(z_j)$ of z_j on U_j with $0<\varepsilon(z_j)\leqq\varepsilon$ such that the subdomain

$$\mathcal{W}_j^*=\{(z_j,s) \,|\, |z_j|<1, |s|<\varepsilon(z_j)\}$$

of $\mathscr{W}_j$ is contained in $g^{-1}(\mathscr{U}_j)$:

$$U_j \subset \mathscr{W}_j^* \subset g^{-1}(\mathscr{U}_j) \cap \mathscr{W}_j.$$

Since the holomorphic map g maps $\mathscr{W}_j^*$ into $\mathscr{U}_j$, g can be written on $\mathscr{W}_j^*$ in the form

$$g: (z_j, s) \to (\zeta_j, t) = (g_j(z_j, s), h(s)), \qquad (z_j, s) \in \mathscr{W}_j^*, \tag{6.6}$$

where each component $g_j^\alpha(z_j, s)$ of $g_j(z_j, s) = (g_j^1(z_j, s), \ldots, g_j^n(z_j, s))$ is a holomorphic function of $z_j^1, \ldots, z_j^n, s_1, \ldots, s_l$ defined on $\mathscr{W}_j^*$. Since

$$g: (z_j, 0) \to (\zeta_j, 0) = (z_j, 0)$$

is the identity, we have

$$g_j(z_j, 0) = z_j, \qquad h(0) = 0. \tag{6.7}$$

Expanding $g_j(z_j, s)$ into power series of $s_1, \ldots, s_l$, we write it in the form

$$g_j(z_j, s) = z_j + \sum_{\nu=1}^{\infty} g_{j|\nu}(z_j, s), \tag{6.8}$$

where, in the notation of § 5.3(b),

$$g_{j|\nu}(z_j, s) = \sum_{\nu_1 + \cdots + \nu_l = \nu} g_{j\nu_1\cdots\nu_l}(z_j) s_1^{\nu_1} \cdots s_l^{\nu_l}$$

is a homogeneous polynomial of degree ν of $s_1, \ldots, s_l$, and each component $g_{j\nu_1\cdots\nu_l}^\alpha(z_j)$, $\alpha = 1, \ldots, n$, of the coefficient

$$g_{j\nu_1\cdots\nu_l}(z_j) = (g_{j\nu_1\cdots\nu_l}^1(z_j), \ldots, g_{j\nu_1\cdots\nu_n}^n(z_j))$$

is *a holomorphic function of* $z_j^1, \ldots, z_j^n$ *defined on* U_j. For simplicity we express such situation as $g_{j\nu_1\cdots\nu_n}(z_j)$ is *a vector-valued holomorphic function defined on* U_j. In general the region of convergence of the power series $g_j(z_j, s)$ of $s_1, \ldots, s_l$ may depend on z_j, but *they all converge for* $|s| < \varepsilon(z_j)$. Expanding $h(s)$ into power series, we put

$$h(s) = \sum_{\nu=1}^{\infty} h_\nu(s). \tag{6.9}$$

For $\mathscr{W}_j^* \cap \mathscr{W}_k^* \neq \varnothing$, if $(z_j, s) \in \mathscr{W}_j^*$ and $(z_k, s) \in \mathscr{W}_k^*$ are the same point, then their images by g, namely, $(\zeta_j, t) = (g_j(z_j, s), h(s))$ and $(\zeta_k, t) = (g_k(z_k, s), h(s))$ are, of course, the same point. Hence by (6.2) and (6.1), if $z_j = f_{jk}(z_k, s)$, then we have $g_j(z_j, s) = g_{jk}(g_k(z_k, s), h(s))$. In other words, for

$(z_k, s) \in \mathcal{W}_k^* \cap \mathcal{W}_j^*$, the equality

$$g_j(f_{jk}(z_k, s), s) = g_{jk}(g_k(z_k, s), h(s)) \tag{6.10}$$

holds. If we expand both sides of this equality into power series of $s_1, \ldots, s_l$, their coefficients are holomorphic functions on $U_j \cap U_k$. Hence, from (6.10) *we obtain equalities between these holomorphic functions defined on* $U_j \cap U_k$. We note that, if

$$g_j(z_j, s) = z_j + \sum_{\nu=1}^{\infty} g_{j|\nu}(z_k, s), \qquad g_k(z_k, s) = z_k + \sum_{\nu=1}^{\infty} g_{k|\nu}(z_k, s),$$

$$\textit{and} \quad h(s) = \sum_{\nu=1}^{\infty} h_\nu(s)$$

are all considered to be formal power series of $s_1, \ldots, s_l$, *the equality* (6.10) *still makes sense.* For, by writing down (6.10) as

$$f_{jk}(z, s) + \sum_{\nu=1}^{\infty} g_{j|\nu}(f_{jk}(z_k, s), s) = g_{jk}\left(z_k + \sum_{\nu=1}^{\infty} g_{k|\nu}(z_k, s), \sum_{\nu=1}^{\infty} h_\nu(s)\right),$$

we can obviously see that, for each $\mu = 1, 2, 3, \ldots$, the terms of degree at most μ of the power series expansions of both sides of (6.10) contains no $g_{j|\nu}(\ , s)$, $g_{k|\nu}(\ , s)$, $h_\nu(s)$ with $\nu > \mu$.

Thus assuming that we obtain holomorphic maps h and g with the required properties, and representing g on each domain $\mathcal{W}_j^*$ in the form (6.6): $(z_j, s) \to (g_j(z_j, s), h(s))$, we have seen that the quality (6.10) holds on each $U_j \cap U_k \neq \emptyset$.

In order to obtain such h and g, *we begin with constructing formal power series* $h(s)$, *as well as formal power series* $g_j(z_j, s)$, *for each* U_j, *whose coefficients are vector-valued holomorphic functions on* U_j *such that the equality* (6.10) *holds.*

In order to avoid confusion, we use $\alpha, \beta, \ldots$ to denote the indices of coordinates and use $\mu, \nu, \ldots$ to denote the degrees. Furthermore, we use the notation of §5.3(b) for power series of $s_1, \ldots, s_l$. In particular, we write

$$h^\nu(s) = h_1(s) + \cdots + h_\nu(s),$$

$$g_j^\nu(s) = z_j + g_{j|1}(z_j, s) + \cdots + g_{j|\nu}(z_j, s).$$

The equality (6.10) is equivalent to the following system of the infinitely many congruences:

$$g_j^\nu(f_{jk}(z_k, s), s) \underset{\nu}{\equiv} g_{jk}(g_k^\nu(z_k, s), h^\nu(s)), \qquad \nu = 0, 1, 2, \ldots, \tag{6.11}_\nu$$

where we indicate by $\equiv_\nu$ that the power series expansions of both sides of $(6.11)_\nu$ coincide up to the term of degree ν. $(6.11)_0$ means $f_{jk}(z_k, 0) = g_{jk}(z_k, 0)$, which clearly holds by (6.3) and (6.4). Now we want to construct $h^\nu(s)$, $g_j^\nu(z_j, s)$ by induction on ν so that $(6.11)_\nu$ hold on each $U_j \cap U_k \neq \varnothing$. Suppose that $h^{\nu-1}(s)$ and $g_j^{\nu-1}(z_j, s)$ are already constructed in such a manner that, for each $U_j \cap U_k \neq \varnothing$,

$$g_j^{\nu-1}(f_{jk}(z_k, s), s) \underset{\nu-1}{\equiv} g_{jk}(g_k^{\nu-1}(z_j, s), h^{\nu-1}(s)) \qquad (6.11)_{\nu-1}$$

holds. Since $g_j^\nu(z_j, s) = g_j^{\nu-1}(z_j, s) + g_{j|\nu}(z_j, s)$, the left-hand side of $(6.11)_\nu$ becomes

$$g_j^\nu(f_{jk}(z_k, s), s) = g_j^{\nu-1}(f_{jk}(z_k, s), s) + g_{j|\nu}(f_{jk}(z_k, s), s),$$

where $g_{j|\nu}(z_j, s)$ is a homogeneous polynomial of degree ν of $s_1, \ldots, s_l$, and $f_{jk}(z_k, 0) = b_{jk}(z_k)$ by (6.3). Therefore we have

$$g_{j|\nu}(f_{jk}(z_k, s), s) \underset{\nu}{\equiv} g_{j|\nu}(b_{jk}(z_k), s),$$

hence, putting $z_j = b_{jk}(z_k)$, we obtain

$$g_j^\nu(f_{jk}(z_k, s), s) \underset{\nu}{\equiv} g_j^{\nu-1}(f_{jk}(z_k, s), s) + g_{j|\nu}(z_j, s). \qquad (6.12)$$

The right-hand side of $(6.11)_\nu$ is given by

$$g_{jk}(g_k^\nu(z_k, s), h^\nu(s)) = g_{jk}(g_k^{\nu-1}(z_k, s) + g_{k|\nu}(z_k, s), h^{\nu-1}(s) + h_\nu(s)).$$

By expanding $g_{jk}(\zeta_k + \xi, t + u)$ into power series of $\xi^1, \ldots, \xi^n, u_1, \ldots, u_m$, we obtain

$$g_{jk}(\zeta_k + \xi, t + u) = g_{jk}(\zeta_k, t) + \sum_{\beta=1}^{n} \frac{\partial g_{jk}}{\partial \zeta_k^\beta}(\zeta_k, t) \cdot \xi^\beta + \sum_{r=1}^{m} \frac{\partial g_{jk}}{\partial t_r}(\zeta_k, t) \cdot u_r + \cdots,$$

where ... denotes the terms of degree $\geqq 2$ in $\xi^1, \ldots, \xi^n, u_1, \ldots, u_m$. Letting $h_\nu(s) = (h_{1|\nu}(s), \ldots, h_{m|\nu}(s))$, and writing $g_k^{\nu-1}(s)$, $g_{k|\nu}(s)$ instead of $g_k^{\nu-1}(z_k, s)$, $g_{k|\nu}(z_k, s)$, respectively, we have

$$\begin{aligned} g_{jk}(g_k^\nu(z_k, s), h^\nu(s)) &- g_{jk}(g_k^{\nu-1}(z_k, s), h^{\nu-1}(s)) \\ &\underset{\nu}{\equiv} \sum_{\beta=1}^{n} \frac{\partial g_{jk}}{\partial \zeta_k^\beta}(g_k^{\nu-1}(s), h^{\nu-1}(s)) \cdot g_{k|\nu}^\beta(s) \\ &\quad + \sum_{r=1}^{m} \frac{\partial g_{jk}}{\partial t_r}(g_k^{\nu-1}(s), h^{\nu-1}(s)) \cdot h_{r|\nu}(s) \end{aligned}$$

$$\equiv_{\nu} \sum_{\beta=1}^{n} \frac{\partial g_{jk}}{\partial z_k^\beta}(g_k^{\nu-1}(0), h^{\nu-1}(0)) \cdot g_{k|\nu}^\beta(s)$$

$$+ \sum_{r=1}^{m} \frac{\partial g_{jk}}{\partial t_r}(g_k^{\nu-1}(0), h^{\nu-1}(0)) \cdot h_{r|\nu}(s)$$

$$= \sum_{\beta=1}^{n} \frac{\partial g_{jk}}{\partial \zeta_k^\beta}(z_k, 0) \cdot g_{k|\nu}^\beta(s) + \sum_{r=1}^{m} \left(\frac{\partial g_{jk}(z_k, t)}{\partial t_r}\right)_{t=0} \cdot h_{r|\nu}(s).$$

Therefore, since $g_{jk}(z_k, 0) = b_{jk}(z_k) = z_j$ by (6.14), we have

$$g_{jk}(g_k^\nu(z_k, s), h^\nu(s)) \equiv_{\nu} g_{jk}(g_k^{\nu-1}(z_k, s), h^{\nu-1}(s))$$

$$+ \sum_{\beta=1}^{n} \frac{\partial z_j}{\partial z_k^\beta} \cdot g_{k|\nu}^\beta(z_k, s) + \sum_{r=1}^{m} \left(\frac{g_{jk}(z_k, t)}{\partial t_r}\right)_{t=0} \cdot h_{r|\nu}(s).$$

From this and (6.12), it follows that the congruence $(6.11)_\nu$ is equivalent to the following:

$$g_j^{\nu-1}(f_{jk}(z_k, s), s) - g_{jk}(g_k^{\nu-1}(z_k, s), h^{\nu-1}(s))$$

$$\equiv_{\nu} \sum_{\beta} \frac{\partial z_j}{\partial z_k^\beta} g_{k|\nu}^\beta(z_k, s) - g_{j|\nu}(z_j, s) + \sum_{r} \left(\frac{\partial g_{jk}(z_k, t)}{\partial t_r}\right)_{t=0} \cdot h_{r|\nu}(s).$$

By the hypothesis of induction, the left-hand side of this congruence $\equiv_{\nu-1} 0$. Hence, if we let $\Gamma_{jk|\nu}$ denote the sum of the terms of degree ν of the left-hand side, we obtain

$$\Gamma_{jk|\nu}(z_j, s) \equiv_{\nu} g_j^{\nu-1}(f_{jk}(z_k, s), s) - g_{jk}(g_k^{\nu-1}(z_k, s), h^{\nu-1}(s)). \tag{6.13}$$

Hence $(6.11)_\nu$ is equivalent to the following:

$$\Gamma_{jk|\nu}(z_j, s) = \sum_{r=1}^{m} \left(\frac{\partial g_{jk}(z_k, t)}{\partial t_r}\right)_{t=0} \cdot h_{r|\nu}(s)$$

$$+ \sum_{\beta=1}^{n} \frac{\partial z_j}{\partial z_k^\beta} \cdot g_{k|\nu}^\beta(z_k, s) - g_{j|\nu}(z_j, s), \qquad (6.14)_\nu$$

where z_k and $z_j = b_{jk}(z_k)$ are the local coordinates of the same point of N_0.

In order to clarify the meaning of these equations, we introduce holomorphic vector fields as follows:

$$\theta_{rjk} = \sum_{\alpha=1}^{n} \left(\frac{\partial g_{jk}^{\alpha}(z_k, t)}{\partial t_r} \right)_{t=0} \frac{\partial}{\partial z_j^{\alpha}}, \qquad z_k = b_{kj}(z_j),$$

$$\Gamma_{jk|\nu}(s) = \sum_{\alpha=1}^{n} \Gamma_{jk|\nu}^{\alpha}(z_j, s) \frac{\partial}{\partial z_j^{\alpha}},$$

$$g_{k|\nu}(s) = \sum_{\beta=1}^{n} g_{k|\nu}^{\beta}(z_k, s) \frac{\partial}{\partial z_k^{\beta}}.$$

Then in these terms $(6.14)_\nu$ is written in the form

$$\Gamma_{jk|\nu}(s) = \sum_{r=1}^{m} h_{r|\nu}(s)\theta_{rjk} + g_{k|\nu}(s) - g_{j|\nu}(s). \qquad (6.15)_\nu$$

By (6.5), $\mathfrak{U} = \{U_j\}$ is a finite covering of $M_0 = N_0$. Since we assume that $\zeta_j^{\alpha} = z_j^{\alpha}$, $\alpha = 1, \ldots, n$, the 1-cocycle $\{\theta_{rjk}\} \in Z^1(\mathfrak{U}, \Theta)$ represents the infinitesimal deformation $\theta_r = \rho_0(\partial/\partial t_r) \in H^1(M_0, \Theta_0)$. Since the coefficients $\Gamma_{jk\nu_1\cdots\nu_l}$ of the homogeneous polynomial

$$\Gamma_{jk|\nu}(s) = \sum_{\nu_1+\cdots+\nu_l=\nu} \Gamma_{jk\nu_1\cdots\nu_l}\, s_1^{\nu_1} \cdots s_l^{\nu_l}$$

are holomorphic vector fields on $U_j \cap U_k$,

$$\{\Gamma_{jk|\nu}(s)\} = \sum_{\nu_1+\cdots+\nu_l=\nu} \{\Gamma_{jk\nu_1\cdots\nu_l}\} s_1^{\nu_1} \cdots s_l^{\nu_l}$$

is a homogeneous polynomial of degree ν whose coefficients are 1-cochains $\{\Gamma_{jk\nu_1\cdots\nu_l}\} \in C^1(\mathfrak{U}, \Theta_0)$. Similarly

$$\{g_{j|\nu}(s)\} = \sum_{\nu_1+\cdots+\nu_l=\nu} \{g_{j\nu_1\cdots\nu_l}\} S_1^{\nu_1} \cdots s_l^{\nu_l}$$

is a homogeneous polynomial of degree ν whose coefficients are 0-cochains $\{g_{j\nu_1\cdots\nu_l}\} \in C^0(\mathfrak{U}, \Theta_0)$. Under these circumstances, the equation $(6.15)_\nu$ is written in the form

$$\{\Gamma_{jk|\nu}(s)\} = \sum_{r=1}^{m} h_{r|\nu}(s)\{\theta_{rjk}\} + \delta\{g_{j|\nu}(s)\}. \qquad (6.16)_\nu$$

Thus in order to construct $h^{\nu}(s) = h^{\nu-1}(s) + h_\nu(s)$, $g_j^{\nu}(z_j, s) = g_j^{\nu-1}(z_j, s) + g_{j|\nu}(z_j, s)$ so that $(6.11)_\nu$ hold, it suffices to obtain solutions $h_{r|\nu}(s)$, $r = 1, \ldots, m$, $\{g_{j|\nu}(s)\}$ of the equations $(6.16)_\nu$.

If solutions $h_{r|\nu}(s)$, $r = 1, \ldots, m$, $\{g_{j|\nu}(s)\}$ exist, then, since the right-hand side of $(6.16)_\nu$ is a 1-cocycle on $\mathfrak{U}$, $\{\Gamma_{jk|\nu}(s)\}$ is also a 1-cocycle, i.e., for each $U_i \cap U_j \cap U_k \neq \emptyset$, we have

$$\Gamma_{jk|\nu}(s) - \Gamma_{ik|\nu}(s) + \Gamma_{ij|\nu}(s) = 0. \tag{6.17}$$

Conversely, *if* $\{\Gamma_{jk|\nu}(s)\}$ *forms a* 1*-cocycle, then* (6.16)$_\nu$ *has solutions* $h_{r|\nu}(s)$, $r = 1, \ldots, m$, $\{g_{j|\nu}(s)\}$. *Proof.* Let

$$h_{r|\nu}(s) = \sum_{\nu_1 + \cdots + \nu_l = \nu} h_{r\nu_1 \cdots \nu_l} s_1^{\nu_1} \cdots s_l^{\nu_l}.$$

Then (6.16)$_\nu$ can be written as

$$\{\Gamma_{jk\nu_1 \cdots \nu_l}\} = \sum_{r=1}^{m} h_{r\nu_1 \cdots \nu_l} \{\theta_{rjk}\} + \delta\{g_{j\nu_1 \cdots \nu_l}\}.$$

Therefore it suffices to prove that any 1-cocycle $\{\Gamma_{jk}\} \in Z^1(\mathfrak{U}, \Theta_0)$ can be written in the form

$$\{\Gamma_{jk}\} = \sum_{r=1}^{m} h_r \{\theta_{rjk}\} + \delta\{g_j\}, \qquad h_r \in \mathbb{C}, \{g_j\} \in C^0(\mathfrak{U}, \Theta_0). \tag{6.18}$$

Let $\gamma \in H^1(M_0, \Theta_0)$ be the cohomology class of the 1-cocycle $\{\Gamma_{jk}\}$. Since by hypothesis, ρ_0: $T_0(B) \to H^1(M_0, \Theta_0)$ is surjective, γ is written in the form of a linear combination of the θ_r as

$$\gamma = \sum_{r=1}^{m} h_r \theta_r, \qquad h_r \in \mathbb{C}.$$

Consequently the cohomology class of the 1-cocycle $\{\Gamma_{jk}\} - \{\sum_{r=1}^{m} h_r \theta_{rjk}\}$ on $\mathfrak{U}$ is 0. On the other hand, by Theorem 3.4 (p. 121),

$$\Pi^U: H^1(\mathfrak{U}, \Theta_0) = Z^1(\mathfrak{U}, \Theta_0)/\delta C^0(\mathfrak{U}, \Theta_0) \to H^1(M_0, \Theta_0)$$

is injective. Therefore $\{\Gamma_{jk}\} - \sum_{r=1}^{m} h_r \{\theta_{rjk}\} \in \delta C^0(\mathfrak{U}, \Theta_0)$, hence $\{\Gamma_{jk}\}$ can be written in the form (6.18). ∎

Next we shall prove that $\{\Gamma_{jk|\nu}(s)\}$ is a 1-cocycle, i.e., (6.17) holds. *Proof.* By representing each term of (6.17) as vector-valued holomorphic functions, we get

$$\Gamma_{ik|\nu}(z_i, s) = \Gamma_{ij|\nu}(z_i, s) + \sum_{\beta=1}^{n} \frac{\partial z_i}{\partial z_j^\beta} \Gamma_{jk|\nu}^\beta(z_j, s), \qquad z_i = b_{ij}(z_j).$$

Here z_j and $z_i = b_{ij}(z_j)$ are the same point of M_0. Using matrix notation, we write this in the form

$$\Gamma_{ik|\nu}(z_i, s) = \Gamma_{ij|\nu}(z_i, s) + B_{ij}(z_j)\Gamma_{jk|\nu}(z_j, s), \tag{6.19}$$

where $B_{ij}(z_j) = (\partial z_i^\alpha / \partial z_j^\beta)_{\alpha,\beta=1,\ldots,n}$. Writing $f_{ik}(s)$ for $f_{ik}(z_k, s)$, and $g_k^{\nu-1}(s)$ for $g_k^{\nu-1}(z_k, s)$ for simplicity, we obtain from (6.13),

$$\Gamma_{ik|\nu}(z_i, s) \underset{\nu}{\equiv} g_i^{\nu-1}(f_{ik}(s), s) - g_{ik}(g_k^{\nu-1}(s), h^{\nu-1}(s)). \tag{6.20}$$

Since on $U_k \cap U_i \cap U_j \neq \varnothing$, $g_{ik}(\zeta_k, t) = g_{ij}((\zeta_k, t), t)$, and since by (6.13)

$$g_{jk}(g_k^{\nu-1}(s), h^{\nu-1}(s)) \underset{\nu}{\equiv} g_j^{\nu-1}(f_{jk}(s), s) - \Gamma_{jk|\nu}(z_j, s),$$

the second term of the right-hand side of (6.20) is written down as

$$\begin{aligned} g_{ik}(g_k^{\nu-1}(s), h^{\nu-1}(s)) &= g_{ij}(g_{jk}(g_k^{\nu-1}(s), h^{\nu-1}(s)), h^{\nu-1}(s)) \\ &\underset{\nu}{\equiv} g_{ij}(g_j^{\nu-1}(f_{jk}(s), s) - \Gamma_{jk|\nu}(z_j, s), h^{\nu-1}(s)) \\ &\underset{\nu}{\equiv} g_{ij}(g_j^{\nu-1}(f_{jk}(s), s), h^{\nu-1}(s)) \\ &\quad - \sum_{\beta=1}^{n} \frac{\partial g_{ij}}{\partial \zeta_j^\beta}(g_j^{\nu-1}(f_{jk}(s), s), h^{\nu-1}(s))\, \Gamma_{jk|\nu}^\beta(z_j, s). \end{aligned}$$

But, since $g_j^{\nu-1}(f_{jk}(z_k, 0), 0) = b_{jk}(z_k) = z_j$, we have

$$\frac{\partial g_{ij}}{\partial \zeta_j^\beta}(g_j^{\nu-1}(f_{jk}(0), 0), 0) = \frac{\partial g_{ij}}{\partial \zeta_j^\beta}(z_j, 0) = \frac{\partial b_{ij}}{\partial z_j^\beta} = \frac{\partial z_i}{\partial z_j^\beta},$$

hence, we have

$$\begin{aligned} g_{ik}(g_k^{\nu-1}(s), h^{\nu-1}(s)) \underset{\nu}{\equiv} & \; g_{ij}(g_j^{\nu-1}(f_{jk}(s), s), h^{\nu-1}(s)) \\ & - \sum_{\beta=1}^{n} \frac{\partial z_i}{\partial z_j^\beta} \Gamma_{jk|\nu}^\beta(z_j, s). \end{aligned}$$

Therefore by (6.20) we obtain

$$\begin{aligned} \Gamma_{ik|\nu}(z_i, s) \underset{\nu}{\equiv} & \; g_i^{\nu-1}(f_{ik}(s), s) - g_{ij}(g_j^{\nu-1}(f_{jk}(s), s), h^{\nu-1}(s)) \\ & + B_{ij}(z_j)\Gamma_{jk|\nu}(z_j, s). \end{aligned}$$

Thus, to prove (6.19) it suffices to verify

$$\Gamma_{ij|\nu}(z_i, s) \underset{\nu}{\equiv} g_i^{\nu-1}(f_{ik}(s), s) - g_{ij}(g_j^{\nu-1}(f_{jk}(s), s), h^{\nu-1}(s)). \tag{6.21}$$

By (6.13) we get

$$\Gamma_{ij|\nu}(b_{ij}(z_j), s) \underset{\nu}{\equiv} g_i^{\nu-1}(f_{ij}(z_j, s), s) - g_{ij}(g_j^{\nu-1}(z_j, s), h^{\nu-1}(s)).$$

Substituting $z_j = f_{jk}(z_k, s)$ and noting $b_{ij}(f_{jk}(z_k, 0)) = b_{ij}(b_{jk}(z_k)) = b_{ij}(z_j) = z_i$, we have

$$\begin{aligned}\Gamma_{ij|\nu}(b_{ij}(f_{jk}(z_k, s)), s) &\underset{\nu}{\equiv} \Gamma_{ij|\nu}(b_{ij}(f_{jk}(z_k, 0)), s)\\ &= \Gamma_{ij|\nu}(z_i, s).\end{aligned}$$

In view of the equality $f_{ij}(f_{jk}(z_k, s)) = f_{ik}(z_k, s)$, (6.21) holds. ∎

Therefore $\{\Gamma_{jk|\nu}(s)\}$ is a 1-cocycle, hence, by solving $(6.16)_\nu$ we can find $h_{r|\nu}(s)$, $r = 1, \ldots, m$, $\{g_{j|\nu}(s)\}$ such that $h^\nu(s)$ and $g^\nu(z_j, s)$ satisfy the congruences $(6.11)_\nu$. In this way, we determine $h^\nu(s)$ and $g_j^\nu(z_j, s)$ successively for $\nu = 1, 2, 3, \ldots$, and obtain the formal power series $h(s)$ and $g_j(z_j, s)$, which obviously satisfy (6.10).

(b) Proof of Convergence

The power series $h(s)$, $g_j(z_j, s)$ constructed in the previous subsection depend on the choice of solutions $h_{r|\nu}(s)$, $r = 1, \ldots, m$, $\{g_{j|\nu}(s)\}$ of $(6.16)_\nu$. In general the equations $(6.16)_\nu$ have infinitely many solutions. In this subsection we prove that, if we choose appropriate solutions $h_{r|\nu}(s)$ and $\{g_{j|\nu}(z_j, s)\}$ of $(6.16)_\nu$ in each step of the above construction, then $h(s)$ and $g_j(z_j, s)$ converge absolutely in $|s| < \varepsilon$ provided $\varepsilon > 0$ is sufficiently small.

The equations $(6.16)_\nu$ are reduced to (6.18). We first prove a lemma concerning the "magnitude" of the solutions $h_1, \ldots, h_m$, $\{g_j\}$ of the equation (6.18):

$$\{\Gamma_{jk}\} = \sum_{r=1}^{m} h_r\{\theta_{rjk}\} + \delta\{g_j\}.$$

We regard holomorphic vector fields $\Gamma_{jk} = \sum_{\alpha=1}^n \Gamma_{jk}^\alpha(z_j)\,\partial/\partial z_j^\alpha$, $\theta_{rjk} = \sum_{\alpha=1}^n \theta_{rjk}^\alpha(z_j)\,\partial/\partial z_j^\alpha$, and $g_j = \sum_{\alpha=1}^n g_j^\alpha(z_j)\,\partial/\partial z_j^\alpha$ as vector-valued holomorphic functions $\Gamma_{jk}(z_j) = (\Gamma_{jk}^1(z_j), \ldots, \Gamma_{jk}^n(z_j))$, $\theta_{rjk}(z_j) = (\theta_{rjk}^1(z_j), \ldots, \theta_{rjk}^n(z_j))$, and $g_j(z_j) = (g_j^1(z_j), \ldots, g_j^n(z_j))$, respectively. Then (6.18) is written in the form

$$\Gamma_{jk}(z_j) = \sum_{r=1}^{m} h_r\theta_{rjk}(z_j) + B_{jk}(z_k)g_k(z_k) - g_j(z_j). \tag{6.22}$$

Since each $U_j = \{z_j \in \mathbb{C}^n \mid |z_j| < 1\}$ is a coordinate polydisk, we may assume that the coordinate function z_j is defined on a domain of M_0 containing $[U_j]$ (see p. 33). Hence each component $\partial z_j^\alpha / \partial z_k^\beta$ of the matrix $B_{jk}(z_k) = (\partial z_j^\alpha / \partial z_k^\beta)_{\alpha,\beta=1,\ldots,n}$ is bounded on $U_j \cap U_k \neq \emptyset$. We define the norm of the matrix $B_{jk}(z_k)$ by

$$|B_{jk}(z_j)| = \sup_\zeta \frac{|B_{jk}(z_k)\zeta|}{|\zeta|} = \max_\alpha \sum_\beta \left| \frac{\partial z_j^\alpha}{\partial z_k^\beta} \right|, \qquad \zeta \in \mathbb{C}^n, \quad \zeta \neq 0. \tag{6.23}$$

Then there exists a constant K_1 such that for all $U_j \cap U_k = \emptyset$

$$|B_{jk}(z_k)| < K_1, \qquad z_k \in U_k \cap U_j. \tag{6.24}$$

Similarly, since the $\theta_{rjk}^\alpha(z_j) = (\partial g_{jk}^\alpha(z_k, t)/\partial t_r)_{t=0}$ are also bounded on $U_j \cap U_k \neq \emptyset$, there exists a constant K_2 such that

$$|\theta_{rjk}(z_j)| < K_2. \tag{6.25}$$

We denote a 1-cocycle $\{\Gamma_{jk}\}$ by Γ, and define its norm by

$$|\Gamma| = \max_{j,k} \sup_{z_j \in U_j \cap U_k} |\Gamma_{jk}(z_j)|. \tag{6.26}$$

Lemma 6.2. *There exist solutions h_r, $r = 1, \ldots, m$, $\{g_j\}$ of (6.18) which satisfy*

$$|h_r| \leqq K_3 |\Gamma|, \qquad |g_j(z_j)| \leqq K_3 |\Gamma|, \tag{6.27}$$

where K_3 is a constant independent of $\Gamma = \{\Gamma_{jk}\}$.

Proof. Let $\delta > 0$, and put, for each $U_j = \{z_j \in \mathbb{C}^n \mid |z_j| < 1\}$,

$$U_j^\delta = \{z_j \in U_j \mid |z_j| < 1 - \delta\}.$$

Then, since $\{U_j\}$ is a finite covering of M_0, and since M_0 is compact, we have

$$M_0 = \bigcup_j U_j^\delta \tag{6.28}$$

for a sufficiently small δ.

We consider solutions h_r, $r = 1, \ldots, m$, $\{g_j\}$ of (6.18) for a 1-cocycle $\Gamma = \{\Gamma_{jk}\}$ with $|\Gamma| < +\infty$. It can be easily checked that, on each U_j, $|g_j(z_j)|$ is bounded. In fact, since $g_j(z_j)$ is holomorphic on U_j, it is obviously bounded on U_j^δ. If $z_j \notin U_j^\delta$, z_j is contained in some U_k^δ with $k \neq j$, that is, $z_j = b_{jk}(z_k)$, $z_k \in U_k^\delta$. Hence by (6.22) we have

$$g_j(z_j) = \sum_{r=1}^m h_r \theta_{rjk}(z_j) + B_{jk}(z_k) g_k(z_k) - \Gamma_{jk}(z_j). \tag{6.29}$$

By (6.24) $|B_{jk}(z_k)| < K_1$, by (6.25) $|\theta_{rjk}(z_j)| < K_2$, and by hypothesis $|\Gamma_{jk}(z_j)| \leqq |\Gamma| < +\infty$. Moreover $|g_k(z_k)|$ is bounded on U_k^δ. Hence by (6.29), $|g_j(z_j)|$ is bounded on $U_j \cap U_k^\delta$, and, since U_j is covered by U_j^δ and a finite number of the U_k^δ, $k \neq j$, it follows that $|g_j(z_j)|$ is bounded on U_j.

For a 1-cocycle $\Gamma = \{\Gamma_{jk}\}$ with $|\Gamma| < +\infty$, we define $\iota(\Gamma)$ by

$$\iota(\Gamma) = \inf \max_{r,j} \{|h_r|, \sup_{z_j \in U_j} |g_j(z_j)|\},$$

where inf is taken with respect to all the solutions h_r, $g_j(z_j)$ of (6.18). *We must show that there exists a constant K such that for all* 1*-cocycles* $\Gamma \in Z^1(\mathfrak{U}, \Theta_0)$, *the following inequality*

$$\iota(\Gamma) \leqq K|\Gamma|$$

holds. Suppose there is no such constant K. Then, for each natural number ν, we can find a 1-cocycle $\Gamma^{(\nu)}$ such that $\iota((\Gamma^{(\nu)}) > \nu|\Gamma^{(\nu)}|$. Replacing $\Gamma^{(\nu)}$ by $\Gamma^{(\nu)}/\iota(\Gamma^{(\nu)})$, we get a sequence of 1-cocycles $\Gamma^{(\nu)} = \{\Gamma_{jk}^{(\nu)}\} \in Z^1(\mathfrak{U}, \Theta_0)$ such that

$$\iota(\Gamma^{(\nu)}) = 1, \qquad |\Gamma^{(\nu)}| < \frac{1}{\nu}. \tag{6.30}$$

The equality $\iota(\Gamma^{(\nu)}) = 1$ implies that there exist solutions $h_r^{(\nu)}$, $\{g_j^{(\nu)}\}$ of (6.18) for $\Gamma = \Gamma^{(\nu)}$ satisfying

$$|h_r^{(\nu)}| < 2, \qquad |g_j^{(\nu)}(z_j)| < 2.$$

Since, for each U_j, $\{g_j^{(\nu)}(z_j)\} = \{g_j^{(1)}(z_j), g_j^{(2)}(z_j), \ldots\}$ are uniformly bounded on U_j, we can choose a subsequence $\{g_j^{(\nu_m)}(z_j)\}$, $\nu_1 < \nu_2 < \cdots < \nu_m < \ldots$, so as to converge uniformly on each compact subset of U_j. Hence we may assume, from the beginning, that $\{g_j^{(\nu}(z_j)\}$ converges uniformly on each compact subset of U_j. Similarly, all the sequences $\{h_r^{(\nu)}\}$, $\nu = 1, 2, \ldots, r$, may be assumed to converge. Then we can deduce that *each* $\{g_j^{(\nu)}(z_j)\}$ *converges uniformly on the whole* U_j. To show this, we note that U_j is covered by U_j^δ and a finite number of U_k^δ, $k \neq j$, and that, since $[U_j^\delta] \subset U_j$ is compact, $\{g_j^{(\nu)}(z_j)\}$ converges uniformly on U_j^δ. For a point $z_j \in U_j \cap U_k^\delta$, i.e., for $z_j = b_{jk}(z_k)$, $z_k \in U_k^\delta$, we have by (6.29)

$$g_j^{(\nu)}(z_j) = \sum_{r=1}^{m} h_r^{(\nu)} \theta_{rjk}(z_j) + B_{jk}(z_k) g_k^{(\nu)}(z_k) - \Gamma_{jk}^{(\nu)}(z_j).$$

Since $\{h_r^{(\nu)}\}$ converges and $\{g_j^{(\nu)}(z_j)\}$ converges uniformly on U_j^δ, and since $|\Gamma_{jk}^{(\nu)}(z_j)| \leqq |\Gamma^{(\nu)}| \to 0$ $(\nu \to \infty)$ by (6.30), in view of (6.24) and (6.25), we see that $\{g_j^{(\nu)}(z_j)\}$ converges uniformly on U_j.

Put $h_r = \lim_{\nu\to\infty} h_r^{(\nu)}$, and $g_j(z_j) = \lim_{\nu\to\infty} g_j^{(\nu)}(z_j)$. Then $g_j(z_j)$ is a vector-valued holomorphic function on U_j. By (6.22)

$$\Gamma_{jk}^{(\nu)}(z_j) = \sum_{r=1}^{m} h_r^{(\nu)}\theta_{rjk}(z_j) + B_{jk}(z_k)g_k^{(\nu)}(z_k) - g_j^{(\nu)}(z_j).$$

Since $|\Gamma_{jk}^{(\nu)}(z_j)| \leqq |\Gamma^{(\nu)}| \to 0$ $(\nu\to\infty)$, taking the limit for $\nu\to\infty$, we obtain

$$0 = \sum_{r=1}^{m} h_r\theta_{rjk}(z_j) + B_{jk}(z_k)g_k(z_k) - g_j(z_j).$$

Hence, putting $\tilde{h}_r^{(\nu)} = h_r^{(\nu)} - h_r$, and $\tilde{g}_j^{(\nu)}(z_j) = g_j^{(\nu)}(z_j) - g_j(z_j)$, we obtain

$$\Gamma_{jk}^{(\nu)}(z_j) = \sum_{r=1}^{m} \tilde{h}_r^{(\nu)}\theta_{rjk}(z_j) + B_{jk}(z_k)\tilde{g}_k^{(\nu)}(z_k) - \tilde{g}_j^{(\nu)}(z_j).$$

Hence $\tilde{h}_r^{(\nu)}$, $r = 1, \ldots, m$, and $\{\tilde{g}_j^{(\nu)}\}$ constitute another set of solutions of (6.18) for $\Gamma = \Gamma^{(\nu)}$. But when ν tends to infinity, we have $\tilde{h}_r^{(\nu)} \to 0$ and $\sup_{z_j\in U_j} |\tilde{g}_j^{(\nu)}(z_j)| \to 0$, which contradicts $\iota(\Gamma) = 1$. ∎

We shall prove, using the method of majorant series, that $h(s)$ and $g_j(z_j, s)$ converge absolutely for $|s| < \varepsilon$ provided that $\varepsilon > 0$ is sufficiently small. In general, if two power series of $s_1, \ldots, s_l$,

$$P(s) = \sum_{\nu_1,\ldots,\nu_l=0}^{\infty} P_{\nu_1\cdots\nu_l} s_1^{\nu_1}\cdots s_l^{\nu_l}, \qquad P_{\nu_1\cdots\nu_l} \in \mathbb{C}^n,$$

and

$$a(s) = \sum_{\nu_1,\ldots,\nu_l=0}^{\infty} a_{\nu_1\cdots\nu_l} s_1^{\nu_1}\cdots s_l^{\nu_l}, \qquad a_{\nu_1\cdots\nu_l} \geqq 0,$$

are given, we indicate by writing

$$P(s) \ll a(s)$$

that

$$|P_{\nu_1\cdots\nu_l}| \leqq a_{\nu_1\cdots\nu_l}, \qquad \nu_1, \ldots, \nu_l = 0, 1, 2 \ldots.$$

As in § 5.3(c), let

$$A(a) = \frac{b}{16c}\sum_{\nu=1}^{\infty} \frac{c^\nu(s_1 + \cdots + s_l)^\nu}{\nu^2}, \qquad b > 0, \quad c > 0.$$

By (5.116), we have

$$A(s)^2 \ll \frac{b}{c}A(s).$$

By induction on ν, we obtain from this inequality the following inequalities:

$$A(s)^\nu \ll \left(\frac{b}{c}\right)^{\nu-1} A(s), \qquad \nu = 2, 3, \ldots. \tag{6.31}$$

In fact, for $\nu \geqq 3$, we have

$$A(s)^\nu = A(s)^2 A(s)^{\nu-2} \ll \frac{b}{c} A(s) A(s)^{\nu-2} = \frac{b}{c} A(s)^{\nu-1}.$$

In order to prove the convergence of $h(s)$, $g_j(z_j, s)$, it suffices to show the estimates

$$h(s) \ll A(s), \qquad g_j(z_j, s) - z_j \ll A(s), \tag{6.32}$$

provided that the constants b and c are properly chosen. For this, it suffices to prove

$$h^\nu(s) \ll A(s), \qquad g_j^\nu(z_j, s) - z_j \ll A(s) \tag*{(6.33)$_\nu$}$$

for $\nu = 1, 2, 3, \ldots$.

We prove (6.33)$_\nu$ by induction on $\nu = 1, 2, 3, \ldots$.

For $\nu = 1$, since the linear term of $A(s)$ is $(b/16)(s_1 + \cdots + s_l)$, the estimate (6.33)$_1$ obviously holds provided that b is sufficiently large.

Let $\nu \geqq 2$ and assume that (6.33)$_{\nu-1}$ are established. To prove (6.33)$_\nu$ we first estimate $\Gamma_{jk|\nu}(z_j, s)$.

In general, for a power series $P(s) = \sum_{\nu=1}^\infty P_\nu(s)$, we denote by $[P(s)]_\nu$ the term $P_\nu(s)$ of homogeneous part of degree ν. Then by (6.13), we have

$$\Gamma_{jk|\nu}(z_j, s) = [g_j^{\nu-1}(f_{jk}(z_k, s), s)]_\nu - [g_{jk}(g_k^{\nu-1}(z_k, s), h^{\nu-1}(s))]_\nu. \tag{6.34}$$

We first estimate $[g_j^{\nu-1}(f_{jk}(z_k, s), s)]_\nu$. Since the $f_{jk}(z_k, s) = b_{jk}(z_k) + \sum_{\nu=1}^\infty f_{jk|\nu}(z_j, s)$ are given vector-valued holomorphic functions, we may assume that

$$f_{jk}(z_k, s) - b_{jk}(z_k) \ll A_0(s), \qquad A_0(s) = \frac{b_0}{16 c_0} \sum_{\nu=1}^\infty \frac{c_0^\nu (s_1 + \cdots + s_l)^\nu}{\nu^2} \tag{6.35}$$

holds for $z_k \in U_k \cap U_j$ with $b_0 > 0$ and $c_0 > 0$. Put

$$G(z_j, s) = g_j^{\nu-1}(z_j, s) - z_j.$$

Then, by hypothesis of induction (6.33)$_{\nu-1}$, we have $G(z_j, s) \ll A(s)$. Namely, letting

$$G(z_j, s) = \sum_{\nu_1+\cdots+\nu_l \geqq 1} G_{\nu_1\cdots\nu_l}(z_j) s_1^{\nu_1} \cdots s_l^{\nu_l},$$

$$A(s) = \sum_{\nu_1+\cdots+\nu_l \geqq 1} A_{\nu_1\cdots\nu_l} s_1^{\nu_1} \cdots s_l^{\nu_l},$$

we have

$$|G_{\nu_1\cdots\nu_l}(z_j)| \leqq A_{\nu_1\cdots\nu_l}, \qquad z_j \in U_j. \tag{6.36}$$

By (6.28), $M_0 = \bigcup_j U_j^\delta$. If $z_j \in U_j^\delta$, then since $G_{\nu_1\cdots\nu_l}(z_j+\zeta)$ is a vector-valued holomorphic function of $\zeta = (\zeta_1, \ldots, \zeta_n)$, $|\zeta| < \delta$, it can be expanded into power series in $\zeta_1, \ldots, \zeta_n$ as

$$G_{\nu_1\cdots\nu_l}(z_j+\zeta) = \sum_{\mu_1,\ldots,\mu_n} G_{\nu_1\cdots\nu_l\mu_1\cdots\mu_n}(z_j)\zeta_1^{\mu_1} \cdots \zeta_n^{\mu_n}.$$

By Cauchy's integral formula, we have

$$G_{\nu_1\cdots\nu_l\mu_1\cdots\mu_n}(z_j) = \left(\frac{1}{2\pi i}\right)^n \int_{|\zeta_1|=\delta} \cdots \int_{|\zeta_n|=\delta} \frac{G_{\nu_1\cdots\nu_l}(z_j+\zeta)\, d\zeta_1 \cdots d\zeta_n}{\zeta_1^{\mu_1+1} \cdots \zeta_n^{\mu_n+1}}.$$

Hence, since $|G_{\nu_1\cdots\nu_l}(z_j+\zeta)| \leqq A_{\nu_1\cdots\nu_l}$ by (6.36), we have

$$|G_{\nu_1\cdots\nu_l\mu_1\cdots\mu_n}(z_j)| \leqq \frac{A_{\nu_1\cdots\nu_l}}{\delta^{\mu_1+\cdots+\mu_n}}.$$

Hence

$$G_{\nu_1\cdots\nu_l}(z_j+\zeta) - G_{\nu_1\cdots\nu_l}(z_j) \ll A_{\nu_1\cdots\nu_l} \sum_{\mu_1+\cdots+\mu_n \geqq 1} \frac{\zeta_1^{\mu_1} \cdots \zeta_n^{\mu_n}}{\delta^{\mu_1+\cdots+\mu_n}}.$$

Therefore

$$G(z_j+\zeta, s) - G(z_j, s) \ll A(s) \sum_{\mu_1+\cdots+\mu_n \geqq 1} \frac{\zeta_1^{\mu_1} \cdots \zeta_n^{\mu_n}}{\delta^{\mu_1+\cdots+\mu_n}}.$$

For $z_j \in U_j^\delta \cap U_k$, *i.e., for* $z_j = b_{jk}(z_k) \in U_j^\delta$, $z_k \in U_k$, we let $\zeta = f_{jk}(z_k, s) - z_j$. Since $\zeta \ll A_0(s)$ by (6.35), we obtain

$$G(f_{jk}(z_k, s), s) - G(z_j, s) \ll A(s) \sum_{\mu_1+\cdots+\mu_n \geqq 1} \frac{A_0(s)^{\mu_1+\cdots+\mu_n}}{\delta^{\mu_1+\cdots+\mu_h}}.$$

We have

$$\sum_{\mu_1+\cdots+\mu_n\geqq 1}\left(\frac{A_0(s)}{\delta}\right)^{\mu_1+\cdots+\mu_n}=\left[\sum_{\mu=0}^{\infty}\left(\frac{A_0(s)}{\delta}\right)^{\mu}\right]^n-1,$$

and by (6.31)

$$\sum_{\mu=0}^{\infty}\left(\frac{A_0(s)}{\delta}\right)^{\mu}=1+\sum_{\mu=1}^{\infty}\frac{A_0(s)^{\mu}}{\delta^{\mu}}\ll 1+\frac{1}{\delta}\sum_{\mu=1}^{\infty}\left(\frac{b_0}{c_0\delta}\right)^{\mu-1}A_0(s).$$

Remark that (6.35) remains valid if we replace c_0 by a larger constant. Therefore, by taking a sufficiently large c_0, we may assume that

$$\frac{b_0}{c_0\delta}<\frac{1}{2}. \tag{6.37}$$

Thus, since $\sum_{\mu=1}^{\infty}(b_0/c_0\delta)^{\mu-1}<2$, we obtain

$$\begin{aligned}\left[\sum_{\mu=0}^{\infty}\left(\frac{A_0(s)}{\delta}\right)^{\mu}\right]^n-1&\ll\left(1+\frac{2}{\delta}A_0(s)\right)^n-1\\&\ll\sum_{k=1}^{n}\binom{n}{k}\left(\frac{2}{\delta}\right)^k A_0(s)^k\ll\frac{2}{\delta}\sum_{k=1}^{n}\binom{n}{k}\left(\frac{2b_0}{c_0\delta}\right)^{k-1}A_0(s)\\&\ll\frac{2}{\delta}\sum_{k=0}^{n}\binom{n}{k}A_0(s)=\frac{2^{n+1}}{\delta}A_0(s).\end{aligned}$$

Hence we get

$$G(f_{jk}(z_k,s),s)-G(z_j,s)\ll\frac{2^{n+1}}{\delta}A(s)A_0(s).$$

If we take b and c such that

$$b>b_0,\qquad c>c_0, \tag{6.38}$$

we have $A_0(s)\ll(b_0/b)A(s)$. Therefore we get

$$A(s)A_0(s)\ll\frac{b_0}{b}A(s)^2\ll\frac{b_0}{c}A(s).$$

Hence

$$G(f_{jk}(z_k,s),s)-G(z_j,s)\ll\frac{2^{n+1}b_0}{c\delta}A(s),$$

namely,

$$g_j^{\nu-1}(f_{jk}(z_k, s), s) - f_{jk}(z_k, s) - g_j^{\nu-1}(z_j, s) + z_j \ll \frac{2^{n+1}b_0}{c\delta} A(s).$$

Since by (6.35), $f_{jk}(z_k, s) - z_j \ll A_0(s) \ll (b_0/b)A(s)$, we have

$$g_j^{\nu-1}(f_{jk}(z_k, s), s) - g_j^{\nu-1}(z_j, s) \ll \left(\frac{2^{n+1}b_0}{c\delta} + \frac{b_0}{b}\right) A(s).$$

Therefore, taking the terms of degree ν of the left-hand side of this equality, we obtain the following estimate on $U_j^\delta \cap U_k$:

$$[g_j^{\nu-1}(f_{jk}(z_k, s), s)]_\nu \ll \left(\frac{2^{n+1}b_0}{c\delta} + \frac{b_0}{b}\right) A(s). \tag{6.39}$$

Next we estimate the second term $[g_{jk}(g_k^{\nu-1}(z_k, s), h^{\nu-1}(s))]_\nu$ of (6.34). We expand $g_{jk}(z_k + \zeta, t)$ into power series in $\zeta_1, \ldots, \zeta_n, t_1, \ldots, t_m$, and let $L(\zeta, t)$ be its linear term. Then, since $g_{jk}(z_k, 0) = b_{jk}(z_k)$, we may assume that

$$g_{jk}(z_k + \zeta, t) - b_{jk}(z_k) - L(\zeta, t)$$
$$\ll \sum_{\mu=2}^{\infty} a_0^\mu (\zeta_1 + \cdots + \zeta_n + t_1 + \cdots + t_m)^\mu, \qquad a_0 > 0.$$

If we set $\zeta = g_k^{\nu-1}(z_k, s) - z_k$, and $t = h^{\nu-1}(s)$, then, since $\zeta \ll A(s)$ and $t \ll A(s)$ by $(6.33)_{\nu-1}$, we obtain

$$g_{jk}(g_k^{\nu-1}(z_k, s), h^{\nu-1}(s)) - b_{jk}(z_k) - L(g_k^{\nu-1}(z_k, s) - z_k, h^{\nu-1}(s))$$
$$\ll \sum_{\mu=2}^{\infty} a_0^\mu (n+m)^\mu A(s)^\mu \ll \sum_{\mu=2}^{\infty} a_0^\mu (m+n)^\mu \left(\frac{b}{c}\right)^{\mu-1} A(s).$$

Hence, by taking the terms of degree ν of the left-hand side, we obtain

$$[g_{jk}(g_k^{\nu-1}(z_k, s), h^{\nu-1}(s)]_\nu \ll \frac{ba_0^2(m+n)^2}{c} \sum_{\mu=0}^{\infty} \left(\frac{ba_0(m+n)}{c}\right)^\mu A(s).$$

Consequently, taking a constant c such that

$$\frac{ba_0(m+n)}{c} < \frac{1}{2}, \tag{6.40}$$

we obtain

$$[g_{jk}(g_k^{\nu-1}(z_k, s), h^{\nu-1}(s))]_\nu \ll \frac{2ba_0^2(m+n)^2}{c} A(s).$$

Combining this with (6.39) we obtain the estimate

$$\Gamma_{jk|\mu}(z_j, s) \ll K^*A(s), \qquad z_j \in U_j^\delta \cap U_k, \tag{6.41}$$

where K^* is given by

$$K^* = \frac{2^{n+1}b_0}{c\delta} + \frac{b_0}{b} + \frac{2ba_0^2(m+n)^2}{c}. \tag{6.42}$$

We now estimate $\Gamma_{jk|\nu}(z_j, s)$ for arbitrary $z_j \in U_j \cap U_k$. For this purpose, recall that $\{\Gamma_{jk|\nu}(s)\}$ is a 1-cocycle. Since $M_0 = \bigcup_j U_j^\delta$, if $z_j \in U_j \cap U_k$, and $z_j \notin U_j^\delta$, then z_j is contained in some U_i^δ $(i \neq j)$: $z_j = b_{ji}(z_i)$, $z_i \in U_i^\delta$. By (6.19)

$$\Gamma_{jk|\nu}(z_j, s) = B_{ji}(z_i)\Gamma_{ik|\nu}(z_i, s) - B_{ji}(z_i)\Gamma_{ij|\nu}(z_i, s),$$

and, since $z_i \in U_i^\delta \cap U_j \cap U_k$, both $\Gamma_{ik|\nu}(z_i, s)$ and $\Gamma_{ij|\nu}(z_i, s)$ are $\ll K^*A(s)$ by (6.41). Since we have $|B_{jk}(z_k)| < K_1$ with $K_1 > 1$, it follows that

$$\Gamma_{jk|\nu}(z_j, s) \ll 2K_1K^*A(s), \qquad z_j \in U_j \cap U_k.$$

Recall that $h_{r|\nu}(s)$, $r = 1, \ldots, m$, $g_{j|\nu}(z_j, s)$ are solutions of the equations $(6.16)_\nu$:

$$\{\Gamma_{jk|\nu}(s)\} = \sum_{r=1}^{m} h_{j|\nu}(s)\{\theta_{rjk}\} + \delta\{g_{j|\nu}(s)\}.$$

Since $\Gamma_{jk|\nu}(s) \ll 2K_1K^*A(s)$, by Lemma 6.2, *we can choose solutions* $h_{r|\nu}(s)$, $r = 1, \ldots, m$, $\{g_{j|\nu}(s)\}$, *such that*

$$h_{r|\nu}(s) \ll K_3 2K_1K^*A(s), \qquad g_{j|\nu}(s) \ll K_3 2K_1K^*A(s).$$

By definition (6.42) of K^*

$$2K_3K_1K^* = 2K_3K_1\left(\frac{2^{n+1}b_0}{c\delta} + \frac{b_0}{b} + \frac{2ba_0^2(m+n)^2}{c}\right)$$

is independent of ν, *and if we first choose a sufficiently large* b, *and then choose* c *so that* c/b *be sufficiently large, then we obtain*

$$2K_3K_1K^* \leqq 1.$$

Note that b and c obviously satisfy (6.38) and (6.40). Hence the above solutions $h_{r|\nu}(s)$, $\{g_{j|\nu}(s)\}$ satisfy the inequalities

$$h_{r|\nu}(s) \ll A(s), \qquad g_{j|\nu}(s) \ll A(s). \tag{6.43}$$

Therefore, putting

$$h^{\nu}(s)=h^{\nu-1}(s)+h_{\nu}(s), \qquad g_j^{\nu}(z_j, s)=g_j^{\nu-1}(z_j, s)+g_{j|\nu}(z_j, s),$$

we see that $(6.33)_\nu$ follows from (6.43) and $(6.33)_{\nu-1}$. This completes the induction, and the inequalities (6.32):

$$h(s) \ll A(s), \qquad g_j(z_j, s)-z_j \ll A(s)$$

are proved.

These inequalities imply that, *if* $|s|<1/lc$, $h(s)$ *converges absolutely, and* $g_j(z_j, s)$ *converges absolutely and uniformly for* $z_j \in U_j$.

(c) Proof of Theorem of Completeness

In view of $\mathcal{N}=\bigcup_j \mathcal{W}_j$, $\mathcal{W}_j=\{(z_j, s)\,|\,|z_j|<1, |s|<1\}=U_j\times D$, and $U_j=N_0\cap \mathcal{W}_j=U_j=0$, if we put $\Delta=\{s\in\mathbb{C}^l\,|\,|s|<1/lc\}$, then $\mathcal{N}_\Delta=\pi^{-1}(\Delta)=\bigcup_j U_j\times\Delta$. Replacing $\mathcal{W}_j$ by $U_j\times\Delta$, we may assume that

$$\mathcal{N}_\Delta=\bigcup_j \mathcal{W}_j, \qquad \mathcal{W}_j=U_j\times\Delta.$$

On the other hand, we have

$$\mathcal{M}=\bigcup_j \mathcal{U}_j, \qquad \mathcal{U}_j=U_j\times B\subset\mathbb{C}^n\times B.$$

By (6.32) the holomorphic map

$$g_j\colon (z_j, s)\to(\zeta_j, t)=(g_j(z_j, s), h(s))$$

maps $\mathcal{W}_j=U_j\times\Delta$ into $\mathbb{C}^n\times B$, and the restriction $g_j\colon U_j\to U_j\subset M_0$ is nothing but the identity map: $(z_j, 0)\to(\zeta_j, 0)=(z_j, 0)$. We would like to obtain a holomorphic map $g\colon\mathcal{N}_\Delta\to\mathcal{M}$ by glueing up the holomorphic maps $\ldots, g_j, \ldots, g_k, \ldots$. But, in general, we may not have $g_j(\mathcal{W}_j)\subset\mathcal{U}_j\subset\mathbb{C}^n\times B$. Hence we first restrict g_j to $g_j^{-1}(\mathcal{U}_j)$. Since $g_j(U_j)=U_j$, $g_j^{-1}(\mathcal{U}_j)$ is an open subset of $\mathcal{W}_j$ containing U_j:

$$U_j\subset g_j^{-1}(\mathcal{U}_j)\subset\mathcal{W}_j,$$

and g_j maps $g_j^{-1}(\mathcal{U}_j)$ into $\mathcal{U}_j\subset\mathcal{M}$.

For each point $c\in U_k\cap U_j$, *there exists a neighbourhood*

$$\mathcal{V}(c)=\{(z_k, s)\,|\,|z_k-c|<\varepsilon, |s|<\varepsilon\}\subset g_k^{-1}(\mathcal{U}_k)\cap g_j^{-1}(\mathcal{U}_j),$$

such that g_j and g_k coincide there. Proof. The maps g_j and g_k are given by

$$g_j\colon (z_j, s) \to (\zeta_j, t) = (g_j(z_j, s), h(s)),$$

$$g_k\colon (z_k, s) \to (\zeta_k, t) = (g_k(z_k, s), h(s)).$$

Note that (z_j, s) and (z_k, s) are the same point on $\mathcal{N}_\Delta$ if $z_j = f_{jk}(z_k, s)$, while (ζ_j, t) and (ζ_k, t) are the same point of $\mathcal{M}$ if $\zeta_j = g_{jk}(\zeta_k, t)$. Therefore, to prove that g_j and g_k coincide on $\mathcal{V}(c)$, it suffices to show

$$g_j(f_{jk}(z_k, s), s) = g_{jk}(g_k(z_k, s), h(s)) \qquad (z_k, s) \in \mathcal{V}(c). \tag{6.44}$$

But (6.44) is nothing but (6.10). Hence the power series expansions in $s_1, \ldots, s_l$ of both sides of (6.44) are the same. On the other hand, both $g_j(f_{jk}(z_k, s), s)$ and $g_{jk}(g_k(z_k, s), h(s))$ are the holomorphic functions of $z_k^1, \ldots, z_k^n, s_1, \ldots, s_l$ in the polydisk $\mathcal{V}(c)$. Hence (6.44) holds. ∎

For each point $c \in U_j \cap U_k$, choose $\mathcal{V}(c)$ as above and let $\mathcal{V}_{jk} = \bigcup_c \mathcal{V}(c)$. Then we have

$$U_j \cap U_k \subset \mathcal{V}_{jk} \subset g_j^{-1}(\mathcal{U}_j) \cap g_k^{-1}(\mathcal{U}_k),$$

and the holomorphic maps g_j and g_k coincide on the domain $\mathcal{V}_{jk}$. We take a sufficiently small $\delta > 0$, and put $U_j^\delta = \{z_j \in U_j \,|\, |z_j| < 1 - \delta\}$. Then we have

$$\pi^{-1}(0) = N_0 = \bigcup_j U_j^\delta.$$

Therefore, putting $\Delta_\varepsilon = \{s \in \Delta \,|\, |s| < \varepsilon\}$, we obtain

$$\pi^{-1}(\Delta_\varepsilon) = \bigcup_j U_j^\delta \times \Delta_\varepsilon, \qquad U_j^\delta = \Delta_\varepsilon \subset g_j^{-1}(\mathcal{U}_j),$$

$$U_j^\delta \times \Delta_\varepsilon \cap U_k^\delta \times \Delta_\varepsilon \subset \mathcal{V}_{jk},$$

provided that $\varepsilon > 0$ is sufficiently small. Restricting each g_j to $U_j^\delta \times \Delta_\varepsilon$, we see that g_j maps $U_j^\delta \times \Delta_\varepsilon$ into $\mathcal{M}$ and that, if $U_j^\delta \times B \cap U_k^\delta \times \Delta_\varepsilon \neq \varnothing$, g_j and g_k coincide there. Hence we can find a holomorphic map

$$g\colon \pi^{-1}(\Delta_\varepsilon) = \bigcup_j U_j^\delta \times \Delta_\varepsilon \to \mathcal{M}$$

such that g coincides with g_j on each $U_j^\delta \times \Delta_\varepsilon$. From the expression

$$g\colon (z_j, s) \to (\zeta_j, t) = (g_j(z_j, s), h(s)),$$

it follows that $\varpi \circ g = h \circ \pi$, while the equality $g_j(z_j, 0) = z_j$ implies that g is an extension of the identity map $g_0\colon N_0 \to M_0 = N_0$. This completes the proof of the theorem of completeness: Theorem 6.1.

§6.2. Number of Moduli

(a) Number of Moduli

In this section we shall prove that the conjecture stated in §5.3 concerning the number of moduli is true in case $H^2(M, \Theta) = 0$.

Let M be a compact complex manifold and suppose $H^2(M, \Theta) = 0$. Then, by the theorem of existence (Theorem 5.6), there exists a complex analytic family $(\mathcal{M}, B, \varpi)$ with $0 \in B \subset \mathbb{C}^m$ satisfying the following conditions:

(i) $\varpi^{-1}(0) = M$.

(ii) $\rho_0: \partial/\partial t \to (\partial M_t/\partial t_\lambda)_{t=0}$, with $M_t = \varpi^{-1}(t)$, is an isomorphism of $T_0(B)$ onto $H^1(M, \Theta)$: $T_0(B) \overset{\rho_0}{\rightrightarrows} H^1(M, \Theta)$.

By the theorem of completeness (Theorem 6.1), $(\mathcal{M}, B, \varpi)$ is complete at $0 \in B$. We need to prove that $m(M) = \dim H^1(M, \Theta)$ holds provided $m(M)$ is defined. Suppose that $m(M)$ is defined and put $\mu = m(M)$. Then, by Definition 5.4 of the number of moduli (p. 228), there exists a complete complex analytic family $(\mathcal{N}, D, \pi)$ which is effectively parametrized and such that $\pi^{-1}(0) = M$, with D being a domain of $\mathbb{C}^\mu$ containing 0. We denote by $s = (s_1, \ldots, s_\mu)$ a point of D. Since $(\mathcal{N}, D, \pi)$ is complete, it follows that, if we take a sufficiently small domain Δ, $0 \in \Delta \subset B$, then $(\mathcal{M}_\Delta, \Delta, \varpi)$ is a complex analytic family induced from $(\mathcal{N}, D, \pi)$ by a holomorphic map $h: t \to s = h(t)$ with $h(0) = 0$. Hence $M_t = N_{h(t)} = \pi^{-1}(h(t))$. Therefore, by (4.30)

$$\frac{\partial M_t}{\partial t_\lambda} = \sum_{\nu=1}^{\mu} \frac{\partial s_\nu}{\partial t_\lambda} \cdot \frac{\partial N_s}{\partial s_\nu}, \qquad (s_1, \ldots, s_\mu) = s = h(t).$$

Since $\rho_0: T_0(B) \rightrightarrows H^1(M, \Theta)$ is an isomorphism, the cohomology classes

$$\left(\frac{\partial M_t}{\partial t_\lambda}\right)_{t=0} \in H^1(M, \Theta), \qquad \lambda = 1, \ldots, m, \quad m = \dim H^1(M, \Theta)$$

on the left-hand side of this equality are linearly independent. On the other hand, we have $\mu = m(M) \leqq \dim H^1(M, \Theta)$. Hence we obtain the equality $m(M) = \dim H^1(M, \Theta)$. Thus we have obtained

Theorem 6.2. *Suppose $H^2(M, \Theta) = 0$ and that the number of moduli $m(M)$ of M is defined. Then $m(M) = \dim H^1(M, \Theta)$.*

Then what is the condition for $m(M)$ to be defined in case $H^2(M, \Theta) = 0$? If $m(M)$ is defined, $(\mathcal{M}_\Delta, \Delta, \varpi)$ is, as stated above, the complex analytic family induced from an effectively parametrized, complete family $(\mathcal{N}, D, \pi)$ by a holomorphic map $h: \Delta \to D$ with $h(0) = 0$, and $m(M) = \dim H^1(M, \Theta)$.

Since D is an effective parameter space, the linear map $\partial/\partial s \to \partial N_s/\partial s$ is an injection of $T_s(D)$ into $H^1(N_s, \Theta)$. Since $\dim T_s(D) = m(M) = m$, we conclude $\dim H^1(M, \Theta) \geqq m$. On the other hand, $\dim H^1(N_0, \Theta_0) = \dim H^1(N_s, \Theta_s) = m$. Therefore, by the theorem of upper-semicontinuity (Theorem 4.4), $\dim H^1(N_s, \Theta_s) = m$ for any s satisfying $|s| < \varepsilon$ provided that ε is sufficiently small. If we take a sufficiently small domain Δ with $0 \in \Delta \subset B$, we have $M_t = N_{h(t)}$, and $|h(t)| < \varepsilon$ for any $t \in \Delta$. Hence we conclude $\dim H^1(M_t, \Theta_t) = m$. Thus, if $m(M)$ is defined, then $\dim H^1(M_t, \Theta_t) = m$ is a constant independent of t.

Conversely, if $\dim H^1(M_t, \Theta_t) = m$ does not depend on $t \in \Delta$, for a sufficiently small domain Δ, $0 \in \Delta \subset B$, then $m(M)$ is defined. *Proof.* For $t = 0$,

$$\left(\frac{\partial M_t}{\partial t_\lambda}\right)_{t=0} \in H^1(M_0, \Theta_0), \qquad \lambda = 1, \ldots, m$$

are linearly independent. Therefore, if $m(M) = \dim H^1(M_t, \Theta_t)$ is independent of t, it follows that

$$\frac{\partial M_t}{\partial t_\lambda} \in H^1(M_t, \Theta_t), \qquad \lambda = 1, \ldots, m$$

are linearly independent for $|t| < \varepsilon$ provided that ε is sufficiently small. This fact will be proved in the next chapter (Theorem 7.11). If we simply write Δ instead of $\Delta_\varepsilon = \{t \in \Delta \,|\, |t| < \varepsilon\}$, then $(\mathcal{M}_\Delta, \Delta, \varpi)$ is an effectively parametrized complex analytic family, and ρ_t: $T_t(\Delta) \to H^1(M_t, \Theta_t)$ is surjective for $t \in \Delta$. Consequently, by the theorem of completeness (Theorem 6.1), the complex analytic family $(\mathcal{M}_\Delta, \Delta, \varpi)$ is complete. Hence the number of moduli of $M = \varpi^{-1}(0)$ is defined and equals m. ∎

Thus we have proved the following

Theorem 6.3. *Suppose $H^2(M, \Theta) = 0$. Then the number of moduli $m(M)$ is defined if and only if* $\dim H^1(M_t, \Theta_t)$ *is independent of $t \in \Delta$ when Δ, $0 \in \Delta \subset B$, is taken sufficiently small. If this condition is satisfied, then $(\mathcal{M}_\Delta, \Delta, \varpi)$ is an effectively parametrized, complete family.*

In case $H^0(M, \Theta) = 0$ holds, which turns out to be true in many examples, the following theorem holds.

Theorem 6.4. *If $H^0(M, \Theta) = H^2(M, \Theta) = 0$, then $m(M)$ is defined and equals* $\dim H^1(M, \Theta)$: $m(M) = \dim H^1(M, \Theta)$.

Proof. Consider a complex analytic family $(\mathcal{M}, B, \varpi)$ satisfying the above conditions (i) $\pi^{-1}(0) = M$, and (ii) $T_0(B) \overset{\rho_0}{\cong} H^1(M, \Theta)$. Since $H^0(M, \Theta) =$

$H^2(M, \Theta)=0$, we have, by Theorem 4.4, $\dim H^0(M_t, \Theta_t)=\dim H^2(M_t, \Theta_t)=0$ for $t\in\Delta$ provided that Δ with $0\in\Delta\subset B$ is sufficiently small. In general, if $\dim H^{q-1}(M_t, \Theta_t)$ and $\dim H^{q+1}(M_t, \Theta_t)$ are independent of t, so is $\dim H^q(M_t, \Theta_t)$. The proof of this fact will be given in Chapter 7 (Corollary to Theorem 7.13). Therefore, in the present case, $\dim H^1(M_t, \Theta_t)$ is also independent of t. Hence, by Theorem 6.3, $m(M)$ is defined and, by Theorem 6.2, $m(M)=\dim H^1(M, \Theta)$. ∎

Moreover, by Theorem 6.3, $(\mathcal{M}_\Delta, \Delta, \varpi)$ is effectively parametrized and complete. *Thus, if* $H^2(M, \Theta)=H^0(M, \Theta)=0$, *M has an effectively parametrized complete family* $(\mathcal{M}_\Delta, \Delta, \varpi)$ *with* $\pi^{-1}(0)=M$.

By studying several examples of compact complex manifolds, we have observed a phenomenon that $m(M)$ coincides with $\dim H^1(M, \Theta)$ for these examples. It is in order to explain this fact that we have so far developed the theory of deformations, and we finally obtained Theorem 6.4, which makes clear that the equality $m(M)=\dim H^1(M, \Theta)$ holds if $H^0(M, \Theta)=H^2(M, \Theta)=0$. This might be considered a milestone of our theory. But the fact is that there are many examples for which $H^0(M, \Theta)=0$ holds, but $H^2(M, \Theta)\neq 0$. So we have to say that Theorem 6.4 can only apply to the very limited class of compact complex manifolds.

(b) Examples

(i) Quadratic transforms of $\mathbb{P}^2$. We consider the quadratic transforms of $\mathbb{P}^2$:

$$M=Q_{q_\nu}\cdots Q_{q_2}Q_{q_1}(\mathbb{P}^2)$$

as in §5.2(a)(iv). As we have seen in §5.2(a), the number m of apparently effective parameters used in constructing M is equal to $\dim H^1(M, \Theta)$. Since $H^1(M, \Theta)$ is 0 for $\nu\leqq 4$, we suppose $\nu\geqq 5$. Then $H^0(M, \Theta)=H^2(M, \Theta)=0$ (see p. 225). Therefore, by Theorem 6.4, $m(M)$ is defined and $m(M)=\dim H^1(M, \Theta)$. Thus the number of parameters m coincides with the number of moduli $m(M)$.

(ii) Quartic surfaces. Let M_t be a non-singular hypersurface of degree h in $\mathbb{P}^{n+1}$ where we suppose $n\geqq 2$ and $h\geqq 2$ (see p. 219). We have seen in §5.2(c) that, except for the case $n=2$ and $h=4$, $m(M_t)$ is defined and the equality $m(M_t)=\dim H^1(M_t, \Theta_t)$ holds (see p. 247). For the case $n=2$ and $h=4$, that is, for the *quartic surfaces* M_t in $\mathbb{P}^3$, it remains unsettled whether the number of moduli $m(M_t)$ is defined or not. By using Theorem 6.4, we can show as follows that the number of moduli $m(M_t)$ is also defined and equals $\dim H^1(M, \Theta)$ in the case of quartic surfaces.

By (3.157)

$$\dim H^2(M_t, \Theta_t)=\dim H^0(M_t, \Omega^1(K_t)),$$

where K_t denotes the canonical bundle of M_t. Let E be the line bundle over $\mathbb{P}^3$ defined by the transition functions $\zeta_{\beta(k)}/\zeta_{\beta(j)}$ (see p. 175), and let E_t be its restriction to M_t. Then, for a non-singular hypersurface M_t of degree h in $\mathbb{P}^3$, we have

$$K_t = E_t^{h-4} \tag{6.45}$$

(see p. 178). In the present case, since $h=4$, K_t is trivial. Hence we have

$$\dim H^2(M_t, \Theta_t) = \dim H^0(M_t, \Omega_t^1),$$

while $H^0(M_t, \Omega_t^1)$ is 0 (see p. 179). Hence $H^2(M_t, \Theta_t)=0$. On the other hand, $H^0(M_t, \Theta_t)=0$ (p. 245). Hence, by Theorem 6.4, $m(M_t)$ is defined, and since $\dim H^1(M_t, \Theta_t)=20$ (see p. 247), we have

$$m(M_t) = \dim H^1(M_t, \Theta_t) = 20.$$

This justifies what we stated in p. 247.

The equality $\dim H^1(M_t, \Theta_t)=20$ also follows easily from the Riemann-Roch-Hirzebruch theorem (5.24). Since K_t is trivial

$$c_1 = -c(K_t) = 0, \qquad \Omega^2 = \mathcal{O}\left(\bigwedge^2 T^*(M_t)\right) = \mathcal{O}(K_t) = 0,$$

Hence $p_g = \dim H^0(M_t, \Omega^2) = \dim H^1(M_t, \mathcal{O}) = 1$. Since $\dim H^0(M_t, \Theta_t) = \dim H^2(M_t, \Theta_t) = 0$, we have by (5.24) that

$$\dim H^1(M_t, \Theta_t) = \tfrac{5}{6}c_2.$$

As $q = \dim H^0(M_t, \Omega^1) = 0$, we have $p_a = p_g - q = 1$. Consequently, by Noether's formula, we have $c_2 = 24$, and hence $\dim H^1(M_t, \Theta_t) = 20$.

(iii) Surfaces of arbitrary degree. We shall compute $\dim H^2(M_t, \Theta_t)$ for a non-singular surface of degree $h \geqq 3$, $h \neq 4$, in $\mathbb{P}^3$. Since, by (6.45), $K_t = E_t^{h-4}$, $\mathcal{O}(K_t)$ is the restriction of $\mathcal{O}(E^{h-4})$ to M_t. If we denote the restriction map by r_t, then, since $[M_t] = E^h$, we have $\ker r_t = \mathcal{O}(E^{h-4} \otimes [M_t]^{-1}) = \mathcal{O}(E^{-4})$. Hence

$$0 \to \mathcal{O}(E^{-4}) \to \mathcal{O}(E^{h-4}) \xrightarrow{r_t} \mathcal{O}(K_t) \to 0$$

is exact. We note that, in the corresponding exact sequence of the cohomology groups

$$\begin{aligned}0 \to H^0(\mathbb{P}^3, \mathcal{O}(E^{-4})) \to H^0(\mathbb{P}^3, \mathcal{O}(E^{h-4})) \to H^0(M_t, \mathcal{O}(K_t))\\ \to H^1(\mathbb{P}^3, \mathcal{O}(E^{-4})) \to \cdots,\end{aligned}$$

we have, by Bott's theorem (Theorem 5.2), $H^0(\mathbb{P}^3, \mathcal{O}(E^{-4})) = H^1(\mathbb{P}^3, \mathcal{O}(E^{-4})) = 0$. Hence

$$H^0(\mathbb{P}^3, \mathcal{O}(E^{h-4})) \cong H^0(M_t, \mathcal{O}(K_t)). \tag{6.46}$$

As is easily seen from the proof of Lemma 5.4, for any natural number k, $\dim H^0(\mathbb{P}^3, \mathcal{O}(E^k))$ is equal to the number of monomials of degree k in ξ_0, ξ_1, ξ_2, ξ_3, that is $\binom{3+k}{k}$. Hence, by (6.46)

$$p_g = \tfrac{1}{6}(h-1)(h-2)(h-3).$$

Next, since $\dim H^0(M_t, \Theta_t) = 0$, using (5.24) and (5.25), we get

$$\dim H^2(M_t, \Theta_t) - \dim H^1(M_t, \Theta_t) = 2c_1^2 - 10(p_a + 1).$$

Since $c(E_t)^2 = h$ (p. 180), we have $c_1^2 = (h-4)^2 h$ by (6.45), while $p_g = p_a$ because $q = \dim H^0(M_t, \Omega^1) = 0$. Consequently we obtain

$$\dim H^2(M_t, \Theta_t) - \dim H^1(M_t, \Theta_t) = 2(h-4)^2 h - \tfrac{10}{6}(h^3 - 6h^2 + 11h). \tag{6.47}$$

Since we have by (5.21)

$$\dim H^1(M_t, \Theta_t) = \tfrac{1}{6}(h+3)(h+2)(h+1) - 16, \qquad h \neq 4, \tag{6.48}$$

we conclude

$$\dim H^2(M_t, \Theta_t) = \tfrac{1}{2}(h-2)(h-3)(h-5), \qquad h \neq 4. \tag{6.49}$$

In particular, $H^2(M_t, \Theta_t) \neq 0$ for $h \geqq 6$. Thus Theorem 6.4 does not apply to these cases and cannot explain the fact that $m(M_t)$ coincides with $\dim H^1(M_t, \Theta_t)$ in these cases.

As stated before, for a hypersurface M of an Abelian variety A^{n+1}, the number of moduli $m(M)$ is defined and the equality $m(M) = \dim H^1(M, \Theta)$ holds (p. 248). Using the Riemann-Roch-Hirzebruch theorem, we can compute $\dim H^2(M, \Theta)$ for a surface M in A^3. The result is $\dim H^2(M, \Theta) = 3e + 5$, where e is a suitable natural number (see [21]). Hence we always have $H^2(M, \Theta) \neq 0$ in this case.

(iv) Consider the complex analytic family constructed in (2.33) in case $m = 2$ and $h = 1$. Employing the same notation as in Example 2.16, we have

$$M_t = U_1 \times \mathbb{P}^1 \cup U_2 \times \mathbb{P}^1, \qquad U_1 = U_2 = \mathbb{C},$$

where $(z_1, \zeta_1) \in U_1 \times \mathbb{P}^1$ and $(z_2, \zeta_2) \in U_2 \times \mathbb{P}^1$ are the same point of M_t if and only if

$$z_1 z_2 = 1, \qquad \zeta_1 = z_2^2 \zeta_2 + t z_2. \tag{6.50}$$

We have seen that $M_0 = \tilde{M}_2$, and $M_t = \tilde{M}_0 = \mathbb{P}^1 \times \mathbb{P}^1$ for $t \neq 0$ (p. 73). Recall that

$$\tilde{M}_m = U_1 \times \mathbb{P}^1 \cup U_2 \times \mathbb{P}^1,$$

where $(z_1, \zeta_1) \in U_1 \times \mathbb{P}^1$ and $(z_2, \zeta_2) \in U_2 \times \mathbb{P}^1$ are identified if $z_1 z_2 = 1$ and $\zeta_1 = z_2^m \zeta_2$. To compute $\dim H^1(\tilde{M}_m, \Theta)$, we first prove $H^1(\mathbb{C} \times \mathbb{P}^1, \Theta) = 0$.

Proof. Since $\mathbb{P}^1$ is a union of U_1 and U_2, $\mathbb{C} \times \mathbb{P}^1$ is written as

$$\mathbb{C} \times \mathbb{P}^1 = V_1 \cup V_2, \quad \text{where} \quad V_1 = \mathbb{C} \times U_1, \quad V_2 = \mathbb{C} \times U_2,$$

i.e., $\mathfrak{V} = \{V_1, V_2\}$ forms an open covering of $\mathbb{C} \times \mathbb{P}^1$ by two coordinate neighbourhoods V_1 and V_2. Both V_1 and V_2 are biholomorphic to $\mathbb{C}^2 = \mathbb{C} \times \mathbb{C}$, and Dolbeault's lemma (Theorem 3.3) is valid for $\mathbb{C}^2$ regarded as a bidisk of radius ∞. Hence Lemma 5.2 applies to the covering $\mathfrak{V}$, and we obtain

$$H^1(\mathbb{C} \times \mathbb{P}^1, \Theta) = H^1(\mathfrak{V}, \Theta) = Z^1(\mathfrak{V}, \Theta)/\delta C^0(\mathfrak{V}, \Theta).$$

A 1-cocycle $\{\theta_{12}, \theta_{21}\}$ belonging to $Z^1(\mathfrak{V}, \Theta)$ consists of holomorphic vector fields θ_{12} and θ_{21} on $V_1 \cap V_2$ with $\theta_{12} = -\theta_{21}$, while $C^0(\mathfrak{V}, \Theta)$ is the set of all pairs $\{\theta_1, \theta_2\}$, where θ_1 and θ_2 are holomorphic vector fields respectively on V_1 and V_2. Hence, in order to prove $H^1(\mathfrak{V}, \Theta) = 0$, it suffices to show that any 1-cocycle $\{\theta_{12}, \theta_{21}\}$ is represented as $\delta\{\theta_2, \theta_2\}$, i.e.,

$$\theta_{12} = \theta_2 - \theta_1.$$

Let w be the coordinate on $\mathbb{C}$. Then, in terms of the coordinates (w, z_1) on $V_1 = U_1 \times \mathbb{P}^1$, the holomorphic vector field θ_{12} on $V_1 \cap V_2$ is written in the form

$$\theta_{12} = u(w, z_1)\frac{\partial}{\partial w} + v(w, z_1)\frac{\partial}{\partial z_1},$$

where $u(w, z_1)$ and $v(w, z_1)$ are holomorphic functions on $V_1 \cap V_2 = \mathbb{C} \times (U_1 \cap U_2) = \mathbb{C} \times \mathbb{C}^*$. Hence they can be expanded into Laurent series in z_1:

$$u(w, z_1) = \sum_{n=-\infty}^{+\infty} u_n(w) z_1^n, \qquad v(w, z_1) = \sum_{n=-\infty}^{+\infty} v_n(w) z_1^n,$$

where the $u_n(w)$, $v_n(w)$ are entire functions of w. In view of $z_1 = 1/z_2$ and

$\partial/\partial z_1 = -z_2^2(\partial/\partial z_2)$, if we set

$$\theta_1 = -\sum_{n=0}^{+\infty} u_n(w) z_1^n \frac{\partial}{\partial w} - \sum_{n=0}^{+\infty} v_n(w) z_1^n \frac{\partial}{\partial z_1},$$

$$\theta_2 = \sum_{n=1}^{+\infty} u_{-n}(w) z_2^n \frac{\partial}{\partial w} - \sum_{n=1}^{+\infty} v_{-n}(w) z_2^{n+2} \frac{\partial}{\partial z_2},$$

then θ_1 and θ_2 are holomorphic vector fields on V_1 and V_2, respectively, and satisfy $\theta_{12} = \theta_2 - \theta_1$. ∎

Put $W_1 = U_1 \times \mathbb{P}^1$ and $W_2 = U_2 \times \mathbb{P}^1$. Then $\mathfrak{W} = \{W_1, W_2\}$ is an open covering of $\tilde{M}_m$, and both W_1 and W_2 are biholomorphic to $\mathbb{C} \times \mathbb{P}^1$. Hence $H^1(W_1, \Theta) = H^1(W_2, \Theta) = 0$. Therefore, by Theorem 3.4, we have

$$H^1(\tilde{M}_m, \Theta) = H^1(\mathfrak{W}, \Theta) = Z^1(\mathfrak{W}, \Theta)/\delta C^0(\mathfrak{W}, \Theta).$$

We represent a 1-cocycle $\{\theta_{12}, \theta_{21}\} \in Z^1(\mathfrak{W}, \Theta)$ by the holomorphic vector field $\theta_{12} = -\theta_{21}$ on $W_1 \cap W_2$. Then, this 1-cocycle belongs to $\delta C^0(\mathfrak{W}, \Theta)$ if and only if there exist holomorphic vector fields θ_1 and θ_2 respectively on W_1 and W_2 such that

$$\theta_2 - \theta_1 = \theta_{12}. \tag{6.51}$$

By (2.36) θ_1 is written in the form

$$\theta_1 = v_1(z_1)\frac{\partial}{\partial z_1} + (\alpha_1(z_1)\zeta_1^2 + \beta_1(z_1)\zeta_1 + \gamma_1(z_1))\frac{\partial}{\partial \zeta_1},$$

where $v_1(z_1)$, $\alpha_1(z_1)$, $\beta_1(z_1)$, $\gamma_1(z_1)$ are entire functions of z_1. Similarly, by (2.37), θ_2 is written in the form

$$\theta_2 = v_2(z_2)\frac{\partial}{\partial z_2} + (\alpha_2(z_2)\zeta_2^2 + \beta_2(z_2)\zeta_2 + \gamma_2(z_2))\frac{\partial}{\partial \zeta_2},$$

where $v_2(z_2)$, $\alpha_2(z_2)$, $\beta_2(z_2)$, $\gamma_2(z_2)$ are entire functions of z_2. We write θ_{12} as follows in terms of the coordinates (z_1, ζ_1):

$$\theta_{12} = v_{12}(z_1)\frac{\partial}{\partial z_1} + (\alpha_{12}(z_1)\zeta_1^2 + \beta_{12}(z_1)\zeta_1 + \gamma_{12}(z_1))\frac{\partial}{\partial \zeta_1}.$$

Here $v_{12}(z_1)$, $\alpha_{12}(z_1)$, $\beta_{12}(z_1)$, $\gamma_{12}(z_1)$ are holomorphic functions of z_1 on $U_1 \cap U_2 = \mathbb{C}^*$, hence, are expanded into Laurent series in z_1. In terms of the coordinates (z_1, ζ_1) with $z_1 = 1/z_2$, $\zeta_1 = z_2^m \zeta_2$ in place of (z_2, ζ_2), θ_2 is written in the form

$$-z_1^2 v_2(z_2)\frac{\partial}{\partial z_1}+\left(z_1^m \alpha_2(z_2)\zeta_1^2+mz_1 v_2(z_2)\zeta_1+\beta_2(z_2)\zeta_1+\frac{\gamma_2(z_2)}{z_1^m}\right)\frac{\partial}{\partial \zeta_1}$$

(see p. 74). Hence the equation (6.51) is reduced to the following system of equations:

$$\begin{cases} -z_1^2 v_2\left(\dfrac{1}{z_1}\right)-v_1(z_1)=v_{12}(z_1), \\ z_1^m \alpha_2\left(\dfrac{1}{z_1}\right)-\alpha_1(z_1)=\alpha_{12}(z_1), \\ mz_1 v_2\left(\dfrac{1}{z_1}\right)+\beta_2\left(\dfrac{1}{z_1}\right)-\beta_1(z_1)=\beta_{12}(z_1), \\ \dfrac{1}{z_1^m}\gamma_2\left(\dfrac{1}{z_1}\right)-\gamma_1(z_1)=\gamma_{12}(z_1). \end{cases}$$

In case $m=0$ or 1, these equations always have a solution. For $m \geqq 2$, let $\gamma_{12}(z_1)=\sum_{n=-\infty}^{+\infty} c_n z_1^n$ be the Laurent expansion of $\gamma_{12}(z_1)$. Then the above equations have a solution if and only if $c_{-1}=c_{-2}=\cdots=c_{-m+1}=0$. Hence we obtain

$$\dim H^1(\tilde{M}_m, \Theta)=\begin{cases} 0, & m=0,1, \\ m-1, & m=2,3,4,\ldots, \end{cases} \tag{6.52}$$

and for $m \geqq 2$, the 1-cocycles

$$\theta_{12}=z_2^k \frac{\partial}{\partial \zeta_1} \in Z^1(\mathfrak{W}, \Theta), \qquad k=1,2,\ldots,m-1, \tag{6.53}$$

form a basis of $H^1(\tilde{M}_m, \Theta) \cong H^1(\mathfrak{W}, \Theta)$.

Next we shall prove

$$H^2(\tilde{M}_l, \Theta)=0. \tag{6.54}$$

Proof. Since we have $\dim H^2(\tilde{M}_m, \Theta)=\dim H^0(\tilde{M}_m, \Omega^1(K))$ by (3.157), it suffices to prove $H^0(\tilde{M}_m, \Omega^1(K))=0$. Since $\tilde{M}_m=U_1\times\mathbb{P}^1 \cup U_2\times\mathbb{P}^1$, $\mathbb{P}^1=\mathbb{C}\cup\{\infty\}$, any $\psi\in H^0(\tilde{M}_m, \Omega^1(K))$ is written on $U_1\times\mathbb{C}\subset U_1\times\mathbb{P}^1$ in the form

$$\psi=(g(z_1,\zeta_1)\,dz_1+h(z_1,\zeta_1)\,d\zeta_1)\otimes(dz_1\wedge d\zeta_1),$$

where $g(z_1, \zeta_1)$ and $h(z_1, \zeta_1)$ are holomorphic functions of z_1 and $\zeta_1 \neq \infty$. We note that ψ is required to be holomorphic in a neighbourhood of $U_1 \times \infty$. Changing the coordinate ζ_1 to the local coordinate $w = 1/\zeta_1$ at ∞, we obtain

$$\psi = -\left(g\left(z_1, \frac{1}{w}\right) dz_1 - h\left(z_1, \frac{1}{w}\right)\frac{dw}{w^2}\right) \otimes \left(dz_1 \wedge \frac{dw}{w^2}\right).$$

Thus $(1/w^2)g(z_1, 1/w)$ and $(1/w^4)h(z_1, 1/w)$ must be holomorphic in w. Hence $g(z_1, \zeta_1) = h(z_1, \zeta_1) = 0$, i.e., $\psi = 0$. ∎

With these preparations, we return to the study of the complex analytic family defined by (6.50). Since $M_0 = \tilde{M}_2$, and $M_t = \tilde{M}_0$, $t \neq 0$, we have, by (6.52),

$$\dim H^1(M_t, \Theta_t) = \begin{cases} 1, & t = 0, \\ 0, & t \neq 0. \end{cases} \tag{6.55}$$

In order to compute the infinitesimal deformation of M_t, we write (6.50) as

$$z_1 = f_{12}^1(z_2, \zeta_2, t), \qquad \zeta_1 = f_{12}^2(z_2, \zeta_2, t),$$

where $f_{12}^1(z_2, \zeta_2, t) = 1/z_2$ is independent of t, and $f_{12}^2(z_2, \zeta_2, t) = z_2^2\zeta_2 + tz_2$. The infinitesimal deformation $dM_t/dt \in H^1(M_t, \Theta_t)$ is nothing but the cohomology class of the 1-cocycle

$$\theta_{12}(t) = \frac{\partial f_{12}^2(z_2, \zeta_2, t)}{\partial t}\frac{\partial}{\partial \zeta_1} = z_2\frac{\partial}{\partial \zeta_1}.$$

For $t = 0$, $z_2(\partial/\partial\zeta_1)$ forms a basis of $H^1(M_0, \Theta_0) = H^1(\tilde{M}_2, \Theta)$ by (6.53). Hence ρ_0: $T_0(C) \to H^1(M_0, \Theta_0)$ is surjective. Therefore, by the theorem of completeness (Theorem 6.1), the family $\{M_t \mid t \in \mathbb{C}\}$ is complete at $t = 0$. Since $H^2(\tilde{M}_2, \Theta) = 0$ by (6.54), there exists a family $(\mathcal{M}, B, \varpi)$ such that $\varpi^{-1}(0) = \tilde{M}_2$ and that ρ_0: $T_0(B) \cong H^1(\tilde{M}_2, \Theta)$ with $0 \in B \subset C$. This is nothing but our $\{M_t \mid t \in \mathbb{C}\}$. For $t \neq 0$, $H^1(M_t, \Theta_t) = 0$ by (6.55). Hence $dM_t/dt = 0$ and the family $\{M_t \mid t \in \mathbb{C}\}$ is complete at any t. Therefore $\{M_t \mid t \in \mathbb{C}\}$ *is a complete complex analytic family.*

We can also prove $dM_t/dt = 0$ for $t \neq 0$ by direct calculation. By (6.50)

$$\frac{\partial}{\partial z_2} = -\frac{1}{z_2^2}\frac{\partial}{\partial z_1} + (2z_2\zeta_2 + t)\frac{\partial}{\partial \zeta_1}, \qquad \frac{\partial}{\partial \zeta_2} = z_2^2\frac{\partial}{\partial \zeta_1}.$$

Hence

$$tz_2\frac{\partial}{\partial \zeta_1} = z_2\frac{\partial}{\partial z_2} - 2\zeta_2\frac{\partial}{\partial \zeta_2} + z_1\frac{\partial}{\partial z_1}.$$

Therefore, if we put

$$\theta_1(t) = -\frac{z_1}{t}\frac{\partial}{\partial z_1}, \qquad \theta_2(t) = \frac{1}{t}\left(z_2\frac{\partial}{\partial z_2} - 2\zeta_2\frac{\partial}{\partial \zeta_2}\right),$$

$\theta_1(t)$ and $\theta_2(t)$ are holomorphic vector fields on $U_1 \times \mathbb{P}^1$ and $U_2 \times \mathbb{P}^1$ respectively, and we have

$$\theta_{12}(t) = \theta_2(t) - \theta_1(t).$$

Hence, for $t \neq 0$, the cohomology class dM_t/dt of the 1-cocycle $\theta_{12}(t)$ is 0. Note that the limit $\theta_1(0) = \lim_{t\to 0} \theta_1(t)$, $\theta_2(0) = \lim_{t\to 0} \theta_2(t)$ do not exist. This explains the reason why $dM_t/dt \neq 0$ for $t = 0$, even though $\theta_{12}(t) = z_2(\partial/\partial\zeta_1)$ is apparently independent of t.

For the complete family $\{M_t \mid t \in \mathbb{C}\}$, we have $\dim H^1(M_0, \Theta_0) = 1$, and $\dim H^1(M_t, \Theta_t) = 0$ for $t \neq 0$. This implies, by Theorem 6.3, that the number of moduli $m(\tilde{M}_2)$ is not defined. Recall that $\dim H^0(\tilde{M}_2, \Theta) = 7$ (see p. 75). This example shows that even if $H^2(M, \Theta) = 0$, the number of moduli $m(M)$ is not necessarily defined in case $H^0(M, \Theta) \neq 0$.

The complete family $\{M_t \mid t \in \mathbb{C}\}$ contains all sufficiently small deformations of $\tilde{M}_2 = M_0$ (p. 228). Therefore for the study of small deformations of $\tilde{M}_2$, it suffices to investigate $\{M_t \mid t \in \mathbb{C}\}$. For $t \neq 0$, $M_t = \tilde{M}_0 = \mathbb{P}^1 \times \mathbb{P}^1$, so M_t "jumps" from $\tilde{M}_2$ to $\tilde{M}_0$ at the moment t moves away from 0, and M_t does not change as far as $t \neq 0$. This suggests that the number of parameters for $\tilde{M}_2$ is not 0, but "almost" 0. The existence of such examples as $\tilde{M}_2$ above is one of the reasons why we do not define $m(M)$ if there exists no effectively parametrized complete family (M, B, ϖ) with $\varpi^{-1}(0) = M$.

Let $(\mathcal{M}, B, \varpi)$ be a complex analytic (or differentiable) family and let $M_t = \varpi^{-1}(t)$. Suppose that $H^1(M_{t^0}, \Theta_{t^0}) = 0$ at some point $t^0 \in B$. Then, by the Frölicher-Nijenhuis theorem (Theorem 4.5), $M_t = M_{t^0}$ for all t sufficiently near t^0. In other words, the complex structure of M_{t^0} does not vary by a small change of the parameter t. But for a large deviation of t from t^0, M_t is not necessarily biholomorphic to M_{t^0}. The above-mentioned family $\{M_t \mid t \in \mathbb{C}\}$ gives such an example. Namely, if we take $t^0 = 1$, then $M_1 = \mathbb{P}^1 \times \mathbb{P}^1$, $H^1(M_1, \Theta_1) = 0$, but $M_0 = \tilde{M}_2 \neq M_1$.

§6.3. Later Developments

We have explained above how the theory of deformations of compact complex manifolds developed till about 1960. We observed the phenomenon that for several compact complex manifolds M, their number of moduli $m(M)$ coincides with $\dim H^1(M, \Theta)$, and conjectured that the equality $m(M) = \dim H^1(M, \Theta)$ might hold for any compact complex manifolds. Considering this as a working hypothesis, we developed the theory of deformations and proved that the conjecture is true if $H^2(M, \Theta) = H^0(M, \Theta) = 0$ (Theorem 6.4). However, with many examples, *we find that* $H^2(M, \Theta) \neq 0$. For example, as in (b) of the previous section, $H^2(M, \Theta) \neq 0$ for a non-singular surface M of degree at least 6 in $\mathbb{P}^3$. In spite of this, the equality $m(M) = \dim H^1(M, \Theta)$ holds for these surfaces. At that time we could find no example of M for which $m(M) \neq \dim H^1(M, \Theta)$. In order to explain this strange phenomenon, we thought that we need to generalize the theorem of existence (Theorem 5.6) to the case $H^2(M, \Theta) \neq 0$. As stated earlier (p. 259), there exists an M with $\theta \in H^1(M, \Theta)$ such that $[\theta, \theta] \neq 0$. Therefore, in general, there does not exist, for a given M, a complex analytic family $(\mathcal{M}, B, \varpi)$ satisfying the conditions (i) and (ii) of Theorem 5.6. But at that time—about 1958—we could not understand at all what should be the conditions for the complex analytic family that we must seek. We did not know what is to be proved to exist. H. Grauert, who was working very energetically on several complex variables, agreed with us that this was a difficult problem.

The theorem of existence for general case was proved by M. Kuranishi in 1961 ([22]). We will explain the outline of his theory below.

We first recall the equality (5.105):

$$\mathscr{L}^{0,q}(T) = \mathbf{H}^{0,q}(T) \oplus \square \mathscr{L}^{0,q}(T), \qquad T = T(M).$$

Substituting $\mathbf{H}^{0,q}(T) \oplus \square \mathscr{L}^{0,q}(T)$ for $\mathscr{L}^{0,q}(T)$ in the right-hand side, we obtain, in view of $\square \mathbf{H}^{0,q}(T) = 0$,

$$\mathscr{L}^{0,q}(T) = \mathbf{H}^{0,q}(T) \oplus \square \square \mathscr{L}^{0,q}(T). \tag{6.56}$$

Hence any ψ in $\mathscr{L}^{0,q}(T)$ is written in the form

$$\psi = \eta + \square \varphi, \qquad \eta \in \mathscr{H}^{0,q}(T), \qquad \varphi \in \square \mathscr{L}^{0,q}(T).$$

Since $\mathbf{H}^{0,q}(T)$ and $\square \mathscr{L}^{0,q}(T)$ are orthogonal to each other, η and φ are determined uniquely by ψ. Writing $H\psi$ for η and $G\psi$ for φ, we obtain

$$\psi = H\psi + \square G\psi, \qquad H\psi \in \mathbf{H}^{0,q}(T), \qquad G\psi \in \square \mathscr{L}^{0,q}(T). \tag{6.57}$$

Note that $H: \psi \to H\psi$ and $G: \psi \to G\psi$ are linear operators. We call $H\psi$ the harmonic part of ψ, and G the Green operator. The Green operator that was introduced in §5.3(c) is a special case of this one. Since we have $H\bar{\partial}\psi = 0$, it follows by (6.57) that $\bar{\partial}\psi = \square G \bar{\partial}\psi$. On the other hand, applying $\bar{\partial}$ to both sides of (6.57), we obtain

$$\bar{\partial}\psi = \bar{\partial}\square G\psi = \bar{\partial}\mathfrak{d}\bar{\partial} G\psi = \square \bar{\partial} G\psi$$

because $\square = \bar{\partial}\mathfrak{d} + \mathfrak{d}\bar{\partial}$. Hence we obtain

$$G\bar{\partial} = \bar{\partial}G. \tag{6.58}$$

For this G, we have the following estimate similar to (5.111):

$$|G\psi|_{k+\alpha} \leqq c_*|\psi|_{k-2+\alpha}, \qquad \psi \in \mathscr{L}^{0,q}(T), \quad k \geqq 2, \tag{6.59}$$

where c_* is a constant independent of ψ.

Let $\{\beta_1, \ldots, \beta_m\}$ be a basis of $\mathbf{H}^{0,1}(T)$ and put

$$\varphi_1(t) = \beta_1 t_1 + \cdots + \beta_m t_m.$$

We shall determine a power series

$$\varphi(t) = \varphi_1(t) + \sum_{\nu_1+\cdots+\nu_m \geqq 2} \varphi_{\nu_1\cdots\nu_m} t_1^{\nu_1}\cdots t_m^{\nu_m}, \qquad \varphi_{\nu_1\cdots\nu_m} \in \mathscr{L}^{0,1}(T),$$

in $t_1, \ldots, t_m$, by requiring the condition

$$\varphi(t) = \varphi_1(t) + \tfrac{1}{2}\mathfrak{d}G[\varphi(t), \varphi(t)]. \tag{6.60}$$

Writing $\varphi(t) = \varphi_1(t) + \sum_{\nu=2}^{\infty} \varphi_\nu(t)$, where $\varphi_\nu(t)$ is a homogeneous polynomial of degree ν in $t_1, \ldots, t_m$, (6.60) is equivalent to

$$\varphi_\nu(t) = \frac{1}{2}\sum_{\mu=1}^{\nu-1} \mathfrak{d}G[\varphi_\mu(t), \varphi_{\nu-\mu}(t)], \qquad \nu = 2, 3, \ldots.$$

Hence $\varphi(t)$ is uniquely determined by the condition (6.60). By the same method as in §5.3(c), we can prove that $\varphi(t)$ converges with respect to the Hölder norm $|\ |_{k+\alpha}$ for t with $|t| < \varepsilon$, provided $\varepsilon > 0$ is sufficiently small, and that $\varphi(t)$ is C^∞ on $M \times \Delta_\varepsilon$, where $\Delta_\varepsilon = \{t \in \mathbb{C}^m \,|\, |t| < \varepsilon\}$.

Lemma 6.3. *Suppose $\varepsilon > 0$ is sufficiently small and $|t| < \varepsilon$. Then $\varphi(t)$ in (6.60) satisfies the integrability condition* (5.86):

$$\bar{\partial}\varphi(t) = \tfrac{1}{2}[\varphi(t), \varphi(t)]$$

if and only if

$$H[\varphi(t), \varphi(t)] = 0.$$

Proof. For simplicity we put $\varphi = \varphi(t)$. Since $H\bar{\partial} = 0$, the equality $\bar{\partial}\varphi = \frac{1}{2}[\varphi, \varphi]$ implies $H[\varphi, \varphi] = 0$.

Conversely, suppose that $H[\varphi, \varphi]=0$, and put $\psi=\frac{1}{2}[\varphi, \varphi]-\bar{\partial}\varphi$. Since $\bar{\partial}\varphi_1(t)=0$, applying $\bar{\partial}$ to both sides of (6.60), we obtain

$$\bar{\partial}\varphi=\tfrac{1}{2}\mathfrak{d}G[\varphi, \varphi].$$

Since $H[\varphi, \varphi]=0$, $[\varphi, \varphi]=\square G[\varphi, \varphi]$ by (6.57). Therefore, since $\bar{\partial}G=G\bar{\partial}$ by (6.58), it follows

$$\psi=\tfrac{1}{2}(\square-\bar{\partial}\mathfrak{d})G[\varphi, \varphi]=\tfrac{1}{2}\mathfrak{d}\bar{\partial}G[\varphi, \varphi]=\tfrac{1}{2}\mathfrak{d}G\bar{\partial}[\varphi, \varphi].$$

Using (5.88) and (5.87), we obtain

$$\bar{\partial}[\varphi, \varphi]=2[\bar{\partial}\varphi, \varphi]=[[\varphi, \varphi], \varphi]-2[\psi, \varphi],$$

while $[[\varphi, \varphi], \varphi]=0$ by (5.89), hence

$$\psi=-\mathfrak{d}G[\psi, \varphi].$$

By (5.118), (5.119), and (6.59)

$$\begin{aligned} |\mathfrak{d}G[\psi, \varphi]|_{k+\alpha} &\leqq K_1|G[\psi, \varphi]|_{k+1+\alpha} \leqq K_1 c_*|[\psi, \varphi]|_{k-1+\alpha} \\ &\leqq K_1 K_2 c_*|\psi|_{k+\alpha}|\varphi|_{k+\alpha}. \end{aligned}$$

Hence

$$|\psi|_{k+\alpha} \leqq K_1 K_2 c_*|\psi|_{k+\alpha}|\varphi(t)|_{k+\alpha}.$$

If ε is sufficiently small and if $|t|<\varepsilon$, then $K_1 K_2 c_*|\varphi(t)|_{k+\alpha}<\frac{1}{2}$. Hence $|\psi|_{k+\alpha}<\frac{1}{2}|\psi|_{k+\alpha}$, and $\psi=0$. Thus we obtain $\bar{\partial}\varphi=\frac{1}{2}[\varphi, \varphi]$. ∎

Let $\{\gamma_1, \ldots, \gamma_l\}$, $l=\dim H^2(M, \Theta)$, be an orthonormal basis of $\mathbf{H}^{0,2}(T)$. Then we can write

$$H[\varphi(t), \varphi(t)]=\sum_{k=1}^{l}([\varphi(t), \varphi(t)], \gamma_k)\gamma_k.$$

Put $f_k(t)=([\varphi(t), \varphi(t)], \gamma_k)$. Then the $f_k(t)$ are holomorphic functions of t. By Lemma 6.3, $\varphi(t)$ satisfies the integrability condition if and only if $f_1(t)=\cdots=f_l(t)=0$. Therefore, if we set

$$B=\{t\in\Delta_\varepsilon \,|\, f_1(t)=\cdots=f_l(t)=0\},$$

then B is an analytic subset of Δ (p. 33), and, by the Newlander–Nirenberg theorem (Theorem 5.5), $\varphi(t)$, for each $t\in\Delta$, defines a complex structure M_t on the differentiable manifold M. In this way, we obtain a "complex analytic family" $\{M_t \mid t \in B\}$ of deformations of M. But, note that in this case B may have singularities so that we must suitably generalize the definition of "complex analytic family". Furthermore, the family $\{M_t \mid t \in B\}$ can be proved to be complete in a sense similar to Definition 5.3 ([22]), but the proof is rather difficult.

Theorem 6.5 (Theorem of Existence). *For any compact complex manifold M, there exists a complete complex analytic family $\{M_t \mid t \in B\}$, $0 \in B \subset \Delta_\varepsilon$ with $M_0 = M$.*

The above family $\{M_t \mid t \in B\}$ is called the *Kuranishi family.* Here B is an analytic subset of Δ_ε defined by l holomorphic equations $f_1(t) = \cdots = f_l(t) = 0$, where $\dim \Delta_\varepsilon = m = \dim H^1(M, \Theta)$, and $l = \dim H^2(M, \Theta)$. Hence, we have

$$\dim B \geqq \dim H^1(M, \Theta) - \dim H^2(M, \Theta). \tag{6.61}$$

From this inequality, it follows that

$$m(M) \geqq \dim H^1(M, \Theta) - \dim H^2(M, \Theta), \tag{6.62}$$

provided $m(M)$ is defined.

If $H^2(M, \Theta) = 0$, Theorem 6.5 is reduced to Theorem 5.6, hence Theorem 6.5 is an extension of Theorem 5.6. But, if we take the view-point of trying to explain the phenomenon that the equality $m(M) = \dim H^1(M, \Theta)$ holds for various examples of M, then Theorem 6.5 is quite different in nature from Theorem 5.6. For example, let M be a non-singular surface of degree h in $\mathbb{P}^3$. Then, by (6.47)

$$\dim H^1(M, \Theta) - \dim H^2(M, \Theta) = -\frac{h}{3}(h^2 - 18h + 41).$$

Hence the left-hand side is negative when $h \geqq 16$, and in these cases the inequalities (6.61) and (6.62) are trivial. Thus in case $\dim H^2(M, \Theta) \geqq \dim H^1(M, \Theta)$, Theorem 6.5 is an existence theorem for its own sake, and gives us no information about the family $\{M_t \mid t \in B\}$ except that it is complete. In particular, Theorem 6.5 is ineffective in explaining that $m(M)$ coincides with $\dim H^1(M, \Theta)$ for various M. *Nevertheless, it is obvious that Theorem 6.5 is of fundamental importance in the theory of deformations of complex structures.* This reflects a fundamental difference between mathematics and physics. In physics, if a theory is developed for the purpose of explaining certain phenomenon and fails to do so, then that theory must be useless.

Even with Theorem 6.5 in hand, we could not prove our conjecture $m(M) = \dim H^1(M, \Theta)$, so we suspected that there might exist counterexamples to our conjecture. In search of such counterexamples, we calculated $m(M)$ and $\dim H^1(M, \Theta)$ for various algebraic surfaces. We believed that algebraic surfaces were relatively easy to handle, but calculating $m(M)$ and $\dim H^1(M, \Theta)$ was difficult in most cases. Furthermore, to our surprise, $\dim H^1(M, \Theta)$ was much more difficult to calculate than $m(M)$. The dimension of $H^1(M, \Theta)$ is determined only by M itself, while, in order to calculate $m(M)$, we must know an effectively parametrized complete complex analytic family $(\mathcal{M}, B, \varpi)$ with $\varpi^{-1}(0) = M$. So we supposed that $\dim H^1(M, \Theta)$ would be easier to calculate than $m(M)$, but it turned out to be the contrary. For many examples of M, while $m(M)$ was calculated, we could not find $\dim H^1(M, \Theta)$. Still, when we got both of $m(M)$ and $\dim H^1(M, \Theta)$, we found that $m(M) = \dim H^1(M, \Theta)$ holds.

In 1962, D. Mumford constructed a counterexample to this conjecture ([25]). His example is a 3-dimensional complex manifold $M = \mu_C(\mathbb{P}^3)$ obtained from $\mathbb{P}^3$ by a *monoidal transformation* μ_C whose centre is a certain curve $C \subset \mathbb{P}^3$ of genus 24 and degree 14.

Because Mumford's construction cannot be applied to the case of surfaces, we hoped that the conjecture might be true for surfaces, i.e. for 2-dimensional compact complex manifolds. However, in 1967, A. Kas found a 2-dimensional counterexample. A surface M^2 is called an *elliptic surface* if the following two conditions are satisfied: (i) There exists a holomorphic map $\Phi: M^2 \to \Delta$ of M onto an algebraic curve, i.e., a compact Riemann surface Δ. (ii) For a point $u \in \Delta$, the inverse image $C_u = \Phi^{-1}(u)$ is an elliptic curve except for a finite number of points $u = a_1, \ldots, a_\nu$. (For details on elliptic surfaces, see [15].) For a general elliptic surface M^2, the equality $m(M) = \dim H^1(M, \Theta)$ holds, while, for any pair of natural numbers k and g satisfying $g > 24k - 2$, Kas constructed an elliptic surface M^2 with

$$m(M^2) = 3k + 4g - 3,$$

$$\dim H^1(M^2, \Theta) = 11k + 4g - 3$$

([13]). As to this elliptic surface M^2, $\Delta = \Phi(M^2)$ is a curve of genus g, and there exist $2k$ points a_λ, $\lambda = 1, 2, \ldots, 2k$, on Δ such that, for $u \neq a_\lambda$, $C = C_u$ is a fixed elliptic curve not depending on u. Thus M^2 is a certain special elliptic surface with rather simple structure.

In this way, the conjecture $m(M) = \dim H^1(M, \Theta)$ ultimately turned out to be false. However, because of the facts that this conjecture played an important role as a working hypothesis in the development of the theory of deformations and that it was quite difficult to find counterexamples, and so on, we supposed that those M for which $m(M) < \dim H^1(M, \Theta)$ holds might have some special structure, and that for general M, $m(M)$ would coincide with $\dim H^1(M, \Theta)$ even if $H^2(M, \Theta) \neq 0$. But we could not succeed in finding a simple and useful criterion for the equality $m(M) = \dim H^1(M, \Theta)$.

Chapter 7

Theorem of Stability

In this chapter we give proofs of Theorem 4.1, Theorem 4.4, Lemma 4.1, and other results used in the previous chapters. The proofs depend on the theory of differentiable families of strongly elliptic differential operators.

§7.1. Differentiable Family of Strongly Elliptic Differential Operators

(a) Strongly Elliptic Partial Differential Operators

Let X be a compact differentiable manifold and $\{U_j\}$ a sufficiently fine open covering of X where each U_j is a coordinate neighbourhood. Let x_j: $x \to x_j = (x_j^1, \ldots, x_j^n)$ be a system of C^∞ local coordinates of X defined on some domain containing $[U_j]$. We assume that X is oriented and that its orientation is determined by the coordinate system $\{x_j\}$. Let B be a complex vector bundle over X, π its projection, and $(\zeta_j^1, \ldots, \zeta_j^\nu)$ the fibre coordinates of B defined over some domain containing $[U_J]$. If we let $b_{jk}(x) = \{b_{jk\mu}^\lambda(x)\}$ be a system of transition functions of B, then $\pi^{-1}(U_j) \cong U_j \times \mathbb{C}^\nu$, and the points $(x, \zeta_j^1, \ldots, \zeta_j^\nu) \in U_j \times \mathbb{C}^\nu$ and $(x, \zeta_k^1, \ldots, \zeta_k^\nu) \in U_k \times \mathbb{C}^\nu$ are the same point on B if and only if $\zeta_j^\lambda = \sum_\mu b_{jk\mu}^\lambda \zeta_k^\mu$.

A C^∞ section ψ of B over X is represented on each U_j by a vector-valued C^∞ function $\psi = \psi_j(x) = (\psi_j^1(x), \ldots, \psi_j^\nu(x))$, where $\psi_j^\lambda = \sum_\mu b_{jk\mu}^\lambda \psi_k^\mu$ holds on $U_j \cap U_k \neq \emptyset$. Let $L(B)$ be *the vector space of all* C^∞ *sections* ψ *of* B *over* X, and E a *linear partial differential operator of* $L(B)$ into itself. Namely, E: $\psi \to E\psi$ is a linear map of $L(B)$ into itself such that we can write, on each U_j,

$$(E\psi)_j^\lambda(x) = \sum_{\mu=1}^{\nu} E_{j\mu}^\lambda(x, D_j)\psi_j^\mu(x), \qquad \lambda = 1, \ldots, \nu. \tag{7.1}$$

Here $E_{j\mu}^\lambda(x, D_j)$ is a polynomial of $D_{j\alpha} = \partial/\partial x_j^\alpha$, $\alpha = 1, \ldots, n$, whose coefficients are C^∞ functions of x:

$$E_{j\mu}^\lambda(x, D_j) = \sum_{m_1, \ldots, m_n} E_{j\mu m_1 \cdots m_n}^\lambda(x) D_{j1}^{m_1} \cdots D_{jn}^{m_n}.$$

We express $E^{\lambda}_{j\mu}(x, D_j)$ as a sum of homogeneous polynomials $E^{l\lambda}_{j\mu}(x, D_j)$ of degree l in $D_{j\alpha}$, $\alpha = 1, \ldots, n$. Then (7.1) is written as

$$(E\psi)^{\lambda}_{j}(x) = \sum_{l=0}^{m} \sum_{\mu=1}^{\nu} E^{l\lambda}_{j\mu}(x, D_j)\psi^{\mu}_{j}(x), \qquad \lambda = 1, \ldots, \nu. \tag{7.2}$$

Here we may assume that at least one of $E^{m\lambda}_{j\mu}(x, D_j)$ is not identically zero. If so, m is called the *order* of E. In what follows, we assume that *the order of E is even.*

Let $g_{\alpha\beta}$ be a C^{∞} symmetric covariant tensor field of rank 2 on X (p. 106). On each U_j, $g_{\alpha\beta}$ is represented as

$$\sum_{\alpha,\beta=1}^{n} g_{j\alpha\beta}(x)\, dx^{\alpha}_{j}\, dx^{\beta}_{j}, \quad \text{where} \quad dx^{\alpha}_{j}\, dx^{\beta}_{j} = \tfrac{1}{2}(dx^{\alpha}_{j} \otimes dx^{\beta}_{j} + dx^{\beta}_{j} \otimes dx^{\alpha}_{j}),$$

and the $g_{j\alpha\beta}(x) = g_{j\beta\alpha}(x)$ are real C^{∞} functions on U_j. If the corresponding quadratic form $\sum_{\alpha,\beta} g_{j\alpha\beta}(x)\xi^{\alpha}\xi^{\beta}$ in real variables $\xi^1, \ldots, \xi^n$ is positive definite at each point $x \in X$, then $\sum_{\alpha,\beta} g_{j\alpha\beta}(x)\, dx^{\alpha}_{j}\, dx^{\beta}_{j}$ is called a *Riemannian metric* on X. In this case, if we set $g_j = g_j(x) = \det(g_{j\alpha\beta}(x))_{\alpha,\beta=1,\ldots,n}$, we have $g_j(x) > 0$, and

$$\sqrt{g_j(x)}\, dx^1_j \wedge \cdots \wedge dx^n_j = \sqrt{g_k(x)}\, dx^1_k \wedge \cdots \wedge dx^n_k$$

on $U_j \cap U_k \neq \varnothing$. For simplicity, we use the following notation:

$$\sqrt{g}\, dX = \sqrt{g_j}\, dX_j = \sqrt{g_j(x)}\, dx^1_j \wedge \cdots \wedge dx^n_j.$$

Then $\sqrt{g}\, dX$ is a C^{∞} n-form on X, and $\sqrt{g_j(x)} > 0$. Hence $\sqrt{g}\, dX$ is *a volume element of X which is invariant under coordinate change.*

Now we shall introduce an inner product on $L(B)$. For this, we assume that X has a Riemannian metric $\sum_{\alpha,\beta=1}^{n} g_{j\alpha\beta}(x)\, dx^{\alpha}_{j}\, dx^{\beta}_{j}$, and that, on the vector bundle B, a Hermitian metric $\sum_{\lambda,\mu=1}^{\nu} a_{j\lambda\bar{\mu}}(x)\zeta^{\lambda}_{j}\overline{\zeta^{\mu}_{j}}$ is defined on the fibres. For any $\psi, \varphi \in L(B)$, we define their inner product by

$$(\psi, \varphi) = \int_X \sum_{\lambda,\mu=1}^{\nu} a_{j\lambda\bar{\mu}}(x)\psi^{\lambda}_{j}(x)\overline{\varphi^{\mu}_{j}(x)}\sqrt{g}\, dX. \tag{7.3}$$

We denote by $\|\psi\| = \sqrt{(\psi, \psi)}$ the norm of ψ.

Definition 7.1. Let E be a linear partial differential operator. If, for any $\varphi, \psi \in L(B)$, the equality

$$(E\psi, \varphi) = (\psi, E\varphi)$$

holds, then E is said to be *formally self-adjoint.*

If E is formally self-adjoint, then as is easily verified by integrating by part, we have the following equality.

$$\sum_{\sigma=1}^{\nu} a_{j\sigma\bar{\mu}}(x)E_{j\lambda}^{m\sigma}(x, D_j) = \sum_{\sigma=1}^{\nu} a_{j\lambda\bar{\sigma}}(x)\overline{E_{j\mu}^{m\sigma}(x, D_j)}. \tag{7.4}$$

In the sequel, E is not necessarily assumed to be self-adjoint, but we always assume that *the equality* (7.4) *holds for* E.

Consider the polynomials $E_{j\lambda}^{m\sigma}(x, \xi_j)$ in $\xi_{j1}, \ldots, \xi_{jn}$ which are obtained from $E_{j\lambda}^{m\alpha}(x, D_j)$ by replacing $D_{j\alpha} = \partial/\partial x_j^\alpha$ by the real variables $\xi_\alpha = \xi_{j\alpha}$. We regard ξ_α as the components of a covariant vector $\xi \in T_x^*(X)$ at $x \in X$. Put

$$A_j(x, \xi_j, \zeta_j) = (-1)^{m/2} \sum_{j,\mu,\sigma} a_{j\sigma\bar{\mu}}(x)E_{j\lambda}^{m\sigma}(x, \xi_j)\zeta_j^\lambda\overline{\zeta_j^\mu}.$$

Then, by (7.4), $A_j(x, \xi_j, \zeta_j)$ is a *Hermitian form in* $\zeta_j^1, \ldots, \zeta_j^\nu$. Since $E\psi \in L(B)$ for any $\psi \in L(B)$, we have, on $U_j \cap U_k \neq \varnothing$,

$$(E\psi)_j^\lambda(x) = \sum_\mu b_{jk\mu}^\lambda(x)(E\psi)_k^\mu(x),$$

hence

$$\sum_\mu E_{j\mu}^{m\lambda}(x, \xi_j)\zeta_j^\mu = \sum_\sigma b_{jk\sigma}^\lambda(x) \sum_\mu E_{k\mu}^{m\sigma}(x, \xi_k)\zeta_k^\mu.$$

Therefore, we have, on $U_j \cap U_k \neq \varnothing$,

$$A_j(x, \xi_j, \zeta_j) = A_k(x, \xi_k, \zeta_k).$$

Hence, suppressing the subscript j, we may simply write

$$A(x, \xi, \zeta) = A_j(x, \xi_j, \zeta_j). \tag{7.5}$$

Let $(g^{\alpha\beta}(x))_{\alpha,\beta=1,\ldots,n}$ be the inverse of the matrix $(g_{\alpha\beta}(x))_{\alpha,\beta=1,\ldots,n}$. We define the length of a covariant vector $\xi \in T_x^*(X)$ by $|\xi| = \sqrt{\sum_{\alpha,\beta} g^{\alpha\beta}(x)\xi_\alpha\xi_\beta}$.

Definition 7.2. A linear partial differential operator E of even order m is said to be *strongly elliptic* if there exists a constant $\delta > 0$ such that, for any $x \in X$, $\xi \in T_x^*(X)$, and $\zeta \in B_x$, the following inequality

$$A(x, \xi, \zeta) \geqq \delta^2|\xi|^m \sum_{\lambda,\mu=1}^{\nu} a_{j\lambda\bar{\mu}}(x)\zeta_j^\lambda\overline{\zeta_j^\mu} \tag{7.6}$$

holds.

In the rest of this section, we assume that E is a *strongly elliptic, formally self-adjoint linear partial differential operator.*

Let $\mathbb{F}$ be the linear subspace of $L(B)$ consisting of all solutions of $E\psi = 0$. It is known that $\mathbb{F} = \{\psi \in L(B) | E\psi = 0\}$ is *of finite dimension* (for the proof, see Theorem 7.3 in the Appendix). Moreover, $L(B)$ is decomposed as follows into the direct sum of the subspaces orthogonal to each other:

$$L(B) = \mathbb{F} \oplus EL(B) \tag{7.7}$$

(Corollary to Theorem 7.4 in the Appendix). We denote by F the orthogonal projection of $L(B)$ into $\mathbb{F}$. Then there exists a linear map G of $L(B)$ to $EL(B)$ such that, for any $\psi \in L(B)$,

$$EG\psi = GE\psi = \psi - F\psi \tag{7.8}$$

(Theorem 7.4 in the Appendix). The linear map G is called the *Green operator.*

Let $\lambda \in \mathbb{C}$ be a complex number. If there exists $e \in L(B)$ with $e \neq 0$ such that $Ee = \lambda e$, then we call λ an *eigenvalue of* E and e an *eigenfunction of* E. From the assumption that E is formally self-adjoint, it immediately follows that all the eigenvalues of E are real. As regards the eigenvalues and eigenfunctions of E, the following theorem is of fundamental importance.

Theorem 7.1. *We can choose eigenfunctions $e_h \in L(B)$ of E with $Ee_h = \lambda_h e_h$, $h = 1, 2, \ldots$, satisfying the following conditions*:

(i) *$\{e_1, e_2, \ldots, e_h, \ldots\}$ form a complete orthonormal system of $L(B)$. Namely, $(e_h, e_k) = \delta_{hk}$, and any $\psi \in L(B)$ is expanded into the series*

$$\psi = \sum_{h=1}^{\infty} (\psi, e_h) e_h \tag{7.9}$$

which converges with respect to the norm $\| \ \|$.

(ii) *$\lambda_1 \leqq \lambda_2 \leqq \cdots \leqq \lambda_h \leqq \cdots$, and $\lambda_h \to +\infty$ $(h \to +\infty)$.*

Proof of Theorem 7.1 will be given in the Appendix, §7(b).

Since $(E\psi, e_h) = (\psi, Ee_h) = \lambda_h(\psi, e_h)$, we have

$$E\psi = \sum_{h=1}^{\infty} \lambda_h (\psi, e_h) e_h. \tag{7.10}$$

Suppose that $\cdots \leqq \lambda_q < \lambda_{q+1} = \lambda_{q+2} = \cdots = \lambda_p = 0 < \lambda_{p+1} \leqq \cdots$. Then $\{e_{q+1}, \ldots, e_p\}$ form an orthonormal basis of $\mathbb{F}$. Therefore, we have

$$F\psi = \sum_{\lambda_h = 0} (\psi, e_h) e_h. \tag{7.11}$$

As for the Green operator, we have

$$G\psi = \sum_{\lambda_h \neq 0} \frac{1}{\lambda_h}(\psi, e_h)e_h. \tag{7.12}$$

In fact, since $G\psi \in EL(B)$, we have $FG\psi = 0$. Therefore

$$G\psi = \sum_{\lambda_h \neq 0} (G\psi, e_h)e_h.$$

On the other hand, since $EG\psi = \psi - F\psi$ by (7.8), we have

$$EG\psi = \sum_{\lambda_h \neq 0} \lambda_h(G\psi, e_h)e_h = \sum_{\lambda_h \neq 0} (\psi, e_h)e_h.$$

Hence we obtain $(G\psi, e_h) = (1/\lambda_h)(\psi, e_h)$, and (7.12) is proved.

(b) Differentiable Family of Partial Differential Operators

In this subsection, we consider a family of partial differential operators E_t acting on $L(B_t)$, where B_t is a complex vector bundle over X depending on the parameter t. We suppose that the parameter t moves in a small domain Δ of $\mathbb{R}^N$. We first define the notion of a family $\{B_t | t \in \Delta\}$ of complex vector bundles over X being differentiable in t.

Let $\mathscr{B}$ be a complex vector bundle over $X \times \Delta$ and $\pi: \mathscr{B} \to X \times \Delta$ its projection. Then for each $t \in \Delta$, $B_t = \pi^{-1}(X \times t)$ is a complex vector bundle on $X = X \times t$.

Definition 7.3. A family $\{B_t | t \in \Delta\}$ of complex vector bundles B_t over X obtained as above from a complex vector bundle B over $X \times \Delta$ is called a *differentiable family of complex vector bundles over X.*

Since we only consider the case in which Δ is sufficiently small, choosing an open covering $\{U_j\}$ by sufficiently small coordinate neighbourhoods U_j, we may assume

$$\pi^{-1}(U_j) = \mathbb{C}^\nu \times U_j \times \Delta.$$

The points $(\zeta_j, x, t) \in \mathbb{C}^\nu \times U_j \times \Delta$ and $(\zeta_k, x, t) \in \mathbb{C}^\nu \times U_k \times \Delta$ are the same point of $\mathscr{B}$ if and only if

$$\zeta_j^\lambda = \sum_{\mu=1}^{\nu} b_{jk\mu}^\lambda(x, t)\zeta_k^\mu, \tag{7.13}$$

where $b_{jk\mu}^\lambda(x, t)$ is a C^∞ function on $(U_j \cap U_k) \times \Delta$. *For each member B_t of the differentiable family $\{B_t | t \in \Delta\}$, we use $(b_{jk\mu}^\lambda(x, t))$ as transition functions.*

We also use the fibre coordinates $(\zeta_j^1, \ldots, \zeta_j^\nu)$ *of* $\mathscr{B}$ *as those of each* B_t. These fibre coordinates on B_t are called *admissible fibre coordinates* of B_t. For example, if we say that a C^∞ section $\psi_t \in L(B_t)$ of B_t is represented by a vector-valued C^∞ function on U_j as

$$\psi_{tj}(x) = (\psi_{tj}^1(x), \ldots, \psi_{tj}^\nu(x)), \tag{7.14}$$

we always suppose that $(\psi_{tj}^1(x), \ldots, \psi_{tj}^\nu(x))$ are the admissible fibre coordinates of $\psi_t(x) \in B_t$. Hence, on $U_j \cap U_k \neq \emptyset$, we have

$$\psi_{tj}^\lambda(x) = \sum_{\mu=1}^{\nu} b_{jk\mu}^\lambda(x, t)\psi_{tk}^\mu(x). \tag{7.15}$$

Definition 7.4. Suppose we are given a C^∞ section $\psi_t \in L(B_t)$ of B_t for each $t \in \Delta$. If each admissible fibre coordinate $\psi_{tj}^\lambda(x)$ of $\psi_t(x)$ is a C^∞ function of (x, t), then ψ_t is said to *be* C^∞ *differentiable in* t.

Put $\psi_j^\lambda(x, t) = \psi_{tj}^\lambda(x)$. Then, by (7.15)

$$\psi_j^\lambda(x, t) = \sum_{\mu=1}^{\nu} b_{jk\mu}^\lambda(x, t)\psi_k^\mu(x, t),$$

hence, $\psi: (x, t) \to \psi_j(x, t) = (\psi_j^1(x, t), \ldots, \psi_j^\nu(x, t))$ is a section of $\mathscr{B}$ over $X \times \Delta$, and each ψ_t is the restriction of ψ to $X \times t$: $\psi_t = \psi|_{X \times t}$. In other words, to say that the $\psi_t \in L(B_t)$ are C^∞ differentiable in t is equivalent to that each ψ_t is the restriction to $X \times t$ of a C^∞ section ψ of $\mathscr{B}$ over $X \times \Delta$.

Definition 7.5. Suppose we are given a linear operator Λ_t: $L(B_t) \to L(B_t)$ for each $t \in \Delta$. If $\Lambda_t \psi_t$ is C^∞ differentiable in t whenever $\psi_t \in L(B_t)$ is C^∞ differentiable in t, then Λ_t is called C^∞ differentiable in t, and the family $\{\Lambda_t \,|\, t \in \Delta\}$ is called a *differentiable family of linear operators.*

Suppose that a linear differentiable operator E_t: $L(B_t) \to L(B_t)$ is given for each $t \in \Delta$. By (7.1), for arbitrary $\psi_t \in L(B_t)$, we have

$$(E_t\psi_t)_j^\lambda(x) = \sum_{\mu=1}^{\nu} E_{j\mu}^\lambda(x, t, D_j)\psi_{tj}^\mu(x). \tag{7.16}$$

Put $\psi_j^\lambda(x, t) = \psi_{tj}^\lambda(x)$, $\varphi_j^\lambda(x, t) = \varphi_{tj}^\lambda(x)$ where $\varphi_t = E_t\psi_t$. Then (7.16) is written in the form

$$\varphi_j^\lambda(x, t) = \sum_{\mu=1}^{\nu} E_{j\mu}^\lambda(x, t, D_j)\psi_j^\mu(x, t).$$

The C^∞ differentiability of E_t in t means that, if the $\psi_j^\mu(x, t)$, $\mu = 1, \ldots, \nu$ are C^∞ functions of (x, t), then the $\varphi_j^\lambda(x, t)$, $\lambda = 1, \ldots, \nu$, are also C^∞

functions of (x, t). Therefore, E_t is C^∞ *differentiable in* t *if and only if the coefficients of the polynomials* $E^\lambda_{j\mu}(x, t, D_j)$ *in* $D_{j1}, \ldots, D_{jn}$, $\lambda, \mu = 1, \ldots, \nu$, *are all* C^∞ *functions of* (x, t).

Assume that a *Hermitian metric on the fibre* $\sum_{\lambda,\mu=1}^{\nu} a_{j\lambda\bar{\mu}}(x, t)\zeta_j^\lambda \overline{\zeta_j^\mu}$ *is given on* $\mathscr{B}$. Then we define the inner product of $\psi_t, \varphi_t \in L(B_t)$ by

$$(\psi_t, \varphi_t)_t = \int_X \sum_{\lambda,\mu} a_{j\lambda\bar{\mu}}(x, t)\psi^\lambda_{tj}(x)\overline{\varphi^\mu_{tj}(x)}\sqrt{g}\, dX. \tag{7.17}$$

It is important to note that the $a_{j\lambda\bar{\mu}}(x, t)$ are C^∞ *functions of* (x, t). Hereafter, when we say a linear differential operator E_t on $L(B_t)$ is formally self-adjoint or strongly elliptic, *it is so with respect to the inner product* $(\ ,\ )_t$ *defined by* (7.17).

Suppose we are given a differentiable family $\{E_t \mid t \in \Delta\}$ of formally self-adjoint strongly elliptic linear partial differential operators E_t on $L(B_t)$. By virtue of Theorem 7.1, we determine, for each E_t, the eigenvalues $\lambda_h(t)$, and eigenfunctions $e_{th} \in L(B_t)$ with $E_t e_{th} = \lambda_h(t) e_{th}$ satisfying the following conditions:

(i) $\{e_{th} \mid h = 1, 2, \ldots\}$ form a complete orthonormal system of $L(B_t)$.
(ii) $\lambda_1(t) \leqq \cdots \leqq \lambda_h(t) \leqq \cdots$, and $\lambda_h(t) \to \infty\ (h \to \infty)$.

Let $\mathbb{F}_t = \{\psi \in L(B_t) \mid E_t\psi = 0\}$, and let F_t be the orthogonal projection of $L(B_t)$ to $\mathbb{F}_t$. Then by (7.11),

$$F_t\psi = \sum_{\lambda_h(t)=0} (\psi, e_{th})e_{th}, \qquad \psi \in L(B_t).$$

In this section we shall prove the following three theorems ([21]).

Theorem 7.2. $\lambda_h(t)$ *is a continuous function of* $t \in \Delta$.

Theorem 7.3. $\dim \mathbb{F}_t$ *is upper-semicontinuous in* $t \in \Delta$.

Theorem 7.4. *If* $\dim \mathbb{F}_t$ *is independent of* $t \in \Delta$, *then* F_t *is* C^∞ *differentiable in* $t \in \Delta$.

Note that $\lambda_h(t)$ is not necessarily differentiable in t. This can be seen, for example, by the matrix $\begin{pmatrix} 1 & t \\ 1 & 1 \end{pmatrix}$ whose eigenvalues $\lambda_1(t) = 1 + \sqrt{t}$, $\lambda_2(t) = 1 - \sqrt{t}$ are not differentiable at $t = 0$.

We give below proofs of Theorems 7.2, 7.3, and 7.4. Obviously we may assume that the domain Δ is a multi-interval:

$$\Delta_\varepsilon = \{t \in \mathbb{R}^N \mid -\varepsilon < t_\kappa < \varepsilon, \kappa = 1, \ldots, N\}, \qquad \varepsilon > 0.$$

Furthermore, if necessary, we may replace Δ_ε by a smaller $\Delta_{\varepsilon'}$, with $0 < \varepsilon' < \varepsilon$, and prove the theorems for $\Delta = \Delta_{\varepsilon'}$.

Recall that each vector bundle B_t in a differentiable family $\{B_t \mid t\in\Delta_\varepsilon\}$ is the restriction to $X\times t$ of the vector bundle $\mathcal{B}$ over $X\times\Delta_\varepsilon$: $B_t=\mathcal{B}|_{X\times t}=\pi^{-1}(X\times t)$, where π is the projection of $\mathcal{B}$. On the other hand, it is obvious that $B_0\times\Delta_\varepsilon$ is a vector bundle over $X\times\Delta_\varepsilon$.

Lemma 7.1. *The vector bundle $\mathcal{B}$ over $X\times\Delta_\varepsilon$ is C^∞ equivalent to the vector bundle $B_0\times\Delta_\varepsilon$.*

Proof. Let π_0: $B_0\times\Delta_\varepsilon\to X\times\Delta_\varepsilon$ be the projection of $B_0\times\Delta_\varepsilon$. We want to construct a diffeomorphism Ψ of the $B_0\times\Delta_\varepsilon$ onto $\mathcal{B}$ which maps each fibre $\pi_0^{-1}(x,t)$ linearly onto $\pi^{-1}(x,t)$. For this, we apply a method similar to the proof of Theorem 2.4 (pp. 64-66).

Let $\{U_j\}$ be the finite covering of X as mentioned above, and put

$$\mathcal{U}_j=\pi^{-1}(U_j\times\Delta_\varepsilon)=\mathbb{C}^\nu\times U_j\times\Delta_\varepsilon.$$

Then $\mathcal{B}=\bigcup_j\mathcal{U}_j$, where $(\zeta_j,x,t)\in\mathcal{U}_j$ and $(\zeta_k,x,t)\in\mathcal{U}_k$ are the same point on $\mathcal{B}$ if and only if

$$\zeta_j^\lambda=\sum_\mu b_{jk\mu}^\lambda(x,t)\zeta_k^\mu.$$

(1°) The case $N=1$. We denote by $(\partial/\partial t_1)_k$ the vector field $\partial/\partial t_1$ on $\mathcal{U}_k$. Let $\{\rho_k(x)\}$ be a partition of unity subordinate to the covering $\{U_k\}$. Then

$$v=\sum_k\rho_k(x)\left(\frac{\partial}{\partial t_1}\right)_k$$

is a C^∞ vector field on $\mathcal{B}$. On $\mathcal{U}_j\cap\mathcal{U}_k\neq\emptyset$, we have

$$\left(\frac{\partial}{\partial t_1}\right)_k=\sum_\lambda\sum_\sigma\frac{\partial b_{jk\sigma}^\lambda(x,t_1)}{\partial t_1}\zeta_k^\sigma\frac{\partial}{\partial\zeta_j^\lambda}+\left(\frac{\partial}{\partial t_1}\right)_j.$$

In view of $\zeta_k^\sigma=\sum_\mu b_{kj\mu}^\sigma(x,t_1)\zeta_j^\mu$, if we set

$$v_{j\mu}^\lambda(x,t_1)=\sum_k\rho_k(x)\sum_\sigma\frac{\partial b_{jk\sigma}^\lambda(x,t_1)}{\partial t_1}b_{kj\mu}^\sigma(x,t_1),$$

then $v_{j\mu}^\lambda(x,t_1)$ is a C^∞ function on $U_j\times\Delta_\varepsilon$, and we have

$$v=\sum_\mu v_{j\mu}^\lambda(x,t_1)\zeta_j^\mu\frac{\partial}{\partial\zeta_j^\lambda}+\left(\frac{\partial}{\partial t_1}\right)_j \tag{7.18}$$

on $\mathcal{U}_j$. Note that (7.18) is similar to (2.25) except that here *the coefficients $\sum_\mu v_{j\mu}^\lambda(x,t_1)\zeta_j^\mu$ of $\partial/\partial\zeta_j^\lambda$ are linear in $\zeta_j^1,\dots,\zeta_j^\nu$.*

Consider the simultaneous ordinary differential equations for v:

$$\begin{cases} \dfrac{d\zeta_j^\lambda}{dt} = \displaystyle\sum_{\mu=1}^{\nu} v_{j\mu}^\lambda(x, t_1)\zeta_j^\mu, & \lambda = 1, \ldots, n, \\ \dfrac{dt_1}{dt} = 1. \end{cases} \tag{7.19}$$

For each point $(\zeta_{0j}, x, 0)$ of $B_0 = \pi^{-1}(X \times 0)$, we consider the initial conditions

$$\begin{cases} \zeta_j^\lambda(0) = \zeta_{0j}^\lambda, & \lambda = 1, \ldots, n, \\ t_1(0) = 0, \end{cases}$$

and let

$$\begin{cases} \zeta_j^\lambda(t) = \zeta_j^\lambda(\zeta_{0j}, x, t), & \lambda = 1, \ldots, \nu, \\ t_1(t) = t \end{cases}$$

be the solutions of (7.19) satisfying these initial conditions. Then

$$\Psi: (\zeta_{0j}, x, t) \to (\zeta_j, x, t_1) = (\zeta_j(\zeta_{0j}, x, t), x, t)$$

is a diffeomorphism of $B_0 \times \Delta_\varepsilon$ onto $\mathscr{B}$. Since the equations (7.19) are linear in $\zeta_j^1, \ldots, \zeta_j^\nu$, the solutions $\zeta_j^\lambda(\zeta_{0j}, x, t)$ *are also linear in* $\zeta_{0j}^1, \ldots, \zeta_{0j}^\nu$:

$$\zeta_j^\lambda(\zeta_{0j}, x, t) = \sum_{\mu=1}^{\nu} f_{j\mu}^\lambda(x, t)\zeta_{0j}^\mu.$$

Hence Ψ maps each fibre $\pi_0^{-1}(x, t)$ of $B_0 \times \Delta_\varepsilon$ linearly onto $\pi^{-1}(x, t)$.

(2°) General case. We can prove the general case by induction on N just as in the proof of Theorem 2.4. ∎

The diffeomorphism $\Psi: B_0 \times \Delta_\varepsilon \to \mathscr{B}$ is written on each $\pi_0^{-1}(U_j \times \Delta_\varepsilon)$ as

$$\Psi: (\zeta_{0j}, x, t) \to (\zeta_j, x, t), \qquad \zeta_j^\lambda = \sum_\mu f_{j\mu}^\lambda(x, t)\zeta_{0j}^\mu.$$

In other words, $\mathscr{B}$ is a vector bundle obtained from $B_0 \times \Delta_\varepsilon$ by the change of coordinates $(\zeta_{0j}^1, \ldots, \zeta_{0j}^\nu) \to (\zeta_j^1, \ldots, \zeta_j^\nu)$. Therefore $\mathscr{B}$ is the same vector bundle as $B_0 \times \Delta_\varepsilon$, and each B_t in the differentiable family $\{B_t \mid t \in \Delta\}$ is the same as B_0. Consequently, we may assume that the $B_t = B$ are *one and the same vector bundle independent of* t, and consider a differentiable family $\{E_t \mid t \in \Delta\}$ of partial differential operators E_t on $L(B)$.

The Hermitian metric on the fibre $\sum_{\lambda,\mu} a_{j\lambda\bar{\mu}}(x,t)\zeta_j^\lambda\overline{\zeta_j^\mu}$ of the vector bundle $\mathscr{B}=B\times\Delta_\varepsilon$ over $X\times\Delta_\varepsilon$ depends in general on t. Therefore the inner product given by (7.17):

$$(\psi,\varphi)_t=\int_X\sum_{\lambda,\mu} a_{j\lambda\bar{\mu}}(x,t)\psi_j^\lambda(x)\overline{\varphi_j^\mu(x)}\sqrt{g}\,dX,\qquad \psi,\varphi\in L(B)$$

also depends on t. Assume that *each E_t of the differentiable family $\{E_t \mid t\in\Delta\}$ is a strongly elliptic linear partial differential operator of even order m.* Then, if we write E_t in the form

$$(E_t\psi)_j^\lambda(x)=\sum_{l=0}^{m}\sum_{\mu=1}^{\nu} E_{j\mu}^{l\lambda}(x,t,D_j)\psi_j^\mu(x),$$

then the equality (7.4):

$$\sum_{\sigma=1}^{\nu} a_{j\sigma\bar{\mu}}(x,t)E_{j\lambda}^{m\sigma}(x,t,D_j)=\sum_{\sigma=1}^{\nu} a_{j\lambda\bar{\sigma}}(x,t)\overline{E_{j\mu}^{m\sigma}(x,t,D_j)}$$

holds, and if we put

$$A(x,t,\xi,\zeta)=(-1)^{m/2}\sum_{\lambda,\mu,\sigma} a_{j\sigma\bar{\mu}}(x,t)E_{j\lambda}^{m\sigma}(x,t,\xi_j)\zeta_j^\lambda\overline{\zeta_j^\mu},$$

the inequality (7.6):

$$A(x,t,\xi,\zeta)\geqq\delta^2|\xi|^m\sum_{\lambda,\mu} a_{\lambda\bar{\mu}}(x,t)\zeta^\lambda\overline{\zeta^\mu},\qquad \delta>0,\tag{7.20}$$

holds. Restricting the domain of t to $\Delta=\Delta_{\varepsilon'}$ with $\varepsilon'<\varepsilon$, so that $[\Delta]\subset\Delta_\varepsilon$, we may assume that *$\delta$ is a constant independent of t.*

We put $D_{j\alpha}=\partial/\partial x_j^\alpha$ as before. We let D_j^l denote an arbitrary differential of rank l:

$$D_j^l=D_{j1}^{l_1}\cdots D_{jn}^{l_n}=\frac{\partial^l}{\partial x_1^{l_1}\cdots\partial x_n^{l_n}},\qquad l_1+\cdots+l_n=l.$$

We define *the norm* $\|\psi\|_k$, $k=0,1,\ldots,$ of $\psi\in L(B)$ *by*

$$\|\psi\|_k=\left(\sum_{l=0}^{k}\sum_j\sum_{D_j^l}\int_{U_j}\sum_\lambda|D_j^l\psi_j^\lambda(x)|^2\,dX_j\right)^{1/2},$$

where the summation $\sum_{D_j^l}$ is taken over all partial differentials D_j^l of rank l. The norm $\|\psi\|_k$ is equivalent to the Sobolev norm given by (3.7) in the Appendix. Namely, if we denote the Sobolev norm by $\|\ \|_k^{(3.7)}$, then, by Proposition 1.2 in the Appendix, there is a constant $K>1$ depending only on k such that

$$K^{-1}\|\psi\|_k^{(3.7)} \leqq \|\psi\|_k \leqq K\|\psi\|_k^{(3.7)}, \qquad \psi \in L(B).$$

In particular, for $k=0$, we have

$$\|\psi\|_0^2 = \sum_j \int_{U_j} \sum_\lambda |\psi_j^\lambda(x)|^2 \, dX_j,$$

while as for the norm $\|\psi\|_t = \sqrt{(\psi, \psi)_t}$, we have

$$\|\psi\|_t^2 = \int_X \sum_{\lambda,\mu} a_{j\lambda\bar{\mu}}(x, t)\psi_j^\lambda(x)\overline{\psi_j^\mu(x)}\sqrt{g}\, dX.$$

Therefore there is a constant $K_0 > 1$ independent of $t \in \Delta$ such that

$$\|\psi\|_t \leqq K_0\|\psi\|_0 \quad \text{and} \quad \|\psi\|_0 \leqq K_0\|\psi\|_t. \tag{7.21}$$

Lemma 7.2 (Sobolev's Inequality). *For any integer $l \geqq 0$, and any natural number $k > n/2$ with $n = \dim X$, there exists a constant $c_{k,l}$ such that the inequality*

$$|D_j^l \psi_j^\lambda(x)| \leqq c_{k,l}\|\psi\|_{k+l}, \qquad \psi \in L(B) \tag{7.22}$$

holds.

Proof. See (3.33) in the Appendix. ∎

Lemma 7.3 (Friedrichs' Inequality ([6])). *Let k be a non-negative integer. Then there exists a constant c_k independent of t such that the inequality*

$$(\|\psi\|_{k+m})^2 \leqq c_k(\|E_t\psi\|_k^2 + \|\psi\|_0^2), \qquad \psi \in L(B) \tag{7.23$_k$}$$

holds.

Proof. Since $\{E_t \mid t \in \Delta_\varepsilon\}$ is a differentiable family of strongly elliptic operators and since $[\Delta] \subset \Delta_\varepsilon$, by Theorem 4.1 in the Appendix, there is a constant C_k such that

$$(\|\psi\|_{k+m})^2 \leqq C_k(\|E_t\psi\|_k^2 + \|\psi\|_k^2), \qquad \psi \in L(B). \tag{7.24}$$

If $k < h$, then $\|\psi\|_k \leqq \|\psi\|_h$ by definition. Using this fact, $(7.23)_k$ is easily deduced as follows from (7.24) by induction on k. First, $(7.23)_0$ is nothing but (7.24) for $k=0$. We suppose that $(7.23)_{k-1}$ is already proved and prove

$(7.23)_k$. By (7.24)

$$(\|\psi\|_{k+m})^2 \leqq C_k(\|E_t\psi\|_k^2 + \|\psi\|_k^2),$$

while by $(7.23)_{k-1}$

$$\|\psi\|_k^2 \leqq (\|\psi\|_{k-1+m})^2 \leqq c_{k-1}(\|E_t\psi\|_{k-1}^2 + \|\psi\|_0^2).$$

Since $\|E_t\psi\|_{k-1} \leqq \|E_t\psi\|_k$, we get

$$(\|\psi\|_{k+m})^2 \leqq C_k(1 + c_{k-1})(\|E_t\psi\|_k^2 + \|\psi\|_0^2).$$

Hence, putting $c_k = C_k(c_{k-1}+1)$, we obtain $(7.23)_k$. ∎

Theorem 7.5. *Assume that E_t: $L(B) \to L(B)$ is bijective for each $t \in \Delta$. If there exists a constant $c > 0$ independent of $t \in \Delta$, such that, for any $\psi \in L(B)$*

$$\|E_t\psi\|_0 \geqq c\|\psi\|_0, \tag{7.25}$$

then E_t^{-1} is C^∞ differentiable in $t \in \Delta$.

Proof. First we extend Definition 7.4, and define the notion that $\psi_t \in L(B)$ is C^r *differentiable in* $t \in \Delta$. On each U_j, we represent ψ_t as

$$\psi_t = (\psi_j^1(x, t), \ldots, \psi_j^\nu(x, t))$$

in terms of the fibre coordinates. If the $D_j^l\psi_j^\lambda(x, t)$, $\lambda = 1, \ldots, \nu$, are C^r functions of (x, t) for any D_j^l, then $\psi_t \in L(B)$ is called C^r differentiable in $t \in \Delta$. Furthermore, a linear operator Λ_t: $L(B) \to L(B)$ is said to be C^r differentiable in t, if, for any $\psi_t \in L(B)$, C^r differentiable in t, $\Lambda_t\psi_t$ is also C^r differentiable in t.

In order to show that E_t^{-1} is C^∞ differentiable in t, it suffices to prove that E_t^{-1} is C^r differentiable in t for each $r = 0, 1, 2, \ldots$.

By hypothesis, we have, for $t \in \Delta$,

$$\|\psi\|_0 \leqq \frac{1}{c}\|E_t\psi\|_0 \leqq \frac{1}{c}\|E_t\psi\|_k, \qquad \psi \in L(B).$$

Hence by $(7.23)_k$

$$\|\psi\|_{k+m} \leqq c_k'\|E_t\psi\|_k, \qquad t \in \Delta,$$

where $c_k' = c_k^{1/2}(1 + 1/c)$ is a constant independent of $t \in \Delta$ and ψ. On the other hand, by Sobolev's inequality, we have for $k + m - l > n/2$,

$$|D_j^l\psi_j^\lambda(x)| \leqq c_{k+m-l,l}\|\psi\|_{k+m}.$$

Hence, for $k>l-m+n/2$, the inequality

$$|D_j^l\psi_j^\lambda(x)| \leqq c'_{k,l}\|E_t\psi\|_k \tag{7.26}$$

holds for any $\psi \in L(B)$, where $c'_{k,l}=c'_k \cdot c_{k+m-l,l}$ is a constant independent of t.

We prove, by induction on r, that, if $\varphi_t \in L(B)$ is C^r differentiable in t, so is $\psi_t = E_t^{-1}\varphi_t$.

(1°) The case $r=0$. By hypothesis $\varphi_t = E_t\psi_t$ is C^0 differentiable, i.e., continuous in $t \in \Delta$. We need to prove that for any D_j^l, $D_j^l\psi_j^\lambda(x,t)$ is continuous in (x,t). Since $D_j^l\psi_j^\lambda(x,t)$ is continuous in x, it suffices to show that, for any $s\in\Delta$, $D_j^l\psi_j^\lambda(x,t)$ converges to $D_j^l\psi_j^\lambda(x,s)$ uniformly in $x\in U_j$ as $t\to s$. For given l, we choose k with $k>l-m+n/2$ once for all and put $c'=c'_{k,l}$. Then, by (7.26)

$$\begin{aligned}|D_j^l\psi_j^\lambda(x,t)-D_j^l\psi_j^\lambda(x,s)| &\leqq c'\|E_t(\psi_t-\psi_s)\|_k\\ &\leqq c'\|E_t\psi_t-E_s\psi_s\|_k+c'\|(E_t-E_s)\psi_s\|_k\\ &= c'\|\varphi_t-\varphi_s\|_k+c'\|E_t\psi_s-E_s\psi_s\|_k.\end{aligned} \tag{7.27}$$

On the other hand

$$\|\varphi_t-\varphi_s\|_k^2=\sum_{l=0}^k\sum\int_{U_j}|D_j^l\varphi_j^\lambda(x,t)-D_j^l\varphi_j^\lambda(x,s)|^2\,dX_j,$$

where the second $\sum$ designates the summation over j, D_j^l and λ. By hypothesis $D_j^l\varphi_j^\lambda(x,t)$ is continuous in (x,t). Hence $\|\varphi_t-\varphi_s\|_k\to 0\ (t\to s)$. For fixed s,

$$(E_t\psi_s)_j^\lambda(x)=\sum_{l=0}^m\sum_{\mu=1}^\nu E_{j\mu}^{l\lambda}(x,t,D_j)\psi_{sj}^\mu(x)$$

is a C^∞ function of (x,t) because, in the right-hand side, the $D_{j1}^{l_1}\cdots D_{jn}^{l_n}\psi_{sj}^\mu(x)$ are C^∞ functions of x, and their coefficients are C^∞ functions of (x,t). Hence $\|E_t\psi_s-E_s\psi_s\|_k\to 0\ (t\to s)$. Therefore, by (7.27), as $t\to s$, we have $D_j^l\psi_j^\lambda(x,t)\to D_j^l\psi_j^\lambda(x,s)$ uniformly in $x\in U_j$.

(2°) The case $r=1$. Put $\varphi_t=E_t\psi_t$. We must show that if φ_t is C^1 differentiable in t, so is ψ_t. We denote by $(t_1,\dots,t_N)$ the coordinates of $t\in\mathbb{R}^N$. Since by hypothesis φ_t is C^1 differentiable in t, each component $\varphi_j^\lambda(x,t)$ of $\varphi_t=(\varphi_j^1(x,t),\dots,\varphi_j^n(x,t))$ and its derivatives $D_j^l\varphi_j^\lambda(x,t)$ are all continuously differentiable in $x^1,\dots,x^n$, $t_1,\dots,t_N$. Define the derivative of φ_t with respect to t_κ by

$$\frac{\partial\varphi_t}{\partial t_\kappa}=\left(\frac{\partial\varphi_j^1(x,t)}{\partial t_\kappa},\dots,\frac{\partial\varphi_j^\nu(x,t)}{\partial t_\kappa}\right).$$

It is obvious that $\partial\varphi_t/\partial t_\kappa\in L(B)$.

Suppose ψ_t is proved to be C^1 differentiable in $t \in \Delta$. Then, differentiating the equation $\varphi_t = E_t\psi_t$ with respect t_κ, we get

$$\frac{\partial \varphi_t}{\partial t_\kappa} = E_t \frac{\partial \psi_t}{\partial t_\kappa} + \frac{\partial E_t}{\partial t_\kappa}\psi_t,$$

hence,

$$\frac{\partial \psi_t}{\partial t_\kappa} = E_t^{-1}\left(\frac{\partial \varphi_t}{\partial t_\kappa} - \frac{\partial E_t}{\partial t_\kappa}\psi_t\right), \tag{7.28}$$

where $\partial E_t/\partial t_\kappa$ is the partial differential operator obtained from E_t by differentiating with respect to t_κ the coefficients of $E^{l\lambda}_{j\mu}(x, t, D_j)$ as polynomials of $D_{j\alpha}$, $\alpha = 1, \ldots, m$. Namely, for any $\psi \in L(B)$,

$$\left(\frac{\partial E_t}{\partial t_\kappa}\psi\right)^\lambda_j (x) = \sum_{l=0}^{m} \sum_{\mu} \frac{\partial}{\partial t_\kappa} E^{l\lambda}_{j\mu}(x, t, D_j)\psi^\mu_j(x). \tag{7.29}$$

We let $\eta_{\kappa t}$ denote the right-hand side of (7.28):

$$\eta_{\kappa t} = E_t^{-1}\left(\frac{\partial \varphi_t}{\partial t_\kappa} - \frac{\partial E_t}{\partial t_\kappa}\psi_t\right). \tag{7.30}$$

By hypothesis $\partial \varphi_t/\partial t_\kappa$ is continuous in $t \in \Delta$. Since φ_t is continuous in $t \in \Delta$, $\psi_t = E_t^{-1}\varphi_t$ is continuous in t by (1°). Hence $\eta_{\kappa t}$ is continuous in t, that is, for any D_j^l, $D_j^l \eta^\lambda_{\kappa j}(x, t)$ is a continuous function of (x, t). Hence, in order to prove that ψ_t is C^1 differentiable in t, it suffices to prove that, for any D_j^l, $D_j^l \psi^\lambda_j(x, t)$ is differentiable in t_κ, $\kappa = 1, \ldots, N$, and that

$$\frac{\partial}{\partial t_\kappa} D_j^l \psi^\lambda_j(x, t) = D_j^l \eta^\lambda_{\kappa j}(x, t).$$

That is, putting

$$t + h = (t_1, \ldots, t_{\kappa-1}, t_\kappa + h, t_{\kappa+1}, \ldots, t_N),$$

we need to prove

$$\lim_{h\to 0} \frac{1}{h}(D_j^l \psi^\lambda_j(x, t+h) - D_j^l \psi^\lambda_j(x, t)) - D_j^l \eta^\lambda_{\kappa j}(x, t) = 0.$$

For this, by (7.26), it suffices to show

$$\lim_{h\to 0} \left\| E_{t+h}\left(\frac{1}{h}(\psi_{t+h} - \psi_t) - \eta_{\kappa t}\right)\right\|_k = 0. \tag{7.31}$$

Using the equalities $E_{t+h}\psi_{t+h}=\varphi_{t+h}$, $E_t\psi_t=\varphi_t$, and $E_t\eta_{\kappa t}=(\partial\varphi_t/\partial t_\kappa)-(\partial E_t/\partial t_\kappa)\psi_t$, we obtain by simple calculation

$$E_{t+h}\left(\frac{1}{h}(\psi_{t+h}-\psi_t)-\eta_{\kappa t}\right)=\frac{1}{h}(\varphi_{t+h}-\varphi_t)-\frac{\partial\varphi_t}{\partial t_\kappa}$$
$$-\frac{1}{h}(E_{t+h}-E_t)\psi_t+\frac{\partial E_t}{\partial t_\kappa}\psi_t-(E_{t+h}-E_t)\eta_{\kappa t}.$$

Therefore, in order to prove (7.31), it is enough to prove

$$\lim_{h\to 0}\left\|\frac{1}{h}(\varphi_{t+h}-\varphi_t)-\frac{\partial\varphi_t}{\partial t_\kappa}\right\|_k=0, \tag{7.32}$$

$$\lim_{h\to 0}\left\|\frac{1}{h}(E_{t+h}-E_t)\psi_t-\frac{\partial E_t}{\partial t_\kappa}\psi_t\right\|_k=0, \tag{7.33}$$

$$\lim_{h\to 0}\|(E_{t+h}-E_t)\eta_{\kappa t}\|_k=0. \tag{7.34}$$

From the definition of the norm $\|\ \|_k$, in order to prove (7.32), we have to show that, for any D_j^l, $l\leqq k$,

$$\lim_{h\to 0}\int_{U_j}\left|\frac{D_j^l\varphi_j^\lambda(x,t+h)-D_j^l\varphi_j^\lambda(x,t)}{h}-\frac{\partial D_j^l\varphi_j^\lambda(x,t)}{\partial t_\kappa}\right|^2 dX_j=0$$

holds. But this is obvious. In fact, since φ_t is C^1 differentiable in t, and since the fibre coordinates $(\zeta_j^1,\dots,\zeta_j^\nu)$ of B are defined on some domain $W_j\supset[U_j]$, it follows that, for a sufficiently small $\delta>0$, the $D_j^l\varphi_j^\lambda(x,t+h)$ are continuously differentiable functions of $(x_j^1,\dots,x_j^n,h)$ in $W_j\times(-\delta,\delta)$, with t being fixed. Therefore

$$\lim_{h\to 0}\frac{D_j^l\varphi_j^\lambda(x,t+h)-D_j^l\varphi_j^\lambda(x,t)}{h}=\frac{\partial D_j^l\varphi_j^\lambda(x,t)}{\partial t_\kappa}$$

uniformly in $x\in U_j$. This proves (7.32).

Put

$$f_j^\lambda(x,t+h)=(E_{t+h}\psi_t)_j^\lambda(x)=\sum_{l=0}^m\sum_\mu E_{j\mu}^{l\lambda}(x,t+h,D_j)\psi_{tj}^\mu(x).$$

For fixed t, the $f_j^\lambda(x,t+h)$ are C^∞ functions of $(x_j^1,\dots,x_j^n,h)$ in $W_j\times(-\delta,\delta)$ because the coefficients of $E_{j\mu}^{l\lambda}(x,t,D_j)$ as polynomials of $D_{j1},\dots,D_{jn}$ are C^∞ functions of (x,t). In order to show (7.33), it is enough to prove

$$\lim_{h\to 0}\int_{U_j}\left|\frac{D_j^lf_j^\lambda(x,t+h)-D_j^lf_j^\lambda(x,t)}{h}-D_j^l\left(\frac{\partial E_t}{\partial t_\kappa}\psi_t\right)_j^\lambda(x)\right|^2 dX_j=0.$$

This is also obvious. For, since by (7.29)

$$D_j^l\left(\frac{\partial E_t}{\partial t_\kappa}\psi_t\right)_j^\lambda(x)=\left(\frac{dD_j^l f_j^\lambda(x,t+h)}{dh}\right)_{h=0},$$

we have

$$\lim_{h\to 0}\frac{D_j^l f_j^\lambda(x,t+h)-D_j^l f_j^\lambda(x,t)}{h}=D_j^l\left(\frac{\partial E_t}{\partial t_\kappa}\psi_t\right)_j^\lambda(x)$$

uniformly in $x\in U_j$.

(7.34) can be proved similarly.

(3°) The case $r\geqq 2$. Put $\varphi_t=E_t\psi_t$ as before. We shall prove that if φ_t is C^r differentiable in $t\in\Delta$, so is ψ_t. By the hypothesis of induction, ψ_t is C^{r-1} differentiable in t. We denote a partial differential operator of order k in $t_1,\ldots,t_N$ by

$$\partial_t^k=\frac{\partial^k}{\partial t_1^{k_1}\cdots\partial t_N^{k_N}},\qquad k=k_1+\cdots+k_N.$$

Since

$$(E_t\psi_t)_j^\lambda(x)=\sum_{l_1+\cdots+l_n\leqq m}\sum_\mu E_{j\mu l_1\cdots l_n}^\lambda(x,t)D_{j1}^{l_1}\cdots D_{jn}^{l_n}\psi_j^\mu(x,t),$$

$\partial_t^{r-1}(E_t\psi_t)$ is a sum of $E_t\partial_t^{r-1}\psi_t$ and the terms of the form $(\partial_t^{r-1-k}E_t)\partial_t^k\psi_t$, $k=0,1,2,\ldots,r-2$:

$$\partial_t^{r-1}(E_t\psi_t)=E_t\partial_t^{r-1}\psi_t+\sum_{k=0}^{r-2}\sum(\partial_t^{r-1-k}E_t)\partial_t^k\psi_t.$$

Hence

$$E_t\partial_t^{r-1}\psi_t=\partial_t^{r-1}\varphi_t-\sum_{k=0}^{r-2}\sum(\partial_t^{r-1-k}E_t)\partial_t^k\psi_t.$$

Since φ_t is C^r differentiable, ψ_t is C^{r-1} differentiable and E_t is C^∞ differentiable in t, it follows that the right-hand side of the equality is C^1 differentiable in t. Hence, by (2°), $\partial_t^{r-1}\psi_t$ is C^1 differentiable in t. This implies that ψ_t is C^r differentiable in t. This completes the induction. ∎

In order to prove Theorems 7.2, 7.3, and 7.4, we assume in the following that each member E_t of the differentiable family $\{E_t|t\in\Delta\}$ is *formally self-adjoint* and *strongly elliptic with respect to the inner product* $(\ ,\ )_t$. Suppose we have chosen the eigenvalues $\lambda_h(t)$ and eigenfunctions $e_{th}\in L(N)$ with $E_te_{th}=\lambda_h(t)e_{th}$ which satisfy the following conditions:

(i) $\{e_{th}|h=1,2,\ldots\}$ form a complete orthonormal system of $L(B)$ with respect to the inner product $(\ ,\ )_t$.

(ii) $\lambda_1(t) \leqq \lambda_2(t) \leqq \cdots \leqq \lambda_n(t) \leqq \cdots, \qquad \lambda_h(t) \to \infty \quad (h \to \infty)$.

For a moment, we fix $t \in \Delta = \Delta_{\varepsilon'}$, $0 < \varepsilon' < \varepsilon$. Any element $\psi \in L(B)$ is expanded into a series

$$\psi = \sum_{h=1}^{\infty} a_h e_{th}, \qquad a_h = (\psi, e_{th})_t$$

which converges with respect to the norm $\| \ \|_t$. In this way, $\psi \in L(B)$ is represented by the sequence $\{a_h\}$. Clearly we have

$$\sum_{h=1}^{\infty} |a_h|^2 = \|\psi\|_t^2. \tag{7.35}$$

Lemma 7.4. *A necessary and sufficient condition for a sequence $\{a_h\}$ to represent a element of $L(B)$, that is, a condition for the existence of $\psi \in L(B)$ satisfying $\psi = \sum_{h=1}^{\infty} a_h e_{th}$, is that the inequalities*

$$\sum_{h=1}^{\infty} |\lambda_h(t)|^{2l} |a_h|^2 < +\infty, \qquad l = 1, 2, \ldots. \tag{7.36}$$

should hold.

Proof. The necessity of (7.36) is easily verified as follows. If $\psi = \sum_{h=1}^{\infty} a_h e_{th} \in L(B)$, then $E_t^l \psi \in L(B)$. On the other hand, since by (7.10), $E_t^l \psi = \sum_h \lambda_h(t)^l a_h e_{th}$, we have

$$\sum_{h=1}^{\infty} |\lambda_h(t)|^{2l} |a_h|^2 = \|E_t^l \psi\|_t^2 < +\infty.$$

To prove the sufficiency of (7.36), we first show that, for any $\psi = \sum_{h=1}^{\infty} a_h e_{th} \in L(B)$ and any $k = 1, 2, \ldots$,

$$\|\psi\|_k^2 \leqq c_k'' \sum_{h=1}^{\infty} \left(1 + \sum_{l=1}^{k} |\lambda_h(t)|^{2l}\right) |a_h|^2 \tag{7.37}$$

holds, where c_k'' is a constant independent of ψ. Since the inequality $(7.23)_k$:

$$(\|\psi\|_{k+m})^2 \leqq c_k(\|\psi\|_0^2 + \|E_t\psi\|_k^2)$$

remains true if we replace c_k by a larger constant, we may assume that $c_k \geqq 1$. Then we have

$$\|\psi\|_m^2 \leqq c_0(\|\psi\|_0^2 + \|E_t\psi\|_0^2),$$

$$\begin{aligned} \|\psi\|_{2m}^2 &\leqq c_m(\|\psi\|_0^2 + \|E_t\psi\|_m^2) \\ &\leqq c_m c_0(\|\psi\|_0^2 + \|E_t\psi\|_0^2 + \|E_t^2\psi\|_0^2). \end{aligned}$$

Similarly, for $q=3,4,5,\ldots$, we have

$$\|\psi\|_{qm}^2 \leqq c_{(q-1)m}\cdots c_m c_0\left(\|\psi\|_0^2+\sum_{l=1}^{q}\|E_t^l\psi\|_0^2\right).$$

For any given natural number k, we choose q so that $qm-m<k\leqq qm$. Then, since $m\geqq 2$, we have $q\leqq k$. Hence, if we put $\hat{c}_k=c_{(q-1)m}\cdots c_m c_0$, we obtain

$$\|\psi\|_k^2 \leqq \|\psi\|_{qm}^2 \leqq \hat{c}_k\left(\|\psi\|_0^2+\sum_{l=1}^{k}\|E_t^l\psi\|_0^2\right).$$

Therefore, in view of (7.21) $\|\ \|_0\leqq K_0\|\ \|_t$, we obtain

$$\|\psi\|_k^2 \leqq \hat{c}_k K_0^2\left(\|\psi\|_t^2+\sum_{l=1}^{k}\|E_t^l\psi\|_t^2\right).$$

Since

$$\|\psi\|_t^2=\sum_{h=1}^{\infty}|a_h|^2 \quad\text{and}\quad \|E_t^l\psi\|_t^2=\sum_{h=1}^{\infty}|\lambda_h(t)|^{2l}|a_h|^2,$$

(7.37) immediately follows from this inequality if we set $c_k''=\hat{c}_k K_0^2$.

Suppose we are given a sequence $\{a_h\}$ satisfying condition (7.36). Since $\lambda_h(t)\to+\infty$ as $h\to\infty$, it follows immediately from (7.36) that $\sum_{h=1}^{\infty}|a_h|^2<+\infty$. Hence, we have

$$\sum_{h=1}^{\infty}\left(1+\sum_{l=1}^{k}|\lambda_h(t)|^{2l}\right)|a_h|^2<+\infty, \qquad k=1,2,\ldots. \tag{7.38}$$

Put $\psi^{(q)}=\sum_{h=1}^{q}a_h e_{th}$. Of course, we have $\psi^{(q)}\in L(B)$. In order to prove the existence of $\psi\in L(B)$ with $\psi=\sum_{h=1}^{\infty}a_h e_{th}$, it suffices to show that for each D_j^l, $l\geqq 0$, the sequence $\{D_j^l\psi_j^{(q)\lambda}(x)\,|\,q=1,2,\ldots\}$ converges uniformly on U_j. Needless to say, $D_j^0\psi_j^{(q)\lambda}(x)=\psi_j^{(q)\lambda}(x)$. For any pair of natural numbers p, q with $p<q$, we have $\psi^{(q)}-\psi^{(p)}=\sum_{h=p+1}^{q}a_h e_{th}$. Therefore, if we choose k, for each l, so that $k>l+n/2$, we obtain from Sobolev's inequality (7.22) and the inequality (7.37)

$$|D_j^l\psi_j^{(q)\lambda}(x)-D_j^l\psi_j^{(p)\lambda}(x)|^2\leqq(c_{k-l,l})^2\|\psi^{(q)}-\psi^{(p)}\|_k^2$$

$$\leqq(c_{k-l,l})^2 c_k''\sum_{h=p+1}^{q}\left(1+\sum_{\alpha=1}^{k}|\lambda_h(t)|^{2\alpha}\right)|a_h|^2.$$

By (7.38) the right-hand side converges to 0 as $p\to\infty$ and $q\to\infty$. This proves that $\{D_j^l\psi_j^{(q)\lambda}(x)\}$ converges uniformly on U_j. ∎

Let ζ be an arbitrary complex number and put

$$E_t(\zeta) = E_t - \zeta.$$

Then $E_t(\zeta)$ is a strongly elliptic differential operator acting on $L(B)$. For any $\psi = \sum_{h=1}^{\infty} a_h e_{th}$, we have

$$E_t(\zeta)\psi = E_t\psi - \zeta\psi = \sum_{h=1}^{\infty} (\lambda_h(t) - \zeta) a_h e_{th}. \tag{7.39}$$

If $\zeta \neq \lambda_h(t)$, $h = 1, 2, \ldots$, then $E_t(\zeta)$: $L(B) \to L(B)$ *is bijective. Proof.* It is clear from (7.39) that $E_t(\zeta)$ is injective. In order to prove that $E_t(\zeta)$ is surjective, take any $\varphi \in L(B)$ with $\varphi = \sum_{h=1}^{\infty} b_h e_{th}$, and put $a_h = b_h/(\lambda_h(t) - \zeta)$. By Lemma 7.4, we have

$$\sum_{h=1}^{\infty} |\lambda_h(t)|^{2l} |b_h|^2 < +\infty, \qquad l = 1, 2, \ldots.$$

Therefore, since $\lambda_h(t) \neq \zeta$ and $\lambda_h(t) \to +\infty$ as $h \to \infty$, we have

$$\sum_{h=1}^{\infty} |\lambda_h(t)|^{2l} |a_h|^2 = \sum_{h=1}^{\infty} \frac{|\lambda_h(t)|^{2l} |b_h|^2}{|\lambda_h(t) - \zeta|^2} < +\infty.$$

Hence, again by Lemma 7.4, there exists $\psi \in L(B)$ with $\psi = \sum_{h=1}^{\infty} a_h e_{th}$. We obviously have $E_t(\zeta)\psi = \varphi$. Hence $E_t(\zeta)$ is surjective. ∎

Since $E_t(\zeta)$ is bijective for $\zeta \neq \lambda_h(t)$, $h = 1, 2, \ldots$, the inverse $E_t(\zeta)^{-1}$ exists. We denote it by $G_t(\zeta)$: $G_t(\zeta) = E_t(\zeta)^{-1}$. The operator $G_t(\zeta)$ is called the Green operator of $E_t(\zeta)$ (cf. Definition 7.2 in the Appendix). On $\Delta_\varepsilon \times \mathbb{C}$, $E_t(\zeta)$ is C^∞ differentiable in (t, ζ). We shall prove, applying Theorem 7.5 to E_t, that $G_t(\zeta)$ is also C^∞ differentiable in (t, ζ) for $t \in \Delta$ and $\zeta \neq \lambda_h(t)$. We need the following lemma.

Lemma 7.5. *Suppose there are given $t_0 \in \Delta$ and $\zeta_0 \in \mathbb{C}$ with $\zeta_0 \neq \lambda_h(t_0)$, $h = 1, 2, 3, \ldots$. If we take a sufficiently small $\delta > 0$, there exists a constant $c > 0$ such that, for $|t - t_0| < \delta$ and $|\zeta - \zeta_0| < \delta$, the following inequality*

$$\|E_t(\zeta)\psi\|_0 \geqq c\|\psi\|_0, \qquad \psi \in L(B) \tag{7.40}$$

holds.

Proof. Suppose that, for any small $\delta > 0$, there is no such constant. Then, for $q = 1, 2, \ldots$, there exist $t_q \in \Delta$, $\zeta_q \in \mathbb{C}$ and $\psi^{(q)} \in L(B)$ such that

$$|t_q - t_0| < \frac{1}{q}, \qquad |\zeta_q - \zeta_0| < \frac{1}{q}, \qquad \|E_{t_q}(\zeta_q)\psi^{(q)}\|_0 < \frac{1}{q}, \qquad \|\psi^{(q)}\|_0 = 1.$$

By $(7.23)_0$

$$\|\psi^{(q)}\|_m^2 \leqq c_0(\|E_{t_q}(\zeta_q)\psi^{(q)}\|_0^2 + \|\psi^{(q)}\|_0^2) < 2c_0.$$

Hence, in the equation

$$(E_t\psi^{(q)})_j^\lambda(x) = \sum_{l_1+\cdots+l_n \leqq m} \sum_\mu E_{j\mu l_1\cdots l_n}^\lambda(x,t) D_{j1}^{l_1}\cdots D_{jn}^{l_n}\psi_j^{(q)\mu}(x),$$

we have

$$\int_{U_j} |D_{j1}^{l_1}\cdots D_{jn}^{l_n}\psi_j^{(q)\mu}(x)|^2\, dX_j \leqq \|\psi^{(q)}\|_m^2 < 2c_0,$$

and the coefficients $E_{j\mu l_1\cdots l_n}^\lambda(x,t)$ are C^∞ functions of (x, t). Hence

$$\begin{aligned} &\|E_{t_q}(\zeta_q)\psi^{(q)} - E_{t_0}(\zeta_0)\psi^{(q)}\|_0 \\ &\qquad = \|E_{t_q}\psi^{(q)} - E_{t_0}\psi^{(q)} - (\zeta_q - \zeta_0)\psi^{(q)}\|_0 \to 0 \qquad (q\to\infty). \end{aligned}$$

Since $\|E_{t_q}(\zeta_q)\psi^{(q)}\|_0 < 1/q \to 0\ (q\to\infty)$, we have

$$\|E_{t_0}(\zeta_0)\psi^{(q)}\|_0 \to 0 \qquad (q\to\infty).$$

On the other hand, by (7.39), we have for any $\psi\in L(B)$

$$\|E_{t_0}(\zeta_0)\psi\|_{t_0} \geqq \mu_0\|\psi\|_{t_0}, \qquad \mu_0 = \min_h |\lambda_h(t_0) - \zeta_0| > 0.$$

Hence, by (7.21) $K_0^2\|E_{t_0}(\zeta_0)\psi\|_0 \geqq \mu_0\|\psi\|_0$. Consequently we have $\|\psi^{(q)}\|_0 \to 0\ (q\to\infty)$, which contradicts $\|\psi^{(q)}\|_0 = 1$. ∎

Suppose we are given $t_0\in\Delta$, and $\zeta_0 \neq \lambda_h(t_0)$, $h = 1, 2, \ldots$. If we take a sufficiently small $\delta>0$, then, by this lemma, $\|E_t(\zeta)\psi\|_0 \geqq c\|\psi\|_0$ for $|t-t_0|<\delta$ and $|\zeta-\zeta_0|<\delta$, hence, it follows that $\zeta\neq\lambda_h(t)$, $h=1,2,\ldots$. Therefore $E_t(\zeta)$ is bijective, and, by Theorem 7.5, $G_t(\zeta) = E_t(\zeta)^{-1}$ is C^∞ differentiable in (t,ζ). If we put

$$\mathscr{W} = \{(t,\zeta)\in\Delta\times\mathbb{C} \mid \zeta\neq\lambda_h(t),\ h=1,2,\ldots\},$$

then, for any point $(t_0,\zeta_0)\in\mathscr{W}$, we can find a constant $\delta>0$ as above. This implies that *$\mathscr{W}$ is an open subset of $\Delta\times\mathbb{C}$, and $G_t(\zeta)$ is C^∞ differentiable in (t,ζ) on $\mathscr{W}$.*

Fix an arbitrary $t_0\in\Delta$, and take a closed Jordan curve C on the complex plane $\mathbb{C}$ which does not pass through any of the $\lambda_h(t_0)$, $h=1,2,\ldots$.

The Jordan curve C divides the complex plane into two domains, the interior and the exterior of C. We denote by $((C))$ the interior of C. Since

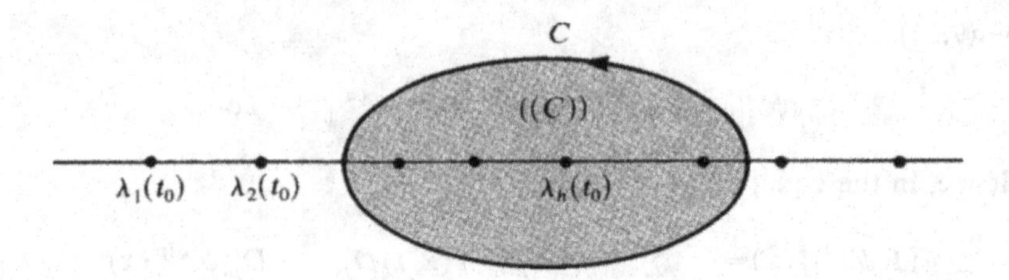

Figure 1

$C\times t_0\subset W$, and since $\mathscr{W}$ is open in $\Delta\times\mathbb{C}$, we have

$$C\times[t_0-\delta,\, t_0+\delta]\subset \mathscr{W}, \tag{7.41}$$

provided that $\delta>0$ is sufficiently small. For any t with $|t-t_0|<\delta$, we define the linear operator $F_t(C)$ of $L(B)$ by

$$F_t(C)\psi=\sum_{\lambda_h(t)\in((C))}(\psi, e_{th})e_{th},\qquad \psi\in L(B),$$

and put $\mathbb{F}_t(C)=F_t(C)L(B)$. Obviously, $\mathbb{F}_t(C)$ is a finite-dimensional linear subspace of $L(B)$, and it consists of all the linear combinations of those e_{th} which belong to the eigenvalues $\lambda_h(t)$ lying in the interior of the Jordan curve C. The linear operator $F_t(C)$ is nothing but the orthogonal projection of $L(B)$ onto $\mathbb{F}_t(C)$ with respect to the inner product $(\ ,\ )_t$.

The operator $F_t(C)$ can be written as follows, using the Green operator $G_t(\zeta)$:

$$F_t(C)\psi=-\frac{1}{2\pi i}\int_C G_t(\zeta)\psi\, d\zeta,\qquad \psi\in L(B). \tag{7.42}$$

Proof. Since, by (7.41), $C\times[t_0-\delta,\, t_0+\delta]\subset\mathscr{W}$, and since $G_t(\zeta)$ is C^∞ differentiable in (t,ζ) in $\mathscr{W}$, the integral $\int_C G_t(\zeta)\psi\, d\zeta$ exists. We let $\psi=\sum_{h=1}^{\infty}a_h e_{th}$. Then, since $G_t(\zeta)=E(t)^{-1}$, we obtain by (7.39)

$$-G_t(\zeta)\psi=\sum_{h=1}^{\infty}\frac{a_h}{\zeta-\lambda_h(t)}e_{th}.$$

Hence, by Cauchy's integral formula,

$$-\frac{1}{2\pi i}\int_C G_t(\zeta)\psi\, d\zeta=\sum_{\lambda_h(t)\in((C))}a_h e_{th}=F_t(C)\psi. \quad \blacksquare$$

Lemma 7.6. *$F_t(C)$ is C^∞ differentiable in t for $|t-t_0|<\delta$.*

Proof. Since $G_t(\zeta)$ is C^∞ differentiable in (t, ζ) in W, it follows that, if $\psi_t \in L(B)$ is C^∞ differentiable in t for $|t-t_0|<\delta$, *then* $F_t(C)\psi_t = -(1/2\pi i)\int_C G_t(\zeta)\,\psi_t\,d\zeta$ is also C^∞ differentiable in t. ∎

Lemma 7.7. $\dim \mathbb{F}_t(C)$ *is independent of* t *for* $|t-t_0|<\delta$ *provided* $\delta>0$ *is sufficiently small*: $\dim \mathbb{F}_t(C) = \dim \mathbb{F}_{t_0}(C)$.

Proof. Put $\dim \mathbb{F}_{t_0}(C) = d$, and let $\{e_1, \ldots, e_r, \ldots, e_d\}$ be a basis of $\mathbb{F}_{t_0}(C)$. By Lemma 7.6, the $F_t(C)e_r$ are C^∞ differentiable in t, and the $F_{t_0}(C)e_r = e_r$, $r = 1, \ldots, d$, are linearly independent. Therefore, the $F_t(C)e_r \in \mathbb{F}_t(C)$, $r = 1, \ldots, d$, are linearly independent for $|t-t_0|<\delta$ provided $\delta>0$ is sufficiently small. Hence $\dim \mathbb{F}_t(C) \geqq d$.

Suppose that for any small $\delta>0$, there exists t with $|t-t_0|<\delta$ such that $\dim \mathbb{F}_t(C) > d$. Then we can find a sequence t_q, $q = 1, 2, \ldots$, with $|t-t_0|<1/q$ such that $\dim \mathbb{F}_{t_q}(C) \geqq d+1$. Then, at least $d+1$ eigenvalues $\lambda_{h_r}(t_q)$, $r = 1, 2, \ldots, d+1$, must lie in the interior of C. Put, for simplicity, $\lambda_r^{(q)} = \lambda_{h_r}(t_q)$ and $e_r^{(q)} = e_{t_q h_r}$. If we take a natural number k with $k > m+1+n/2$, then, by (7.22) and (7.37), for each D_j^l, $l \leqq m+1$,

$$|D_j^l e_r^{(q)\lambda}(x)|^2 \leqq (c_{k-l,l})^2 c_k'' \left(1 + \sum_{\alpha=1}^{k} |\lambda_r^{(q)}|^{2\alpha}\right).$$

Since $\lambda_r^{(q)} \in ((C))$, the sequence $\{D_j^l e_{rj}^{(q)\lambda}(x) \mid q = 1, 2, \ldots\}$ is uniformly bounded in U_j. It follows that, for any $l \leqq m$, $\{D_j^l e_{rj}^{(q)\lambda}(x)\}$ is equicontinuous on U_j. Hence, we can find a uniformly convergent subsequence. Taking *an appropriate subsequence of* $\{t_q\}$, we may assume, from the beginning, that $\{D_j^l e_{rj}^{(q)\lambda}(x)\}$ converges uniformly in U_j for any D_j^l, $l \leqq m$. Put $e_{rj}^\lambda(x) = \lim_{q\to\infty} e_{rj}^{(q)\lambda}(x)$. Then $e_{rj}^\lambda(x)$ is a C^m differentiable function on U_j, and we have

$$\lim_{q\to\infty} D_j^l e_{rj}^{(q)\lambda}(x) = D_j^l e_{rj}^\lambda(x), \qquad l \leqq m.$$

Therefore the $e_r = \lim_{q\to\infty} e_r^{(q)}$ are C^m sections of B and $e_{rj}^\lambda(x)$ are the values of their fibre coordinates. From $(e_r^{(q)}, e_s^{(q)})_{t_q} = \delta_{rs}$, it follows that

$$(e_r, e_s)_{t_0} = \lim_{q\to\infty} (e_r^{(q)}, e_s^{(q)})_{t_q} = \delta_{rs}.$$

Since the $E_{j\mu}^\lambda(x, t, D_j)$ are linear combinations of D_j^l with $l \leqq m$, with the coefficients being C^∞ functions of (x, t), we have

$$\lim_{q\to\infty} E_{t_q} e_r^{(q)} = E_{t_0} e_r.$$

On the other hand, we have $E_{t_q} e_r^{(q)} = \lambda_r^{(q)} e_r^{(q)}$. Therefore, the limit $\lambda_r = \lim_{q\to\infty} \lambda_r^{(q)}$ exists, and we have

$$E_{t_0} e_r = \lambda_r e_r, \qquad \|e_r\|_{t_0} = 1.$$

This means that λ_r is an eigenvalue of E_{t_0}, and e_r is an eigenfunction belonging to λ_r. Since $\lambda_r^{(q)} \in ((C))$, and since there are no eigenvalues of E_{t_0} on C, λ_r must coincide with one of $\lambda_h(t_0) \in ((C))$. Hence $e_r \in \mathbb{F}_{t_0}(C)$, $r = 1, 2, \ldots, d+1$. But, since $(e_r, e_s)_{t_0} = \delta_{rs}$, it follows that $e_1, \ldots, e_{d+1}$ are linearly independent. This contradicts $\dim \mathbb{F}_{t_0}(C) = d$. ∎

Proof of Theorem 7.2. We shall prove by induction on h that $\lim_{t\to t_0} \lambda_h(t) = \lambda_h(t_0)$ for $t_0 \in \Delta$.

(1°) The case $h = 1$. By (7.14) in the Appendix, there is a constant β independent of t such that $\lambda_1(t) > \beta$ for $t \in \Delta$. Taking a sufficiently small $\varepsilon > 0$, we consider the circle C_ε of radius ε with centre $\lambda_1(t_0)$ and a smooth Jordan curve C which intersects transversally with the real axis at β and $\lambda = \lambda_1(t_0) - \varepsilon$. We assume that C does not intersect with the real axis except for at β and λ. Take a sufficiently small $\delta > 0$. Since there are no eigenvalues $\lambda_h(t_0)$ of E_{t_0} in the interior of C or on C, we have $\dim \mathbb{F}_{t_0}(C) = 0$. Hence, by Lemma 7.7, $\dim \mathbb{F}_t(C) = 0$ for $|t - t_0| < \delta$. Therefore, $\lambda_1(t)$ is a real number outside of C. But, since $\lambda_1(t) > \beta$, we have $\lambda_1(t) > \lambda = \lambda_1(t_0) - \varepsilon$.

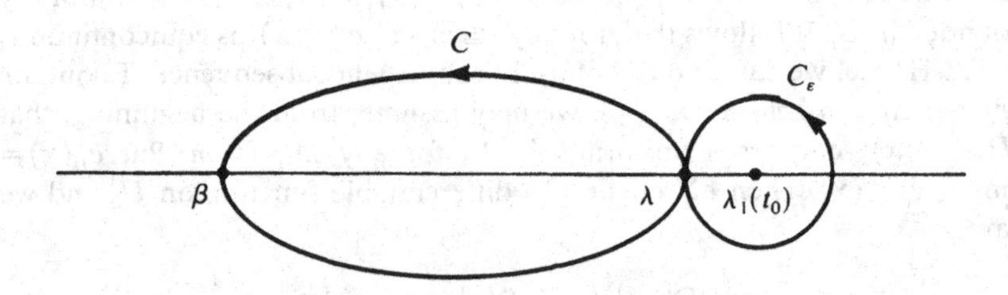

Figure 2

Moreover, by Lemma 7.7, we have $\dim \mathbb{F}_t(C_\varepsilon) = \dim \mathbb{F}_{t_0}(C_\varepsilon) \geqq 1$ for $|t - t_0| < \delta$. Hence there is at least one eigenvalue $\lambda_h(t)$ in the interior of C_ε. This implies that $\lambda_1(t) \leqq \lambda_h(t) < \lambda_1(t_0) + \varepsilon$. Therefore $|\lambda_1(t) - \lambda_1(t_0)| < \varepsilon$ holds if $|t - t_0| < \delta$. This proves that $\lim_{t\to t_0} \lambda_1(t) = \lambda_1(t_0)$.

(2°) The case $h \geqq 2$. We assume that $\lim_{t\to t_0} \lambda_k(t) = \lambda_k(t_0)$ for $k = 1, 2, \ldots, h-1$, and prove $\lim_{t\to t_0} \lambda_h(t) = \lambda_h(t_0)$.

If $\lambda_1(t_0) = \cdots = \lambda_h(t_0)$, then, for the circle C_ε of (1°), we have, by Lemma 7.7, $\dim \mathbb{F}_t(C_\varepsilon) \geqq h$ for $|t - t_0| < \delta$, provided $\delta > 0$ is sufficiently small. Therefore $\lambda_1(t), \ldots, \lambda_h(t)$ lie in the interior of C_ε. Hence $|\lambda_h(t) - \lambda_h(t_0)| < \varepsilon$. This proves $\lim_{t\to t_0} \lambda_h(t) = \lambda_h(t_0)$.

Next we let l be the integer, $2 \leqq l \leqq h$, such that

$$\lambda_1(t) \leqq \lambda_{l-1}(t_0) < \lambda_l(t_0) = \lambda_{l+1}(t_0) = \cdots = \lambda_h(t_0).$$

Let $C_{h\varepsilon}$ be the circle of radius ε with centre $\lambda_h(t_0)$, where ε is assumed to satisfy $0 < \varepsilon < \lambda_l(t_0) - \lambda_{l-1}(t_0)$, and let C_h be a closed Jordan curve which intersects transversally with the real axis only at the two points $\lambda = \lambda_1(t_0) - \varepsilon$ and $\mu = \lambda_h(t_0) - \varepsilon$. We take $\delta > 0$ sufficiently small depending on $C_{h\varepsilon}$ and C_h. Then, since at least $h - l + 1$ eigenvalues $\lambda_l(t_0), \ldots, \lambda_h(t_0)$ are in the

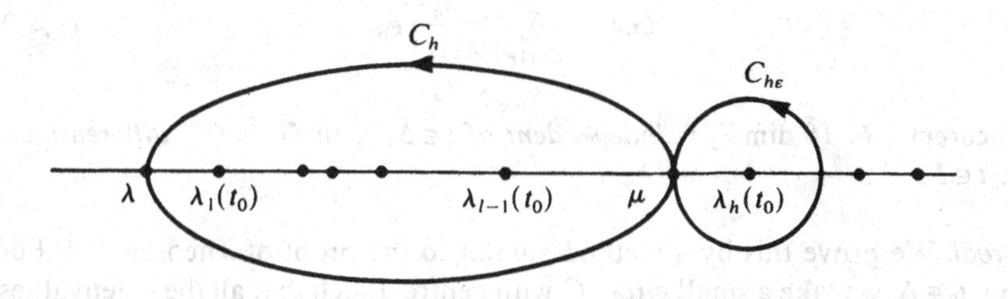

Figure 3

interior of $C_{h\varepsilon}$, we have $\dim \mathbb{F}_{t_0}(C_{h\varepsilon}) \geqq h - l + 1$. Therefore, by Lemma 7.7, at least $h - l + 1$ eigenvalues $\lambda_k(t)$'s are in the interior of $C_{h\varepsilon}$. On the other hand, from $\dim \mathbb{F}_{t_0}(C_h) = l - 1$, it follows that $\dim \mathbb{F}_t(C_h) = l - 1$ for $|t - t_0| < \delta$. Hence, exactly $l - 1$ eigenvalues $\lambda_k(t)$'s are in the interior of C_h. By the hypothesis of induction, $\lim_{t \to t_0} \lambda_k(t) = \lambda_k(t_0)$ for $k = 1, 2, \ldots, l - 1$. Therefore $\lambda_1(t), \ldots, \lambda_{l-1}(t)$ are in the interior of C_h for $|t - t_0| < \delta$. This implies that $\lambda_l(t), \ldots, \lambda_h(t), \ldots$ are all outside of C_h. Hence,

$$\mu < \lambda_l(t) \leqq \lambda_{l+1}(t) \leqq \cdots \leqq \lambda_h(t) \leqq \cdots.$$

Combined with what we have seen above, it follows that $\lambda_l(t)$, $\lambda_{l+1}(t), \ldots, \lambda_h(t)$ must lie inside of $C_{h\varepsilon}$. Hence $|\lambda_h(t) - \lambda_h(t_0)| < \varepsilon$ for $|t - t_0| < \delta$. This proves $\lim_{t \to t_0} \lambda_h(t) = \lambda_h(t_0)$. ∎

Proof of Theorem 7.3. By definition, $\mathbb{F}_t = \{\psi \in L(B) \mid F_t\psi = 0\}$. We need to prove that, for a given $t_0 \in \Delta$, we can find a sufficiently small $\delta > 0$ such that $\dim \mathbb{F}_t \leqq \dim \mathbb{F}_{t_0}$ for $|t - t_0| < \delta$. If we let C_ε denote the circle of radius $\varepsilon > 0$ with centre 0 in the complex plane $\mathbb{C}$, then $\mathbb{F}_{t_0} = \mathbb{F}_{t_0}(C_\varepsilon)$, provided ε is sufficiently small. Therefore, by Lemma 7.7, $\dim \mathbb{F}_t(C_\varepsilon) = \dim \mathbb{F}_{t_0}$ for $|t - t_0| < \delta$, where δ is supposed to be sufficiently small. Since the inclusion $\mathbb{F}_t \subset \mathbb{F}_t(C_\varepsilon)$ is obvious, we obtain $\dim \mathbb{F}_t \leqq \dim \mathbb{F}_t(C_\varepsilon) = \dim \mathbb{F}_{t_0}$. ∎

Proof of Theorem 7.4. By definition F_t is the orthogonal projection of $L(B)$ onto $\mathbb{F}_t$ with respect to the inner product $(\ ,\ )_t$. In general, for any $t_0 \in \Delta$, if we take a sufficiently small δ, then, as we have seen above, the inclusion $\mathbb{F}_t \subset \mathbb{F}_t(C_\varepsilon)$ holds for $|t-t_0|<\delta$. Since we have $\dim \mathbb{F}_t(C_\varepsilon) = \dim \mathbb{F}_{t_0}$, and since $\dim \mathbb{F}_t$ is independent of t by assumption, we conclude $\mathbb{F}_t = \mathbb{F}_t(C_\varepsilon)$. Hence $F_t = F_t(C_\varepsilon)$. But by Lemma 7.6, $F_t(C_\varepsilon)$ is C^∞ differentiable in t for $|t-t_0|<\delta$. Since t_0 is an arbitrary point of Δ, it follows that F_t is C^∞ differentiable in $t \in \Delta$. ∎

Let G_t be the Green operator of E_t. Then, by (7.12), we have, for $\psi = \sum_{h=1}^\infty a_h e_{th}$,

$$G_t\psi = \sum_{\lambda_h(t)\neq 0} \frac{a_h}{\lambda_h(t)} e_{th}. \tag{7.43}$$

Theorem 7.6. *If* $\dim \mathbb{F}_t$ *is independent of* $t \in \Delta$, *then* G_t *is* C^∞ *differentiable in* $t \in \Delta$.

Proof. We prove this by a method similar to the proof of Theorem 7.4. For any $t_0 \in \Delta$, we take a small circle C with centre 0 such that all the eigenvalues $\lambda_h(t_0)\neq 0$ are outside of C. If $\delta>0$ is taken sufficiently small, then, by (7.41), $C\times[t_0-\delta, t_0+\delta] \subset \mathscr{W}$. For t satisfying $|t-t_0|<\delta$, we define the linear operator $G_t(C)$ on $L(B)$ by the formula

$$G_t(C)\psi = \frac{1}{2\pi i}\int_C \frac{1}{\zeta} G_t(\zeta)\psi\, d\zeta.$$

Since $G_t(\zeta)$ is C^∞ differentiable in (t,ζ) in $\mathscr{W}$, it follows that $G_t(C)$ is C^∞ differentiable in t for $|t-t_0|<\delta$. For any $\sum_{h=1}^\infty a_h e_{th}$, we have

$$G_t(\zeta)\psi = \sum_{h=1}^{\infty} \frac{a_h}{\lambda_h(t)-\zeta} e_{th}.$$

By the residue theorem,

$$\frac{1}{2\pi i}\int_C \frac{1}{\zeta(\lambda_h(t)-\zeta)}\, d\zeta = \begin{cases} 0, & \lambda_h(t) \in ((C)), \\ \dfrac{1}{\lambda_h(t)}, & \lambda_h(t) \notin ((C)). \end{cases}$$

Hence for any $\psi = \sum_{h=1}^\infty a_h e_{th}$, we have

$$G_t(C)\psi = \sum_{\lambda_h(t)\notin((C))} \frac{a_h}{\lambda_h(t)} e_{th}. \tag{7.44}$$

Since, by hypothesis, $\dim \mathbb{F}_t$ is independent of t, $\mathbb{F}_t = \mathbb{F}_t(C)$ for $|t-t_0|<\delta$ as was proved in the proof of Theorem 7.4. Therefore, the equality $\lambda_h(t)=0$

amounts to the same as $\lambda_h(t) \in ((C))$. By comparing (7.43) and (7.44), we obtain $G_t = G_t(C)$. Consequently G_t is C^∞ differentiable in t provided $|t - t_0| < \delta$. Since t_0 is arbitrary, G_t is C^∞ differentiable in $t \in \Delta$. ∎

§7.2. Differentiable Family of Compact Complex Manifolds

In this section we prove Theorem 4.1, Theorem 4.4, and other results by applying the results of the preceding section.

(a) Differentiable Family of Compact Complex Manifolds

Consider a differentiable family $M = (\mathcal{M}, I, \varpi) = \{M_t \mid t \in I\}$ of compact complex manifolds over a multi-interval I with $0 \in I \subset \mathbb{R}^m$ (see Definition 4.1), where $M_t = \varpi^{-1}(t)$. By Theorem 2.3, the underlying differentiable manifold X of M_t does not depend on t, and by Theorem 2.4, there is a diffeomorphism Ψ: $X \times I \to \mathcal{M}$ of $X \times I$ onto M such that $\varpi \cdot \Psi$ coincides with the projection $X \times I \to I$. We identify $X \times I$ with $\mathcal{M}$ via Ψ, and *consider* $\mathcal{M} = X \times I$. We denote a point of $X \times I$ by (x, t). By definition, there are a locally finite open covering $\{\mathcal{U}_j \mid j = 1, 2, \ldots\}$ of $X \times I$ and complex-valued C^∞ functions $z_j^1(x, t), \ldots, z_j^n(x, t)$ on $\mathcal{U}_j$ such that for each t,

$$\{x \to (z_j^1(x, t), \ldots, z_j^n(x, t)) \mid \mathcal{U}_j \cap X \times t \neq \varnothing\}$$

forms a system of local complex coordinates on M_t. If we write $z_j(x, t) = (z_j^1(x, t), \ldots, z_j^n(x, t))$, the map $(x, t) \to (z_j(x, t), t)$ is a diffeomorphism of $\mathcal{U}_j$ into $\mathbb{C}^n \times I$ and

$$\{(x, t) \to (z_j(x, t), t) \mid j = 1, 2, \ldots\}$$

is a *system of local coordinates of the differentiable family* $\mathcal{M}$ (see p. 185). On $\mathcal{U}_j \cap \mathcal{U}_k \neq \varnothing$, we have

$$z_j^\alpha(x, t) = f_{jk}^\alpha(z_k(x, t), t), \qquad \alpha = 1, \ldots, n,$$

where $f_{jk}^\alpha(z_k, t) = f_{jk}^\alpha(z_k^1, \ldots, z_k^n, t_1, \ldots, t_m)$ are C^∞ functions of $z_k^1, \ldots, z_k^n, t_1, \ldots, t_m$ which are holomorphic in $z_k^1, \ldots, z_k^n$.

Taking a sufficiently fine finite covering $\{U_j \mid j = 1, 2, \ldots\}$ of X, we may assume that for a sufficiently small multi-interval $\Delta = \{t \in I \mid |t_\kappa| < r, \kappa = 1, \ldots, m\}$ with $0 < r < 1$, each $[U_j \times \Delta]$ is contained in one of $\mathcal{U}_j$'s. Thus, by renumbering $\mathcal{U}_j$, we may assume that $[U_j \times \Delta] \subset \mathcal{U}_j$. Put $\mathcal{M}_\Delta = \varpi^{-1}(\Delta)$. Then $\mathcal{M}_\Delta = (\mathcal{M}_\Delta, \Delta, \varpi)$ is a differentiable family. As a differentiable manifold, we have $\mathcal{M}_\Delta = X \times \Delta$. $\{U_j \times \Delta \mid j = 1, 2, \ldots\}$ forms a finite covering of $X \times \Delta$, and the map

$$(x, t) \to (z_j(x, t), t) = (z_j^1(x, t), \ldots, z_j^n(x, t), t_1, \ldots, t_m)$$

gives local coordinates of the differentiable manifold $\mathcal{M}_\Delta$ on $U_j \times \Delta$. Since the problem is local with respect to the parameter t, it suffices to consider $\mathcal{M}_\Delta$ for a sufficiently small Δ. For simplicity, we write $\mathcal{M} = (\mathcal{M}, \Delta, \varpi)$ for $\mathcal{M}_\Delta = (\mathcal{M}_\Delta, \Delta, \varpi)$ below.

Let $\mathcal{B}$ be a complex vector bundle, and $\pi: \mathcal{B} \to \mathcal{M}$ its projection. Then $B_t = \pi^{-1}(M_t)$ is a complex vector bundle over M_t. We call such $\{B_t | t \in \Delta\}$ a differentiable family of complex vector bundles. In particular, if each B_t is a holomorphic vector bundle over the complex manifold M_t, $\{B_t | t \in \Delta\}$ is called a differentiable family of holomorphic vector bundles. By saying that $\{B_t | t \in \Delta\}$ forms a differentiable family of vector bundles, we mean that $\{B_t | t \in \Delta\}$ is a differentiable family over the underlying differentiable manifold X in the sense of Definition 7.3. Since U_j and Δ are sufficiently small, we have

$$\pi^{-1}(U_j \times \Delta) = \mathbb{C}^\nu \times U_j \times \Delta,$$

and a point of $\pi^{-1}(U_j \times \Delta) \subset \mathcal{B}$ is represented as $(\zeta_j, x, t) = (\zeta_j^1, \ldots, \zeta_j^\nu, x, t)$ where $\zeta_j = (\zeta_j^1, \ldots, \zeta_j^\nu)$ are the fibre coordinates of $\mathcal{B}$ over $U_j \times \Delta$, and on $U_j \times \Delta \cap U_k \times \Delta \neq \varnothing$, we have

$$\zeta_j^\lambda = \sum_{\mu=1}^{\nu} b_{jk\mu}^\lambda(x, t)\zeta_k^\mu, \qquad \lambda = 1, \ldots, \nu.$$

Each component $b_{jk\mu}^\lambda(x, t)$ of the transition function $b_{jk}(x, t) = (b_{jk\mu}^\lambda(x, t))_{\lambda,\mu=1,\ldots,\nu}$ of $\mathcal{B}$ is represented as a C^∞ function of the local coordinates $(z_k, t) = (z_k^1(x, t), \ldots, z_k^n(x, t), t)$ of M by

$$b_{jk\mu}^\lambda(x, t) = b_{jk\mu}^\lambda(z_k, t), \qquad z_k = z_k(x, t).$$

Thus $\{B_t | t \in \Delta\}$ *is a differentiable family of holomorphic vector bundles if and only if all the* $b_{jk\mu}^\lambda(z_k, t) = b_{jk\mu}^\lambda(z_k^1, \ldots, z_k^n, t)$ *are holomorphic in* $z_k^1, \ldots, z_k^n$.

As stated before, we always use admissible fibre coordinates $(\zeta_j^1, \ldots, \zeta_j^\nu)$ for B_t. Let $L(B_t)$ be the linear space of all C^∞ sections of B_t over M_t. $L(B_t)$ is nothing but the linear space of C^∞ sections of B_t over the underlying differentiable manifold X. We write a C^∞ section $\psi_t \in L(B_t)$ of B_t by

$$\psi_t(x, t) = (\psi_j^1(z_j, t), \ldots, \psi_j^\nu(z_j, t)), \qquad z_j = z_j(x, t)$$

on $U_j \times t \subset M$, where $(\psi_j^1(z_j, t), \ldots, \psi_j^\nu(z_j, t))$ are admissible fibre coordinates of $\psi_t(x, t) \in B_t$. Given $\psi_t \in L(B_t)$ for every $t \in \Delta$, we call ψ_t C^∞ differentiable with respect to $t \in \Delta$ if $\psi_j^\lambda(z_j, t)$, $\lambda = 1, \ldots, \nu$, are C^∞ functions of z_j

and t. If each B_t is a holomorphic vector bundle over M_t, and ψ_t is a holomorphic section of B_t, then $\psi_j^\lambda(z_j, t)$ is holomorphic in $z_j = (z_j^1, \ldots, z_j^n)$.

Consider the tangent bundle $T(M_t)$ of M_t. The transition function of $T(M_t)$ is given by

$$\left(\frac{\partial f_{jk}^\alpha(z_k, t)}{\partial z_k^\beta}\right)_{\alpha,\beta=1,\ldots,n},$$

and its component $\partial f_{jk}^\alpha(z_k, t)/\partial z_k^\beta$ is a C^∞ function of (z_k, t) which is holomorphic in z_k. Consequently $\{T(M_t) \mid t \in \Delta\}$ is a differentiable family of holomorphic vector bundles. If we write a C^∞ section $\xi_t \in L(T(M_t))$ of $T(M_t)$ as

$$\xi_t = \sum_{\alpha=1}^n \xi_j^\alpha(z_j, t)\frac{\partial}{\partial z_j^\alpha}$$

on $U_j \times t$, then we have on $U_j \times t \cap U_k \times t$

$$\xi_j^\alpha(z_j, t) = \sum_{\beta=1}^n \frac{\partial f_{jk}^\alpha(z_k, t)}{\partial z_k^\beta}\xi_k^\beta(z_k, t).$$

Namely, $(\xi_j^1(z_j, t), \ldots, \xi_j^n(z_j, t))$ are admissible fibre coordinates. Therefore ξ_t is C^∞ differentiable with respect to $t \in \Delta$ if and only if $\xi_j^\alpha(z_j, t)$ are C^∞ functions of (z_j, t) for $\alpha = 1, \ldots, n$.

From (3.49) the transition function of the dual bundle $T^*(M_t)$ of $T(M_t)$ is given by

$$\left(\frac{\partial f_{kj}^\beta(z_j, t)}{\partial z_j^\alpha}\right)_{\alpha,\beta=1,\ldots,n}, \qquad z_j = f_{jk}(z_k, t).$$

Consequently $\{T^*(M_t) \mid t \in \Delta\}$ is also a differentiable family of holomorphic vector bundles. A C^∞ section $\varphi_t \in L(T^*(M_t))$ of $T^*(M_t)$ is a C^∞ (1, 0)-form on M_t:

$$\varphi_t = \sum_{\alpha=1}^n \varphi_{j\alpha}(z_j, t)\, dz_j^\alpha.$$

$(\varphi_{j1}(z_j, t), \ldots, \varphi_{jn}(z_j, t))$ are clearly admissible fibre coordinates, hence φ_t is C^∞ differentiable with respect to $t \in \Delta$ if and only if each $\varphi_{j\alpha}(z_j, t)$ is a C^∞ function of (z_j, t) for $\alpha = 1, \ldots, n$. $z_j^\alpha = z_j^\alpha(x, t)$ is a C^∞ function of (x, t) on $U_j \times \Delta$, and dz_j^α *is the differential of* z_j^α *as a function of* x *with* t *fixed.* Thus, letting $(x^1, \ldots, x^\gamma, \ldots, x^{2n})$ be arbitrary coordinates of x, we have

$$dz_j^\alpha = \sum_{\gamma=1}^{2n} \frac{\partial z_j^\alpha(x, t)}{\partial x^\gamma}\, dx^\gamma, \tag{7.45}$$

where $\partial z_j^\alpha(x, t)/\partial x^\gamma$ are C^∞ functions of (x, t), which again shows that $(\varphi_{j1}(z_j, t), \ldots, \varphi_{jn}(z_j, t))$ are admissible fibre coordinates.

Suppose given a *differentiable family of holomorphic vector bundles* $\{B_t \mid t \in \Delta\}$. Consider $B_t \otimes \bigwedge^p T^*(M_t) \bigwedge^q \bar{T}^*(M_t)$. Since $\{T^*(M_t) \mid t \in \Delta\}$ is a differentiable family,

$$\left\{B_t \otimes \overset{p}{\bigwedge} T^*(M_t) \overset{q}{\bigwedge} \bar{T}^*(M_t) \,\middle|\, t \in \Delta\right\}$$

is also a differentiable family. A C^∞ section of $B_t \otimes \bigwedge^p T^*(M_t) \bigwedge^q \bar{T}^*(M_t)$ is called a C^∞ (p, q)-form with coefficients in B_t. Let $\mathscr{L}^{p,q}(B_t)$ be the linear space of C^∞ (p, q)-forms with coefficients in B_t:

$$\mathscr{L}^{p,q}(B_t) = L\left(B_t \otimes \overset{p}{\bigwedge} T^*(M_t) \overset{q}{\bigwedge} \bar{T}^*(M_t)\right).$$

$\varphi_t \in \mathscr{L}^{p,q}(B_t)$ is represented on each $U_j \times t \subset M_t$ by a vector whose components are C^∞ (p, q)-forms as

$$\varphi_t = (\varphi_j^1(z_j, t), \ldots, \varphi_j^\nu(z_j, t)) \tag{7.46}$$

where

$$\varphi_j^\lambda(z_j, t) = \sum \varphi_{jA_p\bar{B}_q}^\lambda(z_j, t)\, dz_j^{A_p} \wedge \overline{dz_j^{B_q}}$$

$$= \frac{1}{p!\,q!} \sum \varphi_{j\alpha_1\cdots\alpha_p\bar{\beta}_1\cdots\bar{\beta}_q}^\lambda(z_j, t)\, dz_j^{\alpha_1} \wedge \cdots \wedge dz_j^{\alpha_p} \wedge d\bar{z}_j^{\beta_1} \wedge \cdots \wedge d\bar{z}_j^{\beta_q}.$$

On $U_j \otimes t \cap U_k \times t \neq \varnothing$, we have

$$\varphi_j^\lambda(z_j, t) = \sum_{\mu=1}^{\nu} b_{jk\mu}^\lambda(z_k, t)\varphi_k^\mu(z_k, t).$$

φ_t is called C^∞ differentiable with respect to $t \in \Delta$ if the $\varphi_{j\alpha_1\cdots\alpha_p\bar{\beta}_1\cdots\bar{\beta}_q}^\lambda(z_j, t)$ are C^∞ functions of (z_j, t). In case B_t is a trivial line bundle $\mathbb{C} \times M_t$, $\mathscr{L}^{p,q}(\mathbb{C} \times M_t) = L(\bigwedge^p T^*(M_t) \bigwedge^q \bar{T}^*(M_t))$ is the linear space of all C^∞ (p, q)-forms.

$\bar{\partial}_t\varphi_t$ is defined by

$$\bar{\partial}_t\varphi_t = (\bar{\partial}_t\varphi_j^1(z_j, t), \ldots, \bar{\partial}_t\varphi_j^\nu(z_j, t)),$$

where

$$\bar{\partial}_t\varphi_j^\lambda(z_j, t) = \sum \sum_\alpha \frac{\partial}{\partial \bar{z}_j^\alpha} \varphi_{jA_p\bar{B}_q}^\lambda(z_j, t)\, d\bar{z}_j^\alpha \wedge dz_j^{A_p} \wedge \overline{dz_j^{B_q}}.$$

If $\varphi_t \in \mathscr{L}^{p,q}(B_t)$, then we have $\bar{\partial}_t\varphi_t \in \mathscr{L}^{p,q+1}(B_t)$. *If φ_t is C^∞ differentiable with respect to $t \in \Delta$, then so is $\bar{\partial}_t\varphi_t$, too.* Therefore $\bar{\partial}_t$ is C^∞ *differentiable with respect to $t \in \Delta$* (see Definition 7.5).

As in the proof of Theorem 3.14, we introduce a Hermitian metric on M_t, using a partition of unity $\{\rho_j\}$ subordinate to a finite open covering $\{U_j\}$ of X. We put

$$\omega_t = i\sum_j \rho_j(x)\sum dz_j^\alpha \wedge d\bar{z}_j^\alpha, \quad \text{with} \quad dz_j^\alpha = \sum_{\gamma=1}^{2n} \frac{\partial z_j^\alpha(x,t)}{\partial x^\gamma} dx^\gamma,$$

then ω_t is a real C^∞ (1, 1)-form on M_t, and is represented as

$$\omega_t = i \sum_{\alpha,\beta=1}^n g_{j\alpha\bar{\beta}}(z_j, t)\, dz_j^\alpha \wedge d\bar{z}_j^\beta$$

on each $U_j \times t$, where *the $g_{j\alpha\bar{\beta}}(z_j, t)$ are C^∞ functions of (z_j, t)*. We define a Hermitian metric on M_t by

$$\sum_{\alpha,\beta=1}^n g_{t\alpha\bar{\beta}}(z)\, dz^\alpha \otimes d\bar{z}^\beta = \sum_{\alpha,\beta=1}^n g_{j\alpha\bar{\beta}}(z_j, t)\, dz_j^\alpha \otimes d\bar{z}_j^\beta. \tag{7.47}$$

Similarly we define a Hermitian metric on the fibre of B_t by

$$\sum_{\lambda,\mu=1}^n a_{j\lambda\bar{\mu}}(z_j, t)\zeta_j^\lambda \bar{\zeta}_j^\mu, \tag{7.48}$$

where *the $a_{j\lambda\bar{\mu}}(z_j, t)$ are C^∞ functions of (z_j, t)*.

Now we apply the results of § 3.5(c) to our case. For any local (p, q)-form ψ on M_t, we define $*_t\psi$ by (3.119) with $g_{t\alpha\bar{\beta}}(z)$ instead of $g_{\alpha\bar{\beta}}(z)$. Representing φ_t, $\psi_t \in \mathscr{L}^{p,q}(B_t)$ as in (7.46) on each $U_j \times t \subset M_t$, we have

$$\sum_{\lambda,\mu} a_{j\lambda\bar{\mu}}(z_j, t)\varphi_j^\lambda(z_j, t) \wedge *_t\bar{\psi}_j^\mu(z_j, t)$$

$$= \sum_{\lambda,\mu} a_{k\lambda\bar{\mu}}(z_k, t)\varphi_k^\lambda(z_k, t) \wedge *_t\bar{\psi}_k^\mu(z_k, t)$$

on $U_j \times t \cap U_k \times t \neq \emptyset$. From now on we omit the index j for simplicity, and write $\sum_{\lambda,\mu} a_{\lambda\bar{\mu}}(z, t)\varphi^\lambda(z, t) \wedge *_t\bar{\psi}^\mu(z, t)$ for $\sum_{\lambda,\mu} a_{j\lambda\bar{\mu}}(z_j, t)\varphi_j^\lambda(z_j, t) \wedge *_t\bar{\varphi}_j^\mu(z_j, t)$. We define the inner product of φ_t and ψ_t by (3.139):

$$(\varphi_t, \psi_t)_t = \int_{M_t} \sum_{\lambda,\mu} a_{\lambda\bar{\mu}}(z, t)\varphi^\lambda(z, t) \wedge *_t\bar{\psi}^\mu(z, t).$$

For $\psi_t \in \mathscr{L}^{p,q}(B_t)$, we define $\mathfrak{d}_{at}\psi_t$ by (3.140) as

$$(\mathfrak{d}_{at}\psi)^{\lambda}(z,t) = -\sum_{\tau=1}^{\nu} a^{\bar{\tau}\lambda}(z,t)*_t\partial_t\left(\sum_{\mu=1}^{\nu} a_{\mu\bar{\tau}}(z,t)*_t\psi^{\mu}(z,t)\right),$$

where $(a^{\bar{\lambda}\mu}(z,t))_{\lambda,\mu=1,\dots,\nu} = (a_{\lambda\bar{\mu}}(z,t))^{-1}_{\lambda,\mu=1,\dots,\nu}$. For simplicity we omit the subscript a in $\mathfrak{d}_{at}$ and *write it as* $\mathfrak{d}_t$. $\mathfrak{d}_t$ is the adjoint operator of $\bar{\partial}_t$. Namely, we have

$$(\bar{\partial}_t\varphi, \psi)_t = (\varphi, \mathfrak{d}_t\psi)_t, \qquad \varphi \in \mathscr{L}^{p,q-1}(B_t), \qquad \psi \in \mathscr{L}^{p,q}(B_t).$$

Since $g_{j\alpha\bar{\beta}}(z_j, t)$ are C^∞ functions of (z_j, t), the linear operator $*_t$ *is* C^∞ *differentiable with respect to* $t \in \Delta$. Also since $a_{j\lambda\bar{\mu}}(z_j, t)$ are C^∞, $\mathfrak{d}_t$ *is* C^∞ *differentiable.* By (3.143), we define

$$\Box_t = \bar{\partial}_t\mathfrak{d}_t + \mathfrak{d}_t\bar{\partial}_t.$$

$\Box$ is *a formally self-adjoint strongly elliptic linear partial differential operator on* $\mathscr{L}^{p,q}(B_t)$ *with respect to the inner product* $(\ ,\ )_t$. Since $\bar{\partial}_t$ and $\mathfrak{d}_t$ are C^∞ differentiable with respect to $t \in \Delta$, so is $\Box_t$ also. Thus $\{\Box_t \mid t \in \Delta\}$ *forms a differentiable family.*

$\varphi \in \mathscr{L}^{p,q}(B_t)$ is called a harmonic form with coefficients in B_t if $\mathfrak{d}_t\varphi = \bar{\partial}_t\varphi = 0$. $\varphi \in \mathscr{L}^{p,q}(B_t)$ is a harmonic form with coefficients in B_t if and only if $\Box_t\varphi = 0$. Put

$$\mathbf{H}^{p,q}(B_t) = \{\varphi \in \mathscr{L}^{p,q}(B_t) \mid \Box_t\varphi = 0\}.$$

Then by Theorem 3.19, we have the following orthogonal decomposition:

$$\mathscr{L}^{p,q}(B_t) = \mathbf{H}^{p,q}(B_t) \oplus \Box_t\mathscr{L}^{p,q}(B_t). \tag{7.49}$$

We denote by H_t *the orthogonal projection of* $\mathscr{L}^{p,q}(B_t)$ *onto* $\mathbf{H}^{p,q}(B_t)$. Let G_t be the Green operator of $\Box_t$. Then by (7.8) we have

$$\Box_t G_t\varphi = G_t\Box_t\varphi = \varphi - H_t\varphi, \qquad \varphi \in \mathscr{L}^{p,q}(B_t). \tag{7.50}$$

Consequently any $\varphi \in \mathscr{L}^{p,q}(B_t)$ is written as

$$\varphi = H_t\varphi + \Box_t G_t\varphi = H_t\varphi + \bar{\partial}_t\mathfrak{d}_t G_t\varphi + \mathfrak{d}_t\bar{\partial}_t G_t\varphi. \tag{7.51}$$

Clearly we have

$$H_t\bar{\partial}_t\varphi = \bar{\partial}_t H_t\varphi = 0, \qquad H_t\mathfrak{d}_t\varphi = \mathfrak{d}_t H_t\varphi = 0. \tag{7.52}$$

We claim that G_t *commutes with* $\bar{\partial}_t$. *Proof.* For any $\psi \in \mathscr{L}^{p,q+1}(B_t)$, $\psi = H_t\psi + G_t\Box_t\psi$ by (7.50). Put $\psi = \bar{\partial}_t G_t\varphi$. Then since $\Box_t G_t\varphi = \varphi - H_t\varphi$ by (7.50),

$$\bar{\partial}_t G_t\varphi = G_t\Box_t\bar{\partial}_t G_t\varphi = G_t\bar{\partial}_t\Box_t G_t\varphi = G_t\bar{\partial}_t\varphi. \quad \blacksquare$$

Similarly $\mathfrak{d}_t\mathfrak{d}_t = 0$ implies that $\mathfrak{d}_t G_t\varphi = G_t\mathfrak{d}_t\varphi$. Hence, we have

$$\bar{\partial}_t G_t = G_t\bar{\partial}_t, \qquad \mathfrak{d}_t G_t = G_t\mathfrak{d}_t. \tag{7.53}$$

Let λ be an eigenvalue of $\Box_t$, and e an eigenvector belonging to λ. Then

$$\lambda \|e\|_t^2 = (\Box_t e, e)_t = \|\partial_t e\|_t^2 + \|\mathfrak{d}_t e\|_t^2 \geqq 0,$$

hence $\lambda \geqq 0$. Thus we can arrange all the eigenvalues $\lambda_h^{p,q}(t)$ of $\Box_t$ on $\mathscr{L}^{p,q}(B_t)$ as follows:

$$0 \leqq \lambda_1^{p,q}(t) \leqq \cdots \leqq \lambda_h^{p,q}(t) \leqq \cdots, \qquad \lambda_h^{p,q}(t) \to +\infty.$$

Now we apply the results of the preceding section to the family $\{\Box_t \mid t \in \Delta\}$. First we obtain from Theorem 7.2,

Theorem 7.7. $\lambda_h^{p,q}(t)$ *is continuous in* $t \in \Delta$.

Next we obtain from Theorem 7.3,

Theorem 7.8. $\dim \mathbf{H}^{p,q}(B_t)$ *is an upper-semicontinuous function of* $t \in \Delta$.

Let $\Omega^p(B_t)$ be the sheaf of germs of holomorphic p-forms with coefficients in B_t. Then by (3.148),

$$H^q(M_t, \Omega^p(B_t)) \cong \mathbf{H}^{p,q}(B_t).$$

Hence,

Corollary. $\dim H^q(M_t, \Omega^p(B_t))$ *is an upper-semicontinuous function of* $t \in \Delta$.

From Theorem 7.4 we obtain

Theorem 7.9. *If* $\dim \mathbf{H}^{p,q}(B_t)$ *does not depend on* $t \in \Delta$, *then* H_t *is* C^∞ *differentiable with respect to* $t \in \Delta$.

Finally from Theorem 7.6 we obtain

Theorem 7.10. *If* $\dim \mathbf{H}^{p,q}(B_t)$ *does not depend on* $t \in \Delta$, *then* G_t *is* C^∞ *differentiable with respect to* $t \in \Delta$.

(b) Proofs of Theorems

First we shall prove Theorem 4.4 (p. 200).

Proof of Theorem 4.4. Since the sheaf Θ_t of germs of holomorphic vector fields over M_t is just the sheaf $\mathcal{O}(T(M_t))$ of germs of holomorphic sections of $T(M_t)$, and $\{T(M_t)|t\in\Delta\}$ forms a differentiable family, we see from the Corollary to Theorem 7.8 that $\dim H^1(M_t, \mathcal{O}(T(M_t)))$ *is upper-semicontinuous in* $t\in\Delta$. ∎

Next we proceed to the proof of Theorem 4.1. Put $t=(t_1,\ldots,t_m)$ and consider the infinitesimal deformation $\partial M_t/\partial t_m$. Since $\Theta_t=\mathcal{O}(T(M_t))$, we have

$$H^1(M_t, \Theta_t)=\mathbf{H}^{0,1}(T(M_t)). \tag{7.54}$$

Let $\varphi_t\in\mathbf{H}^{0,1}(T(M_t))$ be the harmonic vector (0, 1)-form corresponding to $\partial M_t/\partial t_m$ by the above isomorphism. Since $z_j^\alpha=f_{jk}^\alpha(z_k, t)$ on $U_j\times t\cap U_k\times t\neq\emptyset$, putting

$$\theta_{jk}(t)=\sum_{\alpha=1}^m \frac{\partial f_{jk}^\alpha(z_k, t)}{\partial t_m}\cdot\frac{\partial}{\partial z_j^\alpha}, \qquad z_k=f_{kj}(z_j, t),$$

we see that $\partial M_t/\partial t_m$ is the cohomology class of the 1-cocycle $\{\theta_{jk}(t)\}$. The corresponding harmonic form φ_t is obtained from $\{\theta_{jk}(t)\}$ as follows: Let $\{\rho_i\}$ be a partition of unity subordinate to a finite open covering $\{U_i\}$ of X. Multiplying the equality

$$\theta_{jk}(t)=\theta_{ik}(t)-\theta_{ij}(t)$$

by ρ_i, and summing up with respect to i, we get

$$\theta_{jk}(t)=\sum_i \rho_i\theta_{ik}(t)-\sum_i \rho_i\theta_{ij}(t).$$

Put $\xi_j(t)=\sum_i \rho_i\theta_{ij}(t)$, and $\xi_k(t)=\sum_i \rho_i\theta_{ik}(t)$. Then

$$\theta_{jk}(t)=\xi_k(t)-\xi_j(t). \tag{7.55}$$

Since $\theta_{ij}(t)=-\theta_{ji}(t)$, we have

$$\xi_j(t)=-\sum_i \rho_i\theta_{ji}(t)=-\sum_i \rho_i(x)\sum_{\alpha=1}^n \frac{\partial f_{ji}^\alpha(z_i, t)}{\partial t_m}\cdot\frac{\partial}{\partial z_j^\alpha}.$$

Consequently if we write $\xi_j(t)$ as

$$\xi_j(t)=\sum_{\alpha=1}^{n}\xi_j^{\alpha}(z_j,t)\frac{\partial}{\partial z_j^{\alpha}},$$

we have

$$\xi_j^{\alpha}(z_j,t)=-\sum_i \rho_i(x)\frac{\partial f_{ji}^{\alpha}(z_i,t)}{\partial t_m},\qquad z_i=f_{ij}(z_j,t).$$

Since $z_j^{\alpha}=z_j^{\alpha}(x,t)$ are C^{∞} functions of (x,t), these $\xi_j^{\alpha}(z_j(x,t),t)$ are C^{∞} functions of (x,t), hence $\xi_j^{\alpha}(z_j,t)$ are C^{∞} functions of (z_j,t).

Since $\theta_{jk}(t)$ is a holomorphic vector field on $U_j\times t\cap U_k\times t\subset M_t$, $\bar\partial_t\theta_{jk}(t)=0$. Therefore by (7.55), $\bar\partial_t\xi_i(t)=\bar\partial_t\xi_k(t)$ on $U_j\times t\cap U_k\times t$. Consequently there is a vector (0, 1)-form $\psi_t\in L^{0,1}(T(M_t))$ on M_t such that on each $U_j\times t\subset M_t$,

$$\psi_t=\bar\partial_t\xi_j(t). \tag{7.56}$$

Clearly $\bar\partial_t\psi_t=0$. If we represent ψ_t on each $U_j\times t$ as

$$\psi_t=\sum_{\alpha}\sum_{\beta}\psi_{j\bar\beta}^{\alpha}(z_j,t)\,d\bar z_j^{\beta}\frac{\partial}{\partial z_j^{\alpha}},\qquad \psi_{j\bar\beta}^{\alpha}(z_j,t)=\frac{\partial\xi_j^{\alpha}(z_j,t)}{\partial\bar z_j^{\beta}},$$

then *the coefficients* $\psi_{j\bar\beta}^{\alpha}(z_j,t)$ *are* C^{∞} *functions of* (z_j,t). *Thus* ψ_t *is* C^{∞} *differentiable with respect to* $t\in\Delta$.

For simplicity we write T_t for $T(M_t)$ below. Let $\mathscr{A}(T_t)$ be the sheaf of germs of C^{∞} sections of T_t. Then since $\Theta_t=\mathcal{O}(T_t)$, we obtain from (3.106),

$$H^1(M_t,\Theta_t)\cong\Gamma(M_t,\bar\partial_t\mathscr{A}(T_t))/\bar\partial_t\Gamma(M_t,\mathscr{A}(T_t)), \tag{7.57}$$

which is induced from the exact cohomology sequence

$$\cdots\to\Gamma(M_t,\mathscr{A}(T_t))\xrightarrow{\bar\partial_t}\Gamma(M_t,\bar\partial_t\mathscr{A}(T_t))\xrightarrow{\delta^*}H^1(M_t,\Theta_t)\to 0.$$

Since by (7.56) and (7.55), $\psi_t=\bar\partial_t\{\xi_j(t)\}$, and $\delta\{\xi_j(t)\}=\{\theta_{jk}(t)\}$ for the 0-cochain $\{\xi_j(t)\}$, we see from the definition of δ^* that the cohomology class $\partial M_t/\partial t_m$ of $\{\theta_{jk}(t)\}$ corresponds to $\psi_t\in\Gamma(M_t,\bar\partial_t\mathscr{A}(T_t))$ by the isomorphism (7.57). Since

$$\Gamma(M_t,\bar\partial_t\mathscr{A}(T_t))=\mathbf{H}^{0,1}(T(M_t))\oplus\bar\partial_t\mathscr{L}^{0,0}(T(M_t)),$$

and

$$\Gamma(M_t,\mathscr{A}(T_t))=\mathscr{L}^{0,0}(T(M_t)),$$

we have

$$\Gamma(M_t,\bar\partial_t\mathscr{A}(T_t))/\bar\partial_t\Gamma(M_t,\mathscr{A}(T_t))\cong\mathbf{H}^{0,1}(T(M_t)), \tag{7.58}$$

where $\psi_t \in \Gamma(M_t, \bar{\partial}_t \mathscr{A}(T_t))$ corresponds to $H_t\psi_t \in \mathbf{H}^{0,1}(T(M_t))$ via this isomorphism. Since the isomorphism (7.54) is the composite of those in (7.57) and (7.58), *letting* φ_t *be the harmonic* $(0,1)$*-form corresponding to the infinitesimal deformation* $\partial M_t/\partial t_m$, *we see that*

$$\varphi_t = H_t\psi_t.$$

Since by (7.51) $\psi_t = H_t\psi_t + \bar{\partial}_t \mathfrak{d}_t G_t \psi_t + \mathfrak{d}_t \bar{\partial}_t G_t \psi_t$, and since $\bar{\partial}_t G_t \psi_t = G_t \bar{\partial}_t \psi_t = 0$, we have

$$\psi_t = \varphi_t + \bar{\partial}_t \eta_t, \qquad \varphi_t = H_t\psi_t, \qquad \eta_t = \mathfrak{d}_t G_t \psi_t. \tag{7.59}$$

Now we return to the proof of Theorem 4.1. Under the assumption that $\partial M_t/\partial t_m = 0$ identically, we must prove that if $\dim H^1(M_t, \Theta_t)$ *is independent of* $t \in \Delta$, *we can find a* 0*-cochain* $\{\theta_j(t)\}$ *which satisfies* $\delta(\theta_j(t)) = \{\theta_{jk}(t)\}$ *with each* $\theta_j(t) = \sum \theta_j^\alpha(z_j, t)(\partial/\partial z_j^\alpha)$ *being a holomorphic vector field on* $U_j \times t \subset M_t$ *such that* $\theta_j^\alpha(z_j, t)$ *are* C^∞ *functions of* (z_j, t). *Proof.* Since $\partial M_t/\partial t_m = 0$, $\varphi_t = 0$, hence $\psi_t = \bar{\partial}_t \eta_t$ and $\eta_t = \mathfrak{d}_t G_t \psi_t$. Since $\psi_t = \bar{\partial}_t \xi_j(t)$ by (7.56), putting

$$\theta_j(t) = \xi_j(t) - \eta_t,$$

we obtain $\bar{\partial}_t \theta_t(t) = 0$. Thus $\theta_j(t)$ is a holomorphic vector field on $U_j \times t \subset M_t$, By (7.55) we have

$$\theta_k(t) - \theta_j(t) = \xi_k(t) - \xi_j(t) = \theta_{jk}(t),$$

that is, $\delta\{\theta_j(t)\} = \{\theta_{jk}(t)\}$. Since $\dim \mathbf{H}^{0,1}(T(M_t)) = \dim H^1(M_t, \Theta_t)$ does not depend on $t \in \Delta$, G_t *is* C^∞ *differentiable with respect to* $t \in \Delta$ by Theorem 7.10. Since ψ_t and $\mathfrak{d}_t$ are C^∞ differentiable with respect to $t \in \Delta$, so is $\eta_t = \mathfrak{d}_t G_t \psi_t$, hence, if we write η_t on $U_j \times t$ as

$$\eta_t = \sum_{\alpha=1}^{n} \eta_j^\alpha(z_j, t) \frac{\partial}{\partial z_j^\alpha},$$

the coefficients $\eta_j^\alpha(z_j, t)$ *are* C^∞ *functions of* (z_j, t). Since $\xi_j^\alpha(z_j, t)$ are C^∞ functions of (z_j, t), we see that

$$\theta_j^\alpha(z_j, t) = \xi_j^\alpha(z_j, t) - \eta_j^\alpha(z_j, t)$$

are also C^∞ functions of (z_j, t). ∎

Theorem 7.11. *Assume that* $\dim H^1(M_t, \Theta_t)$ *is independent of* $t = (t_1, \ldots, t_m) \in \Delta$. *If* $(\partial M_t/\partial t_\kappa)_{t=0} \in H^1(M_0, \Theta_0)$, $\kappa = 1, \ldots, m$, *are linearly independent, then* $\partial M_t/\partial t_\kappa \in H^1(M_t, \Theta_t)$, $\kappa = 1, \ldots, m$, *are linearly independent for* $|t| < \varepsilon$ *provided that* $\varepsilon > 0$ *is sufficiently small.*

Proof. For each $\kappa = 1, \ldots, m$, we construct as above the harmonic (0, 1)-form $\varphi_{\kappa t} = H_t\psi_{\kappa t} \in \mathbf{H}^{0,1}(T(M_t))$ corresponding to $\partial M_t/\partial T_\kappa \in H^1(M_t, \Theta_t)$ via the isomorphism $H^1(M_t, \Theta_t) \cong \mathbf{H}^{0,1}(T(M_t))$. Since by the assumption $\dim \mathbf{H}^{0,1}(T(M_t)) = \dim H^1(M_t, \Theta_t)$ is independent of $t \in \Delta$, *H_t is C^∞ differentiable with respect to t* by Theorem 7.9. Since $\psi_{\kappa t}$ is C^∞ differentiable with respect to *t, so is* $\varphi_{\kappa t} = H_t\psi_{\kappa t}$. If $(\partial M_t/\partial t_\kappa)_{t=0} \in H^1(M_0, \Theta_0)$, $\kappa = 1, \ldots, m$, are linearly independent, the corresponding $\varphi_{\kappa 0} \in \mathbf{H}^{0,1}(T(M_0))$, $\kappa = 1, \ldots, m$, are also linearly independent. Consequently, since each $\varphi_{\kappa t}$ is C^∞ differentiable with respect to t, $\varphi_{\kappa t} \in \mathbf{H}^{0,1}(T(M_t))$, $\kappa = 1, \ldots, m$, are also linearly independent for $|t| < \varepsilon$ provided that $\varepsilon > 0$ is sufficiently small, which in turn implies that the corresponding $\partial M_t/\partial t_\kappa \in H^1(M_t, \Theta_t)$, $\kappa = 1, \ldots, m$, are linearly independent. ∎

As was seen from Example (iv) of §6.2(b), Theorem 7.11 is not necessarily true if $\dim H^1(M_t, \Theta_t)$ varies with t.

In the following we state the outline of the proof of Lemma 4.1. First consider $H^0(M_t, \mathcal{O}(B_t))$, where $\{M_t \mid t \in \Delta\}$ is a differentiable family of compact complex manifolds and $\{B_t \mid t \in \Delta\}$ is a differentiable family of holomorphic vector bundles over M_t.

Lemma 7.8. *If* $\dim H^0(M_t, \mathcal{O}(B_t)) = d$ *is independent of* $t \in \Delta$, *then we can choose a basis* $\{\varphi_{t1}, \ldots, \varphi_{td}\}$ *of* $H^0(M_t, \mathcal{O}(B_t))$ *for* $|t| < \varepsilon$ *such that each* φ_{tq} *is C^∞ differentiable with respect to t for* $q = 1, \ldots, d$ *provided that* $\varepsilon > 0$ *is sufficiently small.*

Proof. Let $\{\varphi_1, \ldots, \varphi_d\}$ be a basis of $H^0(M_0, \mathcal{O}(B_0))$. We can choose C^∞ sections $\psi_{tq} \in \mathcal{L}^{0,0}(B_t)$ of B_t with $q = 1, \ldots, d$ such that ψ_{tq} are C^∞ differentiable with respect to t and that $\psi_{0q} = \varphi_q$. This follows immediately from the fact that if we denote by X the underlying differentiable manifold of B_t, then B_t as a vector bundle over X is the restriction of a vector bundle $\mathcal{B}$ over $X \times \Delta$ to $X = X \times t$, and that $\mathcal{B}$ is C^∞ equivalent to $B_0 \times \Delta$. By (7.49),

$$\mathcal{L}^{0,0}(B_t) = \mathbf{H}^{0,0}(B_t) \oplus \square_t \mathcal{L}^{0,0}(B_t),$$

where $\mathbf{H}^{0,0}(B_t) = \{\varphi \in \mathcal{L}^{0,0}(B_t) \mid \bar\partial_t\varphi = 0\} = H^0(M_t, \mathcal{O}(B_t))$. Put

$$\varphi_{tq} = H_t\psi_{tq}, \qquad q = 1, \ldots, d.$$

Since $\dim \mathbf{H}^{0,0}(B_t) = d$ is independent of t, *H_t is C^∞ differentiable with respect to t.* Since ψ_{tq} are C^∞ differentiable with respect to t, φ_{tq} are also C^∞ differentiable with respect to t for $q = 1, \ldots, d$. Since $\varphi_{0q} = \varphi_q$ are linearly independent, $\varphi_{tq} \in H^0(M_t, \mathcal{O}(B_t))$ are linearly independent for $|t| < \varepsilon$ provided that $\varepsilon > 0$ is sufficiently small, hence $\{\varphi_{t1}, \ldots, \varphi_{td}\}$ forms a basis of $H^0(M_t, \mathcal{O}(B_t))$. ∎

Now consider a complex analytic family $\{M_t \mid t \in \Delta\}$ of compact complex manifolds where $\Delta \subset \mathbb{C}^m$ is a polydisk with centre 0. We assume that $\dim H^0(M_t, \Theta_t) = d$ is independent of $t \in \Delta$. Then by the above Lemma 7.8, we can choose a basis $\{\varphi_{t1}, \ldots, \varphi_{td}\}$ of $H^0(M_t, \Theta_t)$ for $|t| < \varepsilon$ such that φ_{tq} are C^∞ differentiable with respect to t provided that $\varepsilon > 0$ is sufficiently small. Namely, if we represent φ_{tq} on $U_j \times t$ as

$$\varphi_{tq} = \sum_{\alpha=1}^{n} \varphi_{qj}^{\alpha}(z_j, t) \frac{\partial}{\partial z_j^{\alpha}},$$

then the coefficients $\varphi_{qj}^{\alpha}(z_j, t)$ are C^∞ functions of (z_j, t). Since on $U_j \times t \cap U_k \times t \neq \emptyset$,

$$\varphi_{qj}^{\alpha}(z_j, t) = \sum_{\beta=1}^{n} \frac{\partial f_{jk}^{\alpha}(z_k, t)}{\partial z_k^{\beta}} \varphi_{qk}^{\beta}(z_k, t),$$

and since $f_{jk}^{\alpha}(z_k, t)$ are holomorphic in $z_k^1, \ldots, z_k^n, t_1, \ldots, t_m$, we have

$$\frac{\partial \varphi_{qj}^{\alpha}(z_j, t)}{\partial \bar{t}_\kappa} = \sum_{\alpha=1}^{n} \frac{\partial f_{jk}^{\beta}(z_k, t)}{\partial z_k^{\beta}} \cdot \frac{\partial \varphi_{qk}^{\beta}(z_k, t)}{\partial \bar{t}_\kappa}.$$

Therefore

$$\frac{\partial}{\partial \bar{t}_\kappa} \varphi_{tq} = \sum_{\alpha=1}^{n} \frac{\partial \varphi_{qj}^{\alpha}(z_j, t)}{\partial \bar{t}_\kappa} \frac{\partial}{\partial z_j^{\alpha}}, \qquad 1 \leqq \kappa \leqq m,$$

are also holomorphic vector fields on M_t. Hence we have

$$\frac{\partial}{\partial \bar{t}_\kappa} \varphi_{tq} = \sum_{p=1}^{d} a_{\kappa pq}(t) \varphi_{tp}, \qquad 1 \leqq q \leqq d, \tag{7.60}$$

where the coefficients $a_{\kappa pq}(t)$ are C^∞ functions of t on $|t| < \varepsilon$.

Let $c_p(t)$, $p = 1, \ldots, d$, be C^∞ functions of t on $|t| < \varepsilon$, and put $\theta_t = \sum_{p=1}^{d} c_p(t) \varphi_{tp}$. Then $\theta_t \in H^0(M_t, \Theta_t)$ is C^∞ differentiable with respect to t. Thus, if we write θ_t as

$$\theta_t = \sum_{\alpha=1}^{n} \theta_j^{\alpha}(z_j, t) \frac{\partial}{\partial z_j^{\alpha}}$$

on $U_j \times t \subset M_t$, the coefficients $\theta_j^{\alpha}(z_j, t)$ are C^∞ functions of (z_j, t). If $\theta_j^{\alpha}(z_j, t)$ are holomorphic in $z_j^1, \ldots, z_j^n, t_1, \ldots, t_m$, we say that θ_t *is holomorphic with respect to t*. Since $\theta_j^{\alpha}(z_j, t)$ are holomorphic in $z_j^1, \ldots, z_j^n$, θ_t is holomorphic with respect to t if and only if $\partial\theta_t/\partial\bar{t}_\kappa = 0$ for $\kappa = 1, \ldots, m$. Since by (7.60)

$$\frac{\partial}{\partial\bar{t}_\kappa}\theta_t = \sum_p \frac{\partial c_p(t)}{\partial\bar{t}_\kappa}\varphi_{tq} + \sum_q c_q(t)\frac{\partial}{\partial\bar{t}_\kappa}\varphi_{tq}$$

$$= \sum_{p=1}^{d}\left(\frac{\partial}{\partial\bar{t}_\kappa}c_p(t) + \sum_{q=1}^{d} a_{\kappa pq}(t)c_q(t)\right)\varphi_{tp},$$

$\theta_t = \sum_p c_p(t)\varphi_{tp}$ is holomorphic with respect to t if and only if

$$\frac{\partial}{\partial\bar{t}_\kappa}c_p(t) + \sum_{q=1}^{d} a_{\kappa pq}(t)c_q(t) = 0, \qquad p = 1, \ldots, d, \quad \kappa = 1, \ldots, m.$$

Introducing the matrices $A_\kappa(t) = (a_{\kappa pq}(t))_{p,q=1,\ldots,d}$, we can write this condition as

$$\frac{\partial}{\partial\bar{t}_\kappa}\begin{pmatrix} c_1(t) \\ \vdots \\ c_d(t)\end{pmatrix} = A_\kappa(t)\begin{pmatrix} c_1(t) \\ \vdots \\ c_d(t)\end{pmatrix}, \qquad \kappa = 1, \ldots, m. \tag{7.61}$$

We must prove that *taking a sufficiently small* ε' *with* $0 < \varepsilon' < \varepsilon$, *we can choose a basis* $\{\theta_{t1}, \ldots, \theta_{td}\}$ *of* $H^0(M_t, \Theta_t)$ *for* $|t| < \varepsilon'$ *such that each* θ_{tq} *is holomorphic with respect to* t. *Proof.* Let $c_{pq}(t)$ with $p, q = 1, \ldots, d$, be C^∞ functions of t on $|t| < \varepsilon$, and put

$$\theta_{tq} = \sum_{p=1}^{d} c_{pq}(t)\varphi_{tp}, \qquad q = 1, \ldots, d.$$

Since $\dim H^0(M_t, \Theta_t) = d$, if $\det C(t) \neq 0$ with $C(t) = (c_{pq}(t))_{p,q=1,\ldots,d}$, then $\{\theta_{t1}, \ldots, \theta_{td}\}$ forms a basis of $H^0(M_t, \Theta_t)$. Moreover, by (7.61), if $C(t)$ satisfies the system of partial differential equations

$$\frac{\partial}{\partial\bar{t}_\kappa}C(t) + A_\kappa(t)C(t) = 0, \qquad \kappa = 1, \ldots, m, \tag{7.62}$$

each θ_{tq} is holomorphic with respect to t. The system of equations (7.62) has a solution $C(t)$ with $\det C(t) \neq 0$ for $|t| < \varepsilon'$ provided that $\varepsilon' > 0$ is sufficiently small if the integrability conditions

$$\frac{\partial}{\partial\bar{t}_\kappa}A_\lambda(t) - \frac{\partial}{\partial\bar{t}_\lambda}A_\kappa(t) + A_\kappa(t)A_\lambda(t) - A_\lambda(t)A_\kappa(t) = 0 \tag{7.63}$$

are satisfied for $\kappa, \lambda = 1, \ldots, d$ (see [21], §19). It follows immediately from the definition (7.60) that the matrices $A_\kappa(t)$, $\kappa = 1, \ldots, m$, satisfy (7.63).

Consequently there exists a solution $C(t)$ of (7.62) with det $C(t)\neq 0$. Thus, we can find a basis $\{\theta_{t1},\ldots,\theta_{td}\}$ of $H^0(M_t,\Theta_t)$ such that each θ_{tq} is holomorphic with respect to t. ∎

(c) Canonical Basis and Its Applications

In this section we let $\{M_t \mid t\in\Delta\}$ be a differentiable family of compact complex manifolds and $\{B_t \mid t\in\Delta\}$ a differentiable family of holomorphic vector bundles B_t over M_t. Fix a non-negative integer p once for all and consider the linear operator $\square_t=\bar\partial_t\mathfrak{d}_t+\mathfrak{d}_t\bar\partial_t$ acting on the linear space

$$\mathscr{L}^{p,0}(B_t)\oplus\cdots\oplus\mathscr{L}^{p,n}(B_t),$$

where $n=\dim M_t$. Put

$$\mathbf{A}_t^q=\bar\partial_t\mathfrak{d}_t\mathscr{L}^{p,q}(B_t),\qquad \mathbf{D}_t^q=\mathfrak{d}_t\bar\partial_t\mathscr{L}^{p,q}(B_t),$$

where we define $\mathbf{A}_t^0=\mathbf{D}_t^n=0$. Since $\bar\partial_t^2=0$, for any $\varphi,\psi\in\mathscr{L}^{p,q}(B_t)$, we have $(\bar\partial_t\mathfrak{d}_t\varphi,\mathfrak{d}_t\bar\partial_t\psi)_t=(\bar\partial_t^2\mathfrak{d}_t\varphi,\bar\partial_t\psi)_t=0$, hence $\mathbf{A}_t^q$ and $\mathbf{D}_t^q$ are mutually orthogonal. Therefore by (7.49) we have

$$\mathscr{L}^{p,q}(B_t)=\mathbf{H}^{p,q}(B_t)\oplus\mathbf{A}_t^q\oplus\mathbf{D}_t^q. \tag{7.64}$$

Moreover we have

$$\bar\partial_t\mathbf{D}_t^{q-1}=\mathbf{A}_t^q,\qquad \mathfrak{d}_t\mathbf{A}_t^q=\mathbf{D}_t^{q-1},\qquad q=1,\ldots,n. \tag{7.65}$$

In fact, since $\bar\partial_t\mathbf{H}^{p,q-1}(B_t)=0$, and $\bar\partial_t\mathbf{A}_t^{q-1}=0$, we obtain

$$\mathbf{A}_t^q\subset\bar\partial_t\mathscr{L}^{p,q-1}(B_t)=\bar\partial_t\mathbf{D}_t^{q-1}\subset\bar\partial_t\mathfrak{d}_t\mathscr{L}^{p,q}(B_t)=\mathbf{A}_t^q,$$

hence $\bar\partial_t\mathbf{D}_t^{q-1}=\mathbf{A}_t^q$. Therefore

$$\mathfrak{d}_t\mathbf{A}_t^q=\mathfrak{d}_t\bar\partial_t\mathbf{D}_t^{q-1}=\mathfrak{d}_t\bar\partial_t\mathscr{L}^{p,q-1}(B_t)=\mathbf{D}_t^{q-1}.$$

The linear map $\bar\partial_t\colon\mathbf{D}_t^{q-1}\to\mathbf{A}_t^q$ *is bijective.* For, since any $\varphi\in\mathbf{D}_t^{q-1}$ can be written as $\varphi=\mathfrak{d}_t\psi$ with $\psi\in\mathscr{L}^{p,q}(B_t)$, we obtain

$$\|\varphi\|_t^2=(\varphi,\varphi)_t=(\varphi,\mathfrak{d}_t\psi)_t=(\bar\partial_t\varphi,\psi)_t,$$

hence $\bar\partial_t\varphi=0$ implies $\varphi=0$. Similarly $\mathfrak{d}_t\colon\mathbf{A}_t^q\to\mathbf{D}_t^{q-1}$ is also bijective.

Since $\mathfrak{d}_t^2=0$, $\mathfrak{d}_t\mathbf{D}_t^q=0$. Also we have $\mathfrak{d}_t\mathbf{H}^{p,q}(B_t)=0$. Hence,

$$\square_t\mathbf{A}_t^q=\bar\partial_t\mathfrak{d}_t\mathbf{A}_t^q=\bar\partial_t\mathfrak{d}_t\mathscr{L}^{p,q}(B_t)=\mathbf{A}_t^q.$$

Similarly we have $\square_t \mathbf{D}_t^q = \mathbf{D}_t^q$.

$$\square_t \mathbf{A}_t^q = \mathbf{A}_t^q, \qquad \square_t \mathbf{D}_t^q = \mathbf{D}_t^q. \tag{7.66}$$

Let $\lambda_h^{(q)}(t)$ and e_{th}^q be the eigenvalues and eigenfunctions of $\square_t$ on $\mathscr{L}^{p,q}(B_t)$ respectively such that $\{e_{t1}^q, \ldots, e_{th}^q, \ldots\}$ forms an orthonormal basis of $\mathscr{L}^{p,q}(B_t)$ with respect to $(\ .\ )_t$, $\square_t e_{th}^q = \lambda_h^{(q)}(t) e_{th}^q$, and that

$$0 \leqq \lambda_1^{(q)}(t) \leqq \cdots \leqq \lambda_h^{(q)}(t) \leqq \cdots, \qquad \lambda_h^{(q)}(t) \to +\infty \quad (h \to \infty).$$

By (7.66) we may assume that each eigenfunction e_{th}^q *belongs to one of* $\mathbf{H}^{p,q}(B_t)$, $\mathbf{A}_t^q$, and $\mathbf{D}_t^q$. Let $\{e_{th_k}^q\}$ with $h_1 < \cdots < h_k < \cdots$ be the set of all the eigenfunctions belonging to $\mathbf{A}_t^q$ among $\{e_{th}^q\}$, and put $u_{tk}^q = e_{th_k}^q$, $\alpha_k^{(q)}(t) = \lambda_{h_k}^{(q)}(t)$. Then we have

$$\square_t u_{tk}^q = \alpha_k^{(q)}(t) u_{tk}^q, \tag{7.67}$$

and $\{u_{t1}^q, \ldots, u_{tk}^q, \ldots\}$ *forms an orthonormal basis of* $\mathbf{A}_t^q$ with respect to $(\ ,\ )_t$. Clearly we have

$$0 < \alpha_1^{(q)}(t) \leqq \cdots \leqq \alpha_k^{(q)}(t) \leqq \cdots, \qquad \alpha_k^{(q)}(t) \to +\infty \quad (k \to \infty).$$

Since $\bar{\partial}_t u_{tk}^q = 0$, by (7.67) we have

$$\bar{\partial}_t \mathfrak{d}_t u_{tk}^q = \alpha_k^{(q)}(t) u_{tk}^q, \tag{7.68}$$

hence we obtain

$$\begin{aligned}(\mathfrak{d}_t u_{tk}^q, \mathfrak{d}_t u_{th}^q)_t = (\bar{\partial}_t \mathfrak{d}_t u_{tk}^q, u_{th}^q)_t &= \alpha_k^{(q)}(t)(u_{tk}^q, u_{th}^q)_t \\ &= \delta_{kh} \alpha_k^{(q)}(t).\end{aligned}$$

Replacing q by $q+1$ in this equality, we obtain

$$(\mathfrak{d}_t u_{tk}^{q+1}, \mathfrak{d}_t u_{th}^{q+1})_t = \delta_{kh} \alpha_k^{(q+1)}(t).$$

Therefore putting

$$v_{tk}^q = \frac{1}{\sqrt{\alpha_k^{(q+1)}(t)}} \mathfrak{d}_t u_{tk}^{q+1}, \tag{7.69}$$

we have

$$(v_{tk}^q, v_{th}^q) = \delta_{kh}. \tag{7.70}$$

From (7.65) obviously we have $v_{tk}^q \in \mathbf{D}_t^q$. Since by (7.68)

$$\square_t \mathfrak{d}_t u_{tk}^{q+1} = \mathfrak{d}_t \bar{\partial}_t \mathfrak{d}_t u_{tk}^{q+1} = \alpha_k^{(q+1)}(t) \mathfrak{d}_t u_{tk}^{q+1},$$

we have

$$\Box_t v_{tk}^q = \alpha_k^{(q+1)}(t) v_{tk}^q.$$

Thus $v_{tk}^q \in \mathbf{D}_t^q$ is an eigenfunction of $\Box_t$ belonging to $\alpha_k^{(q+1)}(t)$.

$\{v_{t1}^q, v_{t2}^q, \dots\}$ *forms an orthonormal basis of* $\mathbf{D}_t^q$. In order to see this, since $(v_{tk}^q, v_{th}^q)_t = \delta_{kh}$ by (7.70), it suffices to show that any eigenfunction e_{th}^q belonging to $\mathbf{D}_t^q$ is written as a linear combination of v_{tk}^q's. Since $e_{th}^q \in \mathbf{D}_t^q$, $\bar\partial_t e_{th}^q \in \mathbf{A}_t^{q-1}$ by (7.65).

$$\Box_t \bar\partial_t e_{th}^q = \bar\partial_t \Box_t e_{th}^q = \lambda_h^{(q)}(t) \bar\partial_t e_{th}^q.$$

Hence we have

$$\bar\partial_t e_{th}^q = \sum_k^{(h)} a_k u_{tk}^{q+1}, \qquad a_k \in \mathbb{C},$$

where the summation is taken over all k such that $\alpha_k^{(q+1)}(t) = \lambda_h^{(q)}(t)$. Since $\mathfrak{d}_t e_{th}^q = 0$, we have

$$\lambda_h^{(q)}(t) e_{th}^q = \Box_t e_{th}^q = \mathfrak{d}_t \bar\partial_t e_{th}^q = \sum_k^{(h)} a_k \mathfrak{d}_t u_{tk}^{q+1}.$$

Consequently, since by (7.69),

$$\mathfrak{d}_t u_{tk}^{q+1} = \sqrt{\alpha_k^{(q+1)}(t)} \cdot v_{tk}^q,$$

e_{th}^q is written as

$$e_{th}^q = \sum_k^{(h)} c_k v_{tk}^q, \qquad c_k = \frac{a_k}{\sqrt{\alpha_k^{(q+1)}(t)}}.$$

It is clear from (7.68) and (7.69) that

$$\bar\partial_t v_{tk}^q = \sqrt{\alpha_k^{(q+1)}(t)} \cdot u_{tk}^{q+1}.$$

Put $d_q = \dim \mathbf{H}^{p,q}(B_t)$. Then $\{e_{t1}^q, \dots, e_{td_q}^q\}$ forms an orthonormal basis of $\mathbf{H}^{p,q}(B_t)$ where in general d_q depends on t. Since

$$\mathscr{L}^{p,q}(B_t) = \mathbf{H}^{p,q}(B_t) \oplus \mathbf{A}_t^q \oplus \mathbf{D}_t^q,$$

$\{e_{t1}^q, \dots, e_{td_q}^q, u_{t1}^q, \dots, v_{t1}^q, \dots\}$ forms an orthonormal basis of $\mathscr{L}^{p,q}(B_t)$. Note that for $q=0$, we omit $u_{t1}^0, u_{t2}^0, \dots$, and that for $q=n$, we omit $v_{t1}^n, v_{t2}^n, \dots$.

Theorem 7.12. *For each $q = 0, 1, \ldots, n$, we can choose an orthonormal basis*

$$\{e^q_{t1}, \ldots, e^q_{td_q}, u^q_{t1}, \ldots, v^q_{t1}, \ldots\} \quad \text{of } \mathscr{L}^{p,q}(B_t) \tag{7.71}$$

satisfying the following conditions.

(i) $\Box_t e^q_{th} = 0$ *for* $h = 1, \ldots, d_q$ *where* $d_q = \dim \mathbf{H}^{p,q}(B_t)$;

(ii) *For each* $q = 1, \ldots, n$, *there is a sequence* $\{\alpha^{(q)}_k(t)\}$ *with*

$$0 < \alpha^{(q)}_1(t) \leqq \cdots \leqq \alpha^{(q)}_k(t) \leqq \cdots, \qquad \alpha^{(q)}_k(t) \to +\infty \quad (k \to \infty),$$

such that

$$\Box_t u^q_{tk} = \alpha^{(q)}_k(t) u^q_{tk}, \qquad \Box_t v^q_{tk} = \alpha^{(q+1)}_k(t) v^q_{tk};$$

(iii) *The equalities*

$$\mathfrak{d}_t u^q_{tk} = \sqrt{\alpha^{(q)}_k(t)} \cdot v^{q-1}_{tk}, \qquad \bar{\partial}_t v^q_{tk} = \sqrt{\alpha^{(q+1)}_k(t)} \cdot u^{q+1}_{tk}$$

hold.

•

Note that in (7.71) we omit u^0_{tk} and v^n_{tk}. The union of the bases (7.71) for $q = 0, \ldots, n$ forms an orthonormal basis of $\bigoplus_{q=0}^n \mathscr{L}^{p,q}(B_t)$, which we call *the canonical basis.*

By (3.148) we have $H^q(M_t, \Omega^p(B_t)) \cong \mathbf{H}^{p,q}(B_t)$. Put

$$h^{p,q}(t) = \dim H^q(M_t, \Omega^p(B_t)) = \dim \mathbf{H}^{p,q}(B_t).$$

Fix an arbitrary $\varepsilon > 0$, and let $N^q(t)$ be the number of $\lambda^{(q)}_h(t)$ smaller than ε. Thus, if $h \leqq N^q(t)$, then $\lambda^{(q)}_h(t) < \varepsilon$, and if $N^q(t) < h$, then $\lambda^{(q)}_h(t) \geqq \varepsilon$. Similarly let $\nu^q(t)$ be the number of $\alpha^{(q)}_k$ smaller than ε. Since by Theorem 7.12, each $\lambda^{(q)}_h(t)$ *is equal to* 0, $\alpha^{(q)}_k(t)$, *or* $\alpha^{(q+1)}_k(t)$, and since the multiplicity of the eigenvalue 0 is equal to $\dim \mathbf{H}^{p,q}(B_t) = h^{p,q}(t)$, we obtain

$$N^q(t) = h^{p,q}(t) + \nu^q(t) + \nu^{q+1}(t), \qquad q = 0, 1, \ldots, n, \tag{7.72}$$

where we put $\nu^0(t) = \nu^{n+1}(t) = 0$.

For any $t_0 \in \Delta$, choose ε such that

$$0 < \varepsilon < \alpha^{(q)}_k(t_0), \qquad q = 1, \ldots, n, \quad k = 1, 2, \ldots. \tag{7.73}$$

By Theorem 7.7, each $\lambda^{(q)}_h(t) = \lambda^{p,q}_h(t)$ is a continuous function of $t \in \Delta$, while by (7.73), $\lambda^{(q)}_h(t_0) > \varepsilon$ if $\lambda^{(q)}_h(t_0) \neq 0$. Thus $\lambda^{(q)}_h(t) < \varepsilon$ if $\lambda^{(q)}_h(t_0) = 0$, and $\lambda^{(q)}_h(t) > \varepsilon$ if $\lambda^{(q)}_h(t_0) \neq 0$ on $|t - t_0| < \delta$ provided that $\delta > 0$ is sufficiently small. Hence

$$N^q(t) = N^q(t_0) \quad \text{on } |t - t_0| < \delta.$$

Therefore, since by (7.73) $\nu^q(t_0)=0$ for $q=1,2,\ldots,n$, we obtain from (7.72)

$$h^{p,q}(t)+\nu^q(t)+\nu^{q+1}(t)=h^{p,q}(t_0) \quad \text{on } |t-t_0|<\delta. \tag{7.74}$$

Theorem 7.13. If $\dim H^{q-1}(M_t, \Omega^p(B_t))$ *and* $\dim H^{q+1}(M_t, \Omega^p(B_t))$ *are independent of* $t\in\Delta$, *then* $\dim H^q(M_t, \Omega^p(B_t))$ *is also independent of* $t\in\Delta$.

Proof. For any $t_0\in\Delta$, if we take a sufficiently small $\delta>0$, by (7.74) $\nu^q(t)=\nu^{q+1}(t)=0$ on $|t-t_0|<\delta$ since by the assumption $h^{p,q-1}(t)=h^{p,q-1}(t_0)$ and $h^{p,q+1}(t)=h^{p,q+1}(t_0)$. Therefore, again by (7.74), $h^{p,q}(t)=h^{p,q}(t_0)$ on $|t-t_0|<\delta$. Thus $h^{p,q}(t)$ is constant on a neighbourhood of $t_0\in\Delta$. Since t_0 is an arbitrary point of Δ, $\dim H^q(M_t, \Omega^p(B_t))=h^{p,q}(t)$ is independent of $t\in\Delta$. ∎

Applying this theorem to $\Theta_t=\mathcal{O}(T(M_t))$, we obtain the following immediately.

Corollary. *If* $\dim H^{q-1}(M_t, \Theta_t)$ *and* $\dim H^{q+1}(M_t, \Theta_t)$ *are independent of* $t\in\Delta$, *then* $\dim H^q(M_t, \Theta_t)$ *is also independent of* $t\in\Delta$.

Appendix

Elliptic Partial Differential Operators on a Manifold
by Daisuke Fujiwara

§1. Distributions on a Torus

A local coordinate system on an n-dimensional differentiable manifold X gives a diffeomorphism of its domain X_j onto the unit ball U in $\mathbb{R}^n$, while the unit ball in $\mathbb{R}^n$ is diffeomorphic to a domain V in the torus $\mathbb{T}^n = \mathbb{R}^n/2\pi\mathbb{Z}^n$. Thus a section of a vector bundle B with μ-dimensional fibres over X_j can be identified with a vector-valued function with values in $\mathbb{C}^\mu$ over the domain V in $\mathbb{T}^n$. Various function spaces of sections of B over X_j can thus be considered as those of corresponding $\mathbb{C}^\mu$-valued functions over V in $\mathbb{T}^n$. In what follows, we shall always treat the functions defined on V which can be extended to the whole space $\mathbb{T}^n$. Since the torus $\mathbb{T}^n$ is compact, a function space consisting of vector-valued functions on $\mathbb{T}^n$ has much simpler structure than that on $\mathbb{R}^n$.

(a) Definition of Distributions

By $x=(x^1,\ldots,x^n)$ we denote a point in $\mathbb{R}^n$. If we denote by $2\pi\mathbb{Z}^n$ the set of the points in $\mathbb{R}^n$ whose components are 2π times integers, then this becomes an additive group. The n-dimensional torus $\mathbb{T}^n$ is defined as

$$\mathbb{T}^n = \mathbb{R}^n/2\pi\mathbb{Z}^n.$$

A C^∞ function f on $\mathbb{T}^n$ can be regarded as a C^∞ function on $\mathbb{R}^n$, hence we can write it as $f(x)=f(x^1,\ldots,x^n)$. Then we have the multi-periodicity

$$f(x+2\pi\xi)=f(x), \qquad \xi\in\mathbb{Z}^n. \tag{1.1}$$

We write D_j, $j=1,\ldots,n$, for the differential operator $\partial/\partial x^j$. An n-tuple $\alpha=(\alpha_1,\ldots,\alpha_n)$ of non-negative integers will be called a multi-index, and $|\alpha|=\alpha_1+\cdots+\alpha_n$ its *length*. Moreover we introduce the following notation:

$$D^\alpha = D_1^{\alpha_1}\cdots D_n^{\alpha_n}.$$

We denote by $C^\infty(\mathbb{T}^n)$ the totality of complex-valued C^∞ functions on $\mathbb{T}^n$. For $l=0, 1, \ldots,$ the norm $|\ |_l$ is defined by

$$|\varphi|_l = \sum_{|\alpha|\leqq l} \max_{x\in\mathbb{T}^n} |D^\alpha\varphi(x)|. \tag{1.2}$$

With the topology defined by these countably many norms, $C^\infty(\mathbb{T}^n)$ becomes a Fréchet space, which we denote by $\mathscr{D}(\mathbb{T}^n)$.

Definition 1.1. A continuous linear map of $\mathscr{D}(\mathbb{T}^n)$ to $\mathbb{C}$, namely a continuous linear functional on $\mathscr{D}(\mathbb{T}^n)$, is called a *distribution* on $\mathbb{T}^n$. We denote the totality of distributions on $\mathbb{T}^n$ by $\mathscr{D}'(\mathbb{T}^n)$.

In other words, S is a distribution on $\mathbb{T}^n$ if and only if it satisfies the following two conditions:

(1) $S\colon C^\infty(\mathbb{T}^n)\ni\varphi\to S(\varphi)\in\mathbb{C}$ is linear with respect to φ.
(2) There exist a non-negative integer l, and a positive constant C such that for any $\varphi\in C^\infty(\mathbb{T}^n)$

$$|S(\varphi)|\leqq C|\varphi|_l \tag{1.3}$$

holds.

The totality $\mathscr{D}'(\mathbb{T}^n)$ of distributions on $\mathbb{T}^n$ is the dual space of $\mathscr{D}(\mathbb{T}^n)$. We write $\langle S, \varphi\rangle$ for $S(\varphi)$ if $S\in\mathscr{D}'(\mathbb{T}^n)$ and $\varphi\in\mathscr{D}(\mathbb{T}^n)$. $\langle\ ,\ \rangle$ gives a bilinear map $\mathscr{D}'(\mathbb{T}^n)\times\mathscr{D}(\mathbb{T}^n)\to\mathbb{C}$. As the dual space of $\mathscr{D}(\mathbb{T}^n)$, we can introduce a topology into $\mathscr{D}'(\mathbb{T}^n)$. Recall that a sequence $\{S_m\}_{m=1}^\infty$ of distributions converges to S with respect to the weak topology of $\mathscr{D}'(\mathbb{T}^n)$ if and only if

$$\langle S, \varphi\rangle = \lim_{m\to\infty}\langle S_m, \varphi\rangle \tag{1.4}$$

holds for any $\varphi\in\mathscr{D}(\mathbb{T}^n)$.

Let $f(x)$ be an integrable function on $\mathbb{T}^n$, and put

$$\langle f, \varphi\rangle = \int_{\mathbb{T}^n} f(x)\varphi(x)\, \bar{d}x \tag{1.5}$$

for any $\varphi\in\mathscr{D}(\mathbb{T}^n)$, where $\bar{d}x=(1/2\pi)\,dx^1\cdots dx^n$. Then $\langle\ ,\ \rangle$ is linear with respect to φ. By putting $l=0$, and $C=\int_{\mathbb{T}^n}|f(x)|\,dx$, (1.3) holds for $\langle f, \varphi\rangle$, hence by (1.5) f defines a distribution. Thus we have a map

$$L^1(\mathbb{T}^n)\to\mathscr{D}'(\mathbb{T}^n).$$

This map is injective. Indeed, let f and g be elements of $L^1(\mathbb{T}^n)$, and suppose

$$\langle f, \varphi\rangle = \langle g, \varphi\rangle$$

for any $\varphi \in C^\infty(\mathbb{T}^n)$. Then we have

$$\int_{\mathbb{T}^n} (f(x) - g(x))\varphi(x)\, đx = 0.$$

Hence we have the equality

$$f(x) = g(x) = 0$$

for almost every x. Therefore $f = g$ in $L^1(\mathbb{T}^n)$. Consequently the map of $L^1(\mathbb{T}^n)$ to $\mathscr{D}'(\mathbb{T}^n)$ defined above is injective. By this injection $L^1(\mathbb{T}^n)$ can be considered as a subspace of $\mathscr{D}'(\mathbb{T}^n)$: $L^1(\mathbb{T}^n) \subset \mathscr{D}'(\mathbb{T}^n)$.

Let $a(x)$ be a C^∞ function on $\mathbb{T}^n$, and $f(x)$ an integrable function on $\mathbb{T}^n$. Then the product $(af)(x) = a(x)f(x)$ also becomes an integrable function. We want to define a multiplication of any distribution S by a by extending this operation. In a special case $S = f \in L^1(\mathbb{T}^n)$, we have the identity

$$\langle af, \varphi\rangle = \int_{\mathbb{T}^n} a(x)f(x)\varphi(x)\, đx = \int_{\mathbb{T}^n} f(x)a(x)\varphi(x)\, đx = \langle f, a\varphi\rangle$$

holds for any $\varphi \in \mathscr{D}(\mathbb{T}^n)$. In view of this identity we shall define aS as follows:

Definition 1.2. Let $S \in \mathscr{D}'(\mathbb{T}^n)$ and $a \in C^\infty(\mathbb{T}^n)$. Then the *product* $aS \in \mathscr{D}'(\mathbb{T}^n)$ is defined as follows:

$$\langle aS, \varphi\rangle = \langle S, a\varphi\rangle$$

for any $\varphi \in \mathscr{D}(\mathbb{T}^n)$.

To make this definition possible, it should be proved that the linear map

$$\Phi_a : \mathscr{D}(\mathbb{T}^n) \ni \varphi \to a\varphi \in \mathscr{D}(\mathbb{T}^n)$$

is continuous with respect to the topology of $\mathscr{D}(\mathbb{T}^n)$. For any multi-index $\alpha = (\alpha_1, \ldots, \alpha_n)$, we have the following identity (Leibniz' formula):

$$D^\alpha(a(x)\varphi(x)) = \sum_{\beta \leqq \alpha} \binom{\alpha}{\beta} D^\beta a(x) D^{\alpha-\beta}\varphi(x), \tag{1.6}$$

where $\beta = (\beta_1, \ldots, \beta_n)$ is a multi-index, and by $\beta \leqq \alpha$ we mean that $\beta_j \leqq \alpha_j$ for every $j = 1, \ldots, n$. Further we have put

$$\binom{\alpha}{\beta} = \binom{\alpha_1}{\beta_1} \cdots \binom{\alpha_n}{\beta_n}.$$

Using Leibniz' formula, we obtain for $l=0, 1, \ldots,$

$$|a\varphi|_l \leqq n^l |a|_l |\varphi|_l, \tag{1.7}$$

where a $C^\infty(\mathbb{T}^n)$ and φ $C^\infty(\mathbb{T}^n)$. Therefore Φ_a is a continuous linear map.

Since Φ_a is a continuous linear map, we can define its dual

$$'\Phi_a: \mathscr{D}'(\mathbb{T}^n) \to \mathscr{D}'(\mathbb{T}^n),$$

which is continuous with respect to the weak topology. By the definition of the product of distributions given above, we have $'\Phi_a(S) = aS$. Hence the map $\mathscr{D}'(\mathbb{T}^n) \ni \varphi \to aS \in \mathscr{D}'(\mathbb{T}^n)$ is weakly continuous. Its linearity in S is also obvious.

Let $f \in C^1(\mathbb{T}^n)$. Then we obtain the following identity.

$$\left\langle \frac{\partial}{\partial x^j} f, \varphi \right\rangle = \int_{\mathbb{T}^n} \frac{\partial}{\partial x^j} f(x) \varphi(x)\, dx = -\int_{\mathbb{T}^n} f(x) \frac{\partial}{\partial x^j} \varphi(x)\, dx$$

$$= -\left\langle f, \frac{\partial}{\partial x^j} \varphi \right\rangle, \qquad j=1, \ldots, n.$$

This is just the formula of integration by parts, but this characterizes the partial derivative $(\partial/\partial x^j)f$ of f as a distribution. By extending this formula, we define the derivative $(\partial/\partial x_j)S$ of any distribution S as follows. First we note that the map

$$-\frac{\partial}{\partial x^j}: \mathscr{D}(\mathbb{T}^n) \ni \varphi \to -\frac{\partial}{\partial x^j} \varphi \in \mathscr{D}(\mathbb{T}^n), \qquad j=1, \ldots, n, \tag{1.8}$$

is linear and continuous. For, in fact,

$$\left| \frac{\partial}{\partial x^j} \varphi \right|_l \leqq |\varphi|_{l+1}, \qquad l=0, 1, \ldots, \tag{1.9}$$

holds.

Definition 1.3. We denote the dual map of the linear map (1.8) by

$$\frac{\partial}{\partial x^j}: \mathscr{D}'(\mathbb{T}^n) \ni S \to \frac{\partial}{\partial x^j} S \in \mathscr{D}'(\mathbb{T}^n), \qquad j=1, \ldots, n,$$

and call $(\partial/\partial x^j)S$ the *partial derivative* of S in x^j. The map $S \to (\partial/\partial x^j)S$ is continuous in the weak topology of $\mathscr{D}'(\mathbb{T}^n)$.

The above definition of the partial derivative $\partial/\partial x^j$ coincides with that of the usual partial derivative of a C^1 function f. This follows from the above formula of the integration by parts. So we always write

$$D_j = \frac{\partial}{\partial x^j}$$

as an operator acting on distributions.

Leibniz' formula (1.6) can be extended for any $a \in C^\infty(\mathbb{T}^n)$ and $S \in \mathscr{D}'(\mathbb{T}^n)$:

$$D^\alpha(aS) = \sum_{\beta \leqq \alpha} \binom{\alpha}{\beta} D^\beta a D^{\alpha-\beta} S, \tag{1.10}$$

which is called *Leibniz' formula* as well.

Similarly we can define the *translation* of a distribution S in the direction of x_j by h. For $\varphi \in C^\infty(\mathbb{T}^n)$, we put

$$\tau_j^{-h}\varphi(x) = \varphi(x^1, \ldots, x^j + h, \ldots, x^n).$$

Clearly the map thus defined

$$\tau_j^{-h}: \mathscr{D}(\mathbb{T}^n) \to \mathscr{D}(\mathbb{T}^n)$$

is continuous and linear. As its dual map,

$$\tau_j^h: \mathscr{D}'(\mathbb{T}^n) \ni S \to \tau_j^h S \in \mathscr{D}'(\mathbb{T}^n) \tag{1.11}$$

is defined. We call $\tau_j^h S$ the translation of S in the direction of x^j by h.

The *difference quotient operator* Δ_j^h is defined by

$$\Delta_j^h S = h^{-1}(\tau_j^h S - S) \tag{1.12}$$

for any distribution S. Especially in case $S = f \in L^1(\mathbb{T}^n)$, we have

$$\Delta_j^h f(x) = h^{-1}(f(x^1, \ldots, x^j + h, \ldots, x^n) - f(x^1, \ldots, x^n)).$$

The formula $D_k \Delta_j^h = \Delta_j^h D_k$ holds, and for any $a \in \mathscr{D}(\mathbb{T}^n)$ and any $S \in \mathscr{D}'(\mathbb{T}^n)$, we have

$$\Delta_j^h(aS) = \Delta_j^h a \tau_j^{-h} S + a \Delta_j^h S. \tag{1.13}$$

Let S be a distribution on $\mathbb{T}^n$, and K a compact subset of $\mathbb{T}^n$. S is said to vanish on $K^c = T^n - K$ if

$$\langle S, \varphi \rangle = 0$$

holds for any $\varphi \in \mathscr{D}(\mathbb{T}^n)$ vanishing on K. The smallest one of such K is called the *support of* S and is denoted by supp S. If S is a function $f \in L^1(\mathbb{T}^n)$, the support of S coincides with the support of f as a function:

$$\operatorname{supp} S = \operatorname{supp} f.$$

For $a \in C^\infty(\mathbb{T}^n)$,we have

$$\operatorname{supp} aS = \operatorname{supp} a \cap \operatorname{supp} S. \tag{1.14}$$

(b) Vector-Valued Distributions

Let μ be a positive integer. A vector-valued distribution S with values in $\mathbb{C}^\mu$ is defined as a μ-tuple

$$S = (S^1, \ldots, S^\mu) \in \mathcal{D}'(\mathbb{T}^n) \times \cdots \times \mathcal{D}'(\mathbb{T}^n).$$

We denote the space consisting of all such vector-valued distributions by $\mathcal{D}'(\mathbb{T}^n, \mathbb{C}^\mu)$. Since $\mathcal{D}'(\mathbb{T}^n, \mathbb{C}^\mu) = \mathcal{D}'(\mathbb{T}^n) \times \cdots \times \mathcal{D}'(\mathbb{T}^n)$, it is a vector space.

For $a \in C^\infty(\mathbb{T}^n)$ and $S \in \mathcal{D}'(\mathbb{T}^n, \mathbb{C}^\mu)$, we define their product by $aS = (aS^1, \ldots, aS^\mu)$. Similarly we define the partial derivative $D_j S$ and the difference quotient $\Delta_j^h S$ by

$$D_j S = (D_j S^1, \ldots, D_j S^\mu),$$

$$\Delta_j^h S = (\Delta_j^h S^1, \ldots, \Delta_j^h S^\mu).$$

We denote the space of vector-valued C^∞ functions by

$$\mathcal{D}(\mathbb{T}^n, \mathbb{C}^\mu) = \mathcal{D}(\mathbb{T}^n) \times \cdots \times \mathcal{D}(\mathbb{T}^n).$$

For $\varphi = (\varphi^1, \ldots, \varphi^\mu) \in \mathcal{D}(\mathbb{T}^n, \mathbb{C}^\mu)$ and $S = (S^1, \ldots, S^\mu) \in \mathcal{D}'(\mathbb{T}^n, \mathbb{C}^\mu)$, we define the bilinear map

$$\langle S, \varphi \rangle = \sum_\lambda \langle S^\lambda, \varphi^\lambda \rangle,$$

which makes $\mathcal{D}'(\mathbb{T}^n, \mathbb{C}^\mu)$ and $\mathcal{D}(\mathbb{T}^n, \mathbb{C}^\mu)$ dual to each other.

(c) Fourier Series Expansion of Vector-Valued Distributions

We denote by $L^2(\mathbb{T}^n, \mathbb{C}^\mu)$ the totality of vector-valued functions $f(x) = (f^1(x), \ldots, f^\mu(x))$ with values in $\mathbb{C}^\mu$ defined on $\mathbb{T}^n$ such that each component $f^\lambda(x)$ is square-integrable. For $f, g \in L^2(\mathbb{T}^n, \mathbb{C}^\mu)$, we define their inner product by

$$(f, g) = \sum_\lambda \int_{\mathbb{T}^n} f^\lambda(x) \overline{g^\lambda(x)}\, d\!\!\!{}^-x. \tag{1.15}$$

With this inner product $L^2(\mathbb{T}^n, \mathbb{C}^\mu)$ becomes a Hilbert space. The norm of f is given by

$$\|f\| = (f, f)^{1/2} = \left(\sum_\lambda \int_{\mathbb{T}^n} |f^\lambda(x)|^2\, dx\right)^{1/2}.$$

For any $\xi \in \mathbb{Z}^n$, we write $\xi \cdot x = \xi_1 x^1 + \cdots + \xi_n x^n$, and put

$$f_\xi^\lambda = \int_{\mathbb{T}^n} f^\lambda(x) \exp(-i\xi \cdot x)\, dx. \tag{1.16}$$

We call $f_\xi = (f_\xi^1, \ldots, f_\xi^\mu)$ the Fourier coefficient of f. Using these f_ξ^λ, we obtain for each λ a Fourier series expansion

$$f^\lambda(x) = \sum_\xi f_\xi^\lambda \exp(i\xi \cdot x), \qquad \lambda = 1, \ldots, \mu, \tag{1.17}$$

which converges in $L^2(\mathbb{T}^n)$. We denote by $l^2(\mathbb{Z}^n, \mathbb{C}^\mu)$ the totality of μ-tuple of infinite sequences of complex numbers $\{a_\xi^\lambda\}$ with $\xi \in \mathbb{Z}^n$ and $\lambda = 1, \ldots, \mu$, such that

$$\sum_\lambda \sum_\xi |a_\xi^\lambda|^2 < \infty.$$

An element of $l^2(\mathbb{Z}^n, \mathbb{C}^\mu)$ can be considered as a map of $\mathbb{Z}^n$ to $\mathbb{C}^\mu$ which maps $\xi \in \mathbb{Z}^n$ to $a_\xi = (a_\xi^1, \ldots, a_\xi^\mu) \in \mathbb{C}^\mu$. $l^2(\mathbb{Z}^n, \mathbb{C}^\mu)$ is also a Hilbert space. The following is the fundamental theorem for Fourier series.

Theorem 1.1. *The map which associates $f \in L^2(\mathbb{T}^n, \mathbb{C}^\mu)$ to its Fourier coefficients gives an isomorphism of $L^2(\mathbb{T}^n, \mathbb{C}^\mu)$ onto $l^2(\mathbb{Z}^n, \mathbb{C}^\mu)$ as Hilbert spaces.*

$$\|f\|^2 = \sum_\lambda \sum_\xi |f_\xi^\lambda|^2 \quad \text{(Parseval).} \ \blacksquare \tag{1.18}$$

The Fourier series (1.17) converges to f with respect to the norm of $L^2(\mathbb{T}^n, \mathbb{C}^\mu)$.

It is difficult to give the condition under which a square-integrable function on $\mathbb{T}^n$ becomes continuous in terms of the Fourier coefficients. But for our present purpose the following theorem is sufficient.

Theorem 1.2 (Sobolev's Imbedding Theorem). *Let f_ξ^λ be the Fourier coefficients of $f \in L^2(\mathbb{T}^n, \mathbb{C}^\mu)$, and suppose that for some $s > n/2$, we have*

$$\sum_\lambda \sum_\xi (1 + |\xi|^2)^s |f_\xi^\lambda|^2 < \infty,$$

where we put $|\xi|^2 = \xi_1^2 + \cdots + \xi_n^2$. *Then the Fourier series*

$$\tilde{f}^\lambda(x) = \sum_\xi f_\xi^\lambda \exp(i\xi \cdot x), \qquad \lambda = 1, \ldots, \mu,$$
$$\tilde{f}(x) = (\tilde{f}^1(x), \ldots, \tilde{f}^\mu(x)), \tag{1.19}$$

converge absolutely and uniformly on $\mathbb{T}^n$, *hence give continuous functions of* x. *Further*

$$f^\lambda(x) = \tilde{f}^\lambda(x)$$

holds almost everywhere. Moreover there exists a positive constant C_s *depending on* s *such that*

$$\max_{x \in \mathbb{T}^n} |\tilde{f}^\lambda(x)| \leqq C_s \left(\sum_\lambda \sum_\xi (1+|\xi|^2)^s |f_\xi^\lambda|^2 \right)^{1/2}. \tag{1.20}$$

Proof. Putting

$$\|f\|_s = \left(\sum_\lambda \sum_\xi (1+|\xi|^2)^s |f_\xi^\lambda|^2 \right)^{1/2},$$

we have

$$\sum_{\lambda,\xi} |f_\xi^\lambda| \leqq \sum_{\lambda,\xi} (1+|\xi|^2)^{s/2} |f_\xi^\lambda| (1+|\xi|^2)^{-s/2}$$
$$\leqq \|f\|_s \left(\sum_{\lambda,\xi} (1+|\xi|^2)^{-s} \right)^{1/2} < \infty$$

because $s > n/2$. Therefore the infinite series (1.19) whose terms are continuous functions converge absolutely and uniformly in x, hence represent continuous functions of x. On the other hand, by the preceding theorem, f and $\tilde{f}$ coincide as elements of $L^2(\mathbb{T}^n, \mathbb{C}^\mu)$. Consequently on $\mathbb{T}^n$ the equality

$$f^\lambda(x) = \tilde{f}^\lambda(x)$$

holds except possibly for a set of measure 0, which proves the theorem. ∎

Theorem 1.3 (Sobolev's Imbedding Theorem). *Let* f_ξ^λ *be the Fourier coefficients of* $f \in L^2(\mathbb{T}^n, \mathbb{C}^\mu)$, *and* k *a non-negative integer. If for some* $s > n/2 + k$,

$$\sum_\lambda \sum_\xi (1+|\xi|^2)^s |f_\xi^\lambda|^2 < \infty \tag{1.21}$$

holds, then the Fourier series

$$\tilde{f}^\lambda(x) = \sum_\lambda f_\xi^\lambda \exp(i\xi \cdot x), \qquad \lambda = 1, \ldots, \mu, \tag{1.22}$$

and the infinite series obtained by term-wise differentiating them j-times for any $j \leqq k$ all converge absolutely and uniformly. Therefore $\tilde{f}^\lambda(x)$ are C^k. Moreover on $\mathbb{T}^n$, the equality

$$f^\lambda(x) = \tilde{f}^\lambda(x)$$

holds except possibly for a set of measure 0.

Proof. We have already proved that (1.22) converge absolutely and uniformly in the preceding theorem. By differentiating their general term, we have

$$\begin{aligned}|D^\alpha f_\xi^\lambda \exp(i\xi \cdot x)| &= |(i\xi)^\alpha \cdot f_\xi^\lambda \exp(i\xi \cdot x)| \\ &\leqq |\xi|^{|\alpha|} |f_\xi^\lambda|, \qquad \lambda = 1, \ldots, \mu,\end{aligned}$$

where we put $|\alpha| = j \leqq k$, and $\xi^\alpha = \xi_1^{\alpha_1} \cdots \xi_n^{\alpha_n}$. Therefore, similarly as in the proof of Theorem 1.2, we obtain

$$\sum_{\lambda,\xi} |\xi|^{|\alpha|} |f_\xi^\lambda| \leqq \|f\|_s \left(\sum_{\lambda,\xi} (1+|\xi|^2)^{k-s} \right)^{1/2}.$$

Since $s > n/2 + k$, we have $\sum_\xi (1+|\xi|^2)^{k-s} < \infty$, hence

$$\sum_\xi D^\alpha f_\xi^\lambda \exp(i\xi \cdot x), \qquad \lambda = 1, \ldots, \mu,$$

converge absolutely and uniformly, and the $\tilde{f}^\lambda(x)$ are C^k. Also there is a positive constant C_α independent of f such that

$$\max_x |D^\alpha \tilde{f}^\lambda(x)| \leqq C_\alpha \|f\|_s. \tag{1.23}$$

The fact that $\tilde{f}(x) = f(x)$ holds almost everywhere is proved similarly as in the proof of the preceding theorem. ∎

The vector space of all C^k functions defined on $\mathbb{T}^n$ with values in $\mathbb{C}^\mu$ becomes a Banach space if we define the norm of $f = (f^1, \ldots, f^\mu) \in C^k(\mathbb{T}^n, \mathbb{C}^\mu)$ by

$$|f|_k = \sum_\lambda \sum_{|\alpha| \leqq k} \max_{x \in \mathbb{T}^n} |D^\alpha f^\lambda(x)|.$$

Let f_ξ^λ be the Fourier coefficients of f. Then for any multi-index α with $|\alpha| \leqq k$, we have

$$\begin{aligned}(-i)^{|\alpha|} \xi^\alpha f_\xi^\lambda &= \int_{\mathbb{T}^n} f^\lambda(x) D^\alpha \exp(-i\xi \cdot x)\, dx \\ &= -\int_{\mathbb{T}^n} (-D)^\alpha f^\lambda(x) \exp(-i\xi \cdot x)\, dx.\end{aligned}$$

Therefore, recalling $(-D)^\alpha f \in L^2(\mathbb{T}^n, \mathbb{C}^\mu)$, we obtain

$$\sum_\lambda \sum_\xi (1+|\xi|^2)^{2k} |f_\xi^\lambda|^2 = \|(1+D_1^2+\cdots+D_n^2)^k f\| \leqq C_k |f|_{2k}. \tag{1.24}$$

By the deep gap between (1.24) and Sobolev's imbedding theorem, we cannot get the precise characterization of $C^k(\mathbb{T}^n, \mathbb{C}^\mu)$ in this way. But we can characterize $C^\infty(\mathbb{T}^n, \mathbb{C}^\mu)$ as follows.

Theorem 1.4. *$f \in L^2(\mathbb{T}^n, \mathbb{C}^\mu)$ coincides with an element of $C^\infty(\mathbb{T}^n, \mathbb{C}^\mu)$ almost everywhere if and only if its Fourier coefficients satisfy*

$$\sum_\lambda \sum_\xi (1+|\xi|^2)^s |f_\xi^\lambda|^2 < \infty \tag{1.25}$$

for any $s \in \mathbb{R}$.

The proof runs as follows. First by (1.24) we see the necessity of (1.25). Conversely, if (1.25) holds for any s, we see from Sobolev's imbedding theorem that f is equal to an element of $C^\infty(\mathbb{T}^n, \mathbb{C}^\mu)$ almost everywhere. ∎

Definition 1.4. Let φ_ξ^λ be the Fourier coefficients of $\varphi = (\phi^1, \ldots, \varphi^\mu) \in C^\infty(\mathbb{T}^n, \mathbb{C}^\mu)$. For an arbitrary $s \in \mathbb{R}$, we call

$$\|\varphi\|_s = \left(\sum_\lambda \sum_\xi (1+|\xi|^2)^s |\varphi_\xi^\lambda|^2 \right)^{1/2}$$

the Sobolev norm of degree s.

By (1.23) and (1.24), for $k = 0, 1, \ldots$, there exists a positive constant C_k depending only on k such that for any $\varphi \in C^\infty(\mathbb{T}^n, \mathbb{C}^\mu)$,

$$C_k^{-1} \|\varphi\|_{2k} \leqq |\varphi|_{2k} \leqq C_k \|\varphi\|_{2k+[n/2]+1}, \tag{1.26}$$

where $[n/2]$ denotes the integral part of $n/2$. Therefore the topology of the Fréchet space $\mathcal{D}(\mathbb{T}^n, \mathbb{C}^\mu)$ can also be defined by the countable system of norms $\| \ \|_k$, $k = 0, 1, \ldots$. From this fact combined with Theorem 1.4 and the proof of Sobolev's imbedding theorem, we obtain the following theorem.

Theorem 1.5. *Let $\varphi \in \mathcal{D}(\mathbb{T}^n, \mathbb{C}^\mu)$. The Fourier series of φ converges to φ with respect to the topology of $\mathcal{D}(\mathbb{T}^n, \mathbb{C}^\mu)$. Therefore $\{\exp(i\xi \cdot x)\}_{\xi \in \mathbb{Z}^n}$ forms a basis of $\mathcal{D}(\mathbb{T}^n, \mathbb{C}^\mu)$.* ∎

Consider the Fourier series expansion of a vector-valued distribution $S = (S^1, \ldots, S^\mu) \in \mathcal{D}'(\mathbb{T}^n, \mathbb{C}^\mu)$ with values in $\mathbb{C}^\mu$.

Definition 1.5. Put $S_\xi^\lambda = \langle S^\lambda, \exp(-i\xi \cdot x)\rangle$, $\xi \in \mathbb{Z}^n$, $\lambda = 1, \ldots, \mu$. We call $S_\xi = (S_\xi^1, \ldots, S_\xi^\mu)$ the Fourier coefficients of the vector-valued distribution $S =$

$(S^1, \ldots, S^\mu)$. The series

$$\sum_\xi S_\xi \exp(-i\xi \cdot x) = \left(\sum_\xi S_\xi^1 \exp(-i\xi \cdot x), \ldots, \sum_\xi S_\xi^\mu \exp(-i\xi \cdot x) \right)$$

is called the *Fourier series* of S.

Theorem 1.6.

(1°) *The map $\mathscr{D}'(\mathbb{T}^n, C^\mu) \ni S \to (S_\xi^\lambda)$ is injective.*

(2°) *The sequence (S_ξ^λ) is the Fourier coefficient of a vector-valued distribution if and only if there exists an integer $k \geqq 0$ such that*

$$\sum_\lambda \sum_\xi (1 + |\xi|^2)^{-k} |S_\xi^\lambda|^2 < \infty \tag{1.27}$$

holds. This being the case, the Fourier series

$$\sum_\xi S_\xi^\lambda \exp(-i\xi \cdot x)$$

converges weakly in $\mathscr{D}'(\mathbb{T}^n, \mathbb{C}^\mu)$. Let $S \in \mathscr{D}'(\mathbb{T}^n, \mathbb{C}^\mu)$ be its limit. Then the Fourier coefficients of S coincide with S_ξ^λ.

Proof. (1°) It suffices to show that if $S = (S^1, \ldots, S^\mu) \in \mathscr{D}'(\mathbb{T}^n, \mathbb{C}^\mu)$ and for any $\xi \in \mathbb{Z}^n$,

$$S_\xi^\lambda = \langle S^\lambda, \exp(-i\xi \cdot x)\rangle = 0, \qquad \lambda = 1, \ldots, \mu,$$

holds, then $S^\lambda = 0$. But this is clear since $\{\exp(-i\xi \cdot x)\}$ forms a basis of $\mathscr{D}(\mathbb{T}^n, \mathbb{C}^\mu)$.

(2°) By the definition of $\mathscr{D}'(\mathbb{T}^n, \mathbb{C}^\mu)$, there exist a positive integer l and a positive constant C' such that for any $\xi \in \mathbb{Z}^n$,

$$|\langle S^\lambda, \exp(-i\xi \cdot x)\rangle| \leqq C' |\exp(-i\xi \cdot x)|_l$$

holds. Hence by taking a suitable constant C,

$$|S_\xi^\lambda| \leqq C(1 + |\xi|^2)^{l/2}$$

holds. Thus if we choose k so that $k - l > n$, (1.27) holds.

Conversely, suppose that (1.27) holds for some $k \geqq 0$.

If we put $A_\xi^\lambda = (1 + |\xi|^2)^{-k} S_\xi^\lambda$, then we have

$$\sum_\lambda \sum_\xi |A_\xi^\lambda|^2 < \infty.$$

Hence if we put $A^\lambda(x)=\sum_\xi A_\xi^\lambda \exp(-i\xi\cdot x)$, then this becomes a square-integrable function of x. Therefore $A^\lambda(x)$ represents a distribution on $\mathbb{T}^n$, namely, for any $\varphi\in\mathscr{D}(\mathbb{T}^n)$,

$$\langle A^\lambda,\varphi\rangle=\int_{\mathbb{T}^n} A^\lambda(x)\varphi(x)\,dx=\int_{\mathbb{T}^n}\sum_\xi A_\xi^\lambda \exp(-i\xi\cdot x)\varphi(x)\,dx$$

holds. Since the series $\sum_\xi A_\xi^\lambda \exp(-i\xi\cdot x)$ converges strongly in $L^2(\mathbb{T}^n)$, we can exchange the integration with $\sum_\xi$, so we have

$$\langle A^\lambda,\varphi\rangle=\sum_\xi A_\xi^\lambda\int_{\mathbb{T}^n}\exp(-i\xi\cdot x)\phi(x)\,dx=\sum_\xi A_\xi^\lambda\phi_{-\xi},$$

where we denote the Fourier coefficients of $\varphi(x)$ by φ_ξ. Next, we put a partial sum of the series made of $\{S_\xi^\lambda\}$ as

$$S_N^\lambda=\sum_{|\xi|\leqq N} S_\xi^\lambda \exp(-i\xi\cdot x),$$

where $N>0$. This is a C^∞ function, hence gives a distribution on $\mathbb{T}^n$. We shall prove that if $N\to\infty$, S_N^λ converges to the distribution $(1-D_1^2-\cdots-D_n^2)^kA^\lambda$ with respect to the weak topology of $\mathscr{D}'(\mathbb{T}^n)$. In fact, if $\varphi\in\mathscr{D}(\mathbb{T}^n)$, we have

$$\begin{aligned}\langle(1-D_1^2-\cdots-D_n^2)^kA^\lambda,\varphi\rangle&=\langle A^\lambda,(1-D_1^2-\cdots-D_n^2)^k\varphi\rangle\\&=\sum_\xi A_\xi^\lambda\int_{\mathbb{T}^n}\exp(i\xi\cdot x)(1-D_1^2-\cdots-D_n^2)^k\varphi(x)\,dx\\&=\sum_\xi A_\xi^\lambda(1+\xi_1^2+\cdots+\xi_n^2)^k\varphi_{-\xi}=\sum_\xi S_\xi^\lambda\phi_{-\xi}.\end{aligned}$$

Applying (1.25) to $\phi(x)\in C^\infty(\mathbb{T}^n,\mathbb{C}^\mu)$ by putting $s=k$, we have

$$\|\varphi\|_k^2=\sum_\xi(1+|\xi|^2)^k|\varphi_{-\xi}|^2<\infty.$$

Therefore we have

$$\begin{aligned}&|\langle(1-D_1^2-\cdots-D_n^2)^kA^\lambda,\varphi\rangle-\langle S_N^\lambda,\varphi\rangle|\\&\quad=\left|\sum_{|\xi|>N} S_\xi^\lambda\varphi_{-\xi}\right|\\&\quad\leqq\left(\sum_{|\xi|>N}|S_\xi^\lambda|^2(1+|\xi|^2)^{-k}\right)^{1/2}\left(\sum_{|\xi|>N}(1+|\xi|^2)^k|\varphi_{-\xi}|^2\right)^{1/2}\\&\quad\leqq\|\varphi\|_k\left(\sum_{|\xi|>N}|S_\xi^\lambda|^2(1+|\xi|^2)^{-k}\right)^{1/2},\end{aligned}$$

hence by (1.27) we obtain

$$\lim_{N\to\infty}\langle S_N^\lambda, \varphi\rangle = \langle (1-D_1^2-\cdots-D_n^2)^k A^\lambda, \phi\rangle.$$

Thus S_N^λ converges weakly to $(1-D_1^2-\cdots-D_n^2)^k A^\lambda$ in $\mathscr{D}'(\mathbb{T}^n)$. If we put $S^\lambda=(1-D_1^2-\cdots-D_n^2)^k A^\lambda$, then the Fourier coefficients of S^λ are given by

$$\begin{aligned}\langle S^\lambda, \exp(-i\xi\cdot x)\rangle &= \langle A^\lambda, (1-D_1^2-\cdots-D_n^2)^k \exp(-i\xi\cdot x)\rangle\\ &= (1+|\xi|^2)^k\langle A^\lambda, \exp(-i\xi\cdot x)\rangle\\ &= (1+|\xi|^2)^k A_\xi^\lambda = S_\xi^\lambda.\end{aligned}$$

Thus the proof is completed. ▮

Theorem 1.7. *For any $S=(S^1,\ldots,S^\mu)\in\mathscr{D}'(\mathbb{T}^n,\mathbb{C}^\mu)$, there exist a non-negative constant $l\geqq 0$ and $f=(f^1,\ldots,f^\mu)\in L^2(\mathbb{T}^n,\mathbb{C}^\mu)$ such that in the sense of distributions*

$$S^\lambda=(1-D_1^2-\cdots-D_n^2)^l f^\lambda \tag{1.28}$$

holds. (The structure theorem of distributions.)

Proof. Let

$$S^\lambda=\sum_\xi S_\xi^\lambda \exp(i\xi\cdot x).$$

be the Fourier expansion of S^λ. Then (1.27) holds for some $k\geqq 0$. Therefore as in the proof of the preceding theorem, we put

$$f_\xi^\lambda=(1+|\xi|^2)^{-k}S_\xi^\lambda.$$

If we put $f^\lambda(x)=\sum_\xi f_\xi^\lambda \exp(i\xi\cdot x)$, then $f^\lambda(x)\in L^2(\mathbb{T}^n)$, and as is shown in the proof of Theorem 1.6, we have

$$S^\lambda=(1-D_1^2-\cdots-d_n^2)^k f^\lambda$$

in the sense of distributions. By putting $l=k$, the theorem is proved. ▮

Theorem 1.8. *Let S_ξ^λ be the Fourier coefficients of $S\in\mathscr{D}'(\mathbb{T}^n,\mathbb{C}^\mu)$, and φ_ξ^λ the Fourier coefficients of $\varphi\in\mathscr{D}(\mathbb{T}^n,\mathbb{C}^\mu)$. Then*

$$\langle S,\varphi\rangle=\sum_\lambda\sum_\xi S_\xi^\lambda\varphi_{-\xi}^\lambda \tag{1.29}$$

holds. ▮

This formula is already obtained in the proof of the preceding theorem.

(d) Sobolev Space

We introduce a space which lies between $\mathscr{D}(\mathbb{T}^n, \mathbb{C}^\mu)$ and $\mathscr{D}'(\mathbb{T}^n, \mathbb{C}^\mu)$.

Definition 1.6. For any $s \in \mathbb{R}$, we put

$$W^s(\mathbb{T}^n, \mathbb{C}^\mu) = \{S = (S^1, \dots, S^\mu) \in \mathscr{D}'(\mathbb{T}^n, \mathbb{C}^\mu) \mid \sum_\lambda \sum_\xi |S_\xi^\lambda|^2 (1 + |\xi|^2)^s < \infty\},$$

and call it the *vector-valued Sobolev space of degree s with values in* $\mathbb{C}^\mu$, where S_ξ^λ denotes the Fourier coefficient of S^λ.

For $S, T \in W^s(\mathbb{T}^n, \mathbb{C}^\mu)$, we define their inner product by

$$(S, T)_s = \sum_\lambda \sum_\xi (1 + |\xi|^2)^s S_\xi^\lambda \overline{T_\xi^\lambda}, \tag{1.30}$$

which makes $W^s(\mathbb{T}^n, \mathbb{C}^\mu)$ a Hilbert space. Here T_ξ^λ denotes the Fourier coefficient of T^λ. The norm of this space is just the Sobolev norm of degree s

$$\|S\|_s = \left(\sum_\lambda \sum_\xi (1 + |\xi|^2)^s |S_\xi^\lambda|^2 \right)^{1/2}.$$

Theorem 1.9.

(1°) *If* $s' < s$, $W^s(\mathbb{T}^n, \mathbb{C}^\mu) \subset W^{s'}(\mathbb{T}^n, \mathbb{C}^\mu)$.

(2°) $C^\infty(\mathbb{T}^n, \mathbb{C}^\mu) = \bigcap_s W^s(\mathbb{T}^n, \mathbb{C}^\mu)$, $\mathscr{D}'(\mathbb{T}^n, \mathbb{C}^\mu) = \bigcap_s W^s(\mathbb{T}^n, \mathbb{C}^\mu)$.

(3°) *If* $k + n/2 < s$, *there exists a continuous imbedding*

$$W^s(\mathbb{T}^n, \mathbb{C}^\mu) \hookrightarrow C^k(\mathbb{T}^n, \mathbb{C}^\mu), \tag{1.31}$$

namely, there exists a positive constant C_{sk} *such that for any* $S \in W^s(\mathbb{T}^n, \mathbb{C}^\mu)$, *there is an* $S \in C^k(\mathbb{T}^n, \mathbb{C}^\mu)$ *with*

$$|S|_k \leqq C_{sk} \|S\|_s. \tag{1.32}$$

(4°) $\mathscr{D}(\mathbb{T}^n, \mathbb{C}^\mu)$ *is dense in each* $W^s(\mathbb{T}^n, \mathbb{C}^\mu)$.

Proof. (1°) is clear from the definition.

(2°) follows from Theorem 1.4 and Theorem 1.6.

(3°) is just Sobolev's imbedding theorem, Thorem 1.3. Finally, if we put $S = (S^1, \dots, S^\mu) \in W^s(\mathbb{T}^n, \mathbb{C}^\mu)$, and let

$$S^\lambda = \sum_\xi S_\xi^\lambda \exp(i\xi \cdot x)$$

be its Fourier series, then the partial sum $\sum_{|\xi| \leqq N} S_\xi^\lambda \exp(i\xi \cdot x)$ is an element of $\mathscr{D}(\mathbb{T}^n, \mathbb{C}^\mu)$ and converges to S in $W^s(\mathbb{T}^n, \mathbb{C}^\mu)$ for $N \to \infty$.

(4°) follows immediately from this. ∎

The following inequality is often used in the following.

Lemma 1.1. *Let a, b, t be positive numbers. If* $0 \leqq \lambda \leqq 1$,

$$a^{\lambda} b^{1-\lambda} \leqq \lambda t^{1/\lambda} a + (1-\lambda) t^{-1/(1-\lambda)} b \tag{1.33}$$

holds, where the equality holds for $t^{1/\lambda} a = t^{-1/(1-\lambda)} b$.

The proof is given by the comparison of the arithmetic and geometric means. ∎

Proposition 1.1.

(1°) *If* $s'' < s' < s$, *then for any* $f \in W^s(\mathbb{T}^n, \mathbb{C}^\mu)$ *the following interpolation inequality holds*:

$$\|f\|_{s'} \leqq \|f\|_s^{(s'-s'')/(s-s'')} \|f\|_{s''}^{(s-s')/(s-s'')}. \tag{1.34}$$

(2°) *If* $s'' < s' < s$, *for any* $f \in W^s(\mathbb{T}^n, \mathbb{C}^\mu)$ *and any positive number t, the following interpolation inequality holds*:

$$\|f\|_{s'}^2 \leqq \frac{s'-s''}{s-s''} t^{(s-s'')/(s'-s'')} \|f\|_s^2 + \frac{s-s'}{s-s''} t^{-(s-s'')/(s-s')} \|f\|_{s''}^2. \tag{1.35}$$

Proof. By the above lemma, for any $t > 0$, we have

$$(1+|\xi|^2)^{s'} \leqq \frac{s'-s''}{s-s''} t^{(s-s'')/(s'-s'')} (1+|\xi|^2)^s + \frac{s-s'}{s-s''} t^{-(s-s'')/(s-s')} (1+|\xi|^2)^{s''}.$$

Put $f = (f^1, \ldots, f^\mu)$, and let f_ξ^λ be the Fourier coefficients of F^λ. Multiplying the above inequality by $|f_\xi^\lambda|^2$ and summing over $\xi \in \mathbb{Z}^n$, we obtain

$$\|f\|_{s'}^2 \leqq \frac{s'-s''}{s-s''} t^{(s-s'')/(s'-s'')} \|f\|_s^2 + \frac{s-s'}{s-s''} t^{-(s-s'')/(s-s')} \|f\|_{s''}^2.$$

Taking the minimum of the right-hand side for $t > 0$, we obtain

$$\|f\|_{s'}^2 \leqq \|f\|_s^{2(s'-s'')/(s-s'')} \|f\|_{s''}^{2(s-s')/(s-s'')}.$$

Taking the square root, we obtain (1.34).

(2°) follows from this and Lemma 1.1. ∎

Theorem 1.10. *If $s'<s$, the inclusion $W^s(\mathbb{T}^n, \mathbb{C}^\mu) \to W^{s'}(\mathbb{T}^n, \mathbb{C}^\mu)$ is compact.*

Proof. It suffices to show that if a countable sequence $\{f_m\}_m$ in $W^s(\mathbb{T}^n, \mathbb{C}^\mu)$ is bounded, we can choose a subsequence which converges in $W^{s'}(\mathbb{T}^n, \mathbb{C}^\mu)$. Let $f_m = (f_m^1, \ldots, f_m^\mu)$ and $f_{m,\xi}^\lambda$ the Fourier coefficients of f_m^λ where $\xi \in \mathbb{Z}^n$. Since $\{f_m\}_m$ is bounded in $W^s(\mathbb{T}^n, \mathbb{C}^\mu)$, there exists a positive number M such that

$$\|f\|_s^2 = \sum_\lambda \sum_\xi |f_{m,\xi}^\lambda|^2 (1+|\xi|^2)^s < M.$$

Since for each ξ, λ, $\{f_{m,\xi}^\lambda\}_{m=1}^\infty$ is a bounded sequence, we may take a suitable subsequence $\{m'\}$ so that for each ξ, and λ, $\{f_{\xi,m'}^\lambda\}_{m'}$ converges. Then $\{f_{m'}\}$ converges in $W^{s'}(\mathbb{T}^n, \mathbb{C}^\mu)$. In fact, we have the following Fourier expansion:

$$f_{m'}^\lambda = \sum_\xi f_{m',\xi}^\lambda \exp(i\xi \cdot x).$$

For any positive number ε, take a sufficiently large N so that

$$(1+N^2)^{s'-s} < 8^{-1} M^{-1} \varepsilon^2$$

holds. Then for any m', we have the following inequality.

$$\sum_{|\xi| \geqq N} (1+|\xi|^2)^{s'} |f_{m',\xi}^\lambda|^2 \leqq (1+N^2)^{s'-s} \sum_{|\xi| \geqq N} (1+|\xi|^2)^s |f_{m',\xi}^\lambda|^2 < 8^{-1} \varepsilon^2.$$

Fix such N. Since there are only a finite number of lattice points $\xi \in \mathbb{Z}^n$ with $|\xi| \leqq N$, we may choose a sufficiently large m_0 so that for any m_1', $m_2' > m_0$,

$$\sum_\lambda \sum_{|\xi|<N} |f_{m_1',\xi}^\lambda - f_{m_2',\xi}^\lambda|^2 (1+|\xi|^2)^{s'} < 2^{-2} \varepsilon^2$$

holds. Then we have

$$\begin{aligned}
\|f_{m_1'}^\lambda - f_{m_2'}^\lambda\|_{s'}^2 &\leqq \sum_\lambda \sum_{|\xi|<N} |f_{m_1',\xi}^\lambda - f_{m_2',\xi}^\lambda|^2 (1+|\xi|^2)^{s'} \\
&\quad + 2 \sum_\lambda \sum_{|\xi| \geqq N} (|f_{m_1',\xi}^\lambda|^2 + |f_{m_2',\xi}^\lambda|^2)(1+|\xi|^2)^{s'} \\
&< 2^{-2}\varepsilon^2 + 2^{-1}\varepsilon^2 < \varepsilon^2.
\end{aligned}$$

Therefore $\{f_{m'}\}_{m'}$ is a Cauchy sequence in $W^{s'}(\mathbb{T}^n, \mathbb{C}^\mu)$, hence converges in $W^{s'}(\mathbb{T}^n, \mathbb{C}^\mu)$. This completes the proof. ∎

Since the Sobolev space of degree 0 is nothing but $L^2(\mathbb{T}^n, \mathbb{C}^\mu)$, we shall write $\| \ \|$ for $\| \ \|_0$ below. It is difficult to have an intuitive understanding of $W^s(\mathbb{T}^n, \mathbb{C}^\mu)$, but if $s = l$ is a positive integer, it is rather easy to understand. For this we use the following lemma.

Lemma 1.2. *Let $S = (S^1, \ldots, S^\mu) \in D'(\mathbb{T}^n, \mathbb{C}^\mu)$, and S^λ_ξ the Fourier coefficients of S. If we write the Fourier coefficients of $D_jS = (D_jS^1, \ldots, D_jS^\mu)$ as $(D_jS^\lambda)_\xi$ for $j = 1, 2, \ldots, n$, we have*

$$(D_jS^\lambda)_\xi = i\xi_j S^\lambda_\xi, \qquad j = 1, \ldots, n; \quad \lambda = 1, \ldots, \mu. \tag{1.36}$$

Proof.
$$\begin{aligned}(D_jS^\lambda)_\xi &= \langle D_jS^\lambda, \exp(-i\xi \cdot x)\rangle = -\langle S^\lambda, D_j \exp(-i\xi \cdot x)\rangle \\ &= i\xi_j\langle S^\lambda, \exp(-i\xi \cdot x)\rangle = i\xi_j S^\lambda_\xi. \quad ∎\end{aligned}$$

From this lemma we obtain the following proposition.

Proposition 1.2. *Let l be a positive integer. Then*

$$W^l(\mathbb{T}^n, \mathbb{C}^\mu) = \{S \in L^2(\mathbb{T}^n, \mathbb{C}^\mu) \mid D^\alpha S \in L^2(\mathbb{T}^n, \mathbb{C}^\mu) \text{ for any } \alpha \text{ with } |\alpha| \leqq l\}.$$

For any $S \in W^l(\mathbb{T}^n, \mathbb{C}^\mu)$ we have

$$n^{-l/2}\|S\|_l \leqq \left(\sum_{|\alpha| \leqq l} \|D^\alpha S\|\right)^{1/2} \leqq \|S\|_l. \tag{1.37}$$

Proof. It suffices to show (1.37). Letting S^λ_ξ be the Fourier coefficient of S, we have

$$\sum_{|\alpha| \leqq l} \|D^\alpha S\|^2 = \sum_\lambda \sum_\xi \sum_{|\alpha| \leqq l} |\xi|^{2\alpha} |S^\lambda_\xi|^2.$$

For $\xi \in \mathbb{Z}^n$,

$$n^{-l}(1 + |\xi|^2)^l \leqq \sum_{|\alpha| \leqq l} |\xi|^{2\alpha} \leqq (1 + |\xi|^2)^l$$

holds. Therefore, multiplying this by $|S^\lambda_\xi|^2$ and summing over ξ, we obtain

$$n^{-l}\|S\|_l^2 \leqq \sum_{|\alpha| \leqq l} \|D^\alpha S\|^2 \leqq n^l \|S\|_l^2,$$

which proves (1.37). ∎

Theorem 1.11.

(1°) *The restriction of the bilinear form $\langle\ ,\ \rangle$ on $\mathscr{D}'(\mathbb{T}^n, \mathbb{C}^\mu) \times \mathscr{D}(\mathbb{T}^n, \mathbb{C}^\mu)$ to $\mathscr{D}(\mathbb{T}^n, \mathbb{C}^\mu) \times \mathscr{D}(\mathbb{T}^n, \mathbb{C}^\mu)$ gives the following inequality: For any $\phi, \psi \in \mathscr{D}(\mathbb{T}^n, \mathbb{C}^\mu)$ and any $s \in \mathbb{R}$,*

$$|\langle \psi, \varphi\rangle| \leqq \|\psi\|_{-s}\|\varphi\|_s. \tag{1.38}$$

(2°) *The bilinear form* $\mathscr{D}(\mathbb{T}^n, \mathbb{C}^\mu) \times \mathscr{D}(\mathbb{T}^n, \mathbb{C}^\mu) \ni \psi, \varphi \to \langle \psi, \varphi \rangle$ *extends uniquely to a continuous bilinear map of* $W^{-s}(\mathbb{T}^n, \mathbb{C}^\mu) \times W^s(\mathbb{T}^n, \mathbb{C}^\mu)$ *to* $\mathbb{C}$ *which is continuous in* ψ *with respect to the norm* $\| \ \|_{-s}$ *and in* φ *with respect to* $\| \ \|_s$. *By this bilinear form,* $W^{-s}(\mathbb{T}^n, \mathbb{C}^\mu)$ *and* $W^s(\mathbb{T}^n, \mathbb{C}^\mu)$ *become dual to each other.*

(3°) *Let* $s \in \mathbb{R}$. *For* $S \in W^{-s}(\mathbb{T}^n, \mathbb{C}^\mu)$ *we have*

$$\sup_{\varphi} \frac{|\langle S, \varphi \rangle|}{\|\varphi\|_s} \geqq \|S\|_{-s}. \tag{1.39}$$

Here sup *is taken over all non-zero* $\varphi \in \mathscr{D}(\mathbb{T}^n, \mathbb{C}^\mu)$.

Proof. (1°) Let φ_ξ^λ and ψ_ξ^λ be the Fourier coefficients of φ and ψ respectively. Then we have

$$\langle \psi, \varphi \rangle = \sum_\lambda \sum_\xi \psi_\xi^\lambda \varphi_{-\xi}^\lambda.$$

By Schwartz' inequality, we obtain

$$\begin{aligned} |\langle \psi, \varphi \rangle| &\leqq \left(\sum_\lambda \sum_\xi (1+|\xi|^2)^{-s} |\psi_\xi^\lambda|^2 \right)^{1/2} \left(\sum_\lambda \sum_\xi (1+|\xi|^2)^s |\varphi_\xi^\lambda|^2 \right)^{1/2} \\ &= \|\psi\|_{-s} \|\varphi\|_s. \end{aligned}$$

(2°) Since $\mathscr{D}(\mathbb{T}^n, \mathbb{C}^\mu)$ is dense both in $W^{-s}(\mathbb{T}^n, \mathbb{C}^\mu)$ and in $W^s(\mathbb{T}^n, \mathbb{C}^\mu)$, the inequality (1.38) shows that the bilinear form $\langle \ , \ \rangle$ extends uniquely to a continuous bilinear form on $W^{-s}(\mathbb{T}^n, \mathbb{C}^\mu) \times W^s(\mathbb{T}^n, \mathbb{C}^\mu)$ to $\mathbb{C}$. Then (1.39) proves that by this bilinear form, $W^{-s}(\mathbb{T}^n, \mathbb{C}^\mu)$ and $W^s(\mathbb{T}^n, \mathbb{C}^\mu)$ become dual to each other.

(3°) Let S_ξ^λ be the Fourier coefficients of S. For any $N > 0$, we define $\varphi_N = (\varphi_N^1, \dots, \varphi_N^\mu) \in \mathscr{D}(\mathbb{T}^n, \mathbb{C}^\mu)$ by putting

$$\varphi_N^\lambda(x) = \sum_{|\xi| \geqq N} \varphi_{N,\xi}^\lambda \exp(i\xi \cdot x), \qquad \varphi_{N,\xi}^\lambda = (1+|\xi|^2)^{-s} \overline{S_{-\xi}^\lambda}.$$

Then

$$|\langle S, \varphi_N \rangle| = \left| \sum_\lambda \sum_\xi S_\xi^\lambda \varphi_{N,-\xi}^\lambda \right| = \sum_\lambda \sum_{|\xi| \geqq N} |S_\xi^\lambda|^2 (1+|\xi|^2)^{-s}.$$

On the other hand, since $\|\varphi_N\|_s = (\sum_\lambda \sum_{|\xi| \leqq N} (1+|\xi|^2)^{-s} |S_\xi^\lambda|^2)^{1/2}$, we have

$$\frac{|\langle S, \varphi_N \rangle|}{\|\varphi_N\|_s} = \left(\sum_\lambda \sum_{|\xi| \leqq N} (1+|\xi|^2)^{-s} |S_\xi^\lambda|^2 \right)^{1/2}.$$

Taking the supremum with respect to N, we obtain (1.39). ∎

Proposition 1.3. *Let α be a multi-index. If $s > s' + |\alpha|$, then for any $S \in W^s(\mathbb{T}^n, \mathbb{C}^\mu)$, $D^\alpha S \in W^{s'}(\mathbb{T}^n, \mathbb{C}^\mu)$. D^α: $W^s(\mathbb{T}^n, \mathbb{C}^\mu) \to W^{s'}(\mathbb{T}^n, \mathbb{C}^\mu)$ is continuous, and the inequality*

$$\|D^\alpha S\|_{s'} \leqq \|S\|_{s'+|\alpha|} \tag{1.40}$$

holds.

Proof. It suffices to show (1.40). If we let S_ξ^λ be the Fourier coefficients of S, the Fourier coefficients of $D^\alpha S^\lambda$ are given by

$$(D^\alpha S)_\xi^\lambda = (i\xi)^\alpha S_\xi^\lambda.$$

Hence we have

$$\|D^\alpha S\|_{s'}^2 \leqq \sum_\lambda \sum_\xi (1+|\xi|^2)^{s'} |\xi|^{2|\alpha|} |S_\xi^\lambda|^2 \leqq \|S\|_{s'+|\alpha|}^2. \quad \blacksquare$$

Next we consider the product of a function and a distribution.

Lemma 1.3. *Let $f(x) \in C^\infty(\mathbb{T}^n, \mathbb{C}^{\nu\mu})$ be a C^∞ function with values in $(\nu \times \mu)$-matrices, and put $f(x) = (f_1^1(x), \ldots, f_\mu^\nu(x))$. Let $g = (g^1, \ldots, g^\mu) \in W^s(\mathbb{T}^n, \mathbb{C}^\mu)$ be a vector-valued distribution with values in $\mathbb{C}^\mu$. Suppose that*

$$h^\lambda = \sum_{\rho=1}^{\mu} f_\rho^\lambda g^\rho$$

holds in the sense of the product of distributions. Then $h = (h^1, \ldots, h^\nu) \in W^s(\mathbb{T}^n, \mathbb{C}^\mu)$. Moreover there exists a positive constant C depending only on s such that the following inequality holds:

$$\|h\|_s \leqq C|f|_{|s|}\|g\|_s.$$

Proof. The above theorem is true for any $s \in \mathbb{R}$, but its proof is complicated. Therefore we shall give proof only for the case s is an integer here. This is enough for our present purpose.

(1°) The case $s = 0$. The assertion follows from the well-known inequality

$$\int_{\mathbb{T}^n} |f_\rho^\lambda(x) g^\rho(x)|^2 \, dx \leqq \max_x |f_\rho^\lambda(x)| \int_{\mathbb{T}^n} |g^\rho(x)|^2 \, dx$$

with $C = 1$.

The case $s \geqq 1$. For a multi-index α with $|\alpha| \leqq s$, we have by Leibniz' formula

$$D^\alpha(f_\rho^\lambda g^\rho) = \sum_{\beta \geqq \alpha} \binom{\alpha}{\beta} D^\beta f_\rho^\lambda D^{\alpha-\beta} g^\rho.$$

Hence letting $D^\beta f$ be the $(\nu\times\mu)$-matrix function with $D^\beta f^\lambda_\rho$ as its (λ,ρ)-component, and $D^{\alpha-\beta}g$ the $\mathbb{C}^\mu$-valued distribution with the ρth component $D^{\alpha-\beta}g^\rho$, and applying the result of the case $s=0$ to $D^\beta f$ and $D^{\alpha-\beta}g$ instead of f and g respectively, we obtain

$$\|D^\alpha h\|\leqq\sum_{\beta\leqq\alpha}\binom{\alpha}{\beta}|D^\beta f|_0\|D^{\alpha-\beta}g\|$$

by taking the summation with respect to β. Therefore, by Proposition 1.2, there exists a positive constant C_1 such that

$$n^{-s/2}\|h\|_s\leqq\Big(\sum_{|\alpha|\leqq s}\|D^\alpha h\|^2\Big)^{1/2}\leqq C_1|f|_s\|g\|_s.$$

(2°) The case $s=-l<0$. Take an arbitrary $\varphi\neq 0$ with $\varphi\in W^l(\mathbb{T}^n,\mathbb{C}^\nu)$. Using (1.39), we have

$$\|h\|_{-l}=\sup_\varphi\frac{|\langle h,\varphi\rangle|}{\|\varphi\|_l}=\sup_\varphi\frac{|\langle g,\psi\rangle|}{\|\varphi\|_l},$$

where $\psi=(\psi^1,\dots,\psi^\mu)$ and $\psi^\lambda(x)=\sum_{\rho=1}^{\nu}f^\rho_\lambda(x)\varphi^\rho(x)$. By (1°) there exists a positive constant C_1 such that

$$\|\psi\|_l\leqq C_1|f|_l\|\varphi\|_l$$

holds, hence combining with the inequality $|\langle g,\psi\rangle|\leqq\|g\|_{-l}\|\psi\|_l$, we obtain

$$\|h\|_{-l}\leqq C_1|f|_l\|g\|_{-l},$$

which completes the proof. ∎

In particular in case $s\geqq 0$,

$$h^\lambda(x)=\sum_\rho f^\lambda_\rho(x)g^\rho(x),\qquad \lambda=1,\dots,\nu, \tag{1.41}$$

hold for almost every x.

Theorem 1.12. *Let l be a positive integer with $l\geqq n+1$. Then for any $f\in W^l(\mathbb{T}^n,\mathbb{C}^{\mu\nu})$, and $g\in W^l(\mathbb{T}^n,\mathbb{C}^\mu)$, $h=(h^1,\dots,h^\nu)$ given by (1.41) is an element of $W^l(\mathbb{T}^n,\mathbb{C}^\nu)$. Moreover there exists a positive constant C depending only on l, n, μ, ν such that the following estimate holds:*

$$\|h\|_l\leqq C\|f\|_l\|g\|_l.$$

Proof. Using the estimate of $\|D^\alpha h\|$ for the multi-indices α with $|\alpha|\leqq l$, we can find an estimate of $\|h\|_l$. By Leibniz' formula we have

$$D^\alpha h = \sum_{|\beta|\leqq|\alpha|/2}\binom{\alpha}{\beta}D^\beta f^\lambda_\rho D^{\alpha-\beta}g^\rho + \sum_{|\beta|\geqq|\alpha|/2}\binom{\alpha}{\beta}D^\beta f^\lambda_\rho D^{\alpha-\beta}g^\rho.$$

Since $|\alpha|\leqq l$, if $|\beta|\leqq|\alpha|/2$, then $|\beta|+(n+1)/2\leqq l$, hence, by Sobolev's imbedding theorem, there exists a positive constant C independent of f such that on $\mathbb{T}^n$

$$|D^\beta f^\lambda_\rho(x)|\leqq C\|f\|_l$$

holds outside of a set of measure 0. Therefore we have

$$\|\sum_\rho D^\beta f^\lambda_\rho D^{\alpha-\beta}g^\rho\|\leqq C\|f\|_l\|D^{\alpha-\beta}g\|, \qquad |\beta|\leqq|\alpha|/2.$$

Similarly in case $|\beta|>|\alpha|/2$, since $|\alpha-\beta|<|\alpha|/2$, applying Sobolev's imbedding theorem to g^ρ, we obtain

$$\|\sum_\rho D^\beta f^\lambda_\rho D^{\alpha-\beta}g^\rho\|\leqq C\|D^\beta f\|\,\|g\|_l, \qquad |\beta|>|\alpha|/2.$$

Therefore

$$\|D^\alpha h\|\leqq C(\|f\|_l\|g\|_{|\alpha|}+\|f\|_{|\alpha|}\|g\|_l)\leqq C\|f\|_l\|g\|_l$$

holds. Thus by Proposition 1.2, we obtain

$$\|h\|_l\leqq C\|f\|_l\|g\|_l. \quad ∎$$

We define a linear partial differential operator $A(x, D)$ which operates on vector-valued distributions $S=(S^1,\dots,S^\mu)\in\mathcal{D}'(\mathbb{T}^n,\mathbb{C}^\mu)$ by

$$(A(x,D)S)^\lambda=\sum_\rho\sum_{|\alpha|\leqq m}a^\lambda_{\rho\alpha}(x)D^\alpha S^\rho, \qquad \lambda=1,\dots,\nu, \tag{1.42}$$

where the coefficients $a^\lambda_{\rho\alpha}(x)$ are C^∞ functions. If there are some α with $|\alpha|=m$, λ, ρ, and a point x such that $a^\lambda_{\rho\alpha}(x)\neq 0$, we say that $A(x, D)$ is a partial differential operator of order m. If $m=0$, $A(x, D)$ is called a multiplication operator. For a non-negative integer l, we put

$$M_l=\sum_{|\beta|\leqq l}\sum_{\rho,\lambda}\sum_\alpha\max_x|D^\beta a^\lambda_{\rho\alpha}(x)|. \tag{1.43}$$

Proposition 1.4. *Let $A(x, D)$ be a linear partial differential operator of order m with $a^\lambda_{\rho\alpha}(x)$ being C^∞ functions. Then for any $s\in\mathbb{R}$, there exists a positive number C depending only on s, n, m, ν and μ such that for any $S\in W^s(\mathbb{T}^n,\mathbb{C}^\mu)$,* *the following inequality holds*:

$$\|A(x, D)S\|_s \leqq CM_{|s|}\|S\|_{s+m}.$$

This follows from Lemma 1.3 and Proposition 1.3. ∎

Let $\varphi(x) \in C^\infty(\mathbb{T}^n, \mathbb{C})$, and φ the operator of multiplication by $\varphi(x)$: For $S \in \mathscr{D}'(\mathbb{T}^n, \mathbb{C}^\mu)$, $\varphi: S = (S^1, \ldots, S^\mu) \to (\varphi S^1, \ldots, \varphi S^\mu)$. We will often use the commutator of φ and $A(x, D)$ later:

$$[A(x, D), \varphi] = A(x, D)\varphi - \varphi A(x, D). \tag{1.44}$$

In terms of components, this is written as

$$([A(x, D), \varphi])^\lambda = \sum_\rho \sum_{|\alpha| \leqq m} a^\lambda_{\rho\alpha}(x)[D^\alpha, \varphi]S^\rho, \lambda = 1, \ldots, \mu. \tag{1.45}$$

Lemma 1.4.

(1°) *For a multi-index α, the commutator $[D^\alpha, \varphi]$ is a linear partial differential operator of order $|\alpha| - 1$ whose coefficients are derivatives of φ of order up to $|\alpha|$. For any $s \in \mathbb{R}$, there exists a positive constant C depending on s and α such that for any $S \in W^s(\mathbb{T}^n, \mathbb{C}^\mu)$, the following inequality holds*:

$$\|[D^\alpha, \varphi]S\|_s \leqq C|\varphi|_{|s|+|\alpha|}\|S\|_{s+|\alpha|-1}.$$

(2°) *For such a linear partial differential operator $A(x, D)$ as stated in Proposition 1.4, the commutator $[A(x, D), \varphi]$ is a linear partial differential operator of order $m-1$. For any $s \in \mathbb{R}$, there exists a positive number C depending on s, m, ν and μ such that for any $S \in W^s(\mathbb{T}^n, \mathbb{C}^\mu)$,*

$$\|[A(x, D), \varphi]S\|_s \leqq CM_{|s|}|\varphi|_{|s|+m}\|S\|_{s+m-1}$$

holds.

Proof. (1°) We proceed by induction on $|\alpha|$. First let $|\alpha| = 1$. Then

$$([D_j, \varphi]S)^\rho = D_j\varphi S^\rho, \quad j = 1, \ldots, n,$$

hence, there is a positive constant C such that

$$\|[D_j, \varphi]S\|_s \leqq C|D_j\varphi|_{|s|}\|S\|_s \leqq C|\varphi|_{|s|+1}\|S\|_s.$$

Thus in this case (1°) is true. Suppose that (1°) is proved for $|\alpha| \leqq l-1$. For a multi-index α of length $l-1$, and $j=1,\ldots,n$, we have

$$[D_jD^\alpha, \varphi]S = D_j[D^\alpha, \varphi]S + [D_j, \varphi]D^\alpha S,$$

hence, by the hypothesis of induction we obtain

$$\begin{aligned}\|[D_jD^\alpha, \varphi]S\|_s &\leqq \|[D^\alpha, \varphi]S\|_{s+1} + \|[D_j, \varphi]D^\alpha S\|_s \\ &\leqq C|\varphi|_{|s|+|\alpha|+1}\|S\|_{s+|\alpha|} + C|\varphi|_{s+1}\|S\|_{s+|\alpha|} \\ &\leqq C|d|_{s+l}\|S\|_{s+l-1},\end{aligned}$$

which proves (1°).

(2°) follows immediately from (1°). ∎

Next we examine the relation between the difference quotients and the Sobolev spaces. The results obtained will be used in §6.

Theorem 1.13.

(1°) *For $S \in W^s(\mathbb{T}^n, C^\mu)$, the difference quotient $\Delta_i^h S$ is again an element of $W^s(\mathbb{T}^n, \mathbb{C}^\mu)$.*

(2°) *For any h with $|h| \leqq 1$, if $S \in W^{s+1}(\mathbb{T}^n, \mathbb{C}^\mu)$, we have*

$$\|\Delta_j^h S\|_s \leqq \|S\|_{s+1}, \qquad j=1,\ldots,n. \tag{1.46}$$

(3°) *If $S \in W^s(\mathbb{T}^n, \mathbb{C}^\mu)$, and for any h with $0 \neq |h| \leqq 1$ and $j=1,\ldots,n$,*

$$\|\Delta_j^h S\|_s \leqq M \tag{1.47}$$

holds, then $S \in W^{s+1}(\mathbb{T}^n, \mathbb{C}^\mu)$ and

$$\|S\|_{s+1}^2 \leqq nM + \|S\|_s^2 \tag{1.48}$$

holds.

Proof. (1°) Let

$$S^\lambda = \sum_\xi S_\xi^\lambda \exp(i\xi \cdot x)$$

be the Fourier expansion of S. Translating S in the direction of x_j by $-h$, we obtain

$$\tau_j^{-h} S^\lambda = \sum_\xi S_\xi^\lambda \exp(i\xi_j h) \cdot \exp(i\xi \cdot x),$$

hence the Fourier coefficients of $\tau_j^{-h}S$ are given by $S_\xi^\lambda \exp(i\xi_j h)$. Since $|S_\xi^\lambda| = |S_\xi^\lambda \exp(i\xi_j h)|$, we have $\|\tau_j^{-h}S\|_s = \|S\|_s$. On the other hand, the difference quotient is given by

$$\Delta_j^h S = h^{-1}(\tau_j^{-h} S - S),$$

hence for $h \neq 0$, we have $\Delta_j^h S \in W^s(\mathbb{T}^n, \mathbb{C}^\mu)$.

(2°)
$$\Delta_j^h S^\lambda = \sum_\xi S_\xi^\lambda h^{-1}(\exp(ih\xi_j) - 1) \exp(i\xi \cdot x),$$

hence, using the inequality $|\exp(ih\xi_j) - 1| \leqq |h\xi_j|$, we obtain

$$\begin{aligned}\|\Delta_j^h S\|_s^2 &\leqq \sum_\lambda \sum_\xi |S_\xi^\lambda|^2 |\xi_j|^2 (1+|\xi|^2)^s \\ &\leqq \sum_\lambda \sum_\xi |S_\xi^\lambda|^2 (1+|\xi|^2)^{s+1} \\ &\leqq \|S\|_{s+1}^2.\end{aligned}$$

(3°)
$$\Delta_j^h S^\lambda = \sum_\xi S_\xi^\lambda p(h, \xi_j) \exp(i\xi \cdot x),$$

where we put

$$p(h, \xi_j) = h^{-1}(\exp(ih\xi_j) - 1).$$

By the hypothesis (1.47), we have

$$M \geqq \|\Delta_j^h S\|_s^2 = \sum_\lambda \sum_\xi |S_\xi^\lambda|^2 |p(h, \xi_j)|^2 (1+|\xi|^2)^s.$$

By Fatou's lemma,

$$\begin{aligned}M &\geqq \liminf_{h\to 0} \sum_\lambda \sum_\xi |S_\xi^\lambda|^2 |p(h, \xi_j)|^2 (1+|\xi|^2)^s \\ &\geqq \sum_\lambda \sum_\xi \liminf_{h\to 0} |S_\xi^\lambda|^2 |p(h, \xi_j)|^2 (1+|\xi|^2)^s \\ &= \sum_\lambda \sum_\xi |S_\xi^\lambda|^2 |\xi_j|^2 (1+|\xi|^2)^s.\end{aligned}$$

Therefore we obtain

$$\begin{aligned}\|S\|_{s+1}^2 &= \sum_\lambda \sum_\xi |S_\xi^\lambda|^2 (1+|\xi|^2)^s (1+|\xi_1|^2 + \cdots + |\xi_n|^2) \\ &\leqq \|S\|_s + nM^2. \quad \blacksquare\end{aligned}$$

Proposition 1.5. *Set* $0 < s' < 1$, *and let* $S \in W^s(\mathbb{T}^n, \mathbb{C}^\mu)$. *Then* $S \in W^{s+s'}(\mathbb{T}^n, \mathbb{C}^\mu)$ *if and only if there exists a positive number* δ *such that for* *each* $j = 1, \ldots, n$,

$$\int_{-\delta}^{\delta} \{\Delta_j^h S\|_s^2 |h|^{1-2s'} dh < \infty \tag{1.49}$$

holds. This being the case, there exists a positive constant $C = C(s', \delta)$ *depending on* s' *and* δ *such that the following inequality holds*:

$$C^{-1}\|S\|_{s+s'}^2 \leqq \|S\|_s^2 + \sum_j \int_{-\infty}^{\infty} \|\Delta_j^h S\|^2 |h|^{1-2s'} dh \leqq C\|S\|_{s+s'}.$$

Proof. Since $\|\tau_j^h S\|_s = \|S\|_s$, for $h > \delta$, we have

$$\|\Delta_j^h S\|_s^2 \leqq 2h^{-2}\|S\|_s.$$

Hence

$$\int_{|h|>\delta} \|\Delta_j^h S\|_s^2 h^{1-2s'} dh \leqq 2\|S\|_s^2 \int_{|h|>\delta} h^{-1-2s'} dh = \frac{\delta^{-2s'}}{s'}\|S\|_s^2.$$

Therefore the condition (1.49) is equivalent to the following inequality.

$$M = \int_{-\infty}^{\infty} \|\Delta_j^h S\|_s^2 h^{1-2s'} dh < \infty.$$

Let S_ξ^λ be the Fourier coefficients of S. Then we have

$$M \geqq \int_{-\infty}^{\infty} \sum_\lambda \sum_\xi |S_\xi^\lambda|^2 |\exp(ih\xi_j) - 1|^2 (1 + |\xi|^2)^s |h|^{-2s'-1} dh.$$

By Fubini's theorem, changing the order of the integration and $\sum_\lambda \sum_\xi$, we obtain

$$M \geqq \sum_\lambda \sum_\xi |S_\xi^\lambda|^2 (1 + |\xi|^2)^s a(\xi_j),$$

where we put

$$a(\xi_j) = \int_{-\infty}^{\infty} |\exp(ih\xi_j) - 1|^2 |h|^{-2s'-1} dh.$$

Thus $a(\xi_j) > 0$, and for $t \in \mathbb{R}$, we have

$$a(t\xi_j) = \int_{-\infty}^{\infty} |\exp(iht\xi_j) - 1|^2 |h|^{-2s'-1} dh = |t|^{2s'} a(\xi_j).$$

Therefore there is a positive constant $C = C(s')$ such that

$$a(\xi_j) = C(s')|\xi_j|^{2s'}.$$

Consequently there is another positive constant $C=C(s')$ such that

$$C(s')^{-1}(1+|\xi|^2)^{s'}\leqq 1+a(\xi_1)+\cdots+a(\xi_n)\leqq C(s')(1+|\xi|^2)^{s'},$$

$$\begin{aligned} C(s')\|S\|^2_{s+s'} &\leqq \sum_\lambda\sum_\xi |S^\lambda_\xi|^2(1+|\xi|^2)^s(1+a(\xi_1)+\cdots+a(\xi_n)) \\ &= \|S\|_s+\sum_j\int_{-\infty}^{\infty}\|\Delta^h_j S\|^2_s h^{1-2s'}\,dh \\ &\leqq C(s')\|S\|^2_{s+s'}. \end{aligned}$$

This completes the proof. ∎

The following proposition and Lemma 1.5 will be used in §2.

Proposition 1.6. *Let z be a point of $\mathbb{T}^n$, and $B_r(z)$ the ball of radius r with centre z: $B_r(z)=\{x\in\mathbb{T}^n\,|\,\sum_j(x_j-z_j)^2\leqq r\}$. Suppose $s\geqq 1$. For $\varphi\in W^s(\mathbb{T}^n,\mathbb{C}^\mu)$ with* supp $\varphi\subset B_r(z)$, *we put*

$$N_s(z,r)=\sup_{\varphi\neq 0}\frac{\|\varphi\|_{s-1}}{\|\varphi\|_s}.$$

Then $N_s(z,r)$ does not depend on z. If we put $N_s(z,r)=N_s(r)$, we have

$$\lim_{r\to 0}N_s(r)=0. \tag{1.50}$$

Proof. It is clear that $N_s(z,r)$ does not depend on z. Suppose that (1.50) is false. Then there is $\varepsilon_0>0$ such that for any positive integer l, there exists $\varphi_l\in W^s(\mathbb{T}^n,\mathbb{C}^\mu)$ with supp $\varphi_l\subset B_{1/\rho}(z)$ such that

$$\|\varphi_l\|_s=1,\qquad \|\varphi_l\|_{s-1}\geqq\varepsilon_0.$$

Since $\{\varphi_l\}$ is a bounded sequence in $W^s(\mathbb{T}^n,\mathbb{C}^\mu)$, by taking a subsequence if necessary, we may assume that $\{\varphi_l\}$ converges to φ in $W^{s-1}(\mathbb{T}^n,\mathbb{C}^\mu)$. Since $s-1\geqq 0$, we have $\varphi\in L^2(\mathbb{T}^n,\mathbb{C}^\mu)$, but supp $\varphi_l\subset B_{1/l}(z)=\{z\}$. Therefore $\varphi\equiv 0$ as an element of $L^2(\mathbb{T}^n,\mathbb{C}^\mu)$. On the other hand, since $\|\varphi_l\|_{s-1}\geqq\varepsilon_0$ in $W^{s-1}(\mathbb{T}^n,\mathbb{C}^\mu)$, we have $\|\varphi\|_{s-1}=\lim_{l\to\infty}\|\varphi_l\|_{s-1}\geqq\varepsilon_0$, that is, $\varphi\neq 0$ as an element of $W^{s-1}(\mathbb{T}^n,\mathbb{C}^\mu)$. Since $W^{s-1}(\mathbb{T}^n,\mathbb{C}^\mu)\hookrightarrow L^2(\mathbb{T}^n,\mathbb{C}^\mu)$, we must have $\varphi\equiv 0$ in $L^2(\mathbb{T}^n,\mathbb{C}^\mu)$, too. This is a contradiction. Thus we proved (1.50). ∎

Lemma 1.5. *Let $a\in C^\infty(\mathbb{T}^n,\mathbb{C}^\mu)$, and suppose that $a(z)=0$. We denote by $B_r(z)$ the ball of radius r with centre z.*

(i) *Let $l\geqq 0$ be an integer. Then there is a positive function $R(r,l)$ such* *that for any* $\varphi \in W^l(\mathbb{T}^n, \mathbb{C}^\mu)$ *with* $\operatorname{supp}\varphi \subset B_r(z)$,

$$\|a\varphi\|_l \leqq R(r, l)\|\varphi\|_l$$

holds. Here we may choose $R(r, l)$ *so that*

$$R(r, 0) \leqq r|a|_1, \qquad R(r, l) \leqq r|a|_1 + N_{|l|}(r)|a|_{|l|}, \qquad l \neq 0,$$

hold. Moreover we have $\lim_{r\to 0} R(r, l) = 0$.

(ii) *If* $l = -k$ *is a negative integer, there exists a positive-valued function* $C(r, l)$ *such that for any* $\varphi \in W^l(\mathbb{T}^n, \mathbb{C}^\mu)$ *with* $\operatorname{supp}\varphi \subset B_r(z)$,

$$\|a\varphi\|_l \leqq 2r|a|_1\|\varphi\|_l + C(r, l)|a|_{2|l|}\|\varphi\|_{l-1}$$

holds.

Proof. (i) Note that for $x \in B_r(z)$,

$$|a(x)| \leqq r|a|_1$$

holds. First suppose $l \geqq 1$. Then since l is a positive integer, there is a positive constant $C = C(l)$ such that

$$\begin{aligned}\|a\varphi\|_l^2 &\leqq n^{1/2} \sum_{|\alpha|\leqq l} \|D^\alpha a\varphi\|^2 = 2n^{1/2} \sum_{|\alpha|\leqq l} (\|aD^\alpha\varphi\|^2 + \|[a, D^\alpha]\varphi\|^2) \\ &\leqq C(r|a|_1\|\varphi\|_l + |a|_l\|\varphi\|_{l-1}).\end{aligned}$$

Using Proposition 1.6, we obtain

$$\|a\varphi\|_l \leqq C(r|a|_1 + N_l(r)|a|_l)\|\varphi\|_l,$$

hence we have

$$R(r, l) \leqq C(r|a|_1 + N_l(r)|a|_l).$$

In case $l = 0$, $\|\varphi\|_l \leqq r|a|_1\|\varphi\|$, hence we have

$$R(r, 0) = r|a|_1.$$

(ii) First we assume that $l = -k$ is even. Put $\Delta = D_1^2 + \cdots + D_n^2$, and let S_ξ^λ be the Fourier coefficients of $S \in W^{-k}(\mathbb{T}^n, \mathbb{C}^\mu)$. Then since the Fourier coefficients of $(1-\Delta)^{-k/2}S$ are given by $(1+|\xi|^2)^{-k/2}S_\xi^\lambda$, for any integer m, we have

$$(*) \qquad \|S\|_{m-k}^2 = \|(1-\Delta)^{-k/2}S\|_m^2.$$

Take a function $f \in C^\infty(\mathbb{T}^n)$ so that $0 \leqq f(x) \leqq 1$, $f(x) \equiv 1$ on $B_r(z)$ and that $f(x) \equiv 0$ outide of $B_{2r}(z)$. Then $f\varphi \equiv \varphi$. From this and (*), we have

$$\|a\varphi\|_{-k} = \|(1-\Delta)^{-k/2} af\varphi\| = A + B,$$

where $A = \|af(1-\Delta)^{-k/2}\varphi\|$, and $B = \|[af, (1-\Delta)^{-k/2}]\varphi\|$. Since $\operatorname{supp} f \subset B_{2r}(z)$, using (*) we obtain

$$A \leqq 2r|a|_1 \|(1-\Delta)^{-k/2}\varphi\| \leqq 2r|a|_1 \|\varphi\|_{-k}.$$

On the other hand, since

$$[af, (1-\Delta)^{-k/2}] = (1-\Delta)^{-k/2}[af, (1-\Delta)^{k/2}](1-\Delta)^{-k/2},$$

we obtain

$$B \leqq \|[af, (1-\Delta)^{k/2}](1-\Delta)^{-k/2}\varphi\|_{-k} \leqq C|af|_{2k} \|(1-\Delta)^{-k/2}\varphi\|_{-1},$$

where we use Lemma 1.4. By (*) and Theorem 1.12, we have

$$B \leqq C(r)|a|_{2k} \|\varphi\|_{-k-1}, \qquad C(r) = C|f|_{2k}.$$

Combining the estimate of A, we complete the proof.

If $l = -k$ is odd, we have only to use

$$\|S\|_{-k}^2 = \|(1-\Delta)^{-(k+1)/2} S\|^2 + \sum_{j=1}^{n} \|D_j(1-\Delta)^{-(k+1)/2} S\|^2$$

instead of (*). ∎

Finally we prove the following proposition.

Proposition 1.7. *Let K be a compact set of $\mathbb{T}^n$, and let $\{U_j\}_{j=1}^J$ a finite open covering of K. Take $\omega_j(x) \in C^\infty(\mathbb{T}^n, \mathbb{C})$ with $\operatorname{supp} \omega_j \subset U_j$ so that on some open set G containing K,*

$$\Omega(x) = \sum_j^J \omega_j(x) \neq 0.$$

Then for any s, there exists a positive constant C such that for any $S \in W^s(\mathbb{T}^n, \mathbb{C}^\mu)$ with $\operatorname{supp} S \subset K$,

$$C^{-1}\|S\|_s \leqq \sum_{j=1}^{J} \|\omega_j S\|_s \leqq C\|S\|_s$$

holds. $C = C(s)$ may depend on the choice of G and ω.

Proof. Since there is a constant C with $\|\omega_j S\|_s \leqq C\|S\|_s$, the second inequality is obvious. The first inequality follows from the existence of positive numbers C and C' such that

$$\|S\|_s = \|\Omega(x)^{-1}\Omega(x)S\|_s \leqq C\|\Omega S\|_s \leqq C' \sum_j \|\omega_j S\|_s. \quad \blacksquare$$

§2. Elliptic Partial Differential Operators on a Torus

(A) Estimates by the Sobolev Norm

(a) Elliptic Operators with Constant Coefficients

We set

$$(A(D), \varphi)^\lambda = \sum_{|\alpha| \leqq l,\ \rho=1}^{\mu} a_{\rho\alpha}^\lambda D^\alpha \varphi^\rho, \qquad \lambda = 1, \ldots, \mu. \tag{2.1}$$

$A(D)$ is called a linear partial differential operator with constant coefficients, operating on vector-valued distributions with values in $\mathbb{C}^\mu$. When $a_{\rho\alpha}^\lambda \neq 0$ for some α with $|\alpha| = l$, we say that $A(D)$ is of order l. The sum of the terms with $|\alpha| = l$

$$A_l(D) = \sum_{|\alpha|=l} \sum_\rho a_{\rho\alpha}^\lambda D^\alpha \tag{2.2}$$

is called the principal part of $A(D)$. For any $\xi \in \mathbb{Z}^n$ and $w = (w^1, \ldots, w^\mu) \in \mathbb{C}^\mu$, we have $\varphi = (\exp(i\xi \cdot x))w \in D(\mathbb{T}^n, \mathbb{C}^\mu)$. Let $w' = (w'^1, \ldots, w'^\mu)$ be defined as

$$w' = \exp(-i\xi \cdot x)A(D)\{\exp(i\xi \cdot x)w\}. \tag{2.3}$$

Then we have

$$w'^\lambda = \sum_{|\alpha| \leqq l} \sum_\rho a_{\rho\alpha}^\lambda (i\xi)^\alpha w^\rho.$$

The correspondence $w \to w'$ gives a $(\mu \times \mu)$-matrix valued polynomial $A(i\xi)$ in ξ of degree l. Let $A(i\xi) = (a(i\xi)_\rho^\lambda)$. Then

$$w'^\lambda = (A(i\xi)w)^\lambda = \sum_{|\alpha| \leqq l} \sum_\rho a_{\rho\alpha}^\lambda i^{|\alpha|} \xi^\alpha w^\rho, \tag{2.4}$$

namely,

$$a(i\xi)_\rho^\lambda = \sum_{|\alpha| \leqq l} a_{\rho\alpha}^\lambda (i\xi)^\alpha.$$

$A(i\xi)$ is called the characteristic polynomial of $A(D)$.

We obtain from $A_l(D)$ the matrix $A_l(i\xi)$ of homogeneous polynomials of degree l, whose (ρ, λ)-component is equal to $\sum_{|\alpha|=l} a^{\lambda}_{\rho\alpha}(i\xi)^{\alpha}$. $A_l(i\xi)$ is called *the principal symbol* of $A(D)$.

Definition 2.1. When the principal symbol $A_l(i\xi)$ is a regular $(\mu\times\mu)$-matrix for every $s\in\mathbb{Z}^n$, we say that $A(D)$ is *of elliptic type.* Besides the supremum of positive constants satisfying

$$\left(\sum_{\lambda}|w'^{\lambda}|^2\right)^{1/2}\geqq\delta|\xi|^l\left(\sum_{\lambda}|w^{\lambda}|^2\right)^{1/2} \tag{2.5}$$

is called *the constant of ellipticity of* $A(D)$, where we put $w'^{\lambda}=(A_i(i\xi)w)^{\lambda}$.

Lemma 2.1. *Suppose that $A_l(D)$ is an elliptic differential operator of order l which consists of only its principal part, and let δ_0 be its constant of ellipticity. Then for any $s\in\mathbb{R}$ and any $\varphi\in W^{s+l}(\mathbb{T}^n,\mathbb{C}^{\mu})$, the following inequality holds:*

$$\|A_l(D)\varphi\|_s^2+\delta_0^2\|\varphi\|_s^2\geqq 2^{-l}\delta_0^2\|\varphi\|_{s+l}^2. \tag{2.6}$$

Proof. Let

$$\varphi^{\rho}=\sum_{\xi}\varphi^{\rho}_{\xi}\exp(i\xi\cdot x),\qquad \rho=1,\dots,\mu,$$

be the Fourier expansion of $\varphi=(\varphi^1,\dots,\varphi^{\mu})\in W^s(\mathbb{T}^n,\mathbb{C}^{\mu})$. Then, using (2.2), we have

$$\begin{aligned}(A_l(D)\varphi)^{\lambda}&=\sum_{\xi}\sum_{|\alpha|=l}\sum_{\rho}a^{\lambda}_{\rho\alpha}(i\xi)^{\alpha}\varphi^{\rho}_{\xi}\exp(i\xi\cdot x)\\&=i^l\sum_{\xi}(A_l(\xi)\varphi_{\xi})^{\lambda}\exp(i\xi\cdot x),\qquad \lambda=1,\dots,\mu.\end{aligned}$$

Hence, by the definition of the constant of ellipticity, we have

$$\begin{aligned}\|A_l(D)\varphi\|_s^2&=\sum_{\lambda}\sum_{\xi}|(A_l(\xi)\varphi_{\xi})^{\lambda}|^2(1+|\xi|^2)^s\\&\geqq\delta_0^2\sum_{\xi}\sum_{\lambda}|\varphi^{\lambda}_{\xi}|^2(1+|\xi|^2)^s|\xi|^{2l}.\end{aligned}$$

Therefore we have

$$\begin{aligned}\|A_l(D)\varphi\|_s^2+\delta_0^2\|\varphi\|_s^2&\geqq\delta_0^2\sum_{\xi}\sum_{\lambda}|\varphi^{\lambda}_{\xi}|^2(1+|\xi|^2)^s(1+|\xi|^{2l})\\&\geqq 2^{-l}\delta_0^2\|\varphi\|_{s+l}^2.\end{aligned}$$

Thus we have proved Lemma 2.1. ∎

In (2.1) put $l=2m$. Then by Proposition 1.3 we find that for any $\varphi \in W^m(\mathbb{T}^n, \mathbb{C}^\mu)$, $A(D)\varphi \in W^{-m}(\mathbb{T}^n, \mathbb{C}^\mu)$. Therefore by Theorem 1.11 (2°), we can define a Hermitian form on $W^m(\mathbb{T}^n, \mathbb{C}^\mu)$ by putting

$$\langle A(D)\varphi, \psi\rangle = \sum_{|\alpha|\leqq 2m} \sum_\lambda \sum_\rho \langle a^\lambda_{\rho\alpha} D^\alpha \varphi^\rho, \bar{\psi}^\lambda\rangle \tag{2.7}$$

for $\varphi, \psi \in W^m(\mathbb{T}^n, \mathbb{C}^\mu)$. Especially for $\varphi, \psi \in C^\infty(\mathbb{T}^n, \mathbb{C}^\mu)$, we have

$$\langle A(D)\varphi, \bar{\psi}\rangle = (A(D)\varphi, \psi) \tag{2.8}$$

where (,) denotes the inner product of $L^2(\mathbb{T}^n, \mathbb{C}^\mu)$. Since

$$(A(D)\varphi, \psi) = \sum_\lambda \sum_{|\alpha|\leqq 2m} \sum_\rho a^\lambda_{\rho\alpha} \int_{\mathbb{T}^n} D^\alpha \varphi^\rho(x) \bar{\psi}^l(x)\, dx,$$

we can rewrite it by partial integration as

$$(A(D)\varphi, \varphi) = \sum_\lambda \sum_{|\alpha|,|\beta|\leqq m} \sum_\rho b^\lambda_{\rho\alpha\beta} \int_{\mathbb{T}^n} D^\alpha \varphi^\rho(x) \overline{D^\beta \psi(x)}\, dx. \tag{2.9}$$

Here there appear no derivatives of order greater than m in φ nor in ψ. Of course such an expression is not determined uniquely by $A(D)$. We call $\langle A(D)\varphi, \bar{\psi}\rangle$ or $(A(D)\varphi, \psi)$ the bilinear form associated to $A(D)$. This is a continuous bilinear form on $W^m(\mathbb{T}^n, \mathbb{C}^\mu)$, i.e. a bilinear map form $W^m(\mathbb{T}^n, \mathbb{C}^\mu) \times W^m(\mathbb{T}^n, \mathbb{C}^\mu)$ to $\mathbb{C}$.

For $w = (w^1, \ldots, w^\mu) \in \mathbb{C}^\mu$ and $w' \in \mathbb{C}^\mu$, set

$$\varphi = \varphi^{i\xi\cdot x} w = (e^{i\xi\cdot x} w^1, \ldots, e^{i\xi\cdot x} w^\mu),$$

where $\xi \in \mathbb{Z}^n$. We define a bilinear form as follows:

$$\begin{aligned}(A(D)\, e^{i\xi\cdot x} w, e^{i\xi\cdot x} w') &= (e^{-i\xi\cdot x} A(D)\, e^{i\xi\cdot x} w, w') \\ &= \sum_\lambda (A(i\xi)w)^\lambda \bar{w}'^\lambda.\end{aligned} \tag{2.10}$$

This is the Hermitian form on $\mathbb{C}^\mu$ associated to the matrix of characteristic polynomial of $A(D)$. The sum of the terms of the highest order is equal to

$$(-1)^m \sum_x (A_{2m}(\xi)w)^\lambda \bar{w}'^\lambda \tag{2.11}$$

which is the Hermitian form associated to the matrix of the principal symbol of $A(D)$.

Definition 2.2. If the real part of the Hermitian form (2.11) associated to the matrix of the principal symbol of the linear partial differential operator (2.1) is a positive definite form, we say that the differential operator $A(D)$ is *of strongly elliptic type*. The maximum of such positive numbers δ that

$$(-1)^m \operatorname{Re}\sum_{\xi}(A_{2m}(\xi)w)^\lambda \bar{w}^\lambda \geqq \delta^2|\xi|^{2m}\sum_{\lambda}|w^\lambda|^2 \tag{2.12}$$

is called *the constant of strong ellipticity*.

Lemma 2.2. *If a strongly elliptic partial differential operator with constant coefficients $A_{2m}(D)$ of order $2m$ contains no terms of order $\leqq 2m-1$, then we get, for $\varphi \in W^n(\mathbb{T}^n, \mathbb{C}^\mu)$,*

$$\operatorname{Re}(A_{2m}(D)\varphi, \varphi) + \delta_0^2\|\varphi\|^2 \geqq 2^{-m}\delta_0^2\|\varphi\|_m^2, \tag{2.13}$$

where δ_0 is the constant of strong ellipticity.

Proof. For $\varphi = (\varphi^1, \dots, \varphi^\mu) \in W^m(\mathbb{T}^n, \mathbb{C}^\mu)$, let

$$\varphi^\lambda = \sum_{\xi}\varphi_\xi^\lambda \exp(i\xi\cdot x), \qquad \lambda = 1, 2, \dots, \mu$$

be the Fourier expansion of φ. Using (2.10), we have

$$\begin{aligned}\operatorname{Re}(A_{2m}(D)\varphi, \varphi) &= (-1)^m \operatorname{Re}\sum_{\xi}\sum_{\lambda}(A(\xi)\varphi_\xi)^\lambda \bar{\varphi}_\xi^\lambda \\ &\geqq \delta_0^2\sum_{\xi}|\xi|^{2m}|\varphi_\xi^\lambda|^2\end{aligned}$$

since $A(D)$ consists only of the principal part. Therefore we have

$$\begin{aligned}\operatorname{Re}(A_{2m}(D)\varphi, \varphi) + \delta_0\|\varphi\|^2 &\geqq \delta_0^2\sum_{\lambda}\sum_{\xi}(1+|\xi|^{2m})|\varphi_\xi^\lambda|^2 \\ &\geqq 2^{-m}\delta_0^2\|\varphi\|_m^2.\end{aligned}$$

This completes the proof. ∎

(b) Elliptic Linear Partial Differential Operators with Variable Coefficients

Let U be a domain of $\mathbb{T}^n$. We define a differential operator $A(x, D)$ with C^∞ coefficients defined on an open neighbourhood of $[U]$ as follows: For $\varphi \in C^\infty(U, \mathbb{C}^\mu)$,

$$(A(x, D)\varphi)^\lambda(x) = \sum_{\rho}\sum_{|\alpha|\leqq l} a_{\rho\alpha}^\lambda(x) D^\alpha \varphi^\rho(x), \tag{2.14}$$

where $a^{\lambda}_{\rho\alpha}(x)$ is a complex-valued C^{∞} function defined on an open neighbourhood of $[U]$, and for some index α with $|\alpha| = l$, $a^{\lambda}_{\rho\alpha}(x)$ does not vanish identically. l is called the order of $A(x, D)$. The part of the highest order $A_l(x, D)$ of $A(x, D)$ is given by

$$(A_l(x, D)\varphi(x))^{\lambda} = \sum_{\rho} \sum_{|\alpha|=l} a^{\lambda}_{\rho\alpha}(x) D^{\alpha}\varphi^{\rho}(x). \tag{2.15}$$

We call $A_l(x, D)$ the principal part of $A(x, D)$.

Let $w = (w^1, \ldots, w^{\mu}) \in \mathbb{C}^{\mu}$, $f(x)$ a real-valued C^{∞} function defined on U, and t a positive number, and put

$$e^{itf(x)}w = (e^{itf(x)}w^1, \ldots, e^{itf(x)}w^{\mu}).$$

Substituting this for φ, we see that

$$e^{-itf(x)}A(x, D)(e^{itf(x)}w) \tag{2.16}$$

is a polynomial of degree l in t. The term of degree l in t comes from the principal part, and its coefficient is given by

$$(A_l(x, \xi_x)w)^{\lambda} = \sum_{\rho} \sum_{|\alpha|=l} a^{\lambda}_{\rho\alpha}(x)(i\xi_x)^{\alpha} w^{\rho}, \tag{2.17}$$

where $\xi_x = (\xi^1_x, \ldots, \xi^n_x)$ is given by

$$df(x) = \xi^1_x \, dx^1 + \cdots + \xi^n_x \, dx^n, \tag{2.18}$$

which is a cotangent vector of U at x. Namely, $A_l(x, \xi_x)$ is a $(\mu \times \mu)$-matrix which is determined by giving a point $(x, df(x))$ on the cotangent bundle of U. This $(\mu \times \mu)$-matrix-valued C^{∞} function $A_l(x, i\xi_x)$ defined on the cotangent bundle T^*U is called the *principal symbol of* $A(x, D)$. If we take $\xi \cdot x$ as a function $f(x)$, the value of the principal symbol is equal to

$$(A_l(x, i\xi)w)^{\lambda} = \sum_{\rho} \sum_{|\alpha|=l} a^{\lambda}_{\rho\alpha}(x)(i\xi) w^{\rho} \tag{2.19}$$

and this is the part of

$$e^{-ix\cdot\xi}A_l(x, D)(e^{ix\cdot\xi}w) \tag{2.20}$$

of order l with respect to ξ.

Definition 2.3. Suppose a linear partial differential operator $A(x, D)$ of order l, is defined on a neighbourhood of $[U]$. We say that $A(x, D)$ is *of elliptic type*, if there is a positive number δ such that for any point (x, ξ_x) T^*U

with $\xi_x \neq 0$ and for $w \in \mathbb{C}^\mu$, the principal symbol satisfies the inequality

$$\sum_\lambda |A_l(x, \xi_x)w)^\lambda|^2 \geqq \delta^2 |\xi_x|^{2l} \sum_\lambda |w^\lambda|^2. \tag{2.21}$$

The maximum of such δ is called *the constant of ellipticity.*

In other words, $A(x, D)$ is of elliptic type if and only if for every point $z_0 \in [U]$, the differential operator $A(z_0, D)$ derived from $A(x, D)$ by replacing the coefficients $a^\lambda_{\rho\alpha}(x)$ of $A(x, d)$ by $a^\lambda_{\rho\alpha}(z_0)$ is of elliptic type in the sense of Definition 2.1.

We want to show the corresponding fact to Lemma 2.1 concerning elliptic differential operators with variable coefficients. We introduce the following quantity: for $k = 0, 1, 2, \ldots,$

$$M_k = \sum_{\rho,\lambda} \sum_\alpha \sum_{|\beta| \leqq k} \max_{x \in [U]} |D^\beta a^\lambda_{\rho\alpha}(x)|. \tag{2.22}$$

Also we use the following notation:

$$C_0^\infty(U, \mathbb{C}^\mu) = \{\varphi \in C^\infty(\mathbb{T}^n, \mathbb{C}^\mu) \mid \varphi(x) = 0 \text{ for any } x \in U\},$$

$$W_0^s(U, \mathbb{C}^\mu) = \{\text{the closure of } C_0^\infty(U, \mathbb{C}^\mu) \text{ in } W^s(\mathbb{T}^n, \mathbb{C}^\mu)\}.$$

Theorem 2.1 (Local Version of the *a priori* Estimate). *Let $A(x, D)$ be an elliptic linear partial differential operator defined on a neighbourhood of $[U]$ and let δ_0 be its constant of ellipticity. Then for $s \in \mathbb{Z}$ there is a positive constant $C = C(s, \delta, M_{|s|})$ such that for any $\varphi \in W_0^{s+1}(U, \mathbb{C}^\mu)$*

$$\|\varphi\|_{s+l} \leqq C(\|A(x, D)\varphi\|_s + \|\varphi\|_s). \tag{2.23}$$

Proof. In order to prove this theorem, it suffices to show (2.23) for any $\varphi \in C_0^\infty(U, \mathbb{C}^\mu)$. We choose a sufficiently small ε such that

$$2\varepsilon n^l \mu M_1 < 2^{-2-l/2}\delta_0.$$

Cover $[U]$ with open balls $B_1, \ldots, B_J$ of radius ε. Furthermore we choose real-valued C^∞ functions $\omega_j(x), j = 1, \ldots, J,$ defined on $\mathbb{T}^n$ such that $\operatorname{supp} \omega_j \subset B_j$ and that $\sum_j \omega_j(x) \equiv 1$ on $[U]$. For any $\varphi \in C_0^\infty(U, \mathbb{C}^\mu)$, we set $\varphi_j(x) = \omega_j(x)\varphi(x)$. Since

$$\|\varphi\|_{s+l} = \|\sum_j \omega_j \varphi\|_{s+l} = \|\sum_j \varphi_j\|_{s+l} \leqq \sum_j^J \|\varphi\|_{s+l},$$

there exists some j such that the following inequality holds:

$$\|\varphi_j\|_{s+l} \geqq J^{-1} \|\varphi\|_{s+l}. \tag{2.24}$$

Let p_j be the centre of the ball B_j. We write the principal part $A_l(x, D)$ of $A(x, D)$ as in (2.15), and let $A_l(p_j, D)$ be the partial differential operator with constant coefficients obtained from $A_l(x, D)$ by replacing its coefficients $a^\lambda_{\rho\alpha}(x)$ by their values $a^\lambda_{\rho\alpha}(p_j)$ at p_j. This is an elliptic differential operator with δ_0 as the constant of ellipticity. Hence (2.6) holds. Therefore we obtain the following:

$$\|A_l(p_j, D)\varphi_j\|_s + \delta_0\|\varphi_j\|_s \geqq 2^{-l/2}\delta_0\|\varphi_j\|_{s+l}. \tag{2.25}$$

On the other hand, since

$$((A_l(x, D) - A_l(p_j, D))\varphi_j)^\lambda(x) = \sum_{|\alpha|=l} \sum_\rho (a^\lambda_{\rho\alpha}(x) - a^\lambda_{\rho\alpha}(p_j))D^\alpha\varphi^\rho_j(x),$$

by Lemma 1.5 in §1, there exists a positive constant $C(\varepsilon)$ such that

$$\|\sum_\rho (a^\lambda_{\rho\alpha}(x) - a^\lambda_{\rho\alpha}(p_j))D^\alpha\varphi^\rho_j\|_s \leqq 2\varepsilon M_1\|\varphi_j\|_{s+1} + C(\varepsilon)M_{2|s|}\|\varphi_j\|_{s+l-1}.$$

Therefore by the choice of ε,

$$\begin{aligned}
&\|(A_l(x, D) - A_l(p_j, D))\varphi_j\|_s \\
&\quad \leqq n^l\mu(2\varepsilon M_1\|\varphi_j\|_{s+l} + C(\varepsilon)M_{2|s|}\|\varphi_j\|_{s+l-1}) \\
&\quad \leqq 2^{-2-l/2}\delta_0\|\varphi_j\|_{s+l} + n^l\mu C(\varepsilon)M_{2|s|}\|\varphi_j\|_{s+l-1}.
\end{aligned} \tag{2.26}$$

Further, since $A(x, D) - A_l(x, D)$ is a differential operator of order $\leqq l-1$, there is a positive constant C_1 depending only on s, l, μ and n such that

$$\|(A_l(x, D) - A(x, D))\varphi_j\|_s \leqq C_1M_{|s|}\|\varphi\|_{s+l-1}.$$

Combining this with (2.26), we have

$$\begin{aligned}
\|A_l(p_j, D)\varphi\|_s &\leqq \|A(x, D)\varphi_j + (A_l(x, D) - A(x, D))\varphi_j \\
&\qquad - (A_l(x, D) - A_l(p_j, D))\varphi_j\|_s \\
&\leqq \|A(x, D)\varphi_j\|_s + C_1M_{|s|}\|\varphi_j\|_{s+l-1} \\
&\qquad + n^l\mu C(\varepsilon)M_{2|s|}\|\varphi_j\|_{s+l-1} + 2^{-2-l/2}\delta_0\|\varphi_j\|_{s+l}.
\end{aligned} \tag{2.27}$$

Since the commutator $[A(x, D), \omega]$ is a differential operator of order $\leqq l-1$, there is a positive constant C_2 such that

$$\|[A(x, D), \omega_j]\varphi_j\|_s \leqq C_2M_{|s|}\|\varphi\|_{s+l-1}.$$

Hence

$$\begin{aligned}\|A(x, D)\varphi_j\|_s &\leqq \|\omega_j A(x, D)\varphi\|_s + \|[A(x, D), \omega_j]\varphi\|_s \\ &\leqq \|\omega_j A(x, D)\varphi\|_s + C_2 M_{|s|}\|\varphi\|_{s+l-1}.\end{aligned} \tag{2.28}$$

Since there exists a positive constant C_3 depending only on s, l, n, μ such that

$$\|\omega_j A(x, D)\varphi\|_s \leqq C_3\|A(x, D)\varphi\|_s,$$

by (2.25), (2.27), (2.28) and the above inequality, we obtain

$$\begin{aligned}2^{-l/2}\delta_0\|\varphi_j\|_{s+l} \leqq\ & C_3\|A(x, D)\varphi\|_s + \delta_0\|\varphi_j\|_s + C_1 M_{|s|}\|\varphi_j\|_{s+l-1} \\ & + C_2 M_{|s|}\|\varphi\|_{s+l-1} + n^l\mu C(\varepsilon) M_{2|s|}\|\varphi_j\|_{s+l-1} \\ & + 2^{-2-l/2}\delta_0\|\varphi_j\|_{s+l}.\end{aligned}$$

By transposing the last term of the right-hand side, we obtain

$$\begin{aligned}3\cdot 2^{-2-l/2}\delta_0\|\varphi_j\|_{s+l} \leqq\ & C_3\|A(x, D)\varphi\|_s + \delta_0\|\varphi_j\|_s \\ & + C_1 M_{|s|}\|\varphi_j\|_{s+l-1} + C_2 M_{|s|}\|\varphi\|_{s+l-1} \\ & + n^l\mu C(\varepsilon) M_{2|s|}\|\varphi_j\|_{s+l-1}.\end{aligned} \tag{2.29}$$

Since we can take a positive constant C_4 such that

$$\|\varphi_j\|_s \leqq C_4\|\varphi\|_s, \qquad \|\varphi_j\|s_{+l-1} \leqq C_4\|\varphi\|_{s+l-1}$$

we get by (2.24) and (2.29),

$$3\cdot 2^{-2-l/2}\delta_0 J^{-1}\|\varphi\|_{s+l} \leqq C_3\|A(x, D)\varphi\|_s + C_4\delta_0\|\varphi\|_s + C_5\|\varphi\|_{s+l-1} \tag{2.30}$$

where we put $C_5 = C_1 C_4 M_{|s|} + C_2 M_{|s|} + n^l\mu C(\varepsilon) M_{2|s|}$. Taking a suitable t in Proposition 1.1, we obtain a positive constant C_6 depending only on s, l, n and μ such that

$$C_5\|\varphi\|_{s+l-1} \leqq 2^{-2-l/2}\delta_0 J^{-1}\|\varphi\|_{s+l} + C_6\|\varphi\|_s$$

holds. Therefore by this and (2.30), we obtain

$$2^{-1-l/2}\delta_0 J^{-1}\|\varphi\|_{s+l} \leqq C_3\|A(x, D)\varphi\|_s + (C_4\delta_0 + C_6)\|\varphi\|_s,$$

which completes the proof. ∎

Put $l=2m$ and let $w=(w^1,\ldots,w^\mu)$ and $w'=(w'^1,\ldots,w'^\mu)$ be vectors in $\mathbb{C}^\mu$. For each $(x,\xi_x)\in T^*U$, we can define a Hermitian form in w, w'

$$(-1)^m\sum_\lambda (A_{2m}(x,\xi_x)w)^\lambda \bar{w}'^\lambda$$

associated to the $(\mu\times\mu)$-matrix $A_{2m}(x,i\xi_x)$ for the principal symbol of $A(x,D)$.

Definition 2.4. We say that $A(x,D)$ is *of strongly elliptic type* if the above Hermitian form associated to the principal symbol satisfies the following condition: There is a positive number δ such that for any $(x,\xi_x)\in T^*U$, $\xi_x\neq 0$ and for any $w\in\mathbb{C}^\mu$,

$$(-1)^m \operatorname{Re}\sum_\lambda (A_{2m}(x,\xi_x)w)^\lambda \bar{w}^\lambda \geqq \delta^2|\xi_x|^{2m}\sum_\lambda |w^\lambda|^2. \tag{2.31}$$

The supremum of such δ is called *the constant of strong ellipticity.*

Theorem 2.2 (Local Version of Gårding's Inequality). *Let U be a domain of $\mathbb{T}^n$. Suppose that a linear partial differential operator depending on a neighbourhood of $[U]$ is of strongly elliptic type and that δ is its constant of strong ellipticity. Then there exist positive constants δ_1 and δ_2 depending only on δ, m, μ and M_m such that for any $\varphi\in W_0^m(U,\mathbb{C}^\mu)$*

$$\operatorname{Re}(A(x,D)\varphi,\varphi)+\delta_1^2\|\varphi\|^2\geqq \delta_2^2\|\varphi\|_m^2 \tag{2.32}$$

holds.

Proof. For any $\varphi,\psi\in C_{\mathrm{T}}^\infty(U,\mathbb{C}^\mu)$,

$$(A(x,D)\varphi,\psi)=\sum_{|\alpha|\leqq 2m}\sum_{\rho,\lambda}\int_{\mathbb{T}^n} a_{\rho\alpha}^\lambda(x)D^\alpha\varphi^\rho(x)\overline{\psi^\lambda(x)}\,dx.$$

By integrating the terms with $|\alpha|\geqq m+1$ by parts $(|\alpha|-m)$-times, we can rewrite this into the equality containing no partial derivatives of order $\geqq m+1$ as follows:

$$(A(x,D)\varphi,\psi)=\sum_{|\alpha|\leqq m,|\beta|\leqq m}\sum_{\rho,\lambda}\int_{\mathbb{T}^n} b_{\rho\alpha\beta}^\lambda(x)D^\alpha\varphi^\rho(x)\overline{D^\beta\psi^\lambda(x)}\,dx, \tag{2.33}$$

where $b_{\rho\alpha\beta}^\lambda(x)$ is a linear combination of derivatives of order at most m of $a_{\rho'\alpha'}^{\lambda'}(x)$. The sum of the terms with $|\alpha|=|\beta|=m$ is given by

$$(A_{2m}(x,D)\varphi,\psi)=\sum_{|\alpha|=|\beta|=m}\sum_{\rho,\lambda}\int_{t^n} b_{\rho\alpha\beta}^\lambda(x)D^\alpha\varphi^\rho(x)\overline{D^\beta\psi^\lambda(x)}\,dx. \tag{2.34}$$

Let $A_{2m}(x, \xi_x)$ be the matrix of the values of the principal symbol of $A(x, D)$ at $(x, \xi_x) \in T^*U$. For $w, w' \in \mathbb{C}^\mu$, we have

$$(-1)^m \sum_\lambda (A_{2m}(x, \xi_x)w)^\lambda \bar{w}'^\lambda = \sum_{|\alpha|=|\beta|=m} \sum_{\rho,\lambda} b^\lambda_{\rho\alpha\beta}(x)\xi_x^{\alpha+\beta} w^\rho \bar{w}'^\lambda. \tag{2.35}$$

In fact, let $f(x)$ be a real-valued C^∞ function defined on U and $g(x)$ any complex-valued C^∞ function with support in U. Put $\xi_x = df(x)$ with $df(x) \neq 0$. Substituting

$$\varphi(x) = g(x)\, e^{itf(x)} w, \qquad \psi(x) = e^{itf(x)} w'$$

into (2.33), by the definition of $A_{2m}(x, \xi_x)$ we obtain:

$$\begin{aligned}
(-1)^m \sum_\lambda &\int_{\mathbb{T}^n} (A_{2m}(x, \xi_x)w)^\lambda \bar{w}'^\lambda g(x)\, đx \\
&= \lim_{t\to\infty} t^{-2m} (A(x, D)g(x)\, e^{itf(x)} w, e^{itf(x)} w') \\
&= \lim_{t\to\infty} t^{-2m} \sum_{|\alpha|\leq m, |\beta|\leq m} \sum_{\rho,\lambda} \int_{\mathbb{T}^n} b^\lambda_{\rho\alpha\beta}(x) D^\alpha (g(x)\, e^{itf(x)} w^\rho) \overline{D^\beta e^{itf(x)} w'^\lambda}\, đx \\
&= \sum_{|\alpha|=|\beta|=m} \sum_{\rho,\lambda} \int_{\mathbb{T}^n} b^\lambda_{\rho\alpha\beta}(x)(i\xi_x)^\alpha (-i\xi_x)^\beta w^\rho \bar{w}'^\lambda g(x)\, đx.
\end{aligned}$$

Since $g(x)$ is arbitrary,

$$(-1)^m \sum_\lambda (A_{2m}(x, \xi_x)w)^\lambda \bar{w}'^\lambda = \sum_{|\alpha|=|\beta|=m} \sum_{\rho,\lambda} b^\lambda_{\rho\alpha\beta}(x)\xi_x^{\alpha+\beta} w^\rho \bar{w}'^\lambda$$

holds at each x. Since every non-zero element of $T^*_x U$ is written in the form of $df(x)$, we obtain (2.35).

For any $\varepsilon > 0$, we can cover $[U]$ by a finite number of open balls $B_1, \ldots, B_J$ of radius ε. Let $p_1, \ldots, p_J$ be the centres of $B_1, \ldots, B_J$, respectively. We may assume that ε is so small that the inequality

$$\varepsilon n^{2m} M_{m+1} < 2^{-m-2} \delta^2 \tag{2.36}$$

holds. Take real-valued C^∞ functions $\omega_j(x), j = 1, \ldots, J$ on $\mathbb{T}^n$ with $\operatorname{supp} \omega_j \subset B_j$ such that

$$\sum_j \omega_j(x)^2 = 1, \qquad x \in [U]. \tag{2.37}$$

For any $\varphi \in C_0^\infty(U, \mathbb{C}^\mu)$, put $\varphi_j(x) = \omega_j(x)\varphi(x)$. Then $\operatorname{supp} \varphi_j \subset B_j$. Now denote by $A_{2m}(p_j, D)$ the partial differential operator with constant coefficients obtained from (2.34) by replacing $b^\lambda_{\rho\alpha\beta}(x)$ by its value at $x = p_j$, i.e. we put

$$(A_{2m}(p_j, D)\varphi(x))^\lambda = \sum_{|\alpha|=|\beta|=m} \sum_\rho b^\lambda_{\rho\alpha\beta}(p_j) D^{\alpha+\beta}\varphi^\rho(x). \tag{2.38}$$

Then we see by the hypothesis of strong ellipticity and (2.35), that $A_{2m}(p_j, D)$ is a strong elliptic differential operator with the constant of strong ellipticity δ. By Lemma 2.2, we get

$$\mathrm{Re}(A_{2m}(p_j, D)\varphi_j, \varphi_j) + \delta^2\|\varphi_j\|^2 \geqq 2^{-m-1}\delta^2\|\varphi_j\|^2_m. \tag{2.39}$$

By the way, we have

$$\mathrm{Re}(A_{2m}(x, D)\varphi_j, \varphi_j) \geqq \mathrm{Re}(A_{2m}(p_j, D)\varphi_j, \varphi_j) - |(A_{2m}(x, D)\varphi_j, \varphi_j) - (A_{2m}(p_j, D)\varphi_j, \varphi_j)|. \tag{2.40}$$

We shall take an estimate of the second term of the right-hand side.

$$(A_{2m}(x, D)\varphi_j, \varphi_j) - (A_{2m}(p_j, D)\varphi_j, \varphi_j) = \sum_{|\alpha|=|\beta|=m} \sum_{\rho,\lambda} \int_{\mathbb{T}^n} (b^\lambda_{\rho\alpha\beta}(x) - b^\lambda_{\rho\alpha\beta}(p_j)) D^\alpha\varphi^\rho_j(x)\overline{D^\beta\varphi^\lambda_j(x)}\, dx.$$

Since $|x-p_j| \geqq \varepsilon$ for x supp φ_j, we have

$$\|\sum_\rho (b^\lambda_{\rho\alpha\beta}(x) - b^\lambda_{\rho\alpha\beta}(p_j)) D^\alpha\varphi^\rho_j\| \leqq \varepsilon M_{m+1}\|\varphi_j\|_m.$$

Hence we obtain the following estimate:

$$|(A_{2m}(x, D)\varphi_j, \varphi_j) - (A_{2m}(p_j, D)\varphi_j, \varphi_j)| \leqq \varepsilon n^{2m} M_{m+1}\|\varphi_j\|^2_m.$$

By this inequality and by (2.36), (2.39) and (2.40), we obtain

$$\mathrm{Re}(A_{2m}(x, D)\varphi_j, \varphi_j) \geqq 2^{-m-2}\delta^2\|\varphi_j\|^2_m - \delta^2\|\varphi_j\|^2. \tag{2.41}$$

Hence by (2.37),

$$\begin{aligned}(A(x, D)\varphi, \varphi) &= \sum_{\substack{|\alpha|\leqq m\\|\beta|\leqq m}} \sum_{\rho,\lambda}\sum_j \int b^\lambda_{\rho\alpha\beta}(x) D^\alpha\omega_j(x)^2\varphi^\lambda(x)\overline{D^\beta\varphi^\rho(x)}\, dx \\ &= \sum_j (A(x, D)\varphi_j, \varphi_j) \\ &\quad + \sum_{\substack{|\alpha|\leqq m\\|\beta|\leqq m}} \sum_{\rho,\lambda}\sum_j \int b^\lambda_{\rho\alpha\beta}(x)[\omega_j, D^\alpha]\varphi^\lambda_j(x)\overline{D^\beta\varphi^\rho(x)}\, dx \\ &\quad + \sum_{\substack{|\alpha|\leqq m\\|\beta|\leqq m}} \sum_{\rho,\lambda}\sum_j \int b^\lambda_{\rho\alpha\beta}(x) D^\alpha\varphi^\lambda_j(x)\overline{[\omega_j, D^\beta]\varphi^\rho(x)}\, dx.\end{aligned}$$

By Lemma 1.4 there is a positive number C_1 depending only on m, μ, n, and δ such that

$$\operatorname{Re}(A(x, D)\varphi, \varphi) \geqq \sum_j \operatorname{Re}(A(x, D)\varphi_j, \varphi_j) - C_1 M_m \|\varphi\|_m \|\varphi\|_{m-1}. \tag{2.42}$$

On the other hand,

$$\begin{aligned}\operatorname{Re}(A(x, D)\varphi_j, \varphi_j) \geqq \operatorname{Re}(A_{2m}(x, D)\varphi_j, \varphi_j) - |(A(x, D)\varphi_j, \varphi_j)\\ - (A_{2m}(x, D)\varphi_j, \varphi_j)|\end{aligned}$$

Making the difference of (2.33) and (2.34), we have

$$\begin{aligned}&((A(x, D)\varphi_j, \varphi_j) - (A_{2m}(x, D)\varphi_j, \varphi_j)\\ &\quad = \sum_{\substack{|\alpha|\leqq m\\ |\beta|\leqq m}}{}' \sum_{\rho,\lambda} \int_{T^n} b^{\lambda}_{\rho\alpha\beta}(x) D^{\alpha}\varphi^{\rho}(x) \overline{D^{\beta}\varphi^{\lambda}(x)}\, \bar{d}x\end{aligned}$$

where the summation $\sum'$ is taken over all the terms either with $|\alpha| \leqq m-1$ or with $|\beta| \leqq m-1$. There exists a positive number C_2 such that

$$|(A(x, D)\varphi_j, \varphi_j) - (A_{2m}(x, D)\varphi_j, \varphi_j)| \leqq C_2 M_m \|\varphi_j\|_m \|\varphi_j\|_{m-1}.$$

Combining the inequality with (2.41), we can find a positive constant C_3 depending only on m, μ, and δ such that

$$\operatorname{Re}(A(x, D)\varphi_j, \varphi_j) \geqq 2^{-m-2}\delta^2 \|\varphi_j\|_m^2 - \delta^2 \|\varphi_j\|^2 - C_3 M_m \|\varphi_j\|_{m-1} \|\varphi_j\|_m.$$

By the above and (2.42), we have

$$\begin{aligned}\operatorname{Re}(A(x, D)\varphi, \varphi) \geqq 2^{-m-2}\delta^2 \sum_j \|\varphi_j\|_m^2 - \delta^2 \sum_j \|\varphi_j\|^2\\ - C_3 M_m \sum_j \|\varphi_j\|_m \|\varphi_j\|_{m-1}\\ - C_1 M_m \|\varphi\|_m \|\varphi\|_{m-1}.\end{aligned} \tag{2.43}$$

There exists a positive constant C_4 such that

$$\|\varphi_j\|_m \leqq C_4 \|\varphi\|_m, \qquad \|\varphi_j\|_{m-1} \leqq C_4 \|\varphi\|_{m-1}.$$

On the other hand $\sum_j \|\varphi_j\|^2 = \|\varphi\|^2$, and by Lemma 1.7 there is a positive number C_5 such that

$$\|\varphi\|_m^2 \leqq \left(C_5 \sum_{j=1}^{J} \|\varphi_j\|_m \right)^2 \leqq C_5^2 J \sum_j \|\varphi_j\|_m^2.$$

Therefore putting $C_6 = C_3JC_4^2 + C_1$, we get by (2.43)

$$\operatorname{Re}(A(x, D)\varphi, \varphi) \geqq 2^{-m-2}C_5^{-2}J^{-1}\delta^2\|\varphi\|_m^2 - \delta^2\|\varphi\|^2 - C_6M_m\|\varphi\|_m\|\varphi\|_{m-1}. \tag{2.44}$$

By (1.34) in Proposition 1.1, we have

$$\|\varphi\|_m\|\varphi\|_{m-1} \leqq \|\varphi\|_m^{2-1/m}\|\varphi\|^{1/m}.$$

By choosing a sufficiently small t in (1.33), we can find a positive number C_7 such that

$$\|\varphi\|_m\|\varphi\|_{m-1} \leqq (C_6M_m)^{-1}2^{-m-3}C_5^{-2}J^{-1}\delta^2\|\varphi\|_m^2 + C_7\|\varphi\|^2.$$

Therefore by (2.44) and this inequality, we obtain

$$\operatorname{Re}(A(x, D)\varphi, \varphi) \geqq 2^{-m-3}C_5^{-2}J^{-1}\delta^2\|\varphi\|_m^2 - (C_7 + \delta^2)\|\varphi\|^2. \tag{2.45}$$

Hence (2.32) holds for any $\varphi \in C_0^\infty(U, \mathbb{C}^\mu)$. Since $C_0^\infty(U, \mathbb{C}^\mu)$ is dense in $W_0^m(U, \mathbb{C}^\mu)$ and both sides of (2.32) are continuous in $\varphi \in W_0^m(U, \mathbb{C}^\mu)$, (2.30) holds for every $\varphi \in W_0^m(U, \mathbb{C}^\mu)$. ∎

(B) Estimates by the Hölder Norm

(a) The Case of Constant Coefficients

Set $0 < \theta < 1$. We say that a vector-valued function $\varphi(x) = (\varphi^1(x), \ldots, \varphi^\mu(x)) \in C^k(\mathbb{T}^n, \mathbb{C}^\mu)$ is of class $C^{k+\theta}$ if

$$\sum_{\lambda=1}^{\mu} \sum_{|\alpha|=k} \sup_{|x-x'|\leqq 1} \frac{|D^\alpha\varphi^\lambda(x) - D^\alpha\varphi^\lambda(x')|}{|x - x'|^\theta} < \infty.$$

The totality of such functions φ is denoted by $C^{k+\theta}(\mathbb{T}^n, \mathbb{C}^\mu)$. For $\varphi \in C^{k+\theta}(\mathbb{T}^n, \mathbb{C}^\mu)$ we define the norm by

$$|\varphi|_{k+\theta} = |\varphi|_k + \sum_{\lambda=1}^{\mu} \sum_{|\alpha|=k} \sup_{|x-x'|\leqq 1} \frac{|D^\alpha\varphi^\lambda(x) - D^\alpha\varphi^\lambda(x')|}{|x - x'|^\theta}. \tag{2.46}$$

$C^{k+\theta}(\mathbb{T}^n, \mathbb{C}^\mu)$ is then a Banach space.

The same statements as in Lemma 2.1 can be obtained by replacing $\| \ \|_{s-l}$ and $\| \ \|_s$ by $| \ |_{k+\theta}$ and $| \ |_{k+l+\theta}$, respectively. The proof for the general case is, however, somewhat long. So we restrict ourselves to the most simple case. We assume that the order of the elliptic differential operator is 2 and that $A_2(D)$ is *of diagonal type*, namely, we assume that $A_2(D)$ satisfies

$$(A_2(D)\varphi)^\lambda(x) = \sum_{i,j=1}^n a_{ij}^\lambda D_i D_j \varphi^\lambda(x), \qquad \lambda = 1, \ldots, \mu. \tag{2.47}$$

Furthermore, we suppose that a_{ij}^λ is constant with $a_{ij}^\lambda = a_{ji}^\lambda$ for $i, j = 1, 2, \ldots, n$ and that the operator is strongly elliptic. In other words, there is a positive constant δ such that, for any $\xi \in \mathbb{R}^n$

$$-\mathrm{Re} \sum_{i,j=1}^n a_{ij}^\lambda \xi_i \xi_j \geqq \delta^2 \sum_{i=1} \xi_i^2, \qquad \lambda = 1, \ldots, \mu. \tag{2.48}$$

By the assumption that $A_2(D)$ is of diagonal type, the proof is reduced to the case $\mu = 1$. In what follows, putting $\mu = 1$, we shall consider the strongly elliptic partial differential equation $A(D)$ acting on $C^{k+2+\theta}(\mathbb{T}^n, \mathbb{C})$

$$A(D)\varphi(x) = \sum_{i,j} a^{ij} D_i D_j \varphi(x). \tag{2.49}$$

Let $A = (a^{ij})$ be an $(n \times n)$-matrix with $a^{ij} = a^{ji}$. Let $B = (a_{ij})$ be its inverse matrix.

We introduce the quadratic form on $\mathbb{R}^n$ by

$$Q(x) = \sum_{i,j} a_{ij} x^i x^j. \tag{2.50}$$

Define the constant $C(n)$ by

$$C(n)(n-2) \int_{|z|=1} Q(z)^{-n/2}\, d\sigma_1(z) = 1, \tag{2.51}$$

where $d\sigma_1(z)$ stands for the surface element on the unit sphere $|z| = 1$. Set

$$E(x) = C(n) Q(x)^{(2-n)/2}.$$

Then

$$D_i E(x) = -(n-2) C(n) Q(x)^{-n/2} \Big(\sum_j a_{ij} x^j \Big), \tag{2.52}$$

and

$$D_i D_j E(x) = (n-2) C(n) Q(x)^{-n/2} \times \Big(-a_{ij} + n Q(x)^{-1} \Big(\sum_k a_{ik} x^k \Big) \Big(\sum_l a_{jl} x^l \Big) \Big). \tag{2.53}$$

From these equalities, we can easily see that

$$A(D)E(x)=0$$

for $x\in\mathbb{R}^n$ with $x\neq 0$. Let $\rho(t)$ be a C^∞ function if t such that $\rho(t)\equiv 1$ for $|t|<\frac{1}{2}$ and that $\rho(t)\equiv 0$ for $|t|>\frac{3}{4}$. Put $\zeta(x)=\rho(|x|)$. If we put

$$g(x)=\zeta(x)E(x),$$

then we find that $g(x)\equiv 0$ for $|x|\geqq\frac{3}{4}$, and that

$$D_j g(x)=\zeta(x)D_jE(x)+D_j\zeta(x)E(x), \tag{2.54}$$

$$D_iD_jg(x)=\zeta(x)D_iD_jE(x)+\omega_{ij}(x). \tag{2.55}$$

Here $\omega_{ij}(x)$ denotes a C^∞ function with

$$\operatorname{supp}\omega_{ij}\subset\{x\,|\,\tfrac{1}{2}\leqq|x|\leqq\tfrac{3}{4}\}.$$

Thus there is a C^∞ function w such that for $x\in\mathbb{R}^n$ with $x\neq 0$,

$$A(D)g(x)=w(x) \tag{2.56}$$

and that

$$\operatorname{supp}w\subset\{x\,|\,\tfrac{1}{2}\leqq|x|\leqq\tfrac{3}{4}\}.$$

We extend the functions $g(x)$, $\omega_{ij}(x)$ and $w(x)$ to the periodic C^∞ functions on $\mathbb{R}^n$, and identify them with those on $\mathbb{T}^n$. We denote such function on $\mathbb{T}^n$ by the same notation $g(x)$, $\omega_{ij}(x)$ and $w(x)$. $g(x)$ is called a parametrix for the differential operator $A_l(D)$. In the general case, it is rather difficult to construct a parametrix.

Lemma 2.3. *Set $f(x)=A(D)u(x)$ for $u\in C^2(\mathbb{T}^n,\mathbb{C}^\mu)$. Then we have*

$$u(x)=-\int_{\mathbb{T}^n} g(x-y)f(y)\,đy+\int_{\mathbb{T}^n} w(x-y)u(y)\,đy. \tag{2.57}$$

Proof. Set $0<\xi<1$. Put $\Omega_\varepsilon=\{y\in\mathbb{T}^n\,|\,|y-x|>\varepsilon\}$ for a fixed x. By means of Green's formula,

$$\begin{aligned}\int_{\Omega_\varepsilon} A(D_y)u(y)g(x-y)\,đy-\int_{\Omega_\varepsilon} u(y)A(D_y)g(x-y)\,đy\\ =\int_{\partial\Omega_\varepsilon}(A\nu\cdot D_y)u(y)g(x-y)\,d\sigma(y)\\ -\int_{\partial\Omega_\varepsilon}(A\nu\cdot D_y)g(x-y)u(y)\,d\sigma(y),\end{aligned} \tag{2.58}$$

where D_y stands for the differentiation with respect to y, and $(A\nu \cdot D_y) = \sum a^{ij}\nu_i(\partial/\partial y^i)$ with $\nu = (\nu_1, \dots, \nu_n)$ the outer unit normal vector to $\partial\Omega_\varepsilon$. $d\sigma(y)$ denotes the surface element of $\partial\Omega_\varepsilon$, and $A(D_y)g(x-y) = w(x-y)$. Since $|(A\nu \cdot D)u(y)| \leqq C|u|_1$ and $|g(x-y)| \leqq CQ(x-y)^{(2-n)/2}$ on $\partial\Omega_\varepsilon$, we have

$$\lim_{\varepsilon \to 0} \int_{\partial\Omega_\varepsilon} (A\nu \cdot D)u(y)g(x-y)\, d\sigma(y) = 0.$$

On the other hand, since

$$\nu = \frac{x-y}{|x-y|} = \left(\frac{x^1-y^1}{|x-y|}, \dots, \frac{x^n-y^n}{|x-y|}\right),$$

by (2.52), we get

$$(A\nu \cdot D_y)g(x-y) = (n-2)C(n)Q(x-y)^{-n/2}|x-y|$$

for x, y with $0 < |x-y| < 1$. Hence we have by (2.51)

$$\lim_{\varepsilon \to 0} \int_{\partial\Omega_\varepsilon} (A\nu \cdot D)g(x-y)u(y)\, d\sigma(y) = u(x).$$

Therefore putting $\varepsilon \to 0$ in (2.58), we obtain

$$u(x) = -\int_{\mathbb{T}^n} g(x-y)f(y)\, đy + \int_{\mathbb{T}^n} w(x-y)u(y)\, đy. \quad \blacksquare$$

Lemma 2.4. *As for the integral transformation $Hf(x) = \int_{\mathbb{T}^n} h(x-y)f(y)\, đy$, we get the following.*

(1°) *If h is integrable and f is continuous, then Hf is continuous and*

$$Hf(x) = \int_{\mathbb{T}^n} h(y)f(x-y)\, đy.$$

Moreover the following estimate holds:

$$|Hf|_0 \leqq N(h)|f|_0,$$

where we put

$$N(h) = \int_{\mathbb{T}^n} |h(x)|\, đx.$$

(2°) *If for $j=1,2,\dots,n$, $D_j h$ is integrable, then $Hf(x)$ is of class C^1. Set $C_1=\int_{\mathbb{T}^n}|h(x)|\,dx+\sum_j\int_{\mathbb{T}^n}|D_j h(x)|\,dx$, then we have*

$$|Hf|_1 = C_1|f|_0.$$

(3°) *Put $k\geqq 0$. If $D^\alpha h$ is integrable for any α with $|\alpha|\leqq k$ then $Hf\in C^k(\mathbb{T}^n,\mathbb{C})$ and*

$$|Hf|_k\leqq C|f|_0$$

where $C=\sum_{|\alpha|\leqq k}N(D^\alpha h)$.

The proof is easy, hence it is omitted.

By Lemma 2.3 we write

$$u(x)=Gf(x)+Wu(x), \tag{2.59}$$

where

$$Gf(x)=-\int_{\mathbb{T}^n} g(x-y)f(y)\,đy \tag{2.60}$$

and

$$Wu(x)=\int_{\mathbb{T}^n} w(x-y)u(y)\,đy. \tag{2.61}$$

By Lemma 2.4, for any integer $k\geqq 0$, we have

$$|Wu|_k\leqq C_k|u|_0 \tag{2.62}$$

where $C_k=\sum_{|\alpha|\leqq k}N(D^\alpha w)$.

Since $D_j g(x)$ is integrable for any j, we obtain from Lemma 2.3,

$$D_jGf(x)=-\int_{\mathbb{T}^n} D_j g(x-y)f(y)\,đy=\int_{\mathbb{T}^n}\frac{\partial}{\partial y_j}g(x-y)f(y)\,đy.$$

Let $f\in C^1(\mathbb{T}^n,\mathbb{C})$. Then, by integration by parts, we obtain

$$D_jGf(x)=-\int_{\mathbb{T}^n} g(x-y)\frac{\partial}{\partial y_j}f(y)\,đy.$$

Hence, for $f\in C^1(\mathbb{T}^n,\mathbb{C})$, we have, for $i,j=1,\dots,n$,

$$D_iD_jGf(x)=-\int_{\mathbb{T}^n} D_i g(x-y)\frac{\partial}{\partial y_j}f(y)\,đy.$$

Since $(\partial/\partial y_j)f(y) = (\partial/\partial y_j)(f(y)-f(x))$, we obtain

$$D_iD_jGf(x) = -\int_{\mathbb{T}^n} D_ig(x-y)\frac{\partial}{\partial y_j}(f(y)-f(x))\, đy.$$

Put $\Omega_\varepsilon = \{y \in \mathbb{T}^n \,\|x-y| > \varepsilon\}$. Then

$$D_iD_jGf(x) = -\lim_{\varepsilon\to 0}\int_{\Omega_\varepsilon} D_ig(x-y)\frac{\partial}{\partial y_j}(f(y)-f(x))\, đy.$$

By integration by parts, we get

$$D_iD_jGf(x) = \lim_{\varepsilon\to 0}\int_{\Omega_\varepsilon} \frac{\partial}{\partial y_j}D_ig(x-y)(f(y)-f(x))\, đy$$
$$-\lim_{\varepsilon\to 0}\int_{|x-y|=\varepsilon} \frac{x^j-y^j}{|x-y|}D_ig(x-y)(f(y)-f(x))\, d\sigma(g). \quad (2.63)$$

By the way for θ with $0<\theta<1$, there exists a positive number C such that

$$\left|\frac{x^j-y^j}{|x-y|}\right||D_ig(x-y)|\,|f(y)-f(x)| \leqq C|x-y|^{1-n+\theta}|f|_\theta.$$

Therefore the second term of the right-hand side of (2.63) vanishes. Consequently, we have

$$D_iD_jGf(x) = \int_{\mathbb{T}^n} \frac{\partial}{\partial y_j}D_ig(x-y)(f(y)-f(x))\, đy$$
$$= -\int_{\mathbb{T}^n} D_jD_ig(x-y)(f(y)-f(x))\, đy. \quad (2.64)$$

The following fact is important for this integral transformation.

Lemma 2.5. *Suppose that* $0<\varepsilon\leqq\frac{1}{2}$. *Then*

$$\int_{|z|=\varepsilon} D_jD_ig(z)\, d\sigma_\varepsilon(z) = 0, \qquad i,j=1,2,\ldots,n, \quad (2.65)$$

where $d\sigma_\varepsilon(z)$ *denotes the surface element on the sphere* $|z|=\varepsilon$.

Proof. Suppose $\eta(t)$ is a C^∞ function of $t\in\mathbb{R}$ with $\eta(t)\geqq 0$ such that $\eta(t)\equiv 1$ on $|t|<2^{-2}$ and that $\eta(t)\equiv 0$ on $|t|>\frac{1}{2}$. Then we get

$$\infty > M = \int_{\mathbb{T}^n} D_ig(z)D_j\eta(|z|)\, dz = \lim_{\varepsilon\downarrow 0}\int_{|z|\geqq\varepsilon} D_ig(z)D_j\eta(|z|)\, dz,$$

since $D_j\eta(|z|)$ vanishes on $|z|<2^{-2}$. Besides, since $g(z)=E(z)$ on $|z|<2^{-1}$,

$$\begin{aligned}M_\varepsilon &= \int_{|z|>\varepsilon} D_i g(z) D_j \eta(|z|)\,dz\\ &= -\int_{|z|\geqq\varepsilon} D_j D_i g(z)\eta(|z|)\,dz - \int_{|z|=\varepsilon} \frac{z^j}{|z|} D_i g(z)\eta(|z|)\,d\sigma_\varepsilon(z)\\ &= -\int_\varepsilon^{2^{-1}} \eta(t)t^{-1}\,dt \int_{|z|=1} D_j D_i E(z)\,d\sigma_1(z)\\ &\quad -\eta(\varepsilon)\int_{|z|=1}\frac{z^j}{|z|} D_i E(z)\,d\sigma_1(z).\end{aligned}$$

Hence

$$\begin{aligned}M_\varepsilon + \eta(\varepsilon)\int_{|z|=1}\frac{z^j}{|z|} D_i E(z)\,d\sigma_1(z)\\ = -\int_\varepsilon^{2^{-1}} \eta(t)t^{-1}\,dt \int_{|z|=1} D_j D_i E(z)\,d\sigma_1(z).\end{aligned}$$

Putting $\varepsilon\to 0$, we see that the left-hand side converges to a finite value. Since

$$\int_0^{2^{-1}} \eta(t)t^{-1}\,dt = \infty,$$

we must have

$$\int_{|z|=1} D_j D_i E(z)\,d\sigma_1(z)=0.$$

Thus we see that, for any ε with $0<|\varepsilon|<\frac{1}{2}$,

$$\int_{|z|=\varepsilon} D_j D_i g(z)\,d\sigma_\varepsilon(z) = \varepsilon^{-1}\rho(\varepsilon)\int_{|z|=1} D_j D_i E(z)\,d\sigma_1(z)=0,$$

holds, which completes the proof of the lemma. ∎

Lemma 2.6. *Suppose that* $0<\theta<1$. *If* $f\in C^\theta(\mathbb{T}^n,\mathbb{C})$, *then for* $i,j=1,2,\dots,n$ $D_jD_iGF\in C^\theta(\mathbb{T}^n,\mathbb{C})$, *and there exists a positive number* C *such that*

$$|D_jD_iGf|_\theta \leqq CN_3(g)|f|_\theta,\qquad i,j=1,2,\dots,n, \tag{2.66}$$

where we put

$$N_3(g)=\sum_{|\alpha|=2}\sup_{0<|z|\leqq 1}|z|^n|D^\alpha g(z)| + \sum_{|\alpha|=3}\sup_{0<|z|\leqq 1}|z|^{n+1}|D^\alpha g(z)|. \tag{2.67}$$

Proof. For $1>\rho>0$, we set

$$I(x,\rho)=\int_{|x-y|\leqq\rho} D_jD_ig(x-y)(f(y)-f(x))\,đy.$$

Denoting by $\sigma(n)$ the area of the unit sphere, we have

$$I(x,\rho)\leqq N_3(g)|f|_\theta\int_{|z|\leqq\rho}|z|^{\theta-n}\,dz=\theta^{-1}\sigma(n)N_3(g)|f|_\theta\rho^\theta. \tag{2.68}$$

Since $g(z)=0$ on $|z|\geqq 1$, for $\rho>1$, we have

$$I(x,\rho)=I(x,1)=D_jD_iGf(x).$$

Hence by (2.68)

$$|D_jD_iGf|_0\leqq\theta^{-1}\sigma(n)N_3(g)|f|_\theta. \tag{2.69}$$

Next suppose $|x-x'|=\varepsilon<\frac{1}{10}$. We shall estimate $D_jD_iGf(x)-D_jD_iGf(x')$. First we write

$$D_jD_iGf(x)=\int_{\mathbb{T}^n} D_jD_ig(x-y)(f(y)-f(x))\,đy,$$

$$D_jD_iGf(x')=\int_{|y-x'|>5\varepsilon} D_jD_ig(x'-y)(f(y)-f(x'))\,đy+I(x',5\varepsilon).$$

By Lemma 2.5, we can rewrite $D_jD_iGf(x')$ as follows:

$$D_jD_iGf(x')=\int_{|y-x'|>5\varepsilon} D_jD_ig(x'-y)(f(y)-f(x))\,đy+I(x',5\varepsilon)$$
$$+(f(x)-f(x'))\int_{|x'-y|>1/2} D_jD_ig(x'-y)\,đy.$$

Hence

$$D_jD_iGf(x)-D_jD_iGf(x')=J_1+J_2+J_3-I(x',5\varepsilon), \tag{2.70}$$

where

$$J_1=\int_{|y-x'|>5\varepsilon}\{D_jD_ig(x-y)-D_jD_ig(x'-y)\}(f(y)-f(x))\,đy,$$

$$J_2=\int_{|y-x'|<5\varepsilon} D_jD_ig(x-y)(f(y)-f(x))\,đy, \tag{2.71}$$

$$J_3=M(f(x)-f(x')),\ M=\int_{|z|>1/2} D_jD_ig(z)\,dz.$$

If $|y-x'| \geqq 5|x-x'|$, then the line segment connecting $y-x$ with $y-x'$ does not pass through singular points to $g(z)$. Thus, by the mean value theorem, we see that there is a t with $0<t<1$ such that

$$D_jD_ig(x-y)-D_jD_ig(x'-y)=\sum_{k=1}^{n}(x^k-x'^k)D_kD_jD_ig(tx+(1-t)x'-y).$$

$|y-x'|>5\varepsilon=5|x-x'|$ implies $|x-y|>4|x-x'|$, hence

$$|tx+(1-t)x'-y| \geqq |x-y|-|x-x'| \geqq 2^{-1}|x-y|.$$

Therefore we have

$$|D_jD_ig(x-y)-D_jD_ig(x'-y)| \leqq 2^{n+1}N_3(g)\varepsilon|x-y|^{-n-1}.$$

Since $\varepsilon<10^{-1}$, we see that $|x-y| \leqq \frac{3}{4}+\varepsilon<1$ for y with $D_jD_ig(x-y)-D_jD_ig(x'-y) \neq 0$. Hence

$$\begin{aligned} |J_1| &\leqq \int_{|y-x|>4\varepsilon} 2^{n+1}N_3(g)\varepsilon|x-y|^{n-1}|f|_\theta|x-y|^\theta\, \bar{d}y \\ &\leqq 2^{n+1}\sigma(n)N_3(g)|f|_\theta\cdot\varepsilon(1-\theta)^{-1}(4\varepsilon)^{\theta-1} \end{aligned} \tag{2.72}$$

since $|f(x)-f(y)| \leqq |f|_\theta|x-y|^\theta$ there. On the other hand, since $|y-x'|<5\varepsilon$ implies that $|y-x|<6\varepsilon$, we have

$$\begin{aligned} |J_2| &\leqq \int_{|y-x|<6\varepsilon} |D_jD_ig(x-y)|\,|f(y)-f(x)|\, \bar{d}y \\ &\leqq N_3(g)|f|_\theta \int_{|z|<6\varepsilon} |z|^{\theta-n}\, dz=\theta^{-1}N_3(g)|f|_\theta(6\varepsilon)^\theta. \end{aligned} \tag{2.73}$$

Applying (2.68), (2.72), (2.73) and $|J_3| \leqq |M|\,|f|_\theta\varepsilon^\theta$ to (2.70), we get

$$|D_jD_iGf(x)-D_jD_iGf(x')| \leqq C_2N_3(g)|f|_\theta\varepsilon^\theta$$

for $\varepsilon<10^{-1}$. In the case of $\varepsilon \geqq 10^{-1}$, since

$$|D_jD_iGf(x)-D_jD_iGf(x')| \leqq 20|D_jD_iGf|_0$$

we finally see that there exists a positive constant C_3 such that for $|x-x'| \leqq 1$,

$$|D_jD_iGf(x)-D_jD_iGf(x')| \leqq CN_3(g)|f|_\theta|x-x'|^\theta$$

holds. By this inequality and (2.69), we obtain the lemma. ∎

Lemma 2.7. *Let $A_2(D)$ be a second-order strongly elliptic partial linear differential operator with constant coefficients of diagonal type which operates on vector-valued functions with values in $\mathbb{C}^\mu$. Suppose that its lower terms are equal to zero. Assume that $0<\theta<1$. For any integer $k \geqq 0$, there exists a positive constant C depending only on θ, k, n, μ, and the coefficients of $A_2(D)$ such that, for any $u \in C^{k+2+\theta}(\mathbb{T}^n, \mathbb{C}^\mu)$,*

$$|u|_{2+k+\theta} \leqq C(|A_2(D)u|_{k+\theta} + |u|_k).$$

Proof. In case $k=0$, the lemma follows from Lemmas 2.3, 2.4 and 2.6. For general k, since we have

$$A_2(D)D^\alpha u(x) = D^\alpha A_2(D)u(x)$$

for $|\alpha| = k$, by the estimate for the case of $k=0$, we have

$$|D^\alpha u|_{2+\theta} \leqq C(|D^\alpha A_2(D)u|_\theta + |D^\alpha u|_\theta).$$

On the other hand, since for any positive integer k, there is a positive constant C_1 such that

$$C_1^{-1}|u|_{k+\theta} \leqq \sum_{|\alpha| \leqq k} |D^\alpha u|_\theta \leqq C_1|u|_{k+\theta},$$

the lemma is proved for general k. ∎

Remark. In the general elliptic case, we have to construct a $(\mu \times \mu)$-matrix-valued function $E(x)$ corresponding to that for the above case such that $A(D)E(x) = 0$. But is takes a long procedure. The other treatment is quite similar to the above.

(b) The Case of Variable Coefficients

Lemma 2.8. *Suppose that $0<r<s$. Then there are positive constants C_1 and C_2 such that, for any $\varphi \in C^s(\mathbb{T}^n, \mathbb{C}^\mu)$ and for $t>0$, the following interpolation inequalities hold*:

$$|\varphi|_r \leqq t^{s/r}|\varphi|_s + C_1 t^{-s/(s-r)}|\varphi|_0,$$

$$|\varphi|_r \leqq C_2 |\varphi|_s^{r/s} |\varphi|_0^{1-r/s}.$$

Proof. We have only to prove the case $\mu = 1$.

(1°) The case $0<r<s<1$. We set

$$A = \sup_{|x-x'| \leqq 1} |x-x'|^{-r}|\varphi(x) - \varphi(x')|$$

for $\varphi \in C^s(\mathbb{T}^n, \mathbb{C})$. There exist points x and x' with $|x-x'| \leqq 1$ such that

$$|x-x'|^{-r}|\varphi(x)-\varphi(x')| > 2^{-1}A.$$

Hence we obtain

$$|x-x'|^r \leqq 2A^{-1}(|\varphi(x)|+|\varphi(x')|) \leqq 4A^{-1}|\varphi|_0. \tag{2.74}$$

On the other hand, since

$$|x-x'|^{-r}|\varphi(x)-\varphi(x')| \leqq |x-x'|^{s-r}|\varphi|_s,$$

we have

$$2^{-1}A \leqq |x-x'|^{s-r}|\varphi|_s \leqq (4A^{-1}|\varphi|_0)^{(s-r)/r}|\varphi|_s,$$

hence we obtain $A \leqq 4^{(s-r)/s}2^{r/s}|\varphi|_s^{r/s}|\varphi|_0^{1-r/s}$. Combining this with the trivial inequality $|\varphi|_0 \leqq |\varphi|_s^{r/s}|\varphi|_0^{1-r/s}$, we obtain

$$|\varphi|_r \leqq 4^{1-r/s}2^{r/s}|\varphi|_s^{r/s}|\varphi|_0^{1-r/s}. \tag{2.75}$$

(2°) The case $0<r<s=1$. By virtue of the mean value theorem, we have

$$|x-x'|^{-r}|\varphi(x)-\varphi(x')| \leqq |x-x'|^{(1-r)/r}|\varphi|_1.$$

By this and (2.74), we see that

$$2^{-1}A \leqq (4A^{-1}|\varphi|_0)^{1-r}|\varphi|_1.$$

Therefore we obtain $A \leqq 2^r 4^{1-r}|\varphi|_1^r|\varphi|_0^{1-r}$. Thus we have

$$|\varphi|_r \leqq 2^r 4^{1-r}|\varphi|_1^r|\varphi|_0^{1-r}. \tag{2.76}$$

(3°) The case $r \leqq 1 < s < 2$. Put $s = 1+\theta$. There exist a point x_0 and j such that

$$M = \sum_j \max_{x\in\mathbb{T}^n} |D_j\varphi(x)| \leqq 2n|D_j\varphi(x_0)|.$$

Let y be the point obtained from x_0 by the translation in the direction of x^j by $t>0$. By the mean value theorem, we find ξ between x_0 and y such that

$$\varphi(y)-\varphi(x_0) = tD_j\varphi(\xi).$$

This leads to the following estimation:

$$|D_j\varphi(\xi)| \leqq t^{-1}(|\varphi(y)|+|\varphi(x_0)|) \leqq 2t^{-1}|\varphi|_0. \tag{2.77}$$

Since $\varphi \in C^{1+\theta}(\mathbb{T}^n, \mathbb{C})$,

$$|D_j\varphi(x_0) - D_j\varphi(\xi)| \leqq |x_0 - \xi|^\theta |D_j\varphi|_\theta.$$

Hence

$$|D_j\varphi(x_0)| \leqq 2t^{-1}|\varphi|_0 + t^\theta |D_j\varphi|_\theta \leqq 2t^{-1}|\varphi|_0 + t^\theta|\varphi|_{1+\theta}$$

and therefore $M \leqq 2n(2t^{-1}|\varphi|_0 + t^\theta|\varphi|_{1+\theta})$. Taking the minimal value of the right-hand side with respect to t, we see that there exists a positive C_θ such that

$$M \leqq 2nC_\theta|\theta|_{1+\theta}^{1/(1+\theta)}|\varphi|_0^{1-1/(1+\theta)}.$$

So we can choose a positive constant $C'_\theta \geqq 1$ such that

$$|\varphi|_1 \leqq C'_\theta|\varphi|_s^{1/s}|\varphi|_0^{1-1/s}. \tag{2.78}$$

(4°) The case $r = 1 < s = 2$. For $\varphi \in C^2(\mathbb{T}^n, \mathbb{C})$ we have $|D_j(\varphi(x_0) - D_j\varphi(\xi)| \leqq |x_0 - \xi||\varphi|_2$. Hence, combining this with (2.77), we obtain

$$|D_j\varphi(x_0)| \leqq t|\varphi|_2 + 2t^{-1}|\varphi|_0 \leqq 4|\varphi|_2^{1/2}|\varphi|_0^{1/2}.$$

Hence we have $M \leqq 2n \cdot 4|\varphi|_2^{1/2}|\varphi|_0^{1/2}$. Thus we get

$$|\varphi|_1 \leqq 8n|\varphi|_2^{1/2}|\varphi|_0^{1/2}. \tag{2.79}$$

By (1°), (2°), (3°) and (4°), we have proved the lemma for $r \leqq 1$, $s \leqq 2$. The general case shall be proved by applying them repeatedly as follows: Suppose $0 < \theta < 1$. For some positive constants C_1, C_2, C_3 and C_4, we have

$$\begin{aligned}
|D_j\varphi|_\theta &\leqq C_1|D_j\varphi|_1^\theta|D_j\varphi|_0^{1-\theta} \\
&\leqq C_2(|D_j\varphi|_{1+\theta}^{1/(1+\theta)}|D_j\varphi|_0^{\theta/(1+\theta)})^\theta|D_j\varphi|_0^{1-\theta} \\
&= C_2|D_j\varphi|_{1+\theta}^{\theta/(1+\theta)}|\varphi|_1^{1/(1+\theta)} \\
&\leqq C_3|D_j\varphi|_{1+\theta}^{\theta/(1+\theta)}(|\varphi|_{1+\theta}^{1/(1+\theta)}|\varphi|_0^{\theta/(1+\theta)})^{1/(1+\theta)}.
\end{aligned}$$

Hence we see that by choosing a suitable positive constant C_5, we have

$$|\varphi|_{1+\theta} \leqq C_5|\varphi|_{2+\theta}^{(1+\theta)/(2+\theta)}|\varphi|_0^{1-(1+\theta)/(2+\theta)}. \quad \blacksquare \tag{2.80}$$

Lemma 2.9.

(1°) *Suppose that* $0 < \theta < 1$. *For* $a, f \in C^\theta(\mathbb{T}^n, \mathbb{C})$,

$$|af|_\theta \leqq |a|_\theta|f|_0 + |a|_0|f|_\theta.$$

(2°) *Let $A(x, D)$ be a linear partial differential operator of order l which operates on $\mathbb{C}^\mu$-valued functions. $A(x, D)$ is given as follows:*

$$(A(x, D)\varphi)^\lambda(x) = \sum_\lambda \sum_{|\alpha| \leqq l} a^\lambda_{\rho\alpha}(x) D^\alpha \varphi^\rho(x).$$

For θ with $0 < \theta < 1$, and for any integer $k \geqq 0$, we define M_k by (2.22) and put

$$M_{k+\theta} = M_k + \sum_{\lambda,\rho} \sum_{|\alpha| \leqq l} \sum_{|\beta| \leqq k} \sup_{|x-x'|<1} \frac{|D^\beta a^\lambda_{\rho\alpha}(x) - D^\beta a^\lambda_{\rho\alpha}(x')|}{|x - x'|^\theta}.$$

Then for any integer $k \geqq 0$ there exists a positive constant C depending only on l, k, θ, n and μ such that

$$|A(x, D)\varphi|_{k+\theta} \leqq C M_{k+\theta} |\varphi|_{l+k+\theta}.$$

The proof is easy. We leave it to the reader.

Lemma 2.10. *Let z be an arbitrary point of $\mathbb{T}^n$. Let $B_r(z) = \{x \in \mathbb{T}^n \mid \sum_j (x^j - z^j)^2 < r^2\}$ be a ball of radius $r < 1$. Let $0 < \theta < 1$, and k a non-negative integer. Then, for any $\varphi \in C^\infty(\mathbb{T}^n, \mathbb{C}^\mu)$ with $\operatorname{supp} \varphi \subset B_r(z)$, we have*

$$|\varphi|_k \leqq r^\theta |\varphi|_{k+\theta}. \tag{2.81}$$

Proof. Since φ is equal to zero outside of $B_r(z)$, for any α with $|\alpha| = k$ and any $x \in B_r(z)$, we have

$$|D^\alpha \varphi(x)| \leqq r^\theta |D^\alpha \varphi|_\theta.$$

Hence we get (2.81), because $|\alpha| \leqq k-1$ implies $|D^\alpha \varphi(x)| \leqq r |D^\alpha \varphi|_1$. ∎

Lemma 2.11. *Let A be an operator of multiplication, i.e.,*

$$(A\varphi)^\lambda(x) = \sum_{\rho=1}^{\mu} a^\lambda_\rho(x) \varphi^\rho(x).$$

Suppose $0 < \theta < 1$ and let k be a non-negative integer. Suppose that $a^\lambda_\rho \in C^{k+\theta}(\mathbb{T}^n, \mathbb{C})$ and that

$$a^\lambda_\rho(z) = 0, \qquad \lambda, \rho = 1, 2, \ldots, \mu.$$

Then, there exists a positive constant C depending on k, θ and μ such that

$$|A\varphi|_{k+\theta} \leqq C r^\theta K_{k+\theta} |\varphi|_{k+\theta}$$

for all $\varphi \in C^{k+\theta}(\mathbb{T}^n, \mathbb{C}^\mu)$ *with* supp $\varphi \subset B_r(z)$, *where we set* $K_s = \sum_{\lambda,\rho} |a_\rho^\lambda|_s$ *for any* $s \geqq 0$.

Proof. There exists a positive constant C_1 such that

$$|A\varphi|_k \leqq C_1 K_k |\varphi|_k.$$

From (2.81) we get

$$|A\varphi|_k \leqq C_1 r^\theta K_k |\varphi|_{k+\theta}. \tag{2.82}$$

On the other hand, for a multi-index α with $|\alpha| = k$, we have

$$(D^\alpha A\varphi)^\lambda(x) = \sum_\rho a_\rho^\lambda(x) D^\alpha \varphi^\rho(x) + \sum_{\beta<\alpha} \sum_\rho \binom{\alpha}{\beta} D_{a_\rho}^{\alpha-\beta\lambda}(x) D^\beta \varphi^\rho(x).$$

Hence there is a positive constant C_2 such that

$$|D^\alpha A\varphi|_\theta \leqq \sum_{\lambda,\rho} \max_{|x-z|\leqq r} |a_\rho^\lambda(x)| |D^\alpha \varphi|_\theta + K_\rho |D^\alpha \varphi|_0 + C_2 K_{k+\theta} |\varphi|_{k-1+\theta}.$$

On the other hand, since, for $|x-z| < r$,

$$|a_\rho^\lambda(x)| \leqq r^\theta K_\theta$$

holds, we have

$$|D^\alpha A\varphi|_\theta \leqq \mu^2 r^\theta K_\theta|^2 D^\alpha \varphi|_\theta + K_\theta |\varphi|_k + C_2 K_{k+\theta} |\varphi|_{k-1+\theta}.$$

Applying Lemma 2.10, we see that there exists a positive constant C_3 such that

$$|D^\alpha A\varphi|_\theta \leqq C_3 r^\theta K_{k+\theta} |\varphi|_{k+\theta}.$$

Combining (2.82) with this, we obtain Lemma 2.11. ∎

Let U be a domain of $\mathbb{T}^n$. Suppose that a second-order linear partial differential operator $A(x, D)$ with C^∞ coefficients defined on an open neighbourhood of $[U]$ operates on $\varphi \in C^\infty(\mathbb{T}^n, \mathbb{C}^\mu)$ in the following way:

$$(A(x, D)\varphi)^\lambda(x) = \sum_\beta \sum_{|\alpha|\leqq 2} a_{\rho\alpha}^\lambda(x) D^\alpha \varphi^\rho(x). \tag{2.83}$$

We say that this operator is *of diagonal type in the principal part*, if $a_{\rho\alpha}^\lambda(x) \equiv 0$ for $\lambda \neq \rho$ and $|\alpha| = 2$.

Theorem 2.3. *Let U be a domain in $\mathbb{T}^n$. Suppose that the second-order linear partial differential operator $A(x, D)$ with C^∞ coefficients defined on $[U]$ is of diagonal type in the principal part and strongly elliptic. Let δ be its constant of strong ellipticity on $[U]$. Suppose $0<\theta<1$. Then for any integer $k \geqq 0$, there exists a positive constant C depending on n, μ, k, θ, δ and $M_{k+\theta}$ such that for any $\varphi \in C^{k+2+\theta}(\mathbb{T}^n, \mathbb{C}^\mu)$ with $\operatorname{supp} \varphi \subset U$,*

$$|\varphi|_{k+2+\theta} \leqq C(|A(x, D)\varphi|_{k+\theta} + |\varphi|_0).$$

Proof. This is verified in an almost similar way to Theorem 2.1. We have only to use the norms $|\ |_{k+\theta}$ and $|\ |_{k+2+\theta}$ in place of $\|\ \|_s$ and $\|\ \|_{s+l}$, respectively. We cover $[U]$ with open balls $B_1, B_2, \ldots, B_J$ of radius ε. Let C_1 be a positive constant in Lemma 2.11. For any $z \in [U]$, let $C_2(z)$ be a constant in Lemma 2.7 for the elliptic linear partial differential operator obtained from the principal part of $A(x, D)$ by replacing coefficients by their values at z. Take ε so that for any $z \in [U]$, $C_1 C_2(z)\varepsilon^\theta M_{k+\theta} < 2^{-1}$. Choose C^∞ functions $\omega_j(z)$, $j=1, 2, \ldots, J$, on $\mathbb{T}^n$ such that $\operatorname{supp} \omega_j \subset B_j$ and that $\sum_j \omega_j(x) \equiv 1$ on $[U]$. Put $\varphi_j = \omega_j \varphi$ for any $\varphi \in C^{k+2+\theta}(\mathbb{T}^n, \mathbb{C}^\mu)$. Then we have

$$|\varphi|_{k+2+\theta} = \left| \sum_j \varphi_j \right|_{k+2+\theta} \leqq \sum_j^J |\varphi_j|_{k+2+\theta}.$$

We see that for some J, the following inequality holds:

$$|\varphi_j|_{k+2+\theta} \geqq J^{-1} |\varphi|_{k+2+\theta}. \tag{2.84}$$

Let p_j be the centre of B_j, and denote by $A_2(a, D)$ the principal part of $A(x, D)$:

$$(A_2(x, D)\varphi)^\lambda = \sum_{i,k=1}^n a_{\lambda ik}^\lambda(x) D_i D_k \varphi^\lambda(x).$$

We denote by $A_2(p_j, D)$ the partial differential operator with constant coefficients obtained from $A_2(x, D)$ by replacing $a_{\lambda ik}^\lambda(x)$ by their values at $x = p_j$:

$$(A_2(p_j, D)\varphi)^\lambda(x) = \sum_{i,k=1}^n a_{\lambda ik}^\lambda(p_j) D_i D_k \varphi^\lambda(x).$$

Applying Lemma 2.7 to their operator, we can take a positive constant $C_2(p_j)$ depending on n, μ, k, θ, δ and M_0 such that

$$|\varphi_j|_{2+k+\theta} \leqq C_2(p_j)(|A_2(p_j, D)\varphi_j|_{k+\theta} + |\varphi_j|_k). \tag{2.85}$$

Since

$$((A_2(x, D) - A_2(p_j, D))\varphi)^\lambda(x) = \sum_{i,k=1}^{n} (a_{\lambda ik}^\lambda(x) - a_{\lambda ik}^\lambda(p_j))D_iD_k\varphi^\lambda(x),$$

by Lemma 2.11, we get a positive constant C_1 such that

$$|A_2(x, D) - A_2(p_j, D))\varphi_j|_{k+\theta} \leqq C_1\varepsilon^\theta M_{k+\theta}|\varphi_j|_{k+2+\theta}. \tag{2.86}$$

C_1 depends only on n, μ, k, and θ. From (2.85) and (2.86) we obtain

$$|\varphi_j|_{2+k+\theta} \leqq C_2(p_j)(|A_2(x, D)\varphi_j|_{k+\theta} + |\varphi_j|_k + C_1C_2(p_j)\varepsilon^\theta M_{k+\theta}|\varphi_j|_{k+2+\theta}.$$

Since for a sufficiently small ε

$$C_1C_2(p_j)\varepsilon^\theta M_{k+\theta} < 2^{-1},$$

we have

$$|\varphi_j|_{2+k+\theta} \leqq 2C_2(p_j)(|A_2(x, D)\varphi_j|_{k+\theta} + |\varphi_j|_k). \tag{2.87}$$

In the same way as in the proof of Theorem 2.1, we see that there is a positive constant C_3 such that

$$|A_2(x, D)\varphi_j - A(x, D)\varphi_j|_{k+\theta} \leqq C_3|\varphi_j|_{k+1+\theta}.$$

Hence

$$|\varphi_j|_{2+k+\theta} \leqq 2C_2(p_j)(|A(x, D)\varphi_j|_{k+\theta} + |\varphi_j|_k) + 2C_2(p_j)C_3|\varphi_j|_{k+1+\theta}.$$

Since $[A(x, D), \omega_j]$ is a first-order differential operator, there exists a positive constant C_4 such that

$$|[A(x, D], \omega_j]\varphi|_{k+\theta} \leqq C_4M_{k+\theta}|\varphi|_{k+1+\theta}.$$

By the way, there is a positive constant C_5 such that

$$|\varphi_j|_k \leqq C_5|\varphi|_k \quad \text{and} \quad |\varphi_j|_{k+1+\theta} \leqq C_r|\varphi|_{k+1+\theta}.$$

Therefore

$$\begin{aligned} J^{-1}|\varphi|_{k+2+\theta} \leqq{}& 2C_2(p_j)(|A(x, D)\varphi|_{k+\theta} + C_4M_{k+\theta}|\varphi|_{k+1+\theta} + C_5|\varphi|_k) \\ &+ 2C_2(p_j)C_3C_5|\varphi|_{k+1+\theta}. \end{aligned} \tag{2.88}$$

By Lemma 2.8, we can find a positive constant C_6 such that

$$2C_2(p_j)(C_4M_{k+\theta} + C_3C_5)|\varphi|_{k+1+\theta} + 2C_2(p_j)C_5|\varphi|_k \leqq 2^{-1}J^{-1}|\varphi|_{k+2+\theta} + C_6|\varphi|_0.$$

Therefore by (2.86)

$$2^{-1}J^{-1}|\varphi|_{k+2+\theta} \leqq 2C_2(p_j)|A(x, D)\varphi|_{k+\theta} + C_6|\varphi|_0.$$

This completes the proof. ∎

§3. Function Space of Sections of a Vector Bundle

Lemma 3.1. *Let U and V be domains in $\mathbb{T}^n$ which are diffeomorphic to each other, and $\Phi: U \to V$ the diffeomorphism between them. Let K be an arbitrary compact set in U. Put $K' = \Phi(K)$. Then for any integer l, there exists a positive constant C depending on l, Φ and K such that for $f \in C^\infty(V, \mathbb{C}^\mu)$ with $\operatorname{supp} f \subset K'$,*

$$C^{-1}\|f\|_l^2 \leqq \|f \cdot \Phi\|_l^2 \leqq C\|f\|_l^2. \tag{3.1}$$

Proof. For $l \geqq 0$, the lemma follows immediately from Proposition 1.2. In the case of $l < 0$, we set $l = -m$. Let C_0 be a constant with which (3.1) holds for $l = m$. Since $\operatorname{supp} f \subset K'$, f can be extended to a C^∞ function $\mathbb{T}^n$ which is identically zero outside of V. Choose a C^∞ function χ with support in V which is equal to 1 on K'. Then for any $\phi \in C^\infty(\mathbb{T}^n, \mathbb{C}^\mu)$,

$$\langle f, \varphi\rangle = \langle \chi f, \varphi\rangle = \langle f, \chi\varphi\rangle.$$

Since there is a positive constant C_1 such that $\|\chi\varphi\|_m \leqq C_1\|\varphi\|_m$, by Theorem 1.11, we have

$$\begin{aligned}\|f\|_{-m} &\leqq \sup \frac{|\langle f, \phi\rangle|}{\|\varphi\|_m} \leqq C_1^{-1} \sup_{\varphi} \frac{|\langle f, \chi\varphi\rangle|}{\|\chi\varphi\|_m} \\ &\leqq C_1^{-1} \sup_{\psi} \frac{|\langle f, \psi\rangle|}{\|\psi\|_m}.\end{aligned} \tag{3.2}$$

Here ψ runs over $C_0^\infty(V, \mathbb{C}^\mu)\backslash\{0\}$. Let $J(x)$ be the Jacobian of Φ and put $\tilde{\psi}(x) = J(x)(\psi \cdot \Phi)(x)$. Then $\langle f, \psi\rangle = \langle f \cdot \Phi, \tilde{\psi}\rangle$. From Proposition 1.7, we obtain a positive constant C_2 such that $\|\tilde{\psi}\|_m = C_2\|\psi \cdot \Phi\|_m$. By (3.1) for $l = m$, we have $\|\psi \cdot \Phi\|_m \leqq \sqrt{C_0}\|\psi\|_m$. Since for $\psi \in C_0^\infty(V, \mathbb{C}^\mu)\backslash\{0\}$, $\tilde{\psi} \in C_0^\infty(U, \mathbb{C}^\mu)\backslash\{0\}$,

$$\begin{aligned}\sup_{\psi} \frac{|\langle f, \psi\rangle|}{\|\psi\|_m} &\leqq (\sqrt{C_0}) \sup_{\psi} \frac{|\langle f \cdot \Phi, \tilde{\psi}\rangle|}{\|\psi \cdot \Phi\|_m} \leqq (\sqrt{C_0})C_2 \sup_{\psi} \frac{|\langle f \cdot \Phi, \tilde{\psi}\rangle|}{\|\tilde{\psi}\|_m} \\ &\leqq \sqrt{C_0}C_2\|f \cdot \Phi\|_{-m}.\end{aligned} \tag{3.3}$$

Hence by (3.2) and (3.3), we have the first inequality in (3.1):

$$\|f\|_{-m} \leqq C_1\sqrt{C_0}C_2\|f\cdot\Phi\|_{-m}.$$

Using Φ^{-1} instead of Φ, we obtain also the second inequality in (3.1). ▮

By use of Proposition 1.5, we can prove (3.1) for any real l. But we do not need this fact here, hence we omit the proof.

By using Lemma 3.1, the Sobolev space of sections of a vector bundle can be defined. Let X be an n-dimensional compact manifold, and $X = \bigcup_{j=1}^{J} X_j$ an open covering of X by a system of sufficiently small coordinate neighbourhoods. Set $x_j\colon p \to x_j(p) = (X_j^1(p), \ldots, x_j^n(p))$ be C^∞ local coordinates on X_j. This maps X_j diffeomorphically onto the open unit ball of $\mathbb{R}^n$. Since the interior of the unit ball is diffeomorphic to a domain in an n-dimensional torus $\mathbb{T}^n$, we may consider x_j as the diffeomorphism of X_j onto a domain V_j in $\mathbb{T}^n$.

Let (B, X, t) be a complex vector bundle over X. The following discussion is also true for real vector bundles. Let μ be the dimension of fibre of B. We may assume that $\pi^{-1}(X_j) \cong X_j \times \mathbb{C}^\mu$ on each coordinate neighbourhood X_j. Let ψ be a section of B over X_j. Then ψ can be identified with a $\mathbb{C}^\mu$-valued function $\psi_j(x_j)$ defined on the domain V_j in $\mathbb{T}^n$, and it is written as follows:

$$\psi_j(x_j) = (\psi_j^1(x_j), \ldots, \psi_j^\mu(x_j)), \qquad x_j \in V_j. \tag{3.4}$$

Suppose that ψ is defined on X_k, too. Then we have a $\mathbb{C}^\mu$-valued function $\psi_k(x_k)$, with $x_k \in V_k$:

$$\psi_k(x_k) = (\psi_k^1(x_k), \ldots, \psi_k^\mu(x_k)), \qquad x_k \in V_k. \tag{3.4$'$}$$

For $p \in X_j \cap X_k$ with $x_j = x_j(p)$ and $x_k = x_k(p)$, we have the transition relation:

$$\psi_j^\lambda(x_j(p)) = \sum_\rho b_{jk\rho}^\lambda(p)\psi_k^\rho(x_k(p)), \tag{3.5}$$

where $b_{jk\rho}^\lambda(p)$ is a C^∞ function of $p \in X_j \cap X_k$. We denote by $C^\infty(X, B)$ the set of all C^∞ sections of B over X. Choose C^∞ functions $\chi_j(x)$ on X such that

$$\sum_j \chi_j(x)^2 \equiv 1, \qquad \operatorname{supp} \chi_j \subset X_j. \tag{3.6}$$

For any C^∞ section ψ of B, define the product $\chi_j\psi$ by $\chi_j\psi(x) = \chi_j(x)\psi(x)$. Then we can consider it as a C^∞-valued function on V_j, since $\operatorname{supp} \chi_j\psi \subset X_j$. Moreover its support is contained in a compact subset of V_j. Therefore for an integer l, we can define the norm $\|\chi_j\psi\|_{j,l}$ in $W^l(V_j, \mathbb{C}^\mu)$. The suffix j

means that X_j is identified with V_j by x_j. Using this, we define the l-norm $\|\psi\|_l$ on X as follows:

$$\|\psi\|_l^2=\sum_j \|\chi_j\psi\|_{j,l}^2=\sum_j \|\chi_j\cdot\chi_j^{-1}\chi_j(x_j)\|_{j,l}^2. \tag{3.7}$$

This norm depends on the choice of $\{\chi_j\}_j$. Instead of $\{\chi_j\}_j^J$ take $\{\omega_j\}_j^J$ such that

$$\sum_j \omega_j(x)^2\equiv 1, \qquad \operatorname{supp}\omega_j\subset X_j.$$

Then the norm $\sum_j \|\omega_j\psi\|_{j,l}^2$ is equivalent to (3.7). To show this fact, it suffices to see that there exists a positive constant C independent of ψ such that, for any $\psi\in C^\infty(X, B)$,

$$C^{-1}\sum_k \|\chi_k\psi\|_{k,l}\leqq\sum_j \|\omega_j\psi\|_{j,l}\leqq C\sum_k \|\chi_k\psi\|_{k,l}. \tag{3.8}$$

For any k, we get, by (3.6),

$$\|\omega_j\psi\|_{j,l}=\left\|\sum_k \chi_k^2\omega_j\psi\right\|_{j,l}\leqq\sum_k \|\chi_k^2\omega_j\psi\|_{j,l}. \tag{3.9}$$

We denote $(\chi_k^2\omega_j)(x_j^{-1}(y))=a_{(kj)}(y)$, $y\in V_j$. Using the representation (3.4) of ψ on V_j, we have

$$\|\chi_k^2\omega_j\omega\|_{j,l}=\|a_{(kj)}\psi_j\|_{j,l}. \tag{3.10}$$

The support of $\chi_k^2\omega_j$ is contained in $X_j\cap X_k$ so that we can apply Lemma 3.1 to $\Phi=x_j\cdot x_k^{-1}$. For $z\in V_k$ we put $a_{(kj)}\cdot\Phi(z)=\tilde{a}_{(kj)}(z)\psi_j\cdot\Phi=\tilde{\chi}_j(z)$ respectively. Then by Lemma 3.1 we can take a positive C_1 such that

$$\|a_{(kj)}\psi_j\|_{j,l}\leqq C_1\|\tilde{a}_{(kj)}\tilde{\psi}_j\|_{k,l}. \tag{3.11}$$

Putting $\tilde{\psi}_j(z)=(\tilde{\psi}_j^1(z),\ldots,\tilde{\psi}_j^\mu(z))$, we have, by the transition formula (3.5),

$$\tilde{\psi}_j^\lambda(z)=\sum_\rho b_{jk\rho}^\lambda(x_k^{-1}(z))\psi_k^\rho(z).$$

For $b_{jk\rho}^\lambda$ is a C^∞ function, there is a positive C_2 such that

$$\|\tilde{a}_{(kj)}\tilde{\psi}_j\|_{k,l}\leqq C\|\tilde{a}_{(kj)}\psi_k\|_{k,l}. \tag{3.12}$$

Since $\tilde{a}_{(kj)}(z)=a_{(kj)}\cdot\Phi(z)=(\chi_k^2\omega_j)(x_k^{-1}(z))$, we have

$$\|a_{(kj)}\psi_k\|_{k,l}=\|\chi_k^2\omega_j\psi\|_{k,l}. \tag{3.13}$$

Since $\chi_k\omega_j$ is a C^∞ function, there is a positive C_3 such that

$$\|\chi_k^2\omega_j\varphi\|_{k,l} \leqq C_3\|\chi_k\psi\|_{k,l}. \tag{3.14}$$

By (3.10)-(3.14), we get

$$\|\chi_k^2\omega_j\psi\|_{j,l} \leqq C_1C_2C_3\|\chi_k\psi\|_{k,l}.$$

By (3.9) and the above inequality,

$$\|\omega_j\psi\|_{j,l} \leqq C_1C_2C_3\sum_k \|\chi_k\psi\|_{k,l}.$$

Taking the summation over j, we obtain

$$\sum_j \|\omega_j\psi\|_{j,l} \leqq C_1C_2C_3J\sum_k \|\chi_k\psi\|_{k,l}$$

which shows the right-half of (3.8). By the way, the roles of $\{\omega_j\}$ and $\{\chi_k\}$ are symmetrical. Therefore the left-half of (3.8) also holds.

Hereafter, choosing $\{x_j\}$ once and for all, we fix the norm as in (3.7). Clearly $\|\varphi\|_{k'} \leqq \|\varphi\|_k$, for $k' < k$.

By $C^\infty(X, B)$ we denote the totality of C^∞ sections of B.

Proposition 3.1. *Let l, l' and l'' be three integers with $l'' < l' < l$. Then*

(1°) *For any $t > 0$ and any $\psi \in C^\infty(X, B)$, the following inequality holds*

$$\|\psi\|_{l'}^2 \leqq \frac{l'-l''}{l-l''}t^{(l-l'')/(l'-l'')}\|\psi\|_l^2 + \frac{l-l'}{l-l''}t^{-(l-l'')/(l-l')}\|\psi\|_{l''}^2. \tag{3.15}$$

(2°) *For any $\psi \in C^\infty(X, B)$,*

$$\|\psi\|_{l'} \leqq \|\psi\|_l^{(l'-l'')/(l-l'')}\|\psi\|_{l''}^{(l-l')/(l-l'')}. \tag{3.16}$$

Proof. For any j, by Lemma 1.1, we have

$$\|\chi_j\psi\|_{j,l'}^2 \leqq \frac{l'-l''}{l-l''}t^{(l-l'')/(l'-l'')}\|\chi_j\psi\|_{j,l}^2 + \frac{l-l'}{l-l''}t^{-(l-l'')/(l-l')}\|\chi_j\psi\|_{j,l''}^2$$

for any $t > 0$ and any $\psi \in C^\infty(X, B)$. Summing these with respect to j, we obtain (3.15). Taking the minimum with respect to t, we get (3.16). ∎

A linear operator $A: C^\infty(X, B) \to C^\infty(X, B)$ is said to be a *multiplication operator*, if it is represented in the form of (3.4) as

$$(A_j\psi)_j^\lambda(x_j(p)) = \sum_\rho a_{j\rho}^\lambda(x_j(p))\psi_j^\rho(x_j(p)), \qquad p \in X_j, \tag{3.17}$$

where $a_{j\rho}^\lambda(x_j(p))$ are C^∞ functions and for fixed $p \in X_j$ the $(\mu \times \mu)$- matrix $(a_{j\rho}^\lambda(x_j(p)))_{\rho,\lambda=1}^\mu$ is a linear transformatin of the fibre $\pi^{-1}(p)$ at p. Therefore A is considered as a section of the vector bundle $\mathrm{Hom}_{\mathbf{C}}(B) \cong B \otimes B^*$ where B^* is the dual vector bundle of B. We put

$$M_l = \sum_j \sum_{\rho,\lambda} \sum_{|\beta| \leq l} \sup_{x_j \in V_j} |D_{x_j}^\beta a_{j\rho}^\lambda(x_j)|.$$

Lemma 3.2. *Let l be an integer. Then for a multiplication operator A there exists a positive C such that for any $\varphi \in C^\infty(X, B)$,*

$$\|A\varphi\|_l \leqq CM_{|l|}\|\varphi\|_l. \tag{3.18}$$

C depends on n, l, μ, but is independent of φ and representations of A.

Proof. Take $\{\chi_j\}$ as in (3.6). By (3.17) we have $\chi_j A = A\chi_j$,

$$(\chi_j A\varphi)_j^\lambda(x_j(p)) = \sum_\rho a_{j\rho}^\lambda(x_j(p))\chi_j(p)\varphi_j^\rho(x_j(p)).$$

Since $a_{j\rho}^\lambda$ are C^∞ functions, we can apply to A the case $m = 0$ in Proposition 1.4 and we see that there is a positive C such that

$$\|\chi_j A\varphi\|_{j,l} \leqq CM_{|l|}\|\chi_j\varphi\|_{j,l}.$$

Squaring both sides, and summing with respect to j, we get the desired inequality by taking the square roots of both sides. ∎

Corollary. *In the identity* (3.17), *if there is such a positive δ that*

$$\inf_j \min_{x_j \in V_j} |\det(a_{j\rho}^\lambda(x_j))| \geqq \delta, \tag{3.19}$$

then, for each l, we can take a positive C depending on l, δ, μ, n and $M_{|l|}$, such that

$$\|\psi\|_l \leqq C\|A\psi\|_l \tag{3.20}$$

for any $\varphi \in C^\infty(X, B)$.

Proof. This follows from the fact that the inverse mapping A^{-1} of A is also a multiplication operator. ∎

A linear map $A: C^\infty(X, B) \to C^\infty(X, B)$ is said to be a *linear partial differential operator of order m*, if it has the form

$$(A\psi)_j^\lambda(x_j(p)) = \sum_{|\alpha|\leqq m} \sum_{\rho} a_{j\rho\alpha}^\lambda(x_j(p)) D_j^\alpha \psi_j^\rho(x_j(p)) \tag{3.21}$$

in the local representation (3.4), where $a_{j\rho\alpha}^\lambda$ are C^∞ functions and for a multi-index $\alpha = (\alpha_1, \alpha_2, \ldots, \alpha_n)$ D_j^α means

$$D_j^\alpha = D_{j,1}^{\alpha_1} \cdots D_{j,n}^{\alpha_n} = \left(\frac{\partial}{\partial x_j^1}\right)^{\alpha_1} \cdots \left(\frac{\partial}{\partial x_j^n}\right)^{\alpha_n}. \tag{3.22}$$

Note that the meaning of D_j above is somewhat different from the one given in §1. We put for any non-negative integer l

$$M_l = \sum_j \sum_\alpha \sum_{\rho,\lambda} \sum_{|\beta|\leqq l} \sup_{x_j \in V_j} |D_j^\beta a_{j\rho\alpha}^\lambda(x_j)|. \tag{3.23}$$

Lemma 3.3. *Let A be a linear partial differential operator which operates on $C^\infty(X, B)$. Then there exists a positive C such that for any $\varphi \in C^\infty(X, B)$ the following inequality holds.*

$$\|A\varphi\|_l \leqq CM_{|l|} \|\varphi\|_{l+m}, \tag{3.24}$$

where C is determined by l, m, n and μ and independent of representations of A.

Proof. Take χ_j as in (3.6). Fix j, and let ω_k, $k = 1, \ldots, J$, be C^∞ function on X with supp $\omega_k \subset X_k$ and $\sum_k \omega_k(x)^2 = 1$ such that $\omega_j(x) = 1$ identically on a neighbourhood of the support of χ_j. Such $\{\omega_k\}$ do exist. Since ω_j is identically equal to 1 on the support of χ_j, $\chi_j A\varphi = \chi_j A\omega_j\varphi$. Hence

$$\|\chi_j A\varphi\|_{j,l} = \|\chi_j A\omega_j\varphi\|_{j,l} \leqq \|A\chi_j\varphi\|_{j,l} + \|[\chi_j, A]\omega_j\varphi\|_{j,l}. \tag{3.25}$$

By Proposition 1.4 and Lemma 1.4, we have a positive C_1 such that

$$\|A\chi_j\varphi\|_{j,l} \leqq C_1 M_{|l|} \|\chi_j\varphi\|_{j,l+m}, \tag{3.26}$$

and

$$\|[\chi_j, A]\omega_j\varphi\|_{j,l} \leqq C_1 M_{|l|} \|\omega_j\varphi\|_{j,l+m-1}. \tag{3.27}$$

By the choice of $\{\omega_k\}$, (3.8) holds. That is, there is a positive $C_{2,j}$ such that

$$\sum_k \|\omega_k\varphi\|_{k,l+m-1} \leqq C_{2,j} \sum_k \|\chi_k\varphi\|_{k,l+m-1}.$$

By (3.25), (3.26), and (3.27),

$$\|\chi_j A\varphi\|_{j,l} \leqq C_1 M_{|l|}\left(\|\chi_j\varphi\|_{j,l+m} + C_{2,j}\sum_k \|\chi_k\varphi\|_{k,l+m-1}\right).$$

Summing with respect to j and putting $C_3 = \sum_j C_{2,j}$, we have

$$\sum_j \|\chi_j A\varphi\|_{j,l} \leqq C_1 M_{|l|}\left(\sum_j \|\chi_j\varphi\|_{j,l+m} + C_3\sum_k \|\chi_k\varphi\|_{k,l+m-1}\right)$$

$$\leqq C_1 M_{|l|}(1+C_3)\sum_k \|\chi_k\varphi\|_{k,l+m}.$$

From this inequality, it follows

$$\sum_j \|\chi_j A\varphi\|_{j,l}^2 \leqq C_1^2 M_{|l|}^2 (1+C_3)^2 J \sum_k \|\chi_k\varphi\|_{k,l+m}^2.$$

That is,

$$\|A\varphi\|_l^2 \leqq C_1^2 M_{|l|}^2 (1+C_3)^2 J \|\varphi\|_{l+m}^2.$$

This proves (3.24). ∎

For two C^∞ sections φ and ψ of a vector bundle B, the inner product $(\varphi, \psi)_l$ of order l is defined as

$$(\varphi, \psi)_l = \sum_j (\chi_j\varphi, \chi_j\psi)_{j,l}. \tag{3.28}$$

Here the inner product of the right-hand side is understood to be the one of the Sobolev space, for, the supports of $\chi_j\varphi$ and $\chi_j\psi$ being in X_j, $\chi_j\varphi$ and $\chi_j\psi$ can be identified by the coordinate function χ_j with the vector-valued functions on V_j. The suffix j in (3.28) indicates that X_j is identified with V_j. This inner product is positive definite, and

$$\|\phi\|_l^2 = (\varphi, \varphi)_l. \tag{3.29}$$

The completion of $C^\infty(X, B)$ with respect to the norm $\| \ \|_l$ is called the *Sobolev space of order* l, denoted by $W^l(X, B)$. *The inner product* (3.28) extends canonically to the one on $W^l(X, B)$, for which we use the same notation in (3.28). $W^l(X, B)$ becomes a Hilbert space with respect to this inner product. Clearly this inner product depends on the choice of $\{\chi_j\}$. But the norms determined by $\{\chi_j\}$ are all equivalent. Therefore, as a vector space, $W^l(X, B)$ is independent of the choice of $\{\chi_j\}$, and so is its topological structure.

By Lemma 3.3, a linear partial differential operator A can be uniquely extended to the continuous map from $W^l(X, B)$ to $W^{l-m}(X, B)$. In particular, a multiplication operator is a continuous map from $W^l(X, B)$ to $W^l(X, B)$. For $S \in W^l(X, B)$, the support of S denoted by supp S is defined as the smallest compact set $K \subset X$ such that the product aS vanishes for any C^∞ function $a(x)$ whose support is contained in $X \backslash K$. If S is an arbitrary element of $W^l(X, B)$ and $\chi(x)$ is a C^∞ function, then

$$\operatorname{supp} \chi \cdot S \subset \operatorname{supp} \chi \cap \operatorname{supp} S. \tag{3.30}$$

Suppose that the support of χ is contained in a coordinate neighbourhood X_i and take a family of C^∞ functions $\{\omega_j\}$ on X with supp $\omega_j \subset X_j$ and $\sum_j \omega_j(x)^2 \equiv 1$ such that $\omega_i(x) \equiv 1$ on supp χ. Since $S \in W^l(X, B)$, we have a sequence $\{\varphi_k\}_{k=1}^\infty$ in $C^\infty(X, B)$ which converges to S in $W^l(X, B)$. Since the norms are equivalent if we take $\{\omega_j\}$ instead of $\{X_j\}$, there is a positive C_1 such that for any k, k'

$$\|\chi\varphi_k - \chi\varphi_{k'}\|_{i,l} = \|\omega_i(\chi\varphi_k - \chi\varphi_{k'})\|_{i,l} \leqq C_1 \|\chi\varphi_k - \chi\varphi_k\|_l.$$

By Lemma 3.2, we have a positive C_2 such that

$$\|\chi\varphi_k - \chi\varphi_{k'}\|_l \leqq C_2 \|\varphi_k - \varphi_{k'}\|_l.$$

Hence for any k, k'

$$\|\chi\varphi_k - \chi\varphi_{k'}\|_{i,l} \leqq C_1 C_2 \|\varphi_k - \varphi_{k'}\|_l.$$

Therefore regarding $\chi\varphi_k$ as a vector-valued function, we see that $\{\chi\varphi_k\}$ converges in $W^l(V_i, \mathbb{C}^\mu)$. Letting $\tilde{S}_i$ be its limit, we identify χS with $\tilde{S}_i$. We can choose χ arbitrarily so that we may consider an element S of $W^l(X, B)$ as a section of B which can be locally identified with an element of $W^l(V_i, \mathbb{C}^\mu)$. Of course when the support of χ is contained in $X_k \cap X_j$, χS can be identified both with $\tilde{S}_k \in W^l(V_k, \mathbb{C}^\mu)$ and with $\tilde{S}_j \in W^l(V_j, \mathbb{C}^\mu)$, while between $\tilde{S}_k$ and $\tilde{S}_j$, there is a relation which extends (3.5). Conversely, we can prove that the totality of section of B which can be considered locally as an element of $W^l(V, \mathbb{C}^\mu)$ which satisfies this relation coincides with $W^l(X, B)$. But the detail of the proof is so messy that we omit it. In case $l \geqq 0$, however, it is not difficult, because the convergence of $\chi\varphi_k$ in $W^l(V_i, \mathbb{C}^\mu)$ is stronger than that in $L^2(V_i, \mathbb{C}^\mu)$.

In the above, we take χ_i as χ and let S_i be the element of $W^l(V_i, \mathbb{C}^\mu)$ corresponding to $\chi_i S$. Since

$$\|\chi_i\varphi_k\|_{i,l}^2 \leqq C_1^2 C_2^2 \|\varphi_k\|_l^2 \leqq C_1^2 C_2^2 \sum_j \|\chi_i\varphi_k\|_{i,l}^2,$$

putting $k\to\infty$, we get

$$\|S_i\|_{i,l}^2 \leqq C_1^2 C_2^2 \|S\|_l^2 \leqq C_1^2 C_2^2 \sum_j \|S_j\|_{i,l}^2.$$

Thus there exists a constant C such that, for any $S\in W^l(X, B)$,

$$C^{-1}\sum_j \|S_j\|_{j,l}^2 \leqq \|S\|_l^2 \leqq C\sum_j \|S\|_{j,l}^2 \tag{3.31}$$

Theorem 3.1 (Rellich's Theorem). *Let $l'<l$. Then $W^l(X, B)$ is contained in $W^{l'}_{(X,B)}$, and the inclusion map $W^l(X, B)\to W^{l'}(X, B)$ is a compact linear map.*

Proof. It suffices to show that the inclusion map is compact. Let $\{S^k\}_{k=1}^{\infty}$ be a sequence in $W^l(X, B)$ such that there is a positive constant R with

$$\|S^k\|_l < R.$$

By the above argument we take $S_j^k\in W^l(V_i, \mathbb{C}^\mu)$ corresponding to $j=1,\ldots, J$, then by (3.31) we have

$$C^{-1}\sum_j \|S_j^k\|_{j,l}^2 \leqq R^2,$$

in other words, for each j, $\{S_j^k\}_{k=1}^{\infty}$ is a bounded sequence in $W_0^l(V_j, \mathbb{C}^\mu)$. Thus since it is a bounded sequence in $W^l(\mathbb{T}^n, \mathbb{C}^\mu)$, we can choose a subsequence convergent in $W^{l'}(\mathbb{T}^n, \mathbb{C}^\mu)$ (Theorem 1.10). We may assume that these subsequences have the same set of indices for all j. We write them as $\{S_j^K\}$. We can choose a sufficiently large K_0 such that for any $K, K'>K_0$, and for any j,

$$\|S_j^K - S_j^{K'}\|_{j,l'} < J^{-1}C^{-1}\varepsilon$$

holds, where C is the same as in (3.31). Therefore, by (3.31), for $K, K'>K_0$, we have

$$\|S^K - S^{K'}\|_{l'} < \varepsilon.$$

Thus $\{S^K\}_K$ forms a Cauchy sequence in $W^{l'}(X, B)$, converges in $W^{l'}(X, B)$. Therefore the map $W^l(X, B)\to W^{l'}(X, B)$ is compact. ∎

Let ψ be a section of B over X, and $\psi_j(x_j)$ the vector-valued function (3.4) of x_j on V_j, corresponding to the restriction of ψ to X_j. ψ is called *a $C^{k+\theta}$ section of B* if for each j, $\psi_j(x_j)$ is a $C^{k+\theta}$ function of $x_j\in V_j$, where $0\leqq\theta<1$. The set of all such ψ is denoted by $C^{k+\theta}(X, B)$. We take $\{\chi_j\}$ as in (3.6) and define the norm of $C^{k+\theta}(X, B)$ by

$$|\psi|_{k+\theta} = \sum_j |\chi_j\psi|_{j,k+\theta}, \tag{3.32}$$

where $|\ |_{j,k+\theta}$ means the norm of $\chi_j\psi$ as an element of $C^{k+\theta}(V_j, \mathbb{C}^\mu)$. For a different choice of $\{\chi_j\}$, (3.32) defines an equivalent norm, because a similar estimate to (3.8) holds for $|\ |_{j,l+\theta}$.

Theorem 3.2 (Sobolev's Imbedding Theorem). *Let l and k be non-negative integers, and suppose $l > n/2+k$. Then for $f \in W^l(X, B)$, there exists a unique $\tilde{f} \in C^k(X, B)$ such that $f(p) = \tilde{f}(p)$ on X except for a set of measure 0. Furthermore, there exists a positive constant C depending only on k and l such that*

$$|\tilde{f}|_k \leqq C\|f\|_l \quad \text{(Sobolev's inequality)} \tag{3.33}$$

holds.

Proof. Both $W^l(X, B)$ and $C^k(X, B)$ are imbedded continuously into $W^0(X, B)$, and $C^\infty(X, B)$ is dense in $W^l(X, B)$. Therefore it suffices to show the equality (3.33) for $f \in C^\infty(X, B)$. In this case $\tilde{f}(p) = f(p)$ for all $p \in X$.

By (1.23) in the proof of Theorem 1.3, there is a positive constant C_j such that

$$|\chi_j f|_{j,k} \leqq C_j \|\chi_j f\|_{j,l}. \tag{3.34}$$

Putting $\max C_j = C$, and summing (3.34) with respect to j, we have

$$|f|_k \leqq C \sum_j \|\chi_j f\|_{j,l} \leqq C\sqrt{J}\|f\|_l,$$

which proves (3.33). ∎

Assume that X is oriented. By a volume element of X, we mean a C^∞ differential n-form $v(dx)$ which is positive at every point of X. If a subset N of X is of measure 0 with respect to one volume element, so is it with respect to another one, hence, the notion of the sets of measure 0 does not depend on the choice of a volume element.

By a metric g on a vector bundle B, we mean a C^∞ section of the vector bundle $B^* \otimes B^*$ such that the following condition is satisfied: Let $p \in V_j$, and let (p, ζ), (p, ζ') be two points on the fibre $\pi^{-1}(p)$ with fibre coordinates $\zeta_j = (\zeta_j^1, \ldots, \zeta_j^\mu)$, and $\zeta'_j = (\zeta_j'^1, \ldots, \zeta_j'^\mu)$ respectively. Then the Hermitian form on $\pi^{-1}(p)$ defined by

$$g_{pi}(\zeta, \zeta') = \sum_{\lambda,\rho} g_{i\lambda\bar{\rho}}(x_i(p))\zeta_i^\lambda \overline{\zeta_j'^\rho} \tag{3.35}$$

is positive definite.

Let $\varphi, \psi \in C^\infty(X, B)$. In terms of fibre coordinates over X_j, we write $\varphi_i(p) = (\varphi_i^1(x_i(p)), \ldots, \varphi_i^\mu(x_i(p)))$, and $\psi_i(p) = (\psi_i^1(x_i(p)), \ldots, \psi_i^\mu(x_i(p)))$. We define the inner product of φ and ψ with respect to the volume element $v(dx) = v_i(x_i(p))\, dx = v_i(x_i(p))\, dx_i^1 \cdots \wedge dx_i^\mu$ and the metric g by

$$(\varphi, \psi) = \int_X g_p(\varphi(p), \psi(p)) v(dx)$$

$$= \int_X \sum_{\lambda,\rho} g_{i\lambda\bar\rho}(x_i(p)) \varphi_i^\lambda(x_i(p)) \overline{\psi_i^\rho(x_i(p))}\, v_i(x_i(p))\, dx_i. \tag{3.36}$$

This has an intrinsic meaning. With this inner product, $C^\infty(X, B)$ becomes a pre-Hilbert space. The completion with respect to the norm

$$\|\varphi\| = (\varphi, \varphi)^{1/2} \tag{3.37}$$

is a Hilbert space, which we denote by $L^2(X, B)$. This is the same as $W^0(X, B)$ as a topological vector space, but with a different inner product.

Proposition 3.2. *We fix a metric g and a volume element $v(dx)$. The inner product (,) defined on $C^\infty(X, B)$ is uniquely extended to a continuous sesquilinear map of $W^{-l}(X, B) \times W^l(X, B)$ to $\mathbb{C}$ for any integer l, which we denote by the same (,). This is non-degenerate.*

Proof. Take $\{\chi_j\}$ as in (3.6). Then for $\varphi, \psi \in C^\infty(X, B)$, we have

$$(\varphi, \psi) = \sum_j (\chi_j\varphi, \chi_j\psi). \tag{3.38}$$

Since for each j,

$$(\chi_j\varphi, \chi_j\psi) = \int_X \sum_{\lambda,\rho} g_{j\lambda\bar\rho}(x_j(p)) \chi_j\varphi_j^\lambda(x_j(p)) \overline{\chi_j\psi_j^\rho(x_j(p))}\, v_j(x_j(p))\, dx_j(p) \tag{3.39}$$

by Theorem 1.11 in §1, for any integer l, there is a positive constant $C_{l,j}$ such that

$$|(\chi_j\varphi, \chi_j\psi)| \leqq C_{l,j} \|\chi_j\varphi_j\|_{j,-l} \|\chi_j\psi\|_{j,l}.$$

Summing with respect to j, and using Schwartz' inequality we have

$$|(\varphi, \psi)| \leqq C_l \|\varphi\|_{-l} \|\psi\|_l, \tag{3.40}$$

where we put $C_l = \max_j C_{l,j}$. Thus (,) is continuous with respect to the relative topology of $C^\infty(X, B) \times C^\infty(X, B)$ in $W^{-l}(X, B) \times W^l(X, B)$. Since

$C^\infty(X, B) \times C^\infty(X, B)$ is dense in $W^{-l}(X, B) \times W^l(X, B)$, (,) is extended uniquely to a sesquilinear map of $W^{-l}(X, B) \times W^l(X, B)$.

Let $S \in W^{-l}(X, B)$. Then $\chi_j S$ is identified with $S_j \in W_0^{-l}(V_j, B)$. Define $T_j = (T_j^1, \ldots, T_j^\mu)$ by

$$T_j^\rho = \sum_\lambda v_j(x_j) g_{j\lambda\rho}(x) S_j^\lambda, \qquad \rho = 1, \ldots, \mu. \tag{3.41}$$

Then by (3.38) and (3.39), we have

$$(S, \varphi) = \sum_j \langle T_j, \chi_j \bar{\varphi}_j \rangle_j, \tag{3.42}$$

where $\langle\ ,\ \rangle$ is the natural bilinear map of $W_0^{-l}(V_j, \mathbb{C}^\mu) \times W_0^l(V_j, \mathbb{C}^\mu)$ to $\mathbb{C}$ on V_j. The left-hand side is independent of the choice of $\{\chi_j\}$. Suppose that $S \in W^{-l}(X, B)$ satisfies $(S, \varphi) = 0$ for any $\varphi \in C^\infty(X, B)$. For an arbitrary $a \in C^\infty(X)$ with supp $a \subset X_k$, there is a set $\{\chi_j\}_{j=1}^J$ satisfying (3.6) such that $\chi_k(x) \equiv 1$ on supp a. Since (3.42) holds for such $\{\chi_j\}$, we have for any $\varphi \in C^\infty(X, B)$

$$0 = (S, a\varphi) = \langle S_k, a\chi_k\varphi \rangle_k = \langle aS_k, \chi_k\varphi \rangle_k,$$

hence $aS_k = 0$ as an element of $W_0^{-l}(V_k, \mathbb{C}^\mu)$, hence $(aS)_k = 0$. Since $a\chi_j = 0$ for $j \neq k$, we have $aS_j = 0$. Thus $(aS)_j = 0$ for all j, hence $aS = 0$ as an element of $W^{-l}(X, B)$. Let $\{\omega_j\}_j$ be a partition of unity subordinate to the open covering $X = \bigcup_j X_j$. Namely, supp $\omega_j \subset X_j$, and $\sum \omega_j \equiv 1$. Then by the above consideration, we have $\omega_j S = 0$, $j = 1, \ldots, J$. Therefore we obtain

$$S = \sum_j \omega_j S = 0.$$

Thus $S = 0$, which implies the non-degeneracy of (,). ∎

§4. Elliptic Linear Partial Differential Operators

(A) Estimates with Respect to the Sobolev Norm

(a) *A priori* Estimate

Let X be a compact manifold of dimension n, B a complex vector bundle over X with μ-dimensional fibres, and $\pi: B \to X$ its projection. We use the same notation as in §3.

Let $A(p, D): C^\infty(X, B) \to C^\infty(X, B)$ be a linear partial differential operator of order l. In terms of the local coordinates $x_j: X_j \to V_j$ on the open set X_j, a section $\varphi \in C^\infty(X, B)$ of B is written as

$$\varphi_j(x_j(p)) = (\varphi_j^1(x_j(p)), \ldots, \phi_j^\mu(x_j(p))).$$

We write the operation of $A(p, D)$ as

$$(A(p, D)\varphi)_j^\lambda(x_j(p)) = \sum_{|\alpha| \leqq l} \sum_\rho a_{j\rho\alpha}^\lambda(x_j(p)) D_j^\alpha \varphi_j^\rho(x_j(p)), \tag{4.1}$$

where the $a_{j\rho\alpha}^\lambda(x_j(p))$ are C^∞ functions of $x_j(p)$. The operator $A_l(p, D)$ consisting of the terms with $|\alpha| = l$, given by

$$(A_l(p, D)\varphi)_j^\lambda(x_j(p)) = \sum_{|\alpha| = l} \sum_\rho a_{j\rho\alpha}^\lambda(x_j(p)) D_j^\alpha \varphi_j^\rho(x_j(p)) \tag{4.2}$$

is called the *principal part of* $A(p, D)$. Let $f(p)$ be a real-valued C^∞ function on X, and suppose that $\xi_p = df(p) \neq 0$ at p. Then $t \to e^{itf(p)} A(p, D) e^{-itf(p)} \varphi(p))$ is a polynomial of degree l in t, whose coefficient of the term of degree l is given by

$$(A_l(p, i\xi_p)\varphi)_j^\lambda(x_j(p)) = \sum_{|\alpha| = l} \sum_\rho a_{j\rho\alpha}^\lambda(x_j(p)) i^l \xi_p^\alpha \cdot \varphi_j^\rho(x_j(p)), \tag{4.3}$$

where we put $\xi_p \equiv df(p) = \xi_{j1}\, dx_j^1 + \cdots + \xi_{jn}\, dx_j^n$.

For any non-zero element ξ_p in the cotangent space T_p^*X, there is an $f(p)$ with $\xi_p = df(p)$. For any element w in the fibre B_p at $p \in X$, there is a section φ such that $\varphi(p) = w$. Let $\xi_p \in T_p^*X$ be given by $\xi_{p,j} = \xi_{j1}\, dx_j^1 + \cdots + \xi_{jn}\, dx_j^n$ and let $w \in B_j$ be given by $w_j = (w_j^1, \ldots, w_j^n)$. We define the linear map $B_p \ni w \to w' \in B_p$ by

$$w_j'^\lambda \equiv (A_l(p, \xi_p)w)_j^\lambda = \sum_{|\alpha| = l} \sum_\rho a_{j\rho\alpha}^\lambda(x_j(p)) \xi_p^\alpha w_j^\rho, \qquad \lambda = 1, \ldots, \mu \tag{4.4}$$

for $0 \neq \xi_p \in T_p^*X$, which we call *the principal symbol* of $A(p, D)$.

Definition 4.1. $A(p, D)$ is said to be *elliptic* if for any $p \in X$, and any $\xi_p \in T_p^*X$ with $\xi_p \neq 0$, the principal symbol $A_l(p, \xi_p)$ is an isomorphism of $B_p = \pi^{-1}(p)$. In this case, there exists a positive constant δ such that for any $\xi_p \in T_p^*X$ with $\xi_p \neq 0$, and $w \in B_p$, the inequality

$$\sum_\lambda [(A_m(p, \xi_p)w)_j^\lambda]^2 \geqq \delta^2 \sum_\lambda |w_j^\lambda|^2. \tag{4.5}$$

The supremum of such δ is called the *constant of ellipticity of* $A(p, D)$.

We introduce the following constants as in the preceding section. For $k=0,1,\ldots,$ put

$$M_k=\sum_j\sum_\alpha\sum_{p,\lambda}\sum_{|\beta|\leqq k}\sup_{x_jV_j}|D_j^\beta a_{jp\alpha}^\lambda(x_j)|. \tag{4.6}$$

Theorem 4.1 (L^2 *a priori* Estimate). *Let $A(p,D)$ be an elliptic linear partial differential operator of order l. For any integer k, there is a positive constant C depending only on n, l, k, μ, the constant δ of ellipticity of $A(p,D)$ and $M_{|k|}$, such that for any $u\in W^{l+k}(X,B)$, the inequality*

$$\|u\|_{k+l}\leqq C(\|A(p,D)u\|_k+\|u\|_k) \tag{4.7}$$

holds.

Proof. Since both sides of (4.7) are continuous with respect to the topology of $W^{k+1}(X,B)$, and $C^\infty(X,B)$ is dense in $W^{k+l}(X,B)$, it suffices to prove (4.7) for any $u\in C^\infty(X,B)$. Take functions $\{\chi_j\}_{j=1}^J$ as in (3.6) in the preceding section. For any $u\in C^\infty(X,B)$, there is a j such that

$$J^{1/2}\|u\|_{k+1}\leqq\|\chi_ju\|_{j,k+l}. \tag{4.8}$$

Put $u_j=\chi_ju$. Take another family of functions $\{\omega_i\}_{i=1}^J$ on X such that $\operatorname{supp}\omega_j\subset X_j$ and that $\sum_i\omega_i(p)^2\equiv 1$ on X. Moreover we assume that ω_j is identically equal to 1 in some neighbourhood of $\operatorname{supp}\chi_j$. Then we have

$$\begin{aligned}\|A(p,D)u\|_k&\geqq\|\chi_jA(p,D)u\|_{j,k}\\&=\|\chi_jA(p,D)\omega_ju\|_{j,k}\\&=\|A(p,D)u_j\|_{j,k}-\|[A(p,D),\omega_j]\omega_ju\|_{j,k}.\end{aligned} \tag{4.9}$$

Since $[A(p,D),\chi_j]$ is a partial differential operator of order $(l-1)$, by Theorem 1.4, there is a positive constant C_1 such that

$$\|[A(p,D),\chi_j]\omega_ju\|_{j,k}\leqq C_1M_{|k|}\|\omega_ju\|_{j,k+l-1}, \tag{4.10}$$

while there is a positive constant C_2 by (3.8) such that

$$\|\omega_ju\|_{j,k+l-1}\leqq\sum_i\|\omega_iu\|_{i,k+l-1}\leqq C_2\|u\|_{k+l-1}. \tag{4.11}$$

Since the support of $u_j=\chi_ju$ is contained in the coordinate neighbourhood X_j, u_j can be considered via the local coordinates x_j as an element of $C^\infty(\mathbb{T}^n,\mathbb{C}^\mu)$ with support in V_j. Thus we can apply Theorem 2.1 to find a positive constant C_3 depending on k, l, n, μ, δ and $M_{|k|}$ such that

$$\|A(p,D)u_j\|_{j,k}\geqq C_3\|u_j\|_{j,k+l}-\|u_j\|_{j,k}. \tag{4.12}$$

From the inequality $\|u_j\|_{j,k} \leqq \|u\|_k$, and (4.8), (4.9), (4.10), (4.11) and (4.12), we obtain

$$\|A(p, D)u\|_k \geqq C_3 J^{1/2}\|u\|_{k+l} - \|u\|_k - C_2 C_1 M_{|k|}\|u\|_{k+l-1}. \tag{4.13}$$

By Proposition 3.1, there exists a positive constant C_4 such that

$$C_2 C_1 M_{|k|}\|u\|_{k+l-1} \leqq 2^{-1} C_3 J^{-1/2}\|u\|_{k+l} + C_4\|u\|_k,$$

hence from this and (4.13), we obtain

$$\|A(p, D)u\|_k + (C_4+1)\|u\|_k \geqq 2^{-1} C_3 J^{1/2}\|u\|_{k+l}. \quad \blacksquare$$

(b) Gårding's Inequality

As stated at the end of §3, we consider the Hilbert space $L^2(X, B)$ defined by the volume element $v(dx)$ of X and the metric g on the vector bundle B. We denote its inner product by (,).

Lemma 4.1. *Let $A(p, D)$ be a linear partial differential operator of order $(m+l)$. Then there exists a positive constant C determined by m, n and μ such that for any $\varphi, \psi \in C^\infty(X, B)$,*

$$|(A(p, D)\varphi, \psi)| \leqq C M_l \|\varphi\|_m \|\psi\|_l \tag{4.14}$$

holds.

Proof. Take $\{\chi_j\}$ as in (3.6). Then we have

$$(A(p, D)\varphi, \psi) = \sum_j (A(p, D)\varphi, \chi_j^2\psi). \tag{4.15}$$

Fix j, and choose a family of C^∞ function $\{\psi_k\}_{k=1}^J$ with supp $\omega_k \subset X_k$ and $\sum \omega_k^2(x) \equiv 1$ so that ω_j is identically equal to 1 in some neighbourhood of supp χ_j. Then we have

$$(A(p, D)\varphi, \chi_j^2\psi) = (A(p, D)\omega_j\varphi, \chi_j^2\psi).$$

We may assume that $A(p, D)$ has the form (4.1) with $m+l$ instead of l. Then the right-hand side of the above equality is written as

$$(A(p, D)\omega_j\varphi, \chi_j^2\psi) = \sum_{|\alpha| \leqq m+l} \sum_{\rho, \lambda, \sigma} \int_{V_j} g_{j\lambda\bar{\rho}}(x_j(p)) a_{j\sigma\alpha}^{\lambda}(x_j(p))$$
$$\times D_j^\alpha(\omega_j\varphi_j^\sigma(x_j(p)))\overline{(\chi_j^2\psi_j^\epsilon)}(x_j(p))v_j(x_j(p))\, dx_j(p).$$

Integrating the terms with $|\alpha| \geqq m+1$ by parts $(|\alpha|-m)$ times, we have

$$(A(p, D)\omega_j\varphi, \chi_j^2\psi) = \sum_{\substack{|\alpha|\leqq m \\ |\beta|\leqq l}} \sum_{p,\lambda,\sigma} \int_{V_j} b_{j\lambda\bar{\rho}\alpha\beta}(x_j(p)) D_j^\alpha(\omega_j\varphi_j^\lambda(x_j(p)) \times \overline{D_j^\beta(\chi_j\psi_j^\rho)(x_j(p))} v_j(x_j(p))\, dx_j(p),$$

where the $b_{j\lambda\bar{\rho}\alpha\beta}(x_j(p))$ are linear combinations of partial derivatives of $\chi_j g_{j\lambda\bar{\rho}}(x_j(p)) a_{j\sigma\alpha}^\lambda(x_j(p)) v_j(x_j(p))$ of order at most l.

Therefore there exists a positive constant C_1 such that

$$|(A(p, D)\omega_j\varphi, \chi_j^2\psi)| \leqq C_1 M_l \|\omega_j\varphi\|_{j,m} \|\chi_j\psi\|_{j,l}.$$

By (3.8), there is a positive constant $C_{2,j}$ depending on j such that

$$|(A(p, D)\varphi, \chi_j^2\psi)| \leqq C_{2,j} M_l \|\varphi\|_m \|\psi\|_l.$$

Summing with respect to j, and putting $\sum_j C_{2,j} = C_3$, we obtain

$$\sum_j |(A(p, D)\varphi, \chi_j^2\psi)| \leqq C_3 M_l \|\varphi\|_m \|\psi\|_l.$$

(4.14) follows from this and (4.15). ∎

Let $A(p, D)$ be a linear partial differential operator of order $2m$. For $p \in X$, and $\xi_p \in T_p^*X$ with $\xi_p \neq 0$, the principal symbol $A_{2m}(p, \xi_p)$ of $A(p, D)$ is a linear map $B_p \to B_p$. Therefore, by virtue of the metric g_p on B_p, we can define a Hermitian form on $B_p \times B_p$ associated with this linear map by putting

$$g_p(A_{2m}(p, \xi_p)w, w') \tag{4.16}$$

for $w \in B_p$ and $w' \in B_p'$. For $p \in X_j$, if we write $A(p, D)$ in the form of (4.1) with $l = 2m$, then putting $\xi_p = \xi_{j1}\, dx_j^1 + \cdots + \xi_{jn}\, dx_j^n$, $w = (w_j^1, \ldots, w_j^n)$, $w' = (w_j'^1, \ldots, w_j'^n)$, we have

$$g_p(A_{2m}(p, \xi_p)w, w') = \sum_{\sigma,\lambda,\rho} \sum_{|\alpha|=2m} g_{j\lambda\bar{\rho}}(x_j(p)) z_{j\sigma\alpha}^\lambda(x_j(p)) \times \xi_j^\alpha w_j^\sigma \overline{w_j'^\rho}. \tag{4.17}$$

Definition 4.2. A linear partial differential operator $A(p, D)$ of order $2m$ is said to be *strongly elliptic* if there exists a positive constant δ such that for any $p \in X$, any $\xi_p \in T_p^*X$ with $\xi_p \neq 0$, and any $w \in B_p$ with $w \neq 0$,

$$(-1)^m \operatorname{Re} g_p(A_{2m}(p, \xi_p)w, w) \geqq \delta^2 g_p(w, w)|\xi_p|^{2m} \tag{4.18}$$

holds. The supremum of such δ is called the *constant of strong ellipticity of* $A(p, D)$.

Theorem 4.2 (Gårding's Inequality). *Let $A(p, D)$ be a strongly elliptic partial differential operator of order $2m$. Then there are positive constants δ_1, δ_2 depending on m, n, μ, the constant δ of strong ellipticity of $A(p, D)$ and M_m such that for any $\varphi \in C^\infty(X, B)$,*

$$\operatorname{Re}(A(p, D)\varphi, \varphi) + \delta_1(\varphi, \varphi) \geqq \delta_2 \|\varphi\|_m^2 \tag{4.19}$$

holds.

Proof. Take $\{\chi_j\}_{j=1}^J$ as in (3.6). For simplicity we write A for $A(p, D)$. Since $\sum_j \chi_j(s)^2 \equiv 1$, we have

$$(A\varphi, \varphi) = \sum_j (A\varphi, \chi_j^2\varphi) = \sum_j (\chi_j A\varphi, \chi_j\varphi)$$

$$= \sum_j (A\chi_j\varphi, \chi_j\varphi) + \sum_j (\chi_j[\chi_j, A]\varphi, \varphi).$$

Since $\chi_j[\chi_j, A]$ is a linear partial differential operator of order $(2m-1)$ whose coefficients are linear combinations of those of $A(p, D)$, by Lemma 4.1, there exists a positive constant C_1 depending on n, m, μ and M_m such that

$$|(\chi_j[\chi_j, A]\varphi, \varphi)| \leqq C_1 \|\varphi\|_m \|\varphi\|_{m-1}.$$

Since the support of $\chi_j\varphi$ is contained in X_j, $\chi_j\varphi$ can be considered as a $\mathbb{C}^\mu$-valued C^∞ function on $\mathbb{T}^n$ with support in V_j. Therefore by Theorem 2.2, there are positive constants C_2, C_3 such that

$$\operatorname{Re}(A\chi_j\varphi, \chi_j\varphi) + C_2\|\chi_j\varphi\|_{j,0}^2 \geqq C_3\|\chi_j\varphi\|_{j,m}^2.$$

From the above inequalities we obtain

$$\operatorname{Re}(A\varphi, \varphi) \geqq C_3 \sum_j \|\chi_j\varphi\|_{j,m}^2 - C_2 \sum_j \|\chi_j\varphi\|_{j,0}^2 - JC_1\|\varphi\|_m\|\varphi\|_{m-1}$$

$$= C_3\|\varphi\|_m^2 - C_2\|\varphi\|_0^2 - JC_1\|\varphi\|_m\|\varphi\|_{m-1}. \tag{4.20}$$

By Proposition 1.1, there is a positive number C_4 such that

$$JC_1\|\varphi\|_m\|\varphi\|_{m-1} \leqq \tfrac{1}{2}C_3\|\varphi\|_m^2 + C_4\|\varphi\|_0^2.$$

Consequently by (4.20) above, we obtain

$$\operatorname{Re}(A\varphi, \varphi) \geqq \tfrac{1}{2}C_3\|\varphi\|_m^2 - (C_2 + C_4)\|\varphi\|_0^2.$$

Since there is a positive constant C_5 such that for any $\varphi \in C^\infty(X, B)$,

$$\|\varphi\|_0^2 \leqq C_5(\varphi, \varphi),$$

we obtain

$$\operatorname{Re}(A\varphi, \varphi) \geqq \tfrac{1}{2}C_3\|\varphi\|_m^2 - C_5(C_2 + C_4)(\varphi, \varphi),$$

which proves (4.19). ∎

(B) *A priori* Estimate with Respect to the Hölder Norm

Lemma 4.2. *Let* $0 < r < s$. *Then there exist positive constants* C_1, C_2 *such that for any* $\varphi \in C^s(X, B)$,

$$|\varphi|_r \leqq C_1 |\varphi|_s^{2/r} |\varphi|_0^{1-r/s},$$

and that for any $t > 0$

$$|\varphi|_r \leqq t^{s/r} |\varphi|_s + C_2 t^{-2/(s-r)} |\varphi|_0.$$

This follows immediately from Lemma 2.8. ∎

Definition 4.3. Let $A(p, D)$ be a second-order linear partial differential operator acting on sections of a vector bundle B over X. *The principal part of* $A(p, D)$ *is said to be of diagonal type* if we can choose a system of local coordinates $\{(X_j, x_j)\}_{j=1}^J$ such that when we write $A(p, D)$ in terms of these coordinates as

$$(A(p, D)\phi)_j^\lambda(x_j(p)) = \sum_{|\alpha| \leqq 2} \sum_\rho a_{j\rho\alpha}^\lambda(x_j(p)) D_j^\alpha \varphi_j^\rho(x_j(p)), \tag{4.21}$$

then the following conditions is satisfied:

$$a_{j\rho\alpha}^\lambda(x_j(p)) \equiv 0 \quad \text{for } |\alpha| = 2 \quad \text{and} \quad \lambda \neq \rho, \qquad p \in X_j. \tag{4.22}$$

Remark. If in some choice of local coordinates, the principal part of $A(p, D)$ is of diagonal type, and it is written as

$$(A_2(p, D)\varphi)_j^\lambda(x_j(p)) = \sum_{|\alpha|=2} a_{j\alpha}(x_j(p)) D_j^\alpha \varphi_j^\lambda(x_j(p)), \tag{4.23}$$

then the principal part is of diagonal type in another choice of coordinates.

Theorem 4.3. *Let* $A(p, D)$ *be a strongly elliptic partial differential operator of order 2 with* C^∞ *coefficients acting on sections of a vector bundle* B *over* X *whose principal part is of diagonal type. Let* δ *be its constant of strong ellipticity, and put* $0 < \theta < 1$. *Then for any integer* $k \geqq 0$, *there exists a positive constant* C *depending only on* θ, k, n, μ, δ *and* $M_{k+\theta}$ *such that for any* $u \in C^{k+2+\theta}(X, B)$,

$$|u|_{k+2+\theta} \leqq C(|A(p, D)u|_{k+\theta} + |u|_0)$$

holds, where we put

$$M_{k+\theta}=M_k+\sum_j \sum_{|\alpha|\leqq 2} \sum_{|\beta|\leqq k} \sup_{|x-x'|\leqq 1} \frac{|D^\beta a^\lambda_{jp\alpha}(x)-D^\beta a^\lambda_{jp\alpha}(x')|}{|x-x'|^\theta}.$$

Proof. We may assume that the system of local coordinates $\{X_j\}_{j=1}^J$ is so chosen that the principal part is already diagonal. Take $\{\chi_j\}_{j=1}^J$ as in (3.6). For $u\in C^\infty(X, B)$, by the definition of the norm (3.32), there is a j such that

$$J^{-1}|u|_{k+2+\theta}\leqq |\chi_j u|_{j,k+2+\theta}. \tag{4.24}$$

Put $\chi_j u=u_j$. Choose a family of C^∞ function $\{\omega_i\}_{i=1}^J$ on X with supp $\omega_i\subset X_i$ and $\sum_i \omega_i(p)^2\equiv 1$ such that $\omega_j(p)\equiv 1$ on some neighbourhood of supp χ_j. Then we have

$$\begin{aligned}|A(p,D)u|_{k+\theta}&\geqq|\chi_j A(p,D)u|_{j,k+\theta}=|\chi_j A(p,D)\omega_j u|_{j,k+\theta}\\ &\geqq|A(p,D)u_j|_{j,k+\theta}-|[A(p,D),\chi_j]\omega_j u|_{j,k+\theta}.\end{aligned} \tag{4.25}$$

Since $[A(p, D), \chi_j]$ is a first-order partial differential operator, by Lemma 2.8 there is a positive constant C_1 depending only on k, θ, n, μ such that

$$|[A(p,D),\chi_j]\omega_j u|_{j,k+\theta}\leqq C_1 M_{k+\theta}|\omega_j u|_{j,k+1+\theta}, \tag{4.26}$$

while there is a positive constant C_2 such that

$$|\omega_j u|_{j,k+1+\theta}\leqq C_2|u|_{k+1+\theta}. \tag{4.27}$$

Since the support of u_j is contained in X_j, u_j can be considered, via x_j, an element of $C^{k+2+\theta}(\mathbb{T}^n, \mathbb{C}^\mu)$ with support in V_j. Hence by Theorem 2.3, there is a positive constant C_3 depending only on k, θ, n, μ, δ and $M_{k+\theta}$ such that

$$C_3|u_j|_{j,k+2+\theta}-|u_j|_{j,0}\leqq|A(p,D)u_j|_{j,k+\theta}. \tag{4.28}$$

From the trivial inequality $|u_j|_{j,0}\leqq|u|_0$ and (4.24), (4.25), (4.26), (4.27), (4.28), we obtain

$$|A(p,D)u|_{k+\theta}\geqq C_3 J^{-1}|u|_{k+2+\theta}-|u|_0-C_1C_2M_{k+\theta}|u|_{k+1+\theta}. \tag{4.29}$$

By Lemma 4.2, there is a positive constant C_4 such that

$$C_1C_2M_{k+\theta}|u|_{k+1+\theta}\leqq 2^{-1}C_3J^{-1}|u|_{k+2+\theta}+C_4|u|_0.$$

Consequently from this and (4.29), we obtain

$$|A(p, D)u|_{k+\theta}+(1+C_4)|u|_0 \geqq 2^{-1}C_3J^{-1}7|u|_{k+2+\theta},$$

which proves the theorem. ∎

Remark. Let $A(x, D)$ be an elliptic operator of order l with C^∞ coefficients, and δ its constant of ellipticity. Then if $0<\theta<1$, for any integer $k \geqq 0$, there is a positive constant C depending on n, l, μ, k, θ, δ, $M_{k+\theta}$ such that

$$|u|_{k+l+\theta} \leqq C(|A(x, D)u|_{k+\theta}+|u|_0).$$

We omit the proof.

§5. The Existence of Weak Solutions of a Strongly Elliptic Partial Differential Equation

As was stated at the end of §3, we fix once for all a volume element $v(dx)$ of X and a metric g on the vector bundle B. Then we can define $L^2(X, B)$ with respect to (,).

Proposition 5.1. *For a linear partial differential operator $A(p, D)$ of order l, there exists a unique linear partial differential operator $A(p, D)^\#$ such that for any $\varphi, \psi \in C^\infty(X, B)$, the following equality holds:*

$$(A(p, D)\varphi, \psi)=(\varphi, A(p, D)^\# \psi). \tag{5.1}$$

Moreover $A(p, D)^\#$ is also of order l.

Proof. Let $\{\omega_j\}_{j=1}^J$ be a C^∞ partition of unity subordinate to the open covering $\bigcup_{j=1}^J X_j$ of X. Then we have

$$(A(p, D)\varphi, \psi)=\sum_j (A(p, D)\omega_j\varphi, \psi).$$

If $A(p, D)$ is written in the form (4.1) on X_j, we have

$$(A(p, D)\omega_j\varphi, \psi)=\sum_{|\alpha|\leqq l}\sum_{\rho,\lambda,\sigma}\int_{V_j} g_{j\lambda\bar\rho}(x_j(p))a^\lambda_{j\sigma\alpha}(x_j(p))$$
$$\times D_j^\alpha(\omega_j\varphi_j^\sigma(x_j(p)))\overline{\psi_j^\rho(x_j(p))}v_j(x_j(p))\,dx_j(p). \tag{5.2}$$

Since $\omega_j\varphi_j^\alpha$ vanishes near the boundary of V_j, by partial integration, we obtain

$$(A(p, D)\omega_j\varphi, \psi) = \sum_{\nu,k} \int_{V_j} g_{j\nu\bar{\kappa}}(x_j(p))\omega_j\varphi_j^\nu(x_j(p)) \times \overline{\psi_j'^\kappa(x_j(p))} v_j(x_j(p))\, dx_j(p), \tag{5.3}$$

where we put

$$\psi_j'^\kappa(x_j(p)) = \frac{1}{v_j(x_j(p))} \sum_{\lambda,\sigma,\rho} \sum_{|\alpha|\leq l} g_j^{\bar{\kappa}\sigma}(x_j(p)) \times D_j^\alpha(g_{j\lambda\bar{\rho}}(x_j(p))a_{j\sigma\alpha}^\lambda(x_j(p))v_j(x_j(p))\psi_j^\rho(x_j(p))), \tag{5.4}$$

and $(g_j^{\bar{\kappa}\sigma}(x_j(p)))$ is the inverse of the matrix $(g_{j\lambda\bar{\rho}}(x_j(p)))$. Therefore if we put $\psi' = \sum_j \omega_j\psi_j'$, then $A(p, D)^\#: \psi \to \psi'$ is the required differential operator. The uniqueness if obvious. ∎

Definition 5.1. $A(p, D)^\#$ is called *the formal adjoint of* $A(p, D)$. If $A(p, D)$ is of order l, so is $A(p, D)^\#$.

Proposition 5.2. *Let* $p \in X$ *and* $\xi_x \in T_p^*X$ *with* $\xi_x \neq 0$. *If we write the principal symbol of the differential operator* $A(p, D)^\#$ *as* $A_l(p, \xi_x)^\#$, *then it is the adjoint of the linear map of* B_p *determined by the principal symbol* $A_l(p, \xi_x)$ *of* $A(p, D)$ *with respect to the metric* g_p.

Proof. Let $f(p)$ be a real-valued C^∞ function on X with $df(p) = \xi_p$. Then for any $\varphi, \mu \in C^\infty(X, B)$, we have

$$\lim_{t\to\infty} t^{-1}(\varphi, e^{-itf}A(p, D)^\# e^{itf}\psi) = \lim_{t\to\infty} t^{-1}(e^{-itf}A(p, D)e^{itf}\varphi, \psi), \tag{5.5}$$

which proves the assertion. ∎

We want to solve the partial differential equation

$$A(p, D)u(p) = f(p). \tag{5.6}$$

Definition 5.2. Let $f \in W^k(X, B)$. $u \in W^k(X, B)$ is said to be a *weak solution* of the equation (5.6) if for any $\varphi \in C^\infty(X, B)$,

$$(u, A(p, D)^\#\varphi) = (f, \varphi) \tag{5.7}$$

holds, where $A(p, D)^\#$ is the formal adjoint of $A(p, D)$, and (,) denotes the extended one in Proposition 3.2.

Proposition 5.3. *Let* $A(p, D)$ *be a partial differential operator of order* l. *Then* *by Lemma 3.3, it is extended to a continuous map of* $W^k(X, B)$ *to* $W^{k-1}(X, B)$. *If* u *is a weak solution of* (5.6), *in this sense, we have*

$$A(p, D)u = f \tag{5.8}$$

in $W^{k-l}(X, B)$.

Proof. Let $\{\varphi_n\}_{n=1}^{\infty}$ be a sequence in $C^\infty(X, B)$ converging to u in $W^k(X, B)$. Then $\{A(p, D)\varphi_n\}_{n=1}^{\infty}$ converges to $A(p, D)u$ in $W^{k-l}(X, B)$. For an arbitrary $\varphi \in C^\infty(X, B)$, by Proposition 3.2, we have

$$(A(p, D)u, \varphi) = \lim_{n\to\infty} (A(p, D)\varphi_n, \varphi) = \lim_{n\to\infty} (\varphi_n, A(p, D)^{\#}\varphi)$$

$$= (u, A(p, D)^{\#}\varphi) = (F, \varphi).$$

Hence $A(p, D)u = f$ in $W^{k-l}(X, B)$. ∎

In particular, if $f \in C^0(X, B)$, u is a weak solution of (5.6), and $u \in C^1(X, B)$, then by Proposition 5.1, for any $\varphi \in C^\infty(X, B)$, we have

$$(A(p, D)u, \varphi) = (f, \varphi),$$

hence $A(p, D)u = f$ in the sense of $L^2(X, B)$. But since both sides of this equality are continuous, we obtain

$$A(p, D)u(p) = f(p)$$

for all $p \in X$, namely, we obtain a solution of (5.6) in the usual sense.

In order to show the existence of a weak solution, the following Lax-Milgram theorem is useful.

Theorem 5.1. *Let* $\mathcal{H}$ *be a complex Hilbert space,* $(\ ,\ )$ *its inner product, and* $|\ |$ *its norm. Suppose that* $B(x, y)$ *is a Hermitian form on* $\mathcal{H}$ *satisfying the following condition: There exist positive constants* $C_1 \leqq C_2$ *such that for any* $x, y \in \mathcal{H}$,

$$|B(x, y)| \leqq C_2|x||y|, \tag{5.9}$$

$$\operatorname{Re} B(x, x) \geqq C_1|x|^2. \tag{5.10}$$

Then for any continuous conjugate line $f(x)$ *on* $\mathcal{H}$, *there exists a unique element* F_B *of* $\mathcal{H}$ *with*

$$B(F_B, z) = f(z). \tag{5.11}$$

Proof. As will be shown later, there exists a continuous linear isomorphism

B of $\mathscr{H}$ onto $\mathscr{H}$ such that for any $x, y \in \mathscr{H}$

$$B(x, y) = (Bx, y) \tag{5.12}$$

holds. Assuming this for a moment, we see that by the Riesz representation theorem, there exists an element $y \in \mathscr{H}$ such that for any $z \in \mathscr{H}$

$$(y, z) = f(z).$$

Hence it suffices to put $F_B = B^{-1}y$.

We shall prove the existence of B satisfying the above condition. Fix an arbitrary $x \in \mathscr{H}$. The associated linear map $\mathscr{H} \ni y \to B(x, y) \in \mathbb{C}$ is continuous by (5.9), hence, by the Riesz representation theorem, there exists a $z(x) \in \mathscr{H}$ such that $B(x, y) = (z(x), y)$ for any y. The map $x \to z(x)$ is linear. Therefore there is a linear map B with $z(x) = Bx$. By (5.9) and (5.10), we have

$$C_1|x| \leqq \frac{\operatorname{Re} B(x, x)}{|x|} \leqq |Bx| = \sup_{y \neq 0} \frac{|B(x, y)|}{|y|} \leqq C_2|x|. \tag{5.13}$$

Hence Bx is continuous. If $Bx = 0$, by the left inequality of (5.13) we have $x = 0$, hence B is injective. Furthermore the range of B is closed. In fact, if $\{Bx_n\}_n$ converges to y, then by the left inequality of (5.13), $\{x_n\}_n$ is also a Cauchy sequence, hence, converges to some z. Since B is continuous, $Bz = y$. Moreover the range of B is dense. For, if z is orthogonal to the range of B, we have

$$0 = |(Bz, z)| = |B(z, z)| \geqq \operatorname{Re} B(z, z) \geqq C_1|z|^2,$$

which implies $z = 0$. Thus the range of B is dense. Consequently the range of B coincides with the whole $\mathscr{H}$, and B is bijective. The continuity of B^{-1} is also clear. ▮

Let $A(p, D)$ be a strongly elliptic linear partial differential operator of order $2m$. For a sufficiently large $\lambda > 0$, we shall prove the existence of a weak solution of

$$(A(p, D) + \lambda I)u = w. \tag{5.14}$$

For this, take $\{\chi_j\}_{j=1}^J$ as in (3.6). Then with the inner product (3.28), $W^m(X, B)$ becomes a Hilbert space. Put

$$(\varphi, (A(p, D)^* + \bar{\lambda})\psi) = B(\varphi, \psi) \tag{5.15}$$

for $\varphi, \psi \in C^\infty(X, B)$, where the inner product in the left-hand side denotes that in $L^2(X, B)$.

Proposition 5.4. *Let δ_1, δ_2 be the positive constants given in Theorem 4.2. By (5.15), $B(\varphi, \psi)$ defined on $C^\infty(X, B)$ extends uniquely by continuity to a continuous Hermitian form on $W^m(X, B)$, which we denote by the same notation $B(\varphi, \psi)$. Then there exists a positive constant C_1 determined by n, m, μ, and M_m such that if $\operatorname{Re}\lambda > \delta_1$, for any $\varphi, \psi \in W^m(X, B)$, the following inequalities hold:*

$$|B(\varphi, \psi)| \leqq C_2 \|\varphi\|_m \|\psi\|_m, \tag{5.16}$$

$$\operatorname{Re} B(\varphi, \psi) \geqq \delta_2 \|\varphi\|_m^2. \tag{5.17}$$

Proof. If $\varphi, \psi \in C^\infty(X, B)$, since $B(\varphi, \psi) = ((A(p, D) + \lambda)\varphi, \psi)$, we have (5.16) by Lemma 4.1. In other words, $B(\varphi, \psi)$ is continuous with respect to the relative topology of $C^\infty(X, B)$ in $W^m(X, B)$. Since $C^\infty(X, B)$ is dense in $W^m(X, B)$, $B(\varphi, \psi)$ extends uniquely to a continuous Hermitian form on $W^m(X, B)$. (5.16) still holds for the extended $B(\varphi, \psi)$. By Gårding's inequality, (5.17) holds for $\varphi \in C^\infty(X, B)$. Hence, by continuity, (5.17) holds for all $\varphi \in W^m(X, B)$. ∎

Theorem 5.2. *Let $A(p, D)$ be a strongly elliptic linear partial differential operator of order 2m, and δ_1, δ_2 the positive constants given in Theorem 4.2. If $\operatorname{Re}\lambda > \delta_1$, for any $w \in L^2(X, B)$, there exists a weak solution of the equation*

$$A(p, D)u + \lambda u = w \tag{5.18}$$

contained in $W^m(X, B)$. Moreover the weak solution of this equation contained in $W^m(X, B)$ is unique.

Proof. Let $A(p, D)^\#$ be the formal adjoint of $A(p, D)$, and $B(\varphi, \psi)$ the Hermitian form on $W^m(X, B)$ in Proposition 5.4. Since $W^m(X, B)$ is imbedded in $L^2(X, B)$, putting

$$f(\varphi) = (w, \varphi)$$

for any $\varphi \in W^m(X, B)$, we obtain a conjugate linear form on $W^m(X, B)$. Take $W^m(X, B)$ as $\mathscr{H}$ in Theorem 5.1, and let $B(\varphi, \psi)$ and $f(\varphi)$ be the Hermitian form and the conjugate linear form for it respectively. Then by Theorem 5.1, there exists a $u \in W^m(X, B)$ such that for any $\varphi W^m(X, B)$, the following holds:

$$B(u, \varphi) = (w, \varphi), \tag{5.19}$$

where the left-hand side means B in the extended sense. We may assume that there are $\psi_k \in C^\infty(X, B)$, $k = 1, 2, \ldots$, such that $\{\psi_k\}_k$ converges to u in $W^m(X, B)$. Then $\{\psi_k\}_k$ converges to u also in $L^2(X, B)$. Therefore, for

$\varphi \in C^\infty(X, B)$, we can write

$$B(u, \varphi) = \lim_{k\to\infty} B(\psi_k, \varphi) = \lim_{k\to\infty} (\psi_k, (A(p, D)^\# + \bar{\lambda})\varphi)$$
$$= (u, (A(p, D)^\# + \bar{\lambda})\varphi). \tag{5.20}$$

Hence, u is a weak solution of (5.18). The uniqueness follows from the following inequality by putting $w = 0$.

$$0 = |(w, u)| \geqq \operatorname{Re} B(u, u) \geqq \delta_2 \|u\|_m^2. \quad \blacksquare$$

§6. Regularity of Weak Solution of Elliptic Linear Partial Differential Equations

Theorem 6.1. *Let $A(p, D)$ be an elliptic linear partial differential operator of order l with C^∞ coefficients. Suppose that for some integers s and k, $f \in W^{s-l-k}(X, B)$. Then for a weak solution $u \in W^s(X, B)$ of the equation*

$$A(p, D)u = f, \tag{6.1}$$

there exists a positive constant C such that

$$\|u\|_{s+k} \leqq C(\|f\|_{s+k-l} + \|u\|_s) \tag{6.2}$$

holds, where C depends on l, n, k, μ, the constant of ellipticity δ of $A(p, D)$ and $M_{|s|}$, but is independent of f and u.

Proof. First we prove the case $k = 1$. Take C^∞ functions ω_j, $j = 1, \ldots, J$ on X with $\omega_j(p) \geqq 0$ and $\operatorname{supp} \omega_j \subset X_j$ so that

$$\sum_j \omega_j(p) \equiv 1, \qquad p \in X. \tag{6.3}$$

Put

$$u_j = \omega_j u, \qquad f_j = \omega_j f, \qquad j = 1, \ldots, J.$$

Then in $W^{s-l}(X, B)$, we have

$$A(p, D)u_j = f_j + [A(p, D), \omega_j]u, \qquad j = 1, \ldots, J,$$

while $f_j = \omega_j f \in W^{s-l+1}(X, B)$. Since $[A(p, D), \omega_j]$ is a differential operator of order $(l-1)$, we have $[A(p, D), \omega_j]u \in W^{s-l+1}(X, B)$. Furthermore, putting $g_j = f_j + [A(p, D), \omega_j]u$, we have $g_j \in W^{s-l+1}(X, B)$, and

$$A(p, D)u_j = g_j. \tag{6.4}$$

Since the supports of u_j and g_j are contained in X_j, u_j and g_j can be identified with $\mathbb{C}^\mu$-valued distributions $\tilde{u}_j \in W_0^s(V_j, \mathbb{C}^\mu)$ and $\tilde{g}_j \in W_0^{s-l+1}(V_j, \mathbb{C}^\mu)$ on the open set V_j in $\mathbb{T}^n$ respectively. The supports of these $\tilde{u}_j$ and $\tilde{g}_j$ are contained in a compact set K. Writing $\tilde{u}_j = (u_j^1, \ldots, u_j^n)$ and $\tilde{g}_j = (g_j^1, \ldots, g_j^n)$ in terms of their components, we have

$$g_j^\lambda = \sum_{|\alpha| \leqq l} \sum_\rho a_{j\rho\alpha}^\lambda(x_j) D_j^\alpha u_j^\rho = (A_j(p, D)u_j)^\lambda, \qquad x_j = x_j(p). \tag{6.5}$$

Since K is a compact set in V_j, there is a positive number h_0 such that if $0 < |h| < h_0$, the translation of K in the direction of x_j^k by h is contained in V_j for any $k = 1, \ldots, n$. Let $\tau_{j,k}^h$ be the translation in the direction of x_j by h, and $\Delta_{j,k}^h = h^{-1}(\tau_{j,k}^{-h} - I)$. Then we obtain

$$\Delta_{j,k}^h \tilde{g}_j = A_j(x_j, D)(\Delta_{j,k}^h \tilde{u}_j) + v_{j,k}, \qquad k = 1, \ldots, n, \tag{6.6}$$

where we put

$$\Delta_{j,k}^h \tilde{g}_j = \sum_{|\alpha| \leqq l} \sum_\rho (\Delta_{j,k}^h a_{j\rho\alpha}^\lambda)(x) D_j^\alpha \tau_{j,k}^h u_j^\rho. \tag{6.7}$$

By Theorem 1.3, there exists a positive constant C_1 independent of h such that

$$\| v_{j,k} \|_{j,s-l} \leqq C_1 \| \tau_{j,k}^h \tilde{u}_j \|_{j,s} = C_1 \| \tilde{u}_j \|_{j,s}. \tag{6.8}$$

For $0 < |h| < h_0$, we have $\Delta_{j,k}^h \tilde{u}_j \in W_0^s(V_j, \mathbb{C}^\mu)$. Since $\Delta_{j,k}^h \tilde{u}_j$ satisfies (6.6), by Theorem 2.1, there exists a positive constant C_2 independent of h such that

$$\| \Delta_{j,k}^h \tilde{u}_j \|_{j,s} \leqq C_2(\| \delta_{j,k}^h \tilde{g} \|_{j,s-l} + \| v_{j,k} \|_{j,s-l} + \| \Delta_{j,k}^h \tilde{u}_j \|_{j,s-l}). \tag{6.9}$$

On the other hand by Theorem 1.13, there is a positive constant C_3 independent of h such that for $0 < |h| < h_0$, the following inequalities hold:

$$\| \Delta_{j,k}^h \tilde{g}_j \|_{j,s-l} \leqq C_3 \| \tilde{g}_j \|_{j,s-l+1}, \qquad k = 1, \ldots, n,$$

$$\| \Delta_{j,k}^h \tilde{u}_j \|_{j,s-l} \leqq C_3 \| \tilde{u}_j \|_{j,s-l+1}.$$

Using these inequalities and (6.8), we have the following estimate: For $0 < |h| < h_0$, and for any $k = 1, \ldots, n$,

$$\| \Delta_{j,k}^h \tilde{u}_j \|_{j,s} \leqq C_2 C_3 (\| \tilde{g}_j \|_{j,s-l+1} + \| \tilde{u}_j \|_{j,s-l+1}) + C_1 C_2 \| \tilde{u}_j \|_{j,s}.$$

Since the right-hand side does not depend on h, by Theorem 1.13(3°), we see that $\tilde{u}_j \in W_0^{s+1}(V_j, \mathbb{C}^\mu)$. Therefore we obtain $u_j \in W^{s+1}(X, B)$. By the construction of u_j and (6.3), we have $u = \sum u_j \in W^{s+1}(X, B)$. Therefore by (6.1), using Theorem 4.1, we see that there exists a positive constant C_4

such that

$$\|u\|_{s+1} \leqq C_4(\|g\|_{s-l+1} + \|u\|_{s-l}).$$

Thus the theorem is proved for $k=1$. We proceed by induction on $k>1$. Suppose that the assertion is true for $k \leqq j$. We want to show the case $k=j+1$. In this case we have $f \in W^{s-l+j+1}(X, B)$, and $u \in W^s(X, B)$. By induction hypothesis, we have $u \in W^{s+j}(X, B)$. Taking $s'=s+j$ instead of s, we see that (6.1) holds for $f \in W^{s'-l+1}(X, B)$, and $u \in W^{s'}(X, B)$. Since we have proved the theorem for $k=1$, we have $u \in W^{s'+1}(X, B) = W^{s+j+1}(X, B)$. By Theorem 4.1, there is a positive constant C_5 such that the following inequality holds:

$$\|u\|_{s+j+1} \leqq C_5(\|f\|_{s+j+1-l} + \|u\|_{s+j+1-l}).$$

Since this implies (6.2), we showed the case $k=j+1$, which completes the induction. ∎

Corollary. *Let $A(p, D)$ be an elliptic linear partial differential operator of order l, and suppose that for $f \in C^\infty(X, B)$, $u \in W^s(X, B)$ is a weak solution of*

$$A(p, D)u = f.$$

Then there exists $v \in C^\infty(X, B)$ such that in $L^2(X, B)$

$$u - v = 0$$

holds.

Proof. Since $f \in C^\infty(X, B)$, for any integer $k>0$, we have $f \in W^{s-l+k}(X, B)$, hence, by the above theorem, we have $u \in W^{s+k}(X, B)$. If $k>[n/2]+1$, by Sobolev's imbedding theorem (Theorem 3.2), there is a $v \in C^{s+k-[n/2]-1}(X, B)$ such that in $W^0(X, B)$

$$u = v$$

holds. Thus, $u=v$ in $L^2(X, B)$. Since k is arbitrary, we have $v \in C^\infty(X, B)$. ∎

§7. Elliptic Operators in the Hilbert Space $L^2(X, B)$

Let $A(p, D)$ be an elliptic linear partial differential operator of order l with C^∞ coefficients. $A(p, D)$ extends uniquely to a continuous linear map of $W^s(X, B)$ to $W^{s-l}(X, B)$. By restricting the domain of $A(p, D)$ appropriately, we treat $A(p, D)$ as a closed operator of the Hilbert space $L^2(X, B)$.

(a) Closed Extension of Elliptic Partial Differential Operators

Definition 7.1. We define a linear operator A in the Hilbert space $L^2(X, B)$ as follows. The domain $D(A)$ of A is given by

$$D(A)=\{u\in L^2(X, B)\,|\,A(p, D)u\in L^2(X, B)\}, \tag{7.1}$$

and for $u\in D(A)$, we put $Au=A(p, D)u$.

Theorem 7.1.

(1°) *$D(A)=W^l(X, B)$, and the topology of $D(A)$ defined by the graph norm coincides with that of $W^l(X, B)$.*
(2°) *A is a closed operator.*
(3°) *$C^\infty(X, B)$ is dense in $D(A)$ with respect to the graph norm, that is, for $u\in D(A)$, there is a sequence $\{\varphi_k\}_{k=1}^\infty\subset C^\infty(X, B)$ such that φ_k converges to u in $L^2(X, B)$, and that $A(p, D)\varphi_k$ converges to Au.*

Proof. (1°) Since $L^2(X, B)=W^0(X, B)$, by Theorem 6.1, if $u\in D(A)$, then $u\in W^l(X, B)$. Therefore by the *a priori* estimate in Theorem 4.1, there are positive constants C_1, C_2 such that for any $uD(A)$,

$$\|u\|_l\leqq C_1(\|Au\|+\|u\|)\leqq C_2\|u\|_l, \tag{7.2}$$

while the middle part is just the graph norm of u. Thus (1°) is proved.

(2°) Since $D(A)$ endowed with the topology defined by the graph norm coincides with $W^l(X, B)$, this space is complete. Therefore the graph of A is a complete subspace of the direct product $L^2(X, B)\times L^2(X, B)$, hence closed.

(3°) is clear from the density of $C^\infty(X, B)$ in $W^l(X, B)$. ∎

Since the formal adjoint $A(p, D)^\#$ of $A(p, D)$ is also a partial differential operator of order l, $A(p, D)^\#$ has the unique extension to a continuous map of $W^s(X, B)$ to $W^{s-l}(X, B)$. By Proposition 5.2, $A(p, D)^\#$ is also elliptic.

Theorem 7.2. *Let A^* be the adjoint of A in the sense of the operator on the Hilbert space $L^2(X, B)$. Then the domain $D(A^*)$ is given by $D(A^*)=W^l(X, B)$, and for $v\in D(A^*)$, we have*

$$A^*v=A(p, D)^\# v. \tag{7.3}$$

In particular, if $A(p, D)$ is a formally self-adjoint elliptic partial differential operator, A is self-adjoint.

Proof. $v \in D(A^*)$ if and only if there exists an $f \in L^2(X, B)$ such that for any $u \in D(A)$,

$$(v, Au) = (f, u). \tag{7.4}$$

Since $C^\infty(X, B) \subset D(A)$, for any $\varphi \in C^\infty(X, B)$ we have

$$(v, A(p, D)\varphi) = (f, \varphi). \tag{7.5}$$

This implies that v is a weak solution of the equation

$$A(p, D)^* v = f.$$

Since $f \in L^2(X, B)$, by Theorem 6.1, we have $v \in W^l(X, B)$. Conversely, suppose that $v \in W^l(X, B)$ and $f = A(p, D)^* v$. Then $f \in L^2(X, B)$, and for any $\varphi \in C^\infty(X, B)$, (7.5) holds. Since both sides of (7.5) are continuous in φ with respect to the graph norm, and $C^\infty(X, B)$ is dense in $D(A)$ with respect to the graph norm, (7.4) holds for every $u \in D(A)$. Thus the theorem is proved. ∎

Lemma 7.1. *Let $A(p, D)$ be an elliptic linear partial differential operator of order l. We define the operator A as in Definition 7.1. Then*

(1°) $\ker A$ *is a closed subspace of* $L^2(X, B)$.

(2°) *For any integer $k \geqq 0$, there exists a positive constant C_k such that for any $u \in W^{k+l}(X, B) \cap (\ker A)^\perp$, the following estimate holds*:

$$\|u\|_{l+k} \leqq C_k \|Au\|_k. \tag{7.6}$$

Similarly if $0 < \theta < 1$, for any integer $k \geqq 0$, there exists a positive constant C_k such that for $u \in C^{k+l+\theta}(X, B) \cap (\ker A)^\perp$,

$$|u|_{l+k+\theta} \leqq C_k |Au|_{k+\theta} \tag{7.7}$$

holds.

Proof. (1°) If $\{u_n\}_{n=1}^\infty \subset \ker A$, and if $\{u_n\}$ converges to u in $L^2(X, B)$, then $Au_n = 0$ converges to 0, and, since A is a closed operator, $u \in D(A)$ and $Au = 0$. This proves (1°).

(2°) Suppose that for some k, (7.6) does not hold for any C_k. Then for any positive integer j, there exists a $u_j \in W^{l+k}(X, B) \cap (\ker A)^\perp$ such that

$$\|u_j\|_{l+k} = 1 \quad \text{and} \quad \|Au_j\|_k \leqq j^{-1}. \tag{7.8}$$

Since the sequence $\{u_j\}_j$ is bounded in $W^{l+k}(X, B)$, by Theorem 3.1, $\{u_j\}_j$ is relatively compact in $W^k(X, B)$. Taking a subsequence, if necessary, we may assume that $\{u_j\}_j$ converges to u in $W^k(X, B)$, while $\{Au_j\}$ converges to 0 in $W^k(X, B)$. In $L^2(X, B)$ also, $\{u_j\}_j$ and $\{Au_j\}_j$ converges to u and 0 respectively. Since A is a closed operator on $L^2(X, B)$, we have $u \in D(A) \cap \ker A$. On the other hand, since $\{u_j\} \subset (\ker A)^\perp$, and $(\ker A)^\perp$ is a closed subspace of $L^2(X, B)$, we have $u \in (\ker A)^\perp$. Hence $u = 0$ in $L^2(X, B)$. Since we have already known that $u \in W^k(X, B)$, we see that $u = 0$ in $W^k(X, B)$. Since $\{u_j\}_j$ converges to u in $W^k(X, B)$,

$$\lim_{j\to\infty} \|u_j\|_k = 0. \tag{7.9}$$

Using the *a priori* estimate in Theorem 4.1 for $\{u_j\} \subset W^{l+k}(X, B)$, we see that there is a positive constant C such that

$$\|u_j\|_{k+l} \leqq C(\|Au_j\|_k + \|u_j\|_k), \qquad j = 1, 2, \ldots .$$

Taking the limits of both sides for $j \to \infty$, we have $1 \leqq 0$ by (7.8) and (7.9), which is a contradiction. Hence (7.6) is true. (7.7) is proved similarly. ∎

Theorem 7.3. *Let $A(p, D)$ be an elliptic partial differential operator of order l. We define the operator A and its adjoint A^* as in Definition 7.1. Then we have the following.*

(1°) *Both $\ker A$ and $\ker A^*$ are finite-dimensional subspaces of $C^\infty(X, B)$.*
(2°) *The range $R(A)$ of A and the range $R(A^*)$ of A^* are closed subspaces of $L^2(X, B)$.*
(3°) $R(A) = (\ker A^*)^\perp$, $R(A^*) = (\ker A)^\perp$.

Proof. (1°) By the Corollary to Theorem 6.1, we have $\ker A \subset C^\infty(X, B)$. $\ker A$ is a Banach space as a subspace of $L^2(X, B)$. Let $Q = \{u \in \ker A \mid \|u\| \leqq 1\}$ be its unit ball. Since $Q \subset C^\infty(X, B)$, by Theorem 4.1, we see that if $u \in Q$, $\|u\|_l \leqq C\|u\| \leqq C$. Thus Q is bounded in $W^l(X, B)$. Therefore by Theorem 3.1, it is compact in $L^2(X, B)$, hence in $\ker A$. Consequently $\ker A$ is a locally compact space, hence finite dimensional.

(2°) Suppose that a sequence $\{f_j\}_{k=1}^\infty$ in $R(A)$ converges to f in $L^2(X, B)$. By Lemma 7.1, for any j and j', we have the following inequality:

$$\|u_j - u_{j'}\|_l \leqq C_1 \|f_j - f_{j'}\|.$$

Since $\{f_j\}_j$ is a Cauchy sequence, $\{u_j\}_j$ becomes a Cauchy sequence in $W^l(X, B)$, hence converges to some $u \in W^l(X, B)$ in $W^l(X, B)$. On the other hand, $Au_n = f_n$ converges to f in $L^2(X, B)$, hence we have $Au = f$. Thus $R(A)$ is a closed set. The closedness of $R(A^*)$ can be proved similarly.

(3°) Let $u \in R(A)^\perp$. Since for any $v \in D(A)$, $(Av, u) = 0$, we have $u \in D(A^*)$, and $A^*u = 0$. Conversely if $A^*u = 0$, clearly $u \perp R(A)$, hence we have $\ker A^* = R(A)^\perp$. Since $R(A)$ is closed by (2°), we obtain $(\ker A^*)^\perp = R(A)$. $(\ker A)^\perp = R(A^*)$ is proved similarly. ∎

Lemma 7.2. *Let P and Q be the orthogonal projections to* $\ker A$ *and* $\ker A^*$ *in $L^2(X, B)$ respectively. Then for any integer $k \geqq 0$, there exists a positive constant C_k such that for any $u \in L^2(X, B)$, the following estimates hold.*

$$\|Pu\|_k \leqq C_k\|u\|, \qquad \|Qu\|_k \leqq C_k\|u\|. \tag{7.10}$$

$$|Pu|_{k+\theta} \leqq C_k\|u\|, \qquad |Qu|_{k+\theta} \leqq C_k\|u\|. \tag{7.11}$$

Proof. Let $\varphi_1, \ldots, \varphi_L$ be an orthonormal basis of $\ker A$. Then $Pu = \sum_j (u, \varphi_j)\varphi_j$. Therefore, taking into account the fact that $\varphi_j \in C^\infty(X, B)$, and putting $C_k = (\sum_j \|\varphi_j\|_k^2)^{1/2}$, we obtain

$$\|Pu\|_k \leqq \sum_j |(u, \varphi_j)| \, \|\varphi_j\|_k \leqq C_k\|u\|.$$

The other estimates are obtained similarly. ∎

Definition 7.2. Let $D(A) \cap (\ker A)^\perp = \mathscr{H}$. Since A is a bijection of $\mathscr{H}$ onto $R(A)$, let $\tilde{G}$ be its inverse. $\tilde{G}$ is a bijection of $R(A)$ onto $\mathscr{H}$. We define

$$G = \tilde{G}(1+Q),$$

and call G the *Green operator of* A. G is a linear map of $L^2(X, B)$ to $\mathscr{H}$ which coincides with $\tilde{G}$ on $R(A)$ and vanishes on $\ker A^*$.

Theorem 7.4. *The Green operator G has the following properties:*

(1°) *G is defined on $L^2(X, B)$, and its range $R(G)$ is given by $R(G) = W^l(X, B) \cap (\ker A)^\perp$. For any $u \in L^2(X, B)$, $AGu = (I - Q)u$, and for any $v \in W^l(X, B)$, $GAv = (I - P)v$.*

(2°) *For any integer $k \geqq 0$, there is a positive constant C_k such that for any $u \in W^k(X, B)$, $Gu \in W^{k+l}(X, B)$, and*

$$\|Gu\|_{k+\rho} \leqq C_k\|u\|_k \tag{7.12}$$

holds. If $0 < \theta < 1$, for any integer $k \geqq 0$, there exists a positive constant C_k such that for any $u \in C^{k+\theta}(X, B)$, $Gu \in C^{k+l+\theta}(X, B)$, and

$$|Gu|_{k+l+\theta} \leqq C_k|u|_{k+\theta}. \tag{7.13}$$

Proof. (1°) is clear.

(2°) For $u \in W^k(X, B)$, by Lemma 7.2, we have $\|(1-Q)u\|_k \leqq (C_k+1)\|u\|_k$. Let C'_k be a positive constant in (7.6) of Lemma 7.1. Then

$$\|Gu\|_{k+l} = \|\tilde{G}(1+Q)u\|_{k+l} \leqq C'_k(C_k+1)\|u\|_k,$$

which proves (7.12). (7.13) is proved similarly. ∎

Corollary. *Let $A(p, D)$ be a formally self-adjoint elliptic partial differential operator, P the orthogonal projection to* ker A, *and G the Green operator of A. If $u \in C^\infty(X, B)$, then $Gu \in C^\infty(X, B)$, and the following equality holds*

$$u = Pu + AGu.$$

By this the orthogonal decomposition

$$C^\infty(X, B) = \ker A \oplus AC^\infty(X, B)$$

is given, where $AC^\infty(X, B)$ denotes the image of $C^\infty(X, B)$ by $A(p, D)$.

Proof. Let A be the operator on the Hilbert space $L^2(X, B)$ defined from $A(p, D)$ as in Definition 7.1. Since $A(p, D)$ is formally self-adjoint, A is self-adjoint (Theorem 7.2). Therefore, since $\ker A = \ker A^*$, the orthogonal projection P coincides with the orthogonal projection Q to $\ker A^*$. By Theorem 7.4, 1, for any $u \in C^\infty(X, B) \subset W^l(X, B)$,

$$u = Pu + AGu$$

holds. Furthermore by Theorem 7.4(2°), $Gu \in C^\infty(X, B)$, hence $AGu \in AC^\infty(X, B)$. Moreover since A is self-adjoint, by Theorem 7.3 (3°), ker A and $AC^\infty(X, B)$ are orthogonal to each other. ∎

(b) The Spectrum of a Strongly Elliptic Partial Differential Operator

Let $A(p, D)$ be a strongly elliptic linear partial differential operator of order $2m$. We define the operator A as in Definition 7.1. Let δ be the constant of strong ellipticity of $A(p, D)$, and take δ_1, δ_2 as in Theorem 4.1. If $\mathrm{Re}\,\lambda > \delta_1$, then for any $\varphi \in W^{2m}(X, B)$, we have

$$\|(A+\lambda)\varphi\| \geqq (\mathrm{Re}\,\lambda - \delta_1)\|\varphi\|. \tag{7.14}$$

For, indeed, by Theorem 4.2, we have

$$\|(A+\lambda)\varphi\|\,\|\varphi\| \geqq \mathrm{Re}(A\varphi, \varphi) + \mathrm{Re}\,\lambda(\varphi, \varphi) \geqq (\mathrm{Re}\,\lambda - \delta_1)\|\varphi\|^2.$$

Furthermore there exists a positive constant C depending only on n, m, μ,

δ_1 and M_m such that if $\operatorname{Re}\lambda > \delta_1$, for any $\varphi \in W^{2m}(X, B)$,

$$\|\varphi\|_{2m} \leqq C\|(A+\lambda)\varphi\| \tag{7.15}$$

holds. In fact, it suffices to use $A(p, D)+\lambda$ for $A(p, D)$ in (4.7), and make an estimate of $\|\varphi\|_0$ by (7.14).

Theorem 7.5. *Let $A(p, D)$ be a strongly elliptic partial differential operator. If $\operatorname{Re}\lambda > \delta_1$, then $A+\lambda$ is a linear isomorphism of $W^{2m}(X, B)$ onto $L^2(X, B)$.*

Proof. It is clear that $A+\lambda$ is a continuous injection of $W^{2m}(X, B)$ to $L^2(X, B)$. For any $f \in L^2(X, B)$, there is a weak solution of the equation

$$(A+\lambda)u = f$$

in $W^m(X, B)$ (Theorem 5.2). By Theorem 6.1, $u \in W^{2m}(X, B)$. Hence $A+\lambda$ is surjective. Therefore by (7.15) its inverse is also continuous. ∎

Fix μ with $\mu \geqq \delta_1$, and put

$$G_\mu = (A+\mu)^{-1}. \tag{7.16}$$

Since $W^{2m}(X, B) \subset L^2(X, B)$, G_μ can be considered as a map of $L^2(X, B)$ to $L^2(X, B)$. Then the following theorem follows from Rellich's theorem.

Theorem 7.6. *G_μ is a compact liner map of $L^2(X, B)$ to $L^2(X, B)$.* ∎

For an arbitrary complex number λ, consider the equation

$$(A-\lambda)\varphi = f. \tag{7.17}$$

Multiplying by G_μ from the left, and putting $\zeta = \lambda + \mu$, we have

$$(I - \zeta G_\mu)\varphi = G_\mu f.$$

Putting $G_\mu f = g$, we obtain

$$(I - \zeta G_\mu)\varphi = g. \tag{7.18}$$

We consider (7.18) instead of (7.17).

$$(A-\lambda) = (A+\mu)(I-\zeta G_\mu) = (I - \zeta G_\mu)(A+\mu)$$

holds. Since $(A+\mu)$ is a bijection of $W^{2m}(X, B)$ onto $L^2(X, B)$, $(A-\lambda)$ is bijective if and only if $I - \zeta G_\mu$ is bijective. Thus we have the following theorem.

Theorem 7.7. *A complex number λ is contained in the spectrum of A if and only if $\zeta^{-1}=(\lambda-\mu)^{-1}$ is contained in the spectrum of G_μ.* ∎

Since G_μ is a compact operator on $L^2(X, B)$, its spectrum is rather simple. In particular, except for 0, it has only the point spectrum whose only accumulation point is 0, and the generalized eigenspace belonging to each eigenvalue is finite dimensional.

Theorem 7.8. *Let $A(p, D)$ be a strongly elliptic linear partial differential operator. Then the spectrum of A is contained in the half space* $\operatorname{Re}\lambda > -\delta_1$, *and it consists only of the point spectrum which has no finite accumulation point. Furthermore the generalized eigenspace belonging to each eigen value is finite dimensional.* ∎

§8. C^∞ Differentiability of $\varphi(t)$

In this section we shall prove that the vector (0, 1)-form $\varphi(t)$ constructed in Chapter 5, §3 is C^∞.

Let S be the disk of radius r in $\mathbb{C}^m$ with the origin as its centre, namely, we put $S=\{t=(t_1,\dots,t_m)\,|\,|t|^2=\sum_{j=1}^m |t_j|^2<r^2\}$. Let M be a compact complex manifold of dimension n, and X its underlying C^∞ differentiable manifold. The vector (0, 1)-form $\varphi(t)$ on M parametrized by $t\in S$ satisfies the following quasi-linear partial differential equation.

$$\left(-\sum_{\lambda=1}^{m}\frac{\partial^2}{\partial t^\lambda\,\partial \bar{t}^\lambda}+\square\right)\varphi(t)-\tfrac{1}{2}\theta[\varphi(t),\varphi(t)]=0, \tag{8.1}$$

where [,] denotes the Poisson bracket. $\varphi(t)$ is holomorphic in t, and $C^{k+\theta}$ ($k\geqq 2$, $1>\theta>0$) with respect to $p\in X$. Moreover we may assume that there exists a positive constant K such that

$$|\varphi(t)|_{k+\theta}\leqq KA(t), \tag{8.2}$$

where we put

$$A(t)=\frac{b}{16c}\sum_{\mu=1}^{\infty}\frac{c^\mu}{\mu^2}(t_1+\cdots+t_m)^\mu \tag{8.3}$$

with positive constants b, c.

The aim of this section is to show that $\varphi(t)$ is C^∞ on $X\times\{t\in\mathbb{C}^m\,|\,|t|<2^{-1}r\}$ provided that r is a sufficiently small positive number.

We cover M by coordinate neighbourhoods X_j, $j=1,\dots,J$. Let $z_j=(z_j^1,\dots,z_j^n)$ be the local complex coordinates on X_j with $z_j^k=x_j^k+\sqrt{-1}x_j^{k+n}$. The vector-valued (0, 1)-form $\varphi(t)$ is represented on X_j in terms of these local coordinates as

$$\varphi(t)=\sum_{\lambda}\varphi^{\lambda}(z,t)\partial_{\lambda}=\sum_{\lambda,\alpha}\varphi^{\lambda}_{\bar{\alpha}}(z,t)\,d\bar{z}^{\alpha}\,\partial_{\lambda}. \tag{8.4}$$

Let $2g_{\alpha\bar{\beta}}\,dz^{\alpha}\,d\bar{z}^{\beta}$ be the Hermitian metric tensor, and $(g^{\bar{\beta}\alpha})=(g_{\alpha\bar{\beta}})^{-1}$. Further we denote the covariant differentiation by ∇. Then if we put

$$\theta[\varphi,\varphi]=\sum_{\lambda,\alpha}\Phi^{\lambda}_{\bar{\alpha}}\,d\bar{z}^{\alpha}\,\partial_{\lambda}, \tag{8.5}$$

we have

$$\begin{aligned}\Phi^{\lambda}_{\bar{\alpha}}=-\sum_{\sigma,\rho,\nu}g^{\bar{\rho}\sigma}(&\nabla_{\sigma}\varphi^{\nu}_{\bar{\alpha}}\,\partial_{\nu}\varphi^{\lambda}_{\bar{\rho}}+\varphi^{\nu}_{\bar{\alpha}}\nabla_{\sigma}\,\partial_{\nu}\varphi^{\lambda}_{\bar{\rho}}\\&-\nabla_{\sigma}\varphi^{\nu}_{\bar{\rho}}\,\partial_{\nu}\varphi^{\lambda}_{\bar{\alpha}}-\varphi^{\nu}_{\bar{\rho}}\nabla_{\sigma}\,\partial_{\nu}\varphi^{\lambda}_{\bar{\alpha}}).\end{aligned} \tag{8.6}$$

Since the problem is local, we can use the results from the theory of elliptic partial differential equations. Choose a C^∞ function $\omega_j(p)$ on X with supp $\omega_j\subset X_j$ so that for any $p\in X$,

$$\sum_{j=1}^{J}\omega_j(p)\equiv 1. \tag{8.7}$$

Next, for each $l=1,2,\ldots$, we choose a function $\eta^l(t)$ as follows.

$$\begin{aligned}\eta^l(t)&\equiv 1\quad\text{on}\quad |t|\leqq(2^{-1}+2^{-l-1})r,\\ \eta^l(t)&\equiv 0\quad\text{on}\quad |t|\geqq(2^{-1}+2^{-l})r.\end{aligned} \tag{8.8}$$

We assume that the $\eta^l(t)$ are C^∞. Put

$$\omega_j^l(p,t)=\omega_j(p)\eta^l\eta(t). \tag{8.9}$$

Furthermore, we choose a C^∞ function $\chi_j(p)$ with supp $\chi_j\subset X_j$ which is identically equal to 1 on some neighbourhood of the support of ω_j. We put

$$\chi_j^l(p,t)=\chi_j(p)\eta^l(t). \tag{8.10}$$

Since $\eta^l(t)$ is identically equal to 1 on a neighbourhood of the support of $\eta^{l+2}(t)$, χ_j^l is identically equal to 1 on some neighbourhood of the support of ω_j^{l+2}.

First we shall prove that $\eta^3\varphi$ is $C^{k+1+\theta}$. $\omega_j^3\varphi$ satisfies the equation

$$\left(-\sum_{j=1}^{m}\frac{\partial^2}{\partial t^{\lambda}\,\partial\bar{t}^{\lambda}}+\Box\right)\omega_j^3\varphi=F_0, \tag{8.11}$$

where we put

$$F_0=\tfrac{1}{2}\omega_j^3\theta[\varphi,\varphi]-\{\omega_j^3\square\chi_j^1\varphi-\square\omega_j^3\chi_j^1\varphi\}. \tag{8.12}$$

Here we use the fact that χ_j^1 is identically equal to 1 in a neighbourhood of the support of ω_j^3.

The support of $\omega_j^3\varphi(p,t)$ is contained in $X_j\times S$. In terms of local coordinates, we have

$$\begin{aligned}\omega_j^3\varphi(p,t)&=\sum_\lambda \omega_j^3(z,t)\varphi^\lambda(z,t)\,\partial_\lambda\\&=\sum_{\lambda,\alpha}\omega_j^3(z,t)\varphi_{\bar\alpha}^\lambda(z,t)\,d\bar z^\alpha\,\partial_\lambda.\end{aligned} \tag{8.13}$$

We introduce real coordinates x^α, $\alpha=1,2,\dots,2n+2m$, by putting

$$z^l(p)=x^l+\sqrt{-1}x^{l+n},\qquad l=1,\dots,n,$$
$$t^l=x^{l+2n}+\sqrt{-1}x^{l+2n+m},\qquad l=1,\dots,m.$$

By these $2n+2m$ real coordinates, $X_j\times S$ is identified with an open set U_j of a $(2n+2m)$-dimensional torus $\mathbb{T}^{2n+2m}$.

A complex-valued function $f(p,t)$ defined on $X\times S$ can be considered as a function on $\mathbb{T}^{2n+2m}$ if $\operatorname{supp} f\subset X_j\times S$. In this case we can consider the difference quotient $\Delta_\alpha^h f$ in the direction of x^α, $\alpha=1,\dots,2n+2m$. Since the support of the vector-valued (0, 1)-form $\omega_j^3\varphi$ is also contained in $X_j\times S$, we define its difference quotient $\Delta_\alpha^h\omega_j^3\varphi$ as

$$\Delta_\alpha^h\omega_j^3\varphi(z,t)=\sum_{\lambda,\bar\rho}\Delta_\alpha^h(\omega_j^3\varphi_{\bar\rho}^\lambda)(z,t)\,d\bar z^\rho\,\partial_\lambda. \tag{8.14}$$

Since $\omega_j^3\varphi_{\bar\rho}^\lambda$ is a complex-valued function whose support is contained in $X_j\times S$, the right-hand side is well defined.

(8.11) can be considered as the equation on $\mathbb{T}^{2n+2m}$. Hence, by taking the difference quotient of both sides of this, we obtain the following equation.

$$\left(-\sum_{\lambda=1}^m\frac{\partial^2}{\partial t^\lambda\,\partial\bar t^\lambda}+\square\right)\Delta_\alpha^h\omega_j^3\phi=F_1, \tag{8.15}$$

where we put

$$\begin{aligned}F_1=\tfrac{1}{2}\Delta_\alpha^h\omega_j^3\theta[\varphi,\varphi]-\Delta_\alpha^h\{\omega_j^3\square\chi_j^1\varphi-\square\omega_j^3\chi_j^1\varphi\}\\+(\square\Delta_\alpha^h-\Delta_\alpha^h\square)\omega_j^3\varphi.\end{aligned} \tag{8.16}$$

Since $(-\sum_{\lambda=1}^m\partial^2/\partial t^\lambda\,\partial\bar t^\lambda+\square)$ is an elliptic linear partial differential operator whose principal part is of diagonal type, by Theorem 2.3, we obtain the following *a priori* estimate

$$|\Delta_\alpha^h \omega_j^3 \varphi|_{k+\theta} \leqq C_k(|F_1|_{k-2+\theta} + |\Delta_\alpha^h \omega_j^3 \varphi|_0), \tag{8.17}$$

where C_k is a positive constant, which may depend on k. In order to make an estimate of the right-hand side of this inequality, we use the following two lemmata.

Lemma 8.1. *Let u and v be complex-valued $C^{k+\theta}$ functions defined on $\mathbb{T}^l$ with $k \geqq 0$ and $1 > \theta > 0$. Then the product uv is $C^{k+\theta}$, and there exists a positive constant B_k depending only on k and l, but independent of u and v such that*

$$|uv|_{k+\theta} \leqq B_k \sum_{r+s=k} (|u|_{r+\theta}|v|_s + |u|_r|v|_{s+\theta}). \tag{8.18}$$

The proof is easy, hence we omit it. ∎

In order to make an estimate of the norm of $\Delta_\alpha^h \omega_j^3 \varphi$, the following lemma is useful. This corresponds to Theorem 1.13.

Lemma 8.2. *Let $z_1, \ldots, x^l$ be coordinate functions on the l-dimensional torus $\mathbb{T}^l$ and Δ_α^h the difference quotient operator in the direction of x^α by h. Then for any integer $k \geqq 0$ and any θ with $1 > \theta > 0$, we have the following:*

(i) *For $S \in C^{k+\theta}(\mathbb{T}^l, \mathbb{C}^\mu)$, if $|h| > 0$, the difference quotient $\Delta_\alpha^h S$ is again an element of $C^{k+\theta}(\mathbb{T}^l, \mathbb{C}^\mu)$.*

(ii) *If $S \in C^{k+\theta}(\mathbb{T}^l, \mathbb{C}^\mu)$, then for any $\alpha = 1, \ldots, l$, and any h with $0 < |h| < 1$, the following estimate holds:*

$$|\Delta_\alpha^h S|_{k+\theta} \leqq |S|_{k+1+\theta}. \tag{8.19}$$

(iii) *If $S \in C^{k+\theta}(\mathbb{T}^l, \mathbb{C}^\mu)$, and for any $\alpha = 1, \ldots, l$ and any h with $0 < |h| < 1$, there exists a positive constant M such that*

$$|\Delta_\alpha^h S|_{k+\theta} \leqq M, \tag{8.20}$$

then $S \in C^{k+1+\theta}(\mathbb{T}^l, \mathbb{C}^\mu)$.

This lemma is an analogy of Theorem 1.13 with the Sobolev norm replaced by the Hölder norm. The proof of Lemma 8.2 is, however, very easy unlike that of Theorem 1.13, hence we omit it.

We shall make an estimate of $\|F_1\|_{k-2\theta}$. The second term of the right-hand side of (8.12) is a linear combination of partial derivatives of order at most 1 of $\chi_j^1 \varphi$. Also the third term of (8.16) does not contain the difference quotient of second-order partial derivatives of $\omega_j^3 \varphi$ since they are cancelled.

Therefore, only the first term of F_1

$$\tfrac{1}{2}\Delta_\alpha^h\omega_j^3\theta[\varphi,\varphi]=\tfrac{1}{2}\Delta_\alpha^h\omega_j^3\theta[\chi_j^1\varphi,\chi_j^1\varphi] \tag{8.21}$$

contains the difference quotients of second-order partial derivatives of $\omega_j^3\varphi$ or $\chi_j^1\varphi$. Using the local representation (8.5) and (8.6), we can write (8.21) as

$$\tfrac{1}{2}\Delta_\alpha^h\omega_j^3\theta[\chi_j^1\varphi,\chi_j^1\varphi]=-\tfrac{1}{2}\sum_{\tau,\sigma,\rho,\nu,\lambda} g^{\bar\rho\sigma}\{\chi_j^1\varphi_{\bar\tau}^\nu\nabla_\sigma\,\partial_\nu\Delta_\alpha^h(\omega_j^3\varphi_{\bar\rho}^\lambda)$$
$$-\chi_j^1\varphi_{\bar\rho}^\nu\nabla_\sigma\,\partial_\nu\Delta_\alpha^h(\omega_j^3\varphi_{\bar\tau}^\lambda)\}\times d\bar z^\tau\,\partial_\lambda+\cdots, \tag{8.22}$$

where we omit the terms involving no difference quotients of second-order partial derivatives of $\omega_j^3\varphi$ or $\chi_j^1\varphi$.

By Lemma 8.1, there exist positive numbers L, L' such that

$$|\tfrac{1}{2}\Delta_\alpha^h\omega_j^3\theta[\chi_j^1\varphi,\chi_j^1\varphi]|_{k-2+\theta}$$
$$\leqq L\sum_{\beta,\rho,\nu}|\chi_1^1\varphi_{\bar\beta}^\nu|_0|\Delta_\alpha^h\omega_j^3\varphi_{\bar\rho}|_{k+\theta}+L'|\omega_j^3\varphi|_{k+\theta}|\psi_j^1\varphi|_{k+\theta}. \tag{8.23}$$

By Lemma 8.1 and Lemma 8.2, the $C^{k-2+\theta}$ norm of the terms of F_1 other than (8.22) can be estimated by a positive multiple of $|\omega_j^3\varphi|_{k+\theta}|\chi_j^1\varphi|_{k+\theta}$. On the other hand, applying (8.2) to the estimation of the first term $|\chi_j^1\varphi_{\bar\beta}^\nu|_0$ of (8.23), we obtain

$$|F_1|_{k-2+\theta}\leqq M_0A(r)|\Delta_\alpha^h\omega_j^3\varphi|_{k+\theta}+M_k|\omega_j^3\varphi|_{k+\theta}|\chi_j^1\varphi|_{k+\theta}, \tag{8.24}$$

where M_0 and M_k are positive constants.

By (8.17) and (8.24), we obtain

$$|\Delta_\alpha^h\omega_j^3\varphi|_{k+\theta}\leqq C_kM_0A(r)|\Delta_\alpha^h\omega_j^3\varphi|_{k+\theta}+C_k|\omega_j^3\varphi|_1$$
$$+C_kM_k|\omega_j^3\varphi|_{k+\theta}|\chi_j^1\varphi|_{k+\theta}. \tag{8.25}$$

We choose a sufficiently small r so that

$$M_0C_kA(r)<2^{-1} \tag{8.26}$$

holds. Multiplying both sides of (8.25) by 2, transposing the first term of the right-hand side, and using (8.26), we obtain

$$|\Delta_\alpha^h\omega_j^3\varphi|_{k+\theta}\leqq 2C_k|\omega_j^3\varphi|_1+2C_kM_1|\omega_j^3\varphi|_{k+\theta}|\psi_j^1\varphi|_{k+\theta}. \tag{8.27}$$

Since φ is $C^{k+\theta}$, the right-hand side is bounded independently of h. This is true for any $\alpha=1,\ldots,2n+2m$. Therefore by (iii) of Lemma 8.2, $\omega_j^3\varphi$ is proved to be $C^{k+1+\theta}$. Since this is true for any $j=1,\ldots,J$, summing with respect to j, we see from (8.7) that $\eta^3\varphi$ is a $C^{k+1+\theta}$ vector (0, 1)-form.

Next we shall prove that $h^5\varphi$ is $C^{k+2+\theta}$, using the same r determined by (8.26).

Replacing ω_j^3 by ω_j^5 and χ_j^1 by χ_j^3 in (8.15), and differentiating with respect to χ^β, we obtain

$$\left(-\sum_{\lambda=1}^{m}\frac{\partial^2}{\partial t^\lambda\,\partial\bar{t}^\lambda}+\square\right)\Delta_\alpha^h D_\beta\omega_j^5\varphi=F_2, \tag{8.28}$$

where we put $D_\beta=\partial/\partial x^\beta$, and

$$F_2=D_\beta F_1+\{\square\Delta_\alpha^h D_\beta\omega_j^5\varphi-D_\beta\square\Delta_\alpha^h\omega_j^5\varphi\}. \tag{8.29}$$

By (8.28), we obtain the following *a priori* estimate:

$$|\Delta_\alpha^h D_\beta\omega_j^5\varphi|_{k+\theta}\leqq C_k(|F_2|_{k-2+\theta}+|\Delta_\alpha^h D_\beta\omega_j^5\varphi|_0), \tag{8.30}$$

where C_k is the same as in (8.17). $|\Delta_\alpha^h D_\beta\omega_j^5\varphi|_0$ is estimated by $|\omega_j^5\varphi|_2$. We want to obtain an estimate of $|F_2|_{k-2+\theta}$. Since the second term of the right-hand side of (8.29) does not contain the difference quotients of partial derivatives of order 3 of $\omega_j^5\varphi$, the only term of F_2 which involves difference quotients of the partial derivatives of order 3 of $\omega_j^5\varphi$ or $\chi_j^3\varphi$ is

$$\begin{aligned}\tfrac{1}{2}\Delta_\alpha^h D_\beta\omega_j^5\theta[\chi_j^3\varphi,\chi_j^3\varphi]=-\tfrac{1}{2}\sum_{\tau,\sigma,\rho,\nu,\lambda}g^{\bar\rho\sigma}\{&\chi_j^3\varphi_{\bar\tau}^{\nu}\nabla_\sigma\,\partial_\nu\Delta_\alpha^h(D_\beta\omega_j^5\varphi_{\bar\rho}^{\lambda})\\&-\chi_j^3\varphi_{\bar\rho}^{\nu}\nabla_\sigma\,\partial_\nu\Delta_\alpha^h(D_\beta\omega_j^5\varphi_{\bar\tau}^{\lambda})\}\,d\bar z^\tau\,\partial_\lambda+\cdots,\end{aligned} \tag{8.31}$$

where we omit the terms not involving difference quotients of partial derivatives of order 3 of $\omega_j^5\varphi$ or $\chi_j^3\varphi$. Comparing (8.31) with (8.22), by the same argument as for (8.24) we obtain

$$|F_2|_{k-2+\theta}\leqq M_0A(r)|\Delta_\alpha^h D_\beta\omega_j^5\varphi|_{k+\theta}+M_{k+1}|\chi_j^3\varphi|_{k+1+\theta}\times|\omega_j^5\varphi|_{k+1+\theta}, \tag{8.32}$$

where M_{k+1} may be different from M_k, but M_0 is the same as in (8.24). By (8.32) and (8.30), we obtain

$$\begin{aligned}|\Delta_\alpha^h D_\beta\omega_j^5\varphi|_{k+\theta}\leqq{}&C_kM_0A(r)|\Delta_\alpha^h D_\beta\omega_j^5\varphi|_{k+\theta}\\&+C_kM_{k+1}|\chi_j^3\varphi|_{k+1+\theta}|\omega_j^5\varphi|_{k+1+\theta}+M'_{k+1}|\omega_j^5\varphi|_2.\end{aligned} \tag{8.33}$$

Multiplying both sides of (8.33) by 2, transposing the first term of the right-hand side, and using (8.26), we obtain

$$|\Delta_\alpha^h D_\beta \omega_j^5 \varphi|_{k+\theta} \leqq 2C_k M_{k+1} |\chi_j^3 \varphi|_{k+1+\theta} |\omega_j^5 \varphi|_{k+1+\theta} + 2M'_{k+1} |\omega_j^5 \varphi|_2. \tag{8.34}$$

Since $\eta^3\varphi$ is $C^{k+1+\theta}$, the right-hand side is bounded independently of h. Since this is true for any $\alpha = 1, \ldots, 2n+2m$, by (iii) of Lemma 8.2, $D_\beta \omega_j^5 \varphi$ is $C^{k+1+\theta}$. Since this is true for any $\beta = 1, \ldots, 2n+2m$, we see that $\omega_j^5\varphi$ is $C^{k+2+\theta}$. Summing with respect to j, we see from (8.7) that $\eta^5\varphi$ is $C^{k+2+\theta}$ vector (0, 1)-form. Note that in the above we need not replace r satisfying (8.26) by a smaller one.

Similarly we can prove that, for any $l = 1, 2, \ldots$, $\eta^{2l+1}\varphi$ is $C^{k+l+\theta}$, where we may choose r independent of l. Since $\eta^{2l+1}(t)$ is identically equal to 1 on $|t| < 2^{-1}r$ which is independent of l, φ is C^∞ on $X \times \{t \in \mathbb{C}^m \,|\, |t| < 2^{-1}r\}$. ∎

Bibliography

[1] Bott, R.: Homogeneous vector bundles, *Annals of Math.* **66** (1957), 203-248.
[2] Chow, W. L.: On compact complex analytic varieties, *Amer. J. Math.* **71** (1949), 893-914.
[3] Douglis, A. and Nirenberg, L.: Interior estimates for elliptic systems of partial differential equations, Comm. Pure Appl. Math. **8** (1955), 503-538.
[4] Fischer, W. and Grauert, H.: *Lokal-trivial Familien kompakter komplexen Mannigfaltigkeiten*, Nachr. Akad. Wiss. Gottingen II. Math.-Phys. Kl. 1965, 88-94.
[5] Forster, O. und Knorr, K.: Ein neuer Beweis des Satzes von Kodaira-Nirenberg-Spencer, *Math. Z.* **139** (1974), 257-291.
[6] Friedrichs, K. O.: On the differentiability of the solutions of linear elliptic differential equations, *Comm. Pure Appl. Math.* **6** (1953), 299-326.
[7] Fröhlicher, A. and Nijenhuis, A.: A theorem on stability of complex structures, *Proc. Nat. Acad. Sci., U.S.A.* **43** (1957), 239-241.
[8] Grauert, H.: Ein Theorem der analytischen Garbentheorie und die Modulräume komplexer Strukturen, *Publ. Math. I.H.E.S.* **5** (1960), 233-291.
[9] ———: Der Satz von Kuranishi für kompakter komplexe Räume, *Invent. Math.* **25** (1974), 107-142.
[10] Hirzebruch, F.: *Topological Methods in Algebraic Geometry*, 3rd edition, Springer-Verlag, Berlin, Heidelberg, New York, 1966.
[11] Hopf, H.: Zur Topologie der komplexen Mannigfaltigkeiten, in *Studies and Essays Presented to R. Courant*, New York, 1948.
[12] Hurwitz, A. and Courant, R.: *Funktionentheorie*, Springer-Verlag, Berlin, 1929.
[13] Kas, A.: On obstructions to deformations of complex analytic surfaces, *Proc. Nat. Acad. Sci. U.S.A.* **58** (1967), 402-404.
[14] Kodaira, K.: On compact analytic surfaces, in *Analytic functions*, Princeton Univ. Press, Princeton, NJ, 1960, 121-135.
[15] ———: On compact complex analytic surfaces, I, II, III, *Annals of Math.* **71** (1960), 111-152; **77** (1963), 563-626; **78** (1963), 1-40.
[16] ———: *Collected Works*, Iwanami Shoten, Tokyo, and Princeton Univ. Press, Princeton, NJ, 1975.
[17] ———: *Introduction to Complex Analysis*, Cambridge Univ. Press, Cambridge, 1984.
[18] ———, Nirenberg, L. and Spencer, D. C.: On the existence of deformations of complex analytic structures, *Annals of Math.* **68** (1958), 450-459.
[19] ———, and Spencer, D. C.: On the variation of almost-complex structure, in *Algebraic Geometry and Topology*, Princeton Univ. Press, Princeton, NJ, 1957, 139-150.
[20] ———, ———: A theorem of completeness for complex analytic fibre spaces, *Acta Math.* **100** (1958), 281-294.
[21] ———, ———: On deformations of compelx analytic structures, I-II, III, *Annals of Math.* **67** (1958), 328-466; **71** (1960), 43-76.
[22] Kuranishi, M.: On the locally complete families of complex analytic structures, *Annals of Math.* **75** (1962), 536-577.
[23] Lefschetz, S.: *L'Analysis Situs et la Geometrie Algébrique*, Gauthier-Villars, Paris, 1924.
[24] Milnor, J.: *Morse Theory*, Annals of Mathematics Studies 51, Princeton Univ. Press, Princeton, NJ, 1963.
[25] Mumford, D.: Further pathologies in algebraic geometry, *Amer. J. Math.* **84** (1962), 642-648.

[26] Newlander, A. and Nirenberg, L.: Complex analytic coordinates in almost-complex manifolds, *Annals of Math.* **65** (1957), 391-404.
[27] Osgood, W. F.: *Funktionentheorie*, Band II, Chelsea, New York, 1932.
[28] Siegel, C. L.: Discontinuous groups, *Annals of Math.* **44** (1946), 674-689.
[29] ———: *Analytic Function of Several Complex Variables*, Lecture notes, Institute for Advanced Study, Princeton NJ, 1948.
[30] Teichmüller, O.: Extremale quaskkonforme Abbildungen und quadratische Differentiale, *Abh. Preuss. Akad. der Wiss. Math.-naturw. Klasse* **22** (1939), 1-97.

Index

M. Aigner Combinatorial Theory ISBN 978-3-540-61787-7
A. L. Besse Einstein Manifolds ISBN 978-3-540-74120-6
N. P. Bhatia, G. P. Szegő Stability Theory of Dynamical Systems ISBN 978-3-540-42748-3
J. W. S. Cassels An Introduction to the Geometry of Numbers ISBN 978-3-540-61788-4
R. Courant, F. John Introduction to Calculus and Analysis I ISBN 978-3-540-65058-4
R. Courant, F. John Introduction to Calculus and Analysis II/1 ISBN 978-3-540-66569-4
R. Courant, F. John Introduction to Calculus and Analysis II/2 ISBN 978-3-540-66570-0
P. Dembowski Finite Geometries ISBN 978-3-540-61786-0
A. Dold Lectures on Algebraic Topology ISBN 978-3-540-58660-9
J. L. Doob Classical Potential Theory and Its Probabilistic Counterpart ISBN 978-3-540-41206-9
R. S. Ellis Entropy, Large Deviations, and Statistical Mechanics ISBN 978-3-540-29059-9
H. Federer Geometric Measure Theory ISBN 978-3-540-60656-7
S. Flügge Practical Quantum Mechanics ISBN 978-3-540-65035-5
L. D. Faddeev, L. A. Takhtajan Hamiltonian Methods in the Theory of Solitons ISBN 978-3-540-69843-2
I. I. Gikhman, A. V. Skorokhod The Theory of Stochastic Processes I ISBN 978-3-540-20284-4
I. I. Gikhman, A. V. Skorokhod The Theory of Stochastic Processes II ISBN 978-3-540-20285-1
I. I. Gikhman, A. V. Skorokhod The Theory of Stochastic Processes III ISBN 978-3-540-49940-4
D. Gilbarg, N. S. Trudinger Elliptic Partial Differential Equations of Second Order ISBN 978-3-540-41160-4
H. Grauert, R. Remmert Theory of Stein Spaces ISBN 978-3-540-00373-1
H. Hasse Number Theory ISBN 978-3-540-42749-0
F. Hirzebruch Topological Methods in Algebraic Geometry ISBN 978-3-540-58663-0
L. Hörmander The Analysis of Linear Partial Differential Operators I – Distribution Theory and Fourier Analysis ISBN 978-3-540-00662-6
L. Hörmander The Analysis of Linear Partial Differential Operators II – Differential Operators with Constant Coefficients ISBN 978-3-540-22516-4
L. Hörmander The Analysis of Linear Partial Differential Operators III – Pseudo-Differential Operators ISBN 978-3-540-49937-4
L. Hörmander The Analysis of Linear Partial Differential Operators IV – Fourier Integral Operators ISBN 978-3-642-00117-8
K. Itô, H. P. McKean, Jr. Diffusion Processes and Their Sample Paths ISBN 978-3-540-60629-1
T. Kato Perturbation Theory for Linear Operators ISBN 978-3-540-58661-6
S. Kobayashi Transformation Groups in Differential Geometry ISBN 978-3-540-58659-3
K. Kodaira Complex Manifolds and Deformation of Complex Structures ISBN 978-3-540-22614-7
Th. M. Liggett Interacting Particle Systems ISBN 978-3-540-22617-8
J. Lindenstrauss, L. Tzafriri Classical Banach Spaces I and II ISBN 978-3-540-60628-4
R. C. Lyndon, P. E Schupp Combinatorial Group Theory ISBN 978-3-540-41158-1
S. Mac Lane Homology ISBN 978-3-540-58662-3
C. B. Morrey Jr. Multiple Integrals in the Calculus of Variations ISBN 978-3-540-69915-6
D. Mumford Algebraic Geometry I – Complex Projective Varieties ISBN 978-3-540-58657-9
O. T. O'Meara Introduction to Quadratic Forms ISBN 978-3-540-66564-9
G. Pólya, G. Szegő Problems and Theorems in Analysis I – Series. Integral Calculus. Theory of Functions ISBN 978-3-540-63640-3
G. Pólya, G. Szegő Problems and Theorems in Analysis II – Theory of Functions. Zeros. Polynomials. Determinants. Number Theory. Geometry ISBN 978-3-540-63686-1
W. Rudin Function Theory in the Unit Ball of $\mathbb{C}^n$ ISBN 978-3-540-68272-1
S. Sakai C*-Algebras and W*-Algebras ISBN 978-3-540-63633-5
C. L. Siegel, J. K. Moser Lectures on Celestial Mechanics ISBN 978-3-540-58656-2
T. A. Springer Jordan Algebras and Algebraic Groups ISBN 978-3-540-63632-8
D. W. Stroock, S. R. S. Varadhan Multidimensional Diffusion Processes ISBN 978-3-540-28998-2
R. R. Switzer Algebraic Topology: Homology and Homotopy ISBN 978-3-540-42750-6
A. Weil Basic Number Theory ISBN 978-3-540-58655-5
A. Weil Elliptic Functions According to Eisenstein and Kronecker ISBN 978-3-540-65036-2
K. Yosida Functional Analysis ISBN 978-3-540-58654-8
O. Zariski Algebraic Surfaces ISBN 978-3-540-58658-6

Printing: Krips bv, Meppel
Binding: Litges & Dopf, Heppenheim